SERIES ON SEMICONDUCTOR SCIENCE AND TECHNOLOGY

Series Editors

SERIES ON SEMICONDUCTOR SCIENCE AND TECHNOLOGY

---

1. M. Jaros: *Physics and applications of semiconductor microstructures*
2. V. N. Dobrovolsky, V. G. Litovchenko: *Surface electronic transport phenomena in semiconductors*
3. M. J. Kelly: *Low-dimensional semiconductors*
4. P. K. Basu: *Theory of optical processes in semiconductors*
5. N. Balkan: *Hot electrons in semiconductors*
6. B. Gil: *Group III nitride semiconductor compounds: physics and applications*
7. M. Sugawara: *Plasma etching*
8. M. Balkanski, R. F. Wallis: *Semiconductor physics and applications*
9. B. Gil: *Low-dimensional nitride semiconductors*
10. L. Challis: *Electron-phonon interactions in low-dimensional structures*
11. V. Ustinov, A. Zhukov, A. Egorov, N. Maleev: *Quantum dot lasers*
12. H. Spieler: *Semiconductor detector systems*
13. S. Maekawa: *Concepts in spin electronics*
14. S. D. Ganichev, W. Prettl: *Intense terahertz excitation of semiconductors*
15. N. Miura: *Physics of semiconductors in high magnetic fields*
16. A. V. Kavokin, J. J. Baumberg, G. Malpuech, F. P. Laussy: *Microcavities*
17. S. Maekawa, S. O. Valenzuela, E. Saitoh, T. Kimura: *Spin current*
18. B. Gil: *III-nitride semiconductors and their modern devices*

III-Nitride Semiconductors and their Modern Devices

Edited by

Bernard Gil

Centre National de la Recherche Scientifique
Université de Montpellier 2

OXFORD
UNIVERSITY PRESS

OXFORD
UNIVERSITY PRESS

Great Clarendon Street, Oxford, OX2 6DP,
United Kingdom

Oxford University Press is a department of the University of Oxford.
It furthers the University's objective of excellence in research, scholarship, and education by publishing worldwide. Oxford is a registered trade mark of Oxford University Press in the UK and in certain other countries

First Edition published in 2013
Impression: 1

Published in the United States of America by Oxford University Press
198 Madison Avenue, New York, NY 10016, United States of America

British Library Cataloguing in Publication Data
Data available

Library of Congress Control Number: 2013940022

ISBN 978–0–19–968172–3

Printed and bound by
CPI Group (UK) Ltd, Croydon, CR0 4YY

Preface

When I met Sönke Adlung, of Oxford University Press, during the annual meeting of the European Materials Research Society in Strasbourg in June 1996, he was taking care of the exhibition booth in his editorial role, while I was co-organizing a symposium dedicated to visible-light emitters.

Nitrides were booming world-wide at that time. During the preceding three years, many researchers, including myself, had reoriented their research activities to nitrides after several Japanese researchers, including Prof Isamu Akasaki and Drs Hiroshi Amano and Shuji Nakamura, had made significant breakthroughs in various scientific areas such as materials science and device physics. These breakthroughs had led to the realization and commercialization of blue light-emitting diodes based on nitride semiconductors. More recently, in January 2006, a blue laser was demonstrated in Japan. Most of the attendees at the visible-light symposium had come to Strasbourg to hear, in a very crowded conference room, the talk by Shuji Nakamura, planned for Friday the 7th, so that they could see with their own eyes the coherent blue light-ray of a nitride-based laser exciting the fluorescence of a paper screen.

During the afternoon of that Friday, Sönke, with whom I was chatting in front of the OUP booth, suggested that I edit a book dedicated to nitrides. *Group III Nitride Semiconductor Compounds: Physics and Applications* was edited one and-a-half years later in 1998, and it was followed by *Low-Dimensional Nitride Semiconductors* in 2002. Today you are holding *III-Nitride Semiconductors and their Modern Devices.*

Ten years have passed since the second volume of this series was published. We believed a few months ago, in light of recent new developments in the field, that it was time to offer a modernized compilation of today's research activities. We kept the initial format: a book accessible to the larger community, including PhD students at the beginning of their studies and not yet familiar with the world of semiconductor physics. Light emission/detection is a newly developing area. Therefore, I asked Professor Amano of Nagoya University, one of the fathers of nitrides, to write a general introductory chapter to provide an overview of the needs and realizations in that area. His chapter combines a review of economic issues and societal needs, together with growth aspects, physics characterization, and device realizations. All other chapters were written by leading researchers in specific fields. Dr Izabella Grzegory has been growing bulk gallium nitride since the beginning of her career. I had in my hands some very tiny pieces of her transparent gallium nitride bulk crystal as early as January 1986! I found it natural to ask her to take care of the chapter dedicated to the homoepitaxial challenge, and she wrote it in collaboration with her colleagues at Warsaw: Michal Bockowski, Piotr Perlin, Czeslaw Skierbiszewski, Tadeusz Suski, Marcin Sarzynski, Stanislaw Krukowski, and Sylwester Porowski. They share it with us here, and they indicate how far they have advanced since their early studies in Poland at the end of the 1970s. Nitrides being often deposited by using modern epitaxial growth techniques (see for instance, the first book of this series), to review the status of the research at the nitride–silicon interface is of value. I asked Professor Alois Krost and Dr Armin Dadgar, of the Otto-von-Guericke-University at Magdeburg, to address this message at the dawn 2013. This constitutes the third chapter, which precedes the chapter dedicated to the growth of bulk aluminium nitride (a very promising substrate for the years to come), written by Zlatko Sitar and Ronny Kirste of the North Carolina State University at Raleigh. Since the beginning of the 1980s, semiconductor quantum dots have been known to be, beyond the scope of promising concepts, efficient adimensional localization centers for electrons and holes, offering the possibility of realizing low-threshold lasing. We now know that quantum dots are useful for realizing single-photon sources and entangled-photon pair-emitters. Andre Strittmatter, of the Technical University of Berlin, agreed to write the fifth chapter, dedicated to the growth of such light-emitters. Band-gap engineering issues are particularly important regarding the growth of heterostructures on lattice mismatched substrates. Chapters 6 and 7 were written by researchers at the Ecole Polytechnique Fédérale de Lausanne and Mie University, respectively. Aluminum indium nitride was initially proposed by Professor Marc Ilegems (now retired) and Dr Jean François Carlin. I thus naturally asked the colleagues that are currently holding Marc's stick to accept the burden of reviewing the status and importance of the AlInN alloy at the composition lattice-matched to GaN. Drs Raphaël Butté, Gatien Cosendey, Lorenzo Lugani, Marlene Glauser, Antonino Castiglia, Guillaume Perillat-Merceroz, Jean-François Carlin, and Professor Nicolas Grandjean did this very nicely. Aluminum-rich AlGaN alloys are also very important for targeting deep-ultraviolet applications. These alloys are treated by an internationally recognized expert in the area of crystal growth of chalcopyrites and nitrides: Professor Hideto Miyake, of Mie University. There are tremendous developments these days regarding the properties of light-emitters grown on non-polar or semipolar orientations. Writing such a chapter, while retaining clarity and a reasonable length, required considerable effort by Professor Michael Kneissl and Dr Tim Wernicke of the Technical University and Ferdinand-Braun-Institute at Berlin. The result is impressive. In Chapter 9, Drs Rudeesun Songmuang and Eva Monroy address the single-nanowire devices, which was quite challenging from the beginning. Dr Jean-Yves Duboz, of the Centre of Research on Hetero-Epitaxy and its Applications, in Valbonne, agreed to write on advanced photonic and nanophotonic devices, the topic of Chapter 10. Dr Yvon Cordier, of the Centre of Research on Hetero-Epitaxy and Applications, Valbonne, and Drs Tatsuya Fujishima, Bin Lu, Elison Matioli, and Professor Tomás Palacios, of the Massachusetts Institute of Technology, Cambridge, next address nitride-based electron devices, widely investigated for the next generation of high-power and high-frequency applications.

The potentialities of having large conduction-band offsets in nitride heterostructures enables targeting of intersubband transitions in low-dimensional nitrides, which are in particular investigated at the Institute of Fundamental Electronics by Drs Maria Tchernycheva and François Julien. They both accepted my invitation to write the chapter dedicated to applications of nitride heterostructures in the mid-infrared. The large oscillator strength of nitrides paves the way to fascinating optical properties, among which is the slow-light propagation phenomenon. Drs Tatiana Shubina and Mikhail Glazov, of the IOFFE Institute in St Petersburg, together with Dr Nikolay Gippius of the General Physics Institute of the Russian Academy of Sciences, in Moscow, placed aside the Cyrillic alphabet in order to write the thirteenth chapter, and I share the responsibility with them. Then follow three chapters dedicated to very prospective issues. Nitride devices and their biofunctionalization for biosensing applications are treated in a fourteenth chapter by my office neighbor Professor Csilla Gergely at Montpellier, and Walter R. L. Lambrecht and Atchara Punya, of the Department of Physics at Case Western Reserve University, Cleveland, address, in a very didactical way, the physics of heterovalent ternary II-IV-N2 compounds and perspectives for a new class of materials that are, *mutatis mutandis*, the analogs of the chalcopyrites. Finally, it has been proposed recently by Professor Alexey Kavokin to produce a radiative relaxation process between a forbidden excitonic transition populated by non-linear two-photon spectroscopy and an allowed one. Two photons are successively emitted—the first by cascading at an energy in the terahertz frequency region, and the second occurring at the host material band-gap energy. This process arm-twists the black-body radiation law, and may lead to a new class of THz emitters. I have asked Drs Oleksandr Kyriienko and Ivan Shelykh, of the University of Iceland, in Reykjavik, and the Division of Physics and Applied Physics, Anyang Technological University of Singapore, and Professor Alexey Kavokin, of the University of Southampton, to provide readers with sufficient elements for them to eventually handle this kind of research. The cover picture represents a photonic billiard designed by Didier Felbacq, realized and shined with a blue laser by Emmanuel Rousseau.

This book obviously departs from perfection, and is not exhaustive. I deeply hope, however, that it fulfils most of the needs of people in academic laboratories and most of the companies. To conclude, I would like to acknowledge all the contributors, and in addition, Jessica White, Sönke Adlung, Gandhimathi Ganesan, and Victoria Mortimer at Oxford University Press.

Bernard Gil
Montpellier, November 2012

Contents

List of Contributors

Hiroshi Amano
Department of Electrical
Engineering and Computer Science
Akasaki Research Center
Nagoya University
Japan
amano@nuee.nagoya-u.ac.jp

Michal Bockowski
Institute of High Pressure Physics
Polish Academy of Sciences
ul. Sokolowska 29/37
01-142 Warsaw
Poland
bocian@unipress.waw.pl

Raphaël Butté
Institute of Condensed
Matter Physics
Ecole Polytechnique Federale
de Lausanne (EPFL)
CH-1015 Lausanne
Switzerland
raphael.butte@epfl.ch

Jean-François Carlin
Institute of Condensed
Matter Physics
Ecole Polytechnique Federale
de Lausanne (EPFL)
CH-1015 Lausanne
Switzerland
jean-francois.carlin@epfl.ch

Antonino Castiglia
Institute of Condensed
Matter Physics
Ecole Polytechnique Federale
de Lausanne (EPFL)
CH-1015 Lausanne
Switzerland
antonino.castiglia@epfl.ch

Yvon Cordier
Centre de Recherche sur
l'Hétéro-Epitaxie et Applications
CNRS
Valbonne
France
yc@crhea.cnrs.fr

Gatien Cosendey
Institute of Condensed
Matter Physics
Ecole Polytechnique Federale
de Lausanne (EPFL)
CH-1015 Lausanne
Switzerland
gatien.cosendey@epfl.ch

Armin Dadgar
Otto-von-Guericke-University
Magdeburg
Universitätsplatz 2
39106 Magdeburg
Germany
armin.dadgar@ovgu.de

Jean-Yves Duboz
Centre de Recherche sur
l'Hétéro-Epitaxie et Applications
CNRS
Valbonne
France
jyd@crhea.cnrs.fr

Tatsuya Fujishima
Massachusetts Institute
of Technology
Cambridge, MA
USA
tatsuyaf@mit.edu

Csilla Gergely
Université Montpellier 2
Laboratoire Charles Coulomb
F-34095 Montpellier
France
csilla.gergely@univ-montp2.fr

Bernard Gil
L2C UMR CNRS 5221
Université Montpellier 2-CNRS
F-34095 Montpellier
France
bernard.gil@univ-montp2.fr

Nikolay A. Gippius
General Physics Institute of RAS
119991, Moscow
Russia
na.gippius@gmail.com

Marlene Glauser
Institute of Condensed
Matter Physics
Ecole Polytechnique Federale
de Lausanne (EPFL)
CH-1015 Lausanne
Switzerland
marlene.glauser@epfl.ch

Mikhail M. Glazov
Ioffe Physical-Technical Institute
of RAS
194021, St. Petersburg
Russia
glazov@coherent.ioffe.ru

Nicolas Grandjean
Institute of Condensed
Matter Physics
Ecole Polytechnique Federale
de Lausanne (EPFL)
CH-1015 Lausanne
Switzerland
Nicolas.grandjean@epfl.ch

Izabella Grzegory
Institute of High Pressure Physics
Polish Academy of Sciences
ul. Sokolowska 29/37
01-142 Warsaw
Poland
izabella@unipress.waw.pl

François H. Julien
Institut d'Electronique
Fondamentale
University of Paris-Sud
UMR 8622 CNRS
91405 Orsay
France
francois.julien@u-psud.fr

Alexey V. Kavokin
Physics and Astronomy School
University of Southampton
Highfield
Southampton SO17 1BJ
United Kingdom
alexey@phys.soton.ac.uk

Ronny Kirste
Department of Materials Science
and Engineering
North Carolina State University
Raleigh, NC 27695-7919
USA
rkirste@ncsu.edu

Michael Kneissl
Institute of Solid State Physics
Technische Universität Berlin
10623 Berlin
Germany
kneissl@physik.tu-berlin.de

Alois Krost
Otto-von-Guericke-University
Magdeburg
Universitätsplatz 2
39106 Magdeburg
Germany
alois.krost@ovgu.de

Stanislaw Krukowski
Institute of High Pressure Physics
Polish Academy of Sciences
ul. Sokolowska 29/37
01-142 Warsaw
Poland
stach@unipress.waw.pl

Oleksandr Kyriienko
Science Institute
University of Iceland
Dunhagi-3, IS-107
Reykjavik
Iceland
kyriienko@ukr.net

Walter R. L. Lambrecht
Department of Physics
Case Western Reserve University
10900 Euclid Avenue
Cleveland, OH 44106-7079
USA
wxl2@cwru.edu

Bin Lu
Massachusetts Institute
of Technology
Cambridge, MA
USA
binlu@mit.edu

Lorenzo Lugani
Institute of Condensed
Matter Physics
Ecole Polytechnique Federale
de Lausanne (EPFL)
CH-1015 Lausanne
Switzerland
lorenzo.lugani@epfl.ch

Elison Matioli
Massachusetts Institute
of Technology
Cambridge, MA
USA
ematioli@mit.edu

Hideto Miyake
Department of Electrical
and Electronic Engineering
Faculty of Engineering
Mie University
Japan
miyake@elec.mie-u.ac.jp

Eva Monroy
CEA-CNRS group "Nanophysique
et Semi-conducteurs"
CEA/INAC and Institut
Néel-CNRS
Grenoble
France
eva.monroy@cea.fr

Tomás Palacios
Massachusetts Institute
of Technology
Cambridge, MA
USA
tpalacios@mit.edu

Guillaume Perillat-Merceroz
Institute of Condensed
Matter Physics
Ecole Polytechnique Federale
de Lausanne (EPFL)
CH-1015 Lausanne
Switzerland
guillaume.perillat@epfl.ch

Piotr Perlin
Institute of High Pressure Physics
Polish Academy of Sciences
ul. Sokolowska 29/37
01-142 Warsaw
Poland
piotr@unipress.waw.pl

Sylwester Porowski
Institute of High Pressure Physics
Polish Academy of Sciences
ul. Sokolowska 29/37
01-142 Warsaw
Poland
sylvek@unipress.waw.pl

Atchara Punya
Department of Physics
Case Western Reserve University
10900 Euclid Avenue
Cleveland, OH 44106-7079
USA
atchara.punya@case.edu

Marcin Sarzynski
Institute of High Pressure Physics
Polish Academy of Sciences
ul. Sokolowska 29/37
01-142 Warsaw
Poland
sarzyn@unipress.waw.pl

Czeslaw Skierbiszewski
Institute of High Pressure Physics
Polish Academy of Sciences
ul. Sokolowska 29/37
01-142 Warsaw
Poland
czeslaw@unipress.waw.pl

Ivan A. Shelykh
Science Institute
University of Iceland
Dunhagi-3, IS-107
Reykjavik
Iceland
ivshel@mail.ru

Tatiana V. Shubina
Ioffe Physical-Technical Institute
of RAS
194021, St. Petersburg
Russia
shubina@beam.ioffe.ru

Zlatko Sitar
Department of Materials Science
and Engineering
North Carolina State University
Raleigh, NC 27695-7919
USA
sitar@ncsu.edu

Rudeesun Songmuang
CEA-CNRS group "Nanophysique
et Semi-conducteurs"
CEA/INAC and Institut
Néel-CNRS
Grenoble
France
Rudeesun.Songmuang@cea.fr

André Strittmatter
Department of Physics
Technische Universität Berlin
10623 Berlin
Germany
strittma@sol.physik.tu-berlin.de

Tadeusz Suski
Institute of High Pressure Physics
Polish Academy of Sciences
ul. Sokolowska 29/37
01-142 Warsaw
Poland
tadek@unipress.waw.pl

Maria Tchernycheva
Institut d'Electronique
Fondamentale
University of Paris-Sud
UMR 8622 CNRS
91405 Orsay
France
maria.tchernycheva@u-psud.fr

Tim Wernicke
Institute of Solid State Physics
Technische Universität Berlin
10623 Berlin
Germany
wernicke@physik.tu-berlin.de

1 Development of the nitride-based UV/DUV LEDs

Hiroshi Amano

1.1 Introduction

The commercialization of the blue LED started in 1993 [1], and three primary-color LEDs are now available commercially. Since 1993, the applications of LEDs in various fields have expanded widely. Three years after the commercialization of the blue LED, violet blue LD was developed [2], which is the shortest-wavelength LD to be realized in the visible region. The use of the white LED, which is composed of a blue LED with phosphors, has led to general lighting applications of LEDs in addition to conventional display applications. All these newly developed devices, including blue LEDs and violet blue LDs, are composed of group III nitride semiconductors. Research activity and commercial activity have stimulated each other, forming a highly effective feedback loop. Breakthroughs made through research have rapidly resulted in the placement of new devices on the market. These new devices are applied in ways not always expected by researchers. The demand for new applications is also acting as a motivating force for new research and development. Therefore, the development of the group-III-nitride-related field is progressing very well.

The successful use of the group III nitrides in blue and the white LEDs and the violet blue LD has led to attempts to realize LEDs and LDs in much shorter wavelength regions such as the ultraviolet (UV) and deep-ultraviolet (DUV) regions. UV/DUV LEDs and UV LDs will be useful for sterilization, lithography, laser-induced fluorescence spectroscopy, biomedical applications, chemical sensing, chemical reaction control, prototyping, drilling, and optical storage, and for use as an excitation source for phosphors and catalysts. For example, the application of a UV/DUV LED to the treatment of skin diseases has been reported [3].

Attempts to fabricate UV/DUV LEDs and UV LDs have been made by many researchers, and rapid progress has been attained. The first UV LED was reported by Pankove in 1972 and used colloidal carbon and nominally undoped n-type GaN [4]. Since the first pn-junction UV LED that emitted 375 nm light in 1989 [5], the emission wavelength of nitride-based LEDs has become shorter. In 2006 it reached 210 nm—the target shortest wavelength based on semiconductor light emitters [6]. AlN was used as the active region. The emission wavelength of UV LDs has also become shorter. Figure 1.1 shows the emission wavelength of

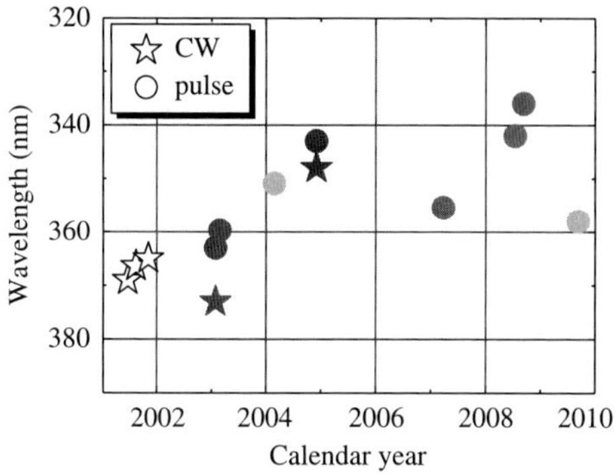

FIG. 1.1. Emission wavelength of laser diodes as a function of calendar year.

short-wavelength UV LDs as a function of calendar year. In 2008 the wavelength reached 336 nm of the shortest-wavelength semiconductor LD, though it was operated upon the pulsed mode [7]. This value has not been improved yet.

However, the performances of UV/DUV LEDs and UV LDs are still inferior to those of blue/white LEDs. Figure 1.2 summarizes the development of the EQE of UV/DUV LEDs as a function of calendar year. In 2012, the EQE reached 14.3% at 2 mA with an emission wavelength of 285 nm, which is still inferior to that of state-of-the-art blue LEDs by a factor of six.

Clarifying the factors that limit the EQE of UV/DUV LEDs and UV LDs is necessary to improve their performance. In general, there are three factors

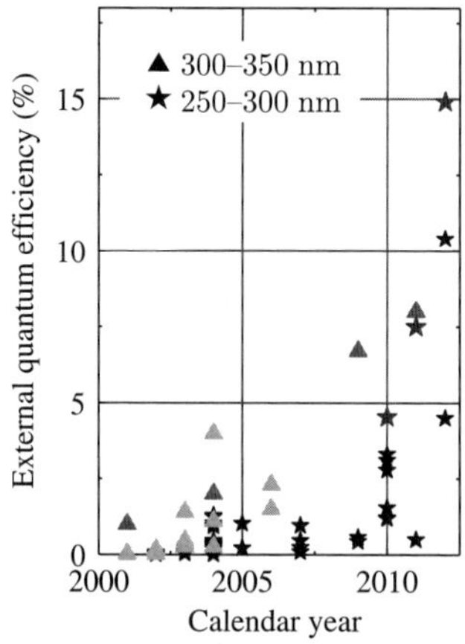

FIG. 1.2. Improvement of the external quantum efficiency of UV (300–350 nm) and DUV (250–300 nm) LEDs as a function of calendar year (as of October 2012).

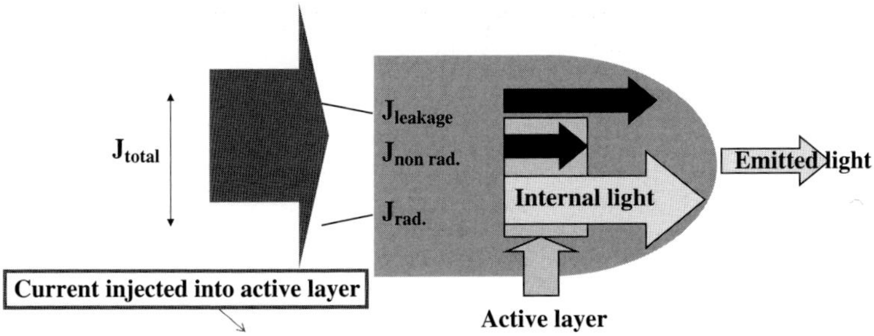

$$J_{total} = J_{rad.} + J_{non\ rad.} + J_{leakage}$$

$$\text{Internal Light Emission Efficiency } (\eta_{ILEE}) = \frac{J_{rad.}}{J_{total}} = \eta_{IQE} \times \eta_{IE}$$

$$\text{Internal Quantum Efficiency } (\eta_{IQE}) = \frac{J_{rad.}}{J_{rad.} + J_{rad.non}}$$

$$\text{Injection Efficiency } (\eta_{IE}) = \frac{J_{total} - J_{leakage}}{J_{total}} = \frac{J_{rad.} + J_{rad.non}}{J_{total}}$$

FIG. 1.3. Definition of each efficiency in this chapter, where J_{total}, $J_{rad.}$, $J_{non\ rad.}$, and $J_{leakage}$ denote total, radiative, non-radiative, and leakage current density, respectively.

that limit the EQE of LEDs: (1) internal quantum efficiency (η_{IQE}), (2) carrier injection efficiency (η_{IE}), and (3) light extraction efficiency (η_{LEE}). Figure 1.3 shows schematically the efficiency-limiting process of LEDs.

Among these factors, IQE is dependent on the radiative recombination rate of each material and the rate of non-radiative recombination, including Auger recombination. Theoretically, the radiative recombination coefficient of free excitons in AlN is larger than that for GaN or InN [8]. Therefore, AlGaN-based QWs should be more radiative or should have a higher IQE than InGaN-based QWs. One of the most important factors for improving IQE is the growth of AlGaN with low defect density. In the following section, the development of the growth of AlGaN and the relationship between IQE and defect density are described.

Carrier injection efficiency is strongly dependent on the hole concentration in the p-type cladding layer as well as the alloy composition. With increasing Al content in p-type AlGaN, the hydrogen-like effective mass acceptor activation energy becomes large. Therefore, the room-temperature hole concentration decreases with increasing Al content. Of course, hole concentration can be increased by increasing the acceptor concentration. However, the heavy doping of Mg causes serious problems. Therefore, there is an optimum Mg concentration for obtaining a high hole concentration. In the following section, the effect of a low hole concentration in a p-type cladding layer on the performance of UV/DUV LEDs is discussed.

The light extraction efficiency is the most important parameter that limits the performance of UV/DUV LEDs. In general, face-up or face-down configurations are used for LEDs. The face-up configuration requires a transparent electrode. Unfortunately, suitable electrode materials that are transparent in the UV/DUV region are uncommon. In contrast, the face-down configuration requires a highly reflective electrode. There are some suitable materials that are reflective in the DUV region. High contact resistivity to p-type AlGaN is one of the most serious issues. In the case of conventional Ni/Au, Indium-Tin-Oxide, or Ag-Pd-Cu electrodes, p-GaN, which is opaque to UV and DUV light, is required to realize low contact resistivity.

In the following section, issues regarding IQE that are related to the crystal growth of AlGaN, IE and LEE are discussed in more detail.

1.2 Efficiency limiting process

1.2.1 *Internal quantum efficiency*

The MOVPE growth of AlGaN alloys was first performed by Matloubian and Gershenzon in 1985 [9]. The parasitic-gas-phase reaction between an Al source such as trimethylaluminum (TMAl) and ammonia (NH_3) [10], the decomposition of TMAl in the gas phase to Al_4C_3 [11], or the very low rate of diffusion of an Al precursor in an NH_3-rich gas flow, causes a reduction in the growth rate and polymerization in the gas phase. Several groups have reported the effectiveness of alternative-source gas supplies or flow-rate modulation using methods such as pulsed-atomic-layer epitaxy [12], the alternative-source feeding technique [13], migration-enhanced MOCVD [14], and the NH_3 pulse-flow multilayer growth method [15]. An alternative-source supply was initiated by atomic-layer epitaxy [16], then modified by gas-flow-rate modulation during growth by MOVPE [17]. The effectiveness of these gas-flow-modulation techniques for the growth of AlN and the fabrication of UV and DUV LEDs is summarized in reviews [18, 19].

Another approach to the growth of high-quality AlN is to grow AlN under a relatively low NH_3 partial-pressure at a high temperature. In the case of GaN growth, the ratio of the molar fraction of group V source gases such as NH_3 to that of group III source gases such as trimethylgallium (TMGa), which is called the V/III ratio, is approximately 1000 or even higher. For InGaN growth, the V/III ratio should be much higher because of the high dissociation pressure of nitrogen in InN. In contrast, the V/III ratio during AlN growth can be lower [20, 21].

Recent experiments have shown that relatively high-quality AlN can be grown at a V/III ratio of as low as 10, which is two orders of magnitude lower than that for the growth of GaN. The growth temperatures of AlN and AlGaN are also an important issue. The sublimation temperature of AlN is much higher than that of GaN [22, 23]. Therefore, the epitaxial growth temperature of AlN is also expected to be higher than that of GaN.

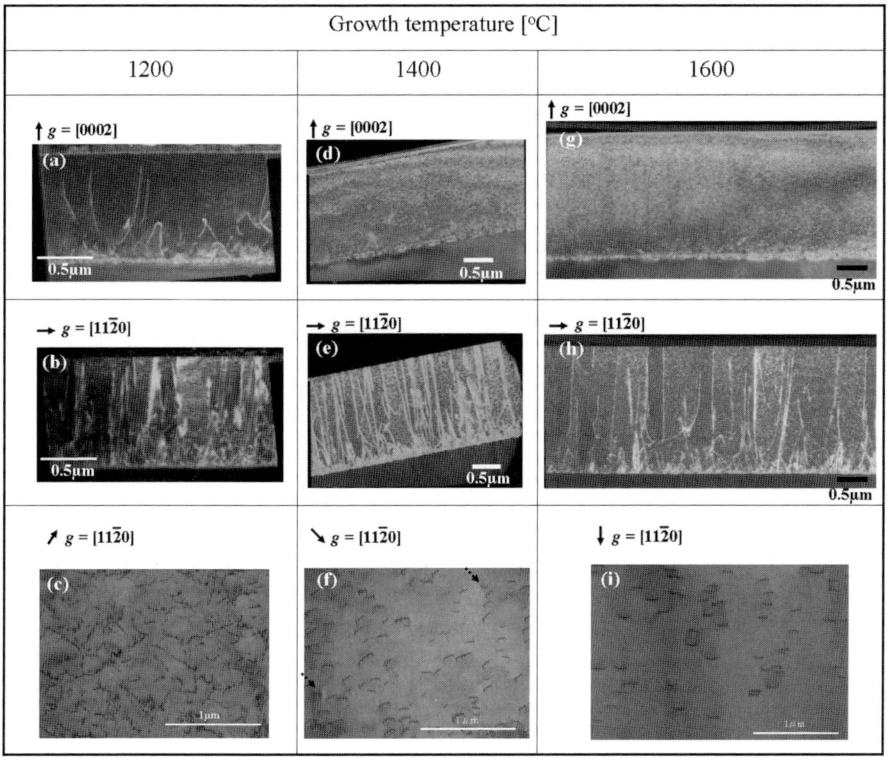

FIG. 1.4. TEM images of AlN grown at different Tg of (a)–(c) 1200°C, (d)–(f) 1400°C, and (g)–(i) 1600°C. The cross-sectional dark-field TEM images in the top row were taken under two-beam conditions with the incident beam $\boldsymbol{g}$// [0002] ((a), (d), and (g)). The TEM images in the middle row were taken with $\boldsymbol{g}$// [11–20] ((b), (e), and (h)). The TEM images on the bottom row show bright-field plan-view TEM images with $\boldsymbol{g}$// [11–20] ((c), (f), and (i)).

Figure 1.4 shows cross-sectional dark-field TEM images observed under two different incident beam conditions and bright-field plan-view images of MOVPE-grown AlN on a 6H-SiC substrate grown at different temperatures [24]. With increasing growth temperature, the screw-type dislocations almost disappeared, as shown in (g), and the density of edge-type dislocations was also much reduced, as in (i).

The AlN molar fraction of AlGaN is considered to be linearly dependent on the ratio of the flow rate of TMAl to the total flow rate of group III alkyls; that, is the sum of the flow rates of TMAl and TMGa. This linear relationship holds only at relatively low-growth temperatures. At high-growth temperatures, GaN desorption leads to the growth of high-AlN-molar-fraction AlGaN. The thermodynamic analysis conducted by Koukitsu *et al.* is very helpful in estimating the

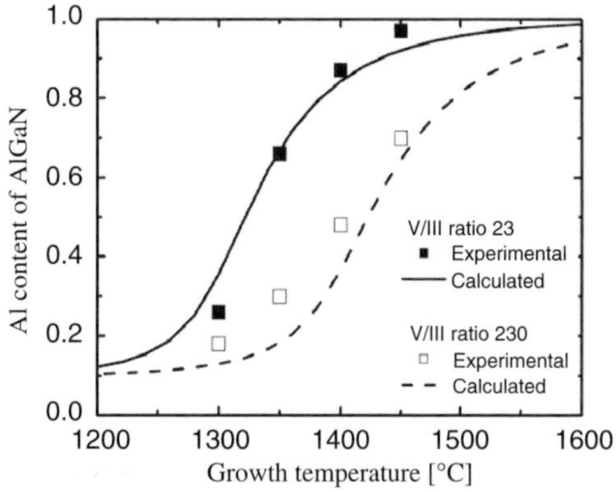

FIG. 1.5. Comparison of the experimental and calculated Al content of AlGaN.

vapor–solid relationship of AlGaN [25, 26]. Figure 1.5 shows the results of experimental and thermodynamic analysis of the relationship between the Al content in AlGaN and the growth temperature for two different V/III ratios under a fixed TMGa/(TMGa+TMAl) supply ratio [26]. The experimental results agree relatively well with the thermodynamic analysis.

To grow low-dislocation-density GaN, epitaxial lateral overgrowth (ELO) using a dielectric mask has been very successful. However, a dielectric mask cannot be applied to the selective growth of AlN-containing materials owing to the deposition of polycrystals on the mask. Therefore, AlGaN growth on a maskless, groove-patterned template has been investigated.

AlGaN growth on a groove-patterned GaN template in combination with a low-temperature-deposited AlN interlayer (LT interlayer) has been found to be very useful for the fabrication of UV LEDs and LDs with emission wavelengths longer than 350 nm [27], whereas AlGaN growth on a patterned AlN template is useful for the fabrication of UV/DUV LEDs with emission wavelengths shorter than 350 nm [28–30]. Figure 1.6 shows a cross-sectional TEM image of AlN grown at 1,450°C on a Si-face 6H SiC substrate [31]. There is a lattice mismatch between the AlGaN and the underlying AlN layer. Therefore, it is not easy to grow low-dislocation-density AlGaN on AlN even though dislocation density in underlying AlN is low.

Figure 1.7 summarizes the compositional dependence of the threading dislocation density (TDD) in AlGaN grown on an AlN template with a dislocation density of 10^7 cm^{-2} [32]. The lattice mismatch increases with increasing Ga content. Therefore, in the initial stage, the dislocation density increases with the Ga content. However, after a certain thickness is reached, island and lateral growth dominate, especially in high-Ga-content AlGaN, so the dislocation density tends to decrease. A grooved template is effective for further reducing the dislocation density.

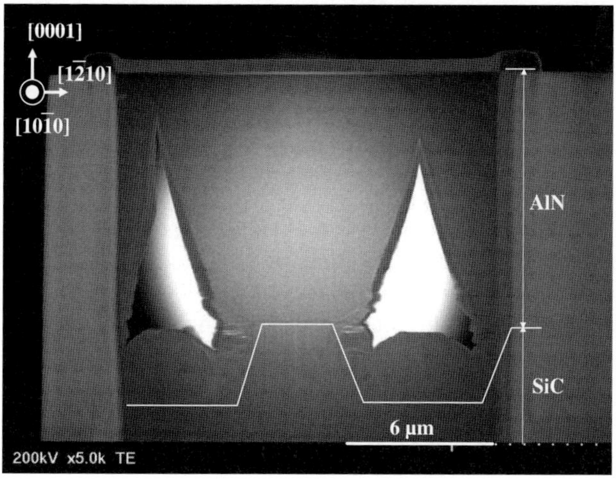

FIG. 1.6. Cross-sectional TEM image of AlN grown at 1,450°C on patterned 6H SiC substrate.

The IQEs of UV/DUV MQWs with different TDDs were characterized by measuring the excitation density dependence of photoluminescence (PL) using an ArF excimer laser with a wavelength of 193 nm as the excitation source, which is known as Shockley–Read–Hall (SRH) analysis [33–36]. In the case of UV/DUV devices, the band structure indicates that the ratio of Auger recombination

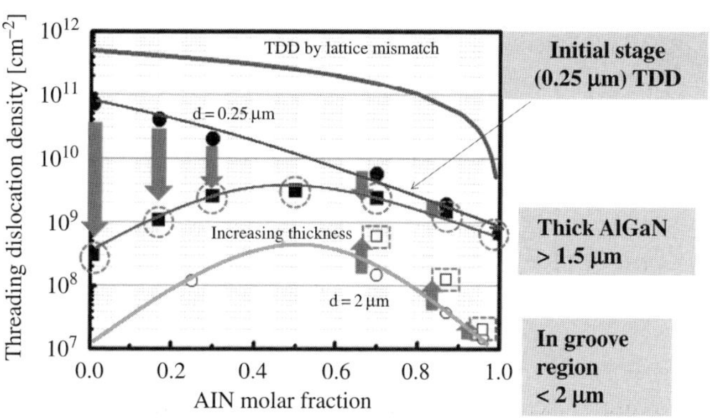

FIG. 1.7. Compositional dependence of AlGaN on planar AlN and grooved AlN. Upper line: TDD estimated from lattice mismatch. Solid circles: TDD in 0.25-μm-thick AlGaN on planar AlN. Solid squares: TDD in AlGaN of thickness greater than 2 μm on planar AlN. Open circles: TDD in 2-μm-thick AlGaN on grooved AlN. Solid squares: TDD in AlGaN thicker than 10 μm on grooved AlN.

should be negligibly small. When the generation rate (***G***) is equal to the total recombination rate (***R***), ***G*** = ***R*** and the PL integrated intensity are given by the following formulae.

$$G = R = An + Bn^2 \tag{1.1}$$

$$I_{PL} = \eta Bn^2 \tag{1.2}$$

Here, n is the excited carrier density, ***A*** is the non-radiative recombination constant, ***B*** is the radiative recombination constant, and η is a constant determined on the basis of the PL collection efficiency. Differences in the ***B*** value cause an error in the absolute IQE value. It is currently unclear which value of ***B*** should be used. However, the relative trend of the TDD dependence of IQE can be discussed regardless of the chosen value of ***B***.

By combining the above two formulas, ***G*** can be rewritten as

$$G = \frac{A}{\sqrt{B\eta}}\sqrt{I_{PL}} + \frac{1}{\eta}I_{PL} \tag{1.3}$$

Then, the non-radiative coefficient ***A*** is deduced by curve fitting, and the IQE can be expressed as

$$IQE = \frac{Bn^2}{An + Bn^2} \tag{1.4}$$

Figures 1.8 and 1.9 show examples of the relationship between the generation rate ***G*** and PL intensity (I_{PL}) (Fig. 1.8) and IQE as a function of n (Fig. 1.9) for QWs emitting 230 nm UV-C with different TDDs [37]. The same procedure was

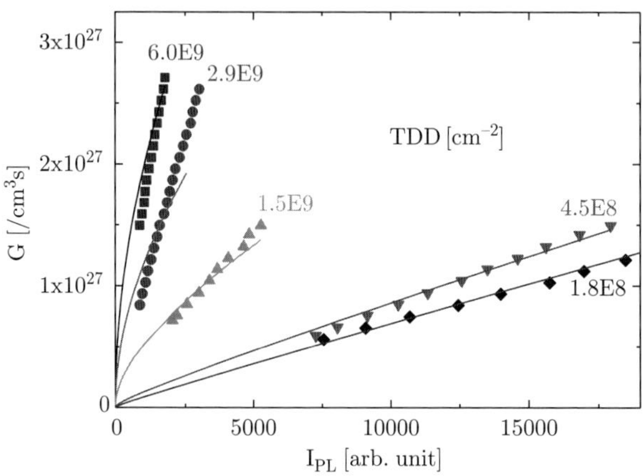

FIG. 1.8. Experimental results and fitting curves of the ***G*** and I_{PL} for 230 nm QWs with different TDDs. The non-radiative coefficient ***A*** is deduced from this fitting curve using eq. (1.3).

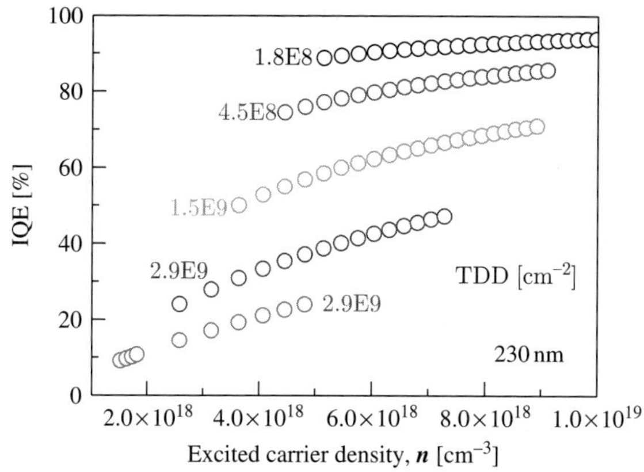

FIG. 1.9. Calculated IQE using eq. (1.4) as a function of $\boldsymbol{n}$ for samples shown in Fig. 1.8.

performed for every sample to deduce the non-radiative recombination coefficient $\boldsymbol{A}$. Figure 1.10 summarizes the TDD dependences of $\boldsymbol{A}$ for 230, 250, 300, and 350 nm QWs. Those of 420 to 495 nm c-plane QWs grown on free-standing GaN with very low dislocation density, and semipolar QWs on a patterned Si substrate are plotted for comparison. From the results, it is clear that $\boldsymbol{A}$ is linearly dependent on TDD, $\boldsymbol{A} \fallingdotseq 0.05 \times \text{TDD}\ (\text{cm}^{-2}) + 8.1 \times 10^6\ (\text{s}^{-1})$, and independent

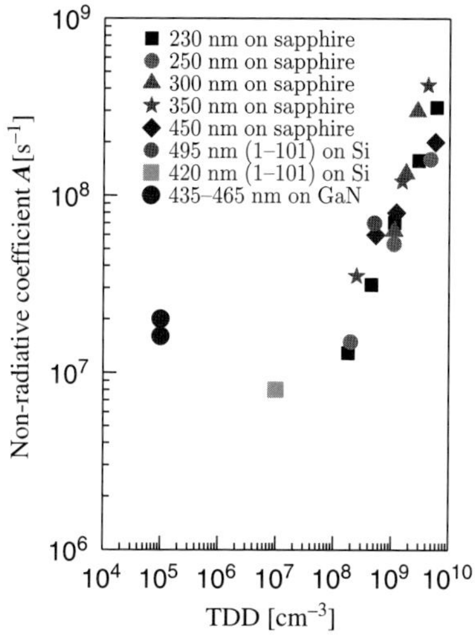

FIG. 1.10. TDD dependences of non-radiative coefficient $\boldsymbol{A}$ for QWs emitting different wavelengths from 230 to 495 nm.

of the emission wavelength. Again, it should be emphasized that this $\boldsymbol{A}$ is deduced assuming $\boldsymbol{B}$ to be $0.2 \times 10^{-10}\,\mathrm{cm^3/s}$. In this assumption, a non-radiative coefficient $\boldsymbol{A}$ of approximately $10^7\,\mathrm{s^{-1}}$ appears to be the limit of accuracy of the measurement system.

1.2.2 *Current injection efficiency*

One of the most important and serious issues in the realization of high-efficiency UV and DUV light emitters is conductivity control—particularly the control of p-type conductivity. Highly conductive p-type layers are necessary to improve the performance of UV/DUV emitters.

However, the control of p-type conductivity in $Al_xGa_{1-x}N$ has proven to be extremely difficult. With increasing Al content in $Al_xGa_{1-x}N$, the activation energy of Mg dopants increases as well as the resistivity of the alloys. In addition, although high-crystalline-quality $Al_xGa_{1-x}N$ films are necessary, it is difficult to grow them owing to the high concentration of dislocations in $Al_xGa_{1-x}N$ alloys. For this reason, it is very important to grow high-crystalline-quality $Al_xGa_{1-x}N$ and control the p-type conductivity of $Al_xGa_{1-x}N$. There have been several reports on the analysis of Mg-doped p-type $Al_xGa_{1-x}N$ by variable-temperature Hall measurements, PL and CL [38–41]. The analysis of Mg-doped p-type AlN is also one of the main challenges for the realization of high-efficiency UV and DUV devices.

To determine the activation energy of Mg in AlGaN, it is necessary to avoid other unwanted effects such as residual impurities and extended defects. High-temperature growth has been found to be effective for reducing the concentration of residual impurities. Figure 1.11 shows the concentrations of hydrogen, oxygen, silicon, and carbon in nominally undoped AlN grown at 1,100°C and 1,600°C [24]. As shown, the concentration of residual impurities is less than the detection limit of SIMS in HT-grown AlN.

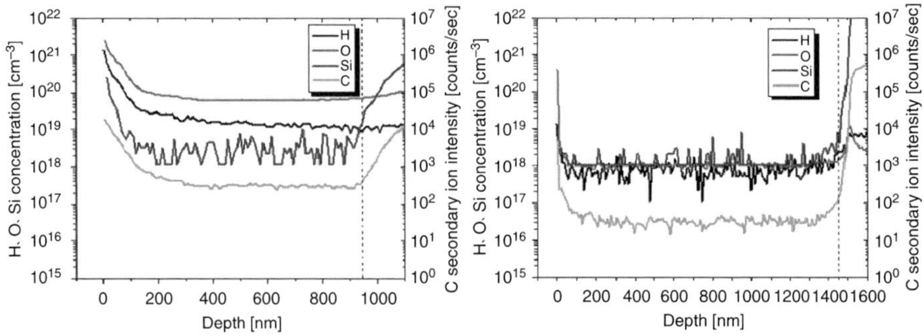

FIG. 1.11. SIMS depth profiles of AlN grown at 1,100°C (left) and 1,600°C (right) on 6H SiC.

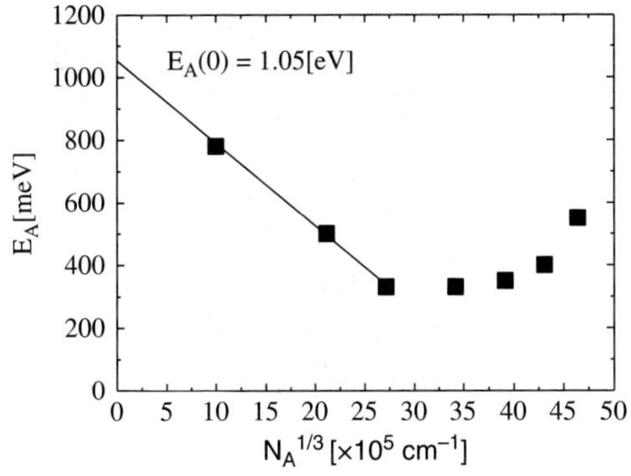

FIG. 1.12. Effective activation energy of Mg acceptor in $Al_{0.5}Ga_{0.5}N$ measured by temperature-dependent Hall effect measurement as a function of Mg concentration to the power of $-1/3$.

For Mg doping, Et-Cp_2Mg or Cp_2Mg is transported into the growth chamber during growth. The as-grown layer becomes highly resistive upon hydrogen passivation [42]. Post-growth annealing is usually conducted to activate the hydrogen-passivated Mg [43].

Figure 1.12 shows the effective acceptor activation energy of Mg in $Al_{0.5}Ga_{0.5}N$ as a function of the Mg concentration to the power of $-1/3$. The effective activation energy decreases with increasing Mg concentration up to $4 \times 10^{19}\,cm^{-3}$. Above this concentration, the activation energy increases again. The reason for the decrease in activation energy with increasing Mg concentration is the Coulomb interaction between negatively charged Mg ions and positively charged valence band holes [44, 45].

In contrast, the reason for the increase in activation energy above a Mg concentration of $4 \times 10^{19}\,cm^{-3}$ has not yet been clarified. Microscopic observation of highly Mg-doped GaN and AlN shows the existence of inversion domain boundaries. Figure 1.13 shows a cross-sectional TEM image of Mg-doped/undoped AlN on a sapphire (0001) substrate. Black dots can only be observed in the Mg-doped layer, which are thought to be inversion domain boundaries caused by Mg segregation at the surface. Similar dots have been reported to exist in heavily Mg-doped GaN [46, 47].

Therefore, the optimum Mg concentration can be deduced and the compositional dependence of the maximum hole concentration at room temperature can be estimated. As shown in Fig. 1.14, the expected hole concentration in AlN is only $3 \times 10^{12}\,cm^{-3}$ at room temperature.

In a p-type AlGaN with such a low hole concentration, current injection efficiency (η_{IE}) should be low. At present, there is no method of directly measuring η_{IE}. In this chapter, our attempts to estimate η_{IE} in a UV LD is explained. The sample discussed was reported in ref. [48] and its structure is depicted in

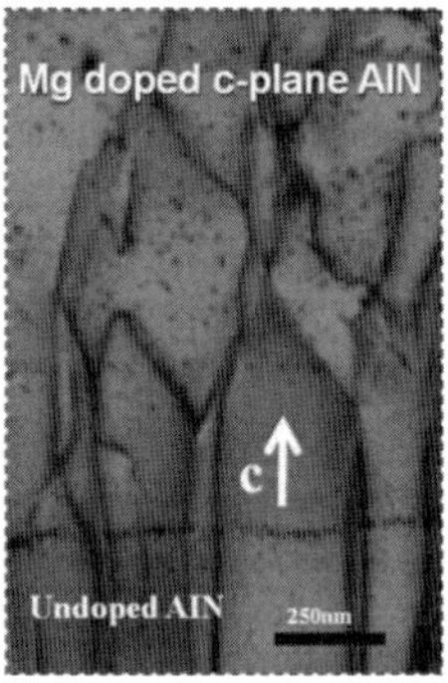

FIG. 1.13. Cross sectional TEM image of AlN:Mg/AlN. Black dots in the AlN:Mg show the inversion domain boundaries caused by Mg precipitation.

Fig. 1.15. LD characteristics are as follows: J_{th}: 8.0 kA/cm^2, n_{th}: 2.6×10^{19} cm^{-3}, $\boldsymbol{A} = 1.2 \times 10^9\,\mathrm{s}^{-1}$, $\boldsymbol{B} = 2.4 \times 10^{-11}\,\mathrm{cm}^3\,\mathrm{s}^{-1}$, η_{IQE} slightly below J_{th}: 34%, λ_{LD}: 354.2 nm. The Auger coefficient is thought to be negligible.

In this LD, a 20 nm p-type $Al_{0.5}Ga_{0.5}N$ electron blocking layer was grown on a p-type $Al_{0.1}Ga_{0.9}N$ waveguide layer to avoid waveguide loss. SRH measurement by current injection was conducted. In the case of PL measurement, the same MQW structure without the p-type layers was measured. Figure 1.16 shows a comparison of the efficiency measured by excitation-density-dependent PL and electroluminescence (EL) current injection. If we assume that η_{IQE} measured by PL gives the true η_{IQE} and that measured by EL gives the product of η_{IQE} and η_{IE}, we can estimate η_{IE}, which is about 30–35%.

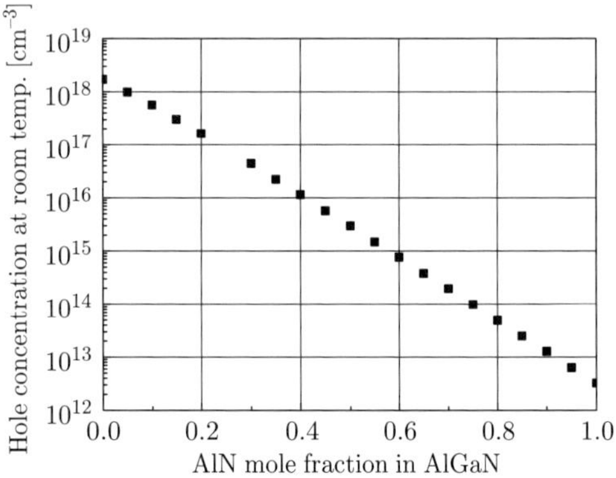

FIG. 1.14. Estimated maximum hole concentration at room temperature as a function of Al content. In this calculation, N_D is assumed to be one tenth of N_A.

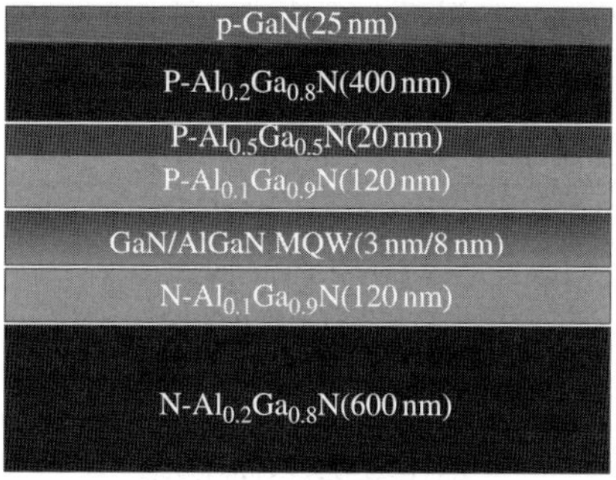

FIG. 1.15. UV LD structure used in this study.

In the case of DUV LEDs, the situation is more serious because the hole concentration in the p-type cladding decreases with increasing Al content. To overcome this problem, several approaches have been considered. One is the use of a multiple quantum barrier structure [49]. This idea was first investigated by Iga *et al.* [50] in 1.5-μm-range InGaAsP/InP and visible AlGaInP lasers. The maximum EQE achieved was 1.2% at an emission wavelength of 250 nm.

Another approach is the use of a polarization field. By using a compositionally graded p-type AlGaN layer on AlN with an Al face, polarization-induced hole doping is possible [51, 52].

Current state-of-the-art DUV LEDs have an EQE of 14.3% at 2 mA with an emission wavelength of 285 nm. In these DUV LEDs, η_{IE} has been estimated to be higher than 70%.

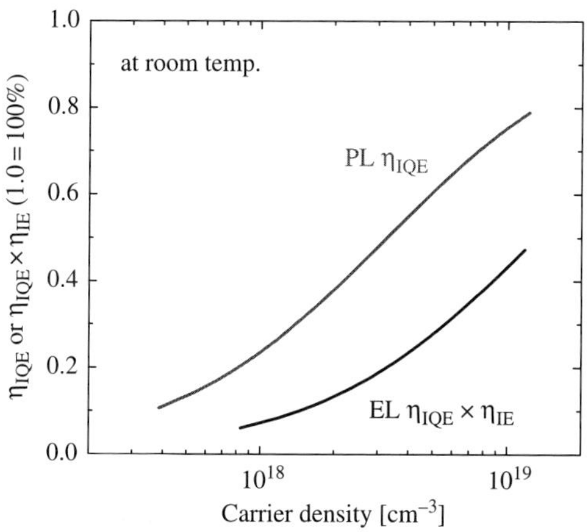

FIG. 1.16. SRH measurement of UV LD characterized by PL and EL.

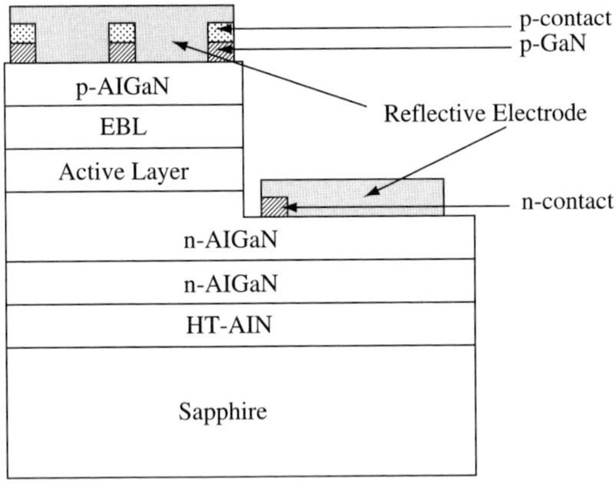

FIG. 1.17. Schematic structure of a DUV LED having a partial reflective electrode.

1.2.3 *Light extraction efficiency*

Compared with η_{IQE} and η_{IE}, η_{LEE} is still limited to a low value, as explained in the introduction. One solution is to use a partially reflective electrode in the face-down configuration [53]. Figure 1.17 shows schematically the structure. The output power was increased by 55% without a serious increase in the operating voltage.

1.3 Summary

The recent development of UV/DUV LEDs and UV LDs and the factors limiting their efficiency have been discussed. The reduction of TDD is essential to improve IQE, while the use of polarization charges is found to be effective in improving IE. The remaining issue is the LEE. The development of new DUV transparent electrode materials and/or DUV reflective materials with a large work function and a new design rule and concept for devices using these polarized wide-band-gap semiconductors are highly anticipated.

Acknowledgments

The author would like to thank Prof. M. Yamaguchi and Dr. Y. Honda, Nagoya University, and Profs. I. Akasaki, S. Kamiyama, T. Takeuchi, and M. Iwaya, Meijo University for fruitful discussions. The results presented in this chapter are based on a Grant-in-Aid for Scientific Research (A) "Super high temperature MOVPE of low dislocation density AlN-based semiconductors for novel device applications" conducted from FY2003 to FY2005, and a Grant-in-Aid for Scientific Research on Priority Areas entitled "Optoelectronics Frontier by

Nitride Semiconductor -Ultimate Utilization of Nitride Semiconductor Material Potential" sponsored by MEXT, conducted from FY2006 to FY2010 by Professor Yasushi Nanishi of Ritsumeikan University, as a delegate.

References

[1] S. Nakamura, S. Senoh, and T. Mukai, *Jpn. J. Appl. Phys.*, 32 (1993) L8.
[2] S. Nakamura, M. Senoh, S. Nagahama, N. Iwasa, T. Yamada, T. Matsushita, H. Kiyoku, and Y. Sugimoto, *Jpn. J. Appl. Phys.*, 35 (1996) L74.
[3] S. A. Inada, H. Amano, I. Akasaki, A. Morita, and K. Kobayashi, *Meijo Daigaku Sogo Kenkyusho Sogo Gakujutsu Kenkyu Ronbunshu*, 5 (2006) 73 (in Japanese).
[4] J. I. Pankove, *J. Lumin.*, 5 (1972) 482.
[5] H. Amano, M. Kito, K. Hiramatsu, and I. Akasaki, *Jpn. J. Appl. Phys.*, 28 (1989) L2112.
[6] Y. Taniyasu, M. Kasu, and T. Makimoto, *Nature*, 441 (2006) 325.
[7] H. Yoshida, Y. Yamashita, M. Kuwabara, and H. Kan, *Appl. Phys. Lett.*, 93 (2008) 241106.
[8] A. Dmitriev and A. Oruzheinikov, *J. Appl. Phys.*, 86 (1999) 3241.
[9] M. Matloubian and M. Gershenzon, *J. Electron. Mater.*, 14 (1985) 633.
[10] M. Morita, N. Uesugi, S. Isogami, K. Tsubouchi, and N. Mikoshiba, *Jpn. J. Appl. Phys.*, 20 (1981) 17.
[11] L. M. Yeddanapalli and C. C. Schubert, *J. Chem. Phys.*, 14 (1946) 1.
[12] M. A. Khan, R. A. Skogman, J. M. Van Hove, D. T. Olson, and J. N. Kuznia, *Appl. Phys. Lett.*, 60 (1992) 1366.
[13] Y. Ishihara, J. Yamamoto, M. Kurimoto, T. Takano, T. Honda, and H. Kawanishi, *Jpn. J. Appl. Phys.*, 38 (1999) L1296.
[14] Q. Fareed, R. Gaska, and M. S. Shur, U.S. Patent, 7 192849, 20 March, 2007.
[15] H. Hirayama, T. Yatabe, N. Noguchi, T. Ohashi, and N. Kamata, *Appl. Phys. Lett.*, 91 (2007) 071901.
[16] M. Ahonen, M. Pessa, and T. Suntola, *Thin Solid Films*, 65 (1980) 301.
[17] N. Kobayashi, T. Makimoto, and Y. Horikoshi, *Jpn. J. Appl. Phys.*, 24 (1985) L962.
[18] A. Khan, K. Balakrishnan, and T. Katona, *Nature Photonics*, 2 (2008) 77.
[19] M. S. Shur, *IEEE Trans. Elec. Devices*, 57 (2010) 12.
[20] Y. Ohba and A. Hatano, *Jpn. J. Appl. Phys.*, 35 (1996) L1013.
[21] M. Imura, K. Nakamo, N. Fujimoto, N. Okada, K. Balakrishnan, M. Iwaya, S. Kamiyama, H. Amano, I. Akasaki, T. Noro, T. Takagi, and A. Bandoh, *Jpn. J. Appl. Phys.*, 45 (2006) 8639.
[22] I. Grzegory, J. Jun, St. Krukowski, M. Bockowski, and S. Porowski, *Physica B*, 185 (1993) 99.
[23] O. Ambacher, M. S. Brandt, R. Dimitrov, T. Metzger, M. Stutzmann, R. A. Fischer, A. Meihr, A. Bergmaier, and G. Dollinger, *J. Vac. Sci. and Technol.*, B14 (1996) 3532.
[24] M. Imura, H. Sugimura, N. Okada, M. Iwaya, S. Kamiyama, H. Amano, I. Akasaki, and A. Bandoh, *J. Crystal Growth*, 310 (2008) 2308.

[25] A. Koukitu, N. Takahashi, and H. Seki, *Jpn. J. Appl. Phys.*, 36 (1997) L1136.

[26] N. Okada, N. Fujimoto, T. Kitano, G. Narita, M. Imura, K. Balakrishnan, M. Iwaya, S. Kamiyama, H. Amano, I. Akasaki, K. Shimono, T. Noro, T. Takagi, and A. Bandoh, *Jpn. J. Appl. Phys.*, 45 (2006) 2502.

[27] S. Kamiyama, M. Iwaya, S. Takanami, S. Terao, A. Miyazaki, H. Amano, and I. Akasaki, *Phys. Stat. Sol. (a)*, 192 (2002) 296.

[28] K. Nakano, M. Imura, G. Narita, T. Kitano, Y. Hirose, N. Fujimoto, N. Okada, T. Kawashima, K. Iida, K. Balakrishnan, M. Tsuda, M. Iwaya, S. Kamiyama, H. Amano, and I. Akasaki, *Phys. Stat. Sol. (a)*, 203 (2006) 1632.

[29] K. Iida, T. Kawashima, M. Iwaya, S. Kamiyama, H. Amano, I. Akasaki, and A. Bandoh, *J. Crystal Growth*, 298 (2007) 265.

[30] R. Jain, W. Sun, J. Yang, M. Shatalov, X. Hu, A. Sattu, A. Lunev, J. Deng, I. Shturm, Y. Bilenko, M. Shur, and R. Gaska, *Appl. Phys. Lett.*, 93 (2008) 051113.

[31] K. Balakrishnan, A. Bandoh, M. Iwaya, S. Kamiyama, H. Amano, and I. Akasaki, *Jpn. J. Appl. Phys.*, 46 (2007) L307.

[32] T. Mori, K. Nagamatsu, K. Nonaka, K. Takeda, M. Iwaya, S. Kamiyama, H. Amano, and I. Akasaki, *Phys. Stat. Sol.*, C6 (2009) 2621.

[33] W. Shockley and W. T. Read Jr., *Phys. Rev.*, 87 (1952) 835.

[34] H.-Y. Ryu, H.-S. Kim, and J.-I. Shim, *Appl. Phys. Lett.*, 95 (2009) 081114.

[35] Y. C. Shen, G. O. Müller, S. Watanabe, N. F. Gardner, A. Munkholm, and M. R. Krames, *Appl. Phys. Lett.*, 91 (2007) 141101.

[36] Q. Dai, M. F. Schubert, M. H. Kim, J. K. Kim, E. F. Schubert, D. D. Koleske, M. H. Crawford, S. R. Lee, A. J. Fischer, G. Thaler, and M. A. Banas, *Appl. Phys. Lett.*, 94 (2009) 111109.

[37] K. Ban, J. Yamamoto, K. Takeda, K. Ide, M. Iwaya, T. Takeuchi, S. Kamiyama, I. Akasaki, and H. Amano, *Appl. Phys. Exp.*, 4 (2011) 052101.

[38] J. Li, T. N. Oder, M. L. Nakarmi, J. Y. Lin, and H. X. Jiang, *Appl. Phys. Lett.*, 80 (2002) 1210.

[39] N. Nepal, M. L. Nakarmi, K. B. Nam, J. Y. Lin, and H. X. Jiang, *Appl. Phys. Lett.*, 85 (2004) 2271.

[40] L. Nakarmi, K. H. Kim, K. Zhu, J. Y. Lin, and H. X. Jiang, *Proc. of SPIE*, 5530 (2004) 7.

[41] H. Yu, E. Ulker, and E. Ozbay, *J. Crystal Growth*, 289 (2006) 419.

[42] J. A. Van Vechten, J. D. Zook, R. D. Horning, and B. Goldenberg, *Jpn. J. Appl. Phys.*, 31 (1992) 3662.

[43] S. Nakamura, T. Mukai, M. Senoh, and N. Iwasa, *Jpn. J. Appl. Phys.*, 31 (1992) L139.

[44] W. Götz, R. S. Kern, C. H. Chen, H. Liu, D. A. Stegerwald, and R. M. Flecher, *Mater. Sci. Eng.*, B59 (1999) 211.

[45] P. Kozodoy, H. Xing, S. P. DenBaars, U. K. Mishra, A. Saxer, R. Perrin, S. Elhamri, and W. C. Mitchel, *J. Appl. Phys.*, 87 (2000) 1832.

[46] Z. Liliental-Weber, M. Benamara, J. Washburn, I. Grzegory, and S. Porowski, *Phys. Rev. Lett.*, 83 (1999) 2370.

[47] P. Vennegues, M. Benaissa, B. Beaumont, E. Feltin, P. De Mierry, S. Dalmasso, M. Leroux, and P. Gibart, *Appl. Phys. Lett.*, 77 (2000) 880.

[48] H. Yoshida, M. Kuwabara, Y. Yamashita, K. Uchiyama, and H. Kan, *Appl. Phys. Lett.*, 96 (2010) 211122.

[49] H. Hirayama, Y. Tsukada, T. Maeda, and N. Kamata, *Appl. Phys. Exp.*, 3 (2010) 031002.

[50] K. Iga, H. Uenohara, and F. Koyama, *Elec. Lett.*, 22 (1986) 1008.

[51] J. Simon, V. Protasenko, C. Lian, H. Xing, and D. Jena, *Science*, 327 (2010) 60.

[52] L. Zhang, K. Ding, J. C. Yan, J. X. Wang, Y. P. Zeng, T. B. Wei, Y. Y. Li, B. J. Sun, R. F. Duan, and J. M. Li, *Appl. Phys. Lett.*, 97 (2010) 062103.

[53] T. Inazu, S. Fukahori, C. Pernot, M. H. Kim, T. Fujita, Y. Nagasawa, A. Hirano, M. Ipponmatsu, M. Iwaya, T. Takeuchi, S. Kamiyama, M. Yamaguchi, Y. Honda, H. Amano, and I. Akasaki, *Jpn. J. Appl. Phys.*, 50 (2011) 122101.

2 The homoepitaxial challenge: GaN crystals grown at high pressure for laser diodes and laser diode arrays

Izabella Grzegory, Michal Bockowski, Piotr Perlin, Czeslaw Skierbiszewski, Tadeusz Suski, Marcin Sarzynski, Stanislaw Krukowski, and Sylwester Porowski

2.1 Introduction

GaN and its alloys with InN and AlN are at present an extremely important group of semiconductors. Due to their outstanding physical properties GaN-related compounds are basic for various applications of fundamental importance. In particular, GaN, AlN, and InN are direct band-gap highly stable crystals and can thus be used as very efficient light sources. Moreover, their energy gaps of 0.65, 3.5, and 6.2 eV for InN, GaN, and GaN respectively, cover an extremely large spectral range including the whole visible part of the spectrum.

On the other hand, technology of this important group of materials started its development only in last two decades of twentieth century, though the first GaN-based light-emitting diode (LED) had already been demonstrated by Pankove *et al.* (1971). A set of fundamental problems were, and still are, related to a lack of single-crystalline high-quality native GaN substrates, p-type doping, lattice mismatch between GaN and its ternaries with In and Al. All these problems were a kind of challenge, and were solved by ways which are not typical for other semiconductors such as Si or GaAs.

The main barrier for obtaining large high-quality bulk crystals of GaN is related to thermodynamic properties of this compound, and in particular its extreme melting conditions and very high decomposition pressure of nitrogen at elevated temperatures necessary for crystallization (see Section 2.2). Therefore, as substrates for epitaxy, highly mismatched sapphire crystals with low-temperature buffer (nucleation) layers of AlN (Amano *et al.*, 1986) or GaN (Nakamura *et al.*, 1992) have been used to obtain device-quality material. The p-type doping was first achieved by electron irradiation of Mg-doped GaN (Amano *et al.*, 1989). Otherwise, the crystal was semi-insulating due to passivation of Mg acceptor with hydrogen.

Sapphire-based heteroepitaxy allowed construction of efficient LEDs (Amano *et al.*, 1989) despite a very high dislocation density of 10^8–10^{10} cm^{-2} in the epitaxial structures. For laser diodes (LDs) a further improvement of the material quality was necessary (Nakamura and Fasol, 1997), and was achieved by epitaxial lateral overgrowth (ELOG), still on patterned sapphire substrates (Usui *et al.*, 1997).

First results confirming quite obvious advantages of the use of low dislocation density bulk GaN crystals came in the late 1990s (for a review see Grzegory and Porowski, 2002 and Section 2.4). The substrate crystals were a few-millimeters-large GaN platelets grown from solution in liquid gallium at very high pressure of nitrogen (1–2 GPa) (Grzegory *et al.*, 1995; Grzegory *et al.*, 2005).

Despite their small size these crystals contained an extremely low number (<100 cm^{-2}) of dislocations, so properties of GaN homoepitaxial layers such as straight atomic step flow morphology, extremely narrow (<0.1 meV) exciton peaks in the PL spectra and very slow decay of the PL due to suppression of the non-radiative recombination, have been reported. The strongly n-type pressure grown GaN crystals have been tested by Nakamura (1999) as substrate for the InGaN 405 nm laser. The cw 30 mW devices with a lifetime exceeding 3000 h were then constructed. The 30 mW power has been achieved at a current of 62 mA. For typical InGaN MQW LDs on sapphire, at a current of 62 mA, the output power was about 15 mW. This increase in efficiency confirmed that dislocations were the main limiting factors for high-power GaN-based lasers.

At present, the great majority of GaN-based laser diodes is being produced on GaN free-standing single crystalline substrates. However, these substrates are not sliced from perfect bulk crystals as in typical cases, but grown by hydride vapor phase epitaxy (HVPE) on foreign (again!) material (mostly GaAs) by a "wafer by wafer" approach, which means a single wafer of GaN obtained on a single GaAs (or sapphire) substrate. Moreover, the method based on GaAs heteroepitaxy developed by Sumitomo Electric Industries (2010; Motoki 2010) includes special procedures (dislocation elimination by epitaxy on inversed pyramids—DEEP or A-DEEP) to obtain areas in the substrate with dislocation densities sufficiently low for high-power laser diodes. The crystals are therefore intentionally non-uniform (they contain domains of inversed polarity), and the LDs have to be positioned in the low-dislocation-density sectors.

The free-standing GaN wafers grown by HVPE and removed from sapphire (i.e. Oshima *et al.*, 2010) are uniform but with a dislocation density of the order of 10^6 cm^{-2} and with crystalline planes deformed (bowed) due to a large lattice and thermal mismatch between sapphire and GaN.

It is worth noticing that the bulk growth of GaN by HVPE on the GaN free-standing substrates is a kind of homoepitaxial challenge in itself. There are no reports of the successful growth of a bulk GaN crystal thicker than 5–6 mm by HVPE, despite the method being highly advanced.

In this chapter the use of GaN crystals grown by both high nitrogen pressure solution (HNPS) and ammonothermal method for laser diodes and laser diode arrays is discussed. The pressure-grown crystals are of high structural perfection, and can be of high electric conductivity and high degree of uniformity, which is especially important for integrated multi-laser devices. In Section 2.2, thermodynamic properties of GaN, including melting conditions, are described. Then, in Section 2.3, the methods of GaN crystallization relevant to this chapter, i.e. the HNPS and the ammonothermal ones, are presented. The HVPE as a leading method is also discussed, for reference.

In Section 2.4 some of the most important features of homoepitaxy by both MOVPE (metal organic vapor phase epitaxy) and MBE (molecular beam epitaxy) on the pressure-grown crystals are shown, with special emphasis on the results obtained by plasma-assisted MBE (PA MBE), which is not common for GaN-based technology dominated by MOVPE, but provides very encouraging results if the pressure-grown crystals are used as substrates.

Section 2.5 is devoted to the development of nitrides epitaxy by PA MBE towards fabrication of LDs and laser diode arrays over a large spectral range. In Section 2.6 an advantage related to the very high free electron concentration possible in the pressure-grown GaN crystals is discussed. The physical basis for this advantage is a refractive index contrast between the very high and lower conductivity GaN. Section 2.7 shows another possibility following from high uniformity and flat crystalline planes in the pressure-grown crystals: the lateral patterning enabling local variation of the substrate miscut, thus the arrays of lasers emitting light with different wavelengths. Section 2.8 presents the state-of the art results on high-power LDs and laser diode arrays based on epitaxial structures grown on both ammonothermal and HNPS GaN substrates.

2.2 Thermodynamics of GaN

2.2.1 *Melting conditions of GaN*

Aluminum, gallium, and indium nitrides are strongly bonded compounds, creating relatively low-density crystalline lattices in which the metallic atoms are surrounded by four nearest neighbors (nitrogen atoms). Such structure is stabilized by interatomic interactions strongly dependent on the mutual orientation of the triples of the metal-nonmetal atoms. Such interactions are covalent in nature, i.e. they are created by the dominant overlap of the wavefunction of the neighboring atoms. In addition, the difference of the energies of the external orbitals of metal and nitrogen atoms is sufficiently large that it contributes to ionic fraction of the bonding energy.

Generally, melting transitions entail disappearance of the long- and short-range order. In order to destroy a strongly bonded lattice, thermal agitation has to be high, therefore melting of these compounds occurs at relatively high temperatures. The demolition of the covalent bonding closes the band gap, therefore molten homogeneous nitrides should be metallic, similar to other III-V semiconductors.

In contrast to the above, the presence of nitrogen shows the nitrides as qualitatively different from other elemental or compound III–V and II–VI semiconductors. Nitrogen has an exceptional tendency to form very strongly bonded diatomic molecules; in fact, the N_2 molecule is the strongest bonded diatomic molecule of all, with a bond energy equal to 9.72 eV (Barin, 1994). It is therefore expected that nitrogen, liberated from the crystalline lattice of the nitrides, would have a strong tendency to form molecules. Nitrogen in molecular form resembles noble gases, and interacts very weakly with other nitrogen molecules (Strak and Krukowski, 2007). In good approximation this interaction may be represented by pair-wise potential (Bohm and Ahlrichs, 1985). Nitrogen critical-point parameters, both temperature (126.19 K) and pressure (3.3978 MPa) (Jacobsen *et al.*, 1986), are much lower than the expected values of these parameters at the melting points of the nitrides, therefore under these conditions pure nitrogen is a supercritical fluid. The molecular structure is very difficult to atomize, as shown by the results of shock-wave compression experiments where at about 33 GPa and 6000 K, nitrogen underwent continuous molecular to non-molecular transition (Nellis *et al.*, 1984). Low-temperature diamond anvil data indicated that at room temperature nitrogen remains in molecular structure up to 130 GPa (Reichlin *et al.*, 1985). These results were confirmed by the demonstration that nitrogen transforms to an amorphous phase, most likely a narrow band-gap semiconductor, at 70–250 GPa and 10–510 K (Gregoryanz *et al.*, 2001; Goncharov *et al.*, 2000; Eremets *et al.*, 2001). DFT calculations confirmed that nitrogen is stable in the molecular form up to pressures of 70 GPa (McMahan and LeSar, 1985; Martin and Needs, 1986). Therefore, it is expected that pure nitrogen remains in the molecular form at pressures and temperatures well above the melting parameters of the nitrides.

Group III metals are different from nitrogen in all possible aspects. In the condensed phase they are metals bonded by uniform attraction of free electrons and positive ions. Their angular dependence of the interaction is weak, so their crystalline lattices are not rigid and their melting temperatures are relatively low, equal to 933.47 K, 302.9146 K, and 429.7485 K for aluminum, gallium, and indium respectively (Glushko, 1979). In contrast, the radial component is very strong, leading to high enthalpies of evaporation, equal to 3.32 eV/atom, 2.81 eV/atom, and 2.50 eV/atom for these substances (Barin, 1994; Glushko, 1979). Accordingly, their vapor pressures are very low and the boiling-point (0.1 MPa) temperatures are high, equal to 2792 K, 2477 K, and 2345 K, for these three metals respectively. It is therefore natural to assume that at the expected nitride melting temperatures, pure Al, Ga, and In are metallic liquids.

The interaction of the group III metals and nitrogen is relatively strong. Nevertheless, little is known about its magnitude and other characteristics. An order of magnitude may be deduced from DFT simulations of nitrogen adsorption on the III group metal surfaces that showed strong catalytic effect, leading to dissociation of molecular nitrogen at adsorption on these metals (Romanowski *et al.*, 2001). The catalytic influence lowers the N_2 dissociation energy barrier from 9.81 eV to much below 4 eV, proving a strong interaction between In, Ga, Al, and nitrogen. At the surface, the nitrogen atoms are attached in stable (Al) or metastable (Ga and In) positions. It has to be noted that these calculations assumed negligible solubility of the nitrogen, as observed during growth from solution in technically accessible nitrogen pressures, i.e. up to 1.5 GPa (Grzegory *et al.*, 2001). From these calculations we could infer that at low concentration nitrogen dissociate, in metal liquids remaining in the atomic form. Unfortunately, no information on the physical properties of the mixture of metals and higher nitrogen contents is available. It is not clear whether nitrogen and metal separate into nitrogen and metal-rich fluids or create a solution of atomic metals and nitrogen. These issues remain to be solved in the future.

The above-described complex scenario was not accounted for in the early attempts to solve the mystery of nitrides melting. Two approaches were used: dynamic and thermodynamic. Dynamic theory was based on the Lindemann criterion of melting, according to which, the melting occurs when the amplitude of thermal lattice vibrations exceeds a specified x_0 fraction of the interatomic spacing (atomic radius) (Lindemann, 1910), which relates the melting temperature T_M, the Debye temperature Θ, and the atomic mass M:

$$T_M = \frac{k_B x_0^2 d^2 M \Theta^2}{9\hbar^2}. \tag{2.1}$$

Usually, the atomic fraction x_0 ranges from 0.2 to 0.25 (Ziman, 1964). The application of the dynamic criterion to the melting of semiconductors was criticized by Van Vechten (1972), who claimed that the approach is deficient because the liquid phase is not involved. For GaN, melting temperature evaluated with eq. (2.1) is 2291 K.

Van Vechten invoked a thermodynamic approach to the melting of semiconductors in which the melting temperature is expressed as the ratio of the enthalpy (ΔH^F) and entropy (ΔS^F) of fusion:

$$T_M = \frac{\Delta H^F}{\Delta S^F}, \tag{2.2}$$

using these parameters for silicon as a base for scaling for other compounds (Van Vechten, 1972). Enthalpy was initially scaled with the Fermi energy, while the entropy, derived as the sum of the contribution from melting of the metallic phase, mixing and dissociation of covalent bonds, was assumed to be constant. In the subsequent more sophisticated approach, the enthalpy of fusion was calculated as the sum of the metallic phase melting contribution, the volume term, and the energy of dissociation of covalent bonds (Van Vechten, 1973). The second term was responsible for the pressure-dependence of the melting temperature. Since it was assumed that a tetrahedrally coordinated semiconductor was less dense, thus following the Clausius–Clapeyron relation, the pressure coefficient of the melting temperatures of all nitrides was determined as negative. The last term assumed sp^3 hybridization of the bonding of the nitrides, which turned out to be justified for AlN only. Recent results obtained from soft X-ray spectroscopy (XPS) and density functional theory (DFT) simulations showed that bonding in GaN proceeds via N 2s–Ga 3d and N 2p–Ga 4s, 4p separately (Magnuson *et al.*, 2010). Thus the basic assumptions of the theory were put into the question. The entropy was taken using an earlier introduced dependence (Van Vechten, 1973). Using this approach, Van Vechten derived the temperature of fusion to be 3487 K, 2791 K, and 831 K for AlN, GaN, and InN respectively. Later, the melting temperature of indium nitride was corrected to 2200 K (Grzegory, 1995).

High-pressure experiments were used for the determination of melting of GaN detected by *in situ* X-ray diffraction (Utsumi *et al.*, 2003). The results of the measurements indicated that GaN was stable at pressures above 6.0 GPa at temperatures up to 2493 K. At higher temperatures the crystal was melted, and crystallized when cooled down, so that small single-crystalline samples were obtained. It was concluded that GaN melts congruently at 2493 K. The pressure dependence of the melting temperature was also determined, showing that the pressure-related variation of the melting temperature was very small, below the precision of the temperature measurements, for pressures increased up to 10 GPa. Another high-pressure experiment brought a different set of data: at pressure of 7.5 GPa GaN was stable up to 2573 K, and unstable at the temperature of 100 K higher (Sokol *et al.*, 2007). Thus it was concluded that the GaN melting pressure is about 7.5 GPa and the temperature is located within 2573—2673 K. In addition, the X-ray measurements reported in Sokol *et al.* (2007) indicated that gallium nitride melts incongruently at this pressure range. It was conjectured that the transition to congruent melting occurs at even higher pressures (Sokol *et al.*, 2007). In summary, these experiments showed that GaN melting occurs at very high pressures above 6.0 GPa and temperatures close to 2500 K.

The nature of the transition is still undetermined, as it is not clear whether molten gallium and nitrogen remain in atomic state or are separated into the atomic gallium liquid and molecular supercritical nitrogen fluid.

Numerical simulations may potentially shed more light on the issue of nitride melting. In absence of extensive experimental data, they are particularly useful, despite formidable potential difficulties. They are related to drastic change of the properties of the coexisting phases and complex interactions between atoms, and possible molecules.

In the molecular dynamics (MD) simulations of GaN melting transition the two main approaches were based either on two-body or three-body interactions. The two-body interactions were used by Harafuji *et al.* (2004). The authors applied extensive MD simulations and obtained two different pressure dependences of the melting temperature:

$$T_M = \begin{cases} 2538 + 177 \times P - 4.62 \times P^2 & P < 8\text{GPa} \\ 2825 + 210 \times P - 5 \times P^2 & P > 9\text{GPa} \end{cases} \tag{2.3}$$

with the discontinuity in the vicinity of 8–9 GPa. The melting point was identified at $T_M = 2500$–2550 K (at normal pressure). Additionally, an extensive metastability region was obtained during cooling. Two-body potential was also used by Moon *et al.* (2007) in MD simulations from which the melting temperature was estimated as 3080 K. Three-body Tersoff empirical potential (Tersoff, 1988, 1989) was applied by Moon and Hwang (2003), who modeled melting of zincblende GaN obtaining $T_M = 2302 \pm 330$ K. Finally, Wang and Zhu (2006) used Stillinger–Weber potential (Stillinger and Weber, 1985) and obtained melting temperature of wurtzite GaN: $T_M = 4200$ K, much above the other estimates.

In summary, melting of nitrides remains one of the great unresolved issues of condensed-matter physics, important for applications and for basic understanding of the matter. Neither the existence or the character of the transition is positively identified. High-pressure experiments are necessary to determine the range of possible parameters. It is also worth noting that MD simulations using many-body potential may not be suitable for verifying the character of the transition, and that massively parallel *ab initio* MD simulation are the best way to solve the issue.

Table 2.1 compares melting temperatures and pressures of most typical semiconductor materials. The melting conditions of GaN are derived from Utsumi *et al.* (2003) where, as already mentioned, they have been determined experimentally by the *in situ* X-ray high-pressure high-temperature measurements. The table shows that both temperature and pressure at melting of GaN are much higher than for typical semiconductors. They are rather similar to the conditions used for high-pressure synthesis of diamond.

2.2.2 *High-pressure thermodynamics of GaN: phase diagrams*

As directly reflected in the high value of melting temperature, gallium nitride is a strongly bonded compound (with a bonding energy of 9.12 eV/atom pair (Melnik *et al.*, 1998)) in comparison with typical III–V semiconductors such as GaAs (bonding energy of 6.5 eV/atom pair). Consequently, the free energy of the crystal is very low in relation to the reference state of free N and Ga atoms. On the other hand, the N_2 molecule is also strongly bonded (4.9 eV/atom (Barin, 1994)). Therefore, the free energy of GaN constituents at their normal states, Ga and

Table 2.1 Melting conditions of semiconductors.

crystal	T_M,° C,	p_M, atm.
Si	1400	<1
GaAs	1250	15
GaP	1465	30
GaN	**2220**	**60000**
diamond (synthesis)	1600	60000

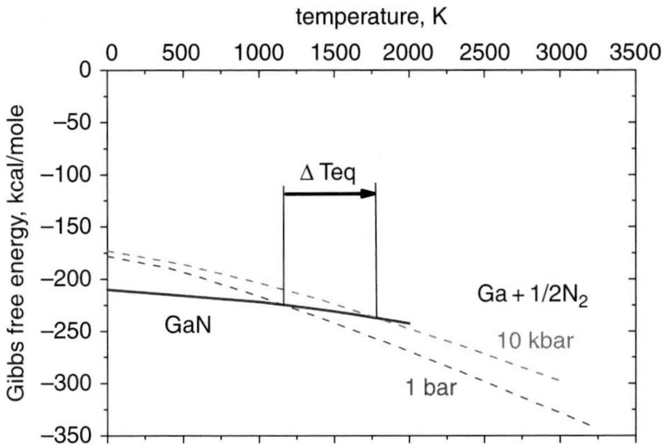

FIG. 2.1. Gibbs free energy of GaN and its constituents.

N_2, becomes quite close to that of the crystal. Figure 2.1 shows where the free energy of GaN (1 mole) and the free energy of the system of its constituents ($Ga + 1/2N_2$) is shown as a function of temperature and N_2 pressure. With increasing temperature, G(T) of the constituents decreases faster than G(T) of the crystal, and at higher temperatures the nitride becomes thermodynamically unstable. The crossing of G(T) curves determines the equilibrium temperature where GaN coexists with its constituents at given N_2 pressure. The application of pressure increases the free energy of the constituents to a much higher degree than G(T) of the crystal. As a consequence, the equilibrium point shifts to higher temperatures and the GaN stability range extends.

The equilibrium p_{N_2}-T conditions for GaN have been studied by several groups (Karpinski and Porowski, 1984; Karpinski *et al.*, 1984; MacChesney *et al.*, 1970; Thurmond and Logan, 1972; Madar *et al.*, 1975; Unland *et al.*, 2003). The most complete and consistent results have been obtained by Karpinski and Porowski (1984) and Karpinski *et al.* (1984) by direct synthesis and by decomposition experiments performed by both the gas-pressure technique (for pressures up to 20 kbar) and high-pressure anvil technique (up to 70 kbar). The curve following from these data is shown in Fig. 2.2. In the temperature range of 1100–1690°C (corresponding to the pressure range of 0.1–20 kbar), the curve was described by the van't Hoff equation:

$$\Delta H_F = \frac{1}{2}R\frac{d\log a_{N_2}}{d(1/T)}, \tag{2.4}$$

where ΔH_F is the formation enthalpy of GaN, and a_{N_2} is the equilibrium activity of N_2 gas with ΔH_F constant and equal to -37.7 kcal/mole.

Figure 2.3 shows the equilibrium p_{N_2}-T curves for AlN, GaN, and InN. The curve for AlN has been calculated by Slack and MacNelly (1976). The curve for InN has been proposed by Grzegory (1996), and follows from Differential

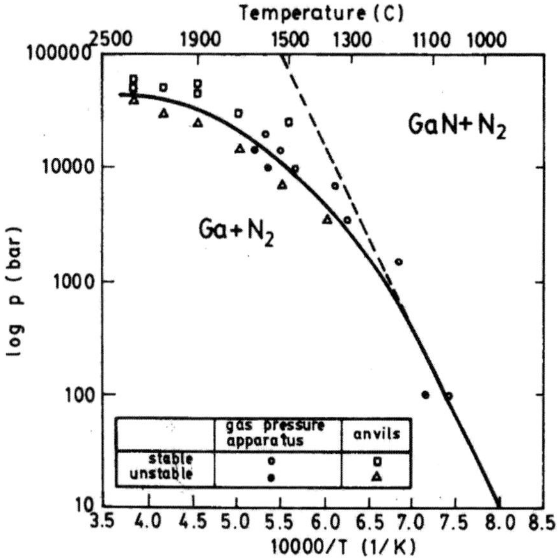

FIG. 2.2. N_2 pressure over GaN in equilibrium with its constituents.

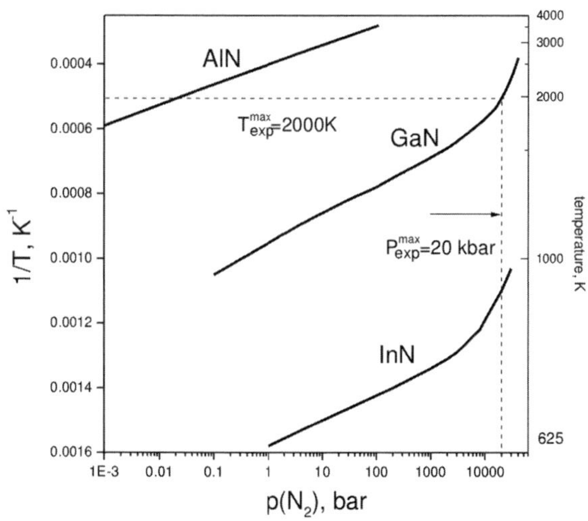

FIG. 2.3. Equilibrium N_2 pressure over III–V nitrides.

Thermal Analysis and annealing of InN at high N_2 pressure. The curves for GaN and InN, at higher pressures deviate from linear dependence due to the non-ideal behavior of N_2 gas.

The diagrams of Fig. 2.3 indicate extremely large differences in thermal stability of III–V nitrides: This is caused mainly by the differences in bonding energy of the crystals. Bonding energy in III–V nitrides is 11.52, 9.12, and 7.72 eV/atom pair for AlN, GaN, and InN, respectively. The consequences of this high bonding energy are high melting temperatures and good thermal stability for the considered compounds, especially AlN and GaN. Indium nitride is much less stable than GaN, and this is the main reason for difficulties in growth of InN and In-rich InGaN epitaxial layers, since low temperatures and high potential of nitrogen are necessary.

The extension of the GaN stability range by the application of pressure allows the growth GaN crystals from the solution in the liquid Ga. Figure 2.4 shows the N solubility data resulting from the annealing of Ga in N_2 atmosphere in the three-phase equilibrium conditions (Grzegory, 1995; Grzegory, 1996). Even the highest temperature available for large-volume gas pressure systems (about 2000 K) is quite far from the melting temperature of GaN (Table 2.1). Therefore, the N concentrations are not high (below 1 at. %), and the growth experiments have to be long to obtain high-quality crystals with dimensions appropriate for research and applications. For this reason, an additional requirement for the experimental system is the extended time for stable work.

The solid line in Fig. 2.4 is the liquidus line for Ga–GaN system calculated in ideal solution approximation (Grzegory, 1995; Grzegory, 1996) with

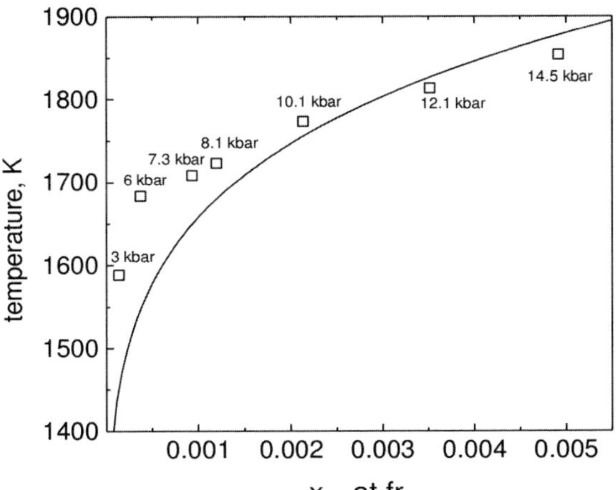

FIG. 2.4. Liquidus line for Ga–GaN system: the solid line was calculated in the ideal solution approximation (Grzegory, 1995; Grzegory, 1996).

Van Vechten's melting temperature of 2790 K (Van Vechten, 1973). For this approximation the solubility can be expressed as follows:

$$n = n_0 \frac{\Delta H_{sol}}{kT}, \tag{2.5}$$

where ΔH_{sol} is the heat of dissolution. For GaN, $\Delta H_{sol} = 44.7$ kcal/mole $=$ 0.49 eV/bond, and expresses the bonding energy in the crystal in relation to its mother phase (the solution). It is equivalent to heat of crystallization, since $\Delta H_{sol} = -\Delta H_{cryst}$.

The analysis of thermodynamical properties of GaN and the system of its constituents explains the role of high pressure as a factor increasing thermodynamical potential of the constituents that makes the crystal stable at temperatures necessary for crystallization. The pressure, however, is important also for kinetics of GaN synthesis. The synthesis of GaN from its constituents is possible due to dissociative chemisorption of N_2 molecule on the Ga surface, as was shown by Romanowski *et al.* (2001) by using the Density Functional Theory calculations. The nitrogen molecule dissociates on the Ga surface only if it overcomes the potential barrier of about 3.5 eV. Therefore, the density of the interacting gas is of crucial importance.

Both the increase of thermodynamical potential of gaseous nitrogen and the enhancement of N_2 dissociation by lowering of the potential barrier can be achieved by other (not compression) methods such as the excitement of N_2 plasma or the use of species containing nitrogen atoms bonded weaker than in

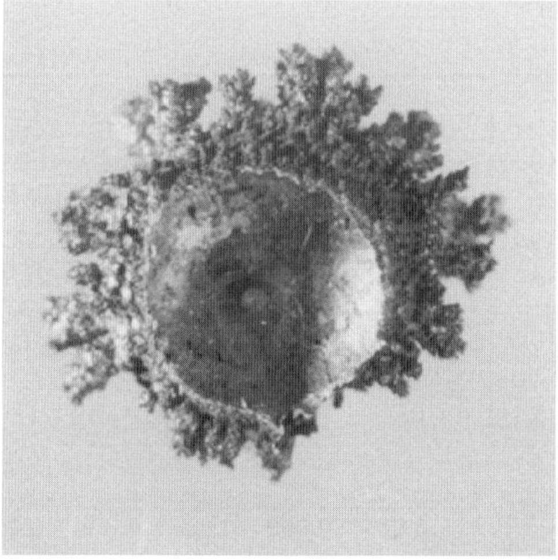

FIG. 2.5. The result of GaN synthesis from the liquid Ga and N_2 plasma. The liquid has been pushed out from the initial droplet covered by GaN crust, by the growing N_2 bubbles.

N_2 molecule (NH_3). These efficient sources of nitrogen are commonly used in epitaxial growth techniques of nitrides: NH_3 in MOVPE and HVPE, and nitrogen plasma in plasma-assisted MBE. It was demonstrated (Porowski and Jun; Lucznik *et al.*) that in microwave nitrogen plasma of 60 Torr and temperature as low as 1100°C, a very efficient GaN (and even InN!) synthesis is possible. The excited N_2 gas was a very efficient source to saturate a Ga droplet with atomic nitrogen and to form a GaN crust on the droplet surface. However, since the pressure in the system was much lower than the equilibrium one (about 100 bar for 1100°C), the N_2 gas phase started to nucleate and grow in the liquid. This resulted in the formation of the structure shown in Fig. 2.5, the empty "Ga droplet" covered with irregular polycrystalline GaN.

2.3 Crystal growth of GaN

As indicated in Section 2.2.1, GaN melts congruently at high nitrogen pressure (>6 GPa) and high temperature (>2497 K) (Utsumi *et al.*, 2003). Due to these extreme melting conditions, GaN cannot be grown from its stoichiometric melt with methods commonly used for other semiconductors such as Si or GaAs. Therefore, GaN single crystals have to be grown using methods combining lower pressures and lower temperatures.

There are several alternative approaches for growing GaN crystals. First is crystallization from the vapor phase, called Hydride Vapor Phase Epitaxy (HVPE), by which GaN is deposited on a foreign substrate (such as sapphire or GaAs), through the reaction of ammonia with gallium chloride at temperatures about 1050°C and at ambient pressure. The foreign substrate is then removed by etching, laser lift-off, or self lift-off techniques, and then a large-diameter free-standing GaN wafer can be obtained. Another important approach producing GaN crystals of the highest quality in terms of their structure and size is crystallization from solution in supercritical NH_3, called the ammonothermal method (Doradzinski *et al.*, 2010). This is a high-pressure method analogous to the hydrothermal crystallization of quartz or oxide crystals such as ZnO. In place of water as a solvent, ammonia is used.

Gallium nitride can be also grown from solution in gallium–sodium mixtures. The growth is carried out at temperatures from 700°C to 900°C and a nitrogen pressure up to 5 MPa (Kawamura *et al.*, 2008). This method has been developed at Osaka University and produces bulk GaN single crystals a few millimeters thick, with a diameter of two or three inches and a defect density of the order of 10^5 cm^{-2}.

The crystals with the lowest dislocation density, 10^2 cm^{-2}, have been obtained by the High Nitrogen Pressure Solution (HNPS) method (Grzegory *et al.*, 1995; Grzegory *et al.*, 2005). This method is based on the direct synthesis reaction between liquid gallium and nitrogen at high temperature (up to 1800 K) and high nitrogen pressure (up to 2 GPa). The spontaneous growth of HNPS-GaN results in crystals having the form of small hexagonal platelets (of about 1 cm^2).

In the following, both high-pressure methods of GaN crystallization HNPS and ammonothermal are described. For comparison, crystallization by HVPE as the most common method of growing free-standing GaN substrates is characterized.

2.3.1 *GaN by HVPE*

The HVPE (Maruska and Titjen, 1969; Seifert *et al.*, 1981; Usui *et al.*, 1997) is the most advanced GaN crystallization method in terms of obtaining large-diameter single-crystalline free-standing substrates. There are two approaches to the fabrication of such GaN wafers: epitaxial growth on foreign substrates, and bulk crystallization. The epitaxial growth on foreign substrates includes deposition of a few hundreds of micrometers thick GaN layer on GaAs or sapphire, and separation of the substrate by chemical etching (GaAs) or by methods based on engineering of strain in a GaN-sapphire system. Usually, the initial substrates are patterned in a special way in order to enforce lateral overgrowth mechanisms allowing significant reduction of dislocation density in the final GaN wafer.

The big advantage of the HVPE method is the relatively fast growth rate of up to 500 μm/h. However, the main disadvantage is the phenomenon of parasitic nucleation. This superfluous GaN nucleation occurs in many places in the HVPE reactor, and often leads to uncontrolled changes in crystal growth conditions during the crystallization run. Thus, the crystallization time must be sufficiently short, and unfortunately this necessitates a thin crystal. This is the main reason why GaN crystals grown by HVPE are generally not thicker than 1 mm.

Some difficulties are also observed with doping to obtain highly n-type or semi-insulating substrates by HVPE. With increasing donor (Si, Ge) or acceptor (Fe) concentrations in HVPE-GaN crystals, the structural quality decreases. Typical free carrier concentration for free-standing GaN material does not exceed 10^{18} cm^{-3}.

The next disadvantage of HVPE technology is lattice bowing of the grown GaN crystals. The absolute value of bowing radii of crystallographic planes of GaN grown by HVPE on sapphire is usually not bigger than 10 meters. This is the reason why free-standing HVPE-GaN crystals are not commonly used as seeds for further HVPE growth.

It is possible to grow crystallographically flat, free-standing HVPE-GaN by changing the growth mode during the HVPE process, from a flat one to a rough one. However, such crystals may contain inversion domains. The best HVPE "free-standing GaN" technology has been developed by SEI (Motoki *et al.*, 2007; Motoki, 2010). In their process, very good quality free-standing HVPE-GaN crystals have been grown by depositing GaN on GaAs wafers. GaAs has a thermal expansion coefficient similar to GaN, therefore the crystallized GaN on GaAs is not so curved. An additional reason why SEIs substrates are crystallographically flat is due to the growth method. It is based on selective growth of the nitride and subsequent regrowth on the surface containing large inverse pyramidal pits (DEEP and A-DEEP methods). Thus, the crystal is grown in the controlled rough-growth mode with the presence of intentionally introduced inversion domains.

The defects in SEIs crystals are collected in an "organized" manner consisting of 400-μm wide stripes, alternating with low and high defect density. In spite of an inhomogeneous threading dislocation density varying from 10^4 cm^{-2} to 5×10^8 cm^{-2}, high-quality laser diode structures have been grown on these substrates, and they are the basis for the manufacture of BluRay laser diodes. Laser chips are positioned on the stripes of low density of defects and without the inversion domains to produce reliable devices. The free carrier concentration in SEIs substrates is about 5×10^{18} cm^{-3}. In the HVPE process the impurities are incorporated into GaN in an anisotropic way (Grzegory *et al.*, 2007). The growth of pits is not the growth in the crystallographic c-direction, but in semi-polar directions. Therefore, the free carrier concentration in SEIs substrates is relatively high. As mentioned, SEI has demonstrated 6-inch GaN substrates, most likely obtained by DEEP or A-DEEP growth methods.

The best method for producing good-quality, free-standing GaN crystals grown on sapphire by HVPE using the flat growth mode has been developed by Hitachi Cable by a technique named "Void Assisted Separation" (VAS) (Oshima *et al.*, 2010). The growth proceeds on a sapphire substrate covered with an ultrathin layer of MOCVD-GaN and nanometric titanium nitride (TiN). The method allows 3-inch GaN substrates with a dislocation density of about 10^6 cm^{-2}, uniformly distributed over the surface of the crystal, a free carrier concentration of about 10^{18} cm^{-3}, and a lattice bowing radius smaller than 10 meters.

An alternative approach—i.e. the real bulk growth with subsequent slicing and polishing of the resulting crystals—is technically more difficult because of parasitic deposition during long HVPE growth runs and exhausting of the gallium ingot. Nevertheless, there are reports on the successful growth of 2-inch GaN boules with a thickness of up to 10 mm. The "bulk approach" has the additional advantage that the crystal can be sliced in various directions. Thus substrates of required orientations including non-polar and semi-polar ones can be obtained. The most important result on bulk crystallization of GaN by HVPE has been reported be Fujito *et al.* (2009) from Mitsubishi Chemical Co. In this work, a beautiful GaN crystal with diameter of 52 mm and thickness of 5.8 mm, free of morphological instability features, with a dislocation density of 10^6 cm^{-2}, was shown. The main problem to solve indicated by the authors was a decrease of the crystal diameter with thickness, due to the formation of inclined facets at the crystal border.

These results strongly supported the view that HVPE can be an effective approach to produce high-quality, thick-bulk GaN crystals. It is very probable that the use of a perfect seed crystal obtained by i.e. ammonothermal method, with no bowing and with very low dislocation density, could significantly accelerate the progress in bulk crystallization of GaN by HVPE. It should, however, be remembered that HVPE of GaN is very sensitive to crystallographic orientation of the crystallization front regarding incorporation of impurities. It can be relevant for the creation of internal stresses in the growing crystals, especially if the growth proceeds with a high rate where three-dimensional growth features can appear. Also, the appearance of semi-polar side facets can introduce sectors of different physical properties than crystal growing on the (0001) polar surface.

2.3.2 *High-nitrogen-pressure solution growth of GaN*

If the concentration of atomic nitrogen in the liquid gallium exceeds the equilibrium value, the excess nitrogen is removed from the liquid by formation of N-rich phases: N_2 if the pressure in the system is lower than required for GaN at given T, or GaN crystals if the pressure is high enough for stability of GaN.

At the p-T conditions corresponding to the GaN stability range, the solubility of nitrogen in gallium is an increasing function of temperature (Fig. 2.4). So, if a temperature gradient is maintained in the system, crystal growth is expected in the cooler part of the solution, which is schematically shown in Fig. 2.6. This is because nitrogen dissolved in the hotter part (larger equilibrium concentration) is transported by diffusion and convection, which leads to the appearance of excess nitrogen in the cooler part (lower equilibrium concentration).

At present, GaN is crystallized in gas-pressure chambers with an internal volume of 1500 cm^3 and 4500 cm^3, allowing crucibles with internal diameters of 25 and 50 mm respectively. The high-pressure–high-temperature reactors consisting of the pressure chambers and the multizone furnaces are equipped with additional systems necessary for *in situ* annealing in vacuum, electronic stabilization and programming of pressure and temperature, and cooling of the pressure chamber. Pressure in the chamber is stabilized with a precision better than 10 bar. The temperature is measured by a set of thermocouples arranged along the furnace and coupled with the standard input power control electronic

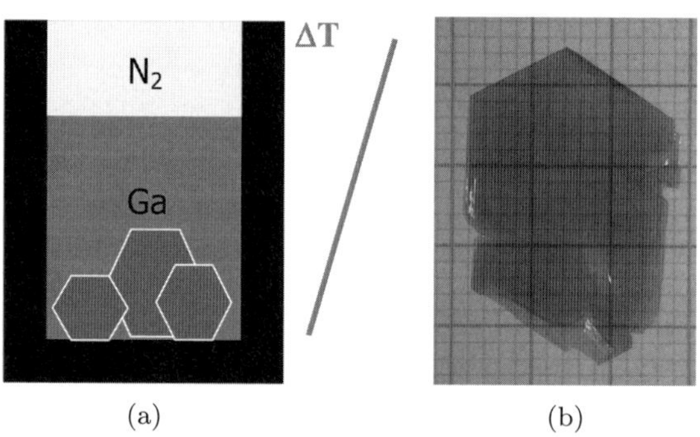

FIG. 2.6. Crystallization of GaN by HNPS method without an intentional seeding: (a) basic configuration, (b) typical crystal.

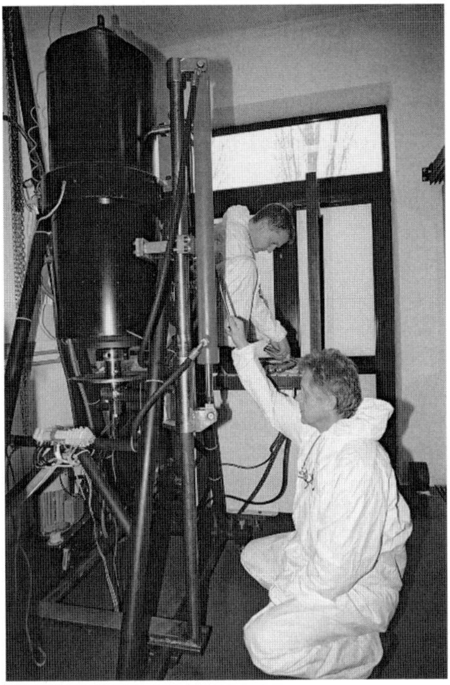

FIG. 2.7. Experimental chamber with an internal diameter of 100 mm.

systems based on Eurotherm units. This allows stabilization of temperature of ±0.2 deg. and programmable changes of temperature distribution in the crucible. The 4500 cm^3-high pressure chamber of internal diameter 100 mm is presented in Fig. 2.7.

The experiments on spontaneous (unseeded) crystallization of GaN were performed mostly in the vertical configuration corresponding to Fig. 2.6 at temperatures of 1700–1850 K (Grzegory *et al.*, 2005). The temperature difference between the hot and the cold zones of the solution was usually 20–50 K. The pressure of N_2 exceeded the equilibrium value corresponding to the highest temperature in the system to keep the solubility relation in agreement with the Ga–GaN liquidus shown in Fig. 2.4. In such a configuration, dissolved nitrogen is transported to the crystallization zone by diffusion and convection. At the typical conditions of our experiments, the convection mechanism plays a dominant role (Grzegory *et al.*, 2002). The crystals nucleated spontaneously on the crucible walls at the cooler zone of the solution, and a typical duration of the processes was 80–150 hours.

The dominating morphological form of GaN crystals grown by the high-pressure method is a thin hexagonal platelet (Fig. 2.6). The large hexagonal surfaces of the platelets correspond to {0001} polar crystallographic planes of the wurzite structure. The side faces of the crystals are mainly the polar $\{10\bar{1}1\}$ and also non-polar $\{10\bar{1}0\}$ planes. The maximum lateral size of the platelets for 100–150 h processes is 10–20 mm, whereas the thickness is 80–120 μm. The crystals in the form of hexagonal platelets grown slowly, with a rate <0.1 mm/h into $\{10\bar{1}0\}$ directions (perpendicular to the c-axis), are usually single crystals of perfect morphology, suggesting stable layer-by-layer growth. They are transparent, with flat mirror-like faces. The habit of the crystals does not change for solutions containing Mg or Be.

As one can deduce from the form of the crystals, the growth is strongly anisotropic, being much faster (about 100 times) in directions perpendicular to the c-axis. This relation is valid at supersaturations corresponding to the average growth rate in $\{10\bar{1}0\}$ directions of 0.05–0.1 mm/h. The supersaturation in the growth solution is determined mainly by the growth temperature, temperature gradients, and mass transport mechanisms in gallium, and also by the local surrounding for a particular crystal (neighboring crystals).

The significant increase of ΔT (i.e. up to 100 K) does not lead to a spectacular increase of the growth rates but rather to growth instabilities on the $\{0001\}$ polar surfaces of the crystals. The most frequently observed features of the unstable growth are cellular structures, accelerated growth at the edges exposed to the N-flow, and in the extreme case the hollow needles elongated into c-directions.

As already mentioned in Section 2.2, the N_2 molecules dissociate at the contact with the Ga surface. However, to approach the surface they have to overcome a quite high (about 3.5 eV) potential barrier which lowers substantially the rate of nitrogen dissociation and its further dissolution in the metal. For oxygen interacting with Ga, there is no potential barrier for dissociation (Krukowski *et al.*, 2001), and therefore even traces of this impurity in the growth system are sources of the unintentional oxygen doping of GaN. Consequently, the crystals are strongly n-type with a free electron concentration of about 5×10^{19} cm^{-3} (metallic conductivity) and mobility of about 60 cm^2/Vs. These free carriers can be fully eliminated by an Mg acceptor added to the growth solution. Then the resistivity of the crystals becomes as high as 10^4–10^6 Ωcm at 300 K. Usually the GaN:Mg crystals become p-type with an activation energy of 150 meV.

The presence of the native point defects in the crystals has been checked by positron annihilation measurements (Saarinen *et al.*, 1997). High concentration of Ga vacancies (V_{Ga}) has been found in the conductive crystals, in contrast to the Mg-doped samples where no Ga-vacancies has been observed. This agreed with the theoretical prediction that the formation energy of V_{Ga} decreases with the increase of the Fermi-level energy (Neugebauer and Van de Walle, 1994), suggesting that the creation of these defects is thermodynamically controlled.

The most important advantage of the pressure-grown GaN is the extremely low density of dislocations (lower than 100 cm^{-2}) as shown by defect selective etching in molten KOH–NaOH eutectics (Weyher *et al.*, 1997) of (0001) Ga-polar surfaces of both n-type and Mg-doped crystals.

As was mentioned earlier in the conditions where the platelets are crystallized, the growth of GaN is strongly anisotropic, being about 100 times faster in directions perpendicular to the c-axis of the crystal than parallel to c. It indicates that the nucleation rates on the polar 0001 surfaces are extremely small. The observed changes in the crystal morphology at the increased supercooling of the solution, suggest that at higher supersaturations the nucleation on these surfaces becomes very fast and its rate is very sensitive to changes of the growth conditions. In this case, the nucleation is difficult to control. In the applied configuration, where a set of crystals is growing simultaneously, it is impossible to control precisely the conditions for a particular crystal or, moreover, for a particular face. The size of the crystals is limited by the available volume of the uniform supersaturation zone in the high-pressure reactors in the temperature gradient configuration used up to now, for this kind of experiment. Therefore, the crystal size is limited to 1–2 cm in their lateral dimension and to 0.1 mm in their thickness. The only solution to enhance the growth on the {0001} surfaces in a stable way is to apply methods of directional crystallization.

Figure 2.8 shows a scheme of the Multi Feed Seed configuration used recently (Bockowski *et al.*, 2012) for directional crystallization of GaN by HNPS method. Several (0001) oriented HVPE-GaN seeds are positioned in a vertical

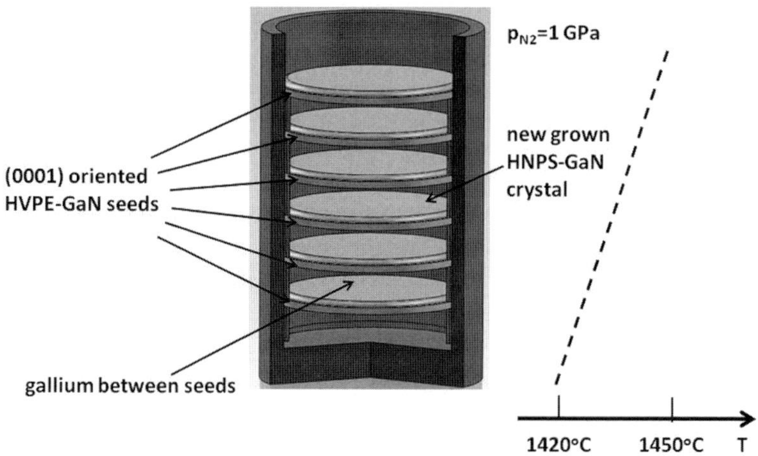

FIG. 2.8. Scheme of the MFS configuration; section of the crucible; the free standing (0001) oriented HVPE-GaN substrates are positioned vertically; liquid gallium discs are between them; due to the temperature gradient applied the bottom sides of HVPE-GaN crystals are dissolved in liquid gallium supplying atomic nitrogen to the solution; nitrogen is then transported in the liquid gallium by convection or diffusion to the underlying seed where the crystallization takes place.

stack configuration with liquid gallium between the seeds. Under high nitrogen pressure (~1 GPa) an axial temperature gradient is applied along the crucible. The seeds are overgrown on their (0001) surface. They dissolve in liquid gallium from their lower $(000\bar{1})$ surface, supplying atomic nitrogen to the solution. The nitrogen is transported to the underlying crystal. Each seed is overgrown and dissolved at the same time, but at a slightly different temperature varying from 1420°C to 1450°C.

A typical crystal grown in the MFS configuration is shown in Fig. 2.9. A stable and macroscopically flat crystallization is observed. The pits visible in the material are situated on the bottom (nitrogen surface) of the seed crystal. No pits are observed on the surface of the pressure-grown material. Two types of surface morphology are noted; macrosteps and hillocks (see Fig. 2.9(b) and (c)). The presence of these two modes of growth depends on the temperature gradient applied. For a larger gradient, a growth rate of 5 μm/h can be obtained and macrostep propagation is detected. In this case some voids and gallium inclusions are observed in the crystals. In comparison, for a low-temperature gradient, the density of voids and inclusions decreases from 10^3 cm^{-2} to 10^0 cm^{-2} and the growth rate is reduced to not higher than 1 μm/h.

The deposition of HNPS-GaN layers and the dissolution of HVPE-GaN improve the structural quality of the obtained crystals. A large improvement is observed between HVPE-GaN and free-standing HNPS-GaN after removing the residual HVPE-GaN from the nitrogen surface of the crystal. The absolute value of the bowing radius can increase from 2 m to 30 m, since removing the HVPE substrate helps to release stress in the HNPS-GaN. For 500 hours of crystallization, one can obtain up to 500 μm thick of MFS-HNPS-GaN (see Fig. 2.10).

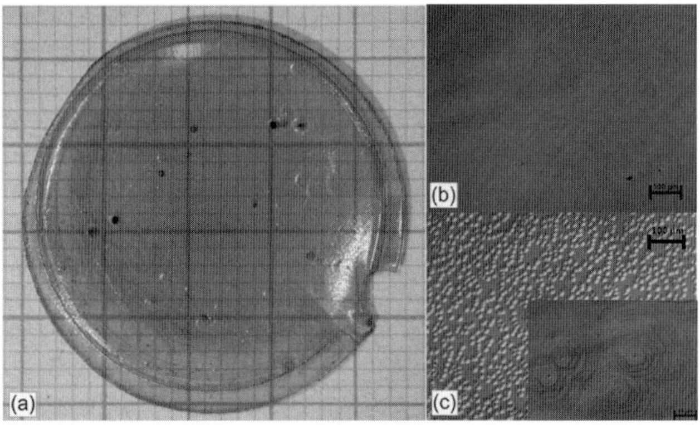

FIG. 2.9. (a) Typical crystal grown in the MFS configuration; stable and macroscopically flat crystallization is observed; two types of morphology are noted: (b) macrosteps; (c) hillocks.

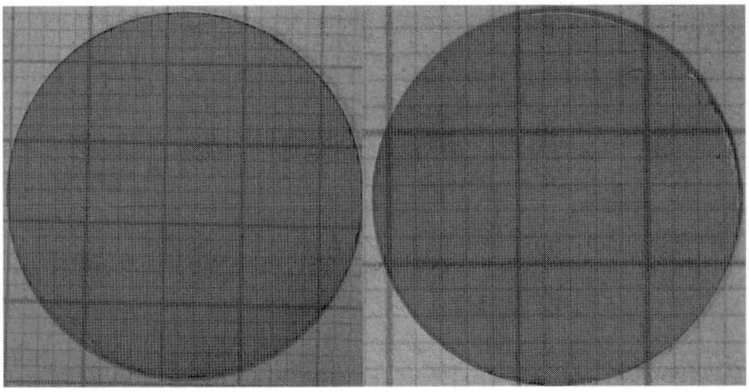

FIG. 2.10. Typical HNPS-GaN crystals used as substrates for making laser diodes.

Laser diodes manufactured using MFS-HNPS-GaN (Sections 2.5–2.8) substrates show a threshold voltage ~4.5V@2.5kA/cm^2, at emission wavelengths from 390 nm to 440 nm and lifetimes exceeding 5000 hours. The average dislocation density revealed by defect selective etching in the laser diode structures does not exceed 5×10^6 cm^{-2}. In turn, the average dislocation density for the MFS-HNPS-GaN is about 1×10^6 cm^{-2}. A "plasmonic" GaN substrate (Section 2.6), with a high free carrier concentration up to 10^{20} cm^{-3}, can be grown at high temperatures of 1440°C. Dependence of free electron concentration in crystals grown in the MFS configuration on growth temperature is shown in Fig. 2.11.

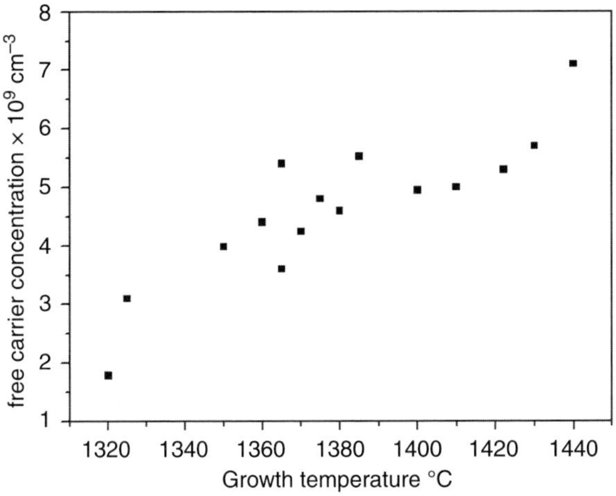

FIG. 2.11. Free electron concentration in GaN crystals grown by HNPS method in MFS configuration on (0001) oriented HVPE-GaN substrates.

A "plasmonic" GaN substrate (Section 2.6) has the advantage of increasing the refractive index contrast between substrate and epitaxial structure, suppressing optical mode leakage into the substrate and optimizing the optical mode distribution in the active region. This feature simplifies the laser diode structure, since otherwise a thick AlGaN layer or an AlGaN/GaN superlattice is required to prevent light propagating from the active (MQW) region into the substrate. Unfortunately, including aluminum increases the strain in the system and often leads to macroscopic bowing, cracking, or misfit dislocations in the epistructure (Gorczyca *et al.*, 2010; Sarzynski *et al.*, 2006).

Today the main disadvantage of the MFS configuration is the slow growth rate. It is planned to optimize the growth conditions by adding impurities such as sodium or lithium to the solution in order to increase the nitrogen solubility in gallium.

2.3.3 *Ammonothermal growth of GaN*

At present, the highest-quality real bulk crystals of GaN in terms of their structural properties and size are grown by ammonothermal method (Doradzinski *et al.*, 2010). The ammonothermal method is based on dissolution of GaN in supercritical ammonia mediated by chemical reactions, with mineralizers transferring GaN into highly soluble molecules. The mineralizers can be acidic (NH_4^+) or basic (NH_2^-) in character, thus the ammonoacidic (Purdy, 1999) and ammonobasic (Dwilinski *et al.*, 1995) methods are distinguished. The most advanced results (Doradzinski *et al.*, 2010) are achieved by the Polish company Ammono S.A. using an ammonobasic approach at a pressure of 0.2–0.3 GPa and temperature of 400–600°C. In the ammonobasic system where pure alkali metals or alkali metal amides ($LiNH_2$, $NaNH_2$, or KNH_2) are used as mineralizers, GaN nutrient (Fig. 2.12) is dissolved in supercritical NH_3 by the formation of soluble metal amide compounds according to the following reactions:

$$KNH_2 + GaN + 2NH_3 \rightarrow KGa(NH_2)_4 \tag{2.6}$$

$$2NaNH_2 + GaN + 2NH_3 \rightarrow Na2Ga(NH_2)_5 \tag{2.7}$$

The crystals are grown in retrograde solubility range (Fig. 2.13), therefore the crystallization zone is hotter than the saturation one. It can lead to higher-purity material because of the regular solubility character of most of the impurities in the system. The solubility data shown in Fig. 2.13 indicate that solubility is an increasing function of pressure, whereas it decreases with temperature.

The autoclaves for ammonothermal growth of GaN are constructed of nickel superalloys characterized by excellent mechanical properties and high resistivity against the ammonobasic environment. This is in contrast to the ammonoacidic systems which require special noble metals liners arranged inside the reaction

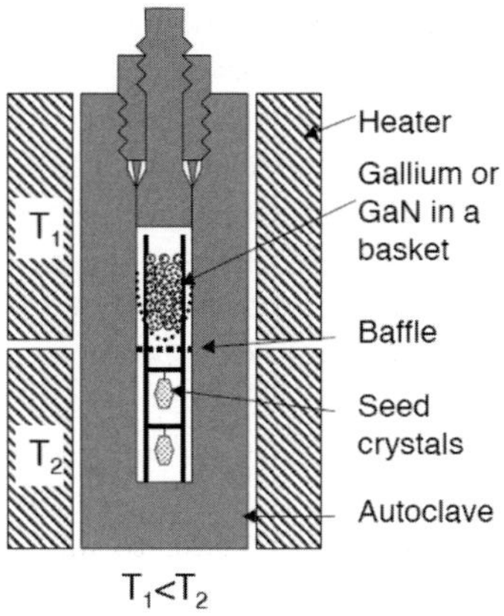

FIG. 2.12. Schematic idea of the ammono-bulk method.

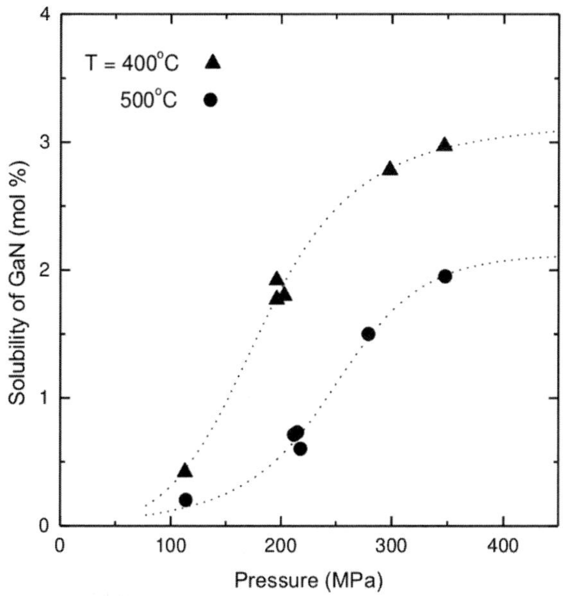

FIG. 2.13. Solubility of GaN in supercritical ammonia as a function of pressure for temperatures of 400°C and 500°C in the KNH_2-NH_3 system with a molar ratio $KNH_2/NH_3 = 0.07$. (Courtesy Ammono S.A.)

chamber. Due to the low viscosity of supercritical ammonia solvent a very efficient mass transport by convection and high degree of supersaturation uniformity in the crystallization zone can be achieved. Therefore, simultaneous use of many seeds in large-volume chambers as well as perfect scalability with the size of the autoclave has been achieved. At present, Ammono S.A. manufactures 1- and 1.5-inch GaN wafers and is on the road to start the production of 2-inch crystals. Figure 2.14(a) shows a bulk GaN crystal grown by Ammono S.A. Such a crystal can be sliced along polar {0001}, non-polar, and semipolar crystallographic planes in order to obtain substrates of different orientations for epitaxial growth. One of the first 2-inch crystals is shown in Fig. 2.14(b).

Ammono crystals are extremely flat with bowing radii of the crystallographic planes reaching up to 1000 m. Threading dislocation density is of the order of 10^4 cm^{-2} which is reflected in extremely narrow X-ray rocking curves being often narrower than 20 arcsec for both symmetric and asymmetric reflections.

Electrical properties of the crystals can be controlled by doping during crystallization process. The free electron concentration as high as 2×10^{19} cm^{-3} has been achieved which is very useful for construction of laser diodes. Native defects in the Ammono-GaN crystals have been studied by positron annihilation spectroscopy (Tuomisto *et al.*, 2010). A high concentration of Ga vacancy related defects in n-type crystals and no such defects in the p-type ones was observed.

High-power laser devices fabricated on Ammono's crystals have already been reported (e.g. by another Polish company, TopGaN (Perlin *et al.*, 2011*b*)) and are discussed in the following parts of this chapter, especially in Sections 2.5 and 2.8.

It seems that the only drawback of the ammonothermal method are extremely low crystallization rates, being as low as about 20 μm/day for the ⟨0001⟩ direction and even lower for the non-polar ones. This slows down development of the

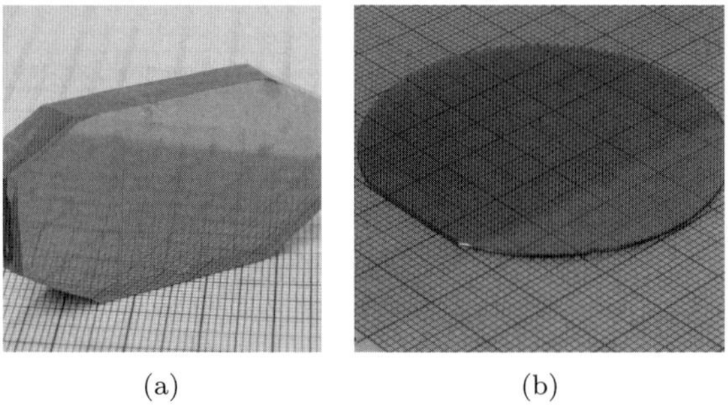

(a) (b)

FIG. 2.14. Single crystals of GaN grown by the ammonothermal method: (a) bulk GaN for fabrication of non-polar and semi polar wafers, (b) 2-inch GaN crystal. (Courtesy Ammono S.A.)

method, especially at its seed-multiplication stage. However, on the other hand, the multi-seed processes in large-volume autoclaves could compensate for this disadvantage.

2.4 Epitaxy of nitrides on single-crystalline HNPS GaN: early results

Homoepitaxy is the deposition of a thin layer of a crystalline material on the crystalline substrate of the same material, having, therefore, exactly the same crystal structure and thermal expansion coefficients. If the substrate is perfect, the homoepitaxial layer should also be a perfect crystal. To approach this result experimentally, a very good coupling between the substrate and the growing layer is necessary. Technically it necessitates excellent preparation of the substrate surface and the choice of the conditions of epitaxy, assuring growth of the continuous layer with a flat surface corresponding usually to one of low-indices crystallographic planes of the grown material. Such a stable growth can proceed by a two-dimensional layer by layer mechanism (mostly at low temperatures and/or on low indices exactly oriented surfaces) or by the atomic step flow (on surfaces vicinal to the low-indices ones). From the point of view of the uniformity of the resultant layer, the step flow growth mode is the optimum one, because the incorporation of unintentional and intentional impurities is often very sensitive to the microscopic situation on the growing surface. The growth by propagation of an array of the well-determined steps should lead, therefore, to the most uniform distribution of the impurities and minimization of the strains in the homoepitaxial layer.

These obvious advantages of homoepitaxial growth of nitrides were demonstrated with the use of pressure-grown HNPS-GaN platelets of extremely low dislocation density. These results are discussed, e.g., in Grzegory and Porowski (2002). In particular, it was shown that for proper surface preparation and optimized conditions of the epitaxial growth, the structure of GaN homoepitaxial layers in terms of dislocation density, follows the structure of the GaN substrates. Figure 2.15 compares surface morphology of GaN epitaxial layers deposited by MOCVD on a GaN crystal grown by HNPS (see Section 2.3.2) and on GaN/sapphire substrate, in the same run (Leszczynski *et al.*, 1999). The atomic step flow on the surface of the homoepitaxial layer is not perturbed as observed for a GaN layer containing threading dislocations.

If a GaN homoepitaxial layer is near-dislocation-free, its optical properties depend mostly on the concentration and distribution of point defects incorporated during the epitaxial growth. It is a function of the purity of the growth system, the growth conditions, and the orientation of the substrate.

Due to the lack of strains and the high degree of homogeneity of the homoepitaxial GaN, the exciton related peaks in low-temperature PL spectra could be extremely narrow. The FWHM of the bound exciton lines less than 0.5 meV was observed for layers grown by both MOCVD (i.e. Stepniewski *et al.*, 1999;

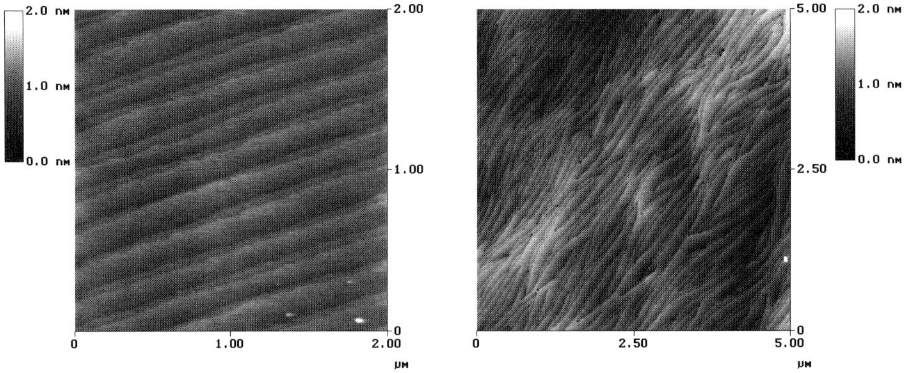

(a) (b)

FIG. 2.15. Surface morphology of GaN epitaxial layers deposited by MOCVD (Leszczynski *et al.*, 1999) on different substrates: (a) GaN pressure-grown crystal, (b) sapphire.

Kirchner *et al.*, 1999) and MBE (i.e. Teisseyre *et al.*, 1996; Grandjean *et al.*, 2000) on the Ga-polar surfaces of the HNPS-grown GaN substrates. For layers deposited on the N-polar surfaces the spectra were usually much wider due to the enhanced and non-uniform incorporation of unintentional impurities on this chemically active side of the GaN crystal. This was suppressed by the use of surfaces vicinal to the $(000\bar{1})$ plane, as was shown by Zauner *et al.* (2000), who obtained narrow excitonic spectra for GaN layers grown on misoriented N-polar surfaces of the GaN pressure-grown substrates.

For GaN layers grown by MBE with NH_3 nitrogen source it was shown (Grandjean *et al.*, 2000) that the band edge PL integrated intensity at RT for the homoepitaxial GaN was about 60 times larger than for GaN grown on sapphire with the same growth conditions. Also, the decay of the PL in these homoepitaxial layers was much slower than for corresponding (grown in similar conditions) heteroepitaxial GaN. This indicated a substantial suppression of the non-radiative recombination by elimination of threading dislocations from the layers.

The PL decay has also been studied at high excitations (Jursenas *et al.*, 2001) for GaN layers grown by MOCVD on sapphire and pressure-grown GaN substrates. The decay time (at RT) of 450 ps measured for homoepitaxial material was five times longer than for corresponding heteroepitaxial layers grown in the same conditions.

The optically pumped LT stimulated emission (Holst *et al.*) experiments confirmed that the near-dislocation-free GaN is a much more efficient source of light than GaN containing dislocations. The stimulated emission in a GaN homoepitaxial layer grown by MOCVD started at the excitation power density lower than 1 MW/cm^2, whereas for GaN grown on SiC only at 10 MW/cm^2. The stimulated emission peak for the homoepitaxial layer was as narrow as about 5 meV (about 40 meV for GaN/SiC).

The growth of GaN/AlGaN structures on GaN pressure-grown crystals has been studied mainly by MBE (Christiansen *et al.*, 1999; Foxon *et al.*, 1999; Frayssinet *et al.*, 2000). Near-dislocation-free GaN/AlGaN multi-quantum wells and structures have been grown on the N-polar surfaces of GaN crystals by RF plasma-assisted MBE (PA MBE) (Christiansen *et al.*, 1999; Foxon *et al.*, 1999) and on the Ga-polar surfaces by MBE with NH_3 nitrogen source (R MBE) (Grandjean *et al.*, 2000; Frayssinet *et al.*, 2000). As a rule, the optical properties of the structures without dislocations were much better than for similar structures grown in the same conditions, on sapphire. For example, the integrated intensities of the PL coming from similar structures (8 ML GaN SQW with the $Al_{0.1}Ga_{0.9}N$ 50 nm barriers) grown by R MBE (Grandjean *et al.*, 2000) on sapphire and bulk pressure-grown GaN substrates were compared. It was shown that for the structure deposited on GaN, the PL was much stronger, especially at RT. For the structure grown on GaN substrate, the PL intensity started to decrease with temperature only at about 100 K, due mainly to the thermal escape of carriers from the quantum well towards the AlGaN barriers, as was observed for classical III-V QW heterostructures with dislocations densities lower than 10^3 cm^{-2} (Lambkin *et al.*, 1990). For heteroepitaxial structure, the presence of dislocations strongly influenced the non-radiative recombination processes. The PL intensity started to decrease at much lower temperatures due to delocalization of excitons and their further interaction with dislocations.

Dislocations in the structures grown on the almost defect-free substrates can obviously appear as a result of the lattice mismatch between GaN and its ternaries InGaN and AlGaN. This was analyzed by Leszczynski *et al.* (2001) by X-ray measurements of lattice parameters of various GaN-based epitaxial layers deposited on GaN substrates as a function of InGaN (AlGaN) composition and thickness. It was shown that the boundary between fully strained and relaxed InGaN (AlGaN) layers is at a safe distance from the mismatch–thickness area for layers necessary for a typical blue laser structure.

As already mentioned in the introduction, strongly n-type pressure-grown GaN crystals have been tested by Nakamura (1999) as substrate for the InGaN 405 nm laser. The cw 30 mW devices with a lifetime exceeding 3000 h were constructed based on epitaxial structures grown by MOCVD. The 30 mW power has been achieved at a current of 62 mA. For typical InGaN MQW LDs, at a current of 62 mA, the output power was about 15 mW. This increase in the efficiency confirmed that dislocations were the main limiting factors for high-power GaN-based lasers.

As appeared later, the pressure-grown GaN substrates were especially important for development of epitaxial growth of quantum nitride structures by Plasma Assisted MBE. Section 2.5 is devoted to this issue.

2.5 Development of nitrides epitaxy by PA MBE

The growth of structures discussed here was performed in a customized V90 Riber MBE reactor equipped with two Veeco RF plasma sources, operating at 240–450 W for 0.8–2 sccm of N_2 flow. The pressure during the growth was $1.5–7 \times 10^{-5}$ Torr. The value of the active nitrogen flux was in the range of 4–13 nm/min. This was quantified by growing GaN in Ga-rich conditions. Three types of substrate were used for the growth from:

(i) High nitrogen pressure solution (HNPS), HNPS-GaN.
(ii) Ammonothermal growth, Ammono-GaN.
(iii) Hydride vapor phase epitaxy (HVPE), HVPE-GaN.

The epi-ready, bulk (0001) substrates, with miscut of about 0.5 degree with respect to c-polar plane, were prepared in a two-step procedure consisting of mechanical and chemo-mechanical polishing. All layers discussed in this section were grown on the (0001) Ga-polarity surface.

2.5.1 *Growth conditions for PAMBE*

The analysis of experimental data for GaAs or Si indicates that the optimum growth temperature for two-dimensional step-flow mode equals about half the melting temperature, T_M (Burton *et al.*, 1951; Ishizaka and Murata, 1994). This rule is also valid for GaN growth, where the GaN melting point lies somewhere between 2540 K (experimentally determined for 6 GPa by Utsumi *et al.*, 2003) and 2800 K (theoretically calculated i.e. Van Vechten, 1973), and the optimum growth temperature used in MOVPE is in the range 1050–1100°C (1320–1370 K). On a microscopic scale, at $T = 0.5\,T_M$ atomic kinks start to become active; i.e. atoms can detach from an atomic kink (Burton *et al.*, 1951; Ishizaka and Murata, 1994) at a rate comparable to typical attachment rates during deposition. This rule is strict for Kossel crystal and is also fulfilled for real crystal structures.

The difficulty for GaN growth in MBE is that for temperature $0.5\,T_M = 1320$ K, a high nitrogen overpressure (about 60 bars for N_2) is required to prevent GaN decomposition (Karpinski and Porowski, 1984; Karpinski *et al.*, 1984). For more chemically active N precursors, such as ammonia, the situation is better, but still an overpressure in the range of 0.1–2 bar must be used. Such overpressures are not compatible with MBE technology, which relies on high vacuum conditions for the delivery of atoms from the effusion cells to the growing layer.

In fact, for typical MBE growth conditions GaN begins to decompose at temperatures in excess of 800°C, restricting the epitaxy to temperatures much below the optimum point. Indeed, due to the arguments given above and the low diffusivity of N adatoms, the early efforts to grow GaN in MBE at temperatures below 800°C, using group V-rich conditions typical for III–V epitaxy, gave unsatisfactory results. These early difficulties, seemingly well grounded in the simple thermodynamics of the processes involved, led many to believe that the only path for successful growth of nitrides in MBE reactors is to push the growth conditions as close as possible to those present in MOVPE reactors.

The breakthrough in the study of growth kinetics in PAMBE came with the finding that in Ga rich conditions it is possible to grow relatively smooth layers at low growth temperatures (Elsass *et al.*, 1999; Heying *et al.*, 2000; Heying *et al.*, 1999; Adelmann *et al.*, 2002; Manfra *et al.*, 2004; Skierbiszewski *et al.*, 2004). However, such growth conditions were prone to the formation of Ga droplets on the GaN surface, and high-quality material was mainly formed in the regions between the droplets. Further study of the Ga auto-surfactant effect revealed that this problem could be avoided provided the Ga:N flux ratio was maintained in a very narrow range of values—low enough to be just below the formation of the droplets, but high enough to ensure the formation of a metallic Ga bilayer on the Ga polarity surface (Elsass *et al.*, 1999; Heying *et al.*, 2000; Heying *et al.*, 1999; Adelmann *et al.*, 2002; Manfra *et al.*, 2004; Skierbiszewski *et al.*, 2004).

This apparent conundrum of finding two-dimensional step-flow growth at temperatures significantly lower than $0.5\,T_M$ can be resolved by considering that the decomposition of GaN (and therefore the activity of atomic kinks) is controlled by a kinetic barrier. Schoonmaker *et al.* (1965) demonstrated more than 40 years ago that the height of this barrier is strongly reduced when GaN is covered by liquid Ga or In. It was also experimentally evidenced and discussed by Groh *et al.* (1974) and Newman (1997) that, for a given temperature, the presence of a metal layer on a GaN surface accelerates etching of GaN in comparison to the exposed surface. Therefore, in the metal-rich regime for PAMBE (we can call it liquid phase epitaxy-like conditions) the formation energies of the kink sites are reduced, substantially lowering the minimum growth temperature necessary to sustain the desired two-dimensional step-flow growth mode.

Another important process necessary for high-quality growth at low temperature, which has also been a subject of intensive theoretical studies, is the adatom diffusivity. Perhaps the most insightful were results of modeling based on the density-functional theory which pointed towards the existence of a very efficient lateral diffusion channel for nitrogen adatoms on a semiconductor surface just below the thin metallic film (Neugebauer *et al.*, 2003). Surprisingly small activation energies for this so-called adlayer enhanced lateral diffusion (AELD) facilitate high-quality step-flow epitaxy at low temperatures. This enhanced surface mobility of nitrogen adatoms, coupled with the reduction in the kink formation energy discussed earlier, promotes two-dimensional growth nucleation and step-flow growth at temperatures far below those needed for "classical" group V-rich conditions.

The importance of the two mechanisms described above—i.e. kink activity and adatoms mobility—is well illustrated by quite recent results demonstrating that high-quality GaN can be grown with PAMBE under "classical" N-rich conditions, in the absence of a metallic layer on the surface, provided that the growth temperature is high enough to cause concurrent decomposition of GaN

(Koblmuller *et al.*, 2007). It should be emphasized nevertheless that such growth conditions are not suitable for the epitaxy of InGaN because of the prohibitively high evaporation rate of In, and are likely to have only limited use for the epitaxy of AlGaN, because of the pinning of the atomic step kinks on Al atoms (due to much higher AlN bond energy) and too-low Al atoms surface mobility. Still, such growth conditions provide a very useful alternative for PAMBE growth of high-quality GaN without the danger of Ga droplet formation.

2.5.2 *The role of threading dislocations and miscut angle on surface morphology in low-temperature PAMBE*

It was found that for PAMBE, in the group III metal-rich regime, spiral growth is always present on substrates containing a high density of threading dislocations (TDs) (Heying *et al.*, 1999). Such morphological features are located at the intersection of a TDs with the crystal surface, and are formed in all crystal growth processes proceeding through two-dimensional nucleation (Swendsen *et al.*, 1976). It is important to realize that such spiral growth leads to the formation of hillocks whose steepness increases monotonically with increasing tightness of the spiral. In Fig. 2.16 we present a comparison of the surface morphology of InGaN layers grown by PAMBE on GaN/sapphire templates and on GaN bulk crystals grown by the high nitrogen pressure solution growth technique (TDs density <100 cm^{-2}). Nearly straight and parallel atomic steps are present on InGaN layers grown on a bulk substrate, while a dislocation mediated step-flow growth mode is observed on the GaN/sapphire substrate.

From our experience, at temperatures higher than 710°C PAMBE growth on high-dislocation-density GaN/sapphire substrates results in nearly flat surfaces (hillocks height below 2 nm). However, as the growth temperature decreases, the tightness of the spirals and thus the height of the hillocks increases rapidly.

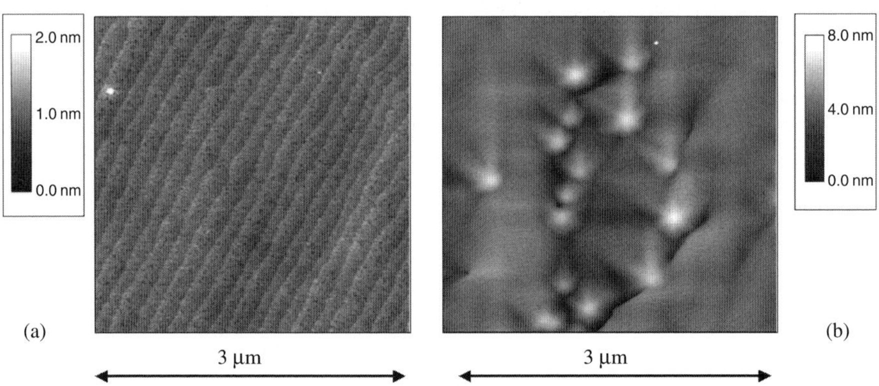

FIG. 2.16. Atomic Force Microscope images of (a) $In_{0.02}Ga_{0.98}N$ layers on a bulk crystal, and (b) on a GaN/sapphire template.

Indeed, at temperatures around 600°C (where InGaN layers are typically grown), hillock heights can exceed 7 nm. As a consequence of the non-uniform dislocation distribution in the substrate, the hillock diameters and sidewall inclinations vary substantially across the surface. Such local disorder on the surface acts as a seed for "catastrophic" degradation of epitaxial layers, particularly when the layers exceed the critical thickness. Thus, low-dislocation substrates should enable the growth of metastable layers, allowing the system to "supercool" considerably before various undesirable crystallographic and morphological transitions are triggered (e.g., inversion domains, InN precipitates, Mg precipitates, dislocation clusters, etc.). Importantly, even though metastable, such layers can be extremely robust once grown, due to the very high melting temperatures for GaN and related compounds. As a consequence of the GaN wurzite symmetry, growth processes are highly anisotropic, impacting on atomic step morphology. In Fig. 2.17 we present a comparison of the morphology of InGaN layers grown under identical conditions on two bulk GaN substrates having different miscut orientation: towards $[1\bar{1}00]$ (Fig. 2.17(a)) and $[11\bar{2}0]$ (Fig. 2.17(b)) directions. For miscut towards $[1\bar{1}00]$ we observe nearly straight parallel atomic steps, while for $[11\bar{2}0]$ the atomic steps are considerably more disordered, showing step meandering, e.g. atomic steps with a characteristic zigzag shape (Shen *et al.*, 2007).

2.5.3 *The growth of high-In-content InGaN layers by PAMBE*

For PAMBE the highest-quality InGaN structures on c-plane (0001) surface are grown in the In-rich growth regime (Neugebauer *et al.*, 2003; Skierbiszewski *et al.*, 2009; Riechert *et al.*, 1996; Siekacz *et al.*, 2008; Siekacz *et al.*, 2009).

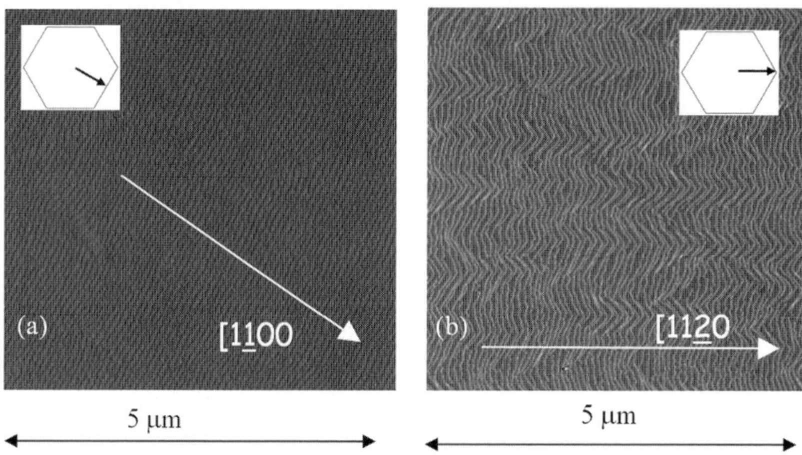

FIG. 2.17. Atomic Force Microscope images of $In_{0.02}Ga_{0.98}N$ layers grown by PAMBE on bulk GaN substrates with miscut orientation towards $[1\bar{1}00]$ (a) and $[11\bar{2}0]$ (b), respectively.

It was already demonstrated (Riechert *et al.*, 1996; Siekacz *et al.*, 2008) that the decrease in the growth temperature T_G, by reducing the decomposition of the InN-fraction, can be used effectively to increase In content in the layers. However, reduction of T_G can deteriorate the layer quality due to the reduction of surface adatom mobility. Therefore, we propose an alternative path to achieve high-In-content layers—by increase of the nitrogen flux, which can improve InN stability at higher temperatures. In Fig. 2.18 we demonstrate the dependence of the indium content for 40-nm thick InGaN layers as a function of the substrate temperature at fixed gallium flux, $\Phi_{Ga} = 2.63$ nm/min, for two nitrogen fluxes Φ_N, equal to 4.1 and 6.5 nm/min. The indium content x was determined after the growth, using XRD reciprocal space mapping (Siekacz *et al.*, 2008).

The expected decrease of the indium content for constant N flux with increasing growth temperatures is clearly seen. Importantly, for higher N flux we obtained higher In content for a given growth temperature. The decrease of the In content for higher growth temperatures is explained by temperature-activated dissociation of In-N bonds, which leads to the nitrogen loss from the growing layer.

In order to investigate the influence of N flux on indium incorporation and the possibility of achieving long-wavelength emission, we have grown a series of 2.2-nm single quantum well (SQW) structures with varying N flux, using a constant growth temperature and Ga flux. The structures were grown on HVPE (0001) GaN crystals. After the 100-nm GaN buffer layer, the 20 nm $In_{0.06}Ga_{0.94}N$ barriers with 2.2 nm $In_xGa_{1-x}N$ well were grown. In Fig. 2.19(a) we present the PL results from these samples. Increasing the N flux from 5.5 nm/min to 12.5 nm/min we were able to shift the PL from 460 nm up to 540 nm, i.e. from blue to green emission wavelength. This result show that change of the N flux

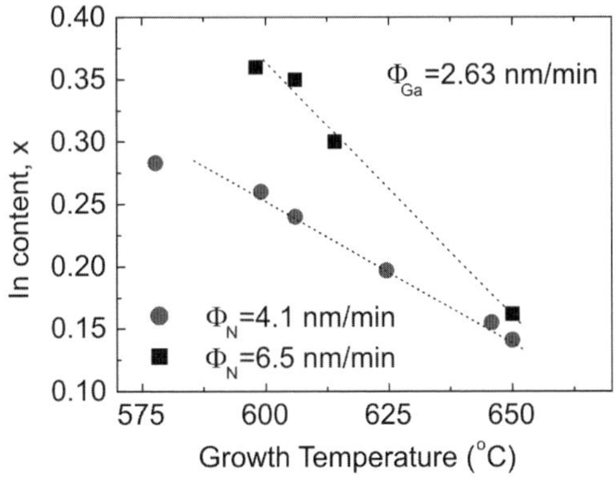

FIG. 2.18. The dependence of the In content in the InGaN layers as a function of the growth temperature for two nitrogen fluxes of 4.1 nm/min and 6.5 nm/min, respectively.

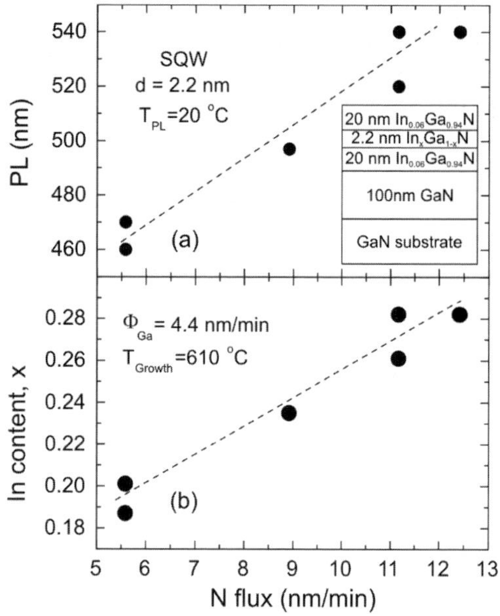

FIG. 2.19. The room-temperature photoluminescence from SQW (a) and the indium content x (b) as a function of the nitrogen flux for the constant Ga flux and growth temperature. The inset shows structure details.

is a very effective method for tuning the light emission wavelength of InGaN QWs grown by PAMBE. As we will demonstrate later, it will be a key factor in achieving green laser action. We used the measured PL energies to estimate the In content in these structures. For that we calculated the expected transition energy from SQW, solving the Schrödinger equation in both conduction and valence bands, following the procedure described in Christmas *et al.* (2005). The results of these calculations are presented in Fig. 2.19, and indicate that by doubling the nitrogen flux we were able to change the In content from approximately 18% to 28% in SQW.

The most important practical message from our experiments is that we can increase the In content at a given growth temperature, increasing the nitrogen flux. As presented in the next Section, the growth of InGaN with high Φ_N is also beneficial for the optical quality of grown layers.

2.5.4 *The influence of the growth conditions on the optical properties of InGaN QWs*

Homogeneous InGaN QWs with a low concentration of dislocations and point defects have to be grown for the achievement of long-wavelength lasing. As we demonstrated, the increase of the In content in InGaN layers can be realized by

i) reduction of the growth temperature or ii) increase of the N flux. The first approach is effective in increasing the In content, but at low growth temperatures the optical quality of the InGaN is dramatically deteriorated, as illustrated by a strong drop in LED efficiency (Siekacz *et al.*, 2009). Here we examined whether an increase of N flux provides a better alternative to decreasing growth temperature for the production of InGaN structures with an optical quality sufficient to sustain laser action at longer wavelengths. For LD development we need to minimize localization potentials and non-radiative recombination processes in QWs. Time-resolved photoluminescence (TRPL) spectroscopy is an ideal tool for monitoring both of these parasitic effects.

The TRPL experiments provide information on the carrier dynamics during and after their pulse laser excitation. The observed PL decay time, τ_{PL}, depends on the radiative (τ_R) and non-radiative (τ_N) recombination times ($\tau_{PL}^{-1} = \tau_R^{-1} + \tau_N^{-1}$). For a perfect QW (i.e., a sample without point defects, etc.) the τ_{PL}^{-1} is determined by the characteristics of the QW confining potential and only weakly changes with temperature (for this situation, $\tau_{PL} = \tau_R$). For practically attainable QWs the factor τ_N^{-1} cannot be neglected because of the presence of non-radiative recombination centers, such as point defects or dislocations. The τ_N is known to decrease with increasing temperature, and as a result, at room temperature, τ_{PL} is strongly reduced for poor-quality QWs. This means that the role of non-radiative recombination centers in InGaN QWs can be compared effectively by measuring the τ_{PL} for QWs at room temperature. In order to perform such a comparison, TRPL in InGaN QWs deposited using different growth parameters was measured at nominally identical conditions. The samples were excited by a 350-nm laser line with a 76 MHz repetition rate and a pulse duration of 150 fs from a mode-locked Ti:Sapphire laser. In order to observe effects associated with point defects, measurements were performed at low excitation conditions (the average excitation power was a few mW). The PL signal was dispersed by a 0.3-m focal length monochromator and detected by a Si streak camera.

In Fig. 2.20 we plot as an example the decay of PL intensity at 425 nm for one of our samples, together with the laser pulse temporal profile. The τ_{PL} has been obtained by fitting the experimental data with a one-exponential decay of PL signal. The τ_{PL} determined in such a way are plotted in Fig. 2.21, together with the PL spectrum taken just after the excitation.

As we pointed out, by decreasing the growth temperature we can shift the PL from 420 nm (see Fig. 2.21(a), sample A) to 520 nm (Fig. 2.21(b), sample B) for the constant nitrogen flux of 4.1 nm/min. For sample A we observe no spectral dispersion of $\tau_{PL} = 0.8$ ns. Such a behaviour of τ_{PL} is typical of homogeneous QWs, i.e., QWs where the carrier localization induced by indium content fluctuations can be neglected. For sample B, the PL decay time is qualitatively different, and τ_{PL} increases from 3 ns at $\lambda = 485$ nm to 13 ns at $\lambda = 570$ nm. Such a spectral dispersion of τ_{PL} implies carrier localizations in this QW, and can be attributed to indium content fluctuations. In general, it is expected that

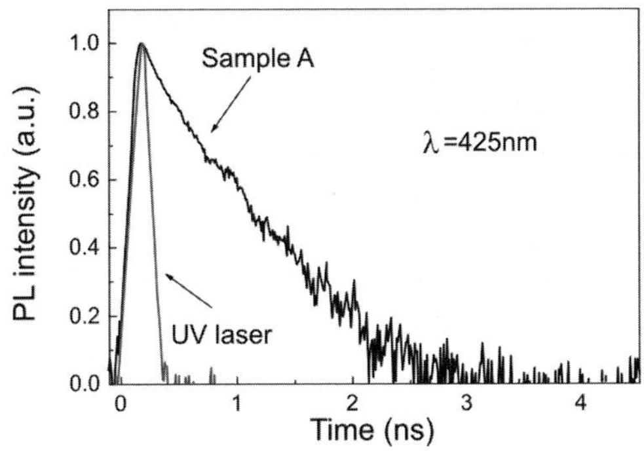

FIG. 2.20. The photoluminescence transient from an MQW structure at 425 nm.

such a localization can effectively reduce the influence of non-radiative recombination channels, but on the other hand it is much harder to obtain the population inversion necessary for lasing in such a system. Therefore, the spectral dispersion of τ_{PL} can be used as a fingerprint of carrier localization, giving valuable feedback for optimizing InGaN QWs growth procedure. As we have already shown, when we increased the N flux we were able to grow InGaN QWs with the same In content (around 26%) at higher growth temperature, as indicated by PL (see Fig. 2.21(c), sample C). For this sample the TRPL measurements show no spectral dispersion of τ_{PL}. This indicates that samples A and C are comparable from the viewpoint of carrier localization. Note that sample A has considerably lower In content in QW: 12%. In addition to the homogeneity improvement, the increased nitrogen flux probably contributes to the reduction in the number of non-radiative recombination centers in the layers by reducing the number of

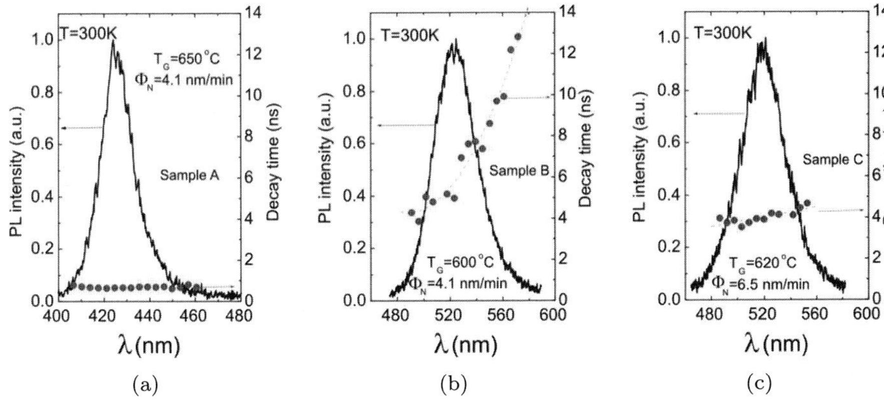

FIG. 2.21. The photoluminescence intensity and its decay time for blue (a) and green (b) and (c) emission, respectively.

nitrogen vacancies. Such a possibility is supported by the observed increase in the absolute value of τ_{PL} for the sample shown in Fig. 2.21(c) in comparison to that in Fig. 2.21(a). Since at room temperature τ_{PL} is expected to be dominated by a non-radiative recombination component τ_N, this increase is not likely to be caused by the larger built-in electric field for the high-In-content QWs of sample C and the related decrease in the oscillator strength for this transition.

2.5.5 *Optical lasing from InGaN QWs*

The laser structures for optical pumping were grown on HVPE bulk substrates with a dislocation density 10^6–10^7 cm^{-2}. The structures consist of 0.5-μm $Al_{0.08}Ga_{0.92}N$ bottom cladding followed by a 400-nm GaN waveguide. The active region located in the middle of the waveguide consists of one or three $In_xGa_{1-x}N$ QW with a thickness from 2.2 nm to 3.5 nm and 10 nm $In_{0.06}Ga_{0.94}N$ barriers. The upper $Al_{0.08}Ga_{0.92}N$ cladding have a thickness of 0.3 μm, and structures were capped by 5-nm GaN layer. The optical lasing was obtained from cleaved laser stripes with length L = 0.4–1.2 mm excited by third harmonics (355 nm) of a Nd:YAG laser beam with an aperture size of 0.25 mm; see inset to Fig. 2.22 for experimental configuration details. The laser pulses had a duration of 5 ns and a repetition rate of 20 Hz.

In Fig. 2.22 we compare optical spectra below and above the lasing threshold for two structures, with QWs optimized for an emission around 420 nm and 495 nm. We achieved optically pumped lasing from these structures at 409 nm and 472 nm, respectively. The structure which lased at 409 nm was grown with N flux 4.1 nm/min, and that lasing at 472 nm with higher N flux 6.5 nm/min, using otherwise nominally identical growth conditions. The lasing emission intensity as a function of the pumping power is shown in Fig. 2.23. It is clear that

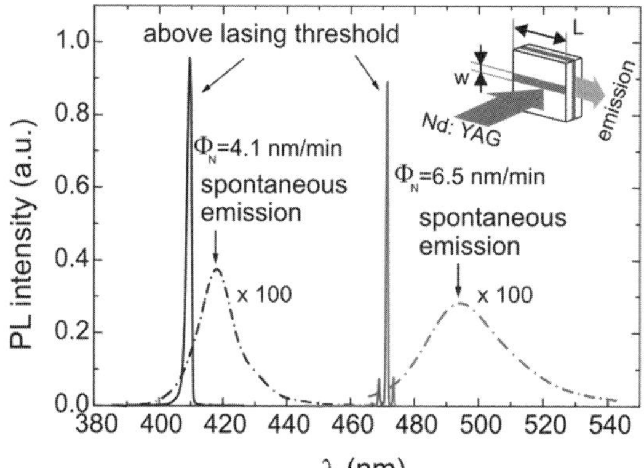

FIG. 2.22. The photoluminescence and lasing spectra of two laser structures optimized for lasing at 409 nm and 472 nm.

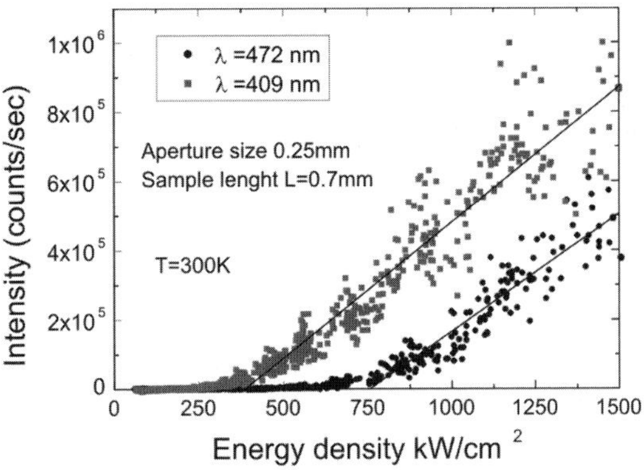

FIG. 2.23. The light-power characteristics for structures lasing at 409 nm and 472 nm.

we are able to achieve lasing at 409 nm much earlier than at 472 nm. Although better optical efficiency of QW for the 409-nm laser cannot be ruled out, we attribute this effect primarily to the superior light confinement in the waveguide at a shorter wavelength (same design for both lasers).

The largest available active nitrogen flux for the nitrogen plasma source initially mounted on our MBE system was 6.5 nm/min, sufficient to grow the structure lasing at 472 nm. After upgrading the MBE system with a second RF-plasma source, the maximum nitrogen flux of 14 nm/min became available. This allowed us to further shift the PL emission towards the green region while maintaining a high-growth temperature of InGaN. In Fig. 2.24 we demonstrate spontaneous and stimulated emission from two SQW laser structures with quantum well widths of 2.2 nm and 3.5 nm, respectively. The growth conditions for both structures were nominally identical, and other than QW thickness the design was the same, with the active region consisting of 10 nm InGaN barriers containing 6% In and SQW with an In content of about 25%. As seen in Fig. 2.24, the maximum of the spontaneous emission is at a longer wavelength for the structure with wider QW: 545 nm vs. 525 nm. We attribute this shift to the stronger QCSE expected for the wider well. Note that for each laser structure we show stimulated emission from two laser bars of different length, and in both cases the stimulated emission from longer bar is red-shifted compared to that from the shorter bar. We were able to shift stimulated emission from 487 nm to 494 nm for the laser structure with the well width $d = 2.2$ nm, and from 488 nm to 501 nm for the one with the well width $d = 3.5$ nm, by increasing the laser bar length from 0.7 nm to 1.2 nm for both cases. This effect is very general. Indeed, for all the investigated laser structures we have found that lasing at longer wavelengths can be realized with a longer resonator length L.

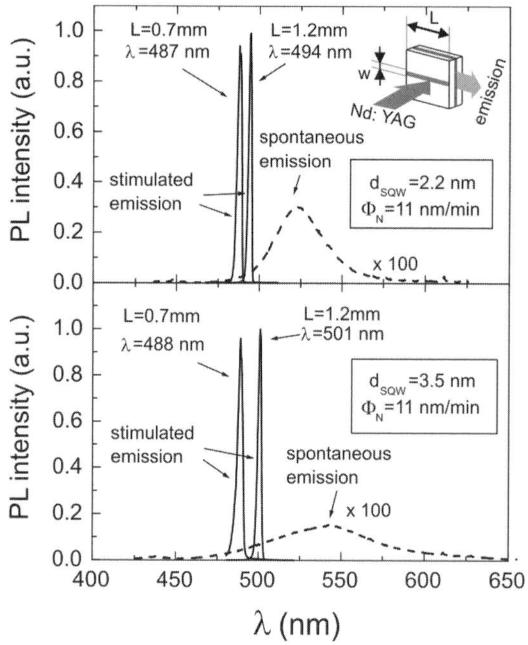

FIG. 2.24. The photoluminescence and lasing spectra around 500 nm for two SQW structures with thickness d = 2.2 nm (a) and d = 3.5 nm (b), respectively.

2.5.6 *Laser diodes*

Laser diode structures were grown on (0001) Ga-polarity, conductive GaN substrates with low dislocation density 10^3–10^4 cm^{-3} (HPNS-GaN, Ammono-GaN), and with a dislocation density of 10^6–10^7 cm^{-2} (HVPE-GaN). For the laser diodes (LDs) operating at 405–420 nm, we used the structure shown in Fig. 2.25 (Skierbiszewski *et al.*, 2005).

The 40-nm GaN:Si buffer layer and 450-nm $Al_{0.08}Ga_{0.92}N$:Si cladding were grown under Ga-rich conditions at 720°C. The bottom waveguide, MQWs, EBL, top waveguide, top cladding, and contact layer were grown in In-rich conditions at 650°C. The active region consisted of three 3-nm $In_{0.1}Ga_{0.9}N$ wells with 7-nm $In_{0.02}Ga_{0.98}N$ barriers. The devices were processed as ridge-waveguide, oxide-isolated lasers. The mesa structures were etched to a depth of 0.3 μm. The 20-μm-wide and 500-μm-long stripes were used as laser resonators. The oxidized Ni/Au ohmic contacts were deposited on the top surface of the devices, and Ti/Au contacts were deposited on the backside of the highly conducting n-GaN substrate crystal. The cleaved laser mirror facets were coated with symmetrically reflecting mirrors. Figure 2.26(a) shows the light current voltage (LIV) characteristics of the CW LDs with lasing threshold current density and voltage

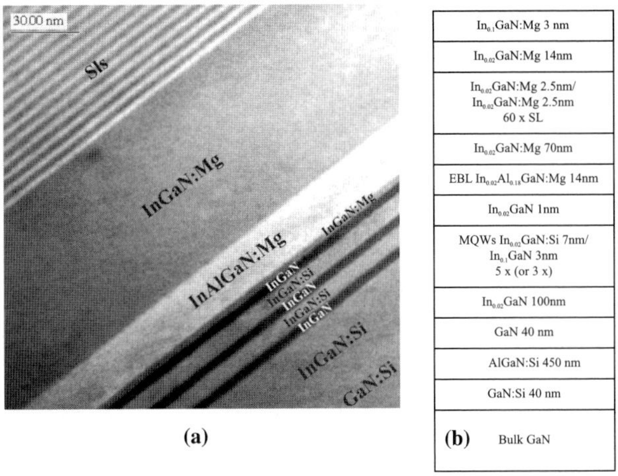

FIG. 2.25. Transmission Electron Microscope image of the active region of a PAMBE laser diode (a), and structure details (b).

of 5.5 kA/cm^2 and 5.7 V, respectively. Lasing was observed up to 60 mW of optical output power (30 mW per facet) at a wavelength of 411 nm, as indicated in Fig. 2.26(b) (Skierbiszewski *et al.*, 2006). This confirms that growth of high-quality layers by PAMBE has been achieved; the smooth interfaces required for LDs can be obtained as shown in Fig. 2.25(a).

The optically pumped stimulated emission at 501 nm was a good starting point for true blue and green laser diodes. It is worth stressing here that not

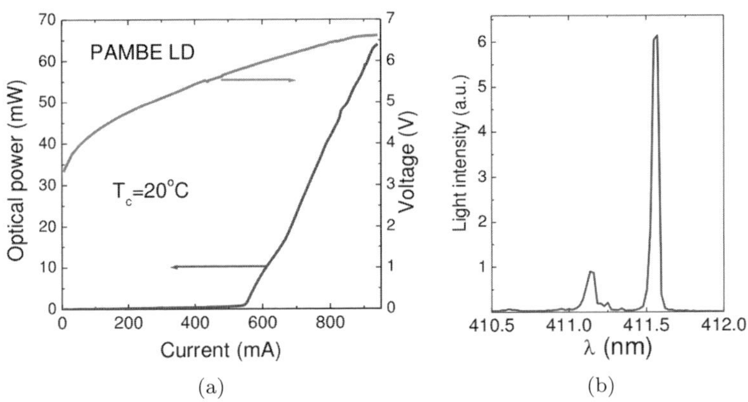

FIG. 2.26. L–I–V characteristics of a PAMBE-grown cw laser diode (a) and lasing spectrum (b).

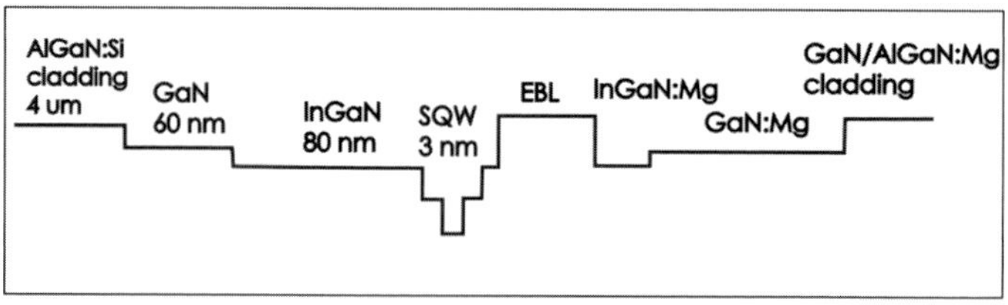

FIG. 2.27. Structural details of an SQW 455-nm laser diode.

only efficient QWs are essential for LDs, but also such design of the LDs ensuring confinement of the optical modes inside the laser waveguide. For the LDs operating at wavelengths longer than 430 nm, we found that the crucial factor for achieving lasing is optimization of the claddings and the waveguide. The details of a single quantum well (SQW) LDs structure lasing at 455 nm are presented in Fig. 2.27 (Skierbiszewski *et al.*, 2012*b*). In order to reduce penetration of the optical modes into the GaN substrate we have grown special claddings comprised of 2-μm heavily doped GaN:Si at the level of 7×10^{19} cm^{-3} and 4 μm GaN/$Al_{0.05}Ga_{0.95}N$ superlattice. For further localization of the optical modes inside the waveguide, and to reduce losses in the gold p-type contact, we also used an "optical confinement layer" (OCL), 80 nm of $In_{0.07}Ga_{0.93}N$, and changed the upper cladding from InGaN/InAlGaN to GaN/AlGaN. As arises from theoretical calculations, the new claddings and waveguide with OCL helped significantly (i) to decrease absorption of the optical modes in the upper gold contact, and (ii) to reduce penetration of the optical modes to the GaN HVPE substrate. The other interesting feature of SQW LDs grown by PAMBE is a relatively small blue shift from spontaneous (at 465 nm) to lasing emission (at 455 nm); Fig. 2.28(b). This is achieved by employing OCL and staggered SQWs which reduce piezoelectric fields inside SQW.

We also introduced a new design of AlGaN-claddings-free separate confinement heterostructure (SCH) LDs grown by PAMBE, and demonstrated LDs operation from 435 nm to 460 nm (Skierbiszewski *et al.*, 2012*a*). The LD structure details are shown in Fig. 2.29. The 0.5-μm GaN:Si buffer layer is followed by undoped 0.3-μm GaN. Next, the asymmetric high-indium-content undoped $In_{0.08}Ga_{0.92}N$ waveguide (80 nm/40 nm) is grown. The active region consist of three 3-nm $In_{0.13}Ga_{0.87}N$ wells with 7-nm $In_{0.08}Ga_{0.92}N$ barriers. The p-type 20 nm $In_{0.01}Al_{0.14}Ga_{0.85}N$ electron-blocking layer (EBL) doped with Mg is located at the top of the upper waveguide. The EBL doping level was 5×10^{19} cm^{-3}. Then 0.5-μm GaN:Mg upper cladding is followed by a 60-nm $In_{0.01}Ga_{0.99}N$:Mg and 5-nm $In_{0.1}Ga_{0.9}N$:Mg contact layer. The devices were processed as ridge-waveguide, oxide-isolated lasers. The mesa structures were etched to a depth of 0.35 μm. The 3-μm-wide and 700-μm-long stripes were used as laser resonators. The oxidized Ni/Au ohmic contacts were deposited on the top surface of the devices, and Ti/Au contacts were deposited on the backside of the conductive

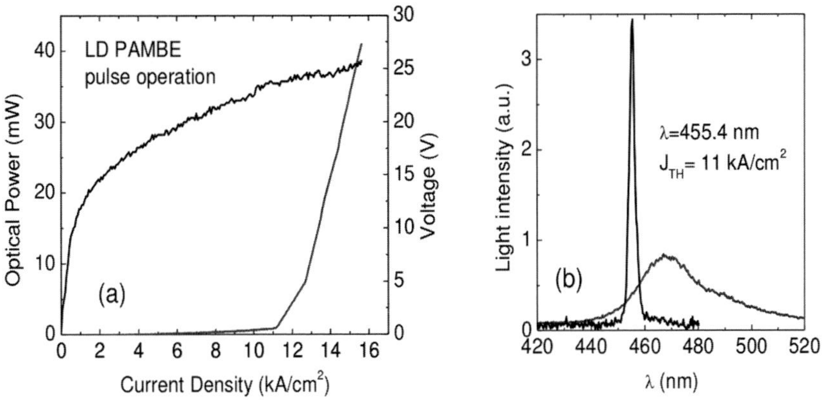

FIG. 2.28. L–I–V characteristics of PAMBE-grown 455 nm SQW laser diode operating in pulse mode (a), and spectrum below and above lasing threshold (b).

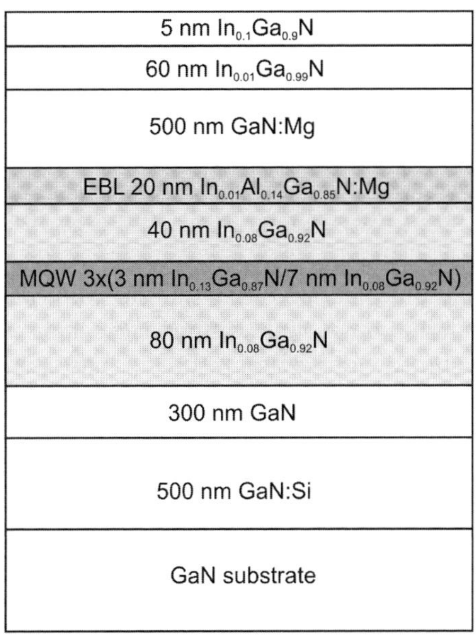

FIG. 2.29. The structure of the AlGaN-cladding free laser diode grown by PAMBE.

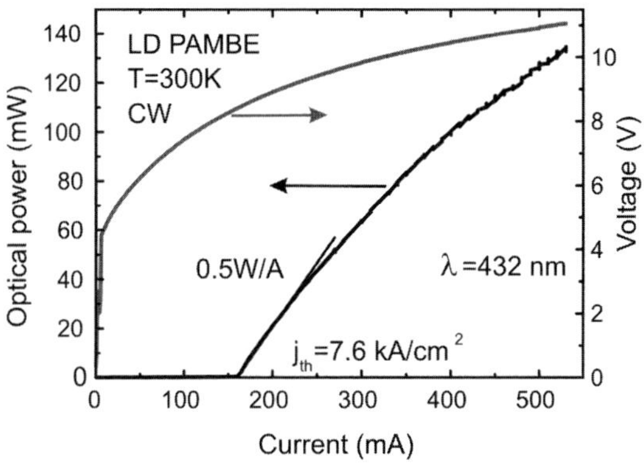

FIG. 2.30. The light–current–voltage characteristic of PAMBE laser diodes operating in continuous wave regime at 432 nm. The mesa for this LD was 3 × 700 μm.

GaN substrate crystal. The LDs have cleaved laser mirror facets covered by reflection (90%) and antireflection dielectric coatings.

The L-I-V characteristics for room-temperature continuous-wave (cw) operation for LDs at 432 nm are shown in Fig. 2.30. The threshold current density is 7.6 kA/cm^2 and threshold voltage is about 8.2 V. The laser diodes are mounted p-down in copper clamps. The maximum optical power of 130 mW is achieved in cw mode with a differential gain of 0.5 W/A.

2.6 "Plasmonic" GaN substrates and their use for lasers

As we have shown earlier (Section 2.3), the crystals of GaN grown by the HNPS method can be strongly n-type doped with free electron concentration as high as 10^{19}–10^{20} cm^{-3}, depending on growth temperature. This can be used as an element of the refractive index engineering in InGaN-based laser diodes (i.e. Perlin *et al.*, 2009; Perlin *et al.*, 2011*a*; Perlin *et al.*, 2012) to achieve efficient transversal confinement of the optical mode. Typically, in nitride-based LDs a wider band gap AlGaN alloys are used as cladding layers providing optical confinement. However in contrast to the lattice-matched GaAs-AlGaAs system, the lattice mismatch between the GaN waveguide and AlGaN cladding makes the AlGaN layer suffer from tensile strain which leads to its cracking and generation of defects. Additionally, the refractive index contrast between GaN and AlGaN decreases with emission wavelength. This limits the possibility of creating a sufficiently strong optical confinement in the device, and causes a significant leakage of the optical mode to the substrate (Smolyakov *et al.*, 2005; Laino *et al.*, 2007), increasing the internal losses in the laser, especially for longer wavelengths. To suppress the mode leakage the AlGaN claddings as thick as 2 μm (Laino *et al.*, 2007) or even 5 μm (Nagahama *et al.*, 2000) are used.

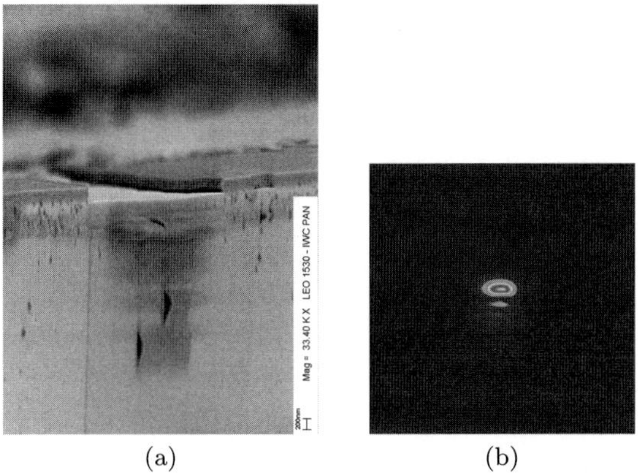

FIG. 2.31. Penetration of light into the substrate of LD: (a) carbon deposit on the LD mirror due to light-induced chemical reactions, (b) near-field image of the beam of LD deposited on a low-conductivity substrate.

The effect of mode penetration into the substrate is sometimes visualized by the formation of a carbon deposit (Package Induced Failure) being the result of light-induced reactions on the facet of the laser. This effect is shown in Fig. 2.31(a).

Perlin *et al.* (2009, 2011*a*, 2012) have shown that the use of a highly conductive HNPS GaN substrate allows suppressing the mode leakage completely in laser diodes with AlGaN lower cladding as thin as 0.5–0.8 μm. This was possible because of a significant decrease of dielectric constant, and consequently a reduction of the refractive index of GaN at frequencies closer to the plasma frequency. This dependence can be well described by the following equation (Perlin *et al.*, 1995):

$$\varepsilon(\omega) = \varepsilon_0 \left(1 - \frac{\omega_p^2}{\omega^2}\right) \tag{2.8}$$

where ω_p is the plasma frequency given by the expression:

$$\omega_p = \frac{N_e e^2}{m^* \varepsilon \varepsilon_0} \tag{2.9}$$

Figure 2.32 shows the dependence of refractive index contrast between strongly n-type and undoped GaN on the wavelength for various electron concentrations. For the blue spectral range (450 nm) the refractive index contrast reaches 0.5% for an electron concentration of 5×10^{19} cm^{-3} and 1% for 1×10^{20} cm^{-3}. This provides the possibility of constructing the laser diode using such a type of substrate as a part of its optical waveguide.

The results of the calculations (Perlin *et al.*, 2009) of the transversal mode distribution in two identical laser structures with 0.6-μm thick 8% AlGaN lower

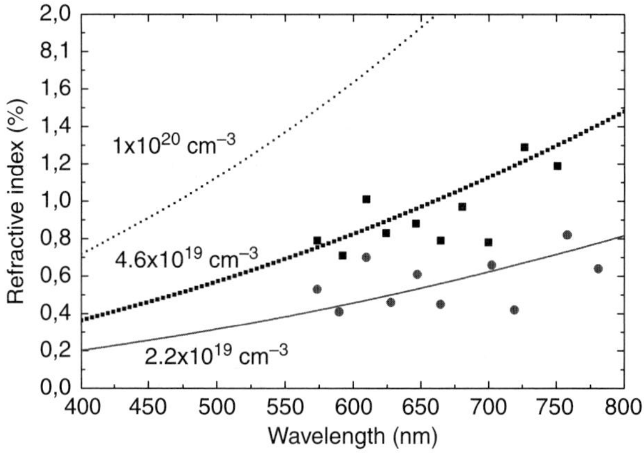

FIG. 2.32. Refractive index contrast versus wavelength for various electron concentrations. Solid line: theoretical curves derived from eq. (2.8); squares and circles: experiment, after Perlin *et al.* (2012).

cladding layers deposited on two different substrates are presented in Fig. 2.33. It is clear that the refractive index contrast induced by free electrons at a concentration of 5×10^{19} cm^{-3} is sufficient to suppress the leakage of the optical mode to the substrate in contrast to the conventional HVPE-grown substrate with an electron concentration of 10^{18} cm^{-3}.

The results of the calculations were confirmed by experimental near-field distributions for laser diodes corresponding to the structures used for modeling. Figure 2.34 presents the result of this experiment. For the conventional

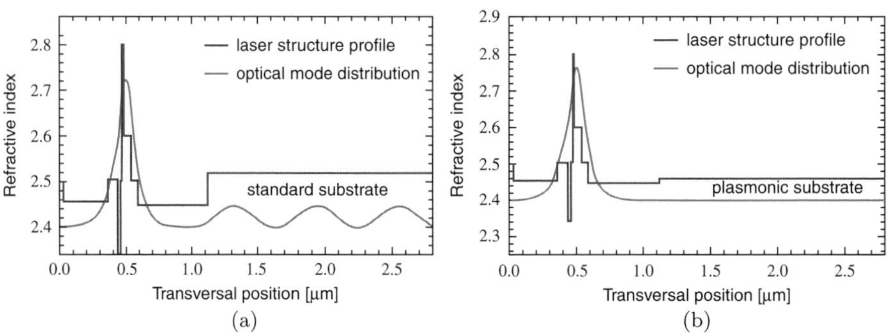

FIG. 2.33. Plane-wave expansion method (Schwarz and Witzigmann, 2007) calculation of the transversal mode distribution in LD: (a) for a standard HVPE-based structure, (b) for the plasmonic substrate.

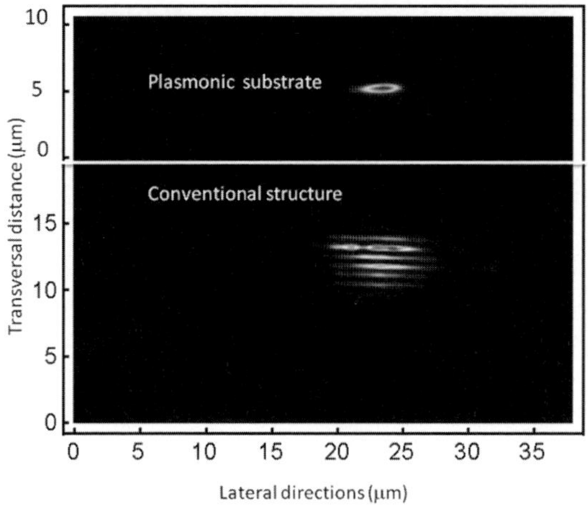

FIG. 2.34. The comparison of the near-field distribution for two structures—the upper one grown on a plasmonic substrate, and the lower one on a standard HVPE substrate.

structure the optical mode extends deep into the substrate, while in the case of the plasmonic substrate the mode is well confined around the active area.

An important and quite obvious effect is a significant reduction in the threshold current density. For considered cases it was from 5.2 kA/cm^2 to 2.3 kA/cm^2 (Perlin *et al.*, 2009).

The threshold current density of the laser structure with an AlGaN cladding of thickness 0.5 μm has been simulated (Perlin *et al.*, 2012) as a dependence on the refractive index contrast of the substrate compared to the undoped GaN. For around 0.5% the threshold current density suddenly dropped. This 0.5% index change corresponds roughly to the electron concentration of 5×10^{19} cm^{-3}, in agreement with the results mentioned previously.

It is therefore shown that the concept of using plasmonic substrates is a way of improving and simplifying the design of blue and green InGaN laser diodes. It can help to design optimum laser structures with minimized strain, and can fully profit from the advantages provided by plasmonic substrates.

2.7 Laser diode arrays on laterally patterned substrates

2.7.1 *Background*

In spite of the growing popularity of semipolar and non-polar substrate orientations, most of the III-nitride optoelectronic devices are still grown on polar c-plane GaN templates or free-standing GaN substrates. The predominant growth technique is metal organic vapor phase epitaxy (MOVPE). In the case of MOVPE, good morphology and chemical uniformity of AlInGaN layers are obtained only in the step-flow growth mode, on vicinal surfaces. On such vicinal surfaces we can distinguish atomic steps, separated by atomically flat terraces. By the vicinal angle δ, we understand the angle between the GaN c-plane and its real surface, also termed as off-cut, off-orientation, miscut, off-axis, or misorientation angle. Let h be the atomic step height and w the atomic terrace width. Then we have $\tan\delta = h/w$. Good morphology of epitaxial layers can be obtained only for δ in a certain range, which for GaN grown by MOVPE is between 0.3 and about 1°.

For $0 < \delta < 0.2°$ the step-flow growth mode with parallel and well-ordered atomic steps by MOVPE is not possible because of the lack of an effective and well-determined source of atomic steps. During growth on exact-oriented substrates the atomic steps tend to flow around the dislocations which are always present, at surface densities of 10^4–10^7 cm^{-3} for free-standing GaN, and 10^8–10^9 cm^{-3} for GaN templates on foreign substrates. As a result, for MOVPE layers grown on exact oriented substrates the density and direction of atomic steps become random. For $\delta > 0.2°$ the density and direction of the atomic steps are uniform.

For GaN/sapphire and GaN/SiC templates used mostly for light-emitting diode and transistor fabrication, the vicinal angle of the GaN epilayer is determined by the miscut angle of the foreign substrate.

In the case of free-standing GaN (made by HVPE, HNPS or ammonothermal methods) used for the production of laser diodes, the desired vicinal angle is realized by polishing the substrate to high precision. Polishing causes, however, a significant loss of the material. This becomes an issue in view of the increasing size of free-standing GaN substrates and their still high price. Starting from exact c-plane-oriented 2-inch GaN wafer it requires as much as 440 μm (51.2 mm × tan(0.5 degree)) of material per side to be polished off to reach 0.5 degree miscut. Therefore, a 300-μm thick 0.5-degree miscut GaN substrate requires initially at least a 1200-μm thick GaN crystal, which accounts for a 25% yield only.

For the step flow growth mode, at a given vertical growth rate, v, we can define the lateral velocity of the atomic steps, s:

$$s = \frac{v}{\sin\delta} \tag{2.10}$$

According to eq. (2.10), the surface atomic steps during epitaxial growth move faster for smaller δ. It is well known that for increasing δ, the amount of indium incorporated into InGaN layers grown by MOVPE in the step-flow mode, at fixed pressure, temperature, and In precursor flow, decreases (Tachibana *et al.*, 2008; Keller *et al.*, 2008; Leszczynski *et al.*, 2011). This effect is shown in Fig. 2.35.

Although all experimental results obtained so far confirm the effect described above, they are focused on how indium content depends on δ, without paying attention to whether the substrate miscut was chosen towards the a or m axis. This issue certainly requires more study, but the first results show that indium content is not sensitive to the azimuth of substrate miscut; see Fig. 2.35(b).

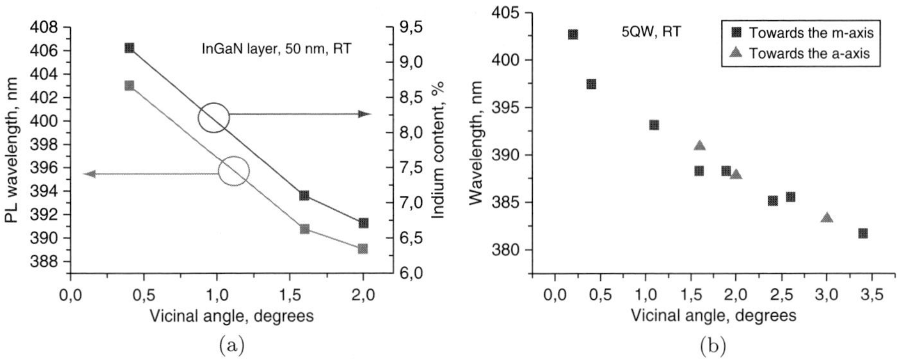

FIG. 2.35. Influence of substrate vicinal angle on photoluminescence wavelength from InGaN: (a) PL wavelength of an InGaN layer grown on patterned GaN/sapphire substrate, and its indium content measured by X-ray diffraction for different substrate vicinal angles. Solid lines are guides to the eye. (b) Room-temperature PL wavelength of a InGaN/GaN MQW structure grown on substrates with different vicinal angles.

The precise control of the quantum-well chemical composition is the key factor in the fabrication of all InGaN-based optoelectronic devices. The band gap and hence the emitted wavelength depends on indium content of QWs, and therefore in the planar light-emitting diodes and laser diodes technology it is very important that the vicinal angle is uniformly constant across the substrate.

Concerning the physics of different indium incorporations on differently angled surfaces, there are two models present in the literature:

i) Indium is incorporated on the atomic terraces at a constant rate and is not incorporated on the atomic steps (Krysko *et al.*, 2007). For larger δ, according to Nakamura and Chichibu (2000), we have a higher density of atomic steps (narrower terraces) and therefore less indium. In this model the indium content should not depend on v for a substrate of a given δ.
ii) Indium incorporation depends on s, because to avoid thermal desorption the indium adatom has to be quickly surrounded by the neighboring gallium adatoms (Keller *et al.*, 2008). In this model, the indium incorporation should increase for smaller δ and constant v and also increase for larger v and constant δ.

Based on Leszczynski *et al.* (2011), for a 50-nm InGaN layer grown on an homogeneous GaN substrates at 820°C and reactor pressure of 300 mbar at $v = 0.7$ Å/s, the indium content increased from 8% to 13% for δ decreasing from 1.8 to 0.3°. For the same growth temperature and pressure, and $\delta = 1°$ the indium content increased from 7% to 12% for v increasing from 0.35 to 1.0 Å/s.

Therefore, it should be concluded that for the step-flow growth mode of InGaN by MOVPE, model ii) is more likely.

2.7.2 *Laser diode arrays on laterally patterned HNPS GaN substrates*

Basing on the findings above we used patterned freestanding GaN for fabrication of monolithic LDs emitting light at different wavelengths. We will mention three applications of such multi-color LDs:

i) Two-wavelength interferometry with the advantage of increased accuracy and elimination of the phase ambiguity (Wang *et al.*, 1995; de Groot and Kishner, 1991).
ii) Down-conversion to generate terahertz frequencies by mixing two laser beams with slightly different wavelengths (Hoffmann and Hofmann, 2007).
iii) Three-dimensional imaging by prismatic projectors, in which the stereoscopic feature works by filtering the two slightly different wavelengths in each eye by special glasses.

For the purpose of illustration we present below the two-wavelength laser diode emitting light in the violet region (about 400 nm). The GaN substrate was grown by the high-pressure nitrogen solution method (Section 2.3) and polished mechanically to obtain zero vicinal angle on the whole surface. Then the substrate was patterned in a special way. Variable orientation of the substrate surface was obtained by special photolithography and ion-etching procedures (Figs. 2.36 and 2.37). The substrate uniformly coated by photoresist was exposed with a variable light dose in a laser writer device. After resist development, a variable-thickness resist layer was obtained. Its profile was transferred onto the substrate surface by ion-etching, since the resist etching rate was the same as for GaN.

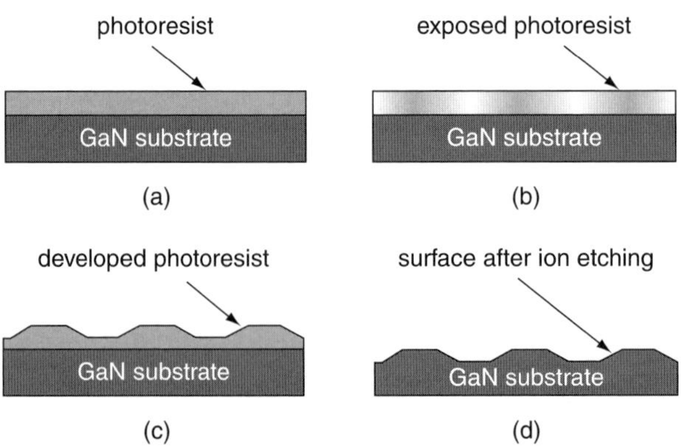

FIG. 2.36. Fabrication of GaN substrate with variable misorientation.

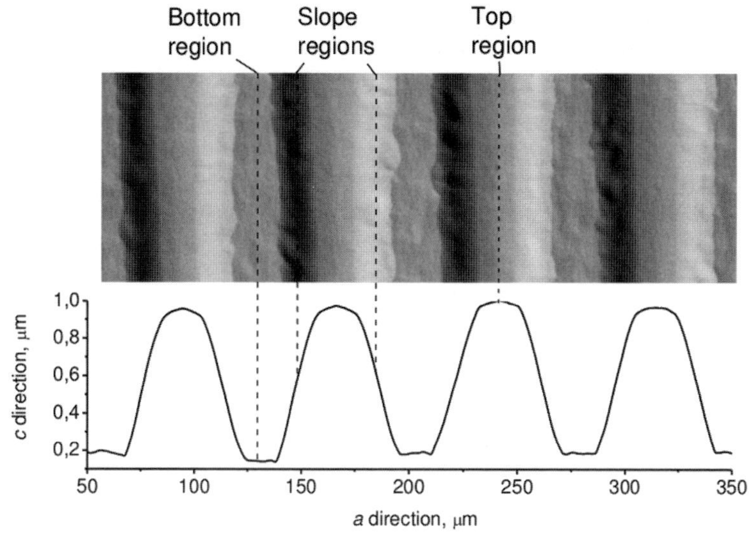

FIG. 2.37. Patterned GaN substrate with flat top and bottom regions and angled slope regions. Optical microscope image with Nomarski contrast is horizontally aligned with a stylus profile scan of the surface. In this pattern the slope angle was 2.0°.

The substrate was patterned to obtain two sets of parallel, differently angled regions aligned along the m-axis: the first (A) had a width of 80 μm, and the second (B) 35 μm (see Fig. 2.38).

The undulations depth was 0.5 μm for both slopes. As a result, the slope angles and their vicinal angles were $0.35 \pm 0.1°$ for A and $0.85 \pm 0.1°$ for B. Then an LD structure was grown by MOVPE in the same reactor as for the

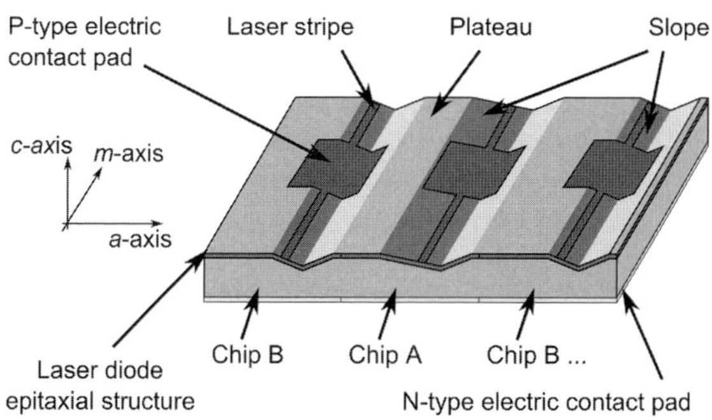

FIG. 2.38. A schematic view of an LD bar grown on a patterned GaN substrate. Note that the slopes for chips A and B are angled differently.

InGaN layers with known dependence of content on the miscut angle. The LD structure consisted of 820 nm $Al_{0.08}Ga_{0.92}N$ lower cladding with the first 20 nm of graded Al-composition (0–8 %), 100 nm GaN and 40 nm InGaN with (intended) 2 % of In lower waveguide, 3 QWs active region with InGaN/InGaN barriers (7 nm)/wells (3 nm) with (intended) 2/10% In respectively, 10-nm-GaN cap layer, a 20-nm graded 12–20% AlGaN electron blocking layer, 150-nm GaN upper waveguide layer, 430-nm AlGaN upper cladding with 5% of Al and 30-nm GaN:Mg subcontact layer. During processing each of the laser stripes was placed exactly in one of the angled regions, also shown schematically in Fig. 2.38.

Again, let us mark LD chips grown in the 0.35 and 0.85-degree regions as A and B, respectively. The laser stripes width was 3 μm and the resonator length 700 μm for both A and B. We tested the LDs in the cw mode and the threshold current was 80 mA for A and 120 mA for B, giving a threshold current density of 3.8 and 5.7 kA/cm^2, respectively, and the threshold voltage was about 5.5 V for both chips. Due to thermal roll-over the maximum optical power was about 70 mW for A and 40 mW for B. Lasing spectra at optical power of 10 mW were acquired, and are shown in Fig. 2.39. The lasing wavelength was 405.8 ± 0.2 nm for set A and 401.0 ± 1 nm for set B. The I–V characteristics for selected A and B are shown in the inset in Fig. 2.39.

The reason for the difference in threshold current for A and B is currently not known and requires further study. Operating voltage at the same current was lower for B, which could be attributed to the fact that the hole concentration

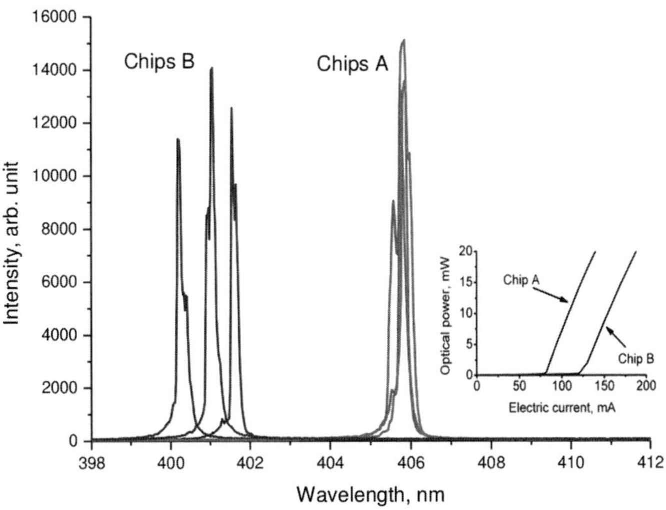

FIG. 2.39. Lasing spectra of six adjacent devices from a single bar. The lasing wavelength for chips A (B) is longer (shorter) due to less (more) angled slopes of the GaN substrate and hence more (less) indium in the active region.

in Mg-doped GaN is higher for a higher miscut, even though Mg concentration is equal (Suski *et al.*, 2008; Suski *et al.*, 2010). The observed difference in lasing wavelength is a consequence of variation in indium content between A and B. We reject the possibility of the wavelength variation due to variation of the QW width (quantum confined Stark effect) because the vertical growth rate is independent of the vicinal angle δ (Franssen *et al.*, 2008). Moreover, due to the high density of carriers injected into the MQWs the electric field is screened (Wang *et al.*, 2004; Kuroda and Takeuchi, 2002). The larger spread of the lasing wavelength observed in B is also currently studied.

To conclude, the indium content of InGaN layers grown by MOVPE can be controlled not only in the growth direction by the growth parameters but also in the growth plane by the local vicinal angle of the substrate. In particular, it is possible to control the wavelength of LDs grown on exact-oriented substrates by patterning and thus avoid polishing the whole substrate to the desired vicinal angle. It should also be possible to fabricate multi-wavelength LD arrays, which would have several applications.

2.8 High-power lasers and arrays on Ammono and HNPS GaN substrates

High optical power, short-wavelength laser diode sources are needed for a number of applications, including material processing, table-top projectors (beamers), and movie projectors. The requirements for these light sources are high optical power in the multi-watt range, compactness, and competitive price. However, the fabrication of high-power single-stripe emitters is challenging. The first reason is that the light-power density at the facet of the laser increases rapidly with the increasing power, approaching the optical damage threshold of the laser mirror. Figure 2.40 shows the power density on the laser facet calculated for 1 W of optical power. For a stripe width of around 4 μm the power density is close to 40 MW/cm^2, a value commonly associated with catastrophic optical mirror damage in nitrides (Ikeda *et al.*, 2004). Unfortunately, the use of broad-area devices was shown to be quite unsuccessful in the case of nitrides structures. For instance, the real broad-stripe laser diode with the stripe width of 100 μm were never demonstrated. The broad-area devices tend to overheat, and fail to operate under a CW regime of current injection (Goto *et al.*, 2003).

Laser diode array is a technical solution which is free of many problems (mentioned earlier) and meeting most of the requirements for high-power light source. However, laser diode arrays have to be large enough to contain from ten up to even thirty emitters, which means a lateral size up to 1 cm. This large chip size requires uniform structures deposited on very-high-quality substrates (low and uniform defect density). We fabricated our laser diode arrays on high-quality gallium nitride substrates synthesized via ammonothermal method (Section 2.3) or by multi-feed seed HNSP method (Section 2.3). An example of the laser structure grown on Ammono substrate is shown in Fig. 2.41.

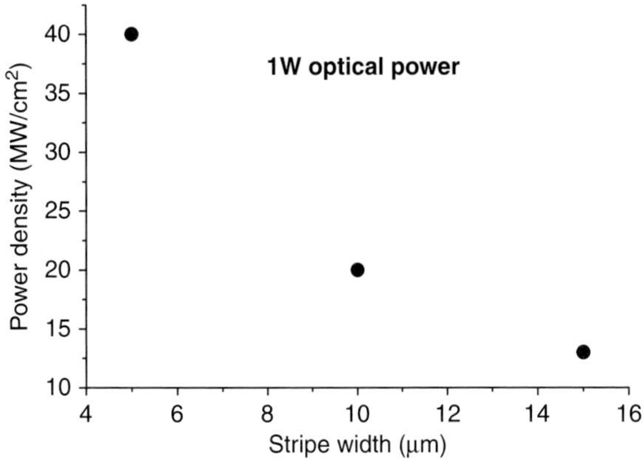

FIG. 2.40. The calculated optical power density on the facet of a typical laser diode at a total optical power of 1 W.

The laser structures presented here were grown by the metalorganic vapor phase epitaxy (MOVPE) method on bulk gallium nitride crystals. GaN c-plane crystals were first cut with a well-defined miscut (between 0.3–0.6°) and then lapped and finally mechanochemically polished to obtain an atomically flat surface. The laser structures were grown in the following way. An initial cladding layer of 800 nm of $Al_{0.08}Ga_{0.92}N$:Si was deposited on the n-type GaN substrate, then a GaN:Si guiding layer of 100 nm and 50 nm $In_{0.02}Ga_{0.98}N$ current injection layer. The active area was formed from 3.5-nm $In_{0.08}Ga_{0.92}N$ quantum wells $\times 3$ with quantum barriers of 10-nm $In_{0.02}Ga_{0.98}N$:Si. The quantum wells

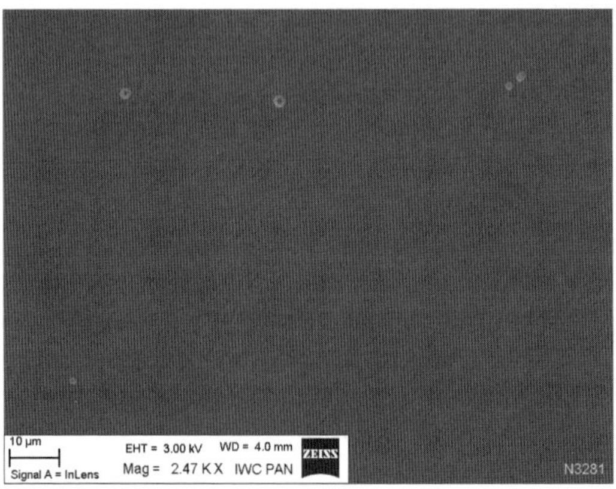

FIG. 2.41. Laser structure fabricated on top of Ammono GaN substrate containing only 6×10^4 cm^{-4} dislocations. The hexagonal pits are dislocations revealed by a selective etching method.

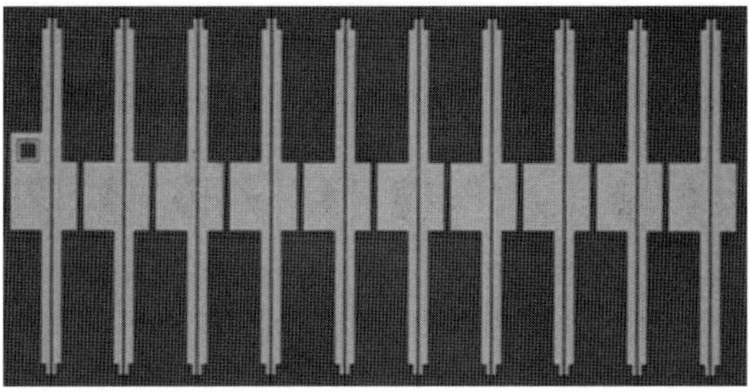

FIG. 2.42. An example of a ten-emitter laser diode array.

were capped with a 10-nm GaN layer and a 20-nm electron blocking layer of $Al_{0.2}Ga_{0.8}N$:Mg. Then a p-type 150-nm guiding layer of GaN was grown, capped with 465-nm $Al_{0.08}Ga_{0.92}N$:Mg p-cladding and a 30-nm GaN:Mg subcontact layer.

The laser diode arrays were processed (Fig. 2.42) in the way similar to conventional single-emitter laser diodes (ridge waveguide, oxide isolated devices). The ridge width was from 3 to 10 μm, the resonator length was 700 μm, and the number of emitters was from three to thirty-four.

Here we want to present the results obtained on two types of laser diode arrays: a wider-stripe (10 μm width) mini array and large sixteen emitters full-size arrays.

The mini-arrays were fabricated as an assembly of three emitters (stripes) with a stripe width of 10 μm and array pitch of 80 μm. We fabricated mini-arrays with three and five emitters in one chip (Perlin *et al.*, 2011*b*). These chips were mounted in a CS-style package having two copper electrodes contacting the bottom and top surfaces of the array.

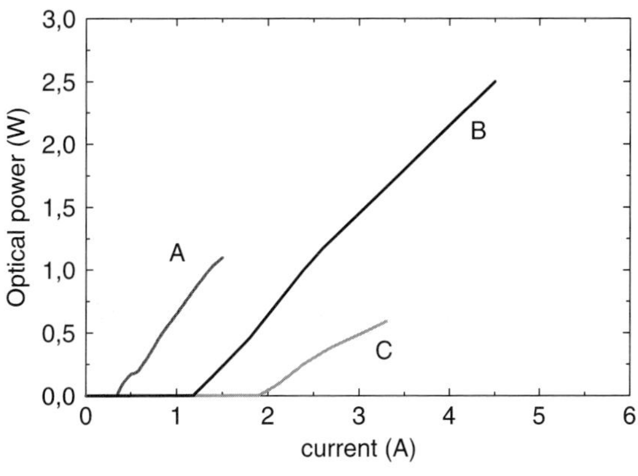

FIG. 2.43. The optical power emitted from three- and five-emitter mini-arrays (B and C respectively) and an identical single emitter (A).

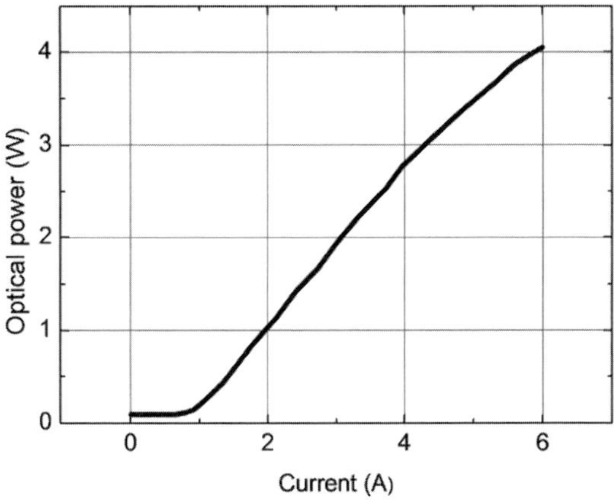

FIG. 2.44. The optical power versus current for the sixteen-emitters device.

We can see in Fig. 2.43 that the maximum optical power (2.5 W) was demonstrated for a three-stripe device, which had similar differential efficiency to the single-emitter laser. What concerns the five-stripe device, the slope efficiency and thermal roll-over is apparently much worse. The overheating of the five-stripe

FIG. 2.45. The CS mount used for the assembly (thirty-four emitters on the right) and the laboratory setup for the array characterization.

device is apparently related with a worse dependence between the total thermal load and chip size which defines the efficiency of the heat dissipation from the device.

The large arrays contained sixteen to thirty-four emitters with stripe width of 3 μm and array pitch of 300 μm. The arrays were mounted in a CS-style package, in a similar way as mentioned in the case of mini-arrays (Fig. 2.45). After packaging, the arrays were mounted on a thermoelectric cooler (TEC) and set on the water-cooled stage.

For the sixteen-emitter laser diode array the maximum optical power demonstrated was 4 W at 6 A, as shown in Fig. 2.44. The optical power was limited by the capability of the external cooling systems. It is worth noticing that the operating current density was only around 18 kA/cm^2, which offers the possibility of obtaining a reasonable lifetime exceeding 5000 h.

The obtained results indicate the great potential and the importance of nitride laser diode arrays for constructing very high optical power emitters especially for large movie projectors.

2.9 Summary and conclusions

Homoepitaxy of GaN is still a challenge in many aspects, mostly because of the still limited availability of large, perfect GaN substrates, especially in non-polar and semi-polar orientations.

The existing methods of GaN crystal growth have their advantages, but also serious limitations slowing down the progress in bulk crystallization. For HVPE the main drawback is bowing of crystallographic planes due to stresses induced by foreign substrates and by growth mechanisms themselves. Therefore, real bulk crystallization by this method is still a challenge.

The crystals grown by ammonothermal and HNPS methods are of high structural perfection, but the low rate of their growth slows down the progress towards mass production and cost effectiveness. On the other hand, GaN crystals grown under pressure are excellent substrates for laser diodes. Due to their uniformity and excellent electrical properties, high-power devices including laser diode arrays were constructed. With the use of the pressure-grown substrates, constant progress is being achieved in the development of the plasma-assisted MBE approach to laser diode fabrication. The results strongly suggest that PA MBE can be at least as efficient as MOVPE in growing epitaxial quantum structures for high-performance GaN-based laser diodes covering a large spectral range.

Further development of GaN bulk crystallization by high-pressure methods is therefore highly desirable, especially for high-power laser diode and laser diode array applications. The GaN crystallization process itself still contains questions of basic importance, including melting conditions of GaN, bulk growth mechanisms by HVPE or the scalability limit of the ammonothermal method. The effort put into solving these issues can be compensated in the near future with new important applications of GaN-related compounds.

Acknowledgments

This work was partially supported by the Polish Ministry of Science and Higher Education Grant No. IT 13426, and the European Union within European Regional Development Fund, through grant Innovative Economy (POIG.01.01. 02-00-008/08).

References

Adelmann, C., Brault, J., Jalabert, D., Gentile, P., Mariette, H., Mula, G., and Daudin, B. (2002). *J. Appl. Phys.*, **91**(12), 9638.

Amano, H., Kito, M., Hiramatsu, K., and Akasaki, I. (1989). *Jpn. J. Appl. Phys.*, **28**, L2112.

Amano, H., Sawaki, N., Akasaki, I., and Toyoda, Y. (1986). *Appl. Phys. Lett.*, **48**, 353.

Barin, I. (1994). *Thermochemical Data of Pure Substances* (3rd edn.). VCH Weinheim, Germany.

Bockowski, M., Grzegory, I., Lucznik, B., Sochacki, T., Nowak, G., Sadovyi, B., Strak, P., Kamler, G., Litwin-Staszewska, E., and Porowski, S. (2012). *J. Cryst. Growth*, **350**(1), 5.

Bohm, H. J. and Ahlrichs, R. (1985). *Mol. Phys.*, **55**, 445.

Burton, W. K., Cabrera, N., and Frank, F. C. (1951). *Phil. Trans. R. Soc. London*, **243**(866), 299.

Christiansen, S. H., Albrecht, M., Strunk, H. P., Foxon, C. T., Korakakis, D., Grzegory, I., and Porowski, S. (1999). *Phys. Status Solidi A*, **176**(1), 285.

Christmas, U. M. E., Andreev, A. D., and Faux, D. A. (2005). *J. Appl. Phys.*, **98**(7), 073522.

de Groot, P. and Kishner, S. (1991). *Appl. Opt.*, **30**(28), 4026.

Doradzinski, R., Dwilinski, R., Garczynski, J., Sierzputowski, L. P., and Kanbara, Y. (2010). In *Technology of Gallium Nitride Crystal Growth* (ed. D. Ehrentraut, E. Meissner, and M. Bockowski), p. 137. Springer-Verlag, Heidelberg.

Dwilinski, R., Wysmolek, A., Baranowski, J., Kaminska, M., Doradzinski, R., Garczynski, J., and Sierzputowski, L. (1995). *Acta Phys. Pol.*, **88**(5), 833.

Elsass, C. R., Smorchkova, I. P., Heying, B., Haus, E., Fini, P., Maranowski, K., Ibbetson, J. P., Keller, S., Petroff, P. M., DenBaars, S. P., Mishra, U. K., and Speck, J. S. (1999). *Appl. Phys. Lett.*, **74**(23), 3528.

Eremets, M. I., Hemley, R. J., Mao, H.-K., and Gregoryanz, E. (2001). *Nature*, **411**, 170.

Foxon, C. T., Davis, C. S., Novikov, S. V., Hughes, O. H., Cheng, T. S., Korakakis, D., Jeffs, N. J., Grzegory, I., and Porowski, S. (1999). *Phys. Status Solidi A*, **176**(1), 723.

Franssen, G., Suski, T., Krysko, M., Lucznik, B., Grzegory, I., Krukowski, S., Khachapuridze, A., Czernecki, R., Grzanka, S., Mensz, P., Leszczynski, M., Porowski, S., and Albrecht, M. (2008). *Phys. Status Solidi C*, **5**(6), 1485.

Frayssinet, E., Knap, W., Lorenzini, P., Grandjean, N., Massies, J., Skierbiszewski, C., Suski, T., Grzegory, I., Porowski, S., Simin, G., Hu, X., Khan, M.A., Shur, M.S., Gaska, R., and Maude, D. (2000). *Appl. Phys. Lett.*, **77**(16), 2551.

Fujito, K., Kubo, S., Nagaoka, H., Mochizuki, T., Namita, H., and Nagao, S. (2009). *J. Cryst. Growth*, **311**(10), 3011.

Glushko, W. P. (1979). *Termodinamiczeskije swojstwa indiwidualnych weszczestw.* Nauka, Moscow, Soviet Union. (In Russian.)

Goncharov, A. F., Gregoryanz, E., Mao, H.-K., Liu, Z., and Hemley, R. J. (2000). *Phys. Rev. Lett.*, **85**, 1262.

Gorczyca, I., Suski, T., Christensen, N. E., and Svane, A. (2010). *Appl. Phys. Lett.*, **96**, 101907.

Goto, S., Ohta, M., Yabuki, Y., Hoshina, Y., Naganuma, K., Tamamura, K., Hashizu, T., and Ikeda, M. (2003). *Phys. Status Solidi A*, **200**(1), 122.

Grandjean, N., Damilano, B., Massies, J., Neu, G., Teissere, M., Grzegory, I., Porowski, S., Gallart, M., Lefebvre, P., Gil, B., and Albrecht, M. (2000). *J. Appl. Phys.*, **88**(1), 183.

Gregoryanz, E., Goncharov, A. F., Hemley, R. J., and Mao, H.-K. (2001). *Phys. Rev. B*, **64**, 052103.

Groh, R., Gerey, G., Bartha, L., and Pankove, J. I. (1974). *Phys. Status Solidi A*, **26**(1), 353.

Grzegory, I. (1995). Ph.D. thesis, Institute of High Pressure Physics, Warsaw, Poland. (In Polish.)

Grzegory, I. (1996). In *Proc. of Joint XV AIRAPT and XXXV EHPRG Conference* (ed. W. Trzeciakowski), Singapore, p. 14. World Scientific.

Grzegory, I., Bockowski, M., Lucznik, B., Krukowski, S., Romanowski, Z., Wroblewski, M., and Porowski, S. (2002). *J. Cryst. Growth*, **246**, 177.

Grzegory, I., Bockowski, M., and Porowski, S. (2005). In *Bulk Crystal Growth of Electronic Optical & Optoelectronic Materials* (ed. P. Capper), p. 174. John Wiley and Sons Ltd.

Grzegory, I., Jun, J., Bockowski, M., Krukowski, S., Wroblewski, M., Lucznik, B., and Porowski, S. (1995). *J. Phys. Chem. Solids*, **56**, 639.

Grzegory, I., Krukowski, S., Leszczynski, M., Perlin, P., Suski, T., and Porowski, S. (2001). *Acta Phys. Pol. A Suppl.*, **100**, 57.

Grzegory, I. and Porowski, S. (2002). In *Low Dimensional Nitride Semiconductors* (ed. B. Gil), pp. 57–78. Oxford University Press.

Grzegory, I., Teisseyre, H., Lucznik, B., Pastuszka, B., Bockowski, M., and Porowski, S. (2007). In *Nitride with Non-polar Surfaces* (ed. T. Paskova), p. 53. Wiley-VCH.

Harafuji, K., Tsuchiya, T., and Kawamura, K. (2004). *J. Appl. Phys.*, **96**, 2501.

Heying, B., Averbeck, R., Chen, L. F., Haus, E., Riechert, H., and Speck, J. S. (2000). *J. Appl. Phys.*, **88**(4), 1855.

Heying, B., Tarsa, E. J., Elsass, C. R., Fini, P., DenBaars, S. P., and Speck, J. S. (1999). *J. Appl. Phys.*, **85**(9), 6470.

Hoffmann, S. and Hofmann, M. R. (2007). *Laser and Photonics Reviews*, **1**(1), 44.

Holst, J. Chr., Hoffman, A., Leszczynski, M., and Porowski, S. Private communication.

Ikeda, M., Mizuno, T., Takeya, M., Goto, S., Ikeda, S., Fujimoto, T., Ohfuji, Y., and Hashizu, T. (2004). *Phys. Status Solidi C*, **1**(6), 1461.

Ishizaka, A. and Murata, Y. (1994). *J. Phys. Cond. Matter*, **6**(45), L693.

Jacobsen, R. T., Stewart, R. B., and Jahangiri, M. (1986). *J. Phys. Chem. Ref. Data*, **15**, 735.

Jursenas, S., Kurilcik, N., Kurilcik, G., Zukauskas, A., Prystawko, P., Leszczynski, M., Suski, T., Perlin, P., Grzegory, I., and Porowski, S. (2001). *Appl. Phys. Lett.*, **78**(24), 3776.

Karpinski, J., Jun, J., and Porowski, S. (1984). *J. Cryst. Growth*, **66**, 1.

Karpinski, J. and Porowski, S. (1984). *J. Cryst. Growth*, **66**, 11.

Kawamura, F., Morishita, M., Tampo, M., Imade, M., Yoshimura, M., Mori, Y., Kitaoka, Y., and Sasaki, T. (2008). *J. Cryst. Growth*, **310**, 3946.

Keller, S., Suh, C. S., Fichtenbaum, N. A., Furukawa, M., Chu, R., Chen, Z., Vijayraghavan, K., Rajan, S., DenBaars, S. P., Speck, J. S., and Mishra, U. K. (2008). *J. Appl. Phys.*, **104**(9), 093510.

Kirchner, C., Schwegler, V., Eberhard, F., Kamp, M., Ebeling, K. J., Kornitzer, K., Ebner, T., Thonke, K., Sauer, R., Prystawko, P., Leszczynski, M., Grzegory, I., and Porowski, S. (1999). *Appl. Phys. Lett.*, **75**(8), 1098.

Koblmuller, G., Wu, F., Mates, T., Speck, J. S., Fernandez-Garrido, S., and Calleja, E. (2007). *Appl. Phys. Lett.*, **91**(22), 221905.

Krukowski, S., Bockowski, M., Grzegory, B., Lucznik, I., Suski, S., Porowski, T., and Romanowski, Z. (2001). *J. Phys. Cond. Matter*, **13**, 8881.

Krysko, M., Franssen, G., Suski, T., Albrecht, M., Lucznik, B., Grzegory, I., Krukowski, S., Czernecki, R., Grzanka, S., Makarowa, I., Leszczynski, M., and Perlin, P. (2007). *Appl. Phys. Lett.*, **91**(21), 211904.

Kuroda, T. and Takeuchi, A. (2002). *J. Appl. Phys.*, **92**(6), 3071–3074.

Laino, V., Roemer, F., Witzigmann, B., Lauterbach, C., Schwarz, U., Rumbolz, C., Schillgalies, M. O., Furtisch, M., Lell, A., and Harle, V. (2007). *IEEE J. Quantum Electron.*, **43**, 16.

Lambkin, J. D., Dunstan, D. J., Homewood, K. P., Howard, L. K., and Emeny, M. T. (1990). *Appl. Phys. Lett.*, **57**(19), 1986.

Leszczynski, M., Czernecki, R., Krukowski, S., Krysko, M., Targowski, G., Prystawko, P., Plesiewicz, J., Perlin, P., and Suski, T. (2011). *J. Cryst. Growth*, **318**(1), 496.

Leszczynski, M., Prystawko, P., Czernecki, R., Lehnert, J., Suski, T., Perlin, P., Wisniewski, P., Grzegory, I., Nowak, G., Porowski, S., and Albrecht, M. (2001). *J. Cryst. Growth*, **231**(3), 352.

Leszczynski, M., Prystawko, P., and Nowak, G. (1999). Private communication.

Lindemann, F. A. (1910). *Z. Phys.*, **11**, 609.

Lucznik, B., Adamczyk, J., Jun, J., Krukowski, S., and Grzegory, I. Unpublished.

MacChesney, J. B., Bridenbaugh, P. M., and O'Connor, P. B. (1970). *Mater. Res. Bull.*, **5**(9), 783.

Madar, R., Jacob, G., Hallais, J., and Fruchard, R. (1975). *J. Cryst. Growth*, **31**, 197.

Magnuson, M., Mattesini, M., Hoglund, C., Birch, J., and Hultman, L. (2010). *Phys. Rev. B*, **81**, 085125.

Manfra, M. J., Baldwin, K. W., Sergent, A. M., West, K. W., Molnar, R. J., and Caissie, J. (2004). *Appl. Phys. Lett.*, **85**(22), 5394.

Martin, R. M. and Needs, R. J. (1986). *Phys. Rev. B*, **34**, 5082.

Maruska, H. P. and Titjen, J. J. (1969). *Appl. Phys. Lett.*, **15**, 327.

McMahan, A. K. and LeSar, R. (1985). *Phys. Rev. Lett.*, **54**, 1929.

Melnik, Y., Nikolaev, A., Nikitina, L., Vassilevski, K., and Dmitriev, V. (1998). *MRS Symp. Proc.*, **482**, 269.

Moon, W. H. and Hwang, H. J. (2003). *Phys. Lett.*, **315**, 319.

Moon, W. H., Kim, H. J., and Choi, C. H. (2007). *Scripta Mater.*, **56**, 345.

Motoki, K. (2010). *SEI Technical Review*, **70**, 28.

Motoki, K., Okahisa, T., Hirota, R., Nakahata, S., and Uematsu, K. (2007). *N. Matsumoto J. Cryst. Growth*, **305**, 377.

Nagahama, S., Iwasa, N., Senoh, M., Matsushita, T., Sugimoto, Y., Kiyoku, H., Kozaki, T., Sano, M., Matsumura, H., Umemoto, H., Chocho, K., and Mukai, T. (2000). *Jpn. J. Appl. Phys.*, **39**(Part 2, No. 7A), L647.

Nakamura, S. (1999). Private communication.

Nakamura, S. and Chichibu, S. F. (ed.) (2000). *Introduction to Nitride Semiconductor Blue Lasers and Light Emitting Diodes*. Taylor and Francis, London.

Nakamura, S. and Fasol, G. (1997). *The Blue Laser Diode*. Springer, Berlin.

Nakamura, S., Mukai, T., and Senoh, M. (1992). *J. Appl. Phys.*, **71**, 5543.

Nellis, W. J., Holmes, N. C., Mitchell, A. C., and van Thiel, M. (1984). *Phys. Rev. Lett.*, **53**, 1661.

Neugebauer, J. and Van de Walle, C. G. (1994). *Phys. Rev. B*, **50**, 8067.

Neugebauer, J., Zywietz, T. K., Scheffler, M., Northrup, J. E., Chen, H., and Feenstra, R. M. (2003). *Phys. Rev. Lett.*, **90**, 056101.

Newman, N. (1997). In *Gallium Nitride (GaN) I* (ed. J. I. Pankove and T. D. Moustakas), Volume 50, Semiconductors and Semimetals, p. 55. Elsevier.

Oshima, Y., Yoshida, T., Eri, T., Watanabe, K., Shibata, M., and Mishima, T. (2010). In *Technology of Gallium Nitride Crystal Growth* (ed. D. Ehrentraut, E. Meissner, and M. Bockowski), p. 79. Springer-Verlag, Heidelberg.

Pankove, J. I., Miller, E. A., Richman, D., and Berkeyheiser, J. E. (1971). *J. Lumin.*, **63**, 4.

Perlin, P., Camassel, J., Knap, W., Taliercio, T., Chervin, J. C., Suski, T., Grzegory, I., and Porowski, S. (1995). *Appl. Phys. Lett.*, **67**(17), 2524.

Perlin, P., Czyszanowski, T., Marona, L., Grzanka, S., Kafar, A., Stanczyk, S., Suski, T., Leszczynski, M., Bockowski, M., Muziol, G., Kuc, M., and Sarzala, R. (2012). *Proc. of SPIE*, **8262**, 826216.

Perlin, P., Holc, K., Sarzynski, M., Leszczynski, M., Czernecki, R., Marona, L., Wisniewski, P., Cywinski, G., Skierbiszewski, C., Bockowski, M., Grzegory, I., and Suski, T. (2011*a*). *Proc. of SPIE*, **7953**, 79530I.

Perlin, P., Holc, K., Sarzynski, M., Scheibenzuber, W., Marona, L., Czernecki, R., Leszczynski, M., Bockowski, M., Grzegory, I., Porowski, S., Cywinski, G., Firek, P., Szmidt, J., Schwarz, U., and Suski, T. (2009). *Appl. Phys. Lett.*, **95**(26), 261108.

Perlin, P., Marona, L., Holc, K., Wisniewski, P., Suski, T., Leszczynski, M., Czernecki, R., Najda, S., Zajac, M., and Kucharski, R. (2011*b*). *Appl. Phys. Express*, **4**, 062103.

Porowski, S. and Jun, J. Unpublished.

Purdy, A. P. (1999). *Chem. Mater.*, **11**, 1648.

Reichlin, R., Schiferl, D., Martin, S., Vanderborgh, C., and Mills, R. L. (1985). *Phys. Rev. Lett.*, **55**, 1464.

Riechert, H., Averbeck, R., Graber, A., Schienle, M., Strauss, U., and Tews, H. (1996). *MRS Symp. Proc.*, **449**, 149.

Romanowski, Z., Krukowski, S., Grzegory, I., and Porowski, S. (2001). *J. Chem. Phys.*, **114**, 6353.

Saarinen, K., Laine, T., Kuisma, S., Hautojarvi, P., Dobrzynski, L., Baranowski, J. M., Pakula, K., Stepniewski, R., Wojdak, M., Wysmolek, A., Suski, T., Leszczynski, M., Grzegory, I., and Porowski, S. (1997). *Phys. Rev Lett.*, **79**, 3030.

Sarzynski, M., Krysko, M., Targowski, G., Czernecki, R., Sarzynska, A., Libura, A., Krupczynski, W., Perlin, P., and Leszczynski, M. (2006). *Appl. Phys. Lett.*, **88**, 121124.

Schoonmaker, R. C., Buhl, A., and Lemley, J. (1965). *J. Phys. Chem.*, **69**(10), 3455.

Schwarz, U. T. and Witzigmann, B. (2007). In *Nitride Semiconductor Devices* (ed. J. Piprek). Wiley-VCH, Berlin.

Seifert, W., Fitzl, G., and Butter, E. (1981). *J. Cryst. Growth*, **52**, 257.

Shen, X. Q., Furuta, K., Nakamura, N., Matsuhata, H., Shimizu, M., and Okumura, H. (2007). *J. Cryst. Growth*, **301–302**(0), 404.

Siekacz, M., Feduniewicz-Zmuda, A., Cywinski, G., Krysko, M., Grzegory, I., Krukowski, S., Waldrip, K. E., Jantsch, W., Wasilewski, Z. R., Porowski, S., and Skierbiszewski, C. (2008). *J. Cryst. Growth*, **310**(17), 3983.

Siekacz, M., Szankowska, M. L., Feduniewicz-Zmuda, A., Smalc-Koziorowska, J., Cywinski, G., Grzanka, S., Wasilewski, Z. R., Grzegory, I., Lucznik, B., Porowski, S., and Skierbiszewski, C. (2009). *Phys. Status Solidi C*, **6**(S2), S917.

Skierbiszewski, C., Siekacz, M., Turski, H., Muziol, G., Sawicka, M., Feduniewicz-Zmuda, A., Cywinski, G., Cheze, C., Grzanka, S., Perlin, P., Wisniewski, P., Wasilewski, Z. R., and Porowski, S. (2012*a*). *Appl. Phys. Express*, **5**(2), 022104.

Skierbiszewski, C., Siekacz, M., Turski, H., Muziol, G., Sawicka, M., Feduniewicz-Zmuda, A., Smalc-Koziorowska, J., Perlin, P., Grzanka, S., Wasilewski, Z. R., Kucharski, R., and Porowski, S. (2012*b*). *J. Vac. Sci. Technol. B*, **30**(2), 02B102.

Skierbiszewski, C., Wasilewski, Z., Siekacz, M., Feduniewicz, A., Pastuszka, B., Grzegory, I., Leszczynski, M., and Porowski, S. (2004). *Phys. Status Solidi A*, **201**(2), 320.

Skierbiszewski, C., Wasilewski, Z. R., Grzegory, I., and Porowski, S. (2009). *J. Cryst. Growth*, **311**(7), 1632.

Skierbiszewski, C., Wasilewski, Z. R., Siekacz, M., Feduniewicz, A., Perlin, P., Wisniewski, P., Borysiuk, J., Grzegory, I., Leszczynski, M., Suski, T., and Porowski, S. (2005). *Appl. Phys. Lett.*, **86**(1), 011114.

Skierbiszewski, C., Wisniewski, P., Siekacz, M., Perlin, P., Feduniewicz-Zmuda, A., Nowak, G., Grzegory, I., Leszczynski, M., and Porowski, S. (2006). *Appl. Phys. Lett.*, **88**(22), 221108.

Slack, G. A. and MacNelly, T. F. (1976). *J. Cryst. Growth*, **34**, 276.

Smolyakov, G. A., Eliseev, P. G., and Osinski, M. (2005). *IEEE J. Quantum Electron.*, **41**, 517.

Sokol, A. G., Polyanov, Yu. N., and Surovtsev, N.V. (2007). *Diam. Rel. Mater.*, **16**, 431.

Stepniewski, R., Potemski, M., Wysmolek, A., Pakula, K., Baranowski, J. M., Lusakowski, J., Grzegory, I., Porowski, S., Martinez, G., and Wyder, P. (1999). *Phys. Rev. B*, **60**, 4438.

Stillinger, F. and Weber, T. A. (1985). *Phys. Rev. B*, **31**, 5262.

Strak, P. and Krukowski, S. (2007). *J. Chem. Phys.*, **126**, 194501.

Sumitomo Electric Industries Ltd. (2010). http://global-sei.com/news/press/10/10_23.html.

Suski, T., Litwin-Staszewska, E., Piotrzkowski, R., Czernecki, R., Krysko, M., Grzanka, S., Nowak, G., Franssen, G., Dmowski, L. H., Leszczynski, M., Perlin, P., Lucznik, B., Grzegory, I., and Jakiela, R. (2008). *Appl. Phys. Lett.*, **93**(17), 172117.

Suski, T., Staszczak, G., Grzanka, S., Czernecki, R., Litwin-Staszewska, E., Piotrzkowski, R., Dmowski, L. H., Khachapuridze, A., Krysko, M., Perlin, P., and Grzegory, I. (2010). *J. Appl. Phys.*, **108**(2), 023516.

Swendsen, R. H., Kortman, P. J., Landau, D. P., and Muller-Krumbhaar, H. (1976). *J. Cryst. Growth*, **35**(1), 73.

Tachibana, K., Nago, H., and Nunoue, S.-Y. (2008). *Phys. Status Solidi C*, **5**(6), 2158.

Teisseyre, H., Nowak, G., Leszczynski, M., Grzegory, I., Bockowski, M., Krukowski, S., Porowski, S., Mayer, M., Pelzmann, A., Kamp, M., Ebeling, K.J., and Karczewski, G. (1996). *MRS Internet J. Nitride Semicond. Res.*, **1**, 13.

Tersoff, J. (1988). *Phys. Rev B*, **27**, 6991.

Tersoff, J. (1989). *Phys. Rev. B*, **39**, 5566.

Thurmond, C. D. and Logan, R. A. (1972). *J. Electrochem. Soc.*, **119**, 622.

Tuomisto, F., Maki, J. M., and Zajac., M. (2010). *J. Cryst. Growth*, **312**, 2620.

Unland, J., Onoderka, B., Davydov, A., and Schmid-Fetzer, R. (2003). *J. Cryst. Growth*, **256**, 33.

Usui, A., Sunakawa, H., Sakai, A., and Yamaguchi, A. (1997). *Jpn. J. Appl. Phys.*, **36**, L899.

Utsumi, W., Saitoh, H., Kaneko, H., Watanuki, T., Aoki, K., and Shimomura, O. (2003). *Nature Mater.*, **2**, 735.

Van Vechten, J. A. (1972). *Phys. Rev. Lett.*, **29**, 769.

Van Vechten, J. A. (1973). *Phys. Rev. B*, **7**, 1479.

Wang, C.-L., Chuang, Y.-H., and Pan, C.-L. (1995). *Opt. Lett.*, **20**(9), 1071.

Wang, T., Parbrook, P.J., Whitehead, M. A., Fan, W. H., and Fox, A. M. (2004). *J. Cryst. Growth*, **273**(1–2), 48.

Wang, Z. and Zhu, X. (2006). *J. Appl. Phys.*, **100**, 063503.

Weyher, J. L., Muller, S., Grzegory, I., and Porowski, S. (1997). *J. Cryst. Growth*, **182**(1–2), 17.

Zauner, A. R. A., Weyher, J. L., Plomp, M., Kirilyuk, V., Grzegory, I., van Enckevort, W. J. P., Schermer, J. J., Hageman, P. R., and Larsen, P. K (2000). *J. Cryst. Growth*, **210**(4), 435.

Ziman, J. M. (1964). *Principles in the Theory of Solids.* Cambridge University Press, Cambridge, UK.

3 Epitaxial growth and benefits of GaN on silicon

Armin Dadgar and Alois Krost

3.1 Introduction

GaN epitaxial growth has a long history reaching back to the late 1960s [1], and even growth on silicon reaches back to the early 1970s [2]. But, apart from a few devices demonstrated, it took more than twenty years to enter mass production. The biggest difficulty for GaN growth at the beginning of GaN research was the lack of suitable homosubstrates for epitaxial growth, leading to low material quality and a high background carrier concentration. Added to this was the difficulty of p-type doping, which required thermal annealing to activate the Mg acceptor. The most commonly used substrate for GaN growth has been, and still is, sapphire, but also SiC is used, mostly for high-power or high-frequency applications. Silicon, which offers the best crystalline properties, large substrate diameters, and lowest substrate cost, was always in principle attractive for GaN growth, but many difficulties appear when it is used as a substrate in GaN epitaxy. Around 2000, GaN-on-silicon technology made some important breakthroughs: the demonstration of an MBE-grown LED structure [3, 4], the development of thicker crack-free GaN layers using an AlGaN intermediate layer [5], and LT-AlN interlayers [6] grown by metalorganic vapor phase epitaxy (MOVPE). Since then it has developed rapidly, and in the meantime it has been proven as a competitive technology to the growth on sapphire in the case of LEDs and even as an advantageous technology in the segment of high-voltage electronics. We discuss the major problems of GaN epitaxy on silicon with a focus on the growth by MOVPE, and discuss solutions for epitaxy and device design to achieve state-of-the-art GaN-on-silicon layers, and demonstrate the benefits of such layers for device applications.

3.2 The GaN-on-silicon challenges

The difficulties in GaN-on-silicon growth have been widely discussed during the last twelve years with a focus on lattice and thermal mismatch, but in our experience this focus ignores other difficulties in GaN-on-Si layer growth. In sum they are:

1. high lattice mismatch, low material quality,
2. high thermal mismatch,

3. meltback etching,
4. plastic substrate deformation,
5. vertical conductivity.

In the following we will briefly discuss the state of the art of these topics.

3.2.1 *Lattice mismatch*

The magnitude of GaN and AlN lattice mismatch on silicon is comparable to lattice mismatch of GaN on sapphire [7]. Using Si as substrate the bond lengths of AlN (1.89 Å) and GaN (1.95 Å) to Si (2.34 Å) differ by –19.2 and –16.7 %, respectively. Thus for AlN on silicon, typically applied as a starting or seeding layer, the situation is comparable to the growth on sapphire. Consequently, the density of misfit dislocations at the substrate/group-III-nitride interface is about the same order of magnitude. The major issue arising from the high misfit is the small coherence length of initial nuclei, leading to small nuclei with no long-range attachment and thus a tendency for high in-plane twist in addition to the usually relatively rough and unstable silicon surface if compared to sapphire. On sapphire the nuclei can be better aligned by applying GaN seeding layers which start as a 2D seed which is then heat-treated to form a few larger 3D islands. Upon further growth they increase in size, coalesce, and again form a 2D layer. By this method, large defect-free islands can be grown. Using AlN seeding layers on sapphire a high material quality probably benefits from the identity of one atom; namely, aluminum. On a microscopic scale most probably an Al_2O_3-AlN transition is formed. Optimizing this by the addition of oxygen, such seeding layers enable an excellent material quality [8] and reduced in-plane twist [9]. Directly-on-silicon GaN seeding layers do not apply, due to the difficulty in growing GaN directly on silicon, as discussed towards the end of this paragraph. The low surface mobility of Al usually applied for growing AlN seeding layers hinders lateral growth as for GaN on sapphire. In addition, the lower stability of the Si surface and the reactivity of Al and Si, enabling the formation of an alloy, most probably inhibits a good alignment if the seeding layer growth conditions are not well optimized. In sum, the small size of the nuclei leading to a relatively high misalignment of the nuclei, the usually imperfect or microscopically rough Si(111) surface, and its reactivity with group-III and group-V elements, are not beneficial for high material quality. Thus special care has to be taken in seeding layer optimization, starting with surface preparation and substrate heating. The seeding layer determines the overall quality that can be achieved, and a poor seeding layer usually inhibits a subsequent high-quality GaN layer. Due to the six-fold symmetry of the GaN crystal the three-fold surface symmetry of silicon (111) is the first choice for growth. On this surface GaN is oriented with its c-axis vertically to the Si surface plane, thus (0001)AlN || (111)Si or [0001]AlN || [111]Si. The in-plane alignment is $[11\bar{2}0]$AlN || [110]Si, along the preferred cleavage direction of both materials, or $[1\bar{1}00]$AlN || Si$[11\bar{2}]$ (Fig. 3.1). Thus cleavage of GaN on Si is more easily applied as on sapphire, where a 30-degree rotation of the GaN layer

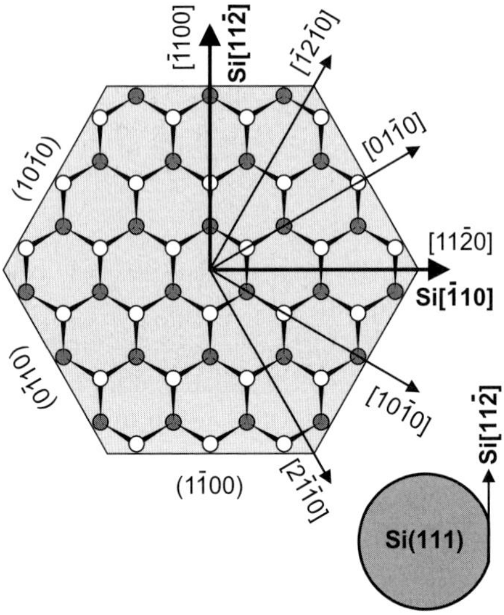

FIG. 3.1. Epitaxial relationship of GaN or AlN on Si(111). In this heteroepitaxial system the preferred cleavage planes of GaN ($\{11\bar{2}0\}$) and Si ($\{110\}$) are oriented along the same direction. (Reprinted from [7] with permission from Elsevier.)

on the substrate also leads to a 30-degree misalignment of the preferred $(1\bar{1}00)$ cleavage plane of GaN to that of sapphire. As a consequence, cleaved layers on sapphire show rugged cleavage planes.

The alignment of the AlN seeding layer on the substrate depends a lot on the growth conditions and the substrate orientation chosen. There are many approaches to grow AlN on Si, e.g., using different temperatures [10], Al [11, 12], ammonia or even silane pre-flow [13, 14]. It has been also demonstrated that under identical growth conditions the growth on Si(110) can be superior to the growth on Si(111) [15]. In this case, TEM studies revealed the arrangement of AlN on Si and showed that for AlN on Si(110) in the [001]Si $\|$ $[1\bar{1}00]$AlN direction a nearly perfect lattice matching is achieved with a 2:1 periodicity (Fig. 3.2) [15–18]. In the perpendicular direction the situation is as on Si(111), usually with a $\approx$4:5 matching which in sum results in a poor overall matching leading to different bond lengths and a higher initial strain anisotropy in the AlN islands forming on the Si surface. For the growth on Si(110) the better alignment in one direction is already leading to better quality in comparison to Si(111).

In contrast to the growth on Si(111) and Si(110), the growth on Si(001)is more demanding [19–22]. Disregarding the situation of two perpendicular surface reconstructions and by this two perpendicular preferred alignments on this surface, even for one type of surface reconstruction lattice mismatch is large and amounts to 19% and 30% with an alignment of [0001]AlN $\|$ [001]Si and in-plane either

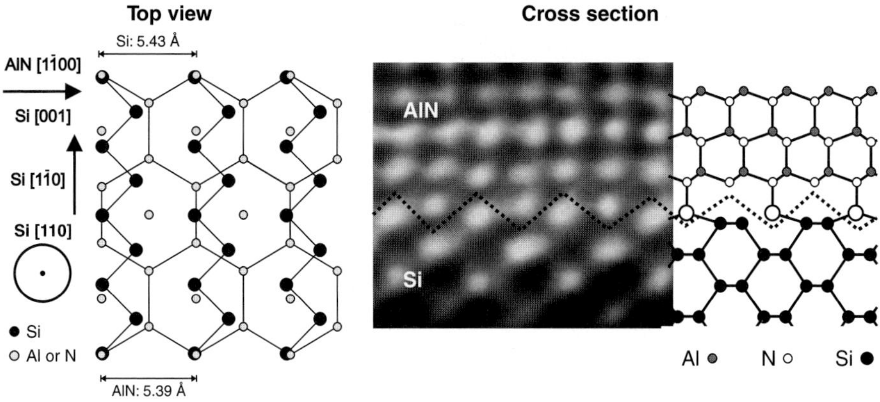

FIG. 3.2. Model of lattice matching of an AlN seeding layer on silicon (110) (left), and TEM cross-section image showing the 2:1 matching of the lattices and the corresponding model. (Left) Reprinted from [17] with permission from IOP. (Right) Reprinted from [16], © 2008 The Japan Society of Applied Physics.

[01$\bar{1}$0]AlN ∥ [110]Si or [$\bar{2}$110]AlN ∥ [110]Si [21, 23, 24]. Thus a good alignment is very difficult to achieve, but functional device layers and devices have been demonstrated on up to 150-mm diameter Si substrates [25–27]. Nevertheless, on all substrate orientations a lot depends on the growth conditions of the seeding layer, and its optimization is the key for high-quality layers. Taking particular care on a proper optimization will yield best layers.

3.2.2 *Thermal mismatch*

Thermal mismatch is the most problematic issue for GaN-on-Si growth. Although the thermal expansion coefficient is not known exactly for GaN up to growth temperature it can be estimated to be roughly twice that of silicon leading to a thermal mismatch above 50 %. Silicon has a thermal expansion coefficient at room temperature of $2.6 \cdot 10^{-6}$ and $4.48 \cdot 10^{-6}$ at 1050°C [28], which averages to $4.04 \cdot 10^{-6}$ between typical GaN growth temperature around 1050°C and room temperature. The difference in thermal expansion is somewhat similar to GaN on sapphire, but, with opposite prefix. Thus GaN on sapphire usually is under compressive stress after cooling from growth temperature, while it is under tensile stress if grown on silicon. As a consequence for MOVPE growth, crack-free layers in excess of 1 μm are not possible to realize as long as no special buffer layer schemes which induce compression in the layer stack as AlN/AlGaN buffers or interlayers are applied [5, 6, 29, 30]. Cracking is the preferential relaxation process of group-III-nitrides if high tensile stress is present. For AlN and GaN under tensile stress, plastic relaxation has not been reported so far.

Today many methods exist which are all either based on pre-straining the growing GaN layer or by growing in reduced areas, small enough to suppress cracking. Attempts to solve thermal mismatch by using silicon on insulator substrates failed. Here it was believed that a thin Si layer can glide upon the Si carrier, and by this relieve stress. However, in reality the glide process would lead to a displacement of the upper Si / GaN structure in the range of tens to hundreds of micrometers, depending on the substrate size. SiO_2 being a brittle material does not support such a long-distance glide process, and either cracking or delamination of the layer will occur [31].

3.2.3 *Meltback etching*

Meltback etching is quite underestimated, and every material grower will face this situation in GaN MOVPE growth on silicon. It was first described by Ishikawa *et al.* [32] and found to lead to SiN and SiGaN deposits [7]. It usually is more problematic at higher growth temperatures and for thin seeding and thin low Al-content buffer layers. AlN and AlGaN help to prevent this reaction, but not all device structures allow sufficiently thick layers to prevent it in any case. Depending on the growth system a high temperature, AlN seeding and/or buffer layer can free significant amounts of gallium which were previously deposited on the reactor or susceptor walls, and promote meltback etching during growth, especially when growing thick layers. This is promoted during AlN growth by the typically low-growth pressure and low V–III ratio for AlN. Ways to circumvent this are conditioning or heating the reactor prior to growth without the substrate, with a substrate dummy, or using a system where parts can be exchanged easily. Alternatively, a proper design of seeding and buffer layer growth with a focus on coalesced smooth layers free from pits is beneficial to hinder meltback etching.

Particles mostly from reactor deposits or from dust prior to loading the substrate can either disturb nucleation and later lead to meltback etching in the direct vicinity of them, or when deposited on the wafer during (Al-rich) buffer layer growth then lead to a locally too-thin buffer and start the reaction from such defects. Typically, meltback etching is propagating along the GaN/Si interface; thus it is not only a local problem, but with advancing growth time it can destroy significant parts of the wafer. Additionally, it forms 3D objects which can be several tens of microns high, leading to trouble in processing. It most probably also produces Si or SiH_4 that can be incorporated in undoped or p-doped layers, changing the electrical properties, and liquid gallium is also sometimes found in these areas. A variety of appearances of meltback etching is shown in Fig. 3.3. The appearance differs, but often meltback etching shows a Ginko leaf-like appearance which represents the propagation of the reaction between layer and substrate. Also, circular or just rough appearances have been observed, depending on growth conditions and growth time.

Unfortunately, meltback etching has not been investigated in detail yet, thus knowledge of the reactions and its impact on the layers grown, as well as possible routes to circumvent it, is scarce.

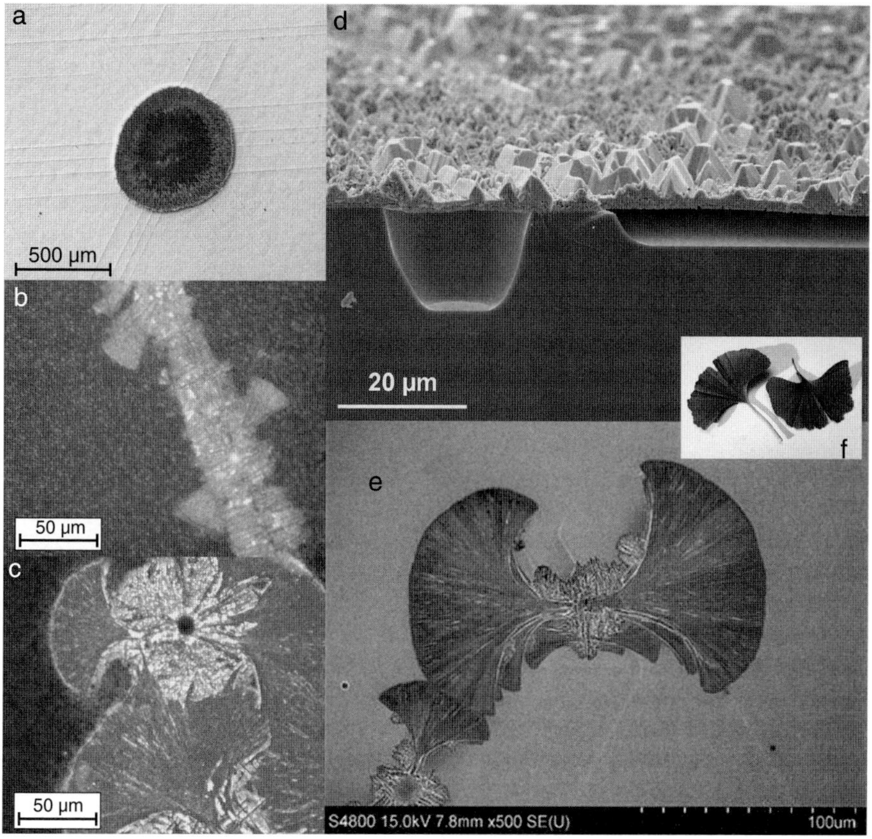

FIG. 3.3. Different appearances of meltback etching. Nomarski microscope images (a–c): A particle is the origin of cracks and meltback etching (a) often visible when too-high tensile stress occurs. Image (b) shows a crack which occurred during growth and acts as origin of meltback etching. Too-thin Al-rich buffer layers and high growth temperatures or simply many particles can cause meltback etching, as in the image (c). Scanning electron microscopy images (d and e) show an extreme form of meltback etching and also the origin of the name, deep hollows etched in the Si substrate (d). Here the nominal layer thickness was only around 3 µm! A typical Ginko leaf (f) appearance is shown in image (e). Typically, such a meltback etching reaction propagates along the Si/AlN/GaN interface.

3.2.4 *Plastic substrate deformation*

Plastic deformation of the substrate is an important topic and often a limitation for the maximum thickness of prestrained layers to achieve crack-free GaN-on-silicon substrates [33–38]. At typical AlN and GaN growth temperatures

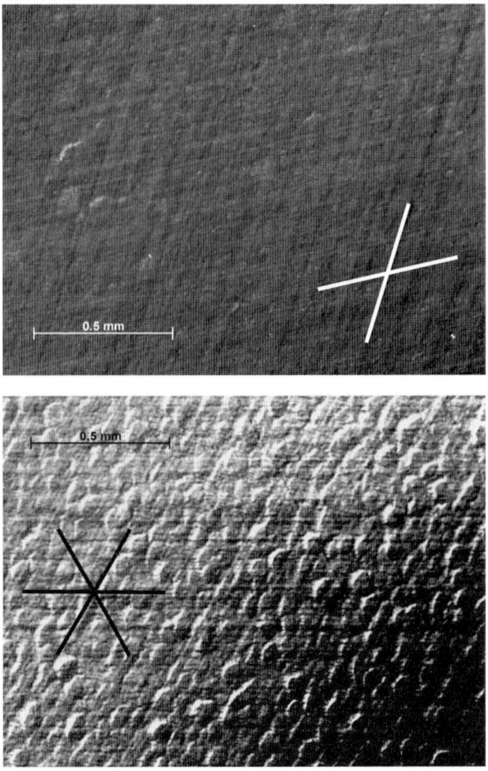

FIG. 3.4. Nomarski microscopy view on two differently plastically deformed samples on Si(111). The slip lines in Si are well visible and, depending on the strength of deformation and position on the substrate, striations are visible in one, two, or all three directions, including an angle of 120°. (Reprinted from [36] with permission.)

(>1000°C) silicon is a relatively soft material and does not withstand high stresses which occur for prestrained layers in excess of several-micron thickness. If plastic deformation occurs, the slip planes of Si are usually well visible by differential interference contrast (Nomarski) microscopy (Fig. 3.4). The deformation of the substrate, though, leads to a non-reproducible high bow, since once deformed the substrate is permanently bowed, even if the layer is removed. Weak plastic deformation, e.g., just the onset, can be acceptable, especially when only part of the wafer, such as the edges, are affected. However, very often this cannot be well controlled. Another effect that was observed for 100 and 150 mm substrates were single cracks with additional slip planes that appear reproducible only in specific regions of the wafer edge. These can be correlated with the impact of the substrate flat region on the substrate symmetry. Typically a substrate flat and to a smaller portion a substrate notch are breaking the substrate rotational symmetry and, when high stresses are applied, lead to a preferred anisotropy orientation of the bow. Therefore, the combination of the substrate flat and crystalline symmetry leads to slip lines and cracks occurring at the same single position of the wafer, but not being repeated at positions one would expect from the crystal symmetry solely, e.g., after a rotation of 120° and 240° for Si(111) substrates.

There are also differences in the stiffness of silicon substrates. These differences were already investigated in the 1980s [39]. Here it was found that Czochralski and float-zone grown substrates differ strongly in their stiffness, which was attributed to their impurity concentration. It has been observed that indeed substrates with high concentrations of dopants such as oxygen or nitrogen, but also typical dopants such as As or P, do decrease the effect of plastic deformation in GaN epitaxy. At the high growth temperatures of GaN on Si and the high compressive stresses that are required during growth to avoid crack formation after cooling, very high impurity concentrations are advisable which are not standard in the silicon industry, where a high purity is typically targeted by the crystal grower. Therefore, at present one is best advised to order special custom-made substrates to efficiently reduce the effect of plastic deformation and avoid differences for different Si substrate batches. Even a Si substrate with a slightly lower quality can be acceptable, since the impact of a lower quality on the AlN/GaN layers is rather small and the benefit of an efficient slip blocker can be bigger.

Due to their high crystalline perfection and typically high purity, float-zone substrates are very sensitive to plastic deformation, and the maximum layer thickness achievable is significantly lower when compared to Czochralski-grown Si.

Another way to postpone the effect of plastic deformation is by using thicker substrates. This not only delays the onset of plastic deformation but also reduces the overall bow during growth, and by this helps to improve layer homogeneity. But also thicker substrates are limited in thickness, due to the demands of the crystal growth and processing equipment. And even when wafer thinning is applied prior to processing, the higher cost also has to be taken into account.

3.2.5 *Vertical conductivity*

A popular argument for LEDs on silicon substrates has often been the possible vertical contacting of devices via the substrate. While LEDs on Si substrates will never be able to compete with thin-film LEDs due to their limitation in light extraction, other devices such as photovoltaic cells or electronics might benefit from vertical contacts or junctions. A major problem with vertical contacts is the usually applied Al-rich seeding layer. It has been shown that using AlGaN or thin seeding layers reduces the contact resistivity [40], but it is still a loss for vertical structures which would limit device performance significantly.

For photovoltaic solar-cell structures based on InGaN the situation is different, and at high In contents around 40% a p-Si/n-InGaN junction is expected to yield low resistivities [41, 42]. However, applying such a layer again leads to an interface reaction, most probably the onset of meltback etching, if grown by MOVPE visible by darker areas in the Si substrate close to the interface in Fig. 3.5. Although the growth temperature of high-In-content InGaN is low, the gases involved in the process and the direct contact of Ga with Si lead to the formation of small hollows in the substrate surface, and layers often show high amounts of In as In droplets on the surface. An alternative is AlInN as a seeding layer, which reduces the amount of holes (Fig. 3.18) and also enables high In contents. Due to its low band-gap energy at high In contents the AlInN band-gap energy is comparable to $In_{0.40}Ga_{0.60}N$ at In contents of 53% [43, 44]. Thus it is most probably also suited as low-resistivity contact material to p-Si.

Apart from the growth of low-resistance Si/group-III-nitride heterostructures one can use standard seeding and buffer layers in combination with standard Si technology, and form via contacts from the Si backside to a conducting GaN layer or from the n-GaN layer into the Si substrate [45].

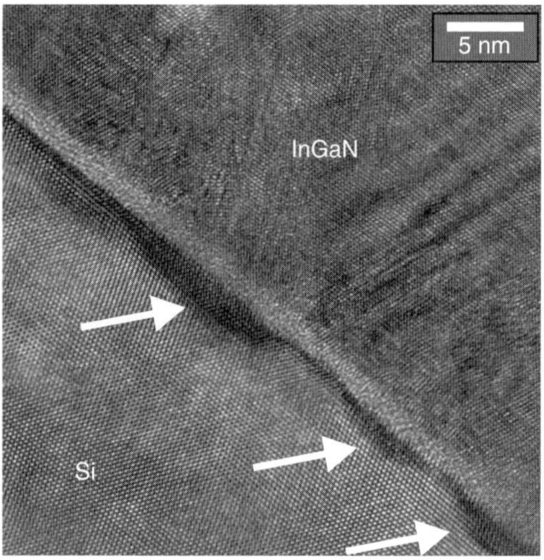

FIG. 3.5. HR-TEM image of an InGaN layer with approx. 40 % In directly grown on Si. The darker areas marked with arrows stem from hollows filled with In and Ga. Typically such interfaces are also observed when meltback etching occurs. Here InGaN was grown around only 600°C in N_2 atmosphere and NH_3. There is also a distorted region visible between the substrate and the layer. (TEM by A. Mogilatenko, HU-Berlin.)

3.3 Seed layer growth

Seed layer growth on Si substrates is the most important growth step. A poor seeding layer quality, from the crystalline standpoint, will never yield in a high layer quality of subsequent layers, although they are usually of better quality than the seeding layer. It is very important to optimize this layer first to achieve best results for the functional layers of a device structure. There are several methods known for seed layer growth as high-temperature (HT) AlN [29], low-temperature (LT) AlN [6], Al-preflow [46–50], no Al or N preflow [51–53], or even nitridation and then AlN growth [13], Al(Ga)N [40], growth on HfN [54], or other buffer layer materials as aluminum oxide [55, 56] or SiC [57, 58]. Today, AlN-based seeding layers are commonly applied due to their good results and reliability, especially with regard to meltback etching protection.

Nevertheless, often each seeding layer growth method can be optimized to a high-quality level. When optimizing, and for later production, one important issue is reproducibility. Here, deposits in the reactor chamber, but also the quality of the Si substrate surface, play an important role. Some groups use Si substrates direct from the vendor's tray and heat treat them under H_2 to clean and deoxidize them, while others use a pre-process chemical etch to form a hydrogen terminated surface [59] and then either heat-treat the substrates or directly heat up and start AlN seed layer growth.

It is very difficult to finally judge on the many different approaches, but there are pros and cons for each method. Long heating and baking always bears the risk of contaminating the Si surface with deposits from the interior of the reactor. This is mainly a reproducibility problem, since these deposits may vary in the rate at which they evaporate during substrate bake. After maintenance a clean reactor will behave differently from one with thick coatings. In contrast to sapphire the silicon surface has a high reactivity, and contaminants usually easily diffuse into the crystal lattice. But the good thing about the baking method is the simplicity if one is able to achieve a high reproducibility. Wet chemical etching prior to growth is different from what is commonly believed: relatively demanding, since standard etching systems often leave a thin trace of contaminants on the surface which leads to a different morphology or even a poor crystal quality at these wafer regions. In most processes this is uncritical, but in epitaxy every surface defect can become decorated. As already mentioned, the seeding layer is also important with regard to meltback etching and a good seeding or Al-rich buffer layer has a fully coalesced surface, and by this seals the Si substrate from Ga which is grown on top of this seed and buffer layer.

Several investigations have been performed to understand the microscopic nature of the AlN/Si interface (Fig. 3.6) which rarely can be observed without an amorphous interlayer, as in the work by Liu *et al.* [47], or with a clean and an amorphous interface, as in the work of Radtke *et al.* [49]. Often a SiN layer or amorphous AlN layer forms at the interface [51, 60], especially for high-temperature-grown AlN [48]. However, such SiN cannot be fully amorphous,

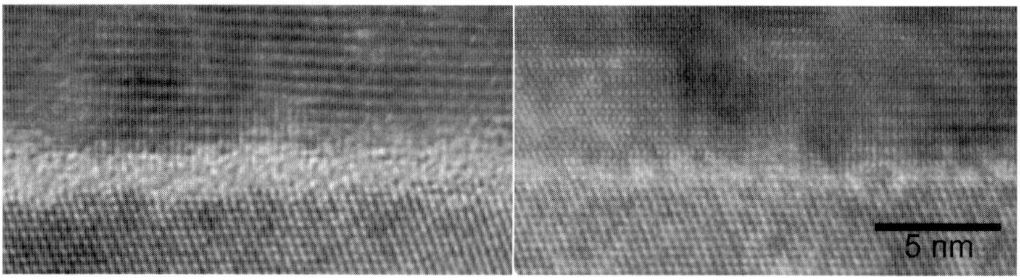

FIG. 3.6. HR-TEM image of the Si/AlN interface grown with a LT-AlN seeding layer with Al-predeposition of one sample at two different sample positions (within less than 1 mm distance). While the left image shows a distorted interface layer, the right image shows only little distortion and at some small areas a perfect transition from Si to AlN. (TEM by P. Veit, Otto-von-Guericke-University Magdeburg.)

as often believed on the basis of TEM images, or it must not be continuous, otherwise no epitaxial growth would be possible. TEM typically applied for these investigations averages through the sample cross-section, thus it is difficult to judge on the quality and details of this SiN layer. There is the additional possibility that some of the SiN forms during AlN growth.

As long as the surface is not continuously covered with AlN, which is usually the case (see Fig. 3.7), Si and N can react, forming SiN. The nitridation process of Si involves Si diffusion to the surface and there the formation of SiN. For a thickness of ≈6 nm this process is self-limiting [61] and the formation of crystalline (0001) oriented β-Si_3N_4 on Si(111) in an ammonia ambient at temperatures above 1075 K has been reported [62, 63]. Also, such a β-Si_3N_4 layer which has a coincidence site lattice of 1:2 to Si(111) and 2:5 to AlN has been used in MBE

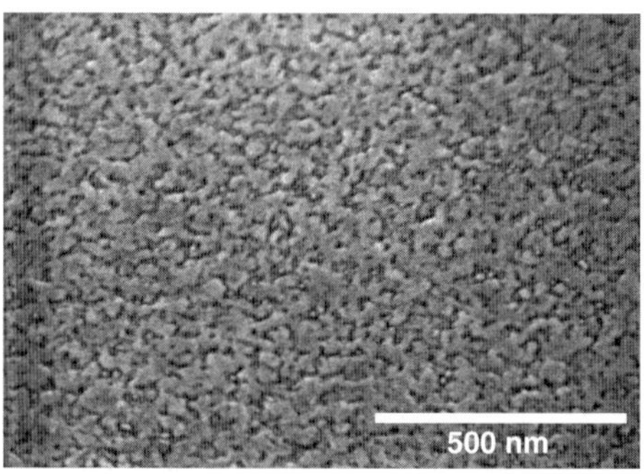

FIG. 3.7. Plan view SEM image of a 20-nm LT-AlN seeding layer on Si(111). The surface is not smooth, and surface pits are a high risk for meltback etching when GaN is grown on such a layer.

to improve the quality of the subsequent layers [64]. From the island shape of AlN during initial epitaxy, such a process might be involved and interrupted by AlN coalescence before thick SiN forms.

Some processes used for seeding layer growth even force the nitridation of the Si surface by a preflow of ammonia [13, 64]. Indeed, AlN can be grown on such treated surfaces, but to our knowledge best layers usually involve Al preflow. Aside the initial growth of the AlN seed, an AlN buffer requires high-quality AlN which can be grown at optimized conditions on the initial seeding layer. As mentioned earlier, one prerequisite to reduce meltback etching is the growth of a continuous layer free from pits. For AlN growth this usually requires high growth temperature and low V–III ratio. But a high growth temperature of the seeding layer can deteriorate or roughen the unprotected surface of the substrate with a melting point of only 1410°C. In addition, the thermal gradient present in cold wall MOVPE systems, as well as the stress induced by the growing high-temperature layer, can be sufficient to initiate plastic substrate deformation already during seed or buffer layer growth at extreme temperatures.

3.4 Stress management

Stress management, often also termed strain engineering, is the key for GaN device growth on silicon. Apart from thermal stress arising upon cooling, other sources of stress are present which often lead to an additional source of tensile stress. In, sum we find the following sources of stress:

- Initial 3D growth: hydrostatic compressive stress [65].
- Island coalescence: tensile stress [66].
- Lattice mismatch: tensile (Al-rich)[67] or compressive (In-rich).
- (Edge) dislocation climb: decrease of compressive stress[66, 68–70].
- Doping: tensile stress (silicon doping) [66, 70–73].
- Cooling: tensile stress when grown on silicon.

Typically, in device structures several of these stress sources are present. Thus the control and manipulation is the key to solve cracking for GaN on Si, but also to reduce the bow to acceptable values for processing. A key instrument for this is *in situ* curvature control. By this, the growth steps introducing stress and the efficiency of countermeasures can be monitored.

Indirectly, it even enables a monitoring of the layer quality. When the material quality is poor, compressive stress is difficult to apply. Additionally, once applied it quickly reduces to low values by stress induced edge dislocation climb but also when doping with Si, both indicating a high edge dislocation density. Layers which do not show this effect or only weakly typically are of high quality with edge dislocation densities below 10^9 cm^{-2}. Monitoring the curvature signal thus enables a fast optimization of the seeding and buffer layers without additional measurements. Details on doping and stress can be found in Section 3.4.1.3.

Stress management for GaN on Si typically involves a layer with a lower in-plane lattice parameter as AlN or AlGaN. As alternative but also an extension of this method selective growth in fields small enough to prevent crack-formation is applicable.

3.4.1 *Stress management by Al(Ga)N layers*

Al-rich layers are commonly applied for stress management of GaN-on-Si. They can either only apply in the lower part of the buffer layer or be distributed in the whole GaN stack. However, usually they are only present in the lower non-functional part of the device grown, because the higher band-gap energy and lower conductivity is usually detrimental for device operation.

3.4.1.1 *Buffer layer* The most common approach is the growth of an AlGaN buffer layer inducing compression on the subsequently grown GaN. This approach has already been used by Ishikawa *et al.*, and led to the basic device structure of Nitronex FETs [29, 32, 74]. To be successful with such a structure a high material quality is prerequisite to maintain the compression built up. Typically, edge-type dislocations in material under high compressive stress incline, and this inclination leads to a shortening of the lattice plane, and they terminate as schematically shown in Fig. 3.8. This reduction of the crystal volume leads to a

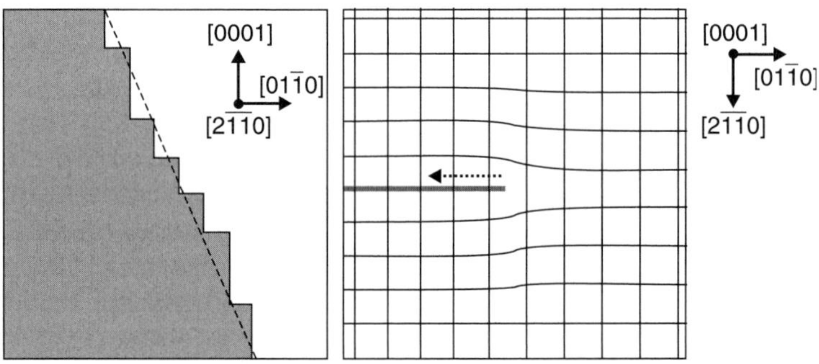

FIG. 3.8. Dislocation climb in group-III-nitrides typically leads to a shortening of the lattice plane which is terminated by an edge-type dislocation. The side view, commonly shown in TEM images as in Fig. 3.12, is shown schematically in the left scheme where the inclination is induced by compressive stress leading to vacancy incorporation or, even for layers under tensile stress, by silicon incorporation at the dislocation probably inducing a similar mechanism as vacancy incorporation. The right image shows the plan view where the additional lattice plane is shown in grey and the direction of its movement by the arrow. Note that in the left part of this plan-view image, lattice planes are more closely spaced due to the additional lattice plane terminated by the dislocation.

reduction in compressive stress. Therefore, to be successful the AlN seeding layer growth must be performed carefully with the aim of minimizing the edge-type dislocation density.

Additionally, the change in AlGaN composition should be well chosen. Huge steps in the lattice constant of the relaxed alloys reduce the critical thickness for fully strained growth, and will lead to the formation of misfit dislocations or cracks at an earlier stage than smaller changes in lattice parameter. On the other hand, high-strain gradients support the inclination and annihilation of dislocations, as shown in Fig. 3.12. Apart from FET growth, such layers are also used as a buffer for other device types to achieve thick GaN layers, e.g., for LED applications. Nevertheless, a high GaN buffer thickness solely using this method is difficult to achieve in combination with low bow values. It has, however, been demonstrated that by a proper optimization of the layers a 4-μm thick GaN buffer can be grown using only this method [75].

3.4.1.2 *Interlayers* The use of interlayers can be manifold. By interlayers we mean the insertion of layers in a GaN or an AlGaN buffer structure typically below the functional device layers. The common basic principle is to use layers with a smaller lattice constant than the main buffer layer, e.g., AlGaN in the case of GaN. Upon relaxation of the narrow lattice material the subsequently grown buffer usually is under compression which can fully or partially compensate tensile thermal stress. Two main approaches can be distinguished: superlattices or single layers or single layer stacks.

Superlattices A combination of AlN and GaN is most often used for superlattice growth. This approach was first successfully demonstrated by Feltin *et al.* for a green LED structure [76]. They analyzed stress development and demonstrated that these SL layers induce compressive stress on the subsequently grown GaN to compensate for tensile thermal stress. Having only a relatively thin GaN main layer this method was not very successful for LEDs, but for FET devices this method gained a lot of interest in the mid-2000s. Sugahara *et al.* and Iwakami *et al.* [30, 77] demonstrated a HV-FET with a thick AlN/GaN SL followed by a 600-nm thick GaN buffer and the active device layer. A benefit of these SL layers is the potentially high resistivity, especially for vertical conduction. However, for such devices it is most likely also important to suppress the formation of pyro- and piezoelectrically-induced 2DEG and 2DEH gases at the interfaces, which could lead to parasitic conduction paths. Thus, compensation doping using C or Fe is most likely necessary for FET applications when applying such buffer structures.

Al-rich AlGaN interlayers At present, the most commonly applied layers for strain engineering are Al-rich interlayers distributed with distances above 100 nm, typically between 500 and 3000 nm, in the GaN matrix. The mode of action of these layers is based on their partial or full relaxation and subsequent GaN growth, whereas this GaN layer grows pseudomorphically or only partially relaxed on top of these interlayers, thus under compressive stress. For relaxation it is not prerequisite to initially form misfit dislocations at the interface. Also micro-cracks or island growth can lead to compression on the subsequent GaN layer; e.g., when AlN is grown at high temperatures it usually cracks when exceeding a critical thickness, but does not necessarily introduce misfit dislocations at the stage of cracking. On the other hand low-temperature AlN grows in an island shape with islands of a lateral size of ≈10–20 nm. In both cases, overgrowth with GaN will fill the crack or the space between the islands, and due to the high strain fields misfit dislocations are preferentially generated at these positions. Surprisingly, with the upper interface GaN does not relax in a manner as does AlN, and therefore fewer misfit dislocations are formed (Fig. 3.9). This has been shown in detail by Markurt *et al.* [78]. With each misfit dislocation at the lower GaN/AlN interface the AlN layer is more and more relaxed, and with each missing misfit dislocation on the upper AlN/GaN interface, thus missing relaxation of GaN on the AlN interlayer, the new lattice parameter of the partially relaxed AlN layer is beneficial in building up compressive stress in the subsequently grown GaN layer.

Such interlayers can be grown in many ways as pure AlN at high or low temperature, as, e.g., AlGaN but also as AlInN, as long as the In content is well below 17%. Even alloying boron, e.g., to compress AlN on Si, might be possible, but until now it has not been demonstrated, and BN itself seems to be difficult to grow in the wurtzitic phase and to be more suited as a masking layer similar to *in situ* deposited SiN. It should be noted that for group-III-nitride growth the hexagonal BN phase has been applied for release and simple layer transfer from a sapphire substrate which should in principle also apply for the growth on Si [79, 80]. In the end, each type of interlayer must be well optimized with regard to composition, thickness, and growth conditions. There is no general rule, and there are even differences for nominally identical layers in different

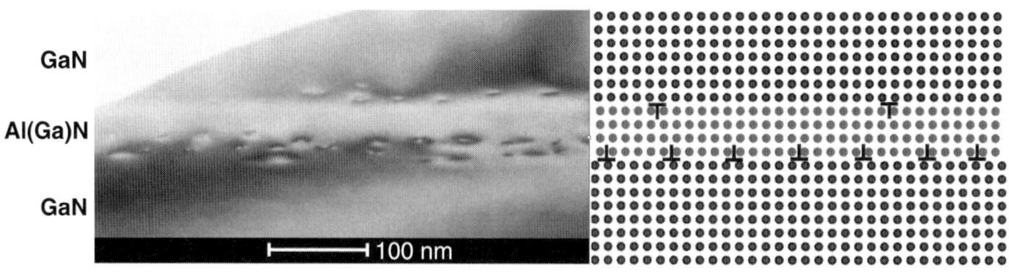

FIG. 3.9. TEM image and schematics of the relaxation process of an Al-rich interlayer sandwiched between GaN. Due to lattice relaxation, misfit dislocations are formed. Their density is higher on the lower interface as compared to the upper. This results in an overall compressive stress in the GaN layer grown on top. (TEM by Toni Markurt and Martin Albrecht, IKZ and HU-Berlin.)

growth systems, most likely due to different parasitic prereactions and effective precursor pressures.

In XRD measurements interlayers are often not identified as pure AlN but as AlGaN. This can be due to the interruption of this layer by cracks or by island formation, thus the layer consist of AlN and GaN. Such cracks or the gaps in the AlN layer fill up with GaN and are also the origin of at least a part of the relaxation process necessary to change the strain state of the subsequently grown GaN layer. Apart from this observation it is also indicated that AlGaN is grown due to a too-low stability of the GaN surface when AlN is grown under unfavorable growth conditions for GaN, leading to an incorporation of Ga in the growing layer. When intentionally applied, indeed growing AlGaN instead of AlN can lead to a better material quality and even to higher compressive stress of the subsequently grown layer (Fig. 3.10); the origin might be in a too-strong deterioration of the lattice (strain fields leading to tilt and twist on a small scale) if highly mismatched AlN is grown, but it might be also correlated with the better mobility of Ga atoms during epitaxial growth improving the overall crystalline quality of the layer.

Fritze *et al.* have demonstrated that in a device buffer structure with three AlGaN interlayers, the three interlayers act differently on strain and quality development [35]. There it has been shown that the first layer is only inducing compression on the subsequently grown GaN, while the second further enhances the degree of compression but also enhances the material quality slightly. The

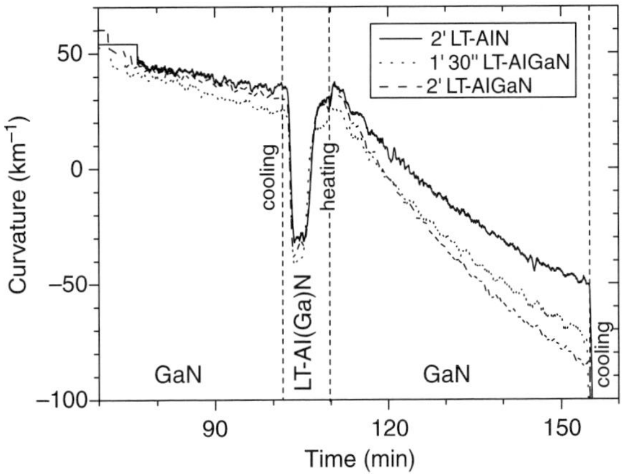

FIG. 3.10. *In situ* curvature measurements of a GaN/Al(Ga)N /GaN structure, demonstrating that the addition of Ga to the interlayer can improve the magnitude of compressive stress in the subsequently grown GaN layer. The amount of Ga was below 30%, and a thickness effect excluded by choosing the AlN interlayer thickness sufficiently thick for maximum compression by nominally pure AlN.

third layer has no impact on strain but significantly improves the material quality. This result indicates that an optimization for each layer is advisable

High compressive stress induced by the interlayers is enhancing edge dislocation climb leading to a reduction of compressive stress [68–70]. Thus, a high dislocation density can be already detected during growth if a high compressive stress is reduced during GaN layer growth, e.g., visible in a change of the *in situ* curvature slope. This method for growth control works well for layers with edge dislocation densities above $\approx 1 \times 10^9$ cm^{-2}. For lower dislocation densities the effect is rather small.

Doping with silicon has been also reported to lead to dislocation climb, here even for layers which are under tension (see also Section 3.4.1.3). We have also found indications that a reduced GaN growth temperatures decreases dislocation climb. But, significantly lowering the growth temperature often leads to a lower material quality, thus it is not a well suited method to reduce this effect.

3.4.1.3 *Doping* A topic not only related to the growth of group-III-nitrides on silicon is the impact of doping on strain and how to circumvent it. Numerous reports exist in the literature on the impact of silicon doping on stress [66, 70–73, 81–84]. In the meantime, many models were developed, either claiming Si-induced surface-roughening inducing edge dislocation climb [70], stress-induced edge-dislocation climb by vacancy formation [68], dislocation masking resulting in edge-dislocation climb [138, 66], Si "atmospheres" blocking-edge dislocation climb [85–87], or Fermi-level induced-edge dislocation climb as the origin of this observation [83, 84]. As has been observed by several authors, compressive stress and/or high Si doping concentrations do lead to an increasing tensile stress. In the latter case, even for a layer already under tensile stress, this tensile stress increases upon Si doping [66], which cannot be explained by a simple energy-balance model. The mechanism seems to be specific to Si doping because, e.g., Ge doping, in concentrations even higher than Si, does not necessarily show this effect, and thus excludes a Fermi-level induced mechanism as claimed by Xie *et al.* [83, 84]. The latter effect can be excluded at typical GaN growth conditions as, e.g., used by machine vendors as AIXTRON for GaN growth ($P_{Reactor} = 200$ mbar, V–III approx. 1000) [138, 88]. It cannot be ruled out yet that at growth conditions with lower growth pressure and likely lower V–III ratio a Fermi-level induced vacancy model might additionally induce dislocation climb.

It should be noted that from all experiments it can be concluded that perfect dislocation-free group-III-nitrides do not suffer from significant tensile stress generation by Si doping. This is only an effect when edge dislocations are present. Therefore, it can be also used to monitor the layer quality already during growth by an *in situ* curvature measurement, since the change of the curvature slope during Si doping is correlated to the edge-type dislocation density [66].

In general, for typical heteroepitaxial material, especially on Si substrates, where dislocation densities are often higher than on sapphire or more difficult to reduce, Ge doping is an interesting alternative to Si [138, 89–91]. For GaN on sapphire even a small amount of tension is acceptable, but for GaN on Si it is often not. Not only the absence or low impact on stress is interesting for layers on Si, but also the higher achievable doping concentration. Often, device layers are limited in thickness on Si and need to be interrupted by strain engineering layers leading to a higher vertical resistivity. A good current spreading and low-contact resistivities can be achieved using Ge dopants, because doping concentrations up to 3×10^{20} cm^{-3} can be achieved without degradation of the layers [88]. But at typical GaN growth temperatures the incorporation of Ge using either GeH_4 or a metalorganic compound as isobutylgermane is driven by the compound's vapor pressure above the GaN surface, and is very ineffective. About three orders of magnitude higher Ge concentrations in the gas phase are required to achieve a comparable doping concentration as for Si doping. In addition, the incorporation depends strongly on the crystal facet. For a-plane GaN a roughly two-orders-of-magnitude higher Ge incorporation has been observed [92] which is not the case for Si doping. Although Ge has been found not to diffuse in GaN, the high amount of Ge in the reactor bears the risk of memory effects, as it can be observed for Mg doping.

3.4.2 *Selective growth*

Alternatively to the usage of interlayers or to expand their potential for thick layers, selective growth by substrate patterning or patterning a buffer layer with a dielectric mask based on, e.g., SiO_2 or Si_3N_4 can achieve thicker layers [31, 93–96]. Here, tensile stress from cooling to room temperature can be partially relieved in the patterned regions, with crack generation often observed in the Si substrate regions between the overgrown areas due to the lower stiffness of the substrate in comparison with the group-III-nitride layer. When using dielectric masks, one difficulty is material transport from the mask to the unstructured region, where an enhanced growth rate is observed at the edge regions. Al-rich compounds tend to form polycrystalline deposits on the mask, which can disturb the growth of thicker layers due to the unavoidable lateral growth from the mask openings. An alternative approach is to etch the substrate prior to growth, also avoiding the usage of any masking material. The trenches formed are required to be deep enough to suppress a coalescence of the layer growing in the trench with the one above. The required suppression of meltback etching at the trench edges is an additional difficulty of this method.

3.5 Dislocation reduction

As mentioned earlier, dislocation reduction starts with an optimized seeding layer. With low initial dislocation densities it is much simpler to achieve an overall satisfying dislocation density in the active layers.

Independent of this, most methods for dislocation reduction in the buffer layer involve lateral or 3D growth. Around the year 2000, methods which commonly involve an interruption of growth as epitaxial lateral overgrowth were very

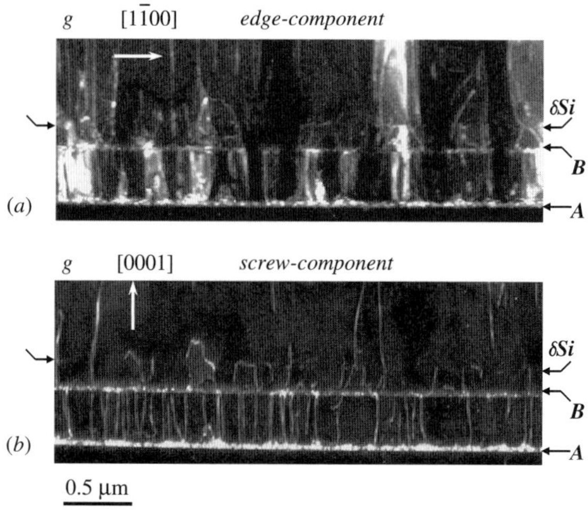

FIG. 3.11. TEM image of a GaN layer on a SiN layer grown on a GaN/AlN buffer on Si. The SiN layer itself cannot be seen, but the impact on dislocations which bend and, especially in the case of screw-type dislocations, undergo enhanced recombination. (Reprinted with permission from [110], © 2002 American Institute of Physics.)

popular [97–101], or pendeo-epitaxy [57, 102–106]. If a buffer layer were required this method is quite time-consuming, requiring two growth steps. Alternatively one can prepattern the substrate either with masked regions or etched trenches. This has been successfully applied by Strittmatter *et al.* [106]. However, this method is difficult to control, and meltback etching is a common problem. One can also transfer ELO methods to the microscale and start the growth on a substrate with, e.g., a nano- or micropatterned surface [107].

Nevertheless, such methods are more time-consuming, and therefore methods which improve the quality by *in situ* methods are more often used. A well-established method on silicon substrates is the application of a SiN masking layer which is believed to form a nearly closed layer hindering GaN nucleation on top of it [10, 108, 109]. Only in small openings of this layer new nuclei form, grow laterally, and coalesce, leading to a reduced dislocation density (Fig. 3.11) [110, 111]. Depending on the layer grown prior to the mask, and its strain state and quality, the GaN grown atop can be still under compression. Indeed, it can even lead to a stronger compression during growth, as is evident from significantly reduced tensile stress at room temperature [112], but usually the 3D growth mode to a larger part relaxes previous compressive stress.

The SiN layer usually is hard to be visualized in TEM directly, but its impact on dislocations can be viewed [110]. Usually, screw-type dislocations can be more easily annihilated which already stems from the much higher statistical probability that two dislocations with opposite burgers vector react, while for edge-type dislocations, with six possible burger vectors, it is less probable. Recently, HR-TEM images and calculations revealed the exact structure of the SiN masking layer, which is found to be a $SiGaN_3$ layer [113].

Apart from this difficulty in visualizing the layers in HR-TEM, in SEM a SiN layer typically can be easily visualized by the contrast it forms due to the enhanced electron concentration it induces as for a doped layer. In principal, any layer acting as anti-surfactant can be used instead of SiN, but only few are known to be suitable as an *in situ* layer, which is the only method suited for mass production. With BN acting as anti-surfactant, small amounts of boron alloyed in AlN or GaN might act similarly to Si, with the benefit that it does not introduce n-doping in the layer. Thus it is potentially better suited for high-frequency and high-power electronics based on silicon substrates. Another approach is the growth of strained layers. Especially edge-type dislocations tend to bend at the interfaces and within strained layers reducing stress by a climb process [68, 70]. For sufficiently high dislocation densities this increases the recombination probability and yields in an overall reduction in dislocation density. Strained layers can be already introduced as buffer layers. A common growth scheme for crack-free GaN on Si involves an AlN seeding layer followed by a graded AlGaN layer. In this graded layer an increasing compressive stress is introduced in the layer, which often leads to an increased dislocation climb (Fig. 3.12). The GaN layer grown atop the AlGaN is further strained. Usually, some misfit dislocations are formed at the interface, but this is not relaxing all compressive stress, and inclined dislocations which annihilate at a high rate can usually be observed [69]. This process is most probably assisted by an initial 3D growth mode originating

FIG. 3.12. Bright-field cross-sectional TEM image of three AlGaN layers ([Al] = 85%, 66%, and 52%) grown on AlN, showing dislocations with a-component (edge and mixed dislocations). The interfaces are marked by the white lines. An increasing compressive stress is induced in the AlGaN layers, first by the AlN seed and buffer, then by the following layers. While dislocations propagate vertically in the AlN which is grown under tensile stress, they bend in the AlGaN layers where they are under an increasing compression. (Reprinted with permission from [68], © 2009 American Institute of Physics.)

in the differing lattice parameters of the AlGaN and GaN layers, leading to dislocation bending, as is observed for 3D growth on, e.g., *in situ* SiN masks.

Also, strain fields, as in superlattices, can be used to force dislocation climb, increase annihilation probability, and reduce the dislocation density. This was first demonstrated by Feltin *et al.* [76], and should also apply for high-voltage buffer layers of AlN and GaN, as demonstrated by Iwakami *et al.* [77].

3.6 Light-emitting diodes

Light-emitting diodes are still the main driver for GaN research. In 2012, the majority of LEDs were still grown on sapphire substrates, Sanken had been selling simple blue and green LED on Si for several years, Latticepower announced that they were ramping up LED on silicon production [114], and Toshiba began selling GaN-on-Si LEDs based on Bridgelux growth development on 200-mm Si substrates. Apart from these three companies which already offer LEDs, others, such as OSRAM OS and Samsung, demonstrated high efficiencies for GaN-on-Si LEDs.

The main driver for this is the expectation that GaN-based LEDs epitaxially grown on silicon substrates enable a significant reduction in manufacturing cost [115]. This stems from the possible use of large-diameter substrates and using depreciated Si fabs to mass produce LEDs. As long as efficiency and yield are competitive for sapphire technology, there are many benefits in growth, processing, and also operation. Until now, light-emitting diodes grown on silicon substrates have been known as laboratory demonstrations for more than ten years [3, 4, 76, 93, 94, 116, 117], but they were long not believed to be a potential alternative to the growth on sapphire and SiC. The main reason for this belief was the poor efficiency due to light absorption by the Si substrate and the common belief that the material quality on silicon cannot compete with that on sapphire and SiC. But, although the material quality has long been inferior to that on sapphire and SiC, the reason for low efficiency mostly originates in absorption by the Si substrate. It is much higher than the often-assumed 50% if compared to sapphire. The reason is that not only back-emitted light is absorbed to about 30–40%, but also laterally-emitted light will be absorbed after a few reflections at the Si/III-nitride interface. Consequently, the poor efficiency is mostly due to optical losses and not necessarily an inferior material quality. After about 0.5 mW output power of a blue LED was achieved in 2003 [118] an output power close to 0.9 mW at 20 mA and 477 nm, reported in 2004, [66] was the record value until 2007. The main reason for the stagnation in improvement was light-absorption by the Si substrate, which will be discussed later in this chapter. But already a poor material quality—namely, dislocations—are significantly reducing the internal quantum efficiency of LEDs [119–121]. Thus, for high-power LEDs it is required to reduce the dislocation density at least to the low 10^8 cm^{-3} range. Lower values are not simply achievable, and usually do not significantly improve quantum efficiency. Dislocation densities in the low 10^8 cm^{-3} range can be typically achieved by a growth scheme employing a high-quality AlN seeding layer followed by an AlGaN transition layer stack, as, e.g., in Fritze *et al.* [35]. The role of this AlGaN layer stack is to induce compression, which results in two effects. First, dislocations incline, increasing their recombination probability (see Fig. 3.12), and compressive prestress prevents cracking. This effect is observed only when a high-quality seeding layer is grown, thus when *in situ* curvature measurements are performed they already enable judgment of the crystalline quality. The GaN layer grown on top of this buffer then usually grows under compressive stress, which prevents cracking upon cooling. Often directly on the AlGaN layer or after a thin GaN layer an *in situ* deposited SiN mask is inserted [117, 118]. This masking layer reduces the compression of the following GaN layer but significantly reduces dislocations by the promotion of 3D island growth. Then a thicker GaN buffer is grown, often with strain-engineering layers based on AlGaN or AlN grown at high or low temperature [6]. For a successful LED device the last functional n-doped GaN and active device layers should not be interrupted by any Al-rich layer for strain engineering. Such strain-engineering layers usually increase series resistance and lead to an increase in operation voltage around 0.5 V for each layer. Typical thicknesses of the final GaN layers are above 2 μm, even when highly doped with Si. A prerequisite for this is low density of edge and mixed dislocations (see also Section 3.4.1.3).

In principle, GaN-based LEDs can be grown in a large variety of other structures. Patterned substrates or selective growth helps in reducing stress and avoids cracking [93, 96], and nanowire LEDs are thought of as a high-brightness alternative due to perfect material and good light extraction [122, 123], but they are difficult to contact and in detail the large surfaces present additional problems. Also light extraction, if looked at in detail, it is not as good as for thin-film LEDs. LED structures can be grown on silicon-on-insulator substrates [124, 125], and also Bragg reflectors were inserted to improve light extraction [126, 127]. Also, AlGaN-based UV-LEDs on Si substrates were investigated [128].

Today, typical GaN-on-Si LEDs are no longer contacted through the Si substrate on which they are grown [129, 130], though Sanken is selling such a device which has a high electrostatic discharge stability of 5000 V. For high external quantum efficiencies, LED structures on Si are processed as thin-film LEDs where light extraction is significantly enhanced. By simple calculations (Figs. 3.13 and 3.14) one already finds that the light of an LED on Si substrate is either directly extracted vertically or, to a much smaller amount, arises from light reflected at the Si substrate (>60% loss) or from a region close to the side facets and laterally emitted. Taking these losses into account, even better reflecting buffer layers than Si as, e.g., HfN [54], will not significantly enhance light emission, because all or at least a larger portion of laterally emitted light is still lost. To significantly prevent this, extremely high reflectivities close to 100% are required. Thus, even for high-brightness LEDs with a silver-based reflector surface, roughening is crucial to minimize reflection losses and extract light with as little loss or as few reflections as possible. Alternatively, one can conclude that vertical nano- or

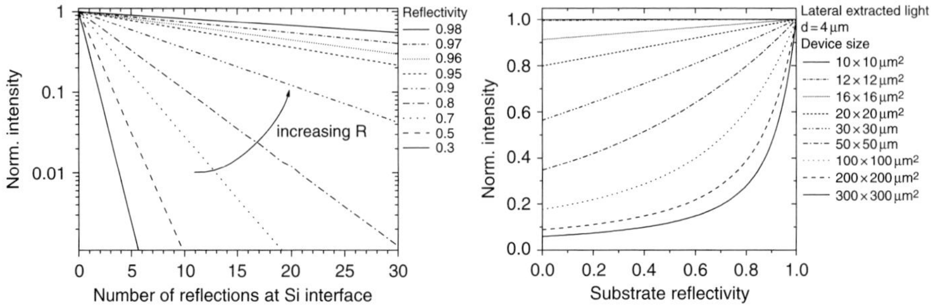

FIG. 3.13. Calculated amount of laterally emitted light and its extraction from LED structures. Left: Impact of multiple reflections on attenuation. Even for a high reflectance of 80% only 1% of light remains after 20 reflections. This strongly reduces laterally emitted light extraction as shown on the right image for 40% substrate reflectivity and different device sizes. Small devices on an absorbing Si substrate enable better light extraction.

microstructures engineered with structures which reduce reabsorption of the laterally emitted light can also be used for high-efficiency and high-power LEDs.

Regarding material quality it is well known that dislocations reduce the internal quantum efficiency [131]. While internal quantum efficiencies are low at high dislocation densities of 10^{10} cm^{-2}, yielding in substantial losses in the quantum efficiency of several 10%, the effect is much lower at lower dislocation densities, and amounts to only a few percent at dislocation densities in the low 10^8 cm^{-2} range.

In the meantime, thin-film LED concepts were developed for layers grown on Si, namely partial or full substrate removal [130–133], where this method significantly enhances light-extraction efficiency and leads to high internal efficiencies [134, 135] and wallplug efficiencies of 58% [136] at high light-extraction efficiencies around 70%, comparable to LEDs grown on sapphire [136].

At present, the commonly applied method of Si substrate removal involves grinding and wet chemical etching of the original III-nitride Si carrier while the layer is glued or bonded to a new carrier as Si or copper [132, 137]. It is a relatively simple and fast method. Recently, a separation method on sapphire was demonstrated which involves a 3-nm thin BN layer as a buffer [79, 80]. This should in principle be also applicable to Si substrates. If it does introduce new difficulties, as delamination during growth of the highly strained film has to be investigated. By today it has been proven that substrate removal applied to GaN on Si yields a comparable optical output power if the device structure on Si is well-optimized [35]. The benefit of growth on Si is the ease of substrate removal by mechanical and chemical methods. Additionally, low-cost Si carriers can be used for thin-film LED bonding, in contrast to Ge-based carriers often used for layers grown on sapphire due to the better thermal match upon bonding.

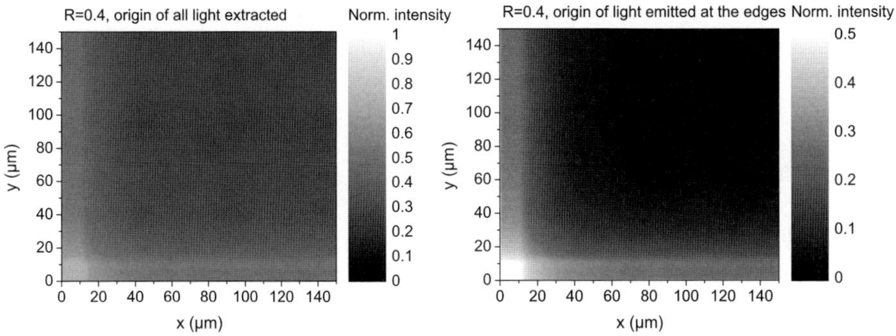

FIG. 3.14. The left image shows the origin of light emitted and extracted laterally at the side facets of a 300 × 300 μ^2m and 4-µm thick LED structure in epoxy resin, of which only the lower left quadrant is shown. This calculation does not take into account roughened surfaces or interfaces but only laterally-emitted light within an opening angle of 38° (e.g., for GaN in an epoxy resin matrix). A significant portion of light extracted laterally originates in a 10-µm wide edge region, while less than 10% and 1% of light from a region >30µm and >120 µm from the edge leaves the structure, respectively. Thus light emitted laterally in the inner part of the structure is mostly absorbed. From this region only vertically-emitted light and a small fraction of backward-emitted and reflected light will be visible, as shown in the right image, where in total 33% of all light emitted in the six 38° light cones is extracted. In principle the limitation to six 38° light cones allows only 63% of all light generated internally to be extracted if 100% substrate reflectivity and no absorption losses are assumed. Therefore, in the example here, 6.1% of all light generated internally is extracted at the side facets (left image), and in total only about 20% can be extracted (right image). Reflection losses and the limitation to a 38% light cone are the most significant limitations for LED light extraction. Surface roughening, reducing multiple reflections, and circumventing the limitation of the light extraction angle, ideally combined with thin-film techniques, will significantly increase LED brightness for layers grown on silicon substrates.

Another benefit of using silicon is the higher thermal conductivity in comparison to germanium.

Although the active part of a GaN-based LED on silicon is quite similar to the one on sapphire, growth needs to be optimized, and the freedom in design is more limited. The latter originates in the limited thickness achievable without any strain-engineering means. While on sapphire one can grow 5-µm thick layers and more without any growth interruption, on silicon, this is usually limited to 2–3 µm, depending on the material quality. Si doping further limits this thickness if a high number of edge-type dislocations are present. In addition to contacting, high doping levels are required, making the choice of an alternative n-type dopant source as Ge interesting to avoid problems, as the limitation in doping to the low 10^{19} cm^{-3} range when using Si and the limitation when the edge-type dislocation density is still too high to neglect its impact on tensile stress generation [88, 138]. A total thickness of the functional layers of an LED around 2–3 μm can be sufficient for current distribution, nevertheless it may limit LED designs in the future.

LED device layer growth on large substrates above 100 mm diameter was first demonstrated in 2006 [139–141], and it was shown that there is even a benefit with regard to bow and homogeneity. In the meantime, Si substrates with diameters as large as 200 mm were demonstrated, with wallplug efficiencies of 43% and IQE of 90% and very narrow (0002) and ($10\bar{1}2$) ω-rocking curves of 220 and 320 arcsec, respectively. The dislocation density was determined to $2\text{–}3 \cdot 10^8$ cm^{-2} [142]. With regard to larger wafer diameters, there is no indication that scaling is significantly limited except by reactor design. Reactor design in MOVPE is usually based on cold-wall reactor concepts. Using large-diameter substrates, wafer bow will lead to temperature and with it thickness and compositional inhomogeneities. This can be reduced and also eliminated with the use of warm- or hot-wall reactor concepts. Alternatively, single-wafer reactors are very capable of achieving homogeneous temperature profiles, even on 300-mm Si substrates, because the temperature profile can usually be adjusted between the different radially arranged heating zones, as, e.g., in Thomas Swan (now AIXTRON Ltd) CCS reactors.

3.7 Electronics

In the beginning, III-nitride electronics was typically connected to RF-electronics, and it was believed that there is a huge market, especially in the communication sector. However, it turned out that high-frequency operation is not as straightforward to achieve, and the market for telecommunication base stations turned out to be rather small. Especially here, device growth cost turned out to be less important than for mass products. But in the mid-2000s another type of device came into the focus: high-power, especially high-voltage, FETs. They do not require operation frequencies in the GHz range, and can even be produced on Si substrates at low cost. Typical operation frequencies are targeted in the kHz to MHz range where the residual conductivity of the substrate does not govern device losses. With high-breakdown field strengths of $3.3 \cdot 10^6$ V/cm^{-3} [143] and high-drift saturation velocities the potential of GaN yields further than that of SiC, but at present the lack of GaN substrates reduces the market to a voltage range above that of Si and below that of SiC. The focus of such devices is voltage transformation as, e.g., in switching power supplies and high current switching. Especially in the voltage range from 600 to 1200 V it is believed that such devices can reach a better price and performance than pure Si-based devices, and it is still possible that they can replace SiC at higher voltages. For

GaN-based switching power supplies high conversion efficiencies well above 90% are expected at low and high load states. In addition, even high-current devices based on Si might be replaced due to the expected higher conversion efficiency and smaller device size of GaN-based switches. HV-FETs on Si are expected to be produced in quantities (substrate area) similar to GaN-based LEDs.

The typical GaN high-voltage device is, as already mentioned, an FET. It is not operated vertically but laterally, and the 2D electron gas of a GaN/AlGaN heterostructure is typically used for current switching. The main disadvantage is the lack of a simple and reliable method to achieve normally-off or enhancement mode FET operation, though there are numerous reports of such devices on Si substrates. This type of device is required to achieve fail-safe transformers.

3.7.1 *RF transistors*

High-frequency devices on silicon are more difficult to realize than on sapphire or SiC due to the residual conductivity of the Si substrate. This leads to parasitic charges below the buffer structure which are even more problematic due to the limited total layer thickness if no interlayers are applied. If applied, interlayers can but must not lead to problems as parasitic charges. In sum, these issues lead to potentially high RC values. Another difficulty is the thermal resistance of silicon, which is about a factor of three higher than for SiC commonly used as a substrate of high-performance RF-FETs. Apart from these problems it has been already demonstrated that GaN-on-Si-based RF-FETs are in principle capable of high-frequency, high-power operation.

At least for operation at mobile telecommunication frequencies around 2 GHz, most problems can be solved as, e.g., NITRONEX is selling RF devices based on an AlN/AlGaN/GaN buffer on silicon. Here no interlayers for strain engineering are used, but the graded AlGaN buffer is applied for stress control yielding in low wafer bow and crack-free layers. Recently also high-frequency operation at 40 GHz with 2.5 W/mm has been demonstrated [144]. Here the buffer on Si consisted of an AlGaN transition layer followed by 1.5 μm $Al_{0.08}Ga_{0.92}N$ and a 150 nm thin GaN layer, followed by 6 nm of AlN and a 3-nm SiN capping layer. More details on FETs in general can be found in Chapters 6 and 11.

3.7.2 *HV transistors*

High-voltage devices require a well-insulating buffer layer which can be achieved by high Al-content or compensation doping of GaN. The theoretically expected values of the breakdown voltage for such layers have, however, not been achieved yet. The main reason are dislocations, and of them, most likely screw-type dislocations. In general, dislocations are potential breakdown paths, since they are charged due to dangling bonds in the dislocation core and by attracting impurities [145]. It has been shown that improving the material quality, namely, reducing the dislocation density, increases the breakdown field strength, enabling higher operation voltages at identical film thicknesses for different layer qualities, as

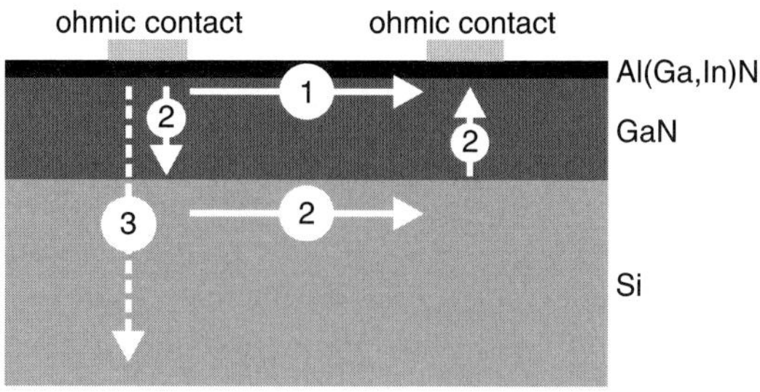

FIG. 3.15. Different breakdown paths for a GaN-on-Si FET structure with two top contacts. In addition to the common breakdown path between the contacts (1) using Si breakdown can also occur through the substrate (2) or even vertically through a grounded substrate (3).

shown in Fig. 3.16, where a maximum breakdown field strength of 150 V/μm or $1.5 \cdot 10^6$ V/cm has been achieved [36].

Apart from the material quality the layer thickness limits the breakdown voltage. Si has an about an order-of-magnitude lower breakdown field strength [37] and is, compared to GaN, highly conductive. For GaN on Si three breakdown paths have been observed (Fig. 3.15). When the device is mounted with an insulating layer below (floating substrate), the substrate breakdown occurs via two paths. Path 1 is between the contacts, which can be along the surface or laterally in the device-layer structure. This is the common breakdown path for small contact distances, also for devices mounted with a grounded substrate. The breakdown voltage along this path is strongly influenced by the material quality. Although dislocations propagate vertically and should not interfere with the device properties they do influence the breakdown field strength. Details of this mechanism are not well understood. One explanation might be a vertical breakdown path (2 or 3), even for small contact distances. In addition, it has not yet been verified if lateral dislocation networks, which appear in the case of lattice relaxation and are part of strain-engineering layers, are possible breakdown paths lowering the breakdown voltage. If this is the case, buffer-layer design has to be reconsidered for many device structures.

Path 2 is the breakdown path observed for contact distances approximately larger than twice the group-III-nitride thickness. Here, breakdown is vertical through the buffer layer into the substrate, which usually is not highly insulating and has a low-breakdown field strength. Thus the substrate can be also treated as a shortening between the GaN regions below the top contacts. This path leads to a saturation of the breakdown voltage even at larger contact distances, as in Fig. 3.16. For insulating substrates this path does not exist as well as the third breakdown path. This is simple vertical breakdown through the group-III-nitride layer and the substrate to ground. For contact distances up to approximately 10 μm, a distance which is still enabling low-resistance devices, the insulation of

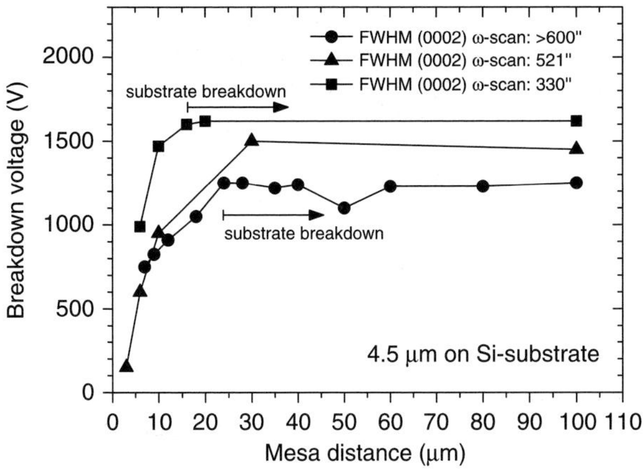

FIG. 3.16. Buffer breakdown of FET structures without an active AlGaN layer and different contact distance. The breakdown voltage increases with increasing distance and saturates at a value depending on buffer quality. In addition to this saturation, the slope for small contact distances differs in dependence of the material properties. Here, high-quality buffer layers enable smaller contact distances and higher breakdown voltages.

the substrate (not grounded) nearly doubles the breakdown voltage for layers up to 5 μm in thickness. Although thicker layers enable high breakdown voltages even with the substrate grounded, they are more costly and more difficult to manufacture due to wafer bow and possible plastic substrate deformation.

In addition to the current paths mentioned previously, additional lateral current paths can originate in the usage of heterostructures for strain compensation in the buffer. AlGaN and GaN are known to lead to 2DEGs at the lower and in principle also 2DHGs at the upper interface for cation terminated (0001) GaN. One possible solution is compensation doping with deep acceptors and deep donors. At present, two dopants are applied for this purpose: carbon [146] and iron. Iron is not very popular for processing in the Si industry due to the fear of cross-contamination, while carbon doping usually requires non-standard growth conditions. It is not yet certain that these growth conditions are well-suited to achieve a high material quality, thus in general Fe doping provides more freedom with regard to the growth parameters. Although Fe does not diffuse significantly in GaN but shows memory effects in many MOVPE reactors. Therefore, doping should be stopped early, before the intentionally undoped region of the buffer layer interferes with the active region of the device.

Most devices presented to date are based on AlGaN/GaN heterostructures. They usually consist of a GaN/AlGaN/GaN heterostructure, of which the last GaN layer is an approximately 3-nm thick capping layer. Also, SiN has been applied as a capping and passivating layer [147]. Simple AlGaN/GaN FETs are normally-on or depletion-mode FETs. Here, without an applied gate voltage the 2DEG is not depleted below the gate, and current flows even at zero voltage applied to the gate. For many applications this is an unwanted property, not being fail-safe and potentially dangerous if a circuit failure enables high current flow through the device. Thus a lot of efforts are nowadays focusing on normally-off or enhancement-mode FETs which can be operated in a fail-safe mode. Several concepts exist to realize such devices. A p-GaN buffer [148] or an insulating buffer [149] was applied by Yoshida *et al.* for high-power operation of an AlGaN/GaN FET on Si. Alternatively, the AlGaN layer can be grown thin [150] or partially or fully removed below the gate contact [151, 152]. As a consequence the missing polarization charges will deplete the 2DEG below the contact, and no current can flow. Also, fluorine implantation was applied [153] or partial removal of an *in situ* grown SiN capping layer [154]. Until now there is no well-established technique for enhancement-mode FETs.

HV devices are predicted to replace Si devices in the voltage range from 300 to 1200 V, and devices with breakdown voltages above 700 V [146, 155, 156] have already been demonstrated. At higher voltages the device cost can be sufficiently low to compete with Si especially being smaller than Si-based devices. By enabling higher operation frequencies in power supplies they enable a higher efficiency, making them attractive for mass production. At present, all devices are usually based on GaN/AlGaN, but with AlInN replacing AlGaN, group-III-nitride-based FETs might become even more attractive for HV-FET applications. AlInN can be grown lattice-matched, potentially reducing device failure during operation. In addition to the lattice-matched case, as well as for higher Al contents, it enables a factor of two higher electron concentrations in the 2DEG [157, 158] nowadays, also at carrier mobilities similar or only slightly lower than for GaN/AlGaN [159], and thus a much lower channel resistance. This can lower the resistance of the device or enable smaller devices, and by this, an even lower cost. GaN/AlN heterostructures offer the best option with regard to high electron concentrations and high mobilities, but they are difficult to grow and highly strained, potentially leading to early device failure.

3.8 Limits of GaN-on-Si MOVPE technology and new developments

Limitations of a technology are very often not real limitations in the long term, even when arguments as basic physics are put up. Quite often it turns out that they are only in the head of the arguer that makes a technology limited. E.g., at the end of the 1990s GaN on Si was commonly believed to be always full of cracks for thicker layers as required for devices, and also that it is highly doped with Si diffusing from the substrate into the layer as known from the group-III arsenides [160]. Both have been proven wrong: cracks can be eliminated by strain engineering, though it was argued that thermal expansion coefficients do not allow this and Si does not diffuse [161, 162] but forms a stable nitride on the non-group-III-nitride covered surface that blocks diffusion as discussed in

Section 3.3. Consequently, limitations are not necessarily general limitations, and scientists always try to push the limit further.

After a short summary of currently known limitations or problems we will show research examples of some of them apart from the current mainstream of research for GaN on Si.

3.8.1 *Limits of GaN-on-Si MOVPE technology*

There are numerous limits of the technology which apply only to GaN on Si, or at least to a larger degree, some of them have been mentioned in the previous chapters:

- Requirement of strain-engineering layers, limitation of uninterrupted GaN thickness, at present 4 μm [75].
- Thick layers require compressive prestress, leading to plastic substrate deformation [33, 36].
- Compressive prestress leads to significant wafer bow which leads to homogeneity problems for large wafer diameters in cold-wall MOVPE systems.
- Al-rich seeding layers obstruct low-resistance Si/Al(Ga,In)N/GaN junctions [40].
- Al-rich seeding layers and Al-rich interlayers lead to higher thermal resistance [163].
- High-resistance silicon substrates usually become conductive during MOVPE growth, which limits RF FET performance.

There are possibly many more limitations of the technology on Si, as there are limitations on all heterosubstrates. In our view these are the most important for GaN on Si technology currently limiting device design and upscaling to large substrate diameters. Nevertheless, many high-performance devices can be realized on silicon substrates, and the initial success of LEDs on Si and also as key technology enabling low-cost GaN HV-FETs demonstrates that the limit of GaN on Si can be pushed much further than first anticipated or even expected to be achievable at all. Thus new developments, some seeming more as academic fun-work, finally might lead to real applications. Some new developments on Si substrates are presented in the next section, from material growth research work closely related to the present mainstream of semi-polar research, to topics which are less intensely studied, such as photovoltaic solar cells and GaN on Si for micro-electro-mechanical systems.

3.8.2 *New developments*

3.8.2.1 *Semipolar GaN* Semi- and non-polar GaN has been an active field of research in recent years. For light emitters in the green-to-orange wavelength region it is thought to be beneficial because high In-content quantum wells are less or not impacted by the quantum-confined Stark effect (QCSE) which is strongest along the piezoelectric c-axis. It has been demonstrated to be especially useful for laser devices—in particular, cladding-free lasers were realized [164]—but with the absence of the QCSE to reach longer wavelengths higher In-content QWs are required, making epitaxial growth more demanding. E.g., it has been demonstrated that high In-content layers are impacted by stacking fault generation due to lattice relaxation [165]. Apart from small-sized high-quality m-plane and semipolar bulk substrates cut from thick c-axis oriented GaN boules, larger-sized non- or semipolar layers are currently usually grown on heterosubstrates. But direct growth on heterosubstrates typically leads to stacking faults which degrade device performance. One way to circumvent this is to use growth methods which apply lateral (usually c-axis oriented) growth where stacking faults are not present or only in the N-polar -c direction. On silicon this has been demonstrated successfully by the group led by Sawaki which shows high-quality layers free or virtually free from stacking faults on low-cost Si. For this the Si substrates are etched prior to growth, so that inclined Si(111) facets occur on which c-axis-oriented GaN is grown. Due to the inclination versus the surface normal semi- or non-polar layers are achieved. The method yields very low stacking fault and dislocation densities, and is very well suited for polarization reduced device growth [166]. With this method one can grow semipolar GaN($1\bar{1}01$) on 7° off-oriented Si(001) [167], GaN($11\bar{2}2$) on Si(113) [168, 169], and non-polar ($11\bar{2}0$) GaN on a (110) Si substrate [170].

As an alternative to this method, Ravash and coworkers have demonstrated that also on planar Si substrates, GaN with inclination angles up to approximately 25° can be grown. This corresponds to a reduction of the polarization in growth direction of approx. 50 % [171]. For this purpose, high-index Si as Si(112) and Si(113) were used [172, 173] and a low-temperature AlN seeding layer applied [174]. When using high-temperature AlN seeding layers only c-axis oriented GaN can be grown [15, 174]. The surface of high-index Si substrates inclined between the Si(001) and Si(111) orientation is thought to consist of Si(001) and Si(111) sections [172, 175]. AlN growth on these high-index surfaces seems to be preferred on the small Si(111) sections as c-axis-oriented AlN which are inclined versus the surface normal. The growth process is difficult to control, because only thin AlN seeding layers lead to the wanted inclination and single crystalline smooth layers. This again puts meltback etching, which is difficult to suppress with such thin seeding layers, into focus. Nevertheless, it is possible to control this by using high-temperature AlN and especially AlGaN following the seeding layer. Stacking faults are also present in these layers, but the use of LT-AlN interlayers, also required for stress control, significantly reduces their density [176, 177]. It is expected that optimized interlayers can in general reduce the density of stacking faults to values well below 10^4 cm^{-1} when an inclination of the c-axis towards the [$10\bar{1}0$] direction is present, since dislocation elimination by this method works also for ($10\bar{1}3$) oriented GaN. Other methods to reduce the stacking fault density have been presented in recent years, ranging from graded AlN/AlGaN/GaN layers to avoid 3D growth, and by this reduce stacking faults on SiC substrates [178] or *ex situ*-deposited Sc metal layers which were subsequently nitridated [179]. However, they are not as efficient, and are more difficult to apply in the case of ScN.

The mechanism of stacking-fault elimination by LT-AlN layers can be viewed as an inverse process to stacking-fault generation in InGaN layers on m-planar GaN, as described by Fischer *et al.* [165]. With AlN a material with a smaller lattice constant is grown on GaN. When stacking faults of type I_1 are present, this AlN layer relaxes, preferably by eliminating I_1 stacking faults. On the upper interface to GaN, less relaxation takes place, leading to compressively strained GaN, and only generating stacking faults of type I_2 which can glide along the basal plane and form small (less than 100 nm long) closed loops, which thus do not impact the following layers (Fig. 3.17).

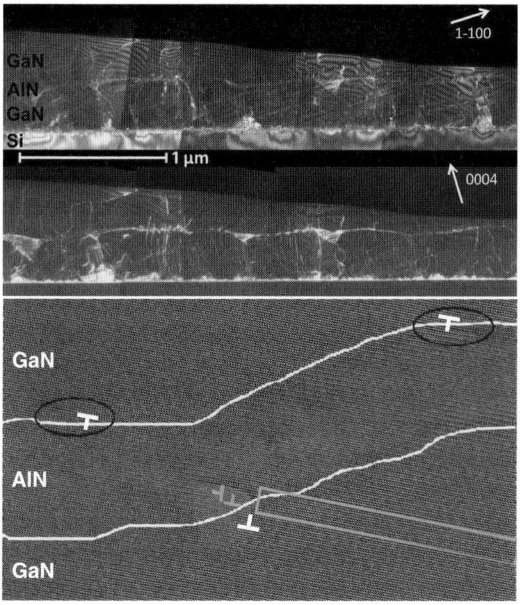

FIG. 3.17. Semipolar nearly $(10\bar{1}6)$ grown GaN on Si(112). The TEM overview (top) shows that at the AlN interlayers, most stacking faults disappear. The waviness of the AlN layer is due to a coalesced but still rough GaN surface when AlN was grown. The HR-TEM image (bottom) shows a detailed view with a-type misfit dislocations marked with white symbols, misfit disloactions with a Burgers vector of $\vec{b}=\frac{1}{6}<2\bar{2}03>$ (grey) terminating stacking faults of type I_1 (within the grey box), and stacking faults of type I_2 (within the dark ovals) with a Burgers vector of $\vec{b}=\frac{1}{3}<1\bar{1}00>$. A-type misfit dislocations and stacking-fault termination and generation are not necessarily appearing together. (TEM by Toni Markurt and Martin Albrecht, IKZ and HU-Berlin.)

3.8.2.2 *Photovoltaic solar cells* GaN-based photovoltaic solar cells have been discussed for quite some time. When alloying In, in principal the whole visible solar spectrum plus near-UV and IR can be accessed for charge-carrier generation. Several concepts have been tested, mostly based on nanowires [180] or quantum wells [181, 182]. QWs have the benefit of possible strained growth but the disadvantage of a low absorption volume. Bulk layers are difficult to grow without relaxation. Nevertheless, it is especially interesting to grow photovoltaic solar cell layers on low-cost substrates. For this, silicon is the ideal cell material, and in principle a simple tandem cell can be realized when growing InGaN with In contents above 40%, which is expected to yield a conversion efficiency above 30% [183].

First, tests of direct InGaN growth on Si(111), which should result in a low-contact resistance for p-Si/n-InGaN [41, 42], revealed that even at low growth temperatures below 700°C, meltback etching occurs. This effect can be reduced by applying a high In content AlInN seeding layer which suppresses a Ga–Si reaction. Nevertheless, InGaN growth is still highly columnar and In droplet formation difficult to avoid (Fig. 3.18). Taking into account the low In precursor efficiency of today's MOVPE growth systems, this approach requires higher precursor efficiency to be competitive with other photovoltaic solar cell approaches.

Until now only InGaN-based cells based on GaN buffer layers or InGaN QWs have been realized. A GaN/Si tandem cell, however, was demonstrated

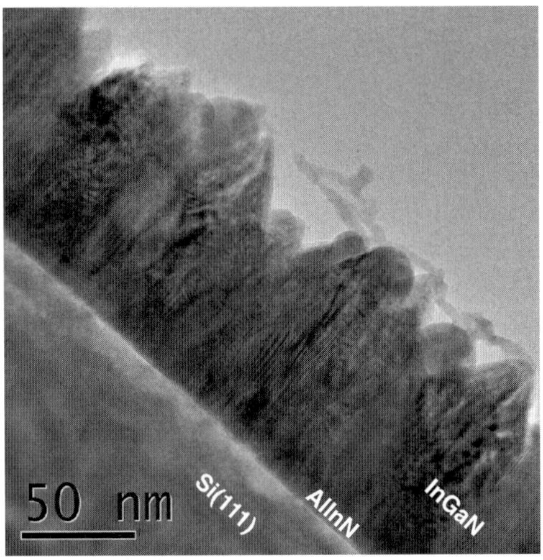

FIG. 3.18. TEM image of an InGaN layer with approximately 40% In on a thin AlInN seeding layer with In-content estimated above 40%. InGaN grows highly columnar with faceted tips. (TEM by A. Mogilatenko, HU-Berlin.)

by Reichertz *et al.* [184], showing that in principal such cells are achievable, but in this case with additional UV illumination to generate electron hole pairs in the GaN top cell.

3.8.2.3 *Micro-electro-mechanical systems (MEMS)* The main applications of GaN-based MEMS are different kinds of oscillators on Si. Here, usually thin bars based on GaN/AlGaN structures are etched and the Si below them removed. Depending on their geometry, these free-standing bars can vibrate at different frequencies and be used, e.g., for frequency filtering [185]. Apart from these applications, GaN is also interesting for other devices where, e.g., high-temperature operation or aggressive media are present. Such a device is a miniature high-pressure sensor as demonstrated by Zimmermann *et al.* [186]. This miniature sensor depends on a Si bar with a FET structure, and bending the bar changes the carrier density and with it the resistance of the 2DEG, which is then measured and is in principal suited for pressure measurements at high temperatures.

Another example of a MEMS device which generates energy and can be tuned electrically is the balance spring of a watch (Fig. 3.19). These balance springs are often manufactured from Si instead of metal by a dry etching process. Growing c-axis-oriented AlN on top of such Si spiral spring surfaces yields additional piezoelectric material which can be used to generate energy. Here the difficulty in MOVPE growth lies in a required homogeneous coverage of the whole Si surface with a c-axis-oriented group-III-nitride layer perpendicular to the surface which cannot be achieved, e.g., by sputtering techniques. With this, piezoelectric layer energy in the range of nanowatts can be generated, e.g., to charge electronics which in return can correct the motion of the balance spring by applying a voltage. This correction of the spring's oscillation frequency by quartz-controlled

FIG. 3.19. SEM images of an AlN-coated Si spiral spring for an application as a balance spring in a wrist watch. AlN was grown with its c-axis-oriented perpendicular to the surface normal by MOVPE (left). A voltage between the inner and outer AlN layers of the spring is induced by the spring motion. This can be used for charging an electronic circuit and a feedback loop to tune the oscillation frequency.

electronics can yield high accuracy in a mechanical watch. One can think of other micro-oscillating devices where energy is gained and used to operate electronics for other functions or oscillation control.

Especially in the field of MEMS, many applications can be thought of which make use of the piezo- and pyroelectric properties of the group-III-nitrides for sensors, filters, or frequency tuning. Here the maturity of the Si technology is of huge benefit when it comes to etching special structures as free-standing group-III-nitride layers or bars or thin Si layers which are subsequently overgrown, as in the case of the spiral spring material.

Acknowledgments

We would like to acknowledge the collaboration and contributions by S. Fritze, J. Bläsing, T. Hempel, A. Diez, and J. Christen of Otto-von-Guericke-University Magdeburg, F. Ponce of Arizona State University, O. E. Contreras of Universidad Nacional Autónoma de México, A. Mogilatenko of HU-Berlin, T. Markurt and M. Albrecht of Institut für Kristallzüchtung Berlin, Peter Stauss, Phillip Drechsel, and Patrick Rode of OSRAM OS, and funding of some of this work by the Deutsche Forschungsgemeinschaft and the Bundesministerium für Bildung und Forschung.

References

[1] H. P. Maruska, *Applied Physics Letters* **15**, 327 (1969).

[2] H. M. Manasevit, F. M. Erdmann, and W. I. Simpson, *Journal of the Electrochemical Society* **118**, 1864 (1971).

[3] S. Guha and N. A. Bojarczuk, *Applied Physics Letters* **72**, 415 (1998).

[4] S. Guha and N. A. Bojarczuk, *Applied Physics Letters* **73**, 1487 (1998).

[5] H. Ishikawa, G.-Y. Zhao, N. Nakada, T. Egawa, T. Jimbo, and M. Umeno, *Japanese Journal of Applied Physics* **38**, L492 (1999).

[6] A. Dadgar, J. Bläsing, A. Diez, A. Alam, M. Heuken, and A. Krost, *Japanese Journal of Applied Physics* **39**, L1183 (2000).

[7] A. Krost and A. Dadgar, *Materials Science and Engineering*, B **93**, 77 (2002).

[8] B. Kuhn and F. Scholz, *physica status solidi (a)* **188**, 629 (2001).

[9] J. Bläsing, A. Krost, J. Hertkorn, F. Scholz, L. Kirste, A. Chuvilin, and U. Kaiser, *Journal of Applied Physics* **105**, 033504 (2009).

[10] P. Hageman, S. Haffouz, V. Kirilyuk, A. Grzegorczyk, and P. Larsen, *physica status solidi (a)* **188**, 523 (2001).

[11] P. Chen, R. Zhang, Z. Zhao, D. Xi, B. Shen, Z. Chen, Y. Zhou, S. Xie, W. Lu, and Y. Zheng, *Journal of Crystal Growth* **225**, 150 (2001).

[12] A. Dadgar and A. Krost, MOVPE growth of GaN on Si, in *Vacuum Science and Technology*, edited by T. Paskova, B. Monemar, and S. Krukowski, Research Signpost, Trivandrum, Kerala, India, 2002.

[13] E. L. Piner, J. C. Roberts, and P. Rajagopal, US020050285142, Gallium nitride materials and methods associated with the same, 1. April 2005.

[14] A. Dadgar, New technology approaches, in *III-V Compound Semiconductor*, edited by T. Li, M. A. Mastro, and A. Dadgar, CRC Press, Boca Raton, FL, 2011.

[15] F. Reiher, A. Dadgar, J. Bläsing, M. Wieneke, and A. Krost, *Journal of Crystal Growth* **312**, 180 (2010).
[16] O. E. Contreras, F. Ruiz-Zepeda, A. Dadgar, A. Krost, and F. A. Ponce, *Applied Physics Express* **1**, 061104 (2008).
[17] A. Dadgar, F. Schulze, M. Wienecke, A. Gadanecz, J. Bläsing, P. Veit, T. Hempel, A. Diez, J. Christen, and A. Krost, *New Journal of Physics* **9**, 389 (2007).
[18] F. Ruiz-Zepeda, O. Contreras, A. Dadgar, and A. Krost, *Applied Physics Letters* **96**, 231908 (2010).
[19] F. Schulze, A. Dadgar, J. Bläsing, and A. Krost, *Journal of Crystal Growth* **272**, 496 (2004).
[20] F. Schulze, A. Dadgar, J. Bläsing, and A. Krost, *Applied Physics Letters* **84**, 4747 (2004).
[21] F. Schulze, A. Dadgar, J. Bläsing, T. Hempel, A. Diez, J. Christen, and A. Krost, *Journal of Crystal Growth* **289**, 485 (2006).
[22] F. Schulze, A. Dadgar, J. Bläsing, A. Diez, and A. Krost, *Applied Physics Letters* **88**, 121114 (2006).
[23] S. Joblot, E. Feltin, E. Beraudo, P. Vennéguès, M. Leroux, F. Omnès, M. Laügt, and Y. Cordier, *Journal of Crystal Growth* **280**, 44 (2005).
[24] V. Lebedev, J. Jinschek, J. Kräußlich, U. Kaiser, B. Schröter, and W. Richter, *Journal of Crystal Growth* **230**, 426 (2001).
[25] F. Schulze, O. Kisel, A. Dadgar, A. Krtschil, J. Bläsing, M. Kunze, I. Daumiller, T. Hempel, A. Diez, R. Clos, J. Christen, and A. Krost, *Journal of Crystal Growth* **299**, 399 (2007).
[26] F. Schulze, A. Dadgar, F. Bertram, J. Bläsing, A. Diez, P. Veit, R. Clos, J. Christen, and A. Krost, *physica status solidi (c)* **4**, 41 (2007).
[27] F. Schulze, A. Dadgar, A. Krtschil, C. Hums, L. Reissmann, A. Diez, J. Christen, and A. Krost, *physica status solidi (c)* **5**, 2238 (2008).
[28] Y. Okada and Y. Tokumaru, *Journal of Applied Physics* **56**, 314 (1984).
[29] H. Marchand, L. Zhao, N. Zhang, B. Moran, R. Coffie, U. K. Mishra, J. S. Speck, S. P. DenBaars, and J. A. Freitas, *Journal of Applied Physics* **89**, 7846 (2001).
[30] T. Sugahara, J.-S. Lee, and K. Ohtsuka, *Japanese Journal of Applied Physics* **43**, L1595 (2004).
[31] S. Zamir, B. Meyler, and J. Salzman, *Journal of Crystal Growth* **243**, 375 (2002).
[32] H. Ishikawa, K. Yamamoto, T. Egawa, T. Soga, T. Jimbo, and M. Umeno, *Journal of Crystal Growth* **189–190**, 178 (1998).
[33] A. Krost, A. Dadgar, G. Strassburger, and R. Clos, *physica status solidi (a)* **200**, 26 (2003).
[34] G. Kipshidze, S. Nikishin, V. Kuryatkov, K. Choi, Ì. Gherasoiu, T. Prokofyeva, M. Holtz, H. Temkin, K. D. Hobart, F. J. Kub, and M. Fatemi, *Journal of Electronic Materials* **30**, 825 (2001).
[35] S. Fritze, P. Drechsel, P. Stauss, P. Rode, T. Markurt, T. Schulz, M. Albrecht, J. Bläsing, A. Dadgar, and A. Krost, *Journal of Applied Physics* **111**, 124505 (2012).
[36] A. Dadgar, T. Hempel, J. Bläsing, O. Schulz, S. Fritze, J. Christen, and A. Krost, *physica status solidi (c)* **8**, 1503 (2011).
[37] K. McKay, *Physical Review* **94**, 877–884 (1954).

[38] A. Dadgar, S. Fritze, O. Schulz, J. Hennig, J. Bläsing, H. Witte, A. Diez, U. Heinle, M. Kunze, I. Daumiller, K. Haberland, and A. Krost, *Journal of Crystal Growth* (2012).

[39] J. Doerschel and F.-G. Kirscht, *physica status solidi (a)* **64**, K85 (1981).

[40] Y. Honda, S. Kato, M. Yamaguchi, and N. Sawaki, *physica status solidi (c)* **4**, 2740 (2007).

[41] J. W. Ager, L. A. Reichertz, Y. Cui, Y. E. Romanyuk, D. Kreier, S. R. Leone, K. M. Yu, W. J. Schaff, and W. Walukiewicz, *physica status solidi (c)* **6**, S413 (2009).

[42] W. Walukiewicz, *Physica B: Condensed Matter* **302–303**, 123 (2001).

[43] E. Sakalauskas, H. Behmenburg, C. Hums, P. Schley, G. Rossbach, C. Giesen, M. Heuken, H. Kalisch, R. H. Jansen, J. Bläsing, A. Dadgar, A. Krost, and R. Goldhahn, *Journal of Physics D: Applied Physics* **43**, 365102 (2010).

[44] E. Sakalauskas, Ö. Tuna, A. Kraus, H. Bremers, U. Rossow, C. Giesen, M. Heuken, A. Hangleiter, G. Gobsch, and R. Goldhahn, *physica status solidi (b)* **249**, 485 (2012).

[45] J. Wei, B. Zhang, G. Wang, B. Fan, Y. Liu, W. Rao, Z. Huang, W. Yang, T. Chen, and T. Egawa, *Japanese Journal of Applied Physics* **49**, 072104 (2010).

[46] S.-J. Lee, S.-H. Jang, S.-S. Lee, and C.-R. Lee, *Journal of Crystal Growth* **249**, 65 (2003).

[47] R. Liu, F. A. Ponce, A. Dadgar, and A. Krost, *Applied Physics Letters* **83**, 860 (2003).

[48] G. Radtke, M. Couillard, G. A. Botton, D. Zhu, and C. J. Humphreys, *Applied Physics Letters* **97**, 251901 (2010).

[49] G. Radtke, M. Couillard, G. A. Botton, D. Zhu, and C. J. Humphreys, *Applied Physics Letters* **100**, 011910 (2012).

[50] D. Xi, Y. Zheng, P. Chen, Z. Zhao, P. Chen, S. Xie, B. Shen, S. Gu, and R. Zhang, *physica status solidi (a)* **191**, 137 (2002).

[51] S. Kaiser, M. Jakob, J. Zweck, W. Gebhardt, O. Ambacher, R. Dimitrov, A. T. Schremer, J. A. Smart, and J. R. Shealy, *Journal of Vacuum Science and Technology B: Microelectronics and Nanometer Structures* **18**, 733 (2000).

[52] M.-H. Kim, Y.-C. Bang, N.-M. Park, C.-J. Choi, T.-Y. Seong, and S.-J. Park, *Applied Physics Letters* **78**, 2858 (2001).

[53] S. Tanaka, Y. Kawaguchi, N. Sawaki, M. Hibino, and K. Hiramatsu, *Applied Physics Letters* **76**, 2701 (2000).

[54] R. Armitage, Q. Yang, H. Feick, J. Gebauer, E. R. Weber, S. Shinkai, and K. Sasaki, *Applied Physics Letters* **81**, 1450 (2002).

[55] N. P. Kobayashi, J. T. Kobayashi, P. D. Dapkus, W.-J. Choi, A. E. Bond, X. Zhang, and D. H. Rich, *Applied Physics Letters* **71**, 3569 (1997).

[56] N. P. Kobayashi, J. T. Kobayashi, X. Zhang, P. D. Dapkus, and D. H. Rich, *Applied Physics Letters* **74**, 2836 (1999).

[57] R. F. Davis, T. Gehrke, K. J. Linthicum, E. Preble, P. Rajagopal, C. Ronning, C. Zorman, and M. Mehregany, *Journal of Crystal Growth* **231**, 335 (2001).

[58] C. Park, J. Kang, K. Kim, E.-K. Suh, K. Lim, and K. Nahm, *Journal of Crystal Growth* **224**, 190 (2001).

[59] M. Grundmann, A. Krost, and D. Bimberg, *Applied Physics Letters* **58**, 284 (1991).

[60] G. W. Auner, F. Jin, V. M. Naik, and R. Naik, *Journal of Applied Physics* **85**, 7879 (1999).

[61] R. d. Almeida and I. Baumvol, *Physical Review B* **62**, R16255 (2000).

[62] X.-S. Wang, G. Zhai, J. Yang, L. Wang, Y. Hu, Z. Li, J. Tang, X. Wang, K. Fung, and N. Cue, *Surface Science* **494**, 83 (2001).

[63] J. Kim and H. Yeom, *Physical Review B* **67** (2003).

[64] C.-L. Wu, J.-C. Wang, M.-H. Chan, T. T. Chen, and S. Gwo, *Applied Physics Letters* **83**, 4530 (2003).

[65] A. Krost, A. Dadgar, J. Bläsing, A. Diez, T. Hempel, S. Petzold, J. Christen, and R. Clos, *Applied Physics Letters* **85**, 3441 (2004).

[66] A. Dadgar, R. Clos, G. Strassburger, F. Schulze, P. Veit, T. Hempel, J. Bläsing, A. Krtschil, I. Daumiller, M. Kunze, A. Kaluza, A. Modlich, M. Kamp, A. Diez, J. Christen, and A. Krost, Strains and Stresses in GaN Heteroepitaxy: Sources and Control, in *Advances in Solid State Physics*, edited by B. Kramer, pp. 313–326, Springer, Berlin and Heidelberg, 2004.

[67] A. Dadgar, F. Schulze, T. Zettler, K. Haberland, R. Clos, G. Straßburger, J. Bläsing, A. Diez, and A. Krost, *Journal of Crystal Growth* **272**, 72 (2004).

[68] D. M. Follstaedt, S. R. Lee, A. A. Allerman, and J. A. Floro, *Journal of Applied Physics* **105**, 083507 (2009).

[69] S. Raghavan, X. Weng, E. Dickey, and J. M. Redwing, *Applied Physics Letters* **88**, 041904 (2006).

[70] A. E. Romanov and J. S. Speck, *Applied Physics Letters* **83**, 2569 (2003).

[71] P. Cantu, F. Wu, P. Waltereit, S. Keller, A. E. Romanov, U. K. Mishra, S. P. DenBaars, and J. S. Speck, *Applied Physics Letters* **83**, 674 (2003).

[72] I. Manning, X. Weng, M. Fanton, D. Snyder, and J. Redwing, *Journal of Crystal Growth* **312**, 1301 (2010).

[73] L. T. Romano, C. G. van de Walle, J. W. Ager, W. Gotz, and R. S. Kern, *Journal of Applied Physics* **87**, 7745 (2000).

[74] T. W. Weeks jr, E. L. Piner, T. Gehrke, and K. J. Linthicum, Patent US6649287, Gallium nitride materials and methods, 18 November 2003.

[75] E. Frayssinet, Y. Cordier, H. P. D. Schenk, and A. Bavard, *physica status solidi (c)* **8**, 1479 (2011).

[76] E. Feltin, B. Beaumont, M. Laügt, P. d. Mierry, P. Venngus, H. Lahrche, M. Leroux, and P. Gibart, *Applied Physics Letters* **79**, 3230 (2001).

[77] S. Iwakami, M. Yanagihara, O. Machida, E. Chino, N. Kaneko, H. Goto, and K. Ohtsuka, *Japanese Journal of Applied Physics* **43**, L831 (2004).

[78] T. Markurt, M. Albrecht, *et al.*, to be published.

[79] Y. Kobayashi, K. Kumakura, T. Akasaka, and T. Makimoto, *Nature* **484**, 223 (2012).

[80] T. Makimoto, K. Kumakura, Y. Kobayashi, T. Akasaka, and H. Yamamoto, *Applied Physics Express* **5**, 072102 (2012).

[81] L. Romano, C. van de Walle, B. Krusor, R. Lau, J. Ho, T. Schmidt, J. Ager III, W. Götz, and R. Kern, *Physica B: Condensed Matter* **273–274**, 50 (1999).

[82] H. Schenk, E. Frayssinet, A. Bavard, D. Rondi, Y. Cordier, and M. Kennard, *Journal of Crystal Growth* **314**, 85 (2011).

[83] J. Xie, S. Mita, L. Hussey, A. Rice, J. Tweedie, J. LeBeau, R. Collazo, and Z. Sitar, *Applied Physics Letters* **99**, 141916 (2011).

[84] J. Xie, S. Mita, A. Rice, J. Tweedie, L. Hussey, R. Collazo, and Z. Sitar, *Applied Physics Letters* **98**, 202101 (2011).

[85] M. A. Moram, M. J. Kappers, F. Massabuau, R. A. Oliver, and C. J. Humphreys, *Journal of Applied Physics* **109**, 073509 (2011).

[86] A. Dadgar and A. Krost, *Journal of Applied Physics* **110**, 096101 (2011).

[87] M. A. Moram, M. J. Kappers, F. Massabuau, R. A. Oliver, and C. J. Humphreys, *Journal of Applied Physics* **110**, 096102 (2011).

[88] S. Fritze, A. Dadgar, H. Witte, M. Bugler, A. Rohrbeck, J. Bläsing, A. Hoffmann, and A. Krost, *Applied Physics Letters* **100**, 122104 (2012).

[89] S. Nakamura, T. Mukai, and M. Senoh, *Japanese Journal of Applied Physics* **31**, 2883 (1992).

[90] J. K. Sheu and G. C. Chi, *Journal of Physics: Condensed Matter* **14**, R657 (2002).

[91] P. Hageman, W. Schaff, J. Janinski, and Z. Liliental-Weber, *Journal of Crystal Growth* **267**, 123 (2004).

[92] Matthias Wieneke, Hartmut Witte, Karsten Lange, Martin Feneberg, Armin Dadgar, Jürgen Bläsing, Rüdiger Goldhahn, and Alois Krost, to be published.

[93] A. Dadgar, A. Alam, T. Riemann, J. Bläsing, A. Diez, M. Poschenrieder, M. Strassburg, M. Heuken, J. Christen, and A. Krost, *physica status solidi (a)* **188**, 155 (2001).

[94] Y. Honda, Y. Kuroiwa, M. Yamaguchi, and N. Sawaki, *Applied Physics Letters* **80**, 222 (2002).

[95] J. W. Yang, A. Lunev, G. Simin, A. Chitnis, M. Shatalov, M. A. Khan, J. E. van Nostrand, and R. Gaska, *Applied Physics Letters* **76**, 273 (2000).

[96] X. Zou, K. M. Wong, N. Yu, P. Chen, and K. M. Lau, *physica status solidi (c)* **9**, 572 (2012).

[97] E. Feltin, B. Beaumont, P. Venngus, T. Riemann, J. Christen, J. Faurie, and P. Gibart, *physica status solidi (a)* **188**, 733 (2001).

[98] E. Feltin, B. Beaumont, M. Laügt, P. d. Mierry, P. Venngus, M. Leroux, and P. Gibart, *physica status solidi (a)* **188**, 531 (2001).

[99] P. Kung, D. Walker, M. Hamilton, J. Diaz, and M. Razeghi, *Applied Physics Letters* **74**, 570 (1999).

[100] H. Marchand, N. Zhang, L. Zhao, Y. Golan, S. Rosner, G. Girolami, P. Fini, J. Ibbetson, S. DenBaars, J. Speck, and U. Mishra, *MRS Internet Journal of Nitride Semiconductor Research* **4.2** (1999).

[101] S. Tanaka, Y. Honda, N. Sawaki, and M. Hibino, *Applied Physics Letters* **79**, 955 (2001).

[102] R. Davis, T. Gehrke, K. Linthicum, T. Zheleva, P. Rajagopal, C. Zorman, and M. Mehregany, *MRS Internet Journal of Nitride Semiconductor Research* **5S2**, W2.1 (2000).

[103] R. Davis, T. Gehrke, K. Linthicum, P. Rajagopal, A. Roskowski, T. Zheleva, E. Preble, C. Zorman, M. Mehregany, U. Schwarz, and J. G. R. Schuck, *MRS Internet Journal of Nitride Semiconductor Research* **6**, 14 (2001).

[104] T. Gehrke, K. Linthicum, P. Rajagopal, E.A. Preble, and R. Davis, *MRS Internet Journal of Nitride Semiconductor Research* **5S1**, W2.4 (2000).

[105] T. M. Katona, M. D. Craven, P. T. Fini, J. S. Speck, and S. P. DenBaars, *Applied Physics Letters* **79**, 2907 (2001).

[106] A. Strittmatter, S. Rodt, L. Reißmann, D. Bimberg, H. Schroder, E. Obermeier, T. Riemann, J. Christen, and A. Krost, *Applied Physics Letters* **78**, 727 (2001).

[107] C.-H. Chiu, C.-C. Lin, D.-M. Deng, D.-W. Lin, J.-C. Li, Z.-Y. Li, G.-W. Shu, T.-C. Lu, J.-L. Shen, H.-C. Kuo, and K.-M. Lau, *IEEE Journal of Quantum Electronics* **47**, 899 (2011).

[108] H. Lahrèche, P. Vennéguès, B. Beaumont, and P. Gibart, *Journal of Crystal Growth* **205**, 245 (1999).

[109] S. Tanaka, M. Takeuchi, and Y. Aoyagi, *Japanese Journal of Applied Physics* **39**, L831 (2000).

[110] O. Contreras, F. A. Ponce, J. Christen, A. Dadgar, and A. Krost, *Applied Physics Letters* **81**, 4712 (2002).

[111] T. Riemann, T. Hempel, J. Christen, P. Veit, R. Clos, A. Dadgar, A. Krost, U. Haboeck, and A. Hoffmann, *Journal of Applied Physics* **99**, 123518 (2006).

[112] A. Dadgar, M. Poschenrieder, A. Reiher, J. Bläsing, J. Christen, A. Krtschil, T. Finger, T. Hempel, A. Diez, and A. Krost, *Applied Physics Letters* **82**, 28 (2003).

[113] T. Markurt, T. Lymperakis, J. Neugebauer, P. Drechsel, P. Stauss, T. Schulz, T. Remmele, V. Grillo, E. Rotunno, and M. Albrecht, *Physical Review Letters* **110**, 036103 (2013).

[114] Lattice Power, Lattice Power–First company to offer volume production of high-power silicon substrate-based LEDs, http://www.latticepower.com/english/newsview.aspx?id=13, 2012.

[115] Eric Virey, *iLED* **3**, 16 (2012).

[116] A. Dadgar, M. Poschenrieder, J. Bläsing, K. Fehse, A. Diez, and A. Krost, *Applied Physics Letters* **80**, 3670 (2002).

[117] A. Dadgar, M. Poschenrieder, O. Contreras, J. Christen, K. Fehse, J. Bläsing, A. Diez, F. Schulze, T. Riemann, F. Ponce, and A. Krost, *physica status solidi (a)* **192**, 308 (2002).

[118] A. Dadgar, M. Poschenrieder, J. Bläsing, O. Contreras, F. Bertram, T. Riemann, A. Reiher, M. Kunze, I. Daumiller, A. Krtschil, A. Diez, A. Kaluza, A. Modlich, M. Kamp, J. Christen, F. Ponce, E. Kohn, and A. Krost, *Journal of Crystal Growth* **248**, 556 (2003).

[119] Q. Dai, M. F. Schubert, M. H. Kim, J. K. Kim, E. F. Schubert, D. D. Koleske, M. H. Crawford, S. R. Lee, A. J. Fischer, G. Thaler, and M. A. Banas, *Applied Physics Letters* **94**, 111109 (2009).

[120] I.-L. Lu, Y.-R. Wu, and J. Singh, *Journal of Applied Physics* **108**, 124508 (2010).

[121] K. Takeda, F. Mori, Y. Ogiso, T. Ichikawa, K. Nonaka, M. Iwaya, S. Kamiyama, H. Amano, and I. Akasaki, *physica status solidi (c)* **7**, 1916 (2010).

[122] F. Li, S. Lee, J. You, T. Kim, K. Lee, J. Lee, Y. Kwon, and T. Kang, *Journal of Crystal Growth* **312**, 2320 (2010).
[123] H. P. T. Nguyen, K. Cui, S. Zhang, S. Fathololoumi, and Z. Mi, *Nanotechnology* **22**, 445202 (2011).
[124] S. Tripathy, V. K. X. Lin, S. L. Teo, A. Dadgar, A. Diez, J. Bläsing, and A. Krost, *Applied Physics Letters* **91**, 231109 (2007).
[125] S. Tripathy, T. E. Sale, A. Dadgar, V. K. X. Lin, K. Y. Zang, S. L. Teo, S. J. Chua, J. Bläsing, and A. Krost, *Journal of Applied Physics* **104**, 053106 (2008).
[126] F. Semond, N. Antoine-Vincent, N. Schnell, G. Malpuech, M. Leroux, J. Massies, P. Disseix, J. Leymarie, and A. Vasson, *physica status solidi (a)* **183**, 163 (2001).
[127] M. B. Charles, Y. Zhang, M. J. Kappers, and C. J. Humphreys, *physica status solidi (a)* **203**, 1618 (2006).
[128] P. Saengkaew, A. Dadgar, J. Blaesing, B. Bastek, F. Bertram, F. Reiher, C. Hums, M. Noltemeyer, T. Hempel, P. Veit, J. Christen, and A. Krost, *physica status solidi (c)* **6**, S455 (2009).
[129] T. Egawa, B. Zhang, and H. Ishikawa, *IEEE Electron Device Letters* **26**, 169 (2005).
[130] T. Egawa and B. A. B. A. Shuhaimi, *Journal of Physics D: Applied Physics* **43**, 354008 (2010).
[131] X. B. Zou, H. Liang, and K. M. Lau, *physica status solidi (c)* **7**, 2171 (2010).
[132] E. Alarcon-Llado, S. Bin-Dolmanan, V. K. X. Lin, S. L. Teo, A. Dadgar, A. Krost, and S. Tripathy, *Journal of Applied Physics* **108**, 114501 (2010).
[133] Ka Ming Wong, Xinbo Zou, Peng Chen, and Kei May Lau, *IEEE Electron Device Letters* **31**, 132 (2010).
[134] J. Lee, Y. Tak, J.-Y. Kim, H.-G. Hong, S. Chae, B. Min, H. Jeong, J. Yoo, J.-R. Kim, and Y. Park, *Journal of Crystal Growth* **315**, 263 (2011).
[135] J.-Y. Kim, Y. Tak, H.-G. Hong, S. Chae, J. W. Lee, H. Choi, J. K. Kim, B. Min, Y. Park, U.-I. Chung, M. Kim, S. Lee, N. Cha, Y. Shin, C. Sone, J.-R. Kim, and J.-I. Shim, *Proceedings of SPIE*, SPIE, 2011.
[136] OSRAM-OS, Success in research: First gallium-nitride LED chips on silicon in pilot stage: OSRAM Opto Semiconductors expands its leading position in high-quality, thin-film LEDs, http://www.osram-os.com/, 12 January 2012.
[137] T. Chen, Y. Wang, P. Xiang, R. Luo, M. Liu, W. Yang, Y. Ren, Z. He, Y. Yang, W. Chen, X. Zhang, Z. Wu, Y. Liu, and B. Zhang, *Applied Physics Letters* **100**, 241112 (2012).
[138] A. Dadgar, J. Bläsing, A. Diez, and A. Krost, *Applied Physics Express* **4**, 011001 (2011).
[139] J. Li, J. Y. Lin, and H. X. Jiang, *Applied Physics Letters* **88**, 171909 (2006).
[140] A. Dadgar, C. Hums, A. Diez, F. Schulze, J. Blasing, and A. Krost, Epitaxy of GaN LEDs on large substrates: Si or sapphire?, in *Proceedings of SPIE*, volume 6355, pp. 63550R–63550R–8, SPIE, 2006.
[141] A. Dadgar, C. Hums, A. Diez, J. Bläsing, and A. Krost, *Journal of Crystal Growth* **297**, 279 (2006).
[142] J.-Y. Kim, Y. Tak, J. Kim, H.-G. Hong, S. Chae, J. W. Lee, H. Choi, Y. Park, U.-I. Chung, J.-R. Kim, and J.-I. Shim, *Proceedings of SPIE*, volume 8262, SPIE, 2012.

[143] T. P. Chow and M. Ghezzo, *MRS Proceedings* **423**, 9 (1996).

[144] F. Medjdoub, M. Zegaoui, B. Grimbert, D. Ducatteau, N. Rolland, and P. A. Rolland, *IEEE Electron Device Letters* **33**, 1168 (2012).

[145] A. Krtschil, A. Dadgar, and A. Krost, *Applied Physics Letters* **82**, 2263 (2003).

[146] S. Kato, Y. Satoh, H. Sasaki, I. Masayuki, and S. Yoshida, *Journal of Crystal Growth* **298**, 831 (2007).

[147] M. Kuroda, T. Murata, S. Nakazawa, T. Takizawa, M. Nishijima, M. Yanagihara, T. Ueda, and T. Tanaka, in *Technical Digest 2008*, p. 4674513, IEEE, Piscataway, NJ, 2008.

[148] S. Yoshida, J. Li, T. Wada, and H. Takehara, *MRS Proceedings* **798**, Y7.3 (2003).

[149] S. Yoshida, J. Li, and N. Ikeda, *physica status solidi (c)* **2**, 2593 (2005).

[150] N. Ikeda, J. Li, H. Takehara, T. Wada, and S. Yoshida, *Journal of Crystal Growth* **275**, e1091 (2005).

[151] M. Kuraguchi, Y. Takada, T. Suzuki, M. Hirose, K. Tsuda, W. Saito, Y. Saito, and I. Omura, *physica status solidi (a)* **204**, 2010 (2007).

[152] W. Saito, Y. Takada, M. Kuraguchi, K. Tsuda, and I. Omura, *IEEE Transactions on Electron Devices* **53**, 356 (2006).

[153] R. Wang, Y. Cai, W. C. W. Tang, K. M. Lau, and K. J. Chen, *physica status solidi (a)* **204**, 2023 (2007).

[154] J. Derluyn, M. van Hove, D. Visalli, A. Lorenz, D. Marcon, P. Srivastava, K. Geens, B. Sijmus, J. Viaene, X. Kang, J. Das, F. Medjdoub, K. Cheng, S. Degroote, M. Leys, G. Borghs, and M. Germain, Low leakage high breakdown e-mode GaN DHFET on Si by selective removal of in-situ grown Si_3N_4, in *2009 IEEE International Electron Devices Meeting (IEDM)*, pp. 1–4, IEEE, 2009.

[155] Bin Lu, E. Piner, and T. Palacios, *IEEE Electron Device Letters* **31**, 302 (2010).

[156] Y. Niiyama, S. Kato, Y. Sato, M. Iwami, J. Li, H. Takehara, H. Kambayashi, N. Ikeda, and S. Yoshida, *MRS Proceedings* **955**, 0955-I16-06 (2006).

[157] A. Dadgar, F. Schulze, J. Bläsing, A. Diez, A. Krost, M. Neuburger, E. Kohn, I. Daumiller, and M. Kunze, *Applied Physics Letters* **85**, 5400 (2004).

[158] A. Dadgar, M. Neuburger, F. Schulze, J. Bläsing, A. Krtschil, I. Daumiller, M. Kunze, K.-M. Günther, H. Witte, A. Diez, E. Kohn, and A. Krost, *physica status solidi (a)* **202**, 832 (2005).

[159] J. Xie, X. Ni, M. Wu, J. H. Leach, U. Ozgur, and H. Morkoc, *Applied Physics Letters* **91**, 132116 (2007).

[160] S. Nozaki, A. T. Wu, J. J. Murray, T. George, T. Egawa, and M. Umeno, *Applied Physics Letters* **57**, 2669 (1990).

[161] A. Strittmatter, A. Krost, M. Straßburg, V. Turck, D. Bimberg, J. Bläsing, and J. Christen, *Applied Physics Letters* **74**, 1242 (1999).

[162] C.-L. Wang, J.-R. Gong, W.-T. Liao, W.-L. Wang, T.-Y. Lin, and C.-K. Lin, *Solid State Communications* **137**, 63 (2006).

[163] J. Kuzmik, S. Bychikhin, M. Neuburger, A. Dadgar, A. Krost, E. Kohn, and D. Pogany, *IEEE Transactions on Electron Devices* **52**, 1698 (2005).

[164] K. M. Kelchner, Y.-D. Lin, M. T. Hardy, C. Y. Huang, P. S. Hsu, R. M. Farrell, D. A. Haeger, H. C. Kuo, F. Wu, K. Fujito, D. A. Cohen, A. Chakraborty, H. Ohta, J. S. Speck, S. Nakamura, and S. P. DenBaars, *Applied Physics Express* **2**, 071003 (2009).

[165] A. M. Fischer, Z. Wu, K. Sun, Q. Wei, Y. Huang, R. Senda, D. Iida, M. Iwaya, H. Amano, and F. A. Ponce, *Applied Physics Express* **2**, 041002 (2009).

[166] C.-H. Chiu, D.-W. Lin, C.-C. Lin, Z.-Y. Li, W.-T. Chang, H.-W. Hsu, H.-C. Kuo, T.-C. Lu, S.-C. Wang, W.-T. Liao, T. Tanikawa, Y. Honda, M. Yamaguchi, and N. Sawaki, *Applied Physics Express* **4**, 012105 (2011).

[167] T. Hikosaka, Y. Honda, M. Yamaguchi, and N. Sawaki, *Journal of Applied Physics* **101**, 103513 (2007).

[168] T. Murase, T. Tanikawa, Y. Honda, M. Yamaguchi, H. Amano, and N. Sawaki, *Japanese Journal of Applied Physics* **50**, 01AD04 (2011).

[169] T. Tanikawa, T. Hikosaka, Y. Honda, M. Yamaguchi, and N. Sawaki, *physica status solidi (c)* **5**, 2966 (2008).

[170] T. Tanikawa, D. Rudolph, T. Hikosaka, Y. Honda, M. Yamaguchi, and N. Sawaki, *Journal of Crystal Growth* **310**, 4999 (2008).

[171] A. E. Romanov, T. J. Baker, S. Nakamura, and J. S. Speck, *Journal of Applied Physics* **100**, 023522 (2006).

[172] R. Ravash, J. Bläsing, T. Hempel, M. Noltemeyer, A. Dadgar, J. Christen, and A. Krost, *Applied Physics Letters* **95**, 242101 (2009).

[173] R. Ravash, J. Blaesing, A. Dadgar, and A. Krost, *Applied Physics Letters* **97**, 142102 (2010).

[174] R. Ravash, J. Blaesing, T. Hempel, M. Noltemeyer, A. Dadgar, J. Christen, and A. Krost, *physica status solidi (b)* **248**, 594 (2011).

[175] A. Baski, S. Erwin, and L. Whitman, *Surface Science* **392**, 69 (1997).

[176] A. Dadgar, R. Ravash, P. Veit, G. Schmidt, M. Muller, A. Dempewolf, F. Bertram, M. Wieneke, J. Christen, and A. Krost, *Applied Physics Letters* **99**, 021905 (2011).

[177] R. Ravash, P. Veit, M. Müller, G. Schmidt, A. Dempewolf, T. Hempel, J. Bläsing, F. Bertram, A. Dadgar, J. Christen, and A. Krost, *physica status solidi (c)* **9**, 507 (2012).

[178] Y. S. Cho, Q. Sun, I.-H. Lee, T.-S. Ko, C. D. Yerino, J. Han, B. H. Kong, H. K. Cho, and S. Wang, *Applied Physics Letters* **93**, 111904 (2008).

[179] M. Moram, C. Johnston, M. Kappers, and C. Humphreys, *Journal of Crystal Growth* **311**, 3239 (2009).

[180] S. Li, S. Fündling, U. Sökmen, S. Merzsch, R. Neumann, P. Hinze, T. Weimann, U. Jahn, A. Trampert, H. Riechert, E. Peiner, H.-H. Wehmann, and A. Waag, *physica status solidi (c)* **7**, 84 (2010).

[181] B. W. Liou, *Japanese Journal of Applied Physics* **48**, 072201 (2009).

[182] B. W. Liou, *Thin Solid Films* **520**, 1084 (2011).

[183] L. Hsu and W. Walukiewicz, *Journal of Applied Physics* **104**, 024507 (2008).

[184] L. A. Reichertz, I. Gherasoiu, K. M. Yu, V. M. Kao, W. Walukiewicz, and J. W. Ager, *Applied Physics Express* **2**, 122202 (2009).

[185] K. Brueckner, F. Niebelschuetz, K. Tonisch, S. Michael, A. Dadgar, A. Krost, V. Cimalla, O. Ambacher, R. Stephan, and M. A. Hein, *Applied Physics Letters* **93**, 173504 (2008).

[186] T. Zimmermann, M. Neuburger, P. Benkart, F. Hernandez-Guillen, C. Pietzka, M. Kunze, I. Daumiller, A. Dadgar, A. Krost, and E. Kohn, *IEEE Electron Device Letters* **27**, 309 (2006).

4 The growth of bulk aluminum nitride

Ronny Kirste and Zlatko Sitar

4.1 Introduction

Tremendous efforts have been made in the growth of bulk GaN and AlN over the last decade. While the most recent results on GaN growth are presented in Chapter 2, this chapter deals with the development of AlN crystals and wafers, as well as AlN-based technology. After a short introduction on AlN bulk growth, which emphasizes the need for high-quality substrates, three AlN growth methods are discussed: (1) sublimation/physical vapor transport (PVT), (2) hydride vapor phase epitaxy (HVPE), and (3) solution growth method. In this context, the influence of different process parameters, growth environment, and growth seeds will be discussed. In addition, the optical and structural properties of bulk AlN crystals obtained by the three methods are described. Since bulk AlN is mainly used as a substrate, the properties of epitaxially overgrown thin AlN and AlGaN layers will be discussed, thus demonstrating their superior quality as compared to layers grown on non-native substrates, i.e., sapphire or SiC. Promising applications based on AlGaN/AlN are presented in the framework of LEDs and lasers emitting in the mid-ultraviolet or UVC wavelength regime (200–280 nm). Finally, a brief outlook on the development of AlN-based technology is presented.

4.2 Bulk AlN: a pathway to high-quality AlGaN

Bulk AlN crystals are envisioned as a source of substrates for the growth of devices using Al-rich $Al_xGa_{1-x}N$ layers ($x \geq 0.7$). These devices include UV solid-state emitters (LEDs and LDs) and detectors in the 200–280 nm range, high-power switches needed in power conversion devices, and high-frequency devices for application in radar and communication technologies.

Figure 4.1 gives an overview of possible applications of UV LEDs and lasers based on AlGaN. These applications include solid-state lighting by using a phosphorous converter, sensing and detection (protein analysis, drug detection, cell imaging, security features for currency and documents), lithography, and disinfection and water purification (based on the absorption peak of DNA at 254 nm). In addition to these optoelectronic devices, electronic devices based on high-Al-content AlGaN, although in the nascent development stages, show much potential:

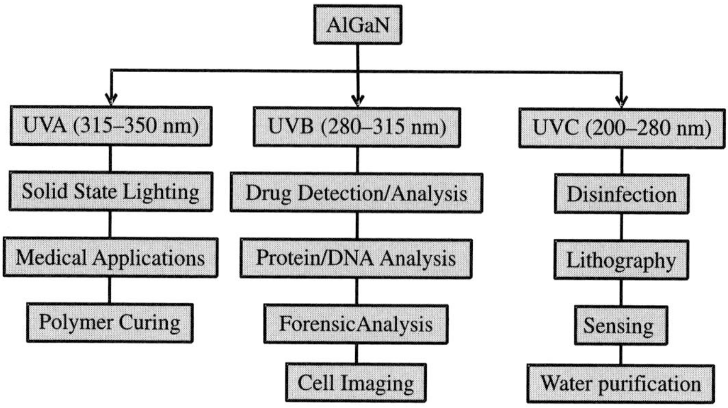

FIG. 4.1. Possible applications for UV LEDs and LDs based on AlGaN.

- High-power electronics. Based on the large bandgap of AlN, this material has an extremely high electric breakdown field of ~20 MV/cm, making it an excellent candidate for electrical devices operating at high voltage/power [1].
- Radiation hard devices and radiation detectors. AlN is extremely radiation hard. Devices based on this material are only marginally harmed by hard radiation [2].
- High-temperature devices. AlN is thermally stable and offers a high thermal conductivity (320 W/mK) [3].

It is well recognized that high-quality AlGaN can be enabled only by large-area native substrates. Technologically feasible solutions are the use of native GaN and AlN substrates for low and high Al-content AlGaN, respectively, since only a low lattice mismatch combined with a low dislocation density in the substrate enables low dislocation densities in epitaxial grown AlGaN layers. However, due to a significant lattice mismatch between AlN and GaN, neither of the substrates can cover the growth of the full spectrum of AlGaN compositions.

An overview of the properties and parameters of typical substrates used for the growth of AlGaN is given in Table 4.1. These include sapphire, SiC, and Si, as well as the native substrates, GaN and AlN. Due to different lattice sizes and thermal expansion coefficients of substrates and AlGaN layers, the layers could be strained, relaxed by forming extended defects, or, in the worst case, even cracked. The lattice mismatch between the layer and substrate is given by [4]:

$$m = \frac{s_{layer} - s_{substrate}}{s_{layer}}, \tag{4.1}$$

where s_{layer} and $s_{substrate}$ are the in-plane lattice parameters of the layer and the substrate, respectively.

Table 4.1 Lattice parameters and properties for substrates typically used in the growth of AlGaN.

	Sapphire [5]	SiC (6H) [5]	Si (111) [5]	AlN [3, 6, 7]	GaN [8, 9]
a-axis [Å]	4.765	3.073	3.840	3.113	3.188
c-axis [Å]	12.982	15.117	3.136	4.982	5.185
RT-thermal expansion coefficient [10^{-6} K^{-1}]	5.0 (a-axis) 9.0 (c-axis)	4.2 (a-axis) 4.68 (c-axis)	2.6	4.2 (a-axis) 5.3 (c-axis)	5.59 (a-axis) 3.17 (c-axis)
thermal conductivity [W/cm K]	0.23	4.9	1.6	3.2	2.1
UV-transparency	excellent	poor	poor	excellent	poor
price	low	high	low	very high	very high
availability	excellent	good	excellent	poor	poor
mismatch to GaN	16.1%	3.5%	17.1%	2.2%	–
mismatch to AlN	13.3%	1.3%	19%	–	2.2%

Note: For Si, lattice parameter in (111) plane is given; mismatch calculations for sapphire account for 30° rotation.

Apart from the low lattice mismatch for high Al-content AlGaN layers, AlN uniquely unites high thermal conductivity, high breakdown voltage, and chemical inertness, which alone would benefit many devices.

4.3 Growth of AlN crystals

Originally, AlN was synthesized in 1862 by Briegleb and Geuther by heating aluminum fillings under nitrogen gas flow. Due to an increase of weight, they concluded that Al and N had reacted to form AlN [10]. First attempts towards real crystal growth of this compound were made much later by Slack *et al.* in the 1970s. They suggested growth via sublimation in a closed tungsten crucible or in an open tube with a gas flow [11]. These experiments resulted in millimeter-sized AlN crystals. Later, these experiments were extended to grow AlN crystals by the Al pellet drop method in a tungsten crucible in an rf-heated tungsten furnace under a nitrogen atmosphere [12]. The resultant crystals were of up to 1 cm in length and 0.3 cm in diameter. Very few new efforts in AlN crystal growth were undertaken in the 1980s and the beginning of the 1990s. However, shortly after the successful p-doping of GaN and the first available GaN-based LEDs, there was increased interest in AlN [13]. This led to a renaissance of layer and later bulk growth of AlN at the end of the 1990s [14].

The three major approaches for the growth of bulk AlN crystals that have been exploited are:

- Sublimation or physical vapor transport (PVT).
- Hydride vapor phase epitaxy (HVPE).
- Growth from solutions.

Currently, the most promising method appears to be PVT in combination with AlN seeding. While solution growth and HVPE basically result in the growth of thick layers, PVT provides voluminous bulk crystals of high quality. Furthermore, only PVT can produce sufficiently large native seeds for subsequent crystal growth, which is crucial for high-quality AlN bulk crystals.

4.3.1 *Physical vapor transport*

For PVT or sublimation growth, AlN-powder (source) is sublimed at temperatures above 2000°C in a nitrogen atmosphere. Due to a temperature gradient between the source and seed, aluminum and nitrogen atoms recondense on the seed and form an AlN bulk crystal. A schematic of a reactor for growth via PVT is shown in Fig. 4.2.

AlN powder is contained in a crucible (source). The crucible is inductively heated and due to the high applied temperature (>2000°C), the AlN is non-congruently sublimed:

$$AlN^{solid} \leftrightarrow Al^{gas} + \frac{1}{2}N_2^{gas} \tag{4.2}$$

On the opposite side of the crucible, a seed is positioned which is heated to temperatures slightly lower than the source. Due to the thermal gradient, the sublimed gas molecules from the source will diffuse towards the seed and lead to the growth of AlN via a reverse reaction. It is crucial that both the source and seed temperatures are controlled and highly stable for the duration of the crystal growth process (50–100 hours).

In general, vaporization of Al could be taken into account as the source for the Al-gas (growth via vaporization). However, attempts using vaporization of Al-metal instead of sublimation of AlN demonstrated that this method was not suitable for long-term growth, since the Al-metal source in these experiments was quickly covered with an AlN crust (due to the presence of nitrogen) which hindered further vaporization of Al and, thus, suppressed the growth rate of AlN [16].

The crucible with the AlN powder (Al-source) is typically heated by induction. The temperature is measured with a pyrometer (Fig. 4.2). Possible temperatures for the growth of AlN close to atmospheric pressure cover a broad range from 1800°C up to 2500°C. The lower limit is determined by the temperature when nitrogen and aluminum start to react significantly. For temperatures below 1800°C, no significant reaction takes place, hindering the growth of any crystals. The upper limit for the AlN crystal growth near the atmospheric pressure is given by the boiling point of Al. At atmospheric pressure and temperatures

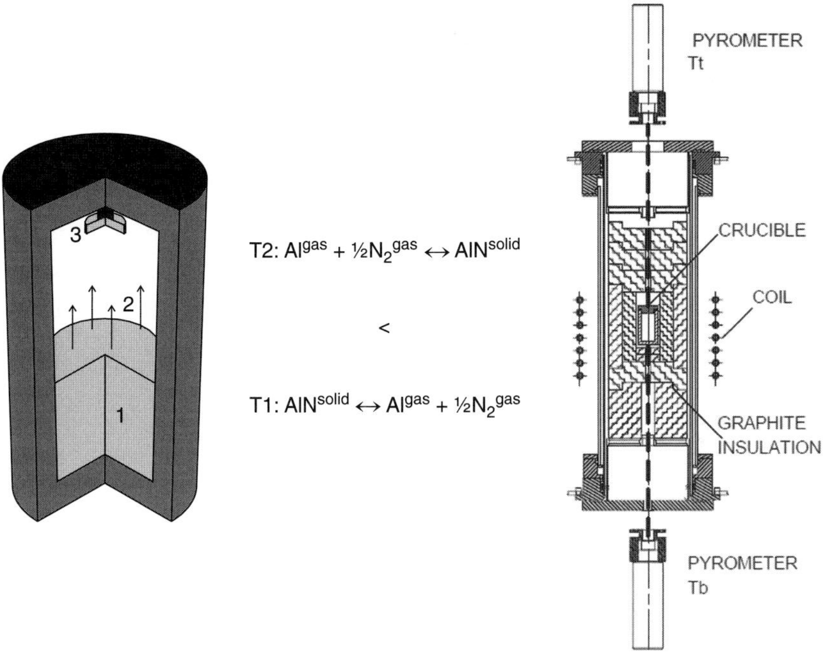

FIG. 4.2. Schematic drawing of a crucible for PVT growth of bulk AlN (left). An AlN source (1) is sublimated leading to vapor (2) which is transported towards the seed (3) due to a temperature gradient (T2 < T1) between the source and seed. Cross sectional drawing of a complete reactor for sublimation growth (right). Note that the size of the crucible and the reactor are disproportional. (Reprinted from [15], © 2004, with permission from Elsevier.)

above 2500°C, AlN is not stable and decomposes. In order to allow for higher temperatures, some attempts were made to grow AlN at elevated nitrogen pressures of up to 1 GPa [17]. Although the first results were promising, little progress has been made over the past few years in the high-pressure growth of AlN. Thus, most current efforts focus on growth at near-atmospheric pressure. Thereby, little influence of the pressure on the growth has been observed [16]. Despite the fact that bulk AlN crystals can be grown at temperatures as low as 1800°C, temperatures around 2200°C are more favorable in order to achieve high growth rates and good crystal quality [11, 16]. In addition to temperature, the temperature gradient between the source and seed is an important parameter for the control of transport and growth rate. Typical values for the temperature gradient are around 50°C/cm. With these reactor parameters, growth rates in the c-direction between 50 μm/h (1800°C), and 1 mm/h (2300°C) can be achieved [15, 16]. However, high growth rates may lead to a reduced crystal quality so that growth rates are usually maintained at several hundred μm/hour. Taking these controllable parameters (pressure, temperature, temperature gradient) into account in a one-dimensional model, the growth rate of an AlN crystal can be described by [18]:

$$v_g = k\frac{\exp\left[\Delta S - \frac{\Delta H}{T}\right]}{RT^{1.2}p^{1.5}}\frac{\Delta T}{\delta} \tag{4.3}$$

Here, v_g is the growth rate, p is the total pressure, T is the temperature, $\Delta\mathrm{T}/\delta$ is the temperature gradient between the AlN source and seed, R is the universal gas constant, S is the entropy, and H the enthalpy. The pre-exponential term k is an adjustable parameter containing the diffusion coefficient of Al in nitrogen.

Not many materials are suitable for crucibles for PVT growth of AlN, as not many materials withstand temperatures >2200°C. In addition, crucibles must be inert to the reaction with the Al vapor at high temperature, not a source of substantial contamination, reusable, and available in the shape and size needed for the growth. These demands drastically decrease the number of possible candidates. Several materials have been tested, including graphite- and boron-based conglomerates, pure tungsten, tantalum, and TaC [14, 16, 19, 20]. Graphite crucibles have been used at low temperatures, but are generally unsuitable for AlN growth at higher temperatures due to the formation of the volatile aluminum carbide, which causes quick deterioration of the crucible. Any B-containing compound introduces enough B into the vapor phase to poison the growth surface, which inhibits the growth in the non-polar directions [16]. So far, the best results have been obtained with crucibles made of tungsten and TaC [21].

Following the reaction schemes presented above, the only source materials needed for the PVT growth of AlN are Al and nitrogen. The nitrogen source typically is ultra-high purity grade molecular nitrogen. AlN powder is used as a source of Al almost exclusively, but the commercially available AlN powder contains several impurities and surface contaminants, mainly oxygen, carbon, silicon, and hydroxides. In order to reduce the impurity levels in the AlN source powder, several groups suggested annealing, recrystallization, or the use of larger AlN blocks [14, 19].

4.3.2 *Hydrid vapor phase epitaxy*

In HVPE, the growth of thick AlN layers is enabled via a chemical reaction of aluminum chloride gas and ammonia, and was demonstrated by various groups [22–24]. This reaction takes place in a temperature-controlled furnace at a relatively low temperature (900–1100°C), leading to high-purity AlN deposits. The growth rate by this method is about one tenth of that obtained by PVT, but the control of unwanted impurities is easier due to a considerably lower deposition temperature. A schematic drawing of the HVPE process is displayed in Fig. 4.3.

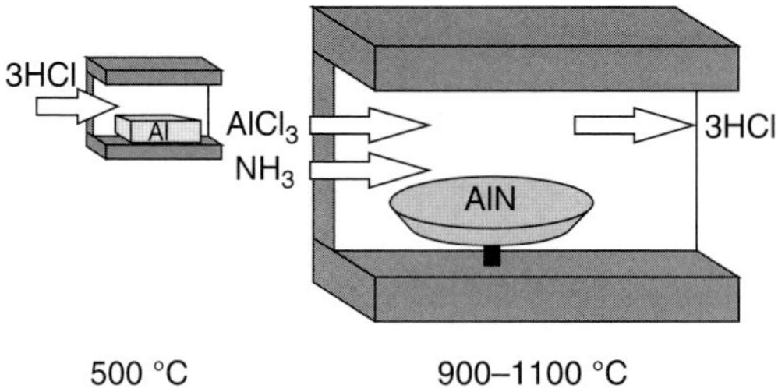

FIG. 4.3. Principle of HVPE growth of bulk AlN. An $AlCl_3$ and ammonia gas mixture is flown through a hot furnace over a substrate on which AlN is grown.

The Al-metal source in the HVPE process is volatilized by the HCl vapor following the reaction:

$$3HCl^{gas} + Al^{liquid/solid} \leftrightarrow AlCl_3^{gas} + \frac{3}{2}H_2^{gas} \tag{4.4}$$

This is flown over a seed or substrate together with ammonia, leading to the growth of AlN [25]:

$$AlCl_3^{gas} + NH_3^{gas} \leftrightarrow AlN^{solid} + 3HCl^{gas} \tag{4.5}$$

HVPE growth is typically performed at temperatures in the range of 900–1100°C and more recently at temperatures up to 1400°C, which is much lower than those used for PVT growth, strongly reducing the demands for the reactor parts. Reactor tubes are often made of quartz glass, and the substrate is resistively heated. It is crucial that Al is supplied as $AlCl_3$, as any unsaturated Al clorides attack quartz hardware. Although cheaper and easier to handle, this reactor construction may introduce some limitations for high-temperature growth with HVPE.

Only a small influence of the temperature on the growth rate of AlN was observed. When the growth temperature was increased from 950°C to 1100°C an increase of the growth rate from 0.9 to 1.7 μm/h was demonstrated [22]. This constant growth rate, practically independent of the temperature, proves that the growth mainly takes place in the mass transport limited regime [26]. It was emphasized that temperatures on the order of 1400°C are necessary to obtain crystals of higher quality [27]. In order to achieve these temperatures, Nagashima *et al.* modified an HVPE reactor by combining a conventional hot-wall type furnace and a heated susceptor with an integrated heating element [26]. Using this modified reactor, growth temperatures between 1200°C and 1400°C were reached. Still, the growth rates were independent of temperature and reached up to 85 μm/h by increasing the supply of precursors. Since the growth of AlN via HVPE is in the mass transport limited regime, the main parameter influencing the growth rate is the $AlCl_3$ input pressure, presenting the supply of Al. Indeed, several works demonstrated a linear correlation between the $AlCl_3$ input pressure and growth rate [22, 26]. Using high input pressures of $2{\cdot}10^{-3}$ atm, growth rates as high as 122 μm/h were achieved [22]. However, growth rates above 30 μm/h led to poor crystal quality, as observed by the FWHM of XRD rocking curves [26, 28].

4.3.3 *Solution growth*

In the solution-growth method, a seed is introduced into an Al/N-containing solution, mostly under a nitrogen atmosphere [29]. Since the growth from the solution method takes place close to the thermal equilibrium, AlN crystals of very high quality can be expected. However, only very few efforts were made for the growth of AlN via this method, which produced small crystals [29–31].

Figure 4.4 shows a schematic of the setup for solution growth. A seed, which is mounted on a rotating shaft, is introduced into a solution containing Al and N. For the seed, high-quality AlN bulk crystals must be used. When the seed is dipped into the solution, Al and nitrogen atoms are expected to deposit onto the seed.

The reaction scheme for the solution growth depends primarily on the chosen solvent. Originally, Ca_3N_2 was chosen as the nitrogen source and AlN powder as the Al-source [30]. In this case, the following reaction scheme can be assumed [32]:

$$Ca_3N_2 + 2Al \leftrightarrow 3Ca + 2AlN \tag{4.6}$$

Other prospective solvents include NaN_3, Be_3N_2, Mg_3N_2, and Li_3N, whereby Kangawa *et al.* concluded from theoretical calculations that Li_3N has a high potential for the solution growth method, since it combines both low melting point and high probability of nitrogen supply [32]. So far, crystals grown by the solution method have been rather small and of relatively poor quality. With growth rates around 0.5 μm/h, up to 5-μm thick AlN layers were grown on AlN/SiC [33]. Higher growth rates of up to 10 μm/h were reported, but the defect density in these crystals drastically increased [32]. Based on general requirements for the solvents, which include sufficient solubility of Al, thermal stability, a low melting point, and the feasibility of separating the solvent from the grown AlN

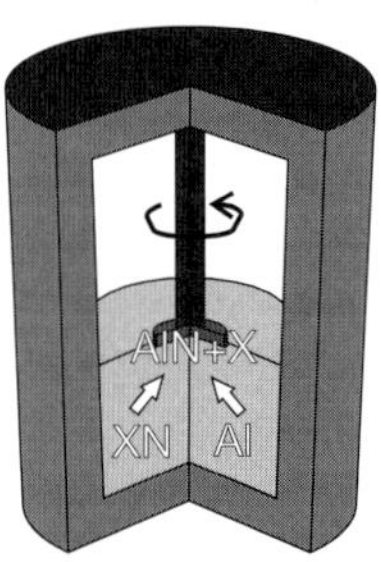

FIG. 4.4. General setup for the solution growth of AlN. A seed is dipped into a nitrogen solvent and melted Al.

crystals [30], in addition to nitrogen-containing compounds, Sn–Ca or Cu–Al–Ti were also proposed [29, 31].

The growth temperature for solution growth is typically much lower than that of sublimation growth (around 1100°C), drastically lowering the demands and stress for the experimental setup. However, for solvents with a high melting point or when a higher temperature was desired in order to improve crystal quality, growth temperatures as high as 1600–1800°C were applied [29, 30].

Despite its appeal from the thermodynamic standpoint, the solution-growth method faces many unresolved challenges. Of the three growth methods discussed in this review, the solution growth is the least developed, and future work will have to find practical solutions to all the challenges and ultimately prove the potential of this growth method.

4.3.4 *Seeding of AlN crystal growth*

The structural quality of the grown AlN crystals obtained by a particular method depends directly on the type and quality of seed crystals. Seed crystals are needed to provide structural information for the condensing Al and N atoms, including polarity and orientation. Over the years, two approaches to seeding have been used in AlN crystal growth: (1) foreign seeds like SiC, and (2) native AlN seeds. The first approach is convenient and provides "instant gratification" as SiC is readily available in different sizes, but severely limits crystal quality due to lattice and thermal mismatch (results in dislocation density in the 10^6–10^7/cm^2 range), while the second approach provides initially small but very-high-quality crystals with dislocation density $<10^3$/cm^2. Since native AlN seeds are not readily available, the second approach necessitates initially two steps: (1) production of seed crystals by spontaneous nucleation, and (2) seeded growth coupled with controlled and gradual expansion of single crystalline boules, which provide increasingly larger and larger seeds and wafers of high quality.

First attempts with seeded growth were made with SiC crystals. 6H-SiC was chosen due to its relatively high-temperature stability, enabling high-temperature growth, and reasonably small differences in the lattice constant and thermal expansion coefficient to those of AlN (see Table 4.1). However, growth on SiC seeds is still highly mismatched heteroepitaxy, generating copious defects and strain at the heteroepitaxial interface. The first results for SiC seeded growth were presented in 1996 by Balkas *et al.* [34]. The crystals grown during these first attempts showed cracks and incorporation of Si and C. An extensive theoretical investigation of the strain in AlN on 6H-SiC revealed that the difference in the thermal expansion coefficients in combination with the high growth temperatures was responsible for cracking [35]. In addition, this study predicted that AlN on SiC needs to be at least 2 mm thick in order to prevent cracking during the cool-down (assuming a growth temperature of 2000°C). So far, the only possibility to avoid cracking seems to be the formation of voids [33, 36, 37]. AlN bulk crystals grown on SiC seeds can reach dislocation densities down to the 10^6/cm^2 range, which is close to the theoretical limit predicted for a mismatched epitaxial system. SiC has been used in both PVT and HVPE [28, 37, 38]. In addition to SiC, sapphire has been widely exploited, especially in HVPE [22, 26, 39]. Thereby, similar problems in defect management as with SiC occurred.

Since native AlN seeds were not available at the onset of AlN crystal growth efforts, initial attempts to grow voluminous crystals were made by self-seeding [12]. For self-seeding, a small growth target was used, typically a conical tube, to limit the number of stable nuclei [40]. Due to the limited space in the target zone and grain competition, a few or even a single crystalline nucleus survived, which acted as a self-seeded base for further crystal growth. However, due to the randomness of the grain selection process, and constrained growth environment, the resultant crystals were small, randomly oriented, and had relatively high dislocation density [40, 41].

Native AlN seeds are paramount for high-quality crystals and wafers. These seeds are typically obtained by spontaneous nucleation and growth of Lely-like, c-oriented platelets close to equilibrium conditions [21]. The first results of bulk growth of AlN on native seeds were presented in 2002 by Schlesser *et al.* [16]. These first attempts resulted in transparent crystals with XRD rocking curve FWHMs as low as 25 arcsec. No cracking was observed. By establishing a procedure which includes the growth of an AlN boule, cutting and polishing a wafer from that boule, and subsequent regrowth on that wafer, Noveski *et al.* established a procedure which led to increased native seed size over time [42]. Using this procedure, commercially available wafers with diameters exceeding 25 mm were grown, whereby a further size increase is in progress [21]. Occasionally, boules with diameters of up to 50 mm have been demonstrated experimentally, but little is known about their quality and they are not available commercially [43].

Very recently, the first results of HVPE-grown AlN on PVT-grown substrates were presented [44]. These experiments tried to combine the advantages of PVT and HVPE by exploiting the high crystalline quality achieved by PVT with the high controllability of point defect incorporation in the HVPE growth. Indeed, using this combination, the growth of UV-transparent crystals with a dislocation density comparable to that of the PVT substrates was demonstrated. Despite the fact that native seeding has proven superior to SiC seeds, this technique is used mainly in the PVT growth. Due to the lack of native seeds on the market, the HVPE and solution growth methods still rely mainly on the growth on SiC or sapphire. This is solely based on the fact that, so far, only a couple of PVT groups world-wide were able to yield large enough, high-quality seeds to make the seeded growth process using native AlN seeds self-sustaining.

4.4 Properties of state-of-the-art bulk AlN

The properties of bulk AlN crystals are determined by the used seeds, purity of the source materials, reactor parts, and growth conditions. Different point defects (native and impurities) and extended defects may be incorporated into the AlN crystals, which influence their structural and optical properties.

4.4.1 *Structural properties*

The structural quality of bulk AlN is strongly dependent on the applied growth conditions. High crystal quality may refer to low defect concentration (with less focus on point defects and more focus on extended defects such as dislocations) and low strain. Methods to investigate the structural properties include X-ray diffraction (XRD) and X-ray topography, Raman spectroscopy, transmission electron microscopy (TEM), and etching experiments in conjunction with optical microscopy.

An easy yet effective method to probe the quality of crystals is X-ray diffraction (XRD) [45]. This method is widely used, since it is fast, non-destructive, and reliable. Typically, a rocking curve of the (002)-diffraction peak is analyzed. Results from asymmetric scans are not often found in literature, as most authors compare only the rocking curves of the (002)-peak to enable a fast assessment of the samples.

For the first AlN bulk crystals grown by PVT for which XRD data were available, the FWHM of the (002)-peak may be estimated to a few hundred arcsec. Following these first results, the FWHM steadily decreased as crystal quality improved, and is more recently quoted in the low teens [21]. However, this value depends greatly on seeding, as discussed previously, and higher values for the FWHM of the (002)-peak exceeding half a degree may still be found in literature [46]. Typical XRD rocking curves of the (002)-peak of PVT-grown AlN are displayed in Fig. 4.5 [21]. The FWHM of the sharpest peak is denoted as 13 arcsec. This spectrum was recorded from a wafer cut from the top part of a boule. The other two spectra were recorded from wafers cut from the same boule but in the middle and at the bottom, respectively. A strong decrease of

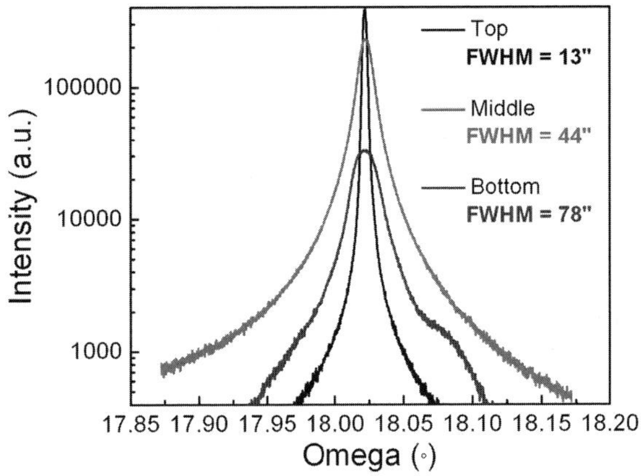

FIG. 4.5. Rocking curves of the (002)-peak from wafers cut from different positions of an AlN boule grown with PVT: top, middle and bottom. (Reprinted from [21], © 2010, with permission from Elsevier.)

the FWHM of the (002)-peak is observed toward the top of the boule. This is interpreted as an increasing crystalline quality of the boule with ongoing growth. The origin of this behavior is presumably due to reduced stress and higher uniformity of the growth conditions during later stages of the growth [21]. Using XRD, the dislocation density of PVT-grown AlN crystals on native seeds can be calculated to $10^6/\text{cm}^2$ [46–48]. For crystals of lower dislocation density, other methods have to be used.

Currently, bulk AlN crystals grown by HVPE on SiC may reach FWHMs of the (002)-peak as low as 40 arcsec [28]. Calculating the dislocation densities for such samples highlights values of 10^6–$10^7/\text{cm}^2$ [22, 28]. These dislocation densities are close to the theoretical limit that can be achieved for foreign seeding. A further reduction of the dislocation density can be expected only if native seeds are used in HVPE [44].

Analyzing the FWHM of XRD peaks is indirect and applicable only to the samples with high dislocation density. Dislocations in AlN have been directly observed by X-ray topography and TEM. Figure 4.6 (left) displays a bright-field image of an AlN crystal with a dislocation density of $10^8/\text{cm}^2$. Furthermore, some dislocation bending is observed. This can be understood as a strain relief mechanism [49].

For dislocation densities below $10^3/\text{cm}^2$, X-ray topography must be used. Furthermore, using this technique, other defects such as cracks, inclusions, and scratches can be easily identified and counted [50, 51]. An X-ray topography image of a PVT-grown AlN crystal with low dislocation density is displayed in Fig. 4.6 (right). In addition, different types of defects are marked. Using this

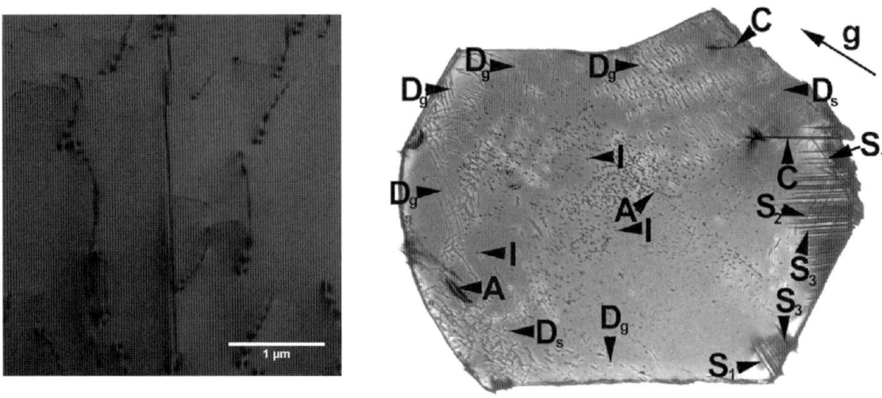

FIG. 4.6. (left) Bright-field TEM image of an AlN crystal with a high dislocation density. (right) Transmission topography of a PVT-grown crystal showing the different defects observed (D_g, growth dislocation; D_s, slip dislocation; S, slip band; C, crack; A, surface artifact/scratch; I, inclusion). (Reprinted from [50], © 2003, with permission from Elsevier.)

technique a dislocation density as low as 800–1000/cm^2 has been determined [50]. In addition, the authors were able to show that the distribution of the dislocations, as well of that of inclusions (density of the order of 10^5/cm^3), was inhomogeneous, with large areas of the wafer nearly free from dislocations. This suggests that TEM has only limited usability if used for the determination of dislocation densities in high-quality AlN crystals.

Another method which is easy to perform is wet KOH-etching and subsequent investigation with an optical microscope or scanning electron microscope (SEM) [52]. This method is based on the fact that AlN is an anisotropic crystal in which different crystallographic planes etch with different etching rates. This makes it possible to distinguish between low-angle grain boundaries, dislocations, polishing scratches, and others [46, 47]. However, this method will not allow a distinction between threading edge and screw dislocations. After the etching, the visible dislocations may be counted (etch pit density—EPD) and projected to the whole wafer to calculate the defect density. Using this method, dislocation densities of PVT-grown crystals of 10^4/cm^2 were determined [46].

4.4.2 *Optical properties and impurities*

In general, the optical properties below the bandgap of bulk AlN crystals are determined by the incorporation of point defects, i.e., vacancies and substitutional atoms. For the latter, oxygen, silicon, carbon, and hydrogen are the dominating species. This finding is explained by the fact that these materials are often involved in the growth process, either as a part of the source material (hydrogen, oxygen, silicon) or in the crucible or reactor parts (silicon, carbon, oxygen). Typical methods used to investigate the optical properties are absorption/transmission measurements and luminescence. Defect concentrations can be determined by secondary ion mass spectroscopy (SIMS) or positron annihilation spectroscopy (vacancies).

Although pure AlN is expected to be transparent down to ~200 nm [53], even high-quality AlN crystals show some degree of coloration which ranges from transparent, yellow, orange, or brown, to even black (opaque), indicating absorption centers around 2.1 eV and higher [54, 55]. A typical crystal is shown in Fig. 4.7 [21].

While the color itself gives a first impression of the optical properties in the visible part of the spectrum, more detailed analyses are needed to determine the location of absorption centers. For this purpose, standard transmission/absorption experiments are often used.

Absorption spectra of bulk AlN crystals grown by PVT in a tungsten furnace are shown in Fig. 4.8 [55]. A few absorption centers below the band gap can be observed. These absorption centers define the color of the crystal. While the main absorption for yellowish or brownish crystals starts around 2.0 eV, crystals transparent in the visible have only a very weak absorption in that region (orange curve in Fig. 4.8). In the UV region, additional absorption bands

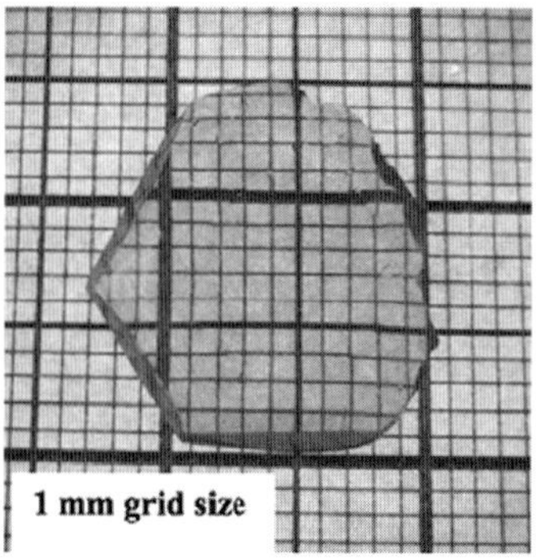

FIG. 4.7. Optical photograph of an AlN wafer with yellow color. (Reprinted from [21], © 2010, with permission from Elsevier.)

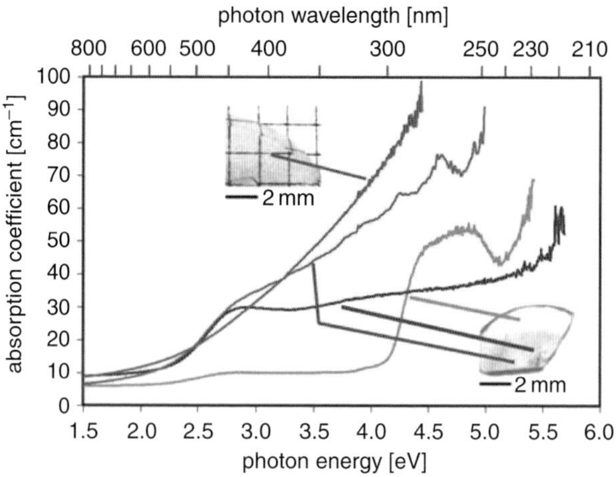

FIG. 4.8. Absorption spectra of PVT-grown AlN. Due to inhomogeneity, the crystals reveal different absorption spectra depending on the position (inset) of the crystal. (Reprinted with permission from [55].)

are observed with an onset around 3.2 eV. These absorption bands may be very broad, and stretch up to an energy around 5.4 eV. The strong absorption at higher energies is related to the band-edge of AlN (~6 eV). In a first approach, four major absorption bands can be identified that are observed in most AlN crystals. An overview of the positions, onset, and the possible origin of these absorption centers is given in Table 4.2.

The center at around 2.8 eV is responsible for absorption in the red-to-green region. One of the first explanations for this center was given by Slack *et al.*, who explained it as either related to a nitrogen vacancy or carbon [12]. These

Table 4.2 Absorption centers typically observed in AlN bulk crystals.

Center energy (eV)	Onset around (eV)	Origin
2.8	< 2.0	V_N [12, 56], C [12, 57], $V_{Al}{}^{3-/2-}$-O [55, 58]
3.6	2.0–2.2	V_{Al}-O_N [59], V_{Al} [55]
4.0	3.2	V_{Al}-O_N [20], C [60], O [12, 56], O-V [55]
4.7	4.2	V_{Al}-2O_N [20], C [60, 61], O_N [55], O [12]

findings were based on chemical analysis and evaluation of the growth conditions. With the beginning of the renaissance of bulk AlN growth in the late 1990s, this explanation was first confirmed by Bickerman *et al.* but later disputed by Sedhain *et al.* [56, 58]. The latter explained the absorption at 2.8 eV as an excitation from the valence band to an electronic state of an aluminum vacancy (V_{Al}). The contribution of Al vacancies to absorption in that region seems to be reasonable, since especially in yellow crystals, very high vacancy concentrations were detected by positron annihilation spectroscopy in bulk AlN crystals [59]. Additionally, the electronic state of such vacancies was calculated to be 2.1–2.5 eV above the valence band [62]. However, the most recent investigations assume that oxygen, which is also one of the most likely impurities in AlN, may be involved [55]. The observations and explanations for the absorption center around 3.6 eV are comparable to that of the center at 2.8 eV. It is typically explained by a transition between an Al vacancy and the conduction band or a shallow donor [55, 59].

The two UV absorption bands with absorption coefficients often well above 100 cm^{-1} are of high technological interest, as AlN crystals are envisaged as substrates for UV LEDs and LDs in the range of 200–280 nm. Since most LEDs are based on light extraction through the back side (substrate), a strong absorption center in this spectral region will hinder successful application. Originally, the occurrence of the UV absorption bands was explained by oxygen [12, 56]. Furthermore, it was assumed that both absorption bands in the UV are connected, since the center shifted from 4.3 to 4.7 eV with increasing oxygen contamination in the samples [63]. Later, this model included an Al vacancy (V_{Al}), so that the UV absorption was explained as a transition from V_{Al} to oxygen [20]. This explanation was based on the observation that AlN with UV absorption centers often had a high amount of oxygen incorporated. But with ongoing exploration of AlN and progress in the control of the growth, it was demonstrated that the oxygen level is relatively high in most samples (10^{18} cm^{-3}, see Fig. 4.10) while these crystals do not necessarily exhibit an UV absorption band. On the other hand, some samples with low oxygen contamination were found to have a strong UV absorption. Recently, more detailed investigations demonstrated that either both, or at least the high-energy UV absorption bands, can be fully explained by the substitutional carbon impurity. Indications for carbon as the origin of the UV absorption were found by Bickermann *et al.*, who showed an unusually high carbon amount in samples with low oxygen contamination and UV luminescence [60]. This finding was later confirmed by Collazo *et al.* [61]. In this work it was demonstrated that the UV absorption strongly declined for samples with a decreasing carbon level. This finding was supported by theoretical DFT calculations, which predicted a carbon-related absorption band above 4 eV.

The counterparts to the absorption measurements are luminescence investigations (photoluminescence—PL or cathodoluminescence—CL). Typically, five below-band-gap luminescence bands are observed in bulk AlN crystals: 320 nm (3.9 eV), 340 nm (3.65 eV), 400 nm (3.1 eV), 460 nm (2.7 eV), and 600 nm (2.1 eV). The most energetic luminescence (3.9 eV) was assigned to the absorption center at 4.7 eV [61]. Thus, its origin is connected to the incorporation of carbon. The emissions at 340 nm and 400 nm are often the strongest defect related luminescence signals and seem to be well understood. For these luminescence lines, a connection to oxygen is discussed [2, 58, 64]. This is based on a broad range of investigations including afterglow and time-resolved behavior. A more detailed model involves an oxygen impurity and Al vacancy (V_{Al}-O), whereby this explanation is widely accepted [65]. For the emission at 460 nm, very little is known. Many authors observed similar behavior of this luminescence band to that centered around 400 nm and some dependence on oxygen incorporation, which provides indications that oxygen might be involved [64]. Others observed a strong connection to the emission at 3.9 eV and the absorption at 4.8 eV, which might be an indicator that carbon is involved [61]. The origin of the red emission is also unclear. This luminescence is often weak and is not observed in all crystals. So far, the most reasonable explanation has been given by Sarua *et al.*, who stated a combination of nitrogen vacancies and excess Al to be responsible for the emission band around 600 nm [66].

Independent of the explanations of the individual absorption and emission bands, most authors agree that the main defects responsible for these features are vacancies (Al/N), oxygen, and carbon. This assignment is supported by measurements with a broad range of methods. For the vacancies, positron annihilation spectroscopy was used to confirm a high level of incorporation [59]. Carbon and oxygen are typically probed with secondary ion mass spectroscopy. High amounts of these atoms were found in different crystals [56, 67]. Additionally, silicon and hydrogen were identified as impurities by SIMS [67]. Si is known to form a shallow donor in AlN. For hydrogen, which is incorporated in various complexes, no influence on the optical properties is known [68, 69].

The origin of the different defects can be found in either the setup used for the growth (carbon, oxygen, silicon) or the source materials used (silicon, hydrogen, oxygen). Point defect incorporation in the AlN crystals is not uniform in all cases. This concerns the time point (regarding the growth) of incorporation in large AlN boules, as well as some sensitivity to different crystal planes. For example, while carbon is incorporated at a comparable amount over the whole boule, the oxygen level appears to decrease over time [12]. However, this different incorporation might strongly differ between the growth setups, reflecting the used crucibles, source materials, and reactor parts. In contrast, the different incorporation of defects depending on the crystal planes appears to be independent of the growth method. As an example, Bickermann *et al.* demonstrated that the incorporation of oxygen is reduced on {1013} facets compared to {0001} facets by around one order of magnitude, while at the same time the carbon concentration rises by a factor of 3–4 for these facets [60].

4.5 Applications and devices

4.5.1 *Properties of AlN and AlGaN epitaxial layers on AlN*

AlN bulk crystals are grown as substrates for the fabrication of AlGaN-based electronic and optoelectronic devices. Epitaxial AlN and AlGaN films were grown on AlN substrates via MOCVD, and it was confirmed that these films have a very high structural quality, far beyond that of AlN layers grown on sapphire and SiC.

The quality of AlN homoepitaxy on substrates fabricated from PVT boules is clearly shown in Fig. 4.9 [67]. No difference between the substrate and the homoepitaxial layers could be observed by TEM or XRD. Thus, the homoepitaxial layers adopted the lattice parameter and orientation of the bulk AlN substrates. Both layer and substrate exhibited a high structural quality and were strain-free. These results demonstrate impressively the advantages of bulk substrates.

In addition to the high structural quality, a low concentration of unintentional impurities in the epilayers is required for the successful fabrication of devices,

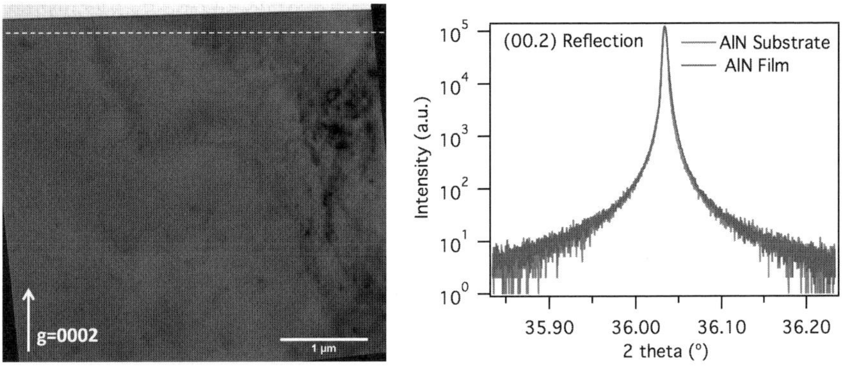

FIG. 4.9. TEM image of a homoepitaxial-grown AlN film (left) and XRD spectra around the (002)-peak of AlN (right). The dashed line marks the substrate to layer interface. (Reprinted from [67] by permission of ECS – The Electrochemical Society.)

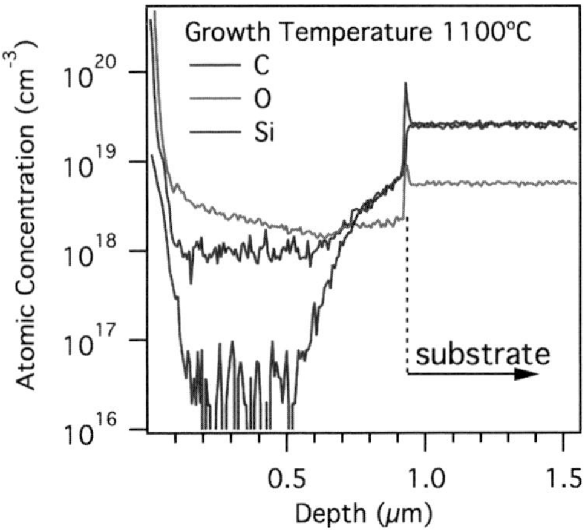

FIG. 4.10. Depth-resolved SIMS measurements of a 1 μm thick undoped AlN layer grown by MOCVD on an AlN substrate. (Reprinted from [67] by permission of ECS – The Electrochemical Society.)

since these impurities may lead to compensation effects and may reduce the carrier lifetime and mobility due to impurity scattering. Figure 4.10 presents a SIMS depth profile of silicon, carbon, and oxygen for a 1 μm thick homoepitaxial AlN layer grown by MOCVD [67]. A high concentration of these impurities is found in the AlN substrate, as discussed previously. The concentration of these impurities decreases for the homoepitaxial layers by one to three orders of magnitude. The strongest decrease is observed for silicon and carbon, while the oxygen concentration remains in the range of 10^{18} cm^{-3}.

The high structural quality and low defect concentration of homoepitaxial AlN layers are also reflected in the PL spectra from these layers [70]. For AlN layers grown on sapphire, the near-band-edge (NBE) luminescence consists mainly of the free A-exciton (X_A) and a silicon-related, donor-bound exciton (D^0X) [68]. Despite the steady increase in the quality of AlN/sapphire layers, this structure of the NBE emission has not changed since the first publication. However, for homoepitaxial AlN layers, a further fine-structure in the bound exciton region was resolved (Fig. 4.11) [71]. Due to the increased quality of these layers, the FWHM of the excitonic emission was reduced to 450 μm. This value is small enough that three different donor-bound excitons (D^0X_1-D^0X_3) can be resolved. An overview of the observed lines and a suggested assignment is given in Table 4.3.

The origin of the narrow FWHMs that lead to the observed fine-structure splitting is the increased crystal quality due to the used AlN bulk substrate. Similar observations were made for the defect luminescence of homoepitaxial layers. In a direct comparison of AlN layers grown on sapphire and AlN, Sedhain *et al.* were able to demonstrate that the defect luminescence around 4.7 eV is much stronger in samples grown heteroepitaxially [72]. Depending on the

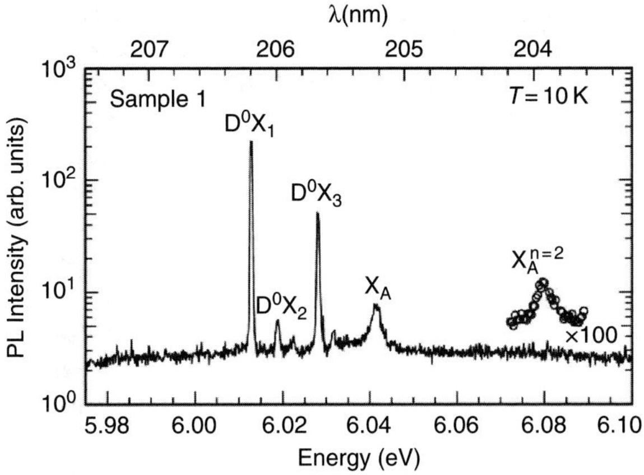

FIG. 4.11. High-resolution PL-spectra of homoepitaxial-grown AlN. (Reprinted with permission from [71].)

Table 4.3 Observed sharp emission lines in homoepitaxial AlN layers [71].

	Center energy (eV)	Localization energy (meV)	Origin
D^0X_1	6.0127	28.7	silicon
D^0X_2	6.0190	22.4	?
D^0X_3	6.0280	13.4	oxygen

growth direction, the defect luminescence of some layers grown on sapphire was even stronger than that of the near-band-edge luminescence, confirming reduced defect incorporation in homoepitaxially-grown AlN samples.

The results discussed for AlN were widely reproduced for AlGaN layers grown on bulk AlN crystals [67, 73, 74]. Similar to the binary system, an improved structural quality, a decreased defect incorporation and better optical properties were observed for layers grown on bulk AlN compared to those grown on sapphire.

4.5.2 *Devices on bulk AlN substrates*

Despite the excellent quality of bulk AlN crystals and epitaxially-grown AlN and AlGaN films thereon, there are only a very few preliminary realizations of electronic and optoelectronic devices on AlN substrates. Devices that have been demonstrated include LEDs and LDs, HFETs, and Schottky diodes.

Among the devices that have been realized on bulk AlN, LEDs can be considered as the most mature. However, due to the limited commercial availability of AlN substrates, these devices are far from perfect and much research needs to be done in the future to fully utilize AlN substrates. The first demonstration of a UV-LED on AlN substrates was made by Nishida *et al.* [75, 76]. Comparing LED structures grown on bulk AlN and on AlN/sapphire, the advantages of AlN substrates were suggested. However, in this work the main advantage of the bulk AlN substrate was solely its higher thermal conductivity (Table 4.1), which was expressed by an increase of the saturation current injection from 150 mA to 300 mA and a reduction of the red shift of the electroluminescence from a few nm down to only 1 nm [75, 76]. The authors did not fully exploit the advantages of AlN substrates, as they used only low Al-content AlGaN (4%—emission at 345 nm) for their active layers, which is highly mismatched with AlN. Thus, no reduction of dislocations could be expected in the active layers of these devices. Later, a comprehensive comparison of LEDs emitting at wavelength around 300 nm grown on sapphire and AlN was presented by Ren *et al.* [77]. The electrical characteristics of the device grown on AlN exceeded those of the device on sapphire (Fig. 4.12). This finding was based on the sharper onset of the LED grown on AlN, a much reduced sheet resistance of 160 Ω/□ (compared to 560–840 Ω/□ for devices on sapphire) and an increased light output. For the latter, an increase up to a factor of 3.5 was observed, directly proving the advantages of bulk AlN substrates. A calculation of the external quantum efficiency (EQE) highlighted a value of 0.14%. Comparing this to typical EQEs achieved by LEDs on sapphire in the same wavelength, an improvement by a factor of 4–5 was achieved.

Mid-UV LEDs on bulk AlN with an emission at 248 nm were demonstrated by Grandusky *et al.* [78]. Despite the high injection currents of up to 300 mA, no

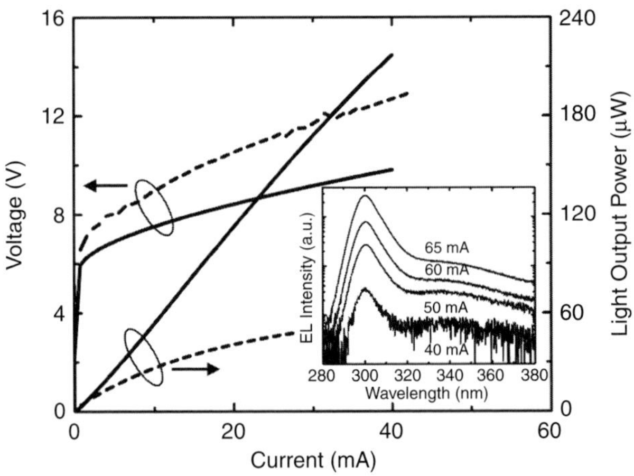

FIG. 4.12. I–V characteristics and light output of an LED grown on sapphire (dashed line) and bulk AlN substrate (solid line). The inset shows related EL luminescence for different currents. (Reprinted with permission from [75], © 2003 Cambridge University Press.)

shift in the emission line was observed, confirming the high thermal stability of the device. A pulsed injection led to a peak EQE of 1.44% with an output power of 1.44 mW. For the cw mode, slightly lower EQE and output power levels were achieved, whereby the drop was attributed to self-heating.

A few examples of laser structures grown on AlN substrates are available in the literature [79–81]. The first demonstration of a 300 mW output laser grown on an AlN bulk substrate was presented by Kneissl *et al.* in 2007 [79]. They reported on electrically driven laser emission around 370 nm using InGaN quantum wells. For this device, a threshold current of 800 mA was determined which corresponded to a current density of 13.3 kA/cm^2. An emission power of 300 mW for pulsed current injection was measured, and an EQE of 6.7% was calculated. Again, these structures were highly mismatched with the AlN substrate. Furthermore, optically pumped devices with emission ranging down to 308 nm were realized. However, due to the use of Ga-rich active layers and not optimized growth conditions, LDs with emission at a lower wavelength revealed a strongly increasing threshold power density reaching only values observed in devices grown on sapphire substrates.

More recently, transverse electric polarized lasing at 280.8 nm and 263.9 nm for AlGaN based multi-quantum-wells and double heterojunction structures fabricated on single crystalline AlN substrates was demonstrated [81]. With cavities shorter than 200 μm, well-defined, equally spaced longitudinal modes with line widths as narrow as 0.014 nm were observed. The low threshold pumping density of 84 kW/cm^2 in these structures suggests that the electrically pumped sub 300 nm ultraviolet laser diodes are imminent.

In addition to optoelectronic devices, (high-power) electronic devices will be one of the most important fields of applications for bulk AlN substrates. Due to the reduced dislocation density and the superior electrical and thermal properties of the AlN crystals (Table 4.1), devices grown on AlN substrates are expected to sustain higher electrical fields while heating up less, thus being more reliable and featuring an extended lifetime. However, only very limited results are available for such devices. Among the electrical devices that have been demonstrated are HFETs [82], X-ray detectors [2], and Schottky diodes [83].

4.6 Outlook

The market for near- and mid-UV light emitting devices is expected to grow dramatically over the next few years. To date, most light-emitters for applications in this spectral range are mercury lamps. There is a high demand to substitute mercury lamps with solid-state emitters, since lower power consumption and drastically enhanced lifetimes are expected from these devices. Without any doubt, AlGaN will be the dominating material for any solid-state UV emitters in the future, and the present body of work shows that for high-quality AlGaN with high Al content, AlN substrates are indispensible.

The structural properties of films grown on AlN wafers are vastly superior to those grown on sapphire or SiC. Dislocation densities down to 10^3 cm^2 have been demonstrated. This value is six orders of magnitude lower than what is typically achieved on SiC or sapphire substrates. The surface preparation of AlN is also at a very advanced level. With mechanical and chemomechnical polishing, a surface roughness with RMS values down to 0.1 nm can be achieved. So, what is hindering the broader use of AlN substrates?

Firstly, AlN substrates are not readily available. There are less than a handful of suppliers of AlN wafers around the world, and the quality of wafers differs dramatically from supplier to supplier. This is primarily a consequence of different seeding techniques, as discussed previously.

Secondly, the size of AlN wafers is small compared to other competing substrates. Since the device production efficiency roughly scales as the square of the wafer diameter, most device manufacturers are taking a wait-and-see approach. Based on current analyses, entry-level wafer sizes for optoelectronics and electronics are 50 mm and 100 mm, respectively.

Thirdly, the cost of AlN substrates is still very high. This is not unusual nor of any real concern, as all other materials technologies have taken a similar path. As crystal growth processes become more efficient and the boules larger, the price eventually reduces. However, current high prices significantly limit the pool of early adaptors.

Fourthly, optical absorption in the UV is high. As discussed, many bulk AlN crystals reveal a relatively strong UV absorption band in the UV region around 265 nm. Since LED emission is typically coupled out through the substrate, absorption in the substrate drastically reduces the efficiency of the light-emitting devices. Therefore, a better understanding of this UV absorption is needed. Recent investigations have shown a connection between carbon and the UV absorption.

Lastly, although great results on AlN and AlGaN epitaxy and on the properties of these layers on AlN substrates have been reported, thus far, most devices have been only first demonstrations and have not fully exploited the properties of the AlN substrates. As a result, the performance of these devices has not been overwhelmingly different from those grown on sapphire. In order to advance this promising technology, much more focused work is needed to develop doping and strain-management schemes that will serve as basic building blocks for future devices.

References

[1] Chu, T. L. and Kelm, R. W. *Journal of The Electrochemical Society* **122**(7), 995–1000 (1975).

[2] Erlbacher, T., Bickermann, M., Kallinger, B., Meissner, E., Bauer, A. J., and Frey, L. *physica status solidi (c)* **9**(3–4), 968–971 (2012).

[3] Slack, G. A., Tanzilli, R., Pohl, R., and Vandersande, J. *Journal of Physics and Chemistry of Solids* **48**(7), 641–647 (1987).

[4] Lahreche, H., Vennegues, P., Tottereau, O., Laugt, M., Lorenzini, P., Leroux, M., Beaumont, B. and Gibart, P. *Journal of Crystal Growth* **217**(12), 13–25 (2000).

[5] Liu, L. and Edgar, J. *Materials Science and Engineering: R: Reports* **37**(3), 61–127 (2002).

[6] Iwama, S., Hayakawa, K., and Arizumi, T. *Journal of Crystal Growth* **56**(2), 265–269 (1982).

[7] Yim, W. M. and Paff, R. J. *Journal of Applied Physics* **45**(3), 1456–1457 (1974).

[8] Lagerstedt, O. and Monemar, B. *Phys. Rev. B* **19**, 3064–3070 Mar (1979).

[9] Florescu, D. I., Asnin, V. M., Pollak, F. H., Jones, A. M., Ramer, J. C., Schurman, M. J., and Ferguson, I. *Applied Physics Letters* **77**(10), 1464–1466 (2000).

[10] Briegleb, F. and Geuther, A. *Justus Liebigs Annalen der Chemie* **123**(2), 228–241 (1862).

[11] Slack, G. A. and McNelly, T. *Journal of Crystal Growth* **34**(2), 263–279 (1976).

[12] Slack, G. A. and McNelly, T. *Journal of Crystal Growth* **42**(0), 560–563 (1977).

[13] Nakamura, S., Mukai, T., and Senoh, M. *Applied Physics Letters* **64**(13), 1687–1689 (1994).

[14] Balkas, C. M., Sitar, Z., Zheleva, T., Bergman, L., Nemanich, R., and Davis, R. *Journal of Crystal Growth* **179**(34), 363–370 (1997).

[15] Noveski, V., Schlesser, R., Mahajan, S., Beaudoin, S., and Sitar, Z. *Journal of Crystal Growth* **264**(13), 369–378 (2004).

[16] Schlesser, R. and Sitar, Z. *Journal of Crystal Growth* **234**(23), 349–353 (2002).

[17] Bockowski, M., Wroblewski, M., Lucznik, B., and Grzegory, I. *Materials Science in Semiconductor Processing* **4**(6), 543–548 (2001).

[18] Ehrentraut, D. and Sitar, Z. *Mrs Bulletin* **34**(4), 259–265 APR (2009).

[19] Helava, H., Davis, S. J., Huminic, G. D., Ramm, M. G., Avdeev, O. V., Barash, I. S., Chemekova, T. Y., Mokhov, E. N., Nagalyuk, S. S., Roenkov, A. D., Segal, A. S., Vodakov, Y. A., and Makarov, Y. N. *physica status solidi (c)* **4**(7), 2281–2284 (2007).

[20] Lu, P., Collazo, R., Dalmau, R. F., Durkaya, G., Dietz, N., and Sitar, Z. *Applied Physics Letters* **93**(13), 131922 (2008).

[21] Herro, Z. G., Zhuang, D., Schlesser, R., and Sitar, Z. *Journal of Crystal Growth* **312**(18), 2519–2521 (2010).

[22] Kumagai, Y., Yamane, T., and Koukitu, A. *Journal of Crystal Growth* **281**(1), 62–67 (2005).

[23] Yim, W. M., Stofko, E. J., Zanzucchi, P. J., Pankove, J. I., Ettenberg, M., and Gilbert, S. L. *Journal of Applied Physics* **44**(1), 292–296 (1973).

[24] Albrecht, M., Nikitina, I. P., Nikolaev, A. E., Melnik, Y. V., Dmitriev, V. A., and Strunk, H. P. *physica status solidi (a)* **176**(1), 453–458 (1999).

[25] Kovalenkov, O., Soukhoveev, V., Ivantsov, V., Usikov, A., and Dmitriev, V. *Journal of Crystal Growth* **281**(1), 87–92 (2005).

[26] Nagashima, T., Harada, M., Yanagi, H., Kumagai, Y., Koukitu, A., and Takada, K. *Journal of Crystal Growth* **300**(1), 42–44 (2007).

[27] Tsujisawa, K., Kishino, S., Liu, Y. H., Miyake, H., Hiramatsu, K., Shibata, T., and Tanaka, M. *physica status solidi (c)* **4**(7), 2252–2255 (2007).

[28] Volkova, A., Ivantsov, V., and Leung, L. *Journal of Crystal Growth* **314**(1), 113–118 (2011).

[29] Kamei, K., Inoue, S., Shirai, Y., Tanaka, T., Okada, N., and Yauchi, A. *physica status solidi (a)* **203**(7), 1720–1723 (2006).

[30] Dugger, C. O. *Materials Research Bulletin* **9**(3), 331–336 (1974).

[31] Isobe, H., Kawamura, F., Kawahara, M., Yoshimura, M., Mori, Y., and Sasaki, T. *Japanese Journal of Applied Physics PART 2: Letters & Express Letters* **44**(16-19), L488–L490 (2005).

[32] Kangawa, Y., Wakigawa, T., and Kakimoto, K. *Japanese Journal of Applied Physics Part 1: Regular Papers Brief Communications & Review Papers* **46**(9A), 5785–5787 SEP (2007).

[33] Kangawa, Y., Toki, R., Yayama, T., Epelbaum, B. M., and Kakimoto, K. *Applied Physics Express* **4**(9) (2011).

[34] Balkas, C. M., Sitar, Z., Zheleva, T., Bergman, L., Shmagin, I. K., Muth, J. F., Kolbas, R., Nemanich, R., and Davis, R. F. In *III-V Nitrides,* Ponce, FA and Moustakas, TD and Akasaki, I and Monemar, BA, editor, volume 449 of *Materials Research Society Symposium Proceedings*, 41–46 (1997).

[35] Liu, L., Liu, B., Edgar, J. H., Rajasingam, S., and Kuball, M. *Journal of Applied Physics* **92**(9), 5183–5188 (2002).

[36] Katagiri, Y., Kishino, S., Okuura, K., Miyake, H., and Hiramatu, K. *Journal of Crystal Growth* **311**(10), 2831–2833 (2009).

[37] Epelbaum, B. M., Bickermann, M., and Winnacker, A. In *Silicon Carbide and Related Materials, 2002,* Bergman, P and Janzen, E, editor, volume 433–4 of *Materials Science Forum*, 983–986, (2002).

[38] Heimann, P., Epelbaum, B. M., Bickermann, M., Nagata, S., and Winnacker, A. *physica status solidi (c)* **3**(6), 1575–1578 (2006).

[39] Sukhoveev, V., Usikov, A., Kovalenkov, O., Ivantsov, V., Syrkin, A., Dmitriev, V., Collins, C., and Wraback, M. In *GaN, AIN, InN and Related Materials,* Kuball, M. and Myers, T.H. and Redwing, J.M. and Mukai, T., editor, volume 892 of *Materials Research Society Symposium Proceedings*, 743–748, (2006).

[40] Rojo, J. C., Slack, G. A., Morgan, K., Raghothamachar, B., Dudley, M., and Schowalter, L. J. *Journal of Crystal Growth* **231**(3), 317–321 (2001).

[41] Tanaka, M., Nakahata, S., Sogabe, K., Nakata, H., and Tobioka, M. *Japanese Journal of Applied Physics Part 2: Letters* **36**(8B), L1062–L1064 AUG 15 (1997).

[42] Noveski, V., Schlesser, R., Raghothamachar, B., Dudley, M., Mahajan, S., Beaudoin, S., and Sitar, Z. *Journal of Crystal Growth* **279**(12), 13–19 (2005).

[43] Bondokov, R. T., Morgan, K. E., Slack, G. A., and Schowalter, L. J. *MRS Online Proceedings Library* **955**, 0955–I0903–0908 (2006).

[44] Kumagai, Y., Kubota, Y., Nagashima, T., Kinoshita, T., Dalmau, R., Schlesser, R., Moody, B., Xie, J., Murakami, H., Koukitu, A., and Sitar, Z. *Applied Physics Express* **5**(5), 055504 (2012).

[45] Moram, M. A. and Vickers, M. E. *Reports on Progress in Physics* **72**(3), 036502 (2009).

[46] Bickermann, M., Epelbaum, B. M., Filip, O., Heimann, P., Nagata, S., and Winnacker, A. *physica status solidi (c)* **5**(6), 1502–1504 (2008).

[47] Dalmau, R., Moody, B., Xie, J., Collazo, R., and Sitar, Z. *physica status solidi (a)* **208**(7), 1545–1547 (2011).

[48] Zhuang, D., Edgar, J., Strojek, B., Chaudhuri, J., and Rek, Z. *Journal of Crystal Growth* **262**(14), 89–94 (2004).

[49] Romanov, A. E. and Speck, J. S. *Applied Physics Letters* **83**(13), 2569–2571 (2003).

[50] Raghothamachar, B., Dudley, M., Rojo, J., Morgan, K., and Schowalter, L. *Journal of Crystal Growth* **250**(12), 244–250 (2003).

[51] Raghothamachar, B., Vetter, W. M., Dudley, M., Dalmau, R., Schlesser, R., Sitar, Z., Michaels, E., and Kolis, J. W. *Journal of Crystal Growth* **246**(34), 271–280 (2002).

[52] Bickermann, M., Schmidt, S., Epelbaum, B., Heimann, P., Nagata, S., and Winnacker, A. *Journal of Crystal Growth* **300**(2), 299–307 (2007).

[53] Li, J., Nam, K. B., Nakarmi, M. L., Lin, J. Y., and Jiang, H. X. *Applied Physics Letters* **81**(18), 3365–3367 (2002).

[54] Han, Q.-F., Duan, C.-H., Qiu, K., Ji, C.-J., Xin-HuanLi, Zhong, F., Yin, Z.-J., Cao, X.-C., Zhou, X.-J., and Wang, Y.-Q. *Chinese Physics Letters* **24**(12), 3555 (2007).

[55] Bickermann, M., Epelbaum, B. M., Filip, O., Heimann, P., Nagata, S., and Winnacker, A. *physica status solidi (c)* **7**(1), 21–24 (2010).

[56] Bickermann, M., Epelbaum, B. M., and Winnacker, A. *physica status solidi (c)* **0**(7), 1993–1996 (2003).

[57] Strassburg, M., Senawiratne, J., Dietz, N., Haboeck, U., Hoffmann, A., Noveski, V., Dalmau, R., Schlesser, R., and Sitar, Z. *Journal of Applied Physics* **96**(10), 5870–5876 (2004).

[58] Sedhain, A., Du, L., Edgar, J. H., Lin, J. Y., and Jiang, H. X. *Applied Physics Letters* **95**(26), 262104 (2009).

[59] Maki, J.-M., Makkonen, I., Tuomisto, F., Karjalainen, A., Suihkonen, S., Raisanen, J., Chemekova, T. Y., and Makarov, Y. N. *Phys. Rev. B* **84**, 081204 (2011).

[60] Bickermann, M., Epelbaum, B. M., Filip, O., Tautz, B., Heimann, P., and Winnacker, A. *physica status solidi (c)* **9**(3–4), 449–452 (2012).

[61] Collazo, R., Xie, J., Gaddy, B. E., Bryan, Z., Kirste, R., Hoffmann, M., Dalmau, R., Moody, B., Kumagai, Y., Nagashima, T., Kubota, Y., Kinoshita, T., Koukitu, A., Irving, D. L., and Sitar, Z. *Applied Physics Letters* **100**(19), 191914 (2012).

[62] Nam, K. B., Nakarmi, M. L., Lin, J. Y., and Jiang, H. X. *Applied Physics Letters* **86**(22), 222108 (2005).

[63] Slack, G. A., Schowalter, L. J., Morelli, D., and Freitas, J. A. *Journal of Crystal Growth* **246**(34), 287–298 (2002).

[64] Berzina, B., Trinkler, L., Jakimovica, D., Korsaks, V., Grabis, J., Steins, I., Palcevskis, E., Bellucci, S., Chen, L.-C., Chattopadhyay, S., and Chen, K.-H. *Journal of Nanophotonics* **3**(1), 031950–031950–16 (2009).

[65] Schweizer, S., Rogulis, U., Spaeth, J.-M., Trinkler, L., and Berzina, B. *physica status solidi (b)* **219**(1), 171–180 (2000).

[66] Sarua, A., Rajasingam, S., Kuball, M., Garro, N., Sancho, O., Cros, A., Cantarero, A., Olguin, D., Liu, B., Zhuang, D., and Edgar, J. In *Gan and Related Alloys, 2003*, Ng, HM and Wraback, M and Hiramatsu, K and Grandjean, N, editor, volume 798 of *Materials Research Society Symposium Proceedings*, 297–302, (2003).

[67] Dalmau, R., Moody, B., Schlesser, R., Mita, S., Xie, J., Feneberg, M., Neuschl, B., Thonke, K., Collazo, R., Rice, A., Tweedie, J., and Sitar, Z. *Journal of The Electrochemical Society* **158**(5), H530–H535 (2011).

[68] Nam, K. B., Nakarmi, M. L., Li, J., Lin, J. Y., and Jiang, H. X. *Applied Physics Letters* **83**(14), 2787–2789 (2003).

[69] Neuschl, B., Thonke, K., Feneberg, M., Mita, S., Xie, J., Dalmau, R., Collazo, R., and Sitar, Z. *physica status solidi (b)* **249**(3), 511–515 (2012).

[70] Feneberg, M., Leute, R. A. R., Neuschl, B., Thonke, K., and Bickermann, M. *Phys. Rev. B* **82**, 075208 Aug (2010).

[71] Feneberg, M., Neuschl, B., Thonke, K., Collazo, R., Rice, A., Sitar, Z., Dalmau, R., Xie, J., Mita, S., and Goldhahn, R. *physica status solidi (a)* **208**(7), 1520–1522 (2011).

[72] Sedhain, A., Nepal, N., Nakarmi, M. L., Tahtamouni, T. M. A., Lin, J. Y., Jiang, H. X., Gu, Z., and Edgar, J. H. *Applied Physics Letters* **93**(4), 041905 (2008).

[73] Grandusky, J., Smart, J., Mendrick, M., Schowalter, L., Chen, K., and Schubert, E. *Journal of Crystal Growth* **311**(10), 2864–2866 (2009).

[74] Ren, Z., Sun, Q., Kwon, S.-Y., Han, J., Davitt, K., Song, Y. K., Nurmikko, A. V., Cho, H.-K., Liu, W., Smart, J. A., and Schowalter, L. J. *Applied Physics Letters* **91**(5), 051116 (2007).

[75] Nishida, T., Ban, T., Saito, H., and Makimoto, T. *MRS Online Proceedings Library* **798** (2003).

[76] Nishida, T., Makimoto, T., Saito, H., and Ban, T. *Applied Physics Letters* **84**(6), 1002–1003 (2004).

[77] Ren, Z., Sun, Q., Kwon, S.-Y., Han, J., Davitt, K., Song, Y. K., Nurmikko, A. V., Liu, W., Smart, J., and Schowalter, L. *physica status solidi (c)* **4**(7), 2482–2485 (2007).

[78] Grandusky, J. R., Gibb, S. R., Mendrick, M. C., and Schowalter, L. J. *Applied Physics Express* **3**(7) (2010).

[79] Kneissl, M., Yang, Z., Teepe, M., Knollenberg, C., Schmidt, O., Kiesel, P., Johnson, N. M., Schujman, S., and Schowalter, L. J. *Journal of Applied Physics* **101**(12), 123103 (2007).

[80] Wunderer, T., Chua, C. L., Yang, Z., Northrup, J. E., Johnson, N. M., Garrett, G. A., Shen, H., and Wraback, M. *Applied Physics Express* **4**(9) (2011).

[81] Xie, J., Guo, W., Bryan, Z., Mita, S., Hussey, L., Moody, B., Schlesser, R., Kirste, R., Gerhold, M., Collazo, R., and Sitar, Z. *Applied Physics Letters* **102**, 171102 (2013).

[82] Hu, X., Deng, J., Pala, N., Gaska, R., Shur, M. S., Chen, C. Q., Yang, J., Simin, G., Khan, M. A., Rojo, J. C., and Schowalter, L. J. *Applied Physics Letters* **82**(8), 1299–1301 (2003).

[83] Xie, J., Mia, S., Dalmau, R., Collazo, R., Rice, A., Tweedie, J., and Sitar, Z. *physica status solidi (c)* **8**(7–8), 2407–2409 (2011).

5 Epitaxial growth of nitride quantum dots

André Strittmatter

5.1 Introduction

The concept of quantum dots (QD) as initially proposed by Arakawa and Sakaki in 1982 was motivated by the quest for semiconductor materials providing higher radiative efficiencies, lower power consumption, and better temperature stability for light emitters such as laser diodes [1]. In the most simple picture quantum dots are three-dimensional boxes out of a semiconductor material confining charge carriers in discretized electronic quantum states. Such confinement is achieved if the dimensions of the boxes are less than the de-Broglie wavelength of the charge carriers. This quantum-mechanically defined quantity, however, is of the order of a few nm up to 20–30 nm for III/V semiconductors. Not only the small sizes are challenging for a materials scientist to realize, but also controlling densities of such tiny objects over several orders of magnitude.

Attempts to engineer quantum dots by means of patterning and etching largely failed, as defect generation deteriorates their optoelectronic properties. A groundbreaking alternative, the self-organized growth of quantum dots, was established around the mid-1990s in GaAs semiconductors and related alloys [2]. Most widely adopted today for the growth of quantum dots is the Stranski–Krastanow mode where coherently strained layers grow initially two-dimensional before nanometer-sized islands are formed upon exceeding a critical layer thickness [3]. Coherently strained islands of 10–30 nm base length, 1–5 nm height, and with densities from 10^6–10^{11} cm^{-2} are controlled by parameters such as temperature, layer thickness, and growth interruption. Another technique, droplet epitaxy, matured recently for site-controlled growth of single QDs and quantum dot molecules [4].

The successful realization of quantum dot-based laser diodes with lower laser threshold current density and improved temperature stability validated their utility for optoelectronic applications [1, 5]. Quantum dots grown by the Stranski–Krastanow method are also demonstrated for InP and related alloys, and in the SiGe/Si material system. Interfacial strain due to lattice mismatch between different compounds is also present in the nitride material system. The limits of 2.5% strain at AlN/GaN interfaces and 11% strain at InN/GaN interfaces makes the SK growth mode likely to be available for the nitrides too. It is also fairly easy to motivate research on nitride quantum dots, as the prospect of increased internal quantum efficiencies will eventually lead to better performances of optoelectronic consumer devices such as lamps, projectors, traffic lights, and the like. Another prospective area is single nitride-based quantum dots for novel applications such as quantum computing and quantum cryptography which require generation of single and entangled photons on demand. For operation of single-photon emitters at room temperature a separation of the quantum-mechanical states of charge carriers well above the kinetic energy of 25 meV at room temperature is a prerequisite. The nitride semiconductor family allows for combination of wide-gap matrix materials and narrow-gap quantum dots which would provide the necessary energy spacing within the quantum dots. Nitride-based quantum dots are therefore considered as potential candidates for room-temperature single-photon emitters.

This chapter is devoted to concepts for realizing nitride quantum dots by epitaxial methods. It starts with GaN quantum dots grown on AlN, as these quantum dots can be grown using the Stranski–Krastanow regime and important parameters controlling their properties are identified. It covers both the molecular beam epitaxy (MBE) as well as the metalorganic vapor phase epitaxy (MOVPE) method, and will also summarize results for non-planar and semi-polar QDs. Next, an attempt is made to present successful growth techniques for InGaN-based QDs. This comprises Stranski–Krastanow growth mode, droplet epitaxy, and surface pre-treatments using various chemicals. Of course, the intensively discussed subject of spontaneous QD formation within InGaN layers will be touched. InN quantum dots are treated separately, as it is probably the most challenging part of nitride growth. The chapter ends with current strategies for site-controlled growth of InGaN QDs, which is mainly done using growth of hexagonal pyramids and subsequent InGaN overgrowth of the apexes.

5.2 GaN quantum dots

5.2.1 *Molecular beam epitaxy*

For strained layer systems the Stranski–Krastanow growth regime has been established as a reliable self-assembling method for coherently strained quantum dots. Typically, the layer growth starts two-dimensional (2D) but upon exceeding a critical layer thickness a transition to three-dimensional (3D) islands is observed. For this transition to occur, the energy balance between surfaces, interfaces, and edges is decisive. When 3D islands form, the surface energy plus the energy of edges is increased as more surface area and edges are formed as compared to 2D growth. At the same time, strain energy is decreased as strained material is removed from the highly strained 2D layer and transferred into less strained quantum dots. In total, the QD formation process succeeds if the energy gain due to strain relaxation surpasses the energy budget required for newly created surfaces and edges. Strain energy increases with lattice mismatch and layer thickness, whereas surface energies depend on crystallographic orientation but also on specific surface terminations.

GaN layers when pseudomorphically grown on AlN(0001) buffer layers are compressively strained by 2.5% due to mismatch of the lattice constants of the basal planes (a_{GaN}=3.189 Å vs. a_{AlN}=3.11 Å). As known from the InGaAs/GaAs system, this is about the lower limit for the Stranski–Krastanow mode [6]. Variations of the surface energy may therefore become significant for the 2D/3D growth mode transition to take place. In MBE growth, this complexity manifests in the sensitive dependency of QD formation on the stabilization of a Ga-bilayer on the growth surface. In such a case, only 2D growth is observed up to the limits of plastic relaxation where dislocations form. This is understood as the consequence of a highly enhanced nitrogen diffusivity on the surface. However, 2D growth is established for Ga/N >1 in the case of plasma-assisted (PA) nitrogen cells and Ga/N <1 in the case of radio-frequency-assisted (rf-assisted) ammonia cells used for the nitrogen supply [7, 8]. If growth conditions are favoring Ga-bilayer formation, the formation of QDs is achieved by applying a growth interruption under vacuum. Thereby, excess Ga atoms evaporate from the surface and stoichiometry between Ga and N is established. Under stoichiometric conditions, however, 3D islands are energetically favored and QDs rapidly appear on the surface [9]. Quantum dots are also obtained if growth conditions are chosen which avoid the Ga-bilayer formation. For PA-MBE such conditions refer to growth using Ga/N-ratio <1. The critical layer thickness at which the 2D/3D transition starts is determined to about 2.5–3 monolayers (ML) GaN [10]. Up to about 8 ML GaN, QD feed on material transfer from the GaN wetting layer (WL), which is consumed until 2.5 ML of GaN remain in the wetting layer. This is illustrated in Fig. 5.1, where the high-resolution transmission electron microscope (HRTEM) image of a 3 QD layer stack is shown. Each QD layer was

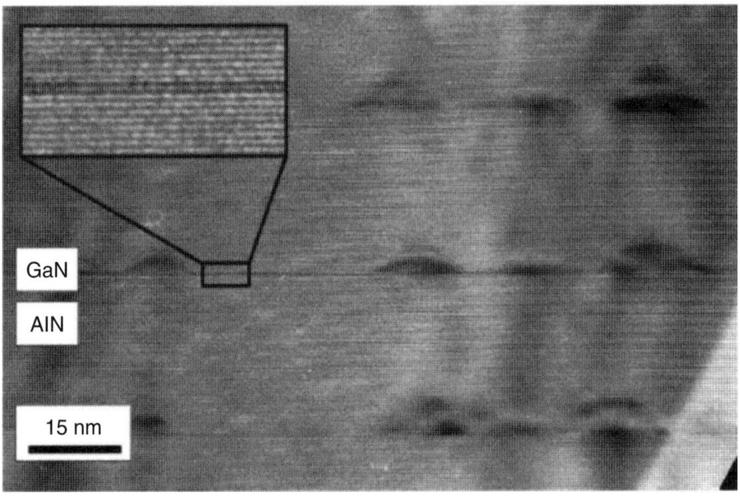

FIG. 5.1. HRTEM image of QDs on AlN(0001)/sapphire. In the inset, the wetting layer thickness of 2.5 MLs is resolved. (Reprinted with permission from [10], © 2002 American Institute of Physics.)

formed during a 120 s long growth interruption after deposition of a 3 ML thick GaN layer under Ga-rich conditions. The thin continuous dark line connecting the QDs is referred to as the remainder of the wetting layer. In the inset the thickness of the WL is resolved to 2.8 MLs. Typical MBE growth temperatures for GaN-QDs are similar to growth temperatures of thick, two-dimensional layers, i.e., 680–800°C. An exponential increase of the density from $3{\cdot}10^{10}$ cm^{-2} to $2{\cdot}10^{11}$ cm^{-2} is observed when the GaN thickness is increased from 2.8 MLs to 6 MLs [9]. Once QDs are formed during growth interruptions the supply of nitrogen stimulates Ostwald ripening, i.e., small dots dissolve and large dots grow in size [7]. Well-defined shapes of truncated hexagonal pyramids with a lateral size between 10–25 nm and 2–5 nm height are found both in TEM and AFM measurements. The side facets of the dots belong to $\{1\bar{1}03\}$ crystallographic planes which make an angle of 32° with the (0001) basal plane of the wurtzite structure in GaN. This shape is preserved during AlN cap layer growth because of the very low intermixing of Ga and Al atoms at the interface between GaN and AlN [12].

Vertical stacking of up to 200 QD planes is reported in the GaN/AlN system [11, 13]. Capping layer growth by AlN results in planarization of the growth front if the layer thickness exceeds the average QD height [12]. Vertically correlated QDs are found for AlN spacer layer thicknesses $\leq$8 nm [14]. An example of vertically correlated dots is displayed in Fig. 5.2. Instead of a random dot distribution in each plane an arrangement into vertical columns is found. One can also notice the lateral size increase of the dots in the upper planes. The vertical correlated growth is due to local strain fields in the AlN matrix surrounding the GaN QDs. For small AlN spacer layer thicknesses the strain fields may protrude to the growth plane of subsequent QDs, thereby reducing locally

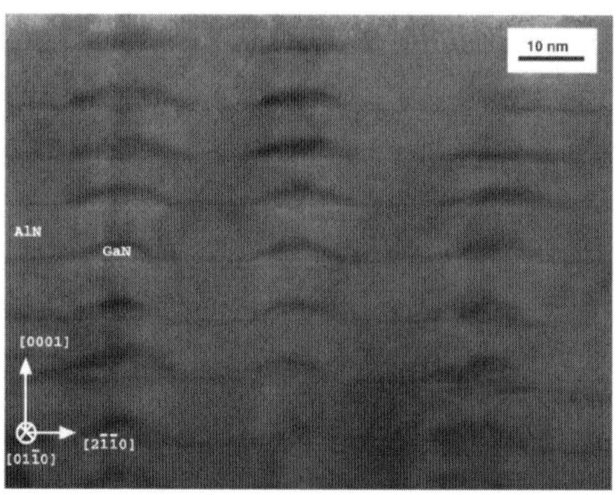

Fig. 5.2. TEM image of vertically correlated QDs on AlN(0001)/sapphire. A threading dislocation runs through the left column of dots. (Reprinted with permission from [11], © 1998 American Institute of Physics.)

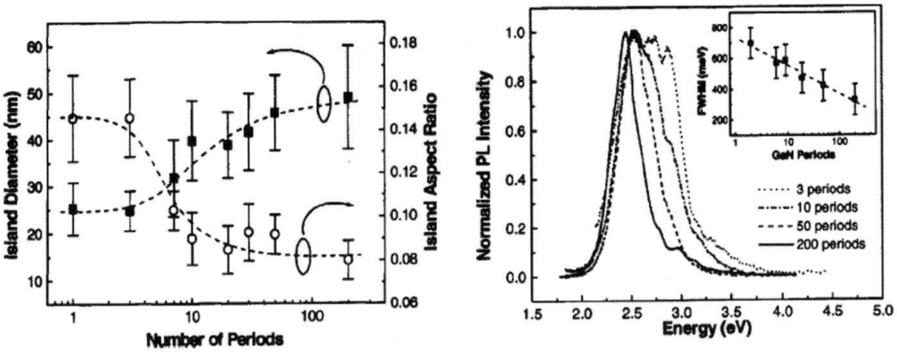

FIG. 5.3. Evolution of shape and emission properties of vertically correlated QDs. Left) Dependence of lateral QD size and aspect ratio on stacking number. Right) Spectra of the ensemble luminescence. The inset shows the narrowing of the spectral width of the luminescence. (Reprinted with permission from [13], © 2004 American Institute of Physics.)

the free energy for QD formation. Uncorrelated stacks of QDs are observed for spacer thicknesses of 20 nm in MBE. The analysis shown in Fig. 5.3 reveals an increase of the lateral QD size from about 25 nm to 50 nm along with a reduction of the statistical size broadening. Thereby, the ground state optical transition undergoes a red-shift and the FWHM of the ensemble luminescence becomes narrower. A detailed survey of the epitaxy of GaN quantum dots by MBE is given by Daudin [15].

5.2.2 *Metalorganic vapor phase epitaxy*

In 1996, Tanaka *et al.* reported the very first result on GaN QD growth employing silicon surface pre-treatment in MOVPE growth [16]. The method allows growth of GaN quantum dots on $Al_xGa_{1-x}N$ ($0.07 \leq x \leq 0.20$) buffer layers by MOVPE. In this case a SiN_x mask may form and the growth of GaN resembles a mode which may be best described by selective area growth (SAG). Self-assembled QD formation via a 2D/3D transition was observed for GaN growth on AlN in a temperature range of 960–970°C using a horizontal reactor geometry. The apparently large difference to MBE growth temperatures is to a great extent caused by the indirect measurement with a thermocouple, reading an inner susceptor temperature which may be offset to the surface temperature by up to 150°C. As shown in Fig. 5.4, the QD density can be effectively controlled between $2{\cdot}10^8$ and $5{\cdot}10^{10}$ cm^{-2} by tuning the GaN coverage from 3.4 and 9 MLs. Besides, control of the dot density is achieved by growth interruptions of 3–30 s and by tuning the V/III ratio during GaN growth. High QD densities of $>10^{10}$ cm^{-2} are obtained at very low V/III ratios of 20–30, whereas for V/III ratios of >10000 one finds densities between 10^7 and 10^9 cm^{-2} [17, 18]. This agrees well with the conclusion

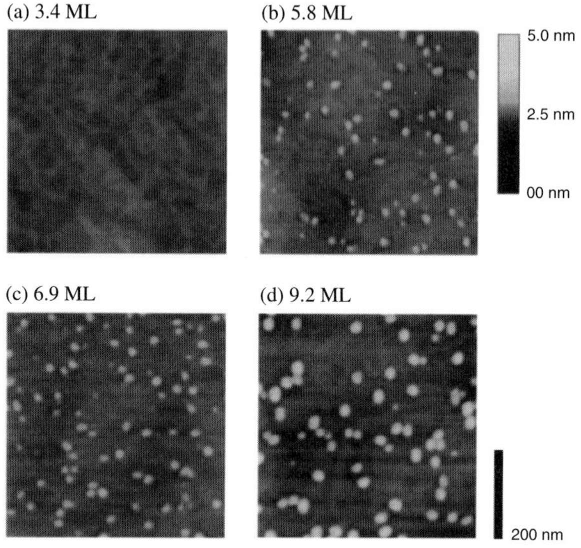

FIG. 5.4. Surfaces for different coverages of GaN (AFM images) grown by MOVPE on AlN(0001)/sapphire. (Reprinted with permission from [17], © 2002 American Institute of Physics.)

from MBE results that the surface diffusion of Ga atoms is regulated by the nitrogen supply. Interestingly, for high V/III ratios the critical layer thickness for QD formation is well below 2.5 MLs and non-exponential growth of the QD density is observed. These discrepancies to MBE results has been ascribed to tolerances on the growth-rate estimations as well as to possibly non-homogeneous wetting layer thicknesses which might be locally exceeding the critical value of about 2 MLs. There is also a noticeable reduction in lateral size of the QD pyramids from 35 nm to 20 nm when the V/III ratio is reduced, and hence the density is increased [17, 18]. Vertically ordered stacks of up to ten QD planes separated by 20 nm AlN spacer layers are also fabricated using MOCVD. The relatively large vertical distance of 20 nm as compared to the upper limit of 8 nm seen for MBE-grown, vertically correlated QD samples is explained by the larger lateral QD size (20 nm to 10 nm) [19]. Similar to the results obtained by MBE, photoluminescence spectra also show reduced size broadening and red-shift of the peak wavelength upon stacking, indicating more homogeneous and larger dot sizes.

5.2.3 *Growth on non-polar and semipolar planes*

Spontaneous as well as piezoelectric polarization charges build up at interfaces normal to the [0001] direction due to the inversion asymmetry of atomic sites for group-A and group-B atoms. The binary nitrides AlN-GaN-InN exhibit strong polarization fields across their {0001} interfaces [20] which separate electron and hole charge carriers along opposite directions. In consequence, the wavefunction overlap of electron and holes is reduced, radiative recombination rates are low, and emission peak wavelength of quantum wells and quantum dots strongly depends on injected carrier density. For light-emitting devices such as LEDs and laser diodes those consequences are detrimental for their performance. Polarization fields can be canceled on non-polar (m-plane, $\{1\bar{1}00\}$ and a-plane, $\{11\bar{2}0\}$) and greatly reduced on semipolar $\{11\bar{2}2\}$ surfaces, and increased performance with respect to light output efficiency is expected [21, 22]. Even though quantum dots confine the charge carriers into small volumes the effects of the polarization fields on radiative recombination rate and emission wavelength are strong [23]. This has motivated studies on QD growth on non-polar and semipolar surfaces.

Non-polar GaN QDs are obtained by MBE growth both on m-plane AlN$(1\bar{1}00)$ and a-plane AlN$(11\bar{2}0)$ buffer layers [24–26]. The Stranski–Krastanow growth regime is achieved using Ga-rich conditions for both surface orientations. This is not easy to understand, given the fact that for Ga-rich conditions surface roughening occurs on m-plane surfaces and layer-by-layer growth is observed for a-plane material. An explanation can be given by assuming different surface energies of the planes covered by Ga atoms similar to what is observed for GaN/AlN(0001) growth. The Stranski–Krastanow regime is accordingly modified by growth interruption strategies to account for the differences in growth. For m-plane QDs, growth interruption are applied without nitrogen stabilization to remove excess Ga atoms from the surface by evaporation. In the case of a-plane QDs, growth interruption is performed under nitrogen to bind excessive Ga atoms on the surface. The 2D/3D transition occurs for GaN layer thicknesses around 2 MLs, which is similar to growth on AlN(0001). In addition, the strain state of the AlN buffer layer is decisive on whether QDs or QW structures result from the growth. For m-plane AlN/SiC$(1\bar{1}00)$ buffers, relaxation of the buffer with increasing thickness (>500 nm) promotes QD formation, since the lattice mismatch between GaN/AlN progressively increases.

Non-polar $In_{0.12}Ga_{0.88}N$-QDs have been realized by MOVPE on m-plane free-standing GaN substrates [27]. In order to achieve the Stranski–Krastanow growth regime for such low In-content, compressive strain was introduced additionally by a low-temperature AlN interlayer and an AlGaN/AlN superlattice structure into a 190-nm thick GaN buffer layer prior to the QD growth.

Symmetric $\{1\bar{1}03\}$ side facets are observed for c-plane (polar) quantum dots resulting in truncated hexagonal pyramids. The lower symmetry of the non-polar a- and m-planes lead to pronounced asymmetries of the shapes of non-polar QDs. As Figs. 5.5a and 5.5b show, a-plane GaN quantum dots are symmetric perpendicular to the [0001] c-axis direction but asymmetric perpendicular to the $[1\bar{1}00]$ direction. The side facets perpendicular to the c-axis are identified as $(02\bar{2}1)$ and $(20\bar{2}1)$ planes, respectively, while the asymmetric side facets are a low-angle $(11\bar{2}1)$ plane and a high-angle $(11\bar{2}\bar{2})$ plane, respectively [28]. Figure 5.6 shows the shape of m-plane QDs in comparison to c-plane QDs when viewed from

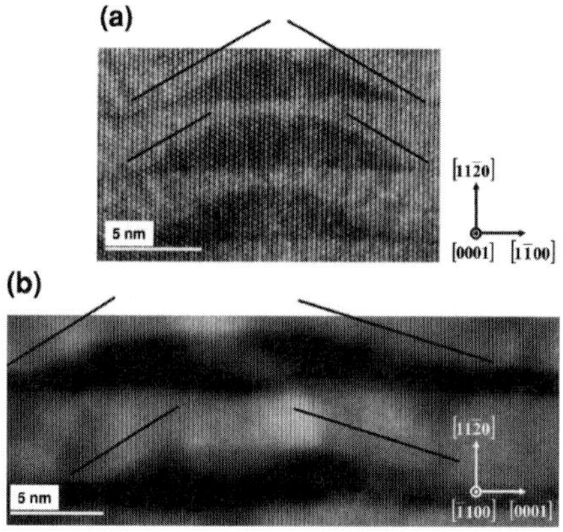

FIG. 5.5. TEM image of vertically correlated of a-plane GaN-QDs on AlN(11$\bar{2}$0). a) Cross-section image along the [0001] direction, symmetric QD shape. b) Cross-section image along the [1$\bar{1}$00] direction, asymmetric QD shape. (Reprinted with permission from [28], © 2007 American Institute of Physics.)

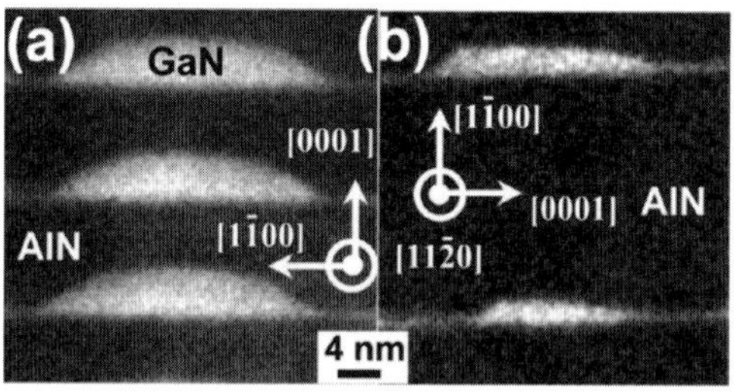

FIG. 5.6. Scanning tunneling microscope (STM) image of a) c-plane and b) m-plane GaN QDs. The asymmetry of m-plane QDs becomes visible when STM is performed on the (11$\bar{2}$0) surface. (Reprinted with permission from [29], © 2008 American Institute of Physics.)

the [11$\bar{2}$0] direction. A strong asymmetry with respect to the [1$\bar{1}$00] direction can be seen [26]. As a rule of thumb, the low-angle facet points in the +c-direction, whereas the high-angle facet points in the opposite −c-direction [29]. A very similar asymmetry of the dot shape is found for InGaN/GaN QDs grown on m-plane GaN [27].

Semipolar GaN{11$\bar{2}$2} and InGaN{11$\bar{2}$2}-QDs are also realized by MBE [30, 31]. Stranski–Krastanow growth is realized under Ga-rich conditions by applying growth interruptions after GaN deposition on AlN(11$\bar{2}$2) buffer layers [30]. Under Ga-rich conditions, the GaN growth remains two-dimensional. However,

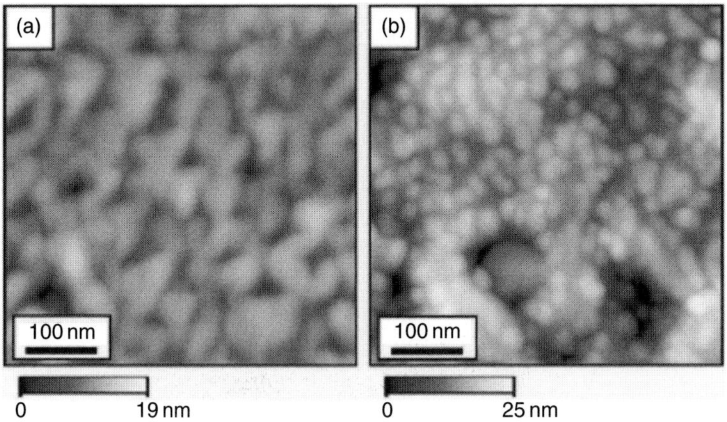

FIG. 5.7. AFM surface images of a) bare AlN($11\bar{2}2$) and b) semipolar QDs after deposition of 5 MLs of GaN on AlN($11\bar{2}2$)/sapphire. (Reprinted with permission from [30], © 2009 American Institute of Physics.)

during the growth interruption, 3D islands are formed if the GaN thickness is larger than about 3 ML. Figure 5.7 shows respective surface images as recorded by AFM prior and after deposition of 5 MLs GaN, with subsequent growth interruption under vacuum of 90s. For the shape of semipolar GaN($11\bar{2}2$) QDs possible variants are deduced from HRTEM analysis [32]. According to this model, the QDs are slightly asymmetric along [$11\overline{23}$] with delimiting ($11\bar{2}6$) and ($11\bar{2}0$) side facets, respectively. For the opposite in-plane direction, [$1\bar{1}00$], symmetric shapes may be observed if either of two equivalent ($10\bar{1}1$) or ($10\bar{1}2$) facets delimit the QDs. However, the model further accounts for the possibility of an asymmetric shape along [$1\bar{1}00$] if both facets coexist in the dot shape. Due to imperfect growth of AlN buffer layers on $\{11\bar{2}2\}$ surfaces, QD nucleation on other than $\{11\bar{2}2\}$ planes is possible [32, 33]. The impact of polarization fields on the optical properties of non-polar and semipolar QDs is greatly reduced as compared to c-plane structures. This is exemplified by photoluminescence experiments shown in Figs. 5.8 and 5.9. As seen in Fig. 5.8, the luminescence peak does not shift with increasing QD size below the bandgap of bulk GaN at 3.47 eV, which indicates reduced polarization fields. In contrast, the emission peak wavelength of polar QDs shifts even into the visible spectral range due to piezoelectric polarization fields [8]. For the same reason, the emission peak of polar QD luminescence shifts towards higher energies with increasing excitation density as the polarization is progressively screened by photoexcited carriers. The comparison of respective excitation-dependent PL spectra shown in Fig. 5.9 demonstrates that this shift of the emission wavelength is not present in a-plane (non-polar) QDs.

Despite the progress on non-polar and semipolar QDs a remaining common problem of non-polar and semipolar nitride heterostructures is an increased

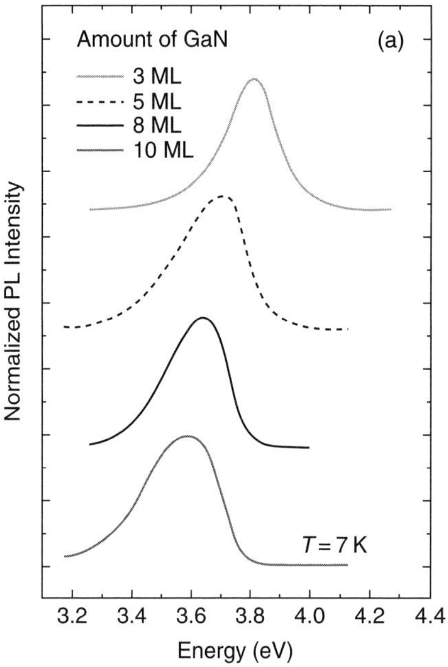

FIG. 5.8. Low-temperature photoluminescence spectra of semipolar GaN/AlN(11$\bar{2}$2) QD stacks for different GaN thicknesses. The QD luminescence is not shifted below the band-gap energy of GaN at 3.47 eV. (Reprinted with permission from [30], © 2009 American Institute of Physics.)

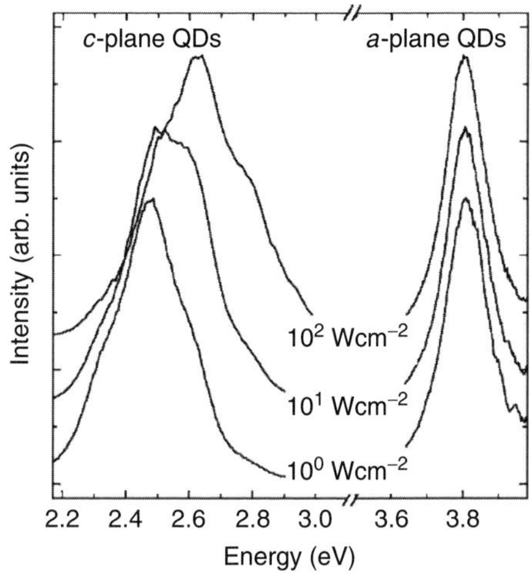

FIG. 5.9. Comparison for polar (c-plane) and non-polar (a-plane) QDs of the spectral position of the photoluminescence at different excitation power densities. (Reprinted with permission from [25], © 2005 American Institute of Physics.)

probability for defect generation within the basal plane of the wurtzite unit cell, as efficient strain relief results from such defects. For instance, basal plane stacking faults terminated by partial dislocations are usually found in heteroepitaxially grown layers [34–36]. In the case of m-plane QDs, dislocations in the [11$\bar{2}$0] direction are identified both at the lower and upper GaN/AlN interface of QD structures [37]. There are also dislocations running in the [1$\bar{1}$00] direction at a-plane AlN/GaN interfaces of the wetting layer [28]. For both orientations, c-axis strain is relieved plastically by dislocation generation within the intersection of the c-plane and the growth plane. However, in-plane strain of the lattice planes perpendicular to the c-axis planes, i.e., parallel to [1$\bar{1}$00] and [11$\bar{2}$0], is relieved elastically through the QD formation.

5.3 $In_xGa_{1-x}N$ quantum dots

5.3.1 *Phase separation and In segregation effects*

Quantum dots in the InGaN material system are achieved in numerous ways, since the instability of the material enables several possibilities for the growth. This section starts with highlighting several issues of the InGaN compound which are important for the understanding of InGaN quantum dot growth. Generally speaking, issues are due to the large differences of the atomic radii of the elemental constituents indium and gallium. In numbers, the Ga-N bond length of 1.95 Å is about 10% smaller than the In-N bond length of 2.15 Å which leads to micro-strains in the unit cell when indium atoms are incorporated on Ga sites. Consequently, the system's energy is increased when In atoms are incorporated on Ga sites. The unit cell reacts by stretching and bending of bonds around In atoms in order to minimize the potential energy of the crystal [38]. Another option for the compound to restore its minimum energy is phase separation by spinodal and binodal decomposition. By calculating the enthalpy of mixing alloy decomposition is expected for temperatures below 1400°C. At a typical growth temperature of 800°C immiscibility of InN and GaN is predicted for $0 \leq x_{In} \leq 0.88$ (see Fig. 5.10). Due to compressive strain exacerbated by GaN barriers onto InGaN the immiscibility region is, however, narrowed and shifted

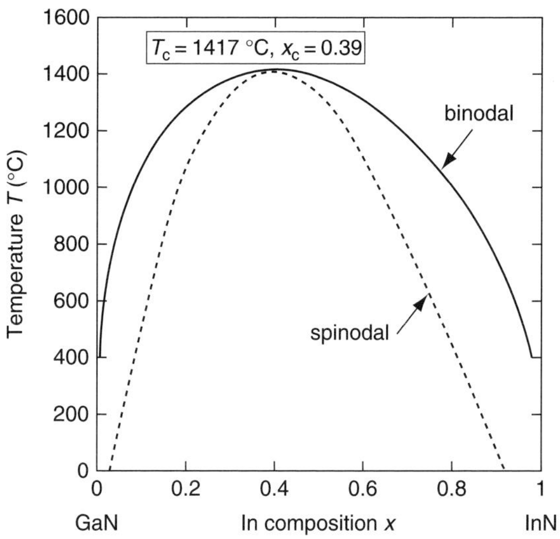

FIG. 5.10. Spinodal and binodal curves within the calculated phase diagram of $In_xGa_{1-x}N$. The alloy is unstable within the binodal curve (solid line). (Reprinted with permission from [38], © 1999 by the American Physical Society.)

towards high-In containing layers, and the critical temperature is lowered below about 700°C [39]. One should also recall that those calculations refer to thermodynamic equilibrium conditions which are not established during epitaxial growth methods like MBE or MOVPE. However, growth interruptions and very low growth rates, which are part of quantum dot formation in the SK regime, are likely to promote equilibration processes.

Important for In incorporation into the growing crystal is also the In-N bond energy (1.89 eV) which is only 80% of the Ga-N bond energy (2.34 eV). Moving an indium atom from one lattice site to another is therefore much more favored than for a gallium atom. Indium atoms sitting on a crystal's surface also make fewer bonds with nitrogen atoms than for the four-fold coordinated group-III lattice sites in the bulk. In addition, indium atoms at crystal surfaces accommodate strain more efficiently by rearranging their bonding to next-neighbors. For these reasons, segregation of In atoms towards the surface is very likely to occur, as the total energy of the system becomes reduced thereby. Segregation phenomena are investigated by Monte Carlo simulation of $In_{0.10}Ga_{0.90}N$ layer growth. Figure 5.11 shows that the surface In coverage exceeds by large amounts (>50%) the nominal bulk value, and less than 5% of indium is found in the bulk [40]. Through such a surface layer the chemical potential of In atoms is raised close to the value where In droplet formation starts. Northrup *et al.* also calculated that In incorporation is favored on $\{11\underline{2}2\}$ surfaces because of In-In bilayer formation promoting In incorporation on such surfaces [41]. They identified the cores of threading dislocations on GaN(0001) as centers for In agglomeration.

Under such circumstances, controlling the self-organized growth of quantum dots in the Stranski–Krastanow regime may deviate from the binary GaN/AlN

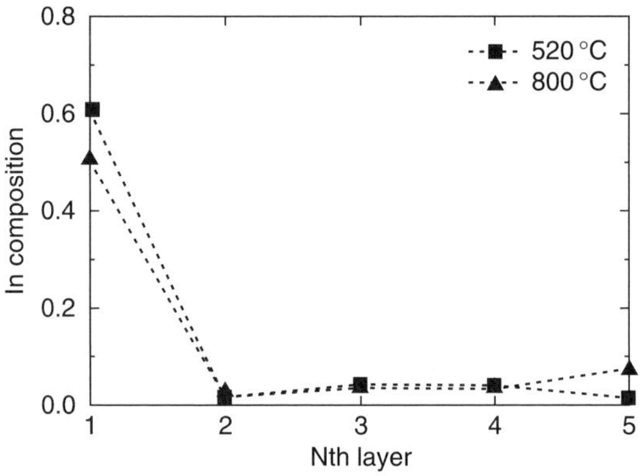

FIG. 5.11. Composition profile near the surface of 15-ML thick $In_{0.10}Ga_{0.90}N(0001)$ films for temperatures of 520°C (triangles) and 800°C (squares) as calculated by the Monte Carlo simulation method. (Reprinted with permission from [40].)

system. Otherwise, miscellaneous techniques are developed for realizing InGaN quantum dots which take advantage of phase separation and In segregation effects.

5.3.2 *Stranski–Krastanow growth mode*

The layer thickness for the transition from two-dimensional to three-dimensional growth as characteristic feature of the SK growth mode depends on lattice mismatch and therefore on In composition for the ternary $In_xGa_{1-x}N$ material. *In situ* growth mode analysis during MBE using Reflection High-Energy Diffraction (RHEED) is used to investigate the 2D/3D critical layer thickness. The results for MBE using ammonia are shown in Fig. 5.12 for $0 < x_{In} < 0.50$. By investigation of the QD formation at 550°C growth temperature, the minimum In composition for SK-QD formation is around 12%, below which only surface roughening occurs [42]. A different onset of the transition at 18% indium content is measured for plasma-assisted MBE using nitrogen. For such an MBE method, SK-QD require stoichiometric or slightly nitrogen-rich conditions. For an $In_{0.35}Ga_{0.65}N$ layer, the critical layer thickness is 1.7 MLs, which is also lower than the value taken from Fig. 5.12. In terms of compositional dependence of the critical layer thickness and lower boundary for the 2D/3D transition, the InGaN material systems resembles very much the behavior of the well-studied InGaAs/GaAs system [6].

Proper MOVPE growth conditions for QD formation are different from those reported for MBE. Foremost, growth temperatures ranging from 600 to 800°C

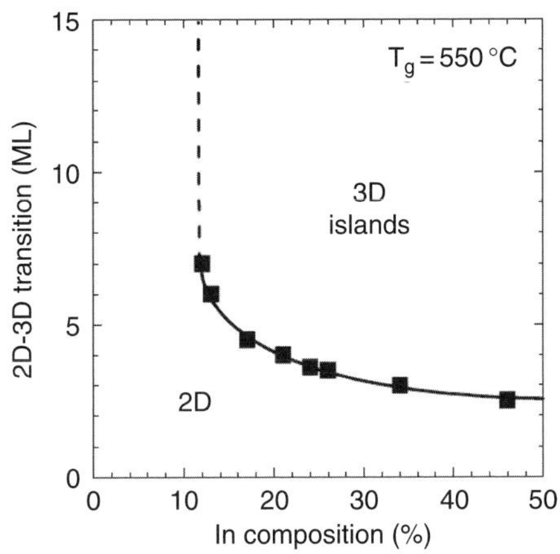

FIG. 5.12. Phase diagram for 2D/3D transition during growth of InGaN by MBE. (Reprinted with permission from [42], © 1998 American Institute of Physics.)

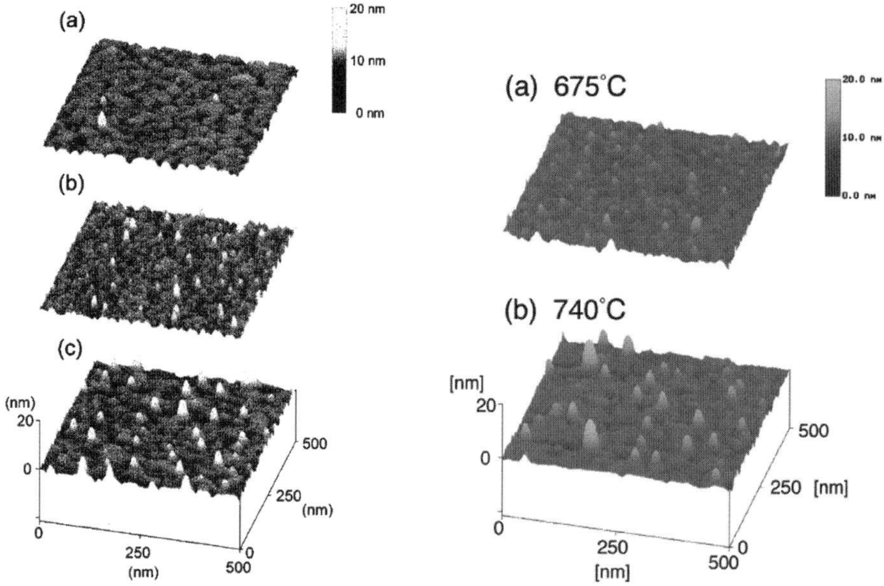

FIG. 5.13. Left) Surface morphologies taken by AFM after MOVPE growth of InGaN on GaN(0001) surfaces for InGaN coverages of a) 5.3, b) 10.4, and c) 18.4 MLs. Right) Reduction of the QD density with increasing temperature for an nominal coverage of 19 ML InGaN. (Reprinted with permission from [45], © 1999 American Institute of Physics.)

and V/III ratios of several thousands are applied owing to the low decomposition rate of ammonia (NH_3). These values are more alike quantum-well growth parameters, which is in contrast to InGaAs/GaAs(001), where lower growth temperatures and low V/III ratios are favored. Also, the critical layer thicknesses for QD formation of 5–10 MLs for In compositions of around 20–40% are significantly larger than for MBE (Fig. 5.13). In part, the latter deviations can be attributed to uncertainties due to *ex situ* calibration of growth rate and composition of InGaN layers for MOVPE. The right panel of Fig. 5.13 shows an example of the decrease of the QD density with increasing temperature. An exponential relation of the QD density with inverse temperature is measured [43, 44].

Typical dimensions of InGaN QDs are between 1 and 5 nm height and 5 and 30 nm base length both for MOVPE and MBE growth. The electronic confinement in the lateral direction may not be sufficient for discretization of energy levels because of the much smaller excitonic Bohr radius of only about 3.4 nm. This may also explain that the 2D/3D transition is not easily detected in the emission spectra as a transition from a spectrally narrow quantum-well emission to an inhomogeneously size-broadened QD ensemble luminescence.

Relatively high ammonia partial pressures are needed to stabilize InGaN surfaces against In droplet formation. The investigation of the impact of different V/III ratios on QD formation at growth temperatures of 755°C using InGaN layers with approximately 20–30% indium indicates that an optimum for the NH_3 flow rate (and hence for the V/III ratio) exists [46]. At very high ammonia flow rates, large three-dimensional objects can show up on the surface which are either relaxed clusters or do not provide three-dimensional quantum confinement for charge carriers. In contrast, very low ammonia flow rates may lead to suppression of the QD formation as In atoms may evaporate from the surface.

Growth interruptions (GRI) after deposition of the wetting layer are commonly used in MOVPE of InGaAs/GaAs(001) QDs to control average QD size and density of QD ensembles. Material transfer from the WL into QDs and from small to large QDs leads to red-shifted emission spectra with increasing growth interruption times. Simultaneously, the QD density gradually decreases due to the dissolution of smaller QDs. The evolution of size and density of InGaN QDs is affected by indium desorption effects as illustrated by Fig. 5.14. During ammonia-stabilized growth interruptions the size distributions at 60s GRI shows a reduction in height and width of the QDs and an increase in QD density as compared to 30s GRI. Indium desorption competes with the QD nucleation during the initial stage of the GRI. Afterwards, the QD density decreases and the QD sizes increase, indicating material transfer from small to large QDs. The continued blue-shift of the emission spectra indicate the desorption of In atoms into the gas phase also at this stage.

In order to provide sufficient optical gain for laser diode devices, stacking of QD planes need to be considered. Thus, overgrowth of QDs with GaN barrier layers is necessary which, due to the volatility of the In-N bond, cannot be performed at temperatures far above the QD growth temperatures. By choosing

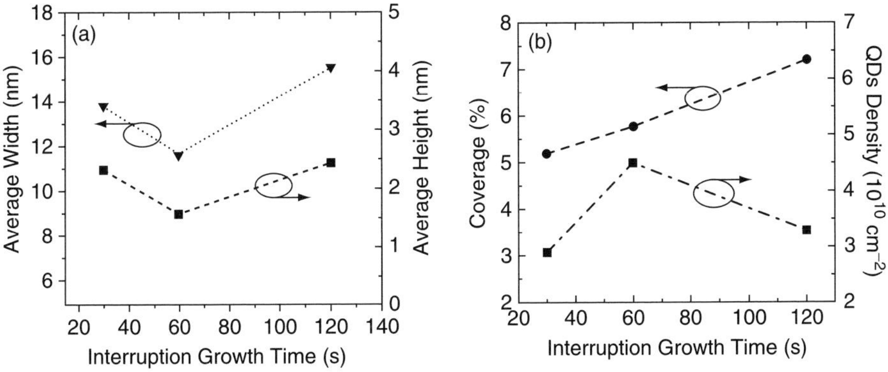

FIG. 5.14. Evolution of InGaN a) QD size and b) QD density for different duration of growth interruption during MOVPE. (Reprinted from [47] with permission from IOP.)

the same temperature for QDs and barrier layers, overgrowth of QDs is achieved for x_{In}<0.5 [42, 48, 49]. At very high In concentrations in the QDs, dissolution during overgrowth takes place both for MBE and for MOVPE [50]. Insight into vertically correlated growth of stacked InGaN quantum dot layers has not yet been gained.

5.3.3 *Spontaneous quantum dot formation in InGaN layers*

The high radiative efficiency of nitride-based light-emitting diodes (LED) despite high dislocation densities in heteroepitaxially grown layers is often attributed to localization centers within InGaN layers. Excitons in InGaN have a Bohr radius of around 3 nm, which approximately sets the length scale on which fluctuating properties of an InGaN layer may lead to quantum-dot-like carrier confinement. As outlined at the beginning of this chapter, compositional fluctuations in InGaN are promoted by In segregation but may also be driven by phase separation. Fluctuations within InGaN layers on the length scale of a few nanometers are often observed by cross-sectional transmission electron microscopy. One has argued that the TEM investigation itself artificially introduces such fluctuations if exposure times are taken too long (>1 min.), but time series show that fluctuations are resolved even with short exposure times (<1 min.) [53, 54]. Figure 5.15 shows a statistical analysis of In fluctuation and thickness deviations within InGaN/GaN samples grown by MOVPE on Si(111) substrates [51]. The analysis is based on cross-sectional TEM images taken at several locations across the sample. The average In composition and mean thickness of the InGaN layers were confirmed by X-ray diffraction measurements. As can be seen, both thickness and In concentration are widely scattered and the length scales are small enough to provide 3D localization.

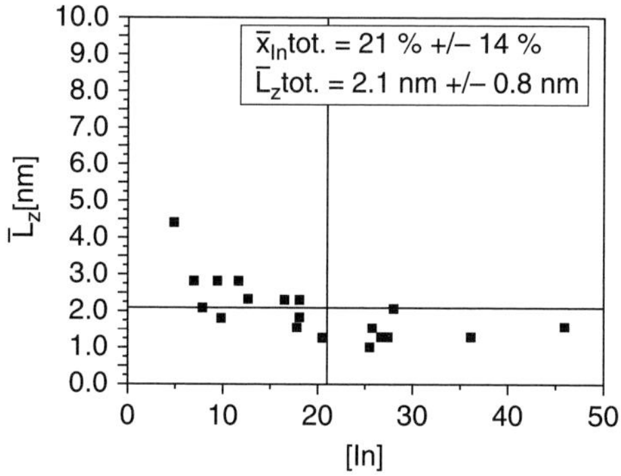

FIG. 5.15. Statistical analysis based on TEM investigations on different sample areas. Compositional fluctuations and thickness deviations from the average values are well resolved. (Reprinted from [51], © 2004, with permission from Elsevier.)

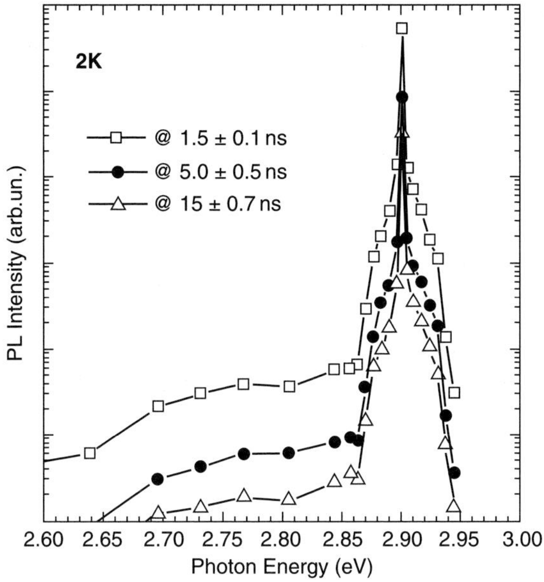

FIG. 5.16. Resonantly excited time-resolved photoluminescence spectra taken at different excitation power densities. The steady spectral distribution of the signal directly proves 3D carrier localization. (Reprinted with permission from [52], © The Japan Society of Applied Physics.)

Complementary results from luminescence experiments exhibiting sharp emission lines with emission linewidths of the order of 500 μeV confirm QD formation [55]. From time-resolved PL spectra shown in Fig. 5.16 an average size of the localization centers of 3–4 nm and a very broad energy distribution are confirmed [56–58].

Significant impact on the localization centers by post-growth annealing and growth interruptions has been identified [59]. Annealing of samples exhibiting pronounced fluctuations after growth at temperatures above 950°C is leading to the disappearance of In-rich clusters in InGaN/GaN-MQW layers [60]. Growth interruptions in an ammonia atmosphere after quantum well growth promote In desorption out of the layers. For short growth interruptions of 5–30s, excess In which is initially incorporated into clusters within the quantum well layers is removed and the in-plane localization of charge carriers is reduced, leading to decreasing photoluminescence intensity [61]. The magnitude of compositional fluctuations also increases with In content in the well layers [62]. Especially for In concentration $x_{In} = 0.3$, quantum dot formation is evidenced by numerous means. Temperature-dependent photoluminescence measurements indicate a tendency towards enhanced in-plane localization with increasing well thickness from 1.2 to 3.9 nm. Corresponding activation energies for carrier delocalization are 27.1 meV, 31 meV, and 35 meV [63].

5.3.4 *Thermal annealing and surface pre-treatment methods*

Annealing of InGaN layers is also used to initiate quantum dot formation. After deposition of a thin $In_{0.20}Ga_{0.80}N$ layer annealing at the quantum well growth temperature in pure nitrogen atmosphere leads to In droplets on the surface. These In droplets may transform into InGaN quantum dots upon capping with GaN. The density of these droplets increases from $5 \cdot 10^9$ to $3 \cdot 10^{10}$ cm^{-2} with decreasing growth temperature of the cap layer between 900°C and 700°C. Using an annealing atmosphere of H_2/NH_3 a meander-like InGaN layer surface is created. These structures are attributed to additional etching effects which hydrogen has on InGaN layers. Similar structures are obtained for layers with very high In-content of x_{In} >90% when annealed at 700°C and 750°C, respectively. Spinodal decomposition is predicted to take place at such high In concentrations [39, 50]. By TEM analysis of uncapped InGaN layers thermally treated at 700°C and 750°C, large truncated pyramids and spherical islands with 20 nm height and 40–50 nm base length are found besides flat structures of 1–1.5 nm height. High In concentrations of 80–90% are measured for the large pyramidal and spherical objects, whereas only 20% In is remaining in the flat structures. Upon overgrowth, the large objects dissolve and the excess In may form a quantum well layer with 6–7% In above the layer containing the flat meandering structures. Quantum dots are identified by sharp emission lines emerging in low-temperature luminescence spectra and by an increased radiative efficiency at room temperature.

Opposite to InGaAs/GaAs-based QDs which are very sensitive to surface recombination, *ex situ* surface patterning methods may be effective to fabricate luminescent InGaN quantum dot structures. This can be inferred from the relatively strong luminescence intensities of InGaN surfaces which point to low non-radiative surface recombination rates.

A very effective way of producing InGaN quantum dots is to pre-treat the growth surface prior to the InGaN growth. By supplying Si to the growth surface the growth mode for subsequent GaN and InGaN layers changes from 2D to 3D growth, thus leading to high-density arrays of quantum dots (see Fig. 5.17) [16, 48]. Originally, an anti-surfactant effect by silicon atoms suppressing In diffusion on the surface has been assumed. More probable is the formation of SiN_x nano-masks on the surface, which lead to selected area growth within the openings of the nano-masks [49, 64]. This is further supported by growth experiments in which Si and ammonia are simultaneously fed into the growth chamber to intentionally allow for SiN_x formation on the growth surface. Similar three-dimensional growth modes are obtained for appropriate exposure times, and thin SiN_x-layers are in such cases identified beneath the InGaN quantum dots. For extended Si and ammonia treatment times InGaN growth can be completely suppressed [49]. Very high quantum dot densities approaching 10^{11} cm^{-2} are realized with Si pre-treatment. A disadvantage of the approach is related to the Si donor properties in the nitride material system. Due to the spatial and temporal proximity of the Si treatment to the quantum dot growth it is very

FIG. 5.17. AFM image of InGaN quantum dots grown by MOVPE on Si pre-treated GaN surfaces. (Reprinted with permission from [16], © 1996 American Institute of Physics.)

likely that Si atoms will incorporate into quantum dots where Si would act as a charged impurity state. Conclusive evidence that Si atoms are not incorporated into QDs are not reported so far.

Another approach of surface pre-treatment is wet-chemical removal of SiO_2 masks completely covering a GaN surface. These masks are deposited *ex situ* without any subsequent patterning. Direct growth of InGaN layers on such surfaces results in 3D growth and QD ensembles with a density of $9{\cdot}10^{10}$ cm^{-2} result. As Fig. 5.18 shows, average lateral QD size is increasing, while density decreases gradually with increasing deposition times, indicating coalescence of the islands.

5.3.5 *InN quantum dots*

InN has become a topic of interest for its band gap, being as small as 0.6–0.7 eV [66]. Potential application as active material for the important 1.3–1.55 μm wavelength region as well as in solar energy conversion using solar cells is expected if InN growth could be sufficiently mastered. Unfortunately, c-plane InN is highly lattice-mismatched by >10% to GaN(0001) and AlN(0001) lattice planes, making it very susceptible to plastic lattice relaxation. Otherwise the Stranski–Krastanow regime benefits from such strain values, and growth of coherently strained optically active quantum dots might be possible. The fairly low dissociation temperature for InN of around 550°C sets upper boundaries on growth temperature making growth of smooth heterostructures even more complicated.

Using MBE high-density ($4{\cdot}10^{11}$cm^{-2}) arrays of InN quantum dots are grown for InN layer thicknesses ≥4 ML and growth temperatures around

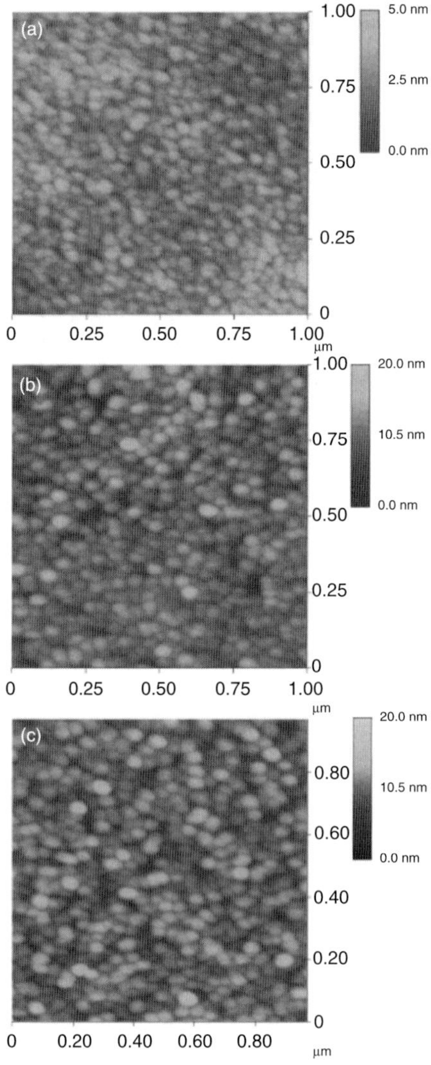

FIG. 5.18. AFM images after InGaN growth for a) 60s, b) 120s, and c) 180s on top of SiO_2 treated GaN. (Reprinted with permission from [65], © 2006 American Institute of Physics.)

350–400°C [67]. Through an *in vacuo* surface characterization by STM a base length and dot height of 4–7 nm and 0.75-1.8 nm is measured, respectively, depending on the InN layer thickness (4–8 ML). Significantly larger dimensions (10 nm height, 30 nm base length) are observed when the material is grown using a plasma-assisted nitrogen source, indicating a sensitive dependence of the

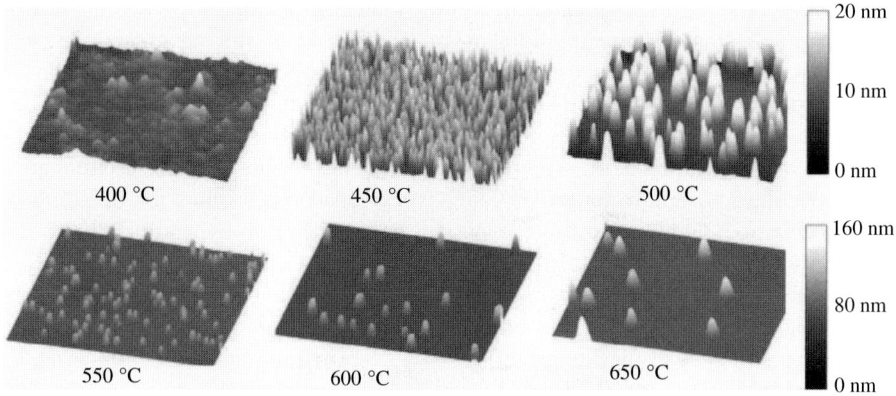

FIG. 5.19. Dependence of InN QD density on growth temperature during MOVPE growth (AFM images). (Reprinted with permission from [70], © 2006 American Institute of Physics.)

growth chemistry on V/III ratio. Another study monitoring oscillations of the characteristic RHEED pattern indicates a critical layer thickness for the 2D/3D transition of about 2 MLs [68]. In the same study, however, plastic relaxation after the first InN monolayer is noticed, i.e., prior to the 2D/3D transition. The case for MOVPE-grown InN quantum dots is a little different, as the precursor pyrolysis of trimethylindium and in particular ammonia requires temperatures well above 400°C. Still, quantum dot structures of roughly 25 nm diameter and 3–6 nm height are realized at 550°C temperature and very high nominal V/III ratios of 15000 [69]. Note that the actual V/III ratio is much lower, as decomposition of ammonia at 550°C amounts to only about 0.01%. In the nitrogen carrier gas, quantum dot densities are of the order of 10^7–10^8 cm^{-2}. An argon carrier gas atmosphere allows reduction of the nominal V/III ratio to 5000, while at the same time the QD density is increased to $5{\cdot}10^{10}$ cm^{-2} [71]. Unfortunately, structural analysis by TEM reveal misfit dislocation networks at the interface between MOCVD-grown InN layers and the underlying GaN buffer layer [72]. Photoluminescence signals detected at energies around 950 meV, besides being weak, show no correlation to characteristic quantum dot properties such as size and barrier material, which underlines the still persisting difficulties to realize the potential of InN for optoelectronic applications [70].

5.4 Site-selective growth

Deterministically addressable individual quantum dots are of major interest for devices operating under the constraints of purely quantum-mechanically defined electronic states like in atoms. Nitride-based quantum dots are promising sources for single photon emission due to large conduction and valence band offsets between AlN-GaN-InN, which may yield large confinement potentials and energy level splittings larger than the kinetic energy at room temperature [73, 74]. Nucleation of single quantum dots at pre-defined sites during growth is possible only if the surface free energy is locally altered, whereby nucleation of atomic species such as In or Ga atoms is attractive at those sites. As the base size of a quantum dot takes on only a few nanometers, the surface free energy has to be changed on about the same length scale. Many technologies exist nowadays to create nanometer-sized patterns on a surface for ordering of nanostructures [75]. For example, electron-beam lithography, nano-imprinting, and focused ion beam implantation were successfully applied for site-controlled growth of InGaAs/GaAs quantum dots [4]. However, one problem of such surface patterning methods is their limited reach of influence being restricted to a few nanometers. In case of nitride growth the short-range impact raises concerns, as growth on lattice-mismatched substrates yields huge densities of threading dislocations which act also as nucleation sites by locally widening the lattice [11]. Also, emission-line broadening by spectral diffusion arising from defects in the vicinity of QDs can be reduced by burying the usually contaminated initial growth surface by a few hundred nm thick buffer layer.

Most approaches for site-controlled quantum dot growth in the nitrides are utilizing, therefore, selective area epitaxy under growth conditions which result in pyramidal cones with very sharp, nanometer-sized apexes and $\{10\bar{1}1\}$ smooth side facets [76, 78–82]. MOVPE is preferred over MBE because of the larger diffusion lengths attainable on the masked areas. After growing nominally 30 nm of GaN onto a SiO_2 masked template with sub-μm openings pyramids, as shown in Fig. 5.20, can be obtained [76]. Suitable growth conditions for the

FIG. 5.20. SEM image of hexagonal pyramids grown on an SiO_2-masked GaN(0001)/sapphire template showing well-defined side facets and sharp apexes. Quantum dots are formed on top of the apexes. (Reprinted with permission from [76], © 2005 American Institute of Physics.)

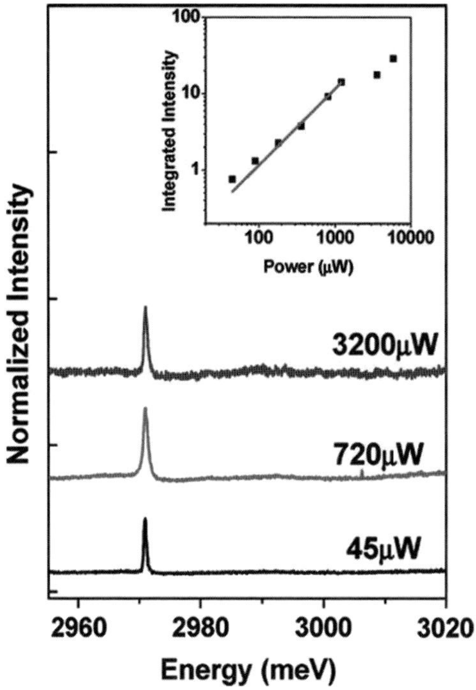

FIG. 5.21. Excitation-power-dependent single QD photoluminescence spectra. The inset reveals a linear relationship on a log–log scale indicating excitonic emission. (Reprinted with permission from [77], © 2011 American Chemical Society.)

formation of GaN hexagonal pyramids favor growth in the [0001] direction and suppress lateral adatom diffusion on the (0001) plane. Growth temperatures of around 900°C yield smooth GaN(0001) surfaces [83, 84]. On top of hexagonal GaN pyramids, quantum dots can be formed by depositing thin layers of InGaN. The nanometer-scaled size of the apexes naturally leads to three-dimensionally confined quantum-dot structures without being forced to fulfill the conditions of the SK regime. Excitonic emission of single quantum dots located at the apexes of the pyramids is demonstrated as shown in Fig. 5.21 [77]. Linewidths as small as 350 μeV at temperatures of 5 K are measured, confirming comparable optical quality to QDs on planar substrates.

5.5 Summary

Many techniques have been developed to engineer quantum dots in the nitride material systems. Strain-driven formation, the realm of the Stranski–Krastanow growth regime, is successfully used for GaN/AlN-, InGaN/GaN-, and InN/GaN-QDs. Other approaches, such as droplet epitaxy or thermal annealing, are also applicable. Still a matter of debate, spontaneous QD formation within InGaN needs further investigation, but signatures of its existence are found in transmission electron microscopy as well as in photoluminescence experiments. As a possible direction towards higher radiative efficiencies, non-polar and semi-polar quantum dots can be realized by Stranski–Krastanow growth, which results in a greatly reduced impact of polarization fields on optical properties of the QDs.

Light-emitting devices (LED) based either on spontaneously formed localization centers or by intentionally employing SK-QDs are demonstrated [85, 86]. Especially in the green spectral region, where the high In content will drive the Stranski–Krastanow growth mode, surface segregation, and decomposition, respectively, quantum-dot formation is important to understand. Laser action under optical pumping is also achieved, but electrical operation is yet to be demonstrated [44].

Future research may be directed towards electrically driven single-photon emitters based on nitrides. First, techniques are developed to control the nucleation site of a single quantum dot on the apex of pyramids. These concepts need to be further developed to integrate those single dots into pn junctions, for example, and attention must be given to achieve narrow emission lines well below 100 μeV. This may require reduction of the size of quantum dots even further than the size obtained by Stranski–Krastanow growth or droplet epitaxy.

References

[1] Y. Arakawa and H. Sakaki, "Multidimensional quantum well laser and temperature dependence of its threshold current," *Applied Physics Letters*, vol. 40, p. 939, June 1982.

[2] D. Bimberg, M. Grundmann, and N. N. Ledentsov, *Quantum Dot Heterostructures.* Wiley-VCH, 1998.

[3] V. Shchukin, N. N. Ledentsov, and D. Bimberg, *Epitaxy of Nanostructures.* Springer, NanoScience and Technology, 2003.

[4] O. G. Schmidt, *Lateral Alignment of Epitaxial Quantum Dots*, vol. 2010. Springer, NanoScience and Technology, 2010.

[5] N. Kirstaedter, N. Ledentsov, M. Grundmann, D. Bimberg, V. Ustinov, S. Ruvimov, M. Maximov, P. Kop'ev, Z. Alferov, U. Richter, P. Werner, U. Gosele, and J. Heydenreich, "Low threshold, large To injection laser emission from (InGa)As quantum dots," *Electronics Letters*, vol. 30, no. 17, p. 1416, 1994.

[6] U. W. Pohl and A. Strittmatter, "Control of Self-Organized In (Ga) As / GaAs Quantum Dot Growth," in *Semiconductor Nanostructures* (D. Bimberg, ed.), ch. 2, pp. 41–65, Springer-Verlag, NanoScience and Technology, 2008.

[7] B. Daudin, F. Widmann, G. Feuillet, Y. Samson, M. Arlery, and J. Rouvière, "Stranski–Krastanov growth mode during the molecular beam epitaxy of highly strained GaN," *Physical Review B*, vol. 56, pp. R7069–R7072, Sept. 1997.

[8] B. Damilano, N. Grandjean, F. Semond, J. Massies, and M. Leroux, "Violet to Orange Room Temperature Luminescence from GaN Quantum Dots on Si(111) Substrates," *physica status solidi (b)*, vol. 216, pp. 451–455, Nov. 1999.

[9] N. Gogneau, D. Jalabert, E. Monroy, T. Shibata, M. Tanaka, and B. Daudin, "Structure of GaN quantum dots grown under modified Stranski–Krastanow conditions on AlN," *Journal of Applied Physics*, vol. 94, p. 2254, Aug. 2003.

[10] C. Adelmann, N. Gogneau, E. Sarigiannidou, J.-L. Rouviere, and B. Daudin, "GaN islanding by spontaneous rearrangement of a strained two-dimensional layer on (0001) AlN," *Applied Physics Letters*, vol. 81, p. 3064, Oct. 2002.

[11] F. Widmann, B. Daudin, G. Feuillet, Y. Samson, J. L. Rouviere, and N. Pelekanos, "Growth kinetics and optical properties of self-organized GaN quantum dots," *Journal of Applied Physics*, vol. 83, p. 7618, June 1998.

[12] N. Gogneau, F. Fossard, E. Monroy, S. Monnoye, H. Mank, and B. Daudin, "Effects of stacking on the structural and optical properties of self-organized GaN/AlN quantum dots," *Applied Physics Letters*, vol. 84, no. 21, p. 4224, 2004.

[13] N. Gogneau, F. Fossard, E. Monroy, S. Monnoye, H. Mank, and B. Daudin, "Effects of stacking on the structural and optical properties of self-organized GaN/AlN quantum dots," *Applied Physics Letters*, vol. 84, p. 4224, May 2004.

[14] J. Coraux, H. Renevier, V. Favre-Nicolin, G. Renaud, and B. Daudin, "In situ resonant x-ray study of vertical correlation and capping effects during GaNAlN quantum dot growth," *Applied Physics Letters*, vol. 88, p. 153125, Apr. 2006.

[15] B. Daudin, "Polar and nonpolar GaN quantum dots," *Journal of Physics: Condensed Matter*, vol. 20, p. 473201, Nov. 2008.

[16] S. Tanaka, S. Iwai, and Y. Aoyagi, "Self-assembling GaN quantum dots on AlxGa1xN surfaces using a surfactant," *Applied Physics Letters*, vol. 69, no. 26, p. 4096, 1996.

[17] M. Miyamura, K. Tachibana, and Y. Arakawa, "High-density and size-controlled GaN self-assembled quantum dots grown by metalorganic chemical vapor deposition," *Applied Physics Letters*, vol. 80, p. 3937, May 2002.

[18] M. Miyamura, K. Tachibana, T. Someya, and Y. Arakawa, "Stranski–Krastanow growth of GaN quantum dots by metalorganic chemical vapor deposition," *Journal of Crystal Growth*, vol. 237–239, pp. 1316–1319, Apr. 2002.

[19] K. Hoshino, S. Kako, and Y. Arakawa, "Formation and optical properties of stacked GaN self-assembled quantum dots grown by metalorganic chemical vapor deposition," *Applied Physics Letters*, vol. 85, p. 1262, Aug. 2004.

[20] F. Bernardini, V. Fiorentini, and D. Vanderbilt, "Spontaneous polarization and piezoelectric constants of III–V nitrides," *Physical Review B*, vol. 56, pp. R10024–R10027, Oct. 1997.

[21] A. E. Romanov, T. J. Baker, S. Nakamura, and J. S. Speck, "Strain-induced polarization in wurtzite III-nitride semipolar layers," *Journal of Applied Physics*, vol. 100, p. 023522, July 2006.

[22] A. Strittmatter, M. Teepe, C. Knollenberg, and N. Johnson, "Coalescence during epitaxial lateral overgrowth of (Al,Ga)N(11.2) layers," *Journal of Crystal Growth*, vol. 314, pp. 1–4, Jan. 2011.

[23] B. Damilano, N. Grandjean, F. Semond, J. Massies, and M. Leroux, "From visible to white light emission by GaN quantum dots on Si(111) substrate," *Applied Physics Letters*, vol. 75, p. 962, Aug. 1999.

[24] S. Founta, F. Rol, E. Bellet-Amalric, J. Bleuse, B. Daudin, B. Gayral, H. Mariette, and C. Moisson, "Optical properties of GaN quantum dots grown on nonpolar (11-20) SiC by molecular-beam epitaxy," *Applied Physics Letters*, vol. 86, p. 171901, Apr. 2005.

[25] N. Garro, A. Cros, J. A. Budagosky, A. Cantarero, A. Vinattieri, M. Gurioli, S. Founta, H. Mariette, and B. Daudin, "Reduction of the internal electric field in wurtzite a-plane GaN self-assembled quantum dots," *Applied Physics Letters*, vol. 87, p. 011101, June 2005.

[26] B. Amstatt, J. Renard, C. Bougerol, E. Bellet-Amalric, B. Gayral, and B. Daudin, "Growth of m-plane GaN quantum wires and quantum dots on m-plane 6H-SiC," *Journal of Applied Physics*, vol. 102, p. 074913, Oct. 2007.

[27] X. Yang, M. Arita, S. Kako, and Y. Arakawa, "Formation of m-plane InGaN/GaN quantum dots using strain engineering of AlGaN/AlN interlayers," *Applied Physics Letters*, vol. 99, no. 6, p. 061914, 2011.

[28] S. Founta, C. Bougerol, H. Mariette, B. Daudin, and P. Vennegues, "Anisotropic morphology of nonpolar a-plane GaN quantum dots and quantum wells," *Journal of Applied Physics*, vol. 102, p. 074304, Oct. 2007.

[29] J.-L. Rouviere, C. Bougerol, B. Amstatt, E. Bellet-Almaric, and B. Daudin, "Measuring local lattice polarity in AlN and GaN by high resolution Z-contrast imaging: The case of (0001) and (1100) GaN quantum dots," *Applied Physics Letters*, vol. 92, p. 201904, May 2008.

[30] L. Lahourcade, S. Valdueza-Felip, T. Kehagias, G. P. Dimitrakopulos, P. Komninou, and E. Monroy, "Stranski–Krastanow growth of (1122)-oriented GaN/AlN quantum dots," *Applied Physics Letters*, vol. 94, p. 111901, Mar. 2009.

[31] A. Das, P. Sinha, Y. Kotsar, P. Kandaswamy, G. Dimitrakopulos, T. Kehagias, P. Komninou, G. Nataf, P. De Mierry, and E. Monroy, "Growth and characterization of polar (0001) and semipolar (1122) InGaN/GaN quantum dots," *Journal of Crystal Growth*, vol. 323, pp. 161–163, May 2011.

[32] G. P. Dimitrakopulos, E. Kalesaki, J. Kioseoglou, T. Kehagias, A. Lotsari, L. Lahourcade, E. Monroy, I. Hausler, H. Kirmse, W. Neumann, G. Jurczak, T. D. Young, P. Duzewski, P. Komninou, and T. Karakostas, "Morphology and strain of self-assembled semipolar GaN quantum dots in (1122) AlN," *Journal of Applied Physics*, vol. 108, p. 104304, Nov. 2010.

[33] A. Kahouli, N. Kriouche, J. Brault, B. Damilano, P. Vennegues, P. de Mierry, M. Leroux, A. Courville, O. Tottereau, and J. Massies, "GaN/Al0.5Ga0.5N (11-22) semipolar nanostructures: A way to get high luminescence efficiency in the near ultraviolet range," *Journal of Applied Physics*, vol. 110, p. 084318, Oct. 2011.

[34] J. Mei, S. Srinivasan, R. Liu, F. A. Ponce, Y. Narukawa, and T. Mukai, "Prismatic stacking faults in epitaxially laterally overgrown GaN," *Applied Physics Letters*, vol. 88, p. 141912, Apr. 2006.

[35] M. D. Craven, S. H. Lim, F. Wu, J. S. Speck, and S. P. DenBaars, "Structural characterization of nonpolar (1120) a-plane GaN thin films grown on (1102) r-plane sapphire," *Applied Physics Letters*, vol. 81, p. 469, July 2002.

[36] A. E. Romanov, E. C. Young, F. Wu, A. Tyagi, C. S. Gallinat, S. Nakamura, S. P. DenBaars, and J. S. Speck, "Basal plane misfit dislocations and stress relaxation in III-nitride semipolar heteroepitaxy," *Journal of Applied Physics*, vol. 109, p. 103522, May 2011.

[37] B. Amstatt, O. Landre, V. F. Nicolin, M. G. Proietti, E. Bellet-Amalric, C. Bougerol, H. Renevier, and B. Daudin, "Anisotropic strain state of the [1100] GaN quantum dots and quantum wires," *Journal of Applied Physics*, vol. 104, p. 063521, Sept. 2008.

[38] T. Saito and Y. Arakawa, "Atomic structure and phase stability of In," *Physical Review B*, vol. 60, no. 3, pp. 1701–1706, 1999.

[39] C. Tessarek, S. Figge, T. Aschenbrenner, S. Bley, A. Rosenauer, M. Seyfried, J. Kalden, K. Sebald, J. Gutowski, and D. Hommel, "Strong phase separation of strained In_{x}Ga_{1-x}N layers due to spinodal and binodal decomposition: Formation of stable quantum dots," *Physical Review B*, vol. 83, Mar. 2011.

[40] S. Inahama, T. Akiyama, K. Nakamura, and T. Ito, "Theoretical Investigation of Indium Surface Segregation in InGaN Thin Films," *e-Journal of Surface Science and Nanotechnology*, vol. 3, no. December, pp. 503–506, 2005.

[41] J. E. Northrup, L. T. Romano, and J. Neugebauer, "Surface energetics, pit formation, and chemical ordering in InGaN alloys," *Applied Physics Letters*, vol. 74, no. 16, pp. 2319–2321, 1999.

[42] N. Grandjean and J. Massies, "Real time control of InxGa1-xN molecular beam epitaxy growth," *Applied Physics Letters*, vol. 72, pp. 1078–1080, March 1998.

[43] Y. Arakawa, "Progress in Growth and Physics of Nitride-Based Quantum Dots," *physica status solidi (a)*, vol. 188, no. 1, pp. 37–45, 2001.

[44] Y. Arakawa, "Progress in GaN-based quantum dots for optoelectronics applications," *IEEE Journal of Selected Topics in Quantum Electronics*, vol. 8, pp. 823–832, July 2002.

[45] K. Tachibana, T. Someya, and Y. Arakawa, "Nanometer-scale InGaN self-assembled quantum dots grown by metalorganic chemical vapor deposition," *Applied Physics Letters*, vol. 74, p. 383, Jan. 1999.

[46] Q. Wang, T. Wang, J. Bai, A. G. Cullis, P. J. Parbrook, and F. Ranalli, "Growth and optical investigation of self-assembled InGaN quantum dots on a GaN surface using a high temperature AlN buffer," *Journal of Applied Physics*, vol. 103, no. 12, p. 123522, 2008.

[47] H. H. Yao, T. C. Lu, G. S. Huang, C. Y. Chen, W. D. Liang, H. C. Kuo, and S. C. Wang, "InGaN self-assembled quantum dots grown by metal–organic chemical vapour deposition with growth interruption," *Nanotechnology*, vol. 17(6), pp. 1713–1716, 2006.

[48] H. Hirayama, S. Tanaka, P. Ramvall, and Y. Aoyagi, "Intense photoluminescence from self-assembling InGaN quantum dots artificially fabricated on AlGaN surfaces," *Applied Physics Letters*, vol. 72, no. 14, p. 1736, 1998.

[49] R.-C. Tu, C.-J. Tun, C.-C. Chuo, B.-C. Lee, C.-E. Tsa, T.-C. Wang, C. Jim, C.-P. Lee, and G.-C. Chi, "Ultra-high-density InGaN quantum dots grown by metalorganic chemical vapor deposition," *Japanese Journal of Applied Physics*, vol. 43, no. 2B, pp. L264–L266, 2004.

[50] S. Figge, C. Tessarek, T. Aschenbrenner, and D. Hommel, "InGaN quantum dot growth in the limits of Stranski–Krastanov and spinodal decomposition," *physica status solidi (b)*, vol. 248, pp. 1765–1776, Aug. 2011.

[51] A. Strittmatter, L. Reißmann, R. Seguin, S. Rodt, A. Brostowski, U. Pohl, D. Bimberg, E. Hahn, and D. Gerthsen, "Influence of the reactor total pressure on optical properties of MOCVD grown InGaN layers," *Journal of Crystal Growth*, vol. 272, pp. 415–419, Dec. 2004.

[52] I. L. Krestnikov, M. Strassburg, A. Strittmatter, N. N. Ledentsov, J. Christen, A. Hoffmann, and D. Bimberg, "Direct Evidence of Nanoscale Carrier Localization in InGaN/GaN Structures Grown on Si Substrates," *Japanese Journal of Applied Physics*, vol. 42, pp. L1057–L1060, Sept. 2003.

[53] T. M. Smeeton, M. J. Kappers, J. S. Barnard, M. E. Vickers, and C. J. Humphreys, "Electron-beam-induced strain within InGaN quantum wells: False indium cluster detection in the transmission electron microscope," *Applied Physics Letters*, vol. 83, p. 5419, Dec. 2003.

[54] T. Li, E. Hahn, D. Gerthsen, A. Rosenauer, A. Strittmatter, L. Reißmann, and D. Bimberg, "Indium redistribution in an InGaN quantum well induced by electron-beam irradiation in a transmission electron microscope," *Applied Physics Letters*, vol. 86, no. 24, p. 241911, 2005.

[55] R. Seguin, A. Schliwa, T. D. Germann, S. Rodt, K. Potschke, A. Strittmatter, U. W. Pohl, D. Bimberg, M. Winkelnkemper, T. Hammerschmidt, and P. Kratzer, "Control of fine-structure splitting and excitonic binding energies in selected individual InAsGaAs quantum dots," *Applied Physics Letters*, vol. 89, no. 26, p. 263109, 2006.

[56] T. Bartel, M. Dworzak, M. Strassburg, A. Hoffmann, A. Strittmatter, and D. Bimberg, "Recombination dynamics of localized excitons in InGaN quantum dots," *Applied Physics Letters*, vol. 85, no. 11, p. 1946, 2004.

[57] M. Dworzak, T. Bartel, M. Strasburg, I. Krestnikov, A. Hoffmann, R. Seguin, S. Rodt, A. Strittmatter, and D. Bimberg, "Optical properties of InGaN quantum dots," *Superlattices and Microstructures*, vol. 36, pp. 763–772, Dec. 2004.

[58] M. Winkelnkemper, R. Seguin, S. Rodt, A. Schliwa, L. Reißmann, A. Strittmatter, A. Hoffmann, and D. Bimberg, "Polarized emission lines from A- and B-type excitonic complexes in single InGaN/GaN quantum dots," *Journal of Applied Physics*, vol. 101, no. 11, p. 113708, 2007.

[59] R. A. Oliver, G. A. D. Briggs, M. J. Kappers, C. J. Humphreys, S. Yasin, J. H. Rice, J. D. Smith, and R. A. Taylor, "InGaN quantum dots grown by metalorganic vapor phase epitaxy employing a post-growth nitrogen anneal," *Applied Physics Letters*, vol. 83, no. 4, p. 755, 2003.

[60] C.-C. Chuo, C.-M. Lee, T.-E. Nee, and J.-I. Chyi, "Effects of thermal annealing on the luminescence and structural properties of high indium-content InGaN/GaN quantum wells," *Applied Physics Letters*, vol. 76, p. 3902, June 2000.

[61] H. K. Cho, J. Y. Lee, N. Sharma, C. J. Humphreys, G. M. Yang, C. S. Kim, J. H. Song, and P. W. Yu, "Effect of growth interruptions on the light emission and indium clustering of InGaN/GaN multiple quantum wells," *Applied Physics Letters*, vol. 79, no. 16, p. 2594, 2001.

[62] M. G. Cheong, E.-K. Suh, H. J. Lee, and M. Dawson, "Growth and properties of InGaN/GaN quantum wells and blue light emitting diodes by metalorganic

chemical vapour deposition," *Semiconductor Science and Technology*, vol. 17, pp. 446–452, May 2002.

[63] S. C. Davies, D. J. Mowbray, Q. Wang, F. Ranalli, and T. Wang, "Influence of crystal quality of underlying GaN buffer on the formation and optical properties of InGaN/GaN quantum dots," *Applied Physics Letters*, vol. 95, no. 10, p. 101909, 2009.

[64] H. L. Tsai, T. Y. Wang, J. R. Yang, T. C. Wang, J. T. Hsu, and M. Shiojiri, "Observation of ultrahigh density InGaN quantum dots," *Journal of Applied Physics*, vol. 102, no. 1, p. 013521, 2007.

[65] P. Chen, S. J. Chua, and J. N. Tan, "High-density InGaN nanodots grown on pretreated GaN surfaces," *Applied Physics Letters*, vol. 89, p. 023114, July 2006.

[66] V. Davydov, A. Klochikhin, R. Seisyan, V. Emtsev, S. Ivanov, F. Bechstedt, J. Furthmller, H. Harima, A. Mudryi, J. Aderhold, O. Semchinova, and J. Graul, "Absorption and Emission of Hexagonal InN. Evidence of Narrow Fundamental Band Gap," *physica status solidi (b)*, vol. 229, pp. r1–r3, Feb. 2002.

[67] C. Nörenberg, R. Oliver, M. Martin, L. Allers, M. Castell, and G. Briggs, "Stranski–Krastanov Growth of InN Nanostructures on GaN Studied by RHEED, STM and AFM," *physica status solidi (a)*, vol. 194, pp. 536–540, Dec. 2002.

[68] Y. F. Ng, Y. G. Cao, M. H. Xie, X. L. Wang, and S. Y. Tong, "Growth mode and strain evolution during InN growth on GaN(0001) by molecular-beam epitaxy," *Applied Physics Letters*, vol. 81, p. 3960, Nov. 2002.

[69] O. Briot, B. Maleyre, and S. Ruffenach, "Indium nitride quantum dots grown by metalorganic vapor phase epitaxy," *Applied Physics Letters*, vol. 83, p. 2919, Oct. 2003.

[70] S. Ruffenach, B. Maleyre, O. Briot, and B. Gil, "Growth of InN quantum dots by MOVPE," *physica status solidi (c)*, vol. 2, pp. 826–832, Feb. 2005.

[71] S. Ruffenach, O. Briot, M. Moret, and B. Gil, "Control of InN quantum dot density using rare gases in metal organic vapor phase epitaxy," *Applied Physics Letters*, vol. 90, p. 153102, Apr. 2007.

[72] J. G. Lozano, A. M. Sánchez, R. García, S. Ruffenach, O. Briot, and D. González, "Strain Relief Analysis of InN Quantum Dots Grown on GaN," *Nanoscale Research Letters*, vol. 2, pp. 442–446, Jan. 2007.

[73] R. Oliver, A. F. Jarjour, R. Taylor, A. Tahraoui, Y. Zhang, M. J. Kappers, and C. Humphreys, "Growth and assessment of InGaN quantum dots in a microcavity: A blue single photon source," *Materials Science and Engineering B*, vol. 147, no. 2-3, pp. 108–113, 2008.

[74] J. Kalden, C. Tessarek, K. Sebald, S. Figge, C. Kruse, D. Hommel, and J. Gutowski, "Electroluminescence from a single InGaN quantum dot in the green spectral region up to 150 K.," *Nanotechnology*, vol. 21, p. 015204, Jan. 2010.

[75] J. V. Barth, G. Costantini, and K. Kern, "Engineering atomic and molecular nanostructures at surfaces," *Nature*, vol. 437, pp. 671–679, Sept. 2005.

[76] V. Perez-Solorzano, A. Groning, M. Jetter, T. Riemann, and J. Christen, "Near-red emission from site-controlled pyramidal InGaN quantum dots," *Applied Physics Letters*, vol. 87, no. 16, p. 163121, 2005.

[77] C.-W. Hsu, A. Lundskog, K. F. Karlsson, U. Forsberg, E. Janzén, and P. O. Holtz, "Single excitons in InGaN quantum dots on GaN pyramid arrays," *Nano Letters*, vol. 11, pp. 2415–2418, June 2011.

[78] K. Tachibana, T. Someya, S. Ishida, and Y. Arakawa, "Selective growth of InGaN quantum dot structures and their microphotoluminescence at room temperature," *Applied Physics Letters*, vol. 76, no. 22, p. 3212, 2000.

[79] J. Wang, M. Nozaki, M. Lachab, Y. Ishikawa, R. S. Qhalid Fareed, T. Wang, M. Hao, and S. Sakai, "Metalorganic chemical vapor deposition selective growth and characterization of InGaN quantum dots," *Applied Physics Letters*, vol. 75, no. 7, pp. 950–952, 1999.

[80] N. Grandjean and M. Ilegems, "Visible InGaN/GaN Quantum-Dot Materials and Devices," *Proceedings of the IEEE*, vol. 95, no. 9, pp. 1853–1865, 2007.

[81] M. Lachab, M. Nozaki, J. Wang, Y. Ishikawa, Q. Fareed, T. Wang, T. Nishikawa, K. Nishino, and S. Sakai, "Selective fabrication of InGaN nanostructures by the focused ion beam/metalorganic chemical vapor deposition process," *Journal of Applied Physics*, vol. 87, p. 1374, Feb. 2000.

[82] P. R. Edwards, R. W. Martin, I. M. Watson, C. Liu, R. A. Taylor, J. H. Rice, J. H. Na, J. W. Robinson, and J. D. Smith, "Quantum dot emission from site-controlled InGaN/GaN micropyramid arrays," *Applied Physics Letters*, vol. 85, no. 19, p. 4281, 2004.

[83] M. Jetter, V. Perez-Solorzano, A. Groening, M. Übl, H. Graebeldinger, and H. Schweizer, "Selective growth of GaInN quantum dot structures," *Journal of Crystal Growth*, vol. 272, pp. 204–210, Dec. 2004.

[84] H. Miyake, A. Motogaito, and K. Hiramatsu, "Effects of Reactor Pressure on Epitaxial Lateral Overgrowth of GaN via Low-Pressure Metalorganic Vapor Phase Epitaxy," *Japanese Journal of Applied Physics*, vol. 38, pp. L1000–L1002, Sept. 1999.

[85] S. F. Chichibu, T. Azuhata, T. Sota, T. Mukai, and S. Nakamura, "Localized quantum well excitons in InGaN single-quantum-well amber light-emitting diodes," *Journal of Applied Physics*, vol. 88, p. 5153, Nov. 2000.

[86] M. Zhang, P. Bhattacharya, and W. Guo, "InGaN/GaN self-organized quantum dot green light emitting diodes with reduced efficiency droop," *Applied Physics Letters*, vol. 97, p. 011103, July 2010.

6 Properties of InAlN layers nearly lattice-matched to GaN and their use for photonics and electronics

Raphaël Butté, Gatien Cosendey, Lorenzo Lugani, Marlene Glauser, Antonino Castiglia, Guillaume Perillat-Merceroz, Jean-François Carlin, and Nicolas Grandjean

6.1 Introduction

The aim of the present chapter is to provide the readers with an update on recent advances regarding the growth, the characterization, and the use of InAlN layers nearly lattice-matched (LM) to GaN. This ternary alloy is the least well known of the III-nitride family, and despite growth attempts dating back to the early 1980s (Starosta, 1981) it was scarcely studied until 2003, when Carlin and Ilegems demonstrated the potential of such layers for photonics through the report of high-reflectivity crack-free InAlN/GaN distributed Bragg reflectors (DBRs) (Carlin and Ilegems, 2003). It was soon followed by several other applications taking advantage of the nearly strain-free character of the InAlN/GaN combination. We can thus cite among others (i) the realization of GaN/InAlN multiple quantum well (MQW) structures suitable for near-infrared intersubband applications (Nicolay *et al.*, 2005), (ii) the selective lateral oxidation of InAlN layers (Dorsaz *et al.*, 2005*a*) and their subsequent selective etching (Simeonov *et al.*, 2008) that can be exploited to achieve an efficient current confinement in III-N light-emitting devices (Dorsaz *et al.*, 2005*a*; Castiglia *et al.*, 2007) or to realize advanced photonic structures such as microdisks (Simeonov *et al.*, 2008) or air-gap Bragg mirrors (Altoukhov *et al.*, 2009), (iii) the characterization of LM GaN/InAlN heterojunctions exhibiting a high two-dimensional carrier density together with a high room-temperature mobility and a low sheet resistance (Gonschorek *et al.*, 2006) that have been successfully processed to demonstrate high electron mobility transistors (HEMTs) grown on SiC featuring a gain cutoff frequency f_t in excess of 200 GHz (Sun *et al.*, 2010*c*), (iv) and the realization of ultraviolet photodiodes (Senda *et al.*, 2008). Beyond those applications, this ternary alloy is still challenging the III-nitride community through several features. Thus, in Section 6.2, relevant information on the structural properties of bulk InAlN layers, including the growth of high-quality InAlN layers on freestanding GaN (FS-GaN) substrates, are given. In particular, the origin of the hillocks and the V-defects apparently inherent to InAlN layers grown on

AlN or GaN templates on *c*-plane sapphire (Aschenbrenner *et al.*, 2010; Butté *et al.*, 2007; Sadler *et al.*, 2009*a*; Sadler *et al.*, 2011*a*; Kehagias *et al.*, 2009; Kehagias *et al.*, 2011; Miao *et al.*, 2009; Miao *et al.*, 2010; Chen *et al.*, 2010; Mouti *et al.*, 2011; Kim-Chauveau *et al.*, 2011; Wei *et al.*, 2012; Chen *et al.*, 2012; Vennégués *et al.*, 2012), but also on FS-GaN, the different morphology of layers grown either by metalorganic vapor phase epitaxy (MOVPE) or by molecular beam epitaxy (MBE) (Zhou *et al.*, 2007; Sahonta *et al.*, 2009; Choi *et al.*, 2012), and the control of residual impurities such as carbon and oxygen (Chung *et al.*, 2011) are discussed. In Section 6.3, optical and electronic properties such as the significant scatter in the reported bowing parameter for the bandgap E_g (Aschenbrenner *et al.*, 2010; Wang *et al.*, 2008; Iliopoulos *et al.*, 2008; Jones *et al.*, 2008; Sakalauskas *et al.*, 2010), the anomalously large Stokes shift (S) that exceeds several hundreds of meV (Butté *et al.*, 2007; Wang *et al.*, 2008; Carlin *et al.*, 2005), the dispersion of the optical refractive index, and the valence-band offset value are discussed. In Section 6.4 we briefly account for the properties of GaN/InAlN quantum wells (QWs) and MQWs. In section 6.5 the successful use of LM InAlN/(Al)GaN bilayers for the realization of DBRs and microcavities is described. In Section 6.6 we focus on InAlN cladding layers employed for edge-emitting laser diodes (LDs). In Section 6.7 we detail the latest advances reported for nearly LM InAlN/GaN HEMTs. Finally, a general conclusion and perspectives for this topic are presented in Section 6.8.

6.2 Growth and structural properties of bulk InAlN layers

6.2.1 *Growth characteristics of InAlN films*

The growth of InAlN layers has been mainly studied using MOVPE, and on several aspects it remains a rather challenging task. First, the huge growth temperature difference between AlN (~1100°C) and InN (~600°C) requires an intermediate growth temperature for InAlN (~800°C), which strongly decreases the surface diffusion length of Al adatoms. Moreover, the large difference between the length and the energy of In-N and Al-N bonds is responsible for indium surface segregation (Massies *et al.*, 1987) and phase separation (Redondo-Cubero *et al.*, 2010), respectively. Whereas the former effect mainly affects superlattices including InAlN layers rather than thick single InAlN layers, as will be discussed hereafter in the DBR section, the latter, related to spinodal decomposition, can strongly affect the crystalline quality and the surface morphology of thick In-rich InAlN layers (Hums *et al.*, 2007).

The indium content fulfilling the LM condition between $In_xAl_{1-x}N$ and GaN is an important parameter which has generated great interest. The latter amounts to about 17% for structures grown on *c*-plane sapphire substrates (Lorenz *et al.*, 2006) and about 18% for those grown on *c*-plane FS-GaN substrates (Darakchieva *et al.*, 2008). This discrepancy is ascribed to the different residual strain state of GaN buffers grown on the two types of substrate. Furthermore, it is worth pointing out that an indium content satisfying the LM condition at room temperature for InAlN layers grown on both types of substrate will result in tensile strain at the InAlN growth temperature because of the difference between InAlN and GaN thermal expansion coefficients. This has to be kept in mind when growing InAlN-based structures, in order to avoid cracks and/or dislocation generation.

InAlN growth parameters have been widely studied during the past few years (Kim-Chauveau *et al.*, 2011; Chung *et al.*, 2011; Lorenz *et al.*, 2010; Sadler *et al.*, 2010; Ichikawa *et al.*, 2012; Liu *et al.*, 2012; Sadler *et al.*, 2012). As can be seen in Fig. 6.1, increasing the growth temperature strongly reduces the indium incorporation because of a large indium desorption rate (Kim-Chauveau *et al.*, 2011; Chung *et al.*, 2011; Lorenz *et al.*, 2010; Liu *et al.*, 2012). For a given temperature, increasing the indium precursor flux also favors indium incorporation, but only up to a given value (Sadler *et al.*, 2009*a*; Sadler *et al.*, 2011*a*; Chung *et al.*, 2011). Surprisingly, when increasing the aluminum precursor flux indium incorporation also increases, but without the saturation effect reported for the trimethylindium flow rate for the range of trimethylaluminum flow rates investigated (Chung *et al.*, 2011). This is ascribed to an increase in the InAlN growth rate, which results in reduced indium desorption. Introducing a small amount of H_2 has proven to be effective in order to improve the InAlN surface morphology, but an increased hydrogen flux also reduces indium incorporation (Kim-Chauveau *et al.*, 2011). Increasing the ammonia flux, and thus the V/III ratio, allows improving the surface roughness and decreasing the pit density (Kim-Chauveau *et al.*, 2011). At first sight, the influence of the nitrogen flux seems identical to that of NH_3 (Kim-Chauveau *et al.*, 2011). When increasing the growth pressure, the indium incorporation is improved, but this is done at

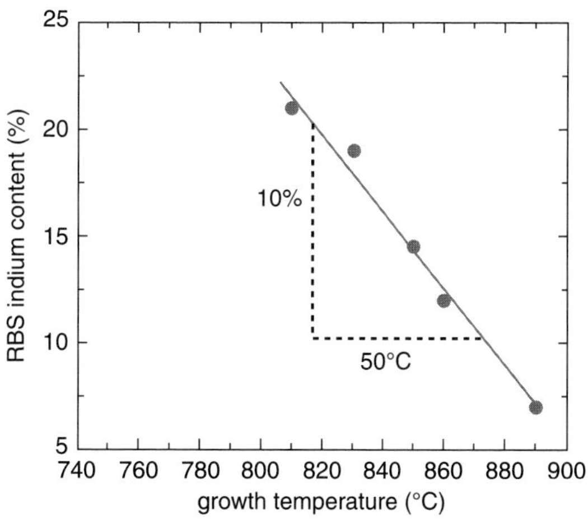

FIG. 6.1. Indium content measured by Rutherford backscattering spectrometry as a function of the InAlN growth temperature.

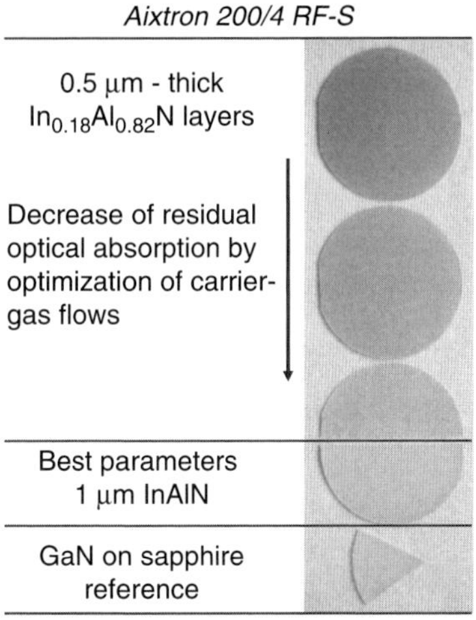

FIG. 6.2. Effect of the optimization of the carrier gas flows on the transparency of InAlN layers.

the expense of the surface roughness (Kim-Chauveau *et al.*, 2011). It has also been shown that no noticeable effect on the growth rate, the indium incorporation, and the surface morphology occurs for a substrate miscut angle varying between 0° and 0.5° (Sadler *et al.*, 2012). As shown in Fig. 6.2, the carrier gas flows have a great impact on the InAlN quality. When performing growth with a low carrier gas flow, InAlN layers LM to GaN, whose absorption edge should lie in the UV range (at ~4.45 eV), exhibit a brownish aspect, indicating that light absorption occurs at visible wavelengths. Eventually the transparency of the layers is improved, and thus the residual absorption is decreased when growth is carried out using a sufficiently high carrier gas flow.

Another aspect of relevant interest deals with the residual doping in nominally undoped InAlN layers, since the latter will strongly impact the ability to intentionally dope such an alloy, which is crucial if we wish to introduce *n*- or *p*-type doped layers in optoelectronic devices. Chung and coworkers have therefore studied the impurity incorporation, namely carbon and oxygen, and its dependence on the growth conditions of optimized InAlN layers, i.e., layers where no relaxation or degradation occurred, by performing secondary ion mass spectrometry measurements (Chung *et al.*, 2011). As far as carbon incorporation is concerned, it appears that [C] values in the 3–4 × 10^{17} cm^{-3} range are obtained for low growth pressures (<100 Torr). It was concluded that carbon mainly originates from the methyl groups from the decomposition of trimethylindium since AlN grown under conditions identical to InAlN layers was characterized by [C] values in the low 10^{17} cm^{-3}. In addition, for a given pressure, the InAlN alloy is more sensitive to carbon contamination compared with the InGaN one, probably due to the stronger binding energy between Al and C atoms vs. that between Ga and C ones. On the other hand, the incorporation of oxygen does not appear to be affected by growth parameters such as the growth pressure, the growth rate, the V/III ratio, and the addition of H_2 as carrier gas to N_2. In any case, InAlN layers exhibit high [O] values (in excess of 1×10^{18} cm^{-3}), which is ascribed to the strong affinity of oxygen atoms to aluminum ones, and the low growth temperatures.

6.2.2 *Structural properties of bulk InAlN layers*

As far as structural properties of InAlN layers are concerned, most of the published works deal with MOVPE-grown layers, but there are also some publications focusing on MBE-grown films. Therefore, we will first briefly describe a few features of InAlN layers grown by MBE before detailing the main structural specificities of MOVPE-grown layers reported in the literature.

MBE-grown InAlN layers LM to GaN. Optimum growth conditions have been described for layers deposited by plasma-assisted MBE (PAMBE) (Ive *et al.*, 2008; Fernández-Garrido *et al.*, 2008). Using transmission electron microcopy (TEM), it has been shown that InAlN layers grown on GaN are characterized by a honeycomb structure, which probably originates from the GaN/InAlN interface (Zhou *et al.*, 2007; Katzer *et al.*, 2005). Those InAlN layers are composed of cells five to ten nanometers in diameter with an indium content close to the LM composition, surrounded by In-rich vertical walls (In content ~30%). Sahonta *et al.* (2009) proposed as an origin for this peculiar structure that Al-rich islands would first form because of the joint low aluminum mobility and the In-desorption occurring under the growth conditions considered. Then, an aluminum adsorbed layer would form, facilitating the incorporation of indium, preferentially at the edges of the islands where the strain is partially relieved and where tensile strain may be present due to the coalescence of islands. This honeycomb structure would be kept during the subsequent growth of the InAlN layer. Note that such a honeycomb structure is inherent to all PAMBE-grown InAlN layers LM to GaN reported so far (Zhou *et al.*, 2007; Sahonta *et al.*, 2009; Choi *et al.*, 2012; Katzer *et al.*, 2005; Mánuel *et al.*, 2010; Gačević *et al.*, 2011; Dasgupta *et al.*, 2011). Interestingly, it was shown that when inserting a 1-nm thick AlN interlayer between the GaN and InAlN layers, the growth of a 60-nm thick $In_{0.18}Al_{0.82}N$ layer with a homogeneous composition is made possible (Mánuel *et al.*, 2010). Concerning ammonia-based MBE, thin homogeneous $In_{0.18}Al_{0.82}N$ layers about 25 nm thick have been obtained (Wong *et al.*, 2012).

MOVPE-grown InAlN layers LM to GaN: structural defects. Various types of defects have been reported for thick MOVPE-grown InAlN films. Cracks are usually observed for an indium composition lower than 17%, i.e., when the InAlN layers are under tensile strain (Sadler *et al.*, 2009*a*; Sadler *et al.*, 2011*a*;

Miao *et al.*, 2010). Metallic indium particles are sometimes seen at the surface, but they can be eliminated by an appropriate tuning of the growth parameters (Sadler *et al.*, 2009*a*; Sadler *et al.*, 2011*a*; Kim-Chauveau *et al.*, 2011; Sadler *et al.*, 2011*b*). The presence of so-called hillocks (whose height and diameter typically range from 1 to 3 nm and from 50 to 200 nm, respectively) is systematic, and it has been the subject of numerous publications in the recent literature (Sadler *et al.*, 2009*a*; Sadler *et al.*, 2011*a*; Miao *et al.*, 2010; Kim-Chauveau *et al.*, 2011; Chung *et al.*, 2011; Lorenz *et al.*, 2008). Their precise origin is still unclear. Indium segregation can also occur around threading dislocations. Finally, V-defects appear either on top of threading dislocations or without any dislocation from a certain critical InAlN thickness.

Hillocks. The formation of hillocks is observed by all the groups growing InAlN layers by MOVPE, whatever the substrate type (Sadler *et al.*, 2009*a*; Sadler *et al.*, 2011*a*; Miao *et al.*, 2010; Kim-Chauveau *et al.*, 2011; Chung *et al.*, 2011; Lorenz *et al.*, 2008). A typical atomic-force microscopy (AFM) image of a 50-nm thick $In_{0.2}Al_{0.8}N$ layer grown on a FS-GaN substrate with a dislocation density $\sim 4 \times 10^7$ cm^{-2} is shown in Fig. 6.3. Hillocks with a diameter of 50–100 nm and a height of $\sim$1 nm are aligned along atomic steps. The origin of these hillocks is not fully understood yet, but they might play a role regarding the optical properties of those layers as reported hereafter (Chen *et al.*, 2010).

Dislocations. In the relevant literature, growth is generally performed on (0001) GaN templates grown on sapphire. Consequently, threading dislocations are inevitably present in the templates, and they propagate in the InAlN layers. Indium segregation has been shown to occur around those dislocations, as confirmed by energy-dispersive X-ray spectrometry (EDX) (Song *et al.*, 2010;

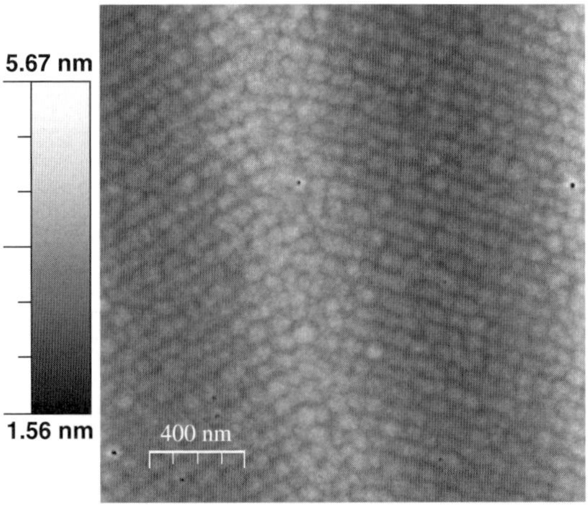

FIG. 6.3. AFM image of a 50 nm thick $In_{0.2}Al_{0.8}N$ layer grown on a FS-GaN substrate showing hillocks.

Kioseoglou *et al.*, 2011). Such In-rich dislocations are known to act as highly conductive channels, which are detrimental to the efficiency of devices (Song *et al.*, 2010; Kioseoglou *et al.*, 2011).

A quantitative study relying on TEM techniques such as Z-contrast imaging, strain measurement, and EDX, and also including a systematic comparison of experimental results with calculations using an elastostatic free-energy model, has been reported (Mouti *et al.*, 2011). Thus, for edge-type threading dislocations this model shows that one half of the (0001) plane is under tensile strain, while the second half is under compressive strain. Consequently, indium segregation should occur in the tensile region, because In atoms are bigger than Al atoms. This was indeed confirmed by high-resolution Z-contrast images showing that indium segregation occurs only on the tensile side of the dislocation core. Concerning screw dislocations, it is theoretically predicted that indium should segregate on the core of the dislocation because of the stress field. Experimentally, such an In-rich region was also observed exactly on the core of the dislocation on a high-resolution Z-contrast image. Using molecular dynamics calculations, it was also shown that indium should segregate around screw dislocations in the InAlN alloy (Lei *et al.*, 2010). Such segregation phenomena occurring on dislocations illustrate the importance of performing growth on FS-GaN substrates to minimize the formation of conductive channels that detrimentally affect the efficiency of optoelectronic devices.

V-defects formed on top of threading dislocations. V-defects (or V-pits) are inverted empty pyramids with a hexagonal base, which are often observed in In-based alloys such as InGaN and InAlN. As shown in numerous studies, dislocations with a Burgers vector with a screw component lead to the formation of a V-defect in InAlN films LM to GaN, whereas pure edge dislocations do not (Miao *et al.*, 2009; Mouti *et al.*, 2011; Song *et al.*, 2010). However, it was also observed that under specific conditions, V-defects can also appear on top of pure edge dislocations (Vennégués *et al.*, 2012). Indium segregation around the core of threading dislocations with a screw component could be responsible for the formation of V-defects as proposed for InGaN layers (Shiojiri *et al.*, 2006). Indium-rich regions would then favor the formation of inclined facets, as shown by first-principles calculations (Northrup and Neugebauer, 1999).

V-defects not due to threading dislocations. In addition to the V-defects which are present on top of threading dislocations, smaller V-defects not connected to any dislocations have also been observed lately (Miao *et al.*, 2010; Vennégués *et al.*, 2012). This is consistent with the fact that the density of V-defects is not systematically correlated to the density of threading dislocations (Sadler *et al.*, 2010). Because these smaller V-defects were first observed in the case of In-rich compressively-strained InAlN layers, it was concluded that they originate from the relaxation of the compressive strain (Miao *et al.*, 2010). However, Vennégués *et al.* (2012) also observed those smaller V-defects not connected to any dislocations in InAlN layers with an indium concentration between 15.5 and

16%, i.e., for layers under tensile strain (Vennéguès *et al.*, 2012). These smaller V-defects appeared from a thickness of 100 nm.

To acquire a deeper understanding of such defects, 100-nm thick $In_{0.2}Al_{0.8}N$ layers grown on FS-GaN substrates with a nominal threading dislocation density of 4×10^7 cm^{-2} (i.e., 1.6 dislocations for a 2×2 μm^2 area) have been characterized. On the AFM image (Fig. 6.4), hillocks and V-defects occurring at the core of bigger hillocks are visible. The density of V-defects is much larger than the dislocation density, thereby indicating that threading dislocations are not the only origin for the formation of V-defects. The reason why such defects can nucleate in InAlN films is not properly understood at the moment, but the literature about GaN and InGaN films can provide some useful clues. Those defects could be due to residual impurities like O, or to residual doping, as it was previously shown that the V-defect density in GaN increased with O, Mg, or Si doping (Liliental-Weber *et al.*, 1997). It could also originate from structural defects other than threading dislocations (Cho *et al.*, 2001) or to In-rich dots (Northrup and Neugebauer, 1999; Watanabe *et al.*, 2003). Eventually, it was shown that V-defects could release the strain in compressed InGaN layers (Song, 2005; Won *et al.*, 2010). It could also be the case in slightly compressed $In_{0.2}Al_{0.8}N$ layers.

Structure and composition of V-defects. Figure 6.5 is a schematic of a typical V-defect. The six facets of the V-defect are generally $\{11\bar{2}x\}$ ones (Kehagias *et al.*, 2009; Kehagias *et al.*, 2011; Mouti *et al.*, 2011; Vennéguès *et al.*, 2012), but for thick samples, $\{1\bar{1}01\}$ facets may also appear (Vennéguès *et al.*, 2012; Zhou *et al.*, 2010*b*). In addition to indium segregation occurring at the core

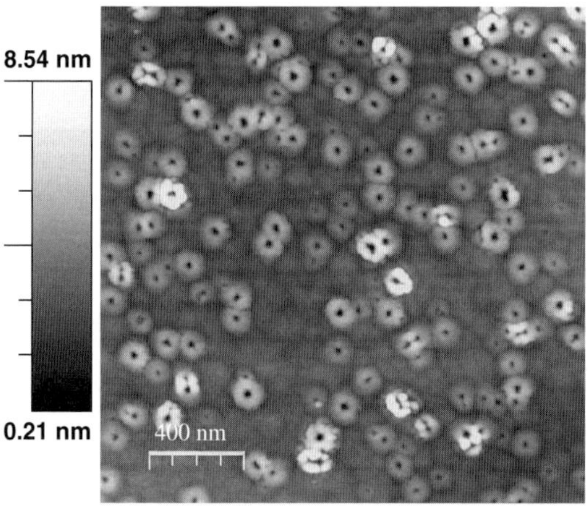

FIG. 6.4. 2×2 μm^2 AFM image of a 100-nm thick $In_{0.2}Al_{0.8}N$ layer grown on a FS-GaN substrate showing hillocks and V-defects surrounded by bigger hillocks. V-defects appear as holes located at the center of the bigger hillocks.

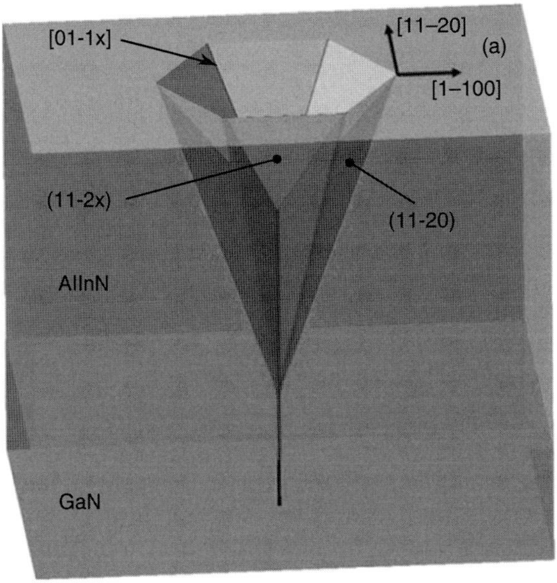

FIG. 6.5. Schematic representation of a V-defect lying on top of a threading dislocation (vertical line). The six facets of the V-defect are $\{11\bar{2}x\}$ ones. The six branches of the V-defect are $\langle 1\bar{1}0x \rangle$ ones and are indium-rich. During growth, triangular indium-rich vertical portions of planes of the $\{11\bar{2}0\}$ type are formed. (Figure adapted from Vennéguès *et al.* (2012), with permission from Elsevier.)

of dislocations, Z-contrast imaging and EDX measurements reveal that indium atoms also segregate on the six $\langle 1\bar{1}0x \rangle$ branches of the V-defects. While V-defects are growing, it gives In-rich $\{11\bar{2}0\}$ triangular portions of planes under them. When observing the samples in the $\langle 0001 \rangle$ direction, it gives six branches clearly visible on TEM images (Kehagias *et al.*, 2009; Kehagias *et al.*, 2011; Mouti *et al.*, 2011; Vennéguès *et al.*, 2012).

A similar V-defect is visible in Fig. 6.6: on the Z-contrast scanning TEM (STEM) image taken in the $\langle 0001 \rangle$ direction, bright branches are visible because of indium segregation occurring along the $\langle 1\bar{1}0x \rangle$ branches of the V-defect.

MOVPE-grown InAlN layers LM to GaN: critical thickness for phase separation. Hums *et al.* (2007) found that InAlN layers with an indium composition ranging between 13% and 31% are single-phase and fully-strained on GaN for the thicknesses considered (16 nm for 31.1% of indium, 43 nm for 16.6% of indium). Similar results were also found for thin InAlN layers (14 nm) with an indium composition ranging between 13% and 18% (Darakchieva *et al.*, 2008). However,

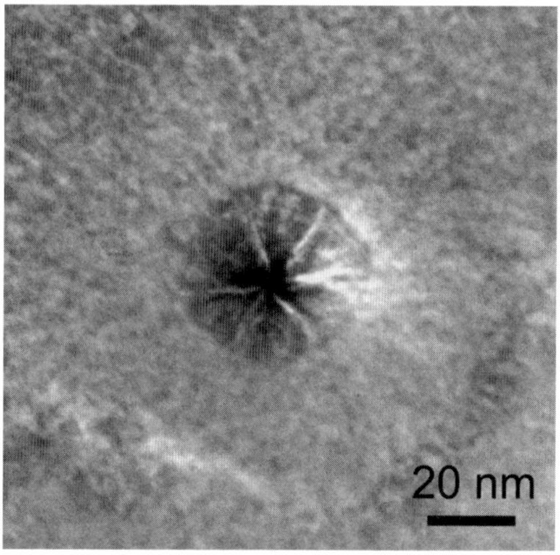

FIG. 6.6. Z-contrast STEM image taken along the ⟨0001⟩ direction of a V-defect in a 100-nm thick $In_{0.2}Al_{0.8}N$ layer grown on a FS-GaN substrate.

thicker layers covering the same range of indium composition (thickness ~100 nm), though single-phase, were partially relaxed (Darakchieva *et al.*, 2008). InAlN layers of identical thickness but with a larger nominal indium content exhibited spontaneous phase separation, with the first 65 nm characterized by an indium content of 22% and the next 35 nm with 26.4% of indium. Another group published similar change of composition for a 120-nm thick layer, with the first sublayer containing 24% of In and the second one 19% (Lorenz *et al.*, 2008). The change in composition was accompanied by a three-dimensional (3D) morphology with an increased roughness rising from 1 to 6 nm. It was attributed to strain relaxation. A change in morphology from 2D to 3D has also been observed and attributed to strain relaxation mediated by the Stranski–Krastanov mechanism in a weakly strained film (Krost *et al.*, 2011), but without unambiguous signature that such a strain-relaxation mechanism indeed occurs. A change in composition was observed for an InAlN layer grown on sapphire with a nominal indium content of 18%, close to the LM composition (Redondo-Cubero *et al.*, 2010). Spontaneously, the 50-nm thick $In_{0.18}Al_{0.82}N$ layer transformed into an $In_{0.15}Al_{0.85}N$ upper layer. It was then deduced that the morphological degradation and the lower indium content were not due to strain relaxation. A degradation occurring beyond a certain critical thickness was also observed by TEM (Kret *et al.*, 2011). It was shown that a homogeneous layer was obtained for an indium content close to LM conditions for thicknesses up to ~100 nm, whereas a columnar structure with phase separation appears on InAlN layers with a nominal indium content of 20% and 27% from a critical thickness of 200 and 10 nm, respectively.

A mechanism has been proposed to explain this change of composition (Chen *et al.*, 2012; Wei *et al.*, 2012). First, V-defects appear on top of threading dislocations, followed by the coalescence of those V-defects. Then, growth occurs on the inclined facets of the V-defects, which leads to a rough upper layer with a lower indium composition. This probably explains why two layers with different compositions have been observed by other groups, the upper one being rougher (Redondo-Cubero *et al.*, 2010; Hums *et al.*, 2007; Darakchieva *et al.*, 2008; Lorenz *et al.*, 2008). However, some V-defects are not linked with threading dislocations on TEM images (Chen *et al.*, 2012; Wei *et al.*, 2012). It is probable that V-defects also appear without the need for threading dislocation as mentioned previously. Then they coalesce, leading to an InAlN upper layer with a different composition and a columnar structure as described by Kret *et al.* (2011). This is further illustrated by TEM observations performed on a 500-nm thick $In_{0.2}Al_{0.8}N$ layer grown on a FS-GaN substrate (see Fig. 6.7) where a similar columnar upper layer is indeed visible from a certain thickness despite the absence of dislocations.

To summarize, the mechanism responsible for the degradation of InAlN layers would proceed as follows. First, hillocks are formed at the beginning of the growth, for thicknesses below 50 nm. Then, V-defects progressively appear, for an InAlN thickness ranging from 50 to 200 nm, on top of threading dislocations or not via a mechanism not yet elucidated. Finally, those V-defects coalesce, for an InAlN thickness about 100–200 nm, and an upper columnar InAlN layer with a different average composition is subsequently formed because of the growth which proceeds on the inclined facets of the V-defects. To conclude this section, it seems crucial to understand the V-defect formation in order to possibly avoid or at least minimize the degradation of InAlN layers with increasing thickness.

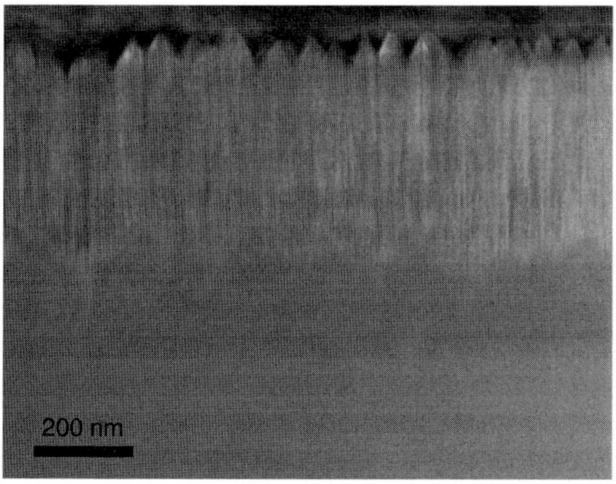

FIG. 6.7. Z-contrast STEM image of a 500-nm thick $In_{0.2}Al_{0.8}N$ layer. After coalescence of the V-defects, a columnar upper layer is obtained.

6.3 Optical and electronic properties of bulk InAlN layers

Following the description of the main structural properties of bulk InAlN layers, it is of relevant interest to focus on their optical and electronic properties. Among the important parameters which are still under debate, we can cite the case of the bandgap bowing parameter b. Indeed, a specific issue with this ternary alloy is that there is no convergence toward a single value for b that would correctly reproduce the variation in the bandgap, depending on whether the epilayers of interest are indium- or aluminum-rich. The reason could be ascribed to various causes that are listed hereafter. First, it is worth pointing out that an accurate determination of b should rely on a consistent set of experimental data, i.e., the position of the band edge should be derived by a given technique. Obviously, mixing photoluminescence (PL) and absorption measurements does not make sense, all the more since the extremely large Stokes shift measured in this ternary alloy would lead to a huge scatter in the data. Anyhow, as far as PL is concerned, a significant variation is expected for the Stokes shift (S), independent of the layer quality, since the degree of disorder is likely to vary across the whole composition range. It has thus been highlighted via *ab initio* calculations that the strong variations reported experimentally for the value of the band gap of the InAlN alloy could originate from the varying degree of indium clustering at the local scale, i.e., depending on whether a perfect random alloy or a clustered one is considered (Gorczyca *et al.*, 2009; Gorczyca *et al.*, 2010). Note in this respect that even for MOVPE-grown layers with the same nominal indium content the two types of structural configuration might occur (Gorczyca *et al.*, 2010). In any case, the determination of b based on absorption measurements appears much more suitable. In addition, another parameter of significant importance to properly consider to derive b is the strain state of the InAlN layer of interest. Indeed, even if the substrate of choice for the growth of such layers is usually c-plane sapphire, the underlying template on which the InAlN layer is grown can be either AlN or GaN. For a given indium content, the InAlN layers will therefore always experience a compressive strain if grown on an AlN template, whereas for layers grown on GaN templates they could be either compressively (In-rich layers) or tensely strained or even be strain-free. To date, the most accurate treatment regarding the dependence of b in the InAlN alloy is probably that given by Sakalauskas *et al.* (2010). In this work, the strain influence on the band-gap energy is readily accounted for via a $k \cdot p$ model which includes the strain-induced energy shift, which then allows derivation of the unstrained band gap for the whole In composition range. Using the conventional empirical expression:

$$E_g^{In_xAl_{1-x}N} = x \cdot E_g^{InN} + (1-x) \cdot E_g^{AlN} - b \cdot x \cdot (1-x) \qquad (6.1)$$

where $E_g^{In_xAl_{1-x}N}$ is the InAlN bandgap, x is the indium content, E_g^{InN} is the InN band gap, and E_g^{AlN} is the AlN bandgap. The best fitted b value describing the evolution of the InAlN bandgap is equal to 5.36 ± 0.36 eV, which is in reasonably good agreement with values reported in the literature

(Wang *et al.*, 2008; Jones *et al.*, 2008; Oh *et al.*, 2008). However, to better account for the experimental data those authors also introduced a modified version of eq. 6.1:

$$E_g^{In_xAl_{1-x}N} = x \cdot E_g^{InN} + (1-x) \cdot E_g^{AlN} - b_2(x) \cdot x \cdot (1-x) \tag{6.2}$$

where $b_2(x)$ is the modified bowing parameter, which is a non-linear function of x:

$$b_2(x) = \frac{A}{1 + Cx^2} \tag{6.3}$$

where the best agreement with experiment is achieved for the values $A = 6.43 \pm 0.12$ eV and $C = 1.21 \pm 0.14$. Note that such a treatment is closely related to that introduced by Iliopoulos *et al.* (2008), where a modified bowing parameter $b_1(x)$ expressed as

$$b_1(x) = \frac{A}{1 + Cx} \tag{6.4}$$

is considered. The justification for the introduction of such a composition-dependent bowing parameter is ascribed to the significant charge transfer occurring in the InAlN alloy when spanning the whole composition range, owing to the large electronegativity difference between aluminum and indium atoms (Iliopoulos *et al.*, 2008; Ferhat and Bechstedt, 2002).

Another puzzling feature of bulk InAlN layers nearly LM to GaN originates from their unusually large Stokes shift, which can reach a value equal to 0.8–1 eV (Butté *et al.*, 2007; Wei *et al.*, 2012; Wang *et al.*, 2008; Carlin *et al.*, 2005). A broad luminescence line is usually monitored in the 330–340 nm range both in PL (Butté *et al.*, 2007; Chen *et al.*, 2010; Wang *et al.*, 2008) and in cathodoluminescence (CL) (Wei *et al.*, 2012; Chung *et al.*, 2011), whose exact origin is presently unknown. Its redshift with increasing temperature remains well below the usual dependence expected for the band edge, thereby underlying strong carrier localization. So far, no clear consensus has been reached to know whether a direct connection could be drawn between composition fluctuations and the origin of the luminescence, since even thin homogeneous layers (~40–50 nm thick) grown by MOVPE with no apparent phase separation reported, e.g., via Z-contrast imaging in STEM or EDX, also exhibit such an anomalously large Stokes shift. The large S value could possibly be intrinsic to the InAlN alloy. It could indeed originate from the numerous possible configurations for the In atoms within atom-supercells, each of them being characterized with different energy gaps leading to different signatures in absorption or emission spectra (de Carvalho *et al.*, 2012). Chen *et al.* (2010) decomposed the broad structureless luminescence lineshape into two components, with a low-energy band ascribed to recombinations occurring in the hillocks that would act as quantum dot-like structures. For PAMBE-grown layers, the lateral phase separation leading to indium-rich walls surrounding honeycomb cells could potentially explain the Stokes shift with respect to the absorption edge (Butté *et al.*, 2007; Zhou *et al.*, 2007). Such luminescence could also originate from deep defects, but no firm demonstration has been reported in the literature. In particular, no clear correlation could be made between the luminescence and the V-defects, which would act as localized exciton recombination centers or deep-level impurities, as has been done for InGaN MQW layers (Wu *et al.*, 1998). Note also that spectra extracted from polarization-resolved PL measurements performed on such layers exhibit only random polarization (Taniyasu, 2012). The latter further support the localized nature of the carriers, whereas for such an Al-rich ternary alloy the polarization selection rules would suggest a stronger edge luminescence with respect to the surface one, i.e., the emitted light with $\mathbf{E}//\mathbf{c}$ would dominate over $\mathbf{E}\perp\mathbf{c}$. (Taniyasu, 2012).

Studies addressing the value of the optical refractive index n_{op} for InAlN layers LM to GaN are also extremely scarce (Carlin and Ilegems, 2003; Aschenbrenner *et al.*, 2010; Sakalauskas *et al.*, 2010; Carlin *et al.*, 2005). The first report made by Carlin and Ilegems when demonstrating the successful fabrication of highly reflective crack-free InAlN/GaN DBRs showed that in the blue–green wavelength range the refractive index contrast to GaN $\Delta n_{op}/n_{op}$ amounts to about 7–8%. It was highlighted that for such InAlN layers with an indium content of 17–18% the dispersion of n_{op} in the green-to-UV range is well matching that of the AlGaN alloy with an aluminum content of 46% (Carlin *et al.*, 2005), as experimentally determined by Brunner *et al.* (1997). This is further confirmed by the good agreement observed between reflectivity measurements performed on Bragg mirrors or vertical cavity structures and corresponding transfer matrix simulations. A more systematic study of the spectral dependence of n_{op} for indium contents in the 10–24% range has been recently performed by variable-angle spectroscopic ellipsometry (Aschenbrenner *et al.*, 2010; Sakalauskas *et al.*, 2010), where the experimental dielectric function is accounted for by a conventional multilayer including surface roughness effects via the Bruggeman effective medium approximation. For InAlN layers LM to GaN, a good qualitative agreement with past studies (Carlin and Ilegems, 2003; Sakalauskas *et al.*, 2010) is reported considering all the above-mentioned issues that can affect the quality and the growth of this ternary alloy. Another parameter of particular relevance is the band alignment at the InAlN/GaN heterointerface. Discarding spontaneous polarization mismatch issues in this LM system leading to the formation of two-dimensional electron gases, it was recently shown by means of X-ray photoelectron spectroscopy (XPS) that the band alignment is probably of type I with a small valence-band offset (VBO) equal to 0.2 ± 0.3 eV (Akazawa *et al.*, 2010). Thus it means that the conduction-band offset (CBO) is as large as $0.9–1.0 \pm 0.3$ eV, thereby indicating a significant deviation with respect to the common anion rule, as the CBO:VBO ratio is much higher than 70:30. Such a small VBO value is in agreement with that computed for the evolution of the conduction and valence band edge positions with respect to the vacuum level for the InAlN alloy, by Jones *et al.* (2008). Note also that such a CBO:VBO ratio does not seem to have a noticeable impact on the interband transition energy of InAlN/GaN QW structures obtained when using the Nextnano[3] software, compared with that derived when assuming that the common anion rule holds.[1]

6.4 Optical features of GaN/InAlN quantum wells

In a previous review on InAlN layers LM to GaN we highlighted the interest of GaN/InAlN single and multiple QWs for optoelectronic applications (Butté *et al.*, 2007). Besides the large CBO making such heterostructures suitable for the fabrication of high-speed mid-infrared intersubband photodetectors and modulators operating in the 2.4–4 μm range (Nicolay *et al.*, 2005), they could represent a promising alternative to the GaN/AlGaN system to realize efficient interband light-emitters. We recall that owing to their strain-free nature, LM GaN/InAlN QWs grown along the *c*-axis will only experience a built-in electric field due to the spontaneous polarization mismatch occurring at the heterointerfaces, which is specific to this layer combination for this crystalline orientation. It was shown experimentally that those single wells experience a built-in field amounting to $\sim$3.64 MV$\cdot$cm^{-1} (Butté *et al.*, 2007). Such a value is in good agreement with that deduced from calculations for GaN/InAlN heterostructures with an indium content of 10–25% pseudomorphically grown on a GaN epilayer which includes the piezoelectric polarization (obviously this component is equal to zero for a perfectly LM heterostructure), and the non-linear dependence of the spontaneous polarization discontinuity with varying indium content.

To illustrate the interesting prospects of GaN/InAlN QWs as light-emitters, we compare in Fig. 6.8(a) the room-temperature PL spectrum of three types of representative 2-nm thick single QWs, namely a GaN/$Al_{0.17}Ga_{0.83}N$ QW, a LM GaN/InAlN QW and an $In_{0.17}Ga_{0.83}N$/GaN one, measured using the UV line ($\lambda = 244$ nm) of a frequency-doubled cw Ar^+ laser operating in the low-density regime (i.e., in absence of screening of the built-in field) and detected by a monochromator UV-enhanced CCD combination. Unsurprisingly, the intensity of the InGaN/GaN QW well exceeds that of the GaN/AlGaN one, as expected for such a ternary QW where localization effects are known to ensure a high radiative efficiency up to room temperature. However, we can notice that for a similar emission wavelength—$\sim$350 nm, i.e., in a spectral region where quantization effects still dominate over the quantum confined Stark effect (QCSE)—the GaN/InAlN QW outperforms the GaN/AlGaN one. As an illustration in Fig. 6.8(b), the ratio of the integrated PL intensity (I) at various temperatures up to 300 K (I_{300K}) to that measured at 10 K (I_{10K}) is displayed for the three above-mentioned single QWs. If we make the usual, optimistic and crude, assumption that the I_{300K}/I_{10K} ratio is a reliable indicator of the internal quantum efficiency (IQE) at RT (thus assuming that at 10 K only radiative recombinations occur), the three systems exhibit an IQE of 32% for the present InGaN/GaN QW vs. 4.6% and 0.4% for the GaN/InAlN and the GaN/AlGaN ones, respectively. It therefore

[1]Nextnano[3] software, http://www.nextnano.de.

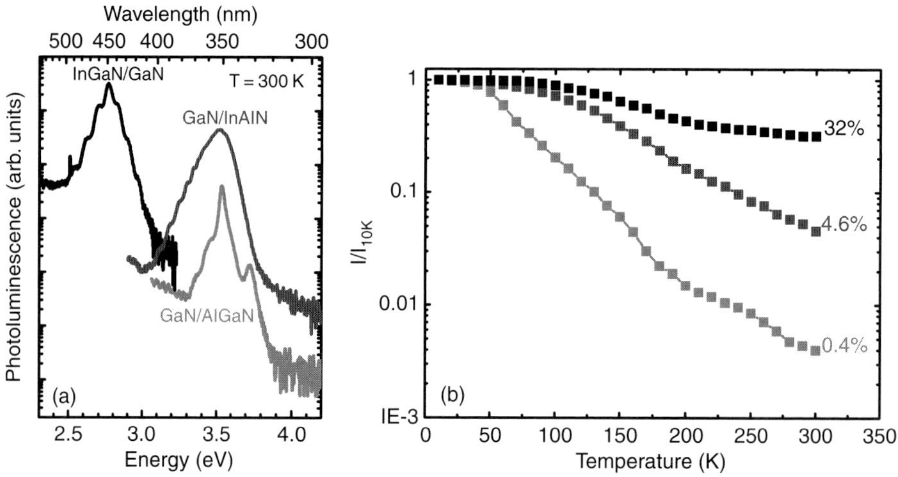

FIG. 6.8. (a) RT PL spectrum of three types of 2-nm thick single QWs (GaN/AlGaN QW; LM GaN/InAlN QW; and InGaN/GaN QW), measured using the UV line ($\lambda = 244$ nm) of a frequency-doubled cw Ar^+ laser operating in the low-density regime. (b) Ratio of the integrated PL intensity (I) at various temperatures up to 300 K to that measured at 10 K (I_{10K}), displayed for the three above-mentioned single QWs (GaN/AlGaN QW; LM GaN/InAlN QW; and InGaN/GaN QW). The RT IQE is quoted in each case (see text for details).

demonstrates the promising potential of LM GaN/InAlN heterostructures for the realization of efficient near-UV light emitters. The reason for the larger IQE of GaN/InAlN QWs compared to the GaN/AlGaN ones could be two-fold. First, the former exhibit a larger localization energy with respect to the latter, for which it is typically smaller than 20 meV. In addition, the band offsets are larger, which leads to a better quantum confinement. Both features could prevent thermalization escape phenomena toward non-radiative centers to occur in the GaN/InAlN QW system.

6.5 Nearly lattice-matched InAlN/(Al)GaN distributed Bragg reflectors

6.5.1 *Growth properties*

As already mentioned, InAlN with an In content of ~18% is lattice-matched to GaN while exhibiting a refractive index contrast to GaN of 7–8% in the blue–green wavelength range (Carlin and Ilegems, 2003). This makes such an alloy of great interest for the realization of III-nitride-based high-quality DBRs. Other approaches usually rely either on the use of AlGaN/GaN (Waldrip *et al.*, 2001;

Natali *et al.*, 2003) or AlN/GaN (Ng *et al.*, 2000) bilayers. The main drawback of such systems is the large in-plane lattice parameter mismatch between the different layers composing the DBR, causing a large build-up of the strain with an increasing number of pairs, which eventually results in cracks and misfit dislocation generation. A strain-compensation technique based on superlattice interlayers has been proposed to improve the quality of AlN/GaN DBRs (Huang *et al.*, 2006), and it has even led to the demonstration of blue cw electrically driven vertical cavity surface-emitting lasers (VCSELs) on sapphire (Lu *et al.*, 2010). However, such a method is not suited for devices grown on low-dislocation-density FS-GaN substrates since the accumulated strain with this bilayer system would inevitably generate threading dislocations, without resulting in a significant gain over growth carried out on a sapphire substrate.

InAlN-based DBRs with a stopband centered either in the UV (Feltin *et al.*, 2006) or in the visible (Butté *et al.*, 2005) range have been demonstrated on *c*-plane sapphire. Reflectivity spectra measured on such DBRs are depicted in Fig. 6.9. For the realization of UV DBRs, AlGaN is used as the second layer together with InAlN in order to avoid GaN absorption below 370 nm. Such UV DBRs have been successfully used to demonstrate RT polariton lasing in nitride-based microcavities (Christopoulos *et al.*, 2007; Christmann *et al.*, 2008), and DBRs with a stopband in the visible region have been used for the realization of electrically driven resonant-cavity light-emitting diodes (Dorsaz *et al.*, 2005*b*) and optically pumped VCSELs (Feltin *et al.*, 2007*a*).

The transfer of such DBRs on FS-GaN substrates would allow a great increase in the device performances, due to the much lower threading dislocation density that would directly benefit the device internal efficiency, and improved thermal management. However, the growth of such DBRs on FS-GaN is not straightforward. Indeed, having fewer threading dislocations also means a reduced residual strain accommodation. In this context, a careful growth optimization of such

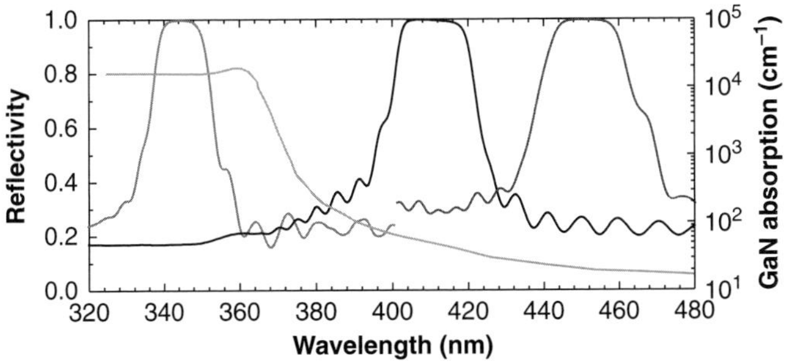

FIG. 6.9. Reflectivity spectra measured on three different InAlN-based DBRs (see text for details). The GaN absorption edge is shown as a reference. (Reprinted, with permission, from Butté *et al.* (2009).)

DBRs on FS-GaN has to be performed. It has been found that the indium-to-aluminum precursor gas flux ratio has to be large in order to ensure flat interfaces and a smooth DBR top surface as well as an optimum peak reflectivity (Cosendey *et al.*, 2011). This has been ascribed to indium surface segregation—a well-known effect occurring in indium-containing ternary alloys of the III-nitride and the III-arsenide family, which is due to the much larger bond length involving indium atoms compared to aluminum or gallium ones (Massies *et al.*, 1987). A possible explanation for the way indium surface segregation affects the quality of InAlN-based DBRs is the following. In the presence of a few InAlN atomic layers with an indium content much lower than the LM value, at the beginning of each InAlN layer the lattice-matched condition in such sublayers is no longer fulfilled, so that a large tensile strain builds up. Then, depending on the indium surface segregation coefficient, the total thickness of those indium-poor sublayers might eventually reach the critical thickness where strain is no longer accommodated by elastic deformation but causes the generation of misfit dislocations. Subsequently, it has been assumed that increasing the indium-to-aluminum precursor gas ratio reduces the indium surface segregation coefficient by increasing the amount of indium adatoms on the sample surface during growth. By doing so, the nominal indium content satisfying the lattice-matched condition is reached before the critical thickness.

Another point that has to be stressed once more concerning the MOVPE growth of InAlN-based DBRs is the large-growth temperature difference between GaN (about 1050°C) and InAlN (about 850°C). This means that when considering the most basic approach for growing such DBRs the surface of each InAlN layer is subject to a temperature ramp up to 1050°C. This is above the thermal stability of lattice-matched InAlN, which is about 960°C (Gadanecz *et al.*, 2007). Indeed, such an annealing step greatly affects the InAlN surface morphology (Brice *et al.*, 2010). Figure 6.10 shows two AFM images, the first one for an as-grown 50-nm thick InAlN layer LM to GaN (Fig. 6.10(a)) and the second one for a similar 50-nm thick InAlN layer that has been annealed *in situ* at 1065°C for two minutes (Fig. 6.10(b)). Both layers have been grown on a FS-GaN substrate under the same conditions than for InAlN layers inserted in the DBRs exhibiting the best optical properties. The AFM images show that after annealing the surface morphology switches from small hillocks to a mesh of small cracks. To illustrate the impact of the temperature ramp, Fig. 6.10(c) shows a Z-contrast STEM image of an InAlN/GaN DBR. InAlN layers appear darker than GaN ones due to their lighter atomic weight. But, close to the InAlN/GaN interfaces, InAlN layers are even darker, suggesting that indium desorption indeed occurs during the temperature ramp. To circumvent this drawback, Sadler *et al.* (2009*b*) proposed two solutions for the growth of GaN layers used in DBRs. They can be grown at low temperature, but it results in a finger-like GaN surface morphology, which is detrimental to the reflectivity of the DBRs. The second approach relies on the low-temperature growth of a thin GaN sublayer followed by a temperature ramp and the subsequent growth of GaN at high temperature.

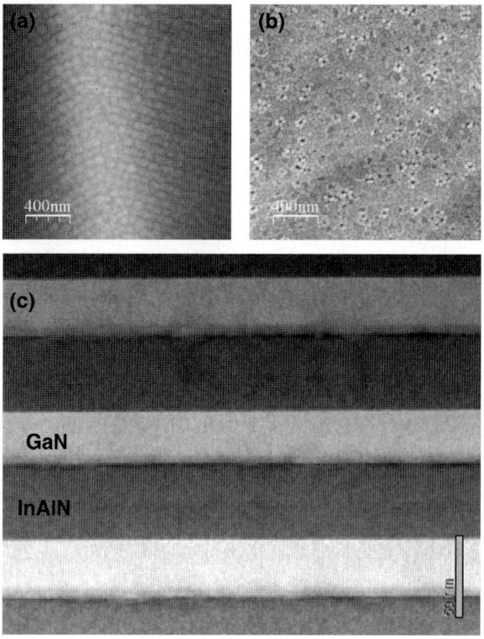

FIG. 6.10. 2×2 µm^2 AFM images of the surface of 50-nm thick InAlN layers (a) as grown and (b) annealed *in situ* at 1065°C. (c) Z-contrast STEM image viewed in cross-section of a DBR.

This latter method seems much more promising, since it should not alter the GaN surface morphology while preventing indium desorption at each InAlN/GaN interface.

6.5.2 *Optical properties*

As discussed previously, the substrate should have a crucial impact on the quality of III-nitride based heterostructures. As an illustration, we briefly discuss the quality of LM InAlN/GaN DBRs grown onto two different types of *c*-plane FS-GaN substrates. The first substrate type exhibits a dislocation density of 1×10^6 cm^{-2}, whereas the second substrate exhibits a larger dislocation density amounting to 4×10^7 cm^{-2}. Identical growth conditions were employed for the two DBRs (see Cosendey *et al.* (2011)) except for a slight variation in the layer thicknesses resulting in a stopband shift of ~20 nm. AFM scans of the final top GaN layer have been performed on both structures to compare the dislocation density. For the DBR grown on the second substrate, a dislocation density similar to that of the substrate is found, in strong contrast with the DBR grown on the high quality FS-GaN substrate for which a dislocation density of 1×10^9 cm^{-2} is extracted. This larger dislocation density might be ascribed to a slight difference in the indium surface segregation coefficient originating from the different thermal conductivity for the two substrates. In addition to standard structural characterization tools (namely AFM, scanning, and transmission electron microscopes) the following characterization method has been used: on top of both DBRs a seven-pair ZrO_2/SiO_2 Bragg mirror has been deposited by electron-beam evaporation, resulting in $\lambda/2$ GaN-ZrO_2 cavities. Afterwards, the in-plane disorder, which is directly linked to the DBR structural and optical properties, has been probed through 2D micro-transmission mappings using the continuous spectrum of a xenon lamp as the incident light source. The light was focused down to a spot size $\sim$2 μm in diameter on the sample, using a long working distance near UV microscope objective ($\times$100) with a numerical aperture of 0.5. The transmitted light was then collected by a UV fiber in far-field configuration ensuring an angular selection of $\sim$0.6°. Transmission spectra of the cavities have been acquired every micron over a 50 $\times$ 50 μm^2 area (Fig. 6.11). For the DBR structure grown onto the low-dislocation-density substrate (Fig. 6.11, left-hand column) a low photonic disorder is evidenced. The cavity mode position is fluctuating by $\pm$0.45 nm ($\sim$0.1%) and can be related to cavity thickness changes of $\pm$1 nm, which is similar to the rms surface roughness determined by AFM. The quality factor (Q) for the microcavity grown onto the high-quality FS-GaN substrate reaches values up to 800, matching those deduced from transfer matrix simulations. On the other hand, for the DBR structure grown onto the intermediate dislocation density substrate (Fig. 6.11, right-hand column), the photonic disorder is considerably increased. The Q factor only amounts to half that of the former one, whereas an identical value was expected *a priori*. Such an analysis clearly demonstrates that the dislocation density measured on the final top GaN layer does not play a crucial role for the achievement of samples characterized with a low photonic disorder. The observed difference might originate from different polishing procedures used for the two kinds of substrate.

6.5.3 *Applications*

As far as UV InAlN/AlGaN DBRs are concerned, they proved instrumental in achieving RT polariton lasing in III-nitride microcavities. In such a case the coupling strength between the cavity photon mode and the excitons in the active material overcomes their damping rates, allowing the system to enter into a non-perturbative regime, the so-called strong-coupling regime (SCR). The new eigenmodes of the system are then the lower and upper polaritons. In the low-density limit these quasiparticles behave as bosons, and polariton condensation is triggered when the ground-state occupancy exceeds unity. The spontaneous decay of this macroscopic quantum state results in a non-linear coherent emission, often referred to as polariton lasing, which is expected to happen at a lower threshold current density than in the case of the conventional stimulated emission ruled by the Bernard–Durrafourg condition, as is the case for VCSELs (Imamoglu *et al.*, 1996).

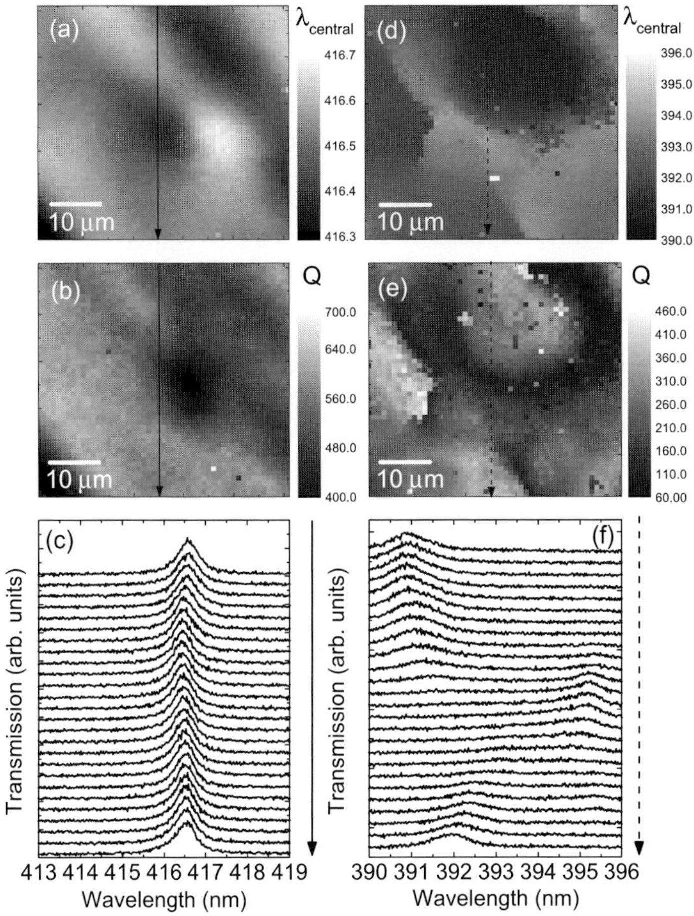

FIG. 6.11. (a) and (d) micro-transmission mappings (50×50 μm^2) of the cavity mode wavelength, (b) and (e) corresponding cavity quality factor Q, and (c) and (f) transmission spectra recorded every 2 μm along the continuous and the dashed arrow on a DBR structure grown on a high-quality FS-GaN substrate (left-hand side column) and on an intermediate-quality FS-GaN substrate (right-hand side column), respectively.

An appropriate choice for the active medium to perform SCR studies is the spectrally narrow excitonic transition of bulk GaN or that of GaN/AlGaN QWs (Feltin *et al.*, 2007*b*). Such an active medium requires UV DBRs. As mentioned previously, AlGaN layers are used together with InAlN layers. Indeed AlGaN can be grown LM to InAlN by decreasing the indium content in the latter ($x_{In} < 18\%$). On a thirty-five-pair $In_{0.15}Al_{0.85}N/Al_{0.2}Ga_{0.8}N$ DBR a peak reflectivity of 99.3% has been measured at 340 nm (Feltin *et al.*, 2006). As the DBR is no longer LM to the GaN buffer layer, a strain-engineering solution relying on the use of short-period GaN/AlN superlattices has been successfully employed to avoid the formation of cracks (Feltin *et al.*, 2006). When completed by a top dielectric DBR, high-quality factor cavities can be formed. In this way, polariton lasing under optical pumping has been observed in such cavities, using either bulk GaN as an active medium (Christopoulos *et al.*, 2007) or GaN/AlGaN MQWs (Christmann *et al.*, 2008), which offer the advantage of a larger exciton binding energy and oscillator strength compared with their bulk counterpart.

InAlN/GaN DBRs grown on *c*-plane FS-GaN substrate have also been used for the realization of optically pumped vertical external cavity surface-emitting lasers (VECSELs) (Zeng *et al.*, 2012). A forty-two-pair DBR has been overgrown by a high-temperature (HT) GaN layer followed by an active region made of twenty-four InGaN (2 nm)/GaN (4.5 nm) QWs capped with another HT GaN layer for a total optical cavity thickness of $5\lambda/2$. The high crystalline quality of the DBR allows obtaining a half-cavity with a very smooth top surface (rms surface roughness of 0.20 nm measured on a 10×10 μm^2 AFM scan). The inset in Fig. 6.12 shows a 2×2 μm^2 AFM scan of the surface of the VECSEL sample, which is characterized with regular atomic steps indicating that the growth proceeds under the step-flow growth mode. A top dielectric DBR consisting of four SiO_2/ZrO_2 pairs with a peak reflectivity of 86.6% has then been deposited on the sample. As reference, a second identical sample has received a top DBR made of seven pairs of SiO_2/ZrO_2, exhibiting a peak reflectivity of 98.1%. This latter sample has been used in a VCSEL configuration in order to evaluate the quality

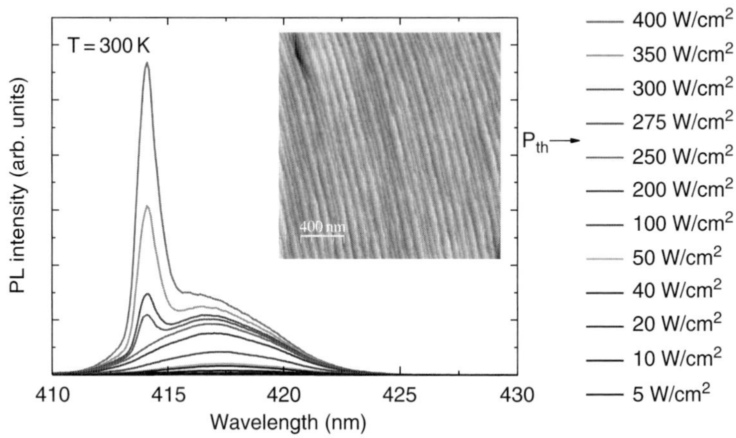

FIG. 6.12. Spectra of an optically pumped VECSEL structure without the external mirror. Inset: 2×2 μm^2 AFM scan of the surface of the VECSEL sample prior to the top DBR deposition.

of the structure. The sample has been optically pumped using the 355 nm line of a frequency-tripled Nd:YAG laser, and exhibits lasing with an average threshold power of about 260 W/cm^2. Spectra measured below and above threshold are shown in Fig. 6.12. Then, both samples have been mounted in an external cavity configuration, using an external concave mirror with a reflectivity of 99.5% and a curvature radius of 50 mm. The threshold pump power and the lasing mode linewidth decrease when the external cavity length is increased from 25 mm to 50 mm. This is ascribed to a decreasing cavity mode size when approaching the semi-confocal configuration, and to reduced external cavity losses for longer VECSEL cavities. The main advantage of a VECSEL with such a large external cavity is the possibility of inserting intracavity elements such as a saturable absorber for the realization of mode-locked VECSELs.

Another application of high-quality highly reflective InAlN-based DBRs is the realization of monolithic VCSELs grown on FS-GaN substrate (Cosendey *et al.*, 2012), which could be a serious alternative to hybrid VCSELs using two dielectric DBRs (Omae *et al.*, 2009; Onishi *et al.*, 2012). The main advantage of the monolithic approach is a much easier process flow, avoiding the requirement for removing the cavity from the substrate and the subsequent wafer bonding step, and which also makes the cavity thickness control much easier. VCSEL structures made of a 7λ cavity comprising a 5 InGaN/InGaN QW active region centered at one of the antinodes of the optical field were grown on a forty-two-pair InAlN/GaN DBR on FS-GaN. Devices were processed using a standard ITO current spreading layer as the top *p*-electrode, and a passivated *p*-GaN surface followed by a top seven-pair TiO_2/SiO_2 DBR. Details about this epitaxial structure and processing are given elsewhere (Cosendey *et al.*, 2012). The electroluminescence (EL) intensity vs. pulsed driving current characteristics of such devices are shown in Fig. 6.13(a) for a device with a 60-μm mesa diameter and

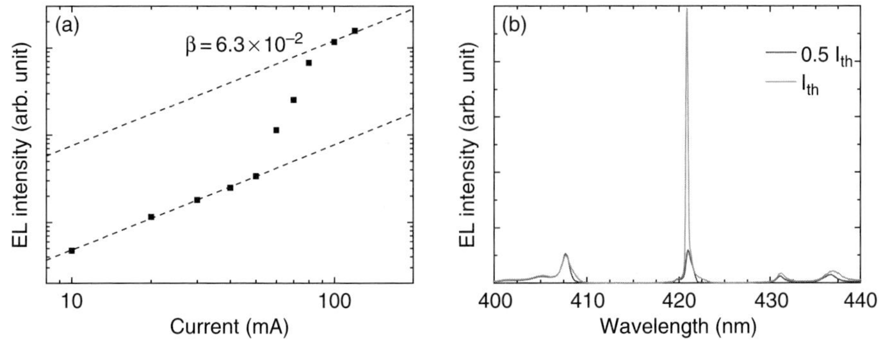

FIG. 6.13. (a) $L-I$ curve of an InAlN-based VCSEL under pulsed current injection with a mesa diameter of 60 μm and a current aperture of 8 μm. (b) Spectra measured from the bottom of the same device for different pulsed currents.

a 8-μm current aperture. The pulse width is 200 ns and the duty cycle 1%. The lasing threshold current is about 70 mA. The measured spontaneous emission coupling factor is as high as 6.3×10^{-2}, but it might be slightly overestimated because of the lack of points for high injection currents. The corresponding EL spectra, measured below and above lasing thresholds, are depicted in Fig. 6.13(b). The cavity mode is well centered with respect to the stopband of the bottom InAlN-based DBR, which can be clearly seen on the spectra.

A possible further improvement for electrically driven devices making use of such InAlN-based DBRs would be to develop conductive Bragg mirrors. It would allow avoiding the requirement for intra-cavity contacts, which would lead to shorter cavities. Concerning n-doped DBRs, the doping of InAlN and GaN layers should not be an issue, but the free charge planes appearing at the interfaces due to the polarization mismatch between InAlN and GaN layers should be properly considered in order to ensure a good vertical current transport through such DBRs. On the other hand, the development of p-doped DBR would be more challenging. It would avoid the use of any transparent conductive oxide as a current-spreading layer inside the VCSEL cavity, which would decrease the losses and thus reduce the threshold current density.

Another route to explore would be the development and study of semipolar and/or non-polar InAlN layers. Devices using such DBRs would have the noticeable property to exhibit a reduced QCSE in their QW-based active region. This would not only increase the oscillator strength of the QWs, but could also result in new device properties. Thus, recently a non-polar VCSEL has been demonstrated which exhibits polarization-locked light emission inherited from the in-plane crystalline anisotropy of the QWs (Holder *et al.*, 2012). However, the in-plane lattice parameter anisotropy along such an orientation would not allow for perfectly LM conditions to GaN contrary to the situation occurring along the c-axis. We would rather refer to strain minimization instead.

6.6 InAlN cladding layers for edge-emitting lasers

One of the most important aspects to keep in mind for the design of a nitride-based LD structure is the optical mode confinement into the waveguide. In this respect, the optical characteristics of InAlN layers with a composition lattice-matched to GaN make this material system a perfect candidate for the fabrication of claddings in GaN-based LDs with an emission wavelength ranging from the UV to the green. They could thus advantageously substitute for the commonly used AlGaN claddings.

As can be deduced from the work of Brunner *et al.* (1997), the AlGaN alloy exhibits a rather low refractive index contrast to GaN in the range of compositions having a small lattice-mismatch with the latter (a few percent in Al composition). Such a low refractive index contrast makes high optical mode confinement in the active region harder to achieve. For the same reason, the transversal optical mode will probably extend into the GaN substrate, giving rise to a poor far-field quality. Thick GaN/AlGaN superlattices or bulk AlGaN cladding layers with low Al content are therefore required in order to decouple the laser structure from the n-GaN substrate and provide at the same time the suitable optical mode confinement (Nakamura, 1997; Sitar *et al.*, 1990). As a matter of fact, $Al_xGa_{1-x}N$ alloys with an Al content well below 10% are widely used as cladding materials for violet and blue LDs. A typical violet LD epitaxial structure emitting at 415 nm is shown in Fig. 6.14(a), together with the corresponding refractive index and simulated near-field intensity profiles (Fig. 6.14(b)). Depending on the bottom cladding thickness, the guided mode can extend into the underlying GaN substrate (Fig. 6.14(b)), giving rise to optical losses and to a poor optical confinement.

The mode leakage into the substrate also affects the beam quality of the LD devices. Modeling of the LD structure shown in Fig. 6.14(a) with bottom and upper $Al_{0.07}Ga_{0.93}N$ claddings shows that although a bottom cladding with a thickness of at least 1 μm would be suitable for the achievement of high optical mode confinement into the waveguide (Fig. 6.15(a)), this thickness would not be enough to provide a high-quality transverse emission pattern (Fig. 6.15(b)). A narrow intensity peak located at negative angle around –10° is observed. It is due to the mode leakage occurring in the GaN substrate. In order to suppress this parasitic peak, the mode leakage in the substrate has to be further reduced. Hence, bottom claddings with a minimum thickness of 2 μm should be grown.

It is then straightforward to conceive that the problem becomes even more critical when moving toward emission wavelengths in the blue–green region ($\lambda > 450$ nm). Different approaches have been proposed to circumvent this issue (Yoshizumi *et al.*, 2009; Tyagi *et al.*, 2010). One solution is to grow AlGaN-free structures on non-polar or semi-polar GaN substrates, including thick InGaN

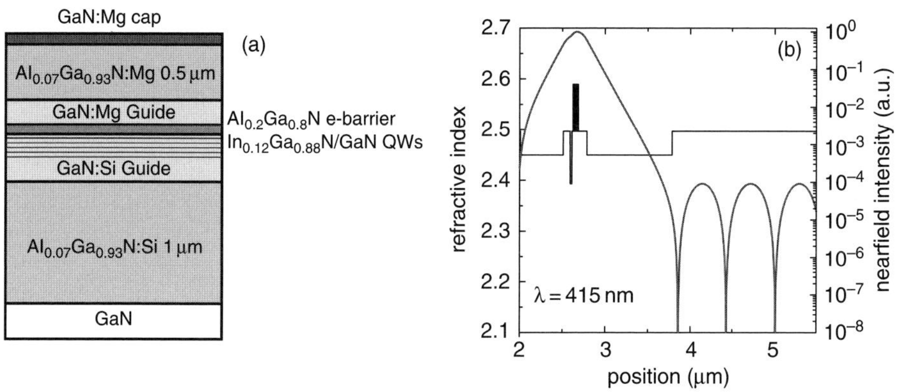

FIG. 6.14. (a) General scheme of a violet LD structure and (b) corresponding simulated near-field mode distribution (light gray curve) and refractive index profile (black curve).

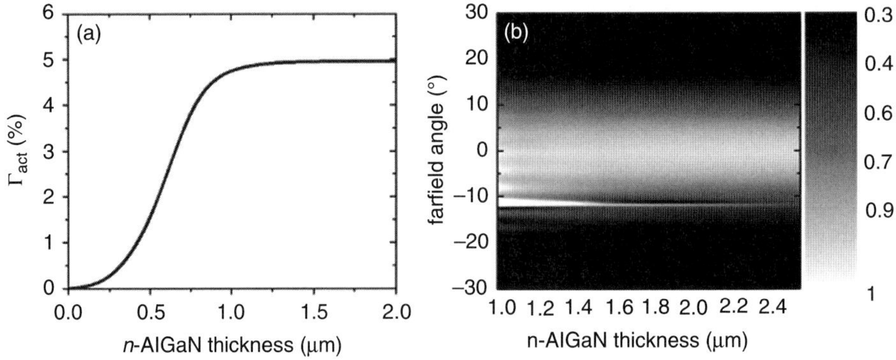

FIG. 6.15. (a) Simulated optical mode confinement factor in a violet LD and (b) far-field distribution as a function of the bottom cladding thickness.

QWs and InGaN inner guides (Yoshizumi *et al.*, 2009). Although the published results are quite promising, the lack of proper substrates could eventually be a major issue in terms of device production.

Recently, several studies on edge-emitting lasers including InAlN LM bottom claddings or optical blocking layers have been carried out (Castiglia *et al.*, 2008; Castiglia *et al.*, 2009; Schenk *et al.*, 2009; Castiglia *et al.*, 2010; Charash *et al.*, 2011; Kim-Chauveau *et al.*, 2012). Lasing has been demonstrated under optical pumping (Schenk *et al.*, 2009) and under electrical injection (Castiglia *et al.*, 2008; Castiglia *et al.*, 2009; Castiglia *et al.*, 2010; Charash *et al.*, 2011; Kim-Chauveau *et al.*, 2012). Improved device optical characteristics can be achieved by implementing InAlN bottom claddings LM to GaN. As mentioned previously, $\Delta n_{op}/n_{op}$ exceeds 6% for the LM InAlN/GaN system in the green wavelength region, whereas this value is limited to less than 2% when using standard low Al content AlGaN claddings. Furthermore, the use of a lattice-matched material allows growing thick layers without introducing additional strain or defects during the growth.

Potential drawbacks when using such material for optoelectronic applications have to be taken into account. The wider energy bandgap, about 4.4–4.5 eV at 300 K for LM InAlN (Butté *et al.*, 2007), compared to GaN, represents an energy barrier for efficient carrier transport at heterointerfaces. Moreover, because of the deep nature of the donors in such material, high dopant concentrations are required to achieve an efficient *n*-type doping. Nonetheless, InAlN optical characteristics can lead to improved laser performances in terms of threshold current density, slope efficiency, and beam shape quality.

Implementing InAlN claddings in GaN-based LDs requires an accurate optical design of the LD structure. Since *p*-type doping has not been demonstrated yet in LM InAlN, the use of this alloy in LD structures under electrical injection is limited to the fabrication of the bottom cladding, while keeping AlGaN with a low Al content for the upper p-type cladding. On such structures, the differences in the index contrast between the lower and the upper part of the LD structure could lead to an asymmetric distribution of the optical mode inside the waveguide.

The influence of the position of the bottom InAlN cladding with respect to the waveguide has been investigated (Castiglia *et al.*, 2009). Three LD structures grown on sapphire substrates are compared. The general LD structure including an InAlN bottom cladding is shown in Fig. 6.16. Here t_1 and t_2 are the n-AlGaN and the n-InAlN thicknesses, respectively. The reference structure consists of a 500-nm thick n-doped $Al_{0.07}Ga_{0.93}N$ layer (structure A; $t_1 = 500$ nm, $t_2 = 0$). The waveguide region consists of a 112-nm thick n-GaN layer, followed by an $In_{0.12}Ga_{0.88}N$ (3 nm)/GaN:Si (12 nm) MQW active region, a 20 nm thick $Al_{0.2}Ga_{0.8}N$:Mg electron-blocking layer and an 80 nm thick p-GaN layer. The LD structures are finally completed by growing a 125-period 2-nm/2-nm $Al_{0.15}Ga_{0.85}N$/GaN p-cladding layer followed by a 10-nm thick p-GaN capping layer for the electrical contact. The second structure is similar to structure A, but it includes a 300-nm thick n-doped $In_{0.17}Al_{0.83}N$ cladding ([Si] $\sim 3 \times 10^{19}$ cm^{-3}) LM to GaN, which is placed below the n-doped AlGaN cladding layer (structure B; $t_1 = 500$ nm, $t_2 = 300$ nm). Finally, a structure with a 300-nm thick $In_{0.17}Al_{0.83}N$ n-cladding ([Si] $\sim 3 \times 10^{19}$ cm^{-3}) instead of the n-doped AlGaN cladding has also been grown (structure C; $t_1 = 0$, $t_2 = 300$ nm).

Gain-guided LD structures with a cavity length of 800 μm and a stripe width of 2 μm were fabricated. The devices were characterized under pulsed current conditions, and lasing action at 415 nm was achieved on the three different structures. Threshold current densities of 39 kA/cm^2, 30 kA/cm^2, and 35 kA/cm^2 and differential quantum efficiencies of 0.33, 0.43, and 0.24 W/A were measured on 800 × 2 μm^2 devices for structures A, B, and C, respectively.

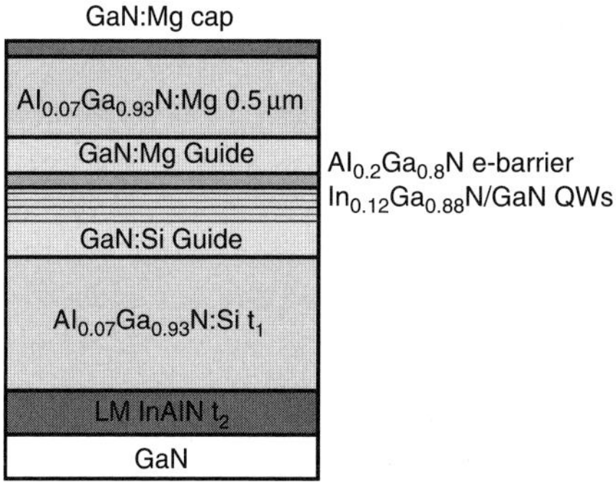

Comparison (gain guided 800 x 2 μm^2)		
Structure	J_{th}(kA/cm^2)	η_{diff}(W/A)
A t_1: 500 nm t_2: 0	39	0.33
B t_1: 500 nm t_2: 300 nm	30	0.43
C t_1: 0 t_2: 300 nm	35	0.24

FIG. 6.16. (a) Schematic cross-section of the LD structures under study, and (b) table summarizing the measured threshold current density and slope efficiency for the different grown structures.

It must be pointed out that the high values found for the threshold current density mainly originate from the weak lateral-mode confinement due to the lack of ridge structure. Devices including a hybrid $In_{0.17}Al_{0.83}N/Al_{0.07}Ga_{0.93}N$ cladding (structure B) show a threshold current 25% lower than standard devices (structure A), whereas the improvement is limited to 10% for devices having only a LM InAlN layer as bottom cladding (structure C). Structure B also exhibits the highest slope efficiency.

AlGaN claddings with an Al content between 4% and 10% are commonly adopted in standard violet LD structures. Here we fixed such composition to 7% and evaluated the influence of the n-AlGaN cladding thickness t_1 on the confinement factor and the transverse far-field quality.

In order to account for the experimental results and to evaluate the impact of the LM InAlN cladding on the mode confinement factor, optical modeling of LD structures on FS-GaN substrates is carried out. The field distribution in the transverse direction is calculated using the CAvity Model FRamework (CAMFR) full-vectorial Maxwell solver.[2] The refractive indices at 415 nm for AlGaN ternary alloys were extracted from experimental data measured by Brunner *et al.* (1997), while for LM InAlN the refractive index was taken from Carlin *et al.* (2005). The 125-period $GaN/Al_{0.15}Ga_{0.85}N$ p-cladding is identified with an AlGaN cladding having an average Al content of 7%. Due to the lack of reliable InGaN refractive index data the QWs refractive index was assumed to be ~2.6.

The optical confinement factor Γ_{act} is defined as the fraction of transverse optical intensity overlapping with the active region:

$$\Gamma_{act} = \frac{\int_{act} E_t^2(y)dy}{\int_{structure} E_t^2(y)dy} \tag{6.5}$$

where E_t is the transverse vectorial electric field. The mode confinement per QW (Γ_{1QW}) can then be defined as the ratio between Γ_{act} and the number of QWs. By extension, the mode overlap of each layer i can be defined as the fraction of optical intensity overlapping with a specific layer i. In this way the mode overlap with the n-GaN substrate (Γ_n) and with the p-GaN (Γ_p) layer can be evaluated in order to account for the mode leakage into the substrate and the optical losses related to Mg doping.

The optical confinement factor per QW Γ_{1QW}, the substrate leakage Γ_n, and the field overlap with the p-type region Γ_p can be evaluated for the structure shown in Fig. 6.16(a) as a function of the thicknesses t_1 (n-AlGaN thickness) and t_2 (n-InAlN thickness). As expected the optical mode confinement is improved when increasing the AlGaN bottom cladding thickness. Moreover, the mode confinement dramatically increases when a thicker n-InAlN thickness is used below the standard AlGaN cladding (Fig. 6.17(a)). From the modeling, a bare n-InAlN cladding having a thickness of 300 nm should be enough to

[2]P. Bienstman, L. Vanholme, M. Ibanescu, P. Dumon, and R. Baets: CAMFR—CAvity Modeling FRamework. Available at: http://camfr.sourceforge.net.

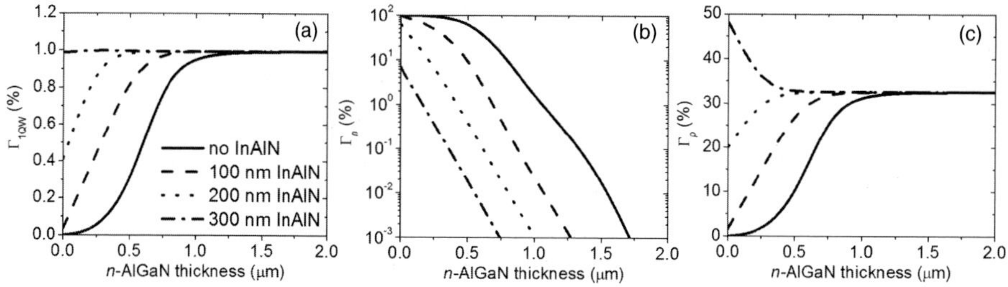

FIG. 6.17. Results derived from the optical modeling of LD structures depicted in Fig. 6.16(a) for different InAlN thicknesses as a function of the AlGaN thickness. (a) Confinement factor; (b) substrate leakage; (c) field overlap with p-type region. (Reprinted, with permission, from Castiglia *et al.* (2009), © 2009 American Institute of Physics.)

maximize the Γ_{1QW} value. Similarly, the implementation of a bottom InAlN cladding leads to a substantial decrease in the mode leakage into the substrate (Fig. 6.17(b)). As observed for the confinement factor, the mode leakage suppression is higher when the n-InAlN layer is placed closer to the waveguide due to the increased refractive index contrast between the waveguide and the underlying cladding. At the same time, the optical mode is pushed toward the p-type region when adding an InAlN cladding layer (Fig. 6.17(c)). The overlap with the p-type region Γ_p increases with t_2 and decreases when increasing the distance of the InAlN layer from the waveguide.

The modeling also suggests that when an InAlN cladding layer LM to GaN is placed directly below the n-type guide (structure C), the mode distribution inside the waveguide becomes strongly asymmetric due to the higher index contrast on the n-side compared to the p-side. As a consequence, the lasing mode is shifted toward the p-type region that may lead to higher absorption in the p-type layers (i.e., to increased internal losses). Thus, a hybrid InAlN/AlGaN cladding offers the best compromise, as it allows keeping a nearly symmetric LD waveguide design while suppressing further optical leakage in the substrate.

Blue LD structures including a thin LM InAlN bottom optical blocking layer (t_1=900 nm; t_2=100 nm) grown on c-plane freestanding GaN substrates have also been reported and compared to standard structures (t_1=1 μm; t_2=0) (Charash *et al.*, 2011).

Index-guided LD structures with a cavity length of 600 μm and a ridge width of 2 μm were fabricated. From the electrical point of view the structures, which include a 100 nm thick InAlN layer, suffer from an increased operating voltage. The $J-V$ characteristics are shifted by 1-2 V to higher voltages when inserting an InAlN optical blocking layer (OBL) below the LD structure (Fig. 6.18(a)). As previously stated, LM InAlN has a wide energy bandgap compared to GaN. The large conduction band offset will affect the electron flow in the direction of the

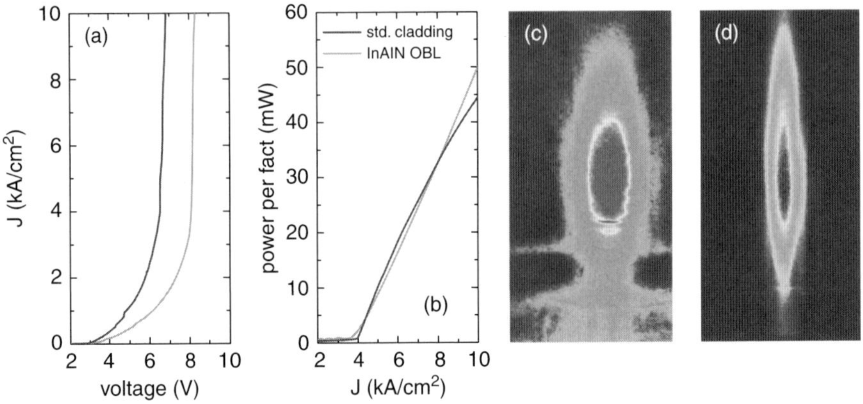

FIG. 6.18. (a) $V - J$ and (b) $L - J$ characteristics of LDs with a standard bottom cladding (dark line) and with an InAlN/AlGaN cladding (light gray line). Far-field patterns of LDs with (c) a standard bottom cladding, and (d) with an InAlN/AlGaN cladding. (Reprinted, with permission, from Castiglia *et al.* (2010), © 2010 American Institute of Physics.)

p-n junction, which likely leads to an increased operating voltage in structures with an InAlN OBL.

Lasing action is achieved at RT for both types of structures under cw electrical injection (Fig. 6.18(b)). They show a comparable threshold current density (around 4 kA/cm^2) and differential efficiency ($\eta_{diff} = 0.55$ W/A per facet without coating). The lack of improvement in the lasing properties of the InAlN OBL LDs can be explained by the presence of a 1-μm thick n-$Al_{0.07}Ga_{0.93}N$ cladding, which is already adequate to provide the required optical mode confinement (Fig. 6.17(a)). In principle, the presence of the InAlN OBL should allow decreasing the n-AlGaN thickness without compromising the optical mode confinement. Nonetheless, the InAlN OBL efficiently suppresses the parasitic mode leakage into the substrate. Measurements of the far-field emission patterns (Figs. 6.18(c) and 6.18(d)) indicate that the structures possess the same vertical divergence angle of 39°. However, the standard devices exhibit intensity dips on the negative part of the fast axis. This characteristic is a clear sign of parasitic mode leakage into the GaN substrate. This undesired feature is completely suppressed in the far-field of the InAlN containing LD, so that mode leakage into the substrate is strongly reduced.

To conclude this section, the InAlN alloy with an In content of 17–18% can be successfully used as a bottom cladding layer in edge-emitting LDs without degrading the device output performances. An improved beam shape is obtained due to the reduced mode leakage into the FS-GaN substrate. The high $\Delta n_{op}/n_{op}$ value together with the LM condition represent a promising approach for the development of green LDs, due to the increased refractive index contrast to GaN at long wavelengths. Note, however, that additional research on n-type InAlN is compulsory for achieving improved electrical characteristics.

6.7 InAlN/GaN high electron mobility transistors

Since the first realization of high-quality InAlN/GaN heterostructures, research on InAlN/GaN high electron mobility transistors (HEMTs) has rapidly progressed, and these devices emerged as an attractive alternative to their AlGaN/GaN counterparts. Indeed InAlN/GaN heterostructures offer various advantages both in terms of power-handling capability and operational frequency, which is inherited from the LM condition of InAlN to GaN. To be more quantitative the huge spontaneous polarization discontinuity at the InAlN/GaN heterointerface gives rise to a two-dimensional electron gas (2DEG) with density as high as 3×10^{13} cm^{-2} (Kuzmik, 2001). The conduction-band profile of a typical InAlN/GaN heterostructure and the corresponding electronic density are shown in Fig. 6.19. For comparison, in AlGaN/GaN heterostructures the important strain due to the lattice-mismatch between AlGaN and GaN limits the Al content of the barrier to ~30% and the 2DEG electron density to $\sim 1.5 \times 10^{13}$ cm^{-2}. The enhanced electron sheet density, combined with the high electron mobility (>1200 cm^2/V·s) provided by the insertion of an AlN spacer

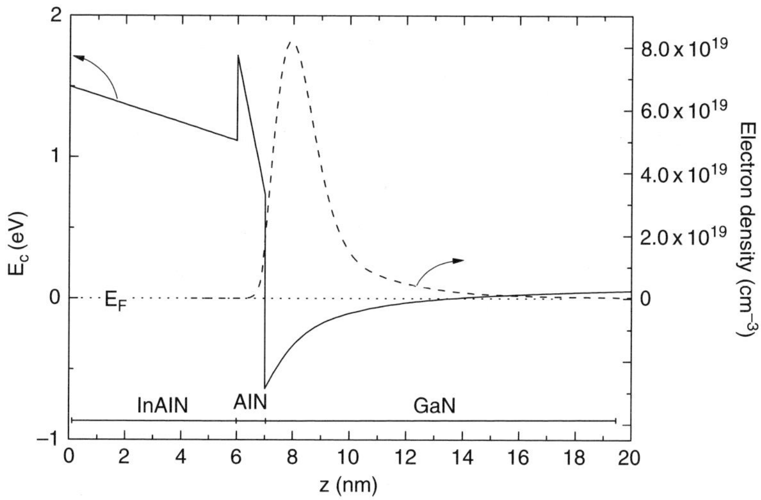

FIG. 6.19. Calculated conduction-band profile and electronic density for an InAlN/GaN heterostructure having a 6-nm thick InAlN barrier. A 1-nm AlN spacer is inserted between GaN and InAlN in order to reduce the penetration of the electronic wavefunction in the InAlN layer, minimizing in this way alloy scattering.

between GaN and InAlN and the high saturation velocity of GaN (3×10^7 cm/s), allows InAlN/GaN HEMTs to reach current densities of >2 A/mm (Sun *et al.*, 2010*c*), which roughly constitutes an improvement by a factor of two over AlGaN/GaN-based technologies. InAlN/GaN HEMTs therefore offer the opportunity of reaching power densities unattainable with other nitride-based systems.

At the same time, due to the reduced surface-related depletion effects in InAlN/GaN heterostructures, high-density 2DEGs can be achieved with barriers as thin as 10 nm or even less (Medjdoub *et al.*, 2008; Gonschorek *et al.*, 2008). This allows keeping a high gate aspect ratio even in deeply scaled-down devices without the need for performing a gate recess, thus avoiding excessive short channel effects in transistors with a sub-100-nm gate footprint. The scaling opportunities offered by InAlN/GaN HEMTs, combined with the high saturation electron velocity of GaN, open up the possibility of reaching cutoff frequencies well above 100 GHz, and therefore compete with the GaAs based technology. Indeed, very high cutoff frequencies in excess of 300 GHz have been obtained (Yue *et al.*, 2012), despite the relatively low maturity of the InAlN/GaN-based technology, demonstrating the impressive potential of this platform. InAlN/GaN HEMTs have also shown impressive temperature stability, with device operation demonstrated up to 900°C for 50 hours without failure (Maier *et al.*, 2010). Interestingly, the cause of failure at these extreme conditions has been identified to be the contact metallization rather than the semiconducting material itself. Indeed, operation at 1000°C has been achieved recently due to the implementation of refractory metal-based contacts (Maier *et al.*, 2012), which nicely shows the extremely robust nature of InAlN/GaN heterostructures. Furthermore, the high chemical stability of these heterostructures makes InAlN/GaN HEMTs the ideal amplifying element for harsh environment sensing, especially if combined with diamond as sensing material (Dipalo *et al.*, 2009). Furthermore, the absence of misfit strain in LM InAlN/GaN heterostructures is thought to be beneficial for the HEMT reliability. There are indeed evidences that in AlGaN/GaN HEMTs the accumulation of misfit strain and voltage-induced strain through the inverse piezoelectric effect can be responsible for device degradation and failure (Park *et al.*, 2009). InAlN/GaN HEMTs therefore combine several unique and exciting properties that make them ideal candidates for high-power and high-frequency applications. The state-of-the-art of the field and the future research axis are described in Section 6.7.1.

6.7.1 *Growth of InAlN/GaN high electron mobility heterostructures*

The main specificities of the growth of high electron mobility InAlN/GaN heterostructures have been already addressed by Butté *et al.* (2007). Here, we just recall that the insertion of a 1-nm AlN spacer between the semi-insulating GaN buffer and the InAlN barrier is necessary for achieving room-temperature mobilities of >1000 $cm^2/V\cdot s$ (Gonschorek *et al.*, 2006), and that smooth heterointerfaces are required in order to reduce interface roughness scattering, which after optical phonon scattering is the main mechanism limiting the mobility at room temperature (Tülek *et al.*, 2009). Variation in the In content of the InAlN barrier changes the spontaneous polarization mismatch with the GaN channel, and at the same time the lattice mismatch introduces a piezoelectric component to the polarization. This affects the 2DEG density, which is enhanced in heterostructures with lower In content (Mikulics *et al.*, 2010). However, lattice-mismatched heterostructures suffer from a lower stability (Gadanecz *et al.*, 2007), and highly strained layers may also present cracks (Liu *et al.*, 2012). Therefore, care must be taken to avoid excessive strain when growing lattice-mismatched heterostructures.

Most of the reports on InAlN/GaN heterostructures published up to now deal with MOVPE-grown materials. Although sapphire has been the first platform for the development of high-quality InAlN/GaN heterostructures (Gonschorek *et al.*, 2006), reports of heterostructures with high electron mobility and low sheet resistivity grown on SiC (Guo *et al.*, 2010) and Si (Cheng *et al.*, 2010) rapidly appeared. Indeed, high-resistivity buffer layers for these different substrates have been extensively optimized for AlGaN/GaN-based applications, and their transfer to InAlN/GaN heterostructures was therefore relatively straightforward. Low sheet resistivities, around 200 $\Omega/\square$, have been reported for all these substrates. The frequency and power performances of MOVPE-grown HEMTs are reviewed in Sections 6.7.2 and 6.7.3. More recently, the growth of InAlN/GaN heterostructures by pulsed MOVPE (PMOVPE) has been reported, both on sapphire (Xue *et al.*, 2011*b*) and SiC (Xue *et al.*, 2011*a*). In PMOVPE ammonia and the metalorganic precursors are supplied as a sequence of separate pulses, which should promote the diffusion of the precursors on the growing surface and therefore improve the InAlN quality. However, despite these potential advantages, the mobility of PMOVPE and MOVPE-grown 2DEGs are up to now quite similar.

Reports on metal-polar MBE-grown InAlN/GaN heterostructures are instead much scarcer. While PAMBE has the serious drawback of producing InAlN layers of poor uniformity and with lateral composition inhomogeneities (Choi *et al.*, 2012), uniform lattice-matched InAlN/GaN heterostructures with high electron mobility (1600 $\text{cm}^2/\text{V}\cdot\text{s}$) and low sheet resistivity (244 $\Omega/\square$) have been obtained recently with ammonia-MBE (Wong *et al.*, 2012), which provides a growth environment closer to the MOVPE one than PAMBE. The research on MBE-grown InAlN/GaN HEMTs is, however, in its early stage, and works really addressing the performances of MBE-grown HEMTs are still lacking.

N-polar heterostructures have also been recently explored. In such heterostructures the inversion of the polarization with respect to metal-polar heterostructures requires the GaN channel to be grown on top of the InAlN barrier. The 2DEG is therefore contacted from the GaN channel and not from the InAlN barrier as in metal-polar HEMTs. This constitutes an interesting strategy for the minimization of the contact resistance (Nidhi *et al.*, 2010) which, as described in Section 6.7.2, is a key requirement for the fabrication of high-frequency devices. N-polar InAlN/GaN heterostructures have been obtained both by MOVPE (Brown *et al.*, 2010) and PAMBE (Dasgupta *et al.*, 2011), but this field is still in its infancy, and reports on processed HEMTs are still limited. However, N-polar InAlN/GaN HEMTs have already shown record transconductance of 1105 mS/mm (Nidhi *et al.*, 2012) and are therefore highly promising for high-frequency applications.

6.7.2 *InAlN/GaN HEMTs for high-frequency applications*

The frequency behavior of transistors is usually described by the current and power gain cutoff frequencies f_t and f_{max}, i.e., the frequencies at which the current or power gain become unitary, respectively. The gain cutoff frequencies of an HEMT depend on many parameters. While the intrinsic properties of the semiconductor material define the ultimate performance attainable in an ideal transistor, the experimental values of f_t and f_{max} are always affected by the presence of parasitic elements that introduce delay times, feedback, and power consumption. A careful device design and processing is therefore as important as the quality of the epitaxial material if high-performance transistors are desired. As a general rule, the current cutoff frequency f_t is proportional to the inverse of the electron transit time under the gate, which gives the following dependence on the gate length L_G:

$$f_t = \frac{v_t}{2\pi L_G} \tag{6.6}$$

where v_t is the electron transit velocity. This ideal equation is valid as long as parasitic delay times and short channel effects are not strong. Parasitic delay times are mainly due to contact and access resistances, and are reduced by minimizing the Ohmic contact resistivity and by decreasing the drain-source and gate-source separation. Ohmic contacts to the 2DEG are usually obtained by deposition of a Ti/Al-based metal stack followed by rapid thermal annealing (Zhou *et al.*, 2010*a*). This allows for a contact resistivity of about 0.3 Ω·mm. For lower contact resistivity, alternative solutions have to be adopted. Regrown Ohmic contacts seem very promising in this sense. The regrown scheme consists in the etching of the barrier and part of the buffer in the contact region, followed by the regrowth, usually by MBE (Guo *et al.*, 2011) of highly doped *n*-GaN, which is eventually contacted with standard Ti/Al contacts. This has allowed for a substantial reduction of the contact resistivity, down to 0.16 Ω·mm (Yue *et al.*, 2012), and the achievement of record f_t of 370 GHz in 30-nm gate devices.

On the other hand, short channel effects become important when the gate lateral dimension becomes comparable to the gate-channel separation. For InAlN/GaN HEMTs, where the InAlN barrier can be made as thin as 10 nm without suffering from surface-related depletion effects, short channel effects become important only at gate lengths of 50 nm or less, which means that sub-100-nm gates can be realized without the need for a gate recess. However, strong short channel effects have been observed with 30-nm long gates (Tirelli *et al.*, 2011; Lee *et al.*, 2011*b*). Therefore, the further scaling down of the gate lateral dimension will require new strategies in order to thin down further the barrier while keeping a high gate aspect ratio without inducing significant surface-related depletion effects. Furthermore, the use of an AlGaN (Lee *et al.*, 2011*c*) or InGaN (Lee *et al.*, 2011*b*) back barrier improves the 2DEG confinement, which looks quite promising for short-channel-effect attenuation.

Concerning f_{max}, its maximization requires both a high f_t and the reduction of the gate resistance and of the parasitic elements that introduce feedback in the transistor, in particular the gate-drain capacitance. The use of T-shaped gates (Fig. 6.20(a)) is therefore particularly important for the fabrication of high-performance HEMTs, as it combines a short gate footprint with a wide cross-section for gate resistance minimization. Rectangular gates (Fig. 6.20(b)), although characterized by lower capacitances allowing for $f_t > 300$ GHz, are generally quite resistive and do not permit high f_{max}. At present, the best f_{max} value reported for an InAlN/GaN HEMT is 220 GHz (Tirelli *et al.*, 2011), obtained in a fully passivated device with a T-shaped gate and a 30-nm footprint.

The values of f_t and f_{max} reported so far for sub-μm gate InAlN/GaN HEMTs as a function of the gate length are summarized in Fig. 6.21, where it can be seen that f_t scales indeed as the reverse of the gate length, with an approximately constant $f_t \cdot L_G$ product of $\sim$ 10 GHz·μm, independent of the substrate material used. It has to be noted, however, that the fastest transistors have been obtained on SiC. f_{max} values are instead more scattered and do not follow this simple trend, which is mainly the consequence of the strong effect of the gate resistance on this parameter.

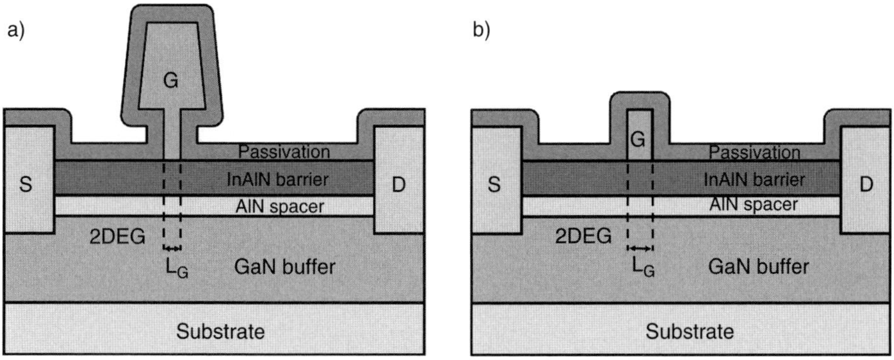

FIG. 6.20. Schematic cross-section of InAlN/GaN HEMTs with (a) T-shaped gate for gate resistance minimization and simultaneous maximization of f_t and f_{max}; (b) rectangular gate for gate capacitance reduction and maximization of f_t.

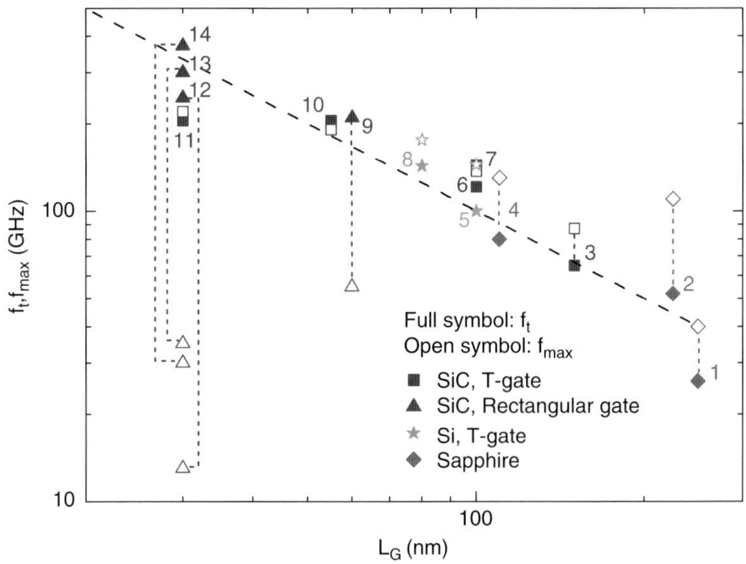

FIG. 6.21. Reported values for the cutoff frequencies f_t and f_{max} of InAlN/GaN HEMTs as a function of the gate length L_G. The data refer to HEMTs grown on sapphire, Si, and SiC substrates. The values of f_t show a good scaling with the reverse of the gate length and are fitted with a constant $f_t \cdot L_G$ product of 10 GHz·µm. The larger scattering of the f_{max} values is due to the dependence of f_{max} on other transistor parameters, in particular the gate resistance. Data are taken from: 1. Medjdoub *et al.* (2007*a*), 2. Lecourt *et al.* (2011*b*), 3. Chung *et al.* (2009), 4. Lecourt *et al.* (2011*a*), 5. Sun *et al.* (2009), 6. Sun *et al.* (2010*a*), 7. Sun *et al.* (2010*b*), 8. Sun *et al.* (2010*d*), 9. Wang *et al.* (2011*a*), 10. Sun *et al.* (2010*c*), 11. Tirelli *et al.* (2011), 12. Lee *et al.* (2011*a*), 13. Lee *et al.* (2011*b*), 14. Yue *et al.* (2012).

It is also worth noting that high-speed InAlN/GaN HEMTs are adapted for low-noise operation. Minimum noise figures of 0.62 and 1.5 dB have been reported at 10 and 20 GHz, respectively, for a device grown on SiC (Sun *et al.*, 2010*a*), showing the great potentials of the InAlN/GaN-based technology for low-noise microwave applications.

6.7.3 *Power performance of InAlN/GaN HEMTs*

If sapphire has been the first substrate material for InAlN/GaN HEMTs, its poor thermal conductivity (42 W/m·K) seriously limits the thermal dissipation, causing strong self-heating effects and limited current densities (Gonschorek *et al.*, 2011). At present, reports on the power performances of InAlN/GaN HEMTs on sapphire are rather scarce. In the early days of InAlN/GaN HEMTs, an output power of 1.5 W/mm at 10 GHz with corresponding power added efficiency (PAE) of 18% was reported (Gaquière *et al.*, 2007), while more recently an output power of 2.9 W/mm at 18 GHz with a PAE of 28% has been obtained (Lecourt *et al.*, 2011*b*). For high-power devices, however, more efficient heat dissipation is required, and other substrate materials have been investigated with the aim of improving thermal management aspects. SiC has been up to now the most successful substrate material, as its high thermal conductivity (390 W/m·K) allows for very efficient heat extraction. Indeed, InAlN/GaN HEMTs grown on SiC have shown excellent power performances, with power densities of 10 W/mm at 10 GHz (Sarazin *et al.*, 2010), 5.8 W/mm at 35 GHz (Crespo *et al.*, 2010) and a PAE of about 50%. Despite the high performances of InAlN/GaN HEMTs on SiC, the high cost of SiC wafers will probably limit their application to fields where performances are the priority, such as highly efficient power amplifiers or the next generation of radars (Pengelly *et al.*, 2012). For widespread applications, cheaper substrate materials are needed. Si combines high thermal conductivity (148 W/m·K) with low prices, and would be therefore the best trade-off between cost and power density. Up to now, the only report on the power performance of InAlN/GaN HEMTs on Si concerns a device displaying a 2.5 W/mm maximum output power at 10 GHz (Sun *et al.*, 2009), but it is expected that the rapid progress made in the InAlN/GaN HEMT technology on Si will lead to improved performances.

At present, the output power of InAlN/GaN HEMTs is limited by the relatively low breakdown voltage of these devices due to parasitic gate leakage currents as high as several mA/mm (Sun *et al.*, 2010*b*), mainly originating from the gate electrode. Research efforts are therefore under way in order to understand the origin of these currents. In analogy with the work of Zhang *et al.* (2006) for AlGaN/GaN heterostructures, Arslan *et al.* (2009) and Chikhaoui *et al.* (2010) attributed the reverse bias current of Ni/Au Schottky contacts on InAlN/GaN heterostructures to a Frenkel–Poole emission process from a trap level located near the metal–semiconductor interface into a continuum of states associated with threading dislocations and located 0.37 eV below the conduction band edge. The activation energy for the emission process was found to be 0.12–0.13 eV. However, the nature of this trap level has not been determined yet. Furthermore, Frenkel–Poole emission is an efficient process only if the emitting trap level can be efficiently populated, i.e., if it lies close to the Fermi level of the metal electrode. This constitutes a weak point, as a Schottky barrier height of >1.47 eV has been reported for Ni/InAlN junctions (Donoval *et al.*, 2011), which is too high to be compatible with the proposed mechanism. The mechanism behind gate leakage is therefore not completely clear at present, and further investigations are needed in order to better understand the transport properties of InAlN Schottky contacts.

A very effective way of reducing gate leakage currents is the insertion of an insulating dielectric layer between the heterostructure and the gate metal. Several oxides have been investigated, such as HfO_2 (Abermann *et al.*, 2009),

ZrO_2 (Abermann *et al.*, 2007; Abermann *et al.*, 2009; Kuzmik *et al.*, 2008), Al_2O_3 (Pozzovivo *et al.*, 2007), and $GdScO_3$ (Čičo *et al.*, 2011), deposited both by MOVPE or atomic layer deposition (ALD) (Abermann *et al.*, 2009). A significant reduction in the gate leakage current, ranging from three to six orders of magnitude depending on the InAlN surface preparation method, has been reported for all the materials (Abermann *et al.*, 2007; Pozzovivo *et al.*, 2007; Čičo *et al.*, 2011). Moreover, the $f_t \cdot L_G$ product was not degraded in HEMTs with gate insulation (Medjdoub *et al.*, 2007*b*), and pulsed measurements revealed that the passivation of the InAlN surface effectively reduces the current collapse observed in Schottky barrier HEMTs (Kuzmik *et al.*, 2008). Interestingly, passivation with Al_2O_3 induces a significant reduction in the sheet resistivity (Chung *et al.*, 2009), probably due to an oxide-induced shift of the InAlN surface potential, which makes Al_2O_3 an attractive passivation dielectric even for Schottky barrier HEMTs, where the gate metal is not isolated from the InAlN surface (Chung *et al.*, 2009). Insulation with Al_2O_3 can also be achieved through the thermal oxidation of the InAlN barrier at elevated temperature (~800°C) in an O_2-rich atmosphere. This diffusion-limited oxidation process (Alomari *et al.*, 2010) allows for the growth of a few nm thick Al_2O_3 passivation layer that effectively reduces the gate leakage currents. The interface trap density of the thermally grown oxide was measured to be 2.5×10^{12} cm^{-2} (Eickelkamp *et al.*, 2011), which corresponds to the values reported for MOVPE and ALD-grown Al_2O_3. From current–voltage measurements, a barrier height of 2.30–2.38 eV was evaluated, which compares well to the value of 2.57 eV obtained from photocurrent measurements (Kováč *et al.*, 2011). An interesting aspect of the thermal oxidation of InAlN is the possibility of performing it on selected regions defined lithographically. This has allowed for a self-aligned process where the InAlN surface is oxidized just in the gate region, allowing for the passivation of the access and output regions of the HEMT with other materials such as Si_3N_4 (Alomari *et al.*, 2009).

6.7.4 *Enhancement-mode InAlN/GaN HEMTs*

The devices presented so far are depletion-mode (D-mode), or normally-on HEMTs, i.e., the threshold gate-source voltage (V_{th}) below which the 2DEG is completely depleted and the transistor is pinched off is negative. However, positive V_{th} values would be of high interest, especially for power applications, as this would allow for failsafe operation and reduced circuit complexity. Research efforts have therefore been devoted to the realization of enhancement-mode (E-mode), or normally-off, InAlN/GaN HEMTs. In a first attempt, a high-In-content InAlN/GaN heterostructure was proposed with the aim of engineering the polarization charge at the heterointerface and the 2DEG density (Choi *et al.*, 2010). This resulted in a device with a threshold voltage of +0.2 V, but the high resistivity of the 2DEG in the access and output regions was responsible for a very low current density and an elevated on-resistance. If this was the first demonstration

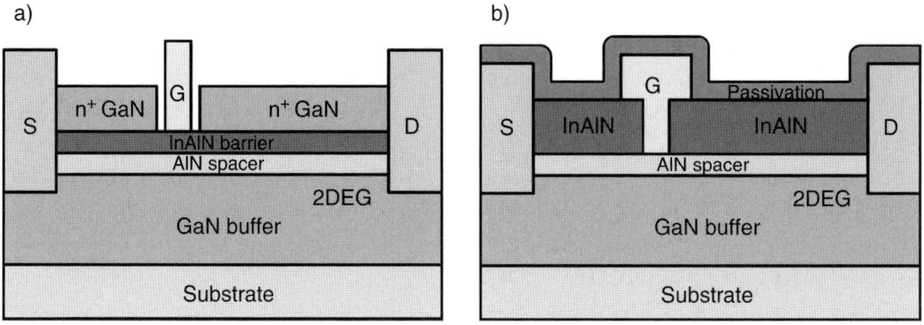

FIG. 6.22. Schematics of the E-mode InAlN/GaN HEMT concepts. Structure (a) has an ultrathin InAlN barrier of 1 nm, lower than the critical thickness for 2DEG formation, and a relatively thick (>5 nm) n^+ GaN cap providing electrons to the 2DEG. The gate is deposited after selective etching of the GaN cap layer, thus depleting the 2DEG in the gate region. Structure (b) is instead characterized by a thick InAlN barrier. The E-mode behavior is obtained by selectively etching the barrier in the gate region.

of a normally-off InAlN/GaN HEMT, it pointed out at the same time the need for solutions able to produce devices with low access and output resistance.

Figure 6.22 illustrates two device concepts, both based on a selective gate recess, that have produced high-performance normally-off InAlN/GaN HEMTs. The first approach (Fig. 6.22(a)) is based on an InAlN/GaN heterostructure with an ultrathin 1-nm thick nearly LM InAlN barrier, well below the critical thickness for 2DEG formation (Ostermaier *et al.*, 2009). The heterostructure is capped with a heavily doped (10^{20} cm^{-3}) 6-nm thick n^+ GaN layer, which provides carriers for the 2DEG formation and at the same time saturates the surface trap levels (Kuzmik *et al.*, 2010), preventing the high-frequency current collapse. A selective fluorine-based plasma etching of the GaN capping layer (Ostermaier *et al.*, 2010*a*) allows placing the gate electrode on the InAlN barrier, thus selectively depleting the 2DEG in the gate region. A positive threshold voltage of +0.7 V is therefore obtained while at the same time keeping low the access and output resistance. Furthermore, a high $f_t \cdot L_G$ product of 16.8 GHz·μm was reported, showing the high potential of E-mode InAlN/GaN HEMTs for high-frequency applications. Later technological optimizations produced a device with a very high transconductance of 640 mS/mm (Ostermaier *et al.*, 2010*b*), while the introduction of a ZrO_2 insulating layer between the gate metal and the InAlN barrier allowed engineering the threshold voltage up to +2.3 V (Alexewicz *et al.*, 2012).

A second possible approach is presented in Fig. 6.22(b). Here a more standard InAlN/GaN heterostructure is adopted with a thickness above the critical one for the 2DEG formation. This gives a low sheet resistivity and consequently keeps low the access and output resistance, while the normally-off operation is obtained by selectively etching the InAlN barrier down to the AlN spacer layer in the gate region (Wang *et al.*, 2010). The implementation of this approach on SiC demonstrated the feasibility of E-mode InAlN/GaN HEMTs with high saturated current density (1.9 A/mm) and transconductance as high as 800 mS/mm. Very low leakage levels and ON/OFF current ratios as high as 10^{12} have been subsequently achieved by means of a postprocessing annealing at 400°C for interface defect reduction (Wang *et al.*, 2011*b*), while relatively high (>1.5 V) threshold voltages have been obtained through the insertion of an insulating layer below the gate (Morgan *et al.*, 2011).

6.8 Conclusion

In conclusion, the main growth parameters affecting the quality of InAlN layers nearly LM to GaN grown by MOVPE, including impurity incorporation, have been first reported with an emphasis on the critical role of the growth temperature as well as those of the precursor flux and the carrier gas flow on the indium incorporation and the layer growth rate. The nature of the main structural defects and their evolution with increasing layer thickness has also been addressed.

Regarding optical properties, it was pointed out that even if a composition-dependent band-gap bowing parameter can better account for the evolution of the unstrained band-gap over the whole indium composition range, a b value of ~5.4 eV will reasonably reproduce experimental values. Following recent theoretical work, it was suggested that the large reported Stokes shift for the InAlN might probably be inherent to the various atomic configurations for indium atoms within the wurtzite phase, leading to regions with strongly varying band-edge energy and thus to a strong difference in absorption and in emission spectra. XPS measurements and theoretical studies also indicate a type I GaN/InAlN heterointerface characterized with a small VBO amounting to ~0.2 eV. PL measurements performed on GaN/InAlN QWs show that such a system could constitute a promising active region for near-UV emitters compared with their GaN/AlGaN counterparts. The larger IQE at RT of the former compared with the latter could be due to an enhanced localization energy and band offsets minimizing thermalization escape phenomena.

Switching to optoelectronic applications, the potential of crack-free highly reflective InAlN/GaN DBRs grown on FS-GaN substrates was illustrated via the achievement of electrically driven VCSELs and optically pumped VECSELs. It was pointed out that the quality of the substrate and its surface preparation are probably playing a major role on the in-plane photonic disorder. It was also demonstrated that the InAlN alloy can be successfully used as a bottom cladding layer in edge-emitting LDs while preserving the device output performances. Thanks to the large refractive index contrast of InAlN with GaN, a reduced mode leakage and an improved far-field pattern over AlGaN-based claddings are reported.

Finally, the present status of InAlN/GaN HEMTs has been summarized. It has been recalled that such structures offer very promising possibilities for high-power and high-frequency applications up to very high temperatures (exceeding several hundreds of degrees Celsius) with a predicted high reliability owing to the absence of misfit strain. High current cutoff frequency operation is inherited from the large electron transit velocity and the possibility of achieving short gate length while using a design avoiding short channel effects. When grown on SiC substrates, power densities of several W/mm (e.g., 10 W/mm at 10 GHz) and a PAE of 50% have been reported for InAlN/GaN HEMTs. For devices grown on Si the figures of merit are lower, but fast progress is expected in this rapidly evolving field. It was also highlighted that such HEMTs offer promising perspectives for the realization of E-mode, or normally-off, devices that would allow for failsafe operation and reduced circuit complexity. All the above-mentioned aspects fully reveal the tremendous potential of the InAlN alloy both for optoelectronic and microelectronic applications. Provided a continuous improvement of the InAlN layers is indeed sustained, it is probable that such a material system will play a crucial role in forthcoming commercial III-nitride-based devices.

Acknowledgments

The contribution of Dr Eric Feltin is acknowledged for the optical measurements carried out on QWs. This work was supported by NCCR Quantum Photonics, research instrument of the Swiss National Science Foundation (SNSF), through Grant No. 200020-113542, the SNSF through Grant No. 200021-132096; the CTI-KTI project "ITO Sputtering Process for GaN-based Solid State Lighting"; the European Commission through the MORGAN project, contract FP7 NMP IP 214610; and the Marie Curie ITN Rainbow project, contract FP7 PITN-GA-2008-213238.

References

Abermann, S., Pozzovivo, G., Kuzmik, J., Ostermaier, C., Henkel, C., Bethge, O., Strasser, G., Pogany, D., Carlin, J.-F., Grandjean, N., and Bertagnolli, E. (2009). *Electronics Letters*, **45**(11), 570–572.

Abermann, S., Pozzovivo, G., Kuzmik, J., Strasser, G., Pogany, D., Carlin, J.-F., Grandjean, N., and Bertagnolli, E. (2007). *Semiconductor Science and Technology*, **22**(12), 1272–1275.

Akazawa, M., Matsuyama, T., Hashizume, T., Hiroki, M., Yamahata, S., and Shigekawa, N. (2010). *Applied Physics Letters*, **96**(13), 132104.

Alexewicz, A., Ostermaier, C., Henkel, C., Bethge, O., Carlin, J.-F., Lugani, L., Grandjean, N., Bertagnolli, E., Pogany, D., and Strasser, G. (2012). *Thin Solid Films*, **520**(19), 6230–6232.

Alomari, M., Chuvilin, A., Toth, L., Pecz, B., Carlin, J.-F., Grandjean, N., Gaquière, C., di Forte-Poisson, M.-A., Delage, S., and Kohn, E. (2010). *physica status solidi (c)*, **7**(1), 13–16.

Alomari, M., Medjdoub, F., Carlin, J.-F., Feltin, E., Grandjean, N., Chuvilin, A., Kaiser, U., Gaquière, C., and Kohn, E. (2009). *Electron Device Letters, IEEE*, **30**(11), 1131–1133.

Altoukhov, A., Levrat, J., Feltin, E., Carlin, J.-F., Castiglia, A., Butté, R., and Grandjean, N. (2009). *Applied Physics Letters*, **95**(19), 191102.

Arslan, E., Bütün, S., and Ozbay, E. (2009). *Applied Physics Letters*, **94**(14), 142106.

Aschenbrenner, T., Dartsch, H., Kruse, C., Anastasescu, M., Stoica, M., Gartner, M., Pretorius, A., Rosenauer, A., Wagner, T., and Hommel, D. (2010). *Journal of Applied Physics*, **108**(6), 063533.

Brice, H. R., Sadler, T. C., Kappers, M. J., and Oliver, R. A. (2010). *Journal of Crystal Growth*, **312**(11), 1800–1806.

Brown, D. F., Keller, S., Mates, T. E., Speck, J. S., DenBaars, S. P., and Mishra, U. K. (2010). *Journal of Applied Physics*, **107**(3), 033509.

Brunner, D., Angerer, H., Bustarret, E., Freudenberg, F., Höpler, R., Dimitrov, R., Ambacher, O., and Stutzmann, M. (1997). *Journal of Applied Physics*, **82**(10), 5090–5096.

Butté, R., Carlin, J.-F., Feltin, E., Gonschorek, M., Nicolay, S., Christmann, G., Simeonov, D., Castiglia, A., Dorsaz, J., Buehlmann, H. J., Christopoulos, S., Baldassarri Höger von Högersthal, G., Grundy, A. J. D., Mosca, M., Pinquier, C., Py, M. A., Demangeot, F., Frandon, J., Lagoudakis, P. G., Baumberg, J. J., and Grandjean, N. (2007). *Journal of Physics D: Applied Physics*, **40**(20), 6328–6344.

Butté, R., Christmann, G., Feltin, E., Castiglia, A., Levrat, J., Cosendey, G., Altoukhov, A., Carlin, J.-F., and Grandjean, N. (2009). *Proc. SPIE 7216, Gallium Nitride and Devices*, IV, 721619.

Butté, R., Feltin, E., Dorsaz, J., Christmann, G., Carlin, J.-F., Grandjean, N., and Ilegems, M. (2005). *Japanese Journal of Applied Physics*, **44**(10), 7207–7216.

Carlin, J.-F. and Ilegems, M. (2003). *Applied Physics Letters*, **83**(4), 668–670.

Carlin, J.-F., Zellweger, C., Dorsaz, J., Nicolay, S., Christmann, G., Feltin, E., Butté, R., and Grandjean, N. (2005). *physica status solidi (b)*, **242**(11), 2326–2344.

Castiglia, A., Carlin, J.-F., Feltin, E., Cosendey, G., Dorsaz, J., and Grandjean, N. (2010). *Applied Physics Letters*, **97**(11), 111104.

Castiglia, A., Feltin, E., Cosendey, G., Altoukhov, A., Carlin, J.-F., Butté, R., and Grandjean, N. (2009). *Applied Physics Letters*, **94**(19), 193506.

Castiglia, A., Feltin, E., Dorsaz, J., Cosendey, G., Carlin, J.-F., Butté, R., and Grandjean, N. (2008). *Electronics Letters*, **44**(8), 521–522.

Castiglia, A., Simeonov, D., Buehlmann, H. J., Carlin, J.-F., Feltin, E., Dorsaz, J., Butté, R., and Grandjean, N. (2007). *Applied Physics Letters*, **90**(3), 033514.

Charash, R., Kim-Chauveau, H., Lamy, J.-M., Akther, M., Maaskant, P. P., Frayssinet, E., de Mierry, P., Dräger, A. D., Duboz, J-Y., Hangleiter, A., and Corbett, B. (2011). *Applied Physics Letters*, **98**(20), 201112.

Chen, Z. T., Fujita, K., Ichikawa, J., and Egawa, T. (2012). *Journal of Applied Physics*, **111**(5), 053535.

Chen, Z. T., Sakai, Y., and Egawa, T. (2010). *Applied Physics Letters*, **96**(19), 191911.

Cheng, K., Degroote, S., Leys, M., Medjdoub, F., Derluyn, J., Sijmus, B., Germain, M., and Borghs, G. (2010). *physica status solidi (c)*, **7**(7–8), 1967–1969.

Chikhaoui, W., Bluet, J.-M., Poisson, M.-A., Sarazin, N., Dua, C., and Bru-Chevallier, C. (2010). *Applied Physics Letters*, **96**(7), 072107.

Cho, H. K., Lee, J. Y., Yang, G. M., and Kim, C. S. (2001). *Applied Physics Letters*, **79**(2), 215–217.

Choi, Suk, Kim, Hee Jin, Lochner, Zachary, Zhang, Yun, Lee, Yi-Che, Shen, Shyh-Chiang, Ryou, Jae-Hyun, and Dupuis, Russell D. (2010). *Applied Physics Letters*, **96**(24), 243506.

Choi, S., Wu, F., Shivaraman, R., Young, E. C., and Speck, J. S. (2012). *Applied Physics Letters*, **100**(23), 232102.

Christmann, G., Butté, R., Feltin, E., Carlin, J.-F., and Grandjean, N. (2008). *Applied Physics Letters*, **93**(5), 051102.

Christopoulos, S., Baldassarri Höger von Högersthal, G., Grundy, A. J. D., Lagoudakis, P. G., Kavokin, A. V., Baumberg, J. J., Christmann, G., Butté, R., Feltin, E., Carlin, J.-F., and Grandjean, N. (2007). *Physical Review Letters*, **98**(12), 126405.

Chung, J. W., Saadat, O. I., Tirado, J. M., Gao, X., Guo, S., and Palacios, T. (2009). *Electron Device Letters, IEEE*, **30**(9), 904–906.

Chung, R. B., Wu, F., Shivaraman, R., Keller, S., DenBaars, S. P., Speck, J. S., and Nakamura, S. (2011). *Journal of Crystal Growth*, **324**(1), 163–167.

Cosendey, G., Carlin, J.-F., Kaufmann, N. A. K., Butté, R., and Grandjean, N. (2011). *Applied Physics Letters*, **98**(18), 181111.

Cosendey, G., Castiglia, A., Rossbach, G., Carlin, J.-F., and Grandjean, N. (2012). *Applied Physics Letters*, **101**(15), 151113.

Crespo, A., Bellot, M. M., Chabak, K. D., Gillespie, J. K., Jessen, G. H., Miller, V., Trejo, M., Via, G. D., Walker, D. E., Winningham, B. W., Smith, H. E., Cooper, T. A., Gao, X., and Guo, S. (2010). *Electron Device Letters, IEEE*, **31**(1), 2–4.

Darakchieva, V., Beckers, M., Xie, M.-Y., Hultman, L., Monemar, B., Carlin, J.-F., Feltin, E., Gonschorek, M., and Grandjean, N. (2008). *Journal of Applied Physics*, **103**(10), 103513.

Dasgupta, S., Nidhi, Choi, S., Wu, F., Speck, J. S., and Mishra, U. K. (2011). *Applied Physics Express*, **4**(4), 045502.

de Carvalho, L. C., Schleife, A., Furthmüller, J., and Bechstedt, F. (2012). *Physical Review B*, **85**(11), 115121.

Dipalo, M., Gao, Z., Scharpf, J., Pietzka, C., Alomari, M., Medjdoub, F., Carlin, J.-F., Grandjean, N., Delage, S., and Kohn, E. (2009). *Diamond and Related Materials*, **18**(5–8), 884–889.

Donoval, D., Chvála, A., Šramatý, R., Kováč, J., Morvan, E., Dua, Ch., DiForte-Poisson, M. A., and Kordoš, P. (2011). *Journal of Applied Physics*, **109**(6), 063711.

Dorsaz, J., Bühlmann, H.-J., Carlin, J.-F., Grandjean, N., and Ilegems, M. (2005*a*). *Applied Physics Letters*, **87**(7), 072102.

Dorsaz, J., Carlin, J.-F., Gradecak, S., and Ilegems, M. (2005*b*). *Journal of Applied Physics*, **97**(8), 084505.

Eickelkamp, M., Weingarten, M., Khoshroo, L. Rahimzadeh, Ketteniss, N., Behmenburg, H., Heuken, M., Donoval, D., Chvála, A., Kordoš, P., Kalisch, H., and Vescan, A. (2011). *Journal of Applied Physics*, **110**(8), 084501.

Feltin, E., Carlin, J.-F., Dorsaz, J., Christmann, G., Butté, R., Laügt, M., Ilegems, M., and Grandjean, N. (2006). *Applied Physics Letters*, **88**(5), 051108.

Feltin, E., Christmann, G., Dorsaz, J., Castiglia, A., Carlin, J.-F., Butté, R., Grandjean, N., Christopoulos, S., Baldassarri Höger von Högersthal, G., Grundy, A. J. D., Lagoudakis, P. G., and Baumberg, J. J. (2007*a*). *Electronics Letters*, **43**(17), 924–926.

Feltin, E., Simeonov, D., Carlin, J.-F., Butté, R., and Grandjean, N. (2007*b*). *Applied Physics Letters*, **90**(2), 021905.

Ferhat, M. and Bechstedt, F. (2002). *Physical Review B*, **65**(07), 075213.

Fernández-Garrido, S., Gačević, Ž., and Calleja, E. (2008). *Applied Physics Letters*, **93**(19), 191907.

Gadanecz, A., Bläsing, J., Dadgar, A., Hums, C., and Krost, A. (2007). *Applied Physics Letters*, **90**(22), 221906.

Gaquière, C., Medjdoub, F., Carlin, J.-F., Vandenbrouck, S., Delos, E., Feltin, E., Grandjean, N., and Kohn, E. (2007). In *Microwave Symposium, 2007. IEEE/MTT-S International*, pp. 2145–2148.

Gačević, Ž., Fernández-Garrido, S., Rebled, J. M., Estradé, S., Peiró, F., and Calleja, E. (2011). *Applied Physics Letters*, **99**(3), 031103.

Gonschorek, M., Carlin, J.-F., Feltin, E., Py, M. A., and Grandjean, N. (2006). *Applied Physics Letters*, **89**(6), 062106.

Gonschorek, M., Carlin, J.-F., Feltin, E., Py, M. A., and Grandjean, N. (2011). *Journal of Applied Physics*, **109**(6), 063720.

Gonschorek, M., Carlin, J.-F., Feltin, E., Py, M. A., Grandjean, N., Darakchieva, V., Monemar, B., Lorenz, M., and Ramm, G. (2008). *Journal of Applied Physics*, **103**(9), 093714.

Gorczyca, I., Kamińska, A., Staszczak, G., Czernecki, R., Łepkowski, S. P., Suski, T., Schenk, H. P. D., Glauser, M., Butté, R., Carlin, J.-F., Feltin, E., Grandjean, N., Christensen, N. E., and Svane, A. (2010). *Physical Review B*, **81**(23), 235206.

Gorczyca, I., Łepkowski, S. P., Suski, T., Christensen, N. E., and Svane, A. (2009). *Physical Review B*, **80**(07), 075202.

Guo, J., Cao, Y., Lian, C., Zimmermann, T., Li, G., Verma, J., Gao, X., Guo, S., Saunier, P., Wistey, M., Jena, D., and Xing, H. (Grace) (2011). *physica status solidi (a)*, **208**(7), 1617–1619.

Guo, S., Gao, X., Gorka, D., Chung, J. W., Wang, H., Palacios, T., Crespo, A., Gillespie, J. K., Chabak, K., Trejo, M., Miller, V., Bellot, M., Via, G., Kossler, M., Smith, H., and Tomich, D. (2010). *physica status solidi (a)*, **207**(6), 1348–1352.

Holder, C., Speck, J. S., DenBaars, S. P., Nakamura, S., and Feezell, D. (2012). *Applied Physics Express*, **5**(9), 092104.

Huang, G. S., Lu, T. C., Yao, H. H., Kuo, H. C., Wang, S. C., Lin, Chih-Wei, and Chang, L. (2006). *Applied Physics Letters*, **88**(6), 061904.

Hums, C., Bläsing, J., Dadgar, A., Diez, A., Hempel, T., Christen, J., Krost, A., Lorenz, K., and Alves, E. (2007). *Applied Physics Letters*, **90**(2), 022105.

Ichikawa, J., Sakai, Y., Chen, Z., Fujita, K., and Egawa, T. (2012). *Japanese Journal of Applied Physics*, **51**(1), 01AF07.

Iliopoulos, E., Adikimenakis, A., Giesen, C., Heuken, M., and Georgakilas, A. (2008). *Applied Physics Letters*, **92**(19), 191907.

Imamoglu, A., Ram, R. J., Pau, S., and Yamamoto, Y. (1996). *Physical Review A*, **53**(06), 4250–4253.

Ive, T., Brandt, O., Kong, X., Trampert, A., and Ploog, K. H. (2008). *Physical Review B*, **78**(03), 035311.

Jones, R. E., Broesler, R., Yu, K. M., Ager, J. W., III, Haller, E. E., Walukiewicz, W., Chen, X., and Schaff, W. J. (2008). *Journal of Applied Physics*, **104**(12), 123501.

Katzer, D. S., Storm, D. F., Binari, S. C., Shanabrook, B. V., Torabi, A., Zhou, Lin, and Smith, David, J. (2005). *Journal of Vacuum Science & Technology B: Microelectronics and Nanometer Structures*, **23**(3), 1204–1208.

Kehagias, Th., Dimitrakopulos, G. P., Kioseoglou, J., Kirmse, H., Giesen, C., Heuken, M., Georgakilas, A., Neumann, W., Karakostas, Th., and Komninou, Ph. (2009). *Applied Physics Letters*, **95**(7), 071905.

Kehagias, Th., Dimitrakopulos, G. P., Kioseoglou, J., Kirmse, H., Giesen, C., Heuken, M., Georgakilas, A., Neumann, W., Karakostas, Th., and Komninou, Ph. (2011). *Applied Physics Letters*, **99**(5), 059902.

Kim-Chauveau, H., de Mierry, P., Chauveau, J.-M., and Duboz, J.-Y. (2011). *Journal of Crystal Growth*, **316**(1), 30–36.

Kim-Chauveau, H., Frayssinet, E., Damilano, B., Mierry, P. De, Bodiou, L., Nguyen, L., Vennéguès, P., Chauveau, J.-M., Cordier, Y., Duboz, J. Y., Charash, R., Vajpeyi, A., Lamy, J.-M., Akhter, M., Maaskant, P. P., Corbett, B., Hangleiter, A., and Wieck, A. (2012). *Journal of Crystal Growth*, **338**(1), 20–29.

Kioseoglou, J., Kalesaki, E., Belabbas, I., Chen, J., Nouet, G., Kirmse, H., Neumann, W., Komninou, Ph., and Karakostas, Th. (2011). *Journal of Applied Physics*, **110**(5), 053715.

Kováč, J., Šramatý, R., Chvála, A., Sibboni, H., Morvan, E., DiForte-Poisson, M. A., Donoval, D., and Kordoš, P. (2011). *Applied Physics Letters*, **98**(16), 162111.

Kret, S., Wolska, A., Klepka, M. T., Letrouit, A., Ivaldi, F., Szczepańska, A., Carlin, J.-F., Kaufmann, N. A. K., and Grandjean, N. (2011). *Journal of Physics: Conference Series*, **326**(1), 012013.

Krost, A., Berger, C., Moser, P., Bläsing, J., Dadgar, A., Hums, C., Hempel, T., Bastek, B., Veit, P., and Christen, J. (2011). *Semiconductor Science and Technology*, **26**(1), 014041.

Kuzmik, J. (2001). *Electron Device Letters, IEEE*, **22**(11), 510–512.

Kuzmik, J., Ostermaier, C., Pozzovivo, G., Basnar, B., Schrenk, W., Carlin, J.-F., Gonschorek, M., Feltin, E., Grandjean, N., Douvry, Y., Gaquière, C., De Jaeger, J.-C., Cico, K., Fröhlich, K., Skriniarova, J., Kovac, J., Strasser, G., Pogany, D., and Gornik, E. (2010). *Electron Devices, IEEE Transactions on*, **57**(9), 2144–2154.

Kuzmik, J., Pozzovivo, G., Abermann, S., Carlin, J.-F., Gonschorek, M., Feltin, E., Grandjean, N., Bertagnolli, E., Strasser, G., and Pogany, D. (2008). *Electron Devices, IEEE Transactions on*, **55**(3), 937–941.

Lecourt, F., Ketteniss, N., Behmenburg, H., Defrance, N., Hoel, V., Eickelkamp, M., Vescan, A., Giesen, C., Heuken, M., and De Jaeger, J. C. (2011*a*). *Electronics Letters*, **47**(3), 212–214.

Lecourt, F., Ketteniss, N., Behmenburg, H., Defrance, N., Hoel, V., Eickelkamp, M., Vescan, A., Giesen, C., Heuken, M., and De Jaeger, J.-C. (2011*b*). *Electron Device Letters, IEEE*, **32**(11), 1537–1539.

Lee, D. S., Chung, J. W., Wang, H., Gao, X., Guo, S., Fay, P., and Palacios, T. (2011*a*). *Electron Device Letters, IEEE*, **32**(6), 755–757.

Lee, D. S., Gao, X., Guo, S., Kopp, D., Fay, P., and Palacios, T. (2011*b*). *Electron Device Letters, IEEE*, **32**(11), 1525–1527.

Lee, D. S., Gao, X., Guo, S., and Palacios, T. (2011*c*). *Electron Device Letters, IEEE*, **32**(5), 617–619.

Lei, H., Chen, J., and Ruterana, P. (2010). *Applied Physics Letters*, **96**(16), 161901.

Liliental-Weber, Z., Chen, Y., Ruvimov, S., and Washburn, J. (1997). *Physical Review Letters*, **79**(15), 2835–2838.

Liu, G., Zhang, J., Li, X.-H., Huang, G. S., Paskova, T., Evans, K. R., Zhao, H., and Tansu, N. (2012). *Journal of Crystal Growth*, **340**(1), 66–73.

Lorenz, K., Franco, N., Alves, E., Pereira, S., Watson, I. M., Martin, R. W., and O'Donnell, K. P. (2008). *Journal of Crystal Growth*, **310**(18), 4058–4064.

Lorenz, K., Franco, N., Alves, E., Watson, I. M., Martin, R. W., and O'Donnell, K. P. (2006). *Physical Review Letters*, **97**(08), 085501.

Lorenz, K., Magalhaes, S., Franco, N., Barradas, N. P., Darakchieva, V., Alves, E., Pereira, S., Correia, M. R., Munnik, F., Martin, R. W., O'Donnell, K. P., and Watson, I. M. (2010). *physica status solidi (b)*, **247**(7), 1740–1746.

Lu, T.-C., Chen, S.-W., Wu, T.-T., Tu, P.-M., Chen, C.-K., Chen, C.-H., Li, Z.-Y., Kuo, H.-C., and Wang, S.-C. (2010). *Applied Physics Letters*, **97**(7), 071114.

Maier, D., Alomari, M., Grandjean, N., Carlin, J.-F., Diforte-Poisson, M.-A., Dua, C., Delage, S., and Kohn, E. (2012). *Electron Device Letters, IEEE*, **33**(7), 985–987.

Maier, D., Alomari, M., Grandjean, N., Carlin, J.-F., Diforte-Poisson, M.-A., Dua, C., Chuvilin, A., Troadec, D., Gaquière, C., Kaiser, U., Delage, S. L., and Kohn, E. (2010). *Device and Materials Reliability, IEEE Transactions on*, **10**(4), 427–436.

Mánuel, J. M., Morales, F. M., Lozano, J. G., González, D., García, R., Lim, T., Kirste, L., Aidam, R., and Ambacher, O. (2010). *Acta Materialia*, **58**(12), 4120–4125.

Massies, J., Turco, F., Saletes, A., and Contour, J. P. (1987). *Journal of Crystal Growth*, **80**(2), 307–314.

Medjdoub, F., Alomari, M., Carlin, J.-F., Gonschorek, M., Feltin, E., Py, M. A., Grandjean, N., and Kohn, E. (2008). *Electron Device Letters, IEEE*, **29**(5), 422–425.

Medjdoub, F., Ducatteau, D., Gaquière, C., Carlin, J.-F., Gonschorek, M., Feltin, E., Py, M. A., Grandjean, N., and Kohn, E. (2007*a*). *Electronics Letters*, **43**(5), 71–72.

Medjdoub, F., Sarazin, N., Tordjman, M., Magis, M., di Forte-Poisson, M. A., Knez, M., Delos, E., Gaquière, C., Delage, S. L., and Kohn, E. (2007*b*). *Electronics Letters*, **43**(12), 691–692.

Miao, Z. L., Yu, T. J., Xu, F. J., Song, J., Huang, C. C., Wang, X. Q., Yang, Z. J., Zhang, G. Y., Zhang, X. P., Yu, D. P., and Shen, B. (2009). *Applied Physics Letters*, **95**(23), 231909.

Miao, Z. L., Yu, T. J., Xu, F. J., Song, J., Lu, L., Huang, C. C., Yang, Z. J., Wang, X. Q., Zhang, G. Y., Zhang, X. P., Yu, D. P., and Shen, B. (2010). *Journal of Applied Physics*, **107**(4), 043515.

Mikulics, M., Stoklas, R., Dadgar, A., Gregušová, D., Novák, J., Grützmacher, D., Krost, A., and Kordoš, P. (2010). *Applied Physics Letters*, **97**(17), 173505.

Morgan, D., Sultana, M., Fatima, H., Sugiyama, S., Fareed, Q., Adivarahan, V., Lachab, M., and Khan, A. (2011). *Applied Physics Express*, **4**(11), 114101.

Mouti, A., Rouvière, J.-L., Cantoni, M., Carlin, J.-F., Feltin, E., Grandjean, N., and Stadelmann, P. (2011). *Physical Review B*, **83**(19), 195309.

Nakamura, S. (1997). *Solid State Communications*, **102**(2–3), 237–248.

Natali, F., Byrne, D., Dussaigne, A., Grandjean, N., Massies, J., and Damilano, B. (2003). *Applied Physics Letters*, **82**(4), 499–501.

Ng, H. M., Moustakas, T. D., and Chu, S. N. G. (2000). *Applied Physics Letters*, **76**(20), 2818–2820.

Nicolay, S., Carlin, J.-F., Feltin, E., Butté, R., Mosca, M., Grandjean, N., Ilegems, M., Tchernycheva, M., Nevou, L., and Julien, F. H. (2005). *Applied Physics Letters*, **87**(11), 111106.

Nidhi, Brown, D. F., Keller, S., and Mishra, U. K. (2010). *Japanese Journal of Applied Physics*, **49**(2), 021005.

Nidhi, Dasgupta, S., Lu, J., Speck, J. S., and Mishra, U. K. (2012). *Electron Device Letters, IEEE*, **33**(6), 794–796.

Northrup, J. E. and Neugebauer, J. (1999). *Physical Review B*, **60**(12), R8473–R8476.

Oh, T. S., Kim, J. O., Jeong, H., Lee, Y. S., Nagarajan, S., Lim, K. Y., Hong, C.-H., and Suh, E.-K. (2008). *Journal of Physics D: Applied Physics*, **41**(9), 095402.

Omae, K., Higuchi, Y., Nakagawa, K., Matsumura, H., and Mukai, T. (2009). *Applied Physics Express*, **2**(5), 052101.

Onishi, T., Imafuji, O., Nagamatsu, K., Kawaguchi, M., Yamanaka, K., and Takigawa, S. (2012). *Quantum Electronics, IEEE Journal of*, **48**(9), 1107–1112.

Ostermaier, C., Pozzovivo, G., Basnar, B., Schrenk, W., Carlin, J.-F., Gonschorek, M., Grandjean, N., Vincze, A., Tóth, L., Pécz, B., Strasser, G., Pogany, D., and Kuzmik, J. (2010*a*). *Japanese Journal of Applied Physics*, **49**(11), 116506.

Ostermaier, C., Pozzovivo, G., Basnar, B., Schrenk, W., Schmid, M., Tóth, L., Pécz, B., Carlin, J.-F., Gonschorek, M., Grandjean, N., Strasser, G., Pogany, D., and Kuzmik, J. (2010*b*). *Applied Physics Letters*, **96**(26), 263515.

Ostermaier, C., Pozzovivo, G., Carlin, J.-F., Basnar, B., Schrenk, W., Douvry, Y., Gaquière, C., DeJaeger, J.-C., Cico, K., Fröhlich, K., Gonschorek, M., Grandjean, N., Strasser, G., Pogany, D., and Kuzmik, J. (2009, Oct.). *Electron Device Letters, IEEE*, **30**(10), 1030–1032.

Park, S. Y., Floresca, C., Chowdhury, U., Jimenez, J. L., Lee, C., Beam, E., Saunier, P., Balistreri, T., and Kim, M. J. (2009). *Microelectronics Reliability*, **49**(5), 478–483.

Pengelly, R. S., Wood, S. M., Milligan, J. W., Sheppard, S. T., and Pribble, W. L. (2012). *Microwave Theory and Techniques, IEEE Transactions on*, **60**(6), 1764–1783.

Pozzovivo, G., Kuzmik, J., Golka, S., Schrenk, W., Strasser, G., Pogany, D., Čičo, K., Ťapajna, M., Fröhlich, K., Carlin, J.-F., Gonschorek, M., Feltin, E., and Grandjean, N. (2007). *Applied Physics Letters*, **91**(4), 043509.

Redondo-Cubero, A., Lorenz, K., Gago, R., Franco, N., di Forte Poisson, M.-A., Alves, E., and Munoz, E. (2010). *Journal of Physics D: Applied Physics*, **43**(5), 055406.

Sadler, T. C., Kappers, M. J., and Oliver, R. A. (2009*a*). *Journal of Crystal Growth*, **311**(13), 3380–3385.

Sadler, T., Massabuau, F., Kappers, M., and Oliver, R. (2012). *physica status solidi (c)*, **9**(3–4), 461–464.

Sadler, T. C., Kappers, M. J., and Oliver, R. A. (2009*b*). *physica status solidi (c)*, **6**(S2), S666–S670.

Sadler, T. C., Kappers, M. J., and Oliver, R. A. (2010). *Journal of Physics: Conference Series*, **209**(1), 012015.

Sadler, T. C., Kappers, M. J., and Oliver, R. A. (2011*a*). *Journal of Crystal Growth*, **314**(1), 13–20.

Sadler, T. C., Kappers, M. J., and Oliver, R. A. (2011*b*). *Journal of Crystal Growth*, **331**(1), 4–7.

Sahonta, S.-L., Dimitrakopulos, G. P., Kehagias, Th., Kioseoglou, J., Adikimenakis, A., Iliopoulos, E., Georgakilas, A., Kirmse, H., Neumann, W., and Komninou, Ph. (2009). *Applied Physics Letters*, **95**(2), 021913.

Sakalauskas, E., Behmenburg, H., Hums, C., Schley, P., Rossbach, G., Giesen, C., Heuken, M., Kalisch, H., Jansen, R. H., Bläsing, J., Dadgar, A., Krost, A., and Goldhahn, R. (2010). *Journal of Physics D: Applied Physics*, **43**(36), 365102.

Sarazin, N., Morvan, E., di Forte Poisson, M. A., Oualli, M., Gaquière, C., Jardel, O., Drisse, O., Tordjman, M., Magis, M., and Delage, S. L. (2010). *Electron Device Letters, IEEE*, **31**(1), 11–13.

Schenk, H. P. D., Nemoz, M., Korytov, M., Vennégués, P., Demolon, P., Dräger, A. D., Hangleiter, A., Charash, R., Maaskant, P. P., Corbett, B., and Duboz, J. Y. (2009). *physica status solidi (c)*, **6**(S2), S897–S901.

Senda, S., Jiang, H., and Egawa, T. (2008). *Applied Physics Letters*, **92**(20), 203507.

Shiojiri, M., Chuo, C. C., Hsu, J. T., Yang, J. R., and Saijo, H. (2006). *Journal of Applied Physics*, **99**(7), 073505.

Simeonov, D., Feltin, E., Altoukhov, A., Castiglia, A., Carlin, J.-F., Butté, R., and Grandjean, N. (2008). *Applied Physics Letters*, **92**(17), 171102.

Sitar, Z., Paisley, M. J., Yan, B., Ruan, J., Choyke, W. J., and Davis, R. F. (1990). *Journal of Vacuum Science & Technology B: Microelectronics and Nanometer Structures*, **8**(2), 316–322.

Song, J., Xu, F. J., Yan, X. D., Lin, F., Huang, C. C., You, L. P., Yu, T. J., Wang, X. Q., Shen, B., Wei, K., and Liu, X. Y. (2010). *Applied Physics Letters*, **97**(23), 232106.

Song, T. L. (2005). *Journal of Applied Physics*, **98**(8), 084906.

Starosta, K. (1981). *physica status solidi (a)*, **68**(1), K55–K57.

Sun, H., Alt, A. R., Benedickter, H., Bolognesi, C. R., Feltin, E., Carlin, J.-F., Gonschorek, M., Grandjean, N., Maier, T., and Quay, R. (2009). *Electron Device Letters, IEEE*, **30**(8), 796–798.

Sun, H., Alt, A. R., Benedickter, H., Feltin, E., Carlin, J., Gonschorek, M., Grandjean, N., and Bolognesi, C. R. (2010*a*). *Microwave and Wireless Components Letters, IEEE*, **20**(8), 453–455.

Sun, H., Alt, A. R., Benedickter, H., Feltin, E., Carlin, J.-F., Gonschorek, M., Grandjean, N., and Bolognesi, C. R. (2010*b*). *Electron Device Letters, IEEE*, **31**(4), 293–295.

Sun, H., Alt, A. R., Benedickter, H., Feltin, E., Carlin, J.-F., Gonschorek, M., Grandjean, N. R., and Bolognesi, C. R. (2010*c*). *Electron Device Letters, IEEE*, **31**(9), 957–959.

Sun, H., Alt, A. R., Benedickter, H., Bolognesi, C. R., Feltin, E., Carlin, J.-F., Gonschorek, M., and Grandjean, N. (2010*d*). *Applied Physics Express*, **3**(9), 094101.

Taniyasu, Y. (2012). Private communication.

Tirelli, S., Marti, D., Sun, H., Alt, A. R., Carlin, J.-F., Grandjean, N., and Bolognesi, C. R. (2011). *Electron Device Letters, IEEE*, **32**(10), 1364–1366.

Tülek, R., Ilgaz, A., Gökden, S., Teke, A., Öztürk, M. K., Kasap, M., Özçelik, S., Arslan, E., and Özbay, E. (2009). *Journal of Applied Physics*, **105**(1), 013707.

Tyagi, A., Farrell, R. M., Kelchner, K. M., Huang, C.-Y., Hsu, P. S., Haeger, D. A., Hardy, M. T., Holder, C., Fujito, K., Cohen, D. A., Ohta, H., Speck, J. S., DenBaars, S. P., and Nakamura, S. (2010). *Applied Physics Express*, **3**(1), 011002.

Čičo, K., Hušeková, K., Ťapajna, M., Gregušová, D., Stoklas, R., Kuzmík, J., Carlin, J.-F., Grandjean, N., Pogany, D., and Fröhlich, K. (2011). *Journal of Vacuum Science & Technology B: Microelectronics and Nanometer Structures*, **29**(1), 01A808.

Vennéguès, P., Diaby, B. S., Kim-Chauveau, H., Bodiou, L., Schenk, H. P. D., Frayssinet, E., Martin, R. W., and Watson, I. M. (2012). *Journal of Crystal Growth*, **353**(1), 108–114.

Waldrip, K. E., Han, J., Figiel, J. J., Zhou, H., Makarona, E., and Nurmikko, A. V. (2001). *Applied Physics Letters*, **78**(21), 3205–3207.

Wang, K., Martin, R. W., Amabile, D., Edwards, P. R., Hernandez, S., Nogales, E., O'Donnell, K. P., Lorenz, K., Alves, E., Matias, V., Vantomme, A., Wolverson, D., and Watson, I. M. (2008). *Journal of Applied Physics*, **103**(7), 073510.

Wang, R., Li, G., Laboutin, O., Cao, Y., Johnson, W., Snider, G., Fay, P., Jena, D., and Xing, H. (2011*a*). *Electron Device Letters, IEEE*, **32**(7), 892–894.

Wang, R., Saunier, P., Tang, Y., Fang, T., Gao, X., Guo, S., Snider, G., Fay, P., Jena, D., and Xing, H. (2011*b*). *Electron Device Letters, IEEE*, **32**(3), 309–311.

Wang, R., Saunier, P., Xing, X., Lian, C., Gao, X., Guo, S., Snider, G., Fay, P., Jena, D., and Xing, H. (2010). *Electron Device Letters, IEEE*, **31**(12), 1383–1385.

Watanabe, K., Yang, J. R., Huang, S. Y., Inoke, K., Hsu, J. T., Tu, R. C., Yamazaki, T., Nakanishi, N., and Shiojiri, M. (2003). *Applied Physics Letters*, **82**(5), 718–720.

Wei, Q. Y., Li, T., Huang, Y., Huang, J. Y., Chen, Z. T., Egawa, T., and Ponce, F. A. (2012). *Applied Physics Letters*, **100**(9), 092101.

Won, D., Weng, X., and Redwing, J. M. (2010). *Journal of Applied Physics*, **108**(9), 093511.

Wong, M. H., Wu, F., Hurni, C. A., Choi, S., Speck, J. S., and Mishra, U. K. (2012). *Applied Physics Letters*, **100**(7), 072107.

Wu, X. H., Elsass, C. R., Abare, A., Mack, M., Keller, S., Petroff, P. M., DenBaars, S. P., Speck, J. S., and Rosner, S. J. (1998). *Applied Physics Letters*, **72**(6), 692–694.

Xue, J., Hao, Y., Zhang, J., Zhou, X., Liu, Z., Ma, J., and Lin, Z. (2011*a*). *Applied Physics Letters*, **98**(11), 113504.

Xue, J., Hao, Y., Zhou, X., Zhang, J., Yang, C., Ou, X., Shi, L., Wang, H., Yang, L., and Zhang, J. (2011*b*). *Journal of Crystal Growth*, **314**(1), 359–364.

Yoshizumi, Y., Adachi, M., Enya, Y., Kyono, T., Tokuyama, S., Sumitomo, T., Akita, K., Ikegami, T., Ueno, M., Katayama, K., and Nakamura, T. (2009). *Applied Physics Express*, **2**(9), 092101.

Yue, Y., Hu, Z., Guo, J., Sensale-Rodriguez, B., Li, G., Wang, R., Faria, F., Fang, T., Song, B., Gao, X., Guo, S., Kosel, T., Snider, G., Fay, P., Jena, D., and Xing, H. (2012). *Electron Device Letters, IEEE*, **33**(7), 988–990.

Zeng, X., Boiko, D. L., Cosendey, G., Glauser, M., Carlin, J.-F., and Grandjean, N. (2012). *Applied Physics Letters*, **101**(14), 141120.

Zhang, H., Miller, E. J., and Yu, E. T. (2006). *Journal of Applied Physics*, **99**(2), 023703.

Zhou, L., Leach, J. H., Ni, X., Morkoç, H., and Smith, D. J. (2010*a*). *Journal of Applied Physics*, **107**(1), 014508.
Zhou, L., McCartney, M. R., Smith, D. J., Mouti, A., Feltin, E., Carlin, J. F., and Grandjean, N. (2010*b*). *Applied Physics Letters*, **97**(16), 161902.
Zhou, L., Smith, D. J., McCartney, M. R., Katzer, D. S., and Storm, D. F. (2007). *Applied Physics Letters*, **90**(8), 081917.

7 Growth and optical properties of aluminum-rich AlGaN heterostructures

Hideto Miyake

7.1 Introduction

$Al_xGa_{1-x}N$ alloys have attracted significant attention owing to their emission covering the wavelength range from 210 to 365 nm upon changing the AlN mole fraction x. This characteristic has been actively applied to the development of deep-ultraviolet light-emitting diodes (LEDs) [1, 2]. However, the emission efficiency of LEDs with wavelengths of less than 280 nm is rather low, partly due to the high resistivity of p-AlGaN and the low quality of the AlGaN template. Many methods have been proposed for improving the crystal quality of AlGaN, such as growth on GaN substrates and the use of low-temperature buffer layer technology [3–5]. However, cracks are generated in AlGaN using these methods in the case of a high AlN mole fraction owing to the large difference in the lattice constants and the thermal expansion mismatch between the substrate and growth layer. Moreover, the crystal quality is generally low because of the high dislocation density [6–8]. Other groups have recently obtained crack-free AlGaN layers with high crystal quality for all AlN mole fractions using an AlN/sapphire substrate [9–17]. However, they did not discuss the growth mechanism of AlGaN on AlN. Moreover, the successful doping of $Al_xGa_{1-x}N$ is a great major requirement for the development of high-efficiency devices, and many groups have studied the doping of $Al_xGa_{1-x}N$ [18, 19].

In this chapter, the growth of Si-doped AlGaN and AlGaN/AlGaN multiple-quantum wells on an AlN/sapphire substrate by low-pressure metalorganic vapor phase epitaxy (LP-MOVPE) with an *in situ* monitoring system have been reported. The quality and growth mechanism of AlGaN on AlN was discussed. GaN and AlGaN with an AlN mole fraction of 0.28 were free from stress because of the occurrence of three-dimensional (3D) growth. The lattice constant a of AlGaN with an AlN mole fraction of 0.4 was larger than that at an AlN mole fraction of 0.28 despite the two-dimensional (2D) growth mode. AlGaN with an AlN mole fraction of more than 0.6 was coherently grown on AlN owing to the fact that the AlN/sapphire template used in this work was subjected to compression. For this reason, in the samples with an AlN mole fraction of more than 0.6, the crystal quality was high because of the coherent growth on the underlying AlN.

7.2 Growth of Si-doped AlGaN on AlN/sapphire templates

A 0.8–1.0-μm thick epitaxial AlN layer on sapphire (0001) was used as a substrate to grow AlxGa1-xN. As a measure of the crystallinity of this AlN substrate, the full width at half maximum (FWHM) of the X-ray rocking curve (XRC) was 50–100 arcsec for (0002) diffraction and 400–550 arcsec for (10–12) diffraction. The AlGaN film was grown by LP-MOVPE on an AlN/sapphire template with an *in situ* monitoring system. Trimethylgallium (TMG), trimethylaluminum (TMA), and ammonia (NH_3) were used as precursors for Ga, Al, and N, respectively. H_2 was used as the carrier gas and CH_3SiH_3 was used as the doping source. The reactor pressure was fixed at 50 Torr. The AlN mole fraction in AlGaN was controlled by varying the growth temperature, and two high-Al-composition layers were inserted at the first stage of growth as a buffer layer.

The crystal quality and AlN mole fraction were determined from the lattice constant measured by X-ray diffraction (XRD). XRD analysis was conducted in accordance with Hooke's law for a wurtzite structure,

$$\varepsilon_{zz} = -2\frac{C_{13}}{C_{33}}\varepsilon_{xx} \tag{7.1}$$

where ε_{zz} and ε_{xx} are the strains in the c-axis and a-axis directions, respectively, in the c-plane, and C_{ij} are the elastic stiffness constants. The Al mole fraction x was determined by comparing the a-axis and c-axis lattice constants evaluated by XRD with those obtained from Vegard's law assuming eq. (7.1). The carrier concentration and mobility were determined by Hall effect measurements.

Figure 7.1 shows the growth rate and AlN mole fraction in AlGaN under different TMG flow rates. The growth rates of AlN and GaN were calculated from

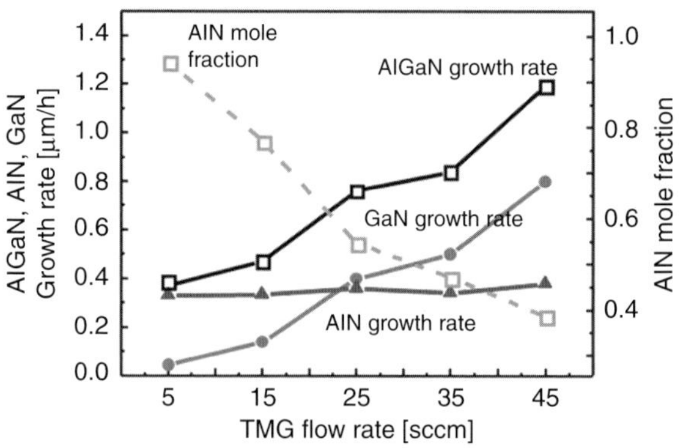

FIG. 7.1. Growth rate and AlN mole fraction in AlGaN under different TMG flow rates. The flow rate of TMA was 15 sccm, and the growth was performed at 1180°C. The growth rates of AlN and GaN were calculated from the growth rate of AlGaN and its mole fraction using the lattice constants of AlN and GaN.

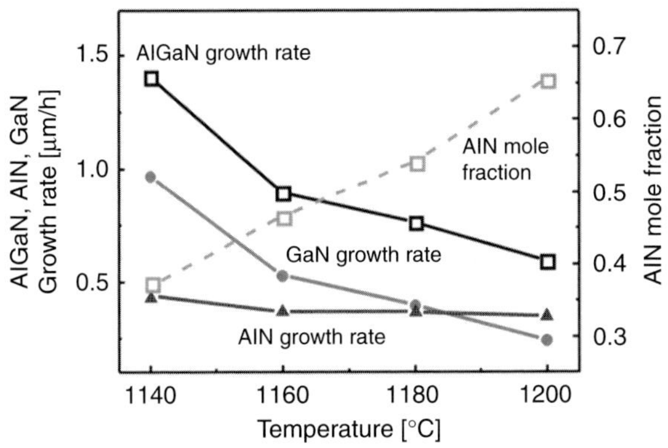

FIG. 7.2. Growth rate and AlN mole fraction in AlGaN at different growth temperatures. The flow rates of TMA and TMG were 15 sccm and 25 sccm, respectively.

the growth rate of AlGaN and its mole fraction using the lattice constants of AlN and GaN. The growth rate of AlN was constant. In contrast, the growth rates of GaN and AlGaN increased with increasing TMG flow rate. In addition, Fig. 7.2 shows the growth rate and AlN mole fraction in AlGaN at different growth temperatures. Similarly to when AlGaN was grown under different TMG flow rates, the growth rate of AlN was constant, regardless of the growth temperature. In contrast, the growth rates of GaN and AlGaN increased with decreasing growth temperature, which is considered to be due to the suppression of Ga desorption at lower growth temperatures. The AlN mole fraction in AlGaN increased linearly with increasing growth temperature. Thus, it is possible to control the growth rate and the AlN molar fraction by controlling the growth temperature.

AlGaN films with AlN mole fractions varying from 0.05 to 0.91 were obtained by increasing the growth temperature from 1100 to 1200°C. Smooth and crack-free AlGaN films were obtained with AlN mole fractions of 0.51 and above, and atomic steps were observed for those surfaces by atomic force microscopy (AFM). Figure 7.3 shows the dependence of the full width at half maximum (FWHM) of the X-ray rocking curve (XRC) peak on the AlN mole fraction. The crystal quality deteriorates as the AlN mole fraction decreases to approximately 0.4, and then improves as the AlN mole fraction decreases further. It is considered that this is due to a different growth mechanism occurring below an AlN mole fraction of 0.4, and that the misfit dislocation increases because of the lattice constant difference between $Al_xGa_{1-x}N$ and AlN as the AlN mole fraction decreases to approximately 0.40, which leads to the deterioration of crystal quality. Figure 7.4 shows *in situ* reflectance intensity spectra for GaN and AlGaN grown with AlN mole fractions of 0.28, 0.40, and 0.58 on an AlN/sapphire substrate. The crystal quality improves as the AlN mole fraction decreases to below 0.4 because of the

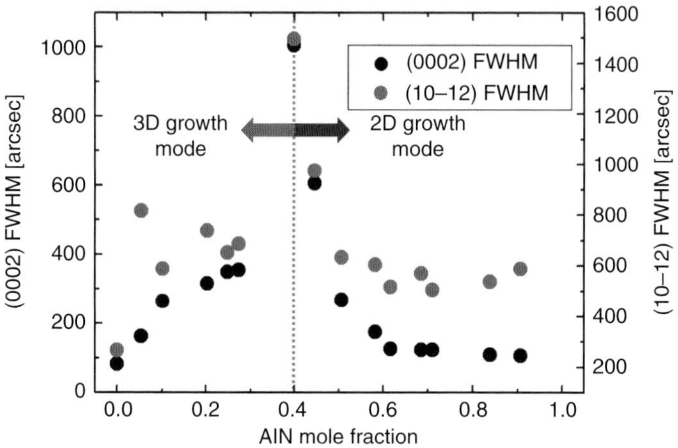

FIG. 7.3. FWHM value of XRC peak obtained from XRD as a function of AlN mole fraction in AlGaN.

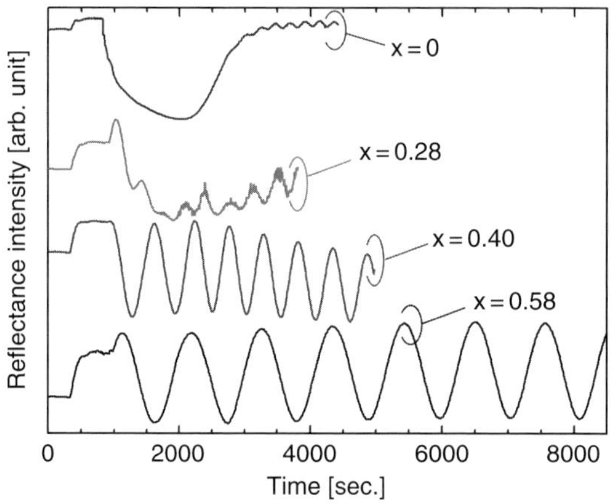

FIG. 7.4. *In situ* reflectance intensity spectra for GaN and AlGaN grown with AlN mole fractions of 0.28, 0.40, and 0.58 on an AlN/sapphire as the substrate.

3D growth that occurs at the early stage of growth owing to the very large difference in the lattice constant between the growth layer and the substrate. The subsequent horizontal growth causes the bending of the propagation direction of the edge dislocations, which is thought to lead to improved crystal quality. At an AlN mole fraction of 0.4 and above, the growth mode changed to the layer-by-layer mode (2D growth mode). Because the difference in the lattice constants

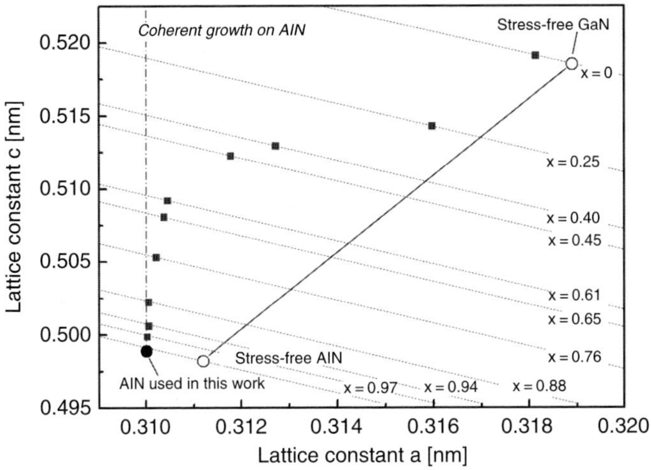

FIG. 7.5. c-axis lattice constant versus a-axis lattice constant for AlGaN. Those of epitaxial AlN used in this work and GaN grown on epitaxial AlN are also plotted for reference. Open circles: lattice constants of stress-free GaN and AlN; dashed line: biaxial strain according to eq. (7.2); solid squares: lattice constants of crack-free AlGaN on AlN.

between the AlGaN layer and the AlN underlying layer decreases with increasing AlN mole fraction, the density of misfit dislocations was reduced; thus, the FWHM value decreased as the AlN mole fraction was increased from 0.4 to 0.6.

Figure 7.5 shows the relationship between the lattice constants a and c for $Al_xGa_{1-x}N$ on the AlN/sapphire templates with various AlN mole fractions. The solid line connecting the values for stress-free GaN and AlN (open circles) gives Vegard's law for the lattice constants of stress-free $Al_xGa_{1-x}N$. The dashed lines show the biaxial strain given by

$$c \approx c_0 \left(1 + \frac{2C_{13}}{C_{33}}\right) - \frac{2C_{13}}{C_{33}} \cdot \frac{c_0}{a_0} \cdot a \tag{7.2}$$

which can be derived from eq. (7.1), implying that the lattice constant c changes with the slope of $-(2C_{13}/C_{33}) \times (c_0/a_0)$ with respect to the lattice constant a [20, 21]. From Fig. 7.6, GaN and AlGaN with an AlN mole fraction of 0.28 are free from stress because of the 3D growth (Fig. 7.4). The value of a for AlGaN with an AlN mole fraction of 0.4 is larger than that for an AlN mole fraction of 0.28, despite the occurrence of 2D growth (Fig. 7.4). This indicates that the lattice mismatch between the AlGaN and underlying AlN has an adverse effect on the crystal quality through a change in the growth mechanism to 2D growth when x is nearly 0.4. On the other hand, AlGaN with an AlN mole fraction of more than 0.6 is coherently grown on AlN owing to the fact that the AlN/sapphire template used in this work is subjected to compression and has excellent crystal quality.

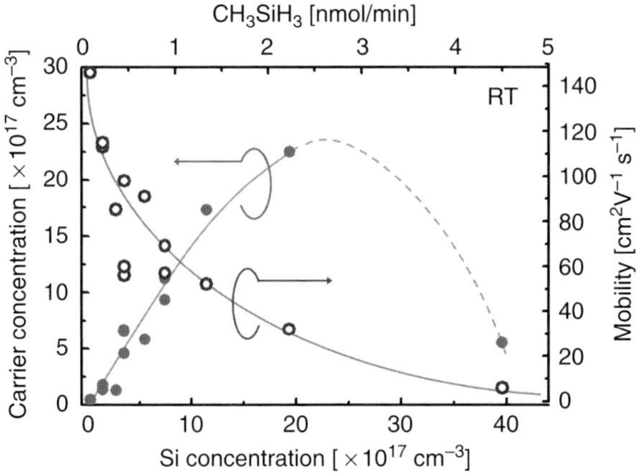

FIG. 7.6. Carrier concentration and mobility of Si-doped AlGaN at room temperature as a function of Si concentration. The Si concentration was evaluated by SIMS.

For this reason, in the samples with an AlN mole fraction of more than 0.6, the crystal quality is high because of the coherent growth on the underlying AlN.

Figure 7.6 shows the carrier concentration and mobility of Si-doped AlGaN as a function of Si concentrations. The Si concentration was evaluated by secondary ion mass spectrometry (SIMS). Here, the AlN mole fraction is fixed at approximately 0.65. The carrier concentration increases linearly with the Si concentration up to an Si concentration of 2×10^{18} cm^{-3}, indicating that the activation rate is approximately 1, while that for a Si concentration of 4×10^{18} cm^{-3} is lower owing to self-compensation. On the other hand, the mobility decreased monotonically with increasing Si concentration.

7.3 Growth of Si-doped AlGaN/AlGaN multiple-quantum wells

Currently, research on AlxGa1-xN-based UV light-emitting diodes (LEDs) is popular. However, their external quantum efficiency is low because the growth of p-type AlxGa1-xN with a high AlN mole fraction and a low resistivity is difficult, and thereby the hole injection efficiency is low. Previously, we studied the application of Si-doped AlxGa1-xN single-layer thin films grown on epitaxial AlN substrates to electron-beam-excited UV light sources [22, 23]. Here, we fabricated multiple quantum well (MQW) structures using Si-doped AlGaN/AlGaN supperlattices and evaluated their light emission characteristics in an attempt to realize deep-UV light emission with a higher output and efficiency.

Figure 7.7 shows the layer structure of AlGaN MQWs on an AlN/sapphire template. Two high-AlN-mole-fraction AlGaN layers with a thickness of 0.1 μm each and an Si-doped AlGaN layer with a thickness of 0.2 μm as buffer layers were

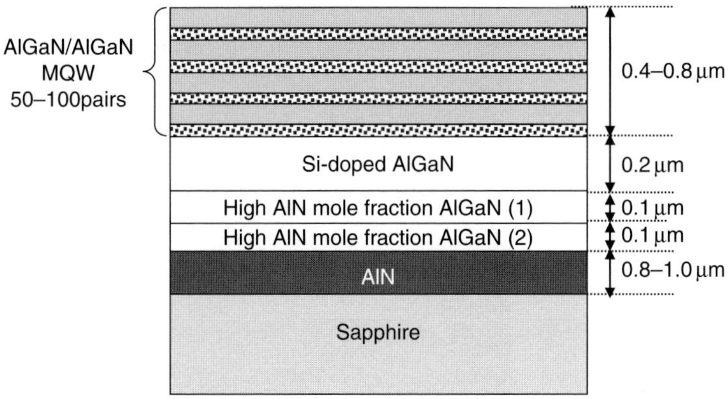

FIG. 7.7. Structure of AlGaN MQWs for application to EB-pumped deep-UV light sources.

grown before the growth of AlGaN MQWs. It is necessary to examine the thickness of well layers to obtain a good carrier confinement effect in quantum well structures. Here, we examined samples with different thicknesses of well layers: the thickness of the well layers was 0.7–2.2 nm with a constant barrier thickness of 7 nm and a total MQW thickness of 600 nm controlled by varying the number of MQW pairs. The AlN mole fractions of the well layers (Si-doped $Al_yGa_{1\text{-}y}N$ layers) and the barrier layers (Si-doped $Al_xGa_{1\text{-}x}N$ layers) were $y = 0.69 \pm 0.01$ and $x = 0.81 \pm 0.01$, respectively. The flow rate of CH_3SiH_3 was 0.45 nmol/min.

Figure 7.8 shows the results of an X-ray diffraction (XRD) (0002) 2θ-ω scan for the AlGaN MQW samples. Clear satellite peaks were observed for all samples,

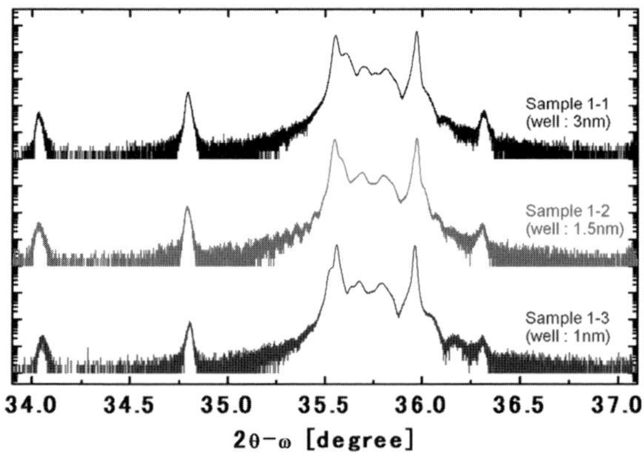

FIG. 7.8. Results of 2θ-ω scan of AlGaN MQWs for different well layer thicknesses.

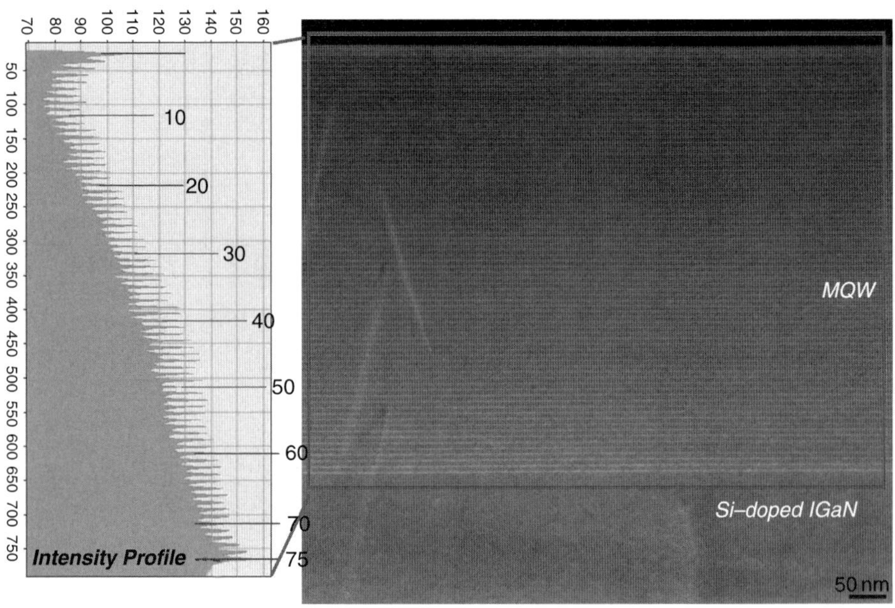

FIG. 7.9. Cross-sectional TEM image of AlGaN MQWs and its intensity histogram.

confirming that the AlGaN MQW structure was almost coherently grown on the AlN substrate. As shown in Fig. 7.9, cross-sectional transmission electron microscopy (TEM) revealed an abrupt interface of the MQWs as well as the film uniformity of the well and barrier layers. The dislocation density calculated from the XRC FWHM was approximately 10^8 cm^{-2} for all the samples. Figure 7.10 shows the results of cathodoluminescence (CL) evaluation. The acceleration voltage of the electron beam was 5 kV. A deep-UV CL peak was observed at a wavelength of approximately 250–260 nm for all samples, and the strongest peak was observed for the sample with a well layer thickness of 1.5 nm. This is considered to be because the carriers generated in the barrier layers were confined in the well layers to enable efficient radiative recombination. The decrease in peak intensity below a well thickness of approximately 1.5 nm might have been caused by the sharpness of the boundary between the well and the barrier, because the lattice constant along the AlGaN c-axis was about 0.5 nm, making it difficult to fabricate a well with a thickness of less than 1 nm. Moreover, the decrease in peak intensity above a well thickness of approximately 1.5 nm is thought to be due to the increase in the degree of wavefunction separation caused by the internal field in the quantum well. The reduction of overlap reduces the probability of recombination between electrons and holes. Thus, we confirmed that the optimum well thickness is approximately 1.5 nm. Moreover,

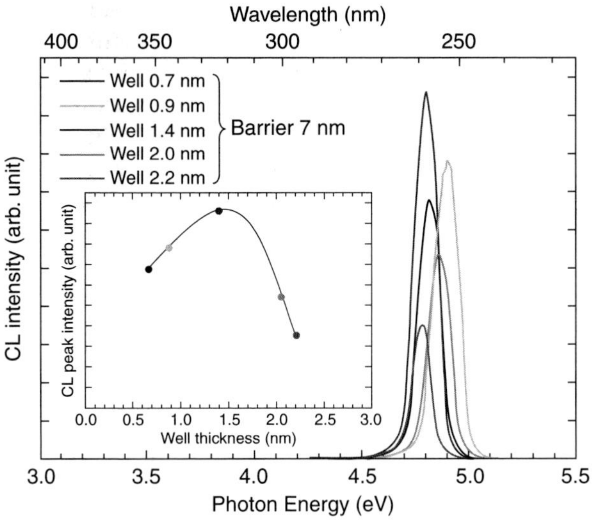

FIG. 7.10. Well thickness dependence of CL intensity. The well thicknesses were 0.7, 0.9, 1.4, 2.0, and 2.2 nm, the barrier thickness was fixed at 7 nm, and the total MQW thickness was 600 nm. The MQW thickness was controlled by varying the number of MQW pairs.

it is considered that light polarized as $E \perp c$ was obtained even for the high-AlN-mole-fraction $Al_yGa_{1-y}N$ because of the significant quantum effect on the valence band structure, and strong radiation was obtained from the (0001) plane [24]. In general, however, light polarized as $E \| c$ is dominant for $Al_xGa_{1-x}N$ thin films coherently grown on AlN substrates when $x > 0.6$, resulting in difficulty in obtaining the radiation from the (0001) plane for wavelengths shorter than approximately 250 nm.

In order to optimize the thickness of barrier layers, we evaluated samples with barrier thicknesses of 3–15 nm with the optimized well thickness of 1.5 nm and a total MQW thickness of 600 nm controlled by varying the number of MQW pairs. The AlN mole fractions of the well layers (Si-doped $Al_yGa_{1-y}N$ layers) and the barrier layers (Si-doped $Al_xGa_{1-x}N$ layers) were $y = 0.59 \pm 0.01$ and $x = 0.70 \pm 0.02$, respectively. The flow rate of CH_3SiH_3 was 0.45 nmol/min. The peak wavelengths for all the samples were in the range of 250–260 nm, as shown in Fig. 7.11. Similarly to the optimization of well thickness, we focused on the peak intensity of the CL spectra. The highest peak intensity was observed at a barrier thickness of 7 nm. The decrease in peak intensity below a barrier thickness of 7 nm is thought to have been caused by the decrease in the number of carriers trapped in the wells. Moreover, the decrease in peak intensity above a barrier thickness of 7 nm was probably caused by the increase in the number of carriers trapped by non-radiative recombination

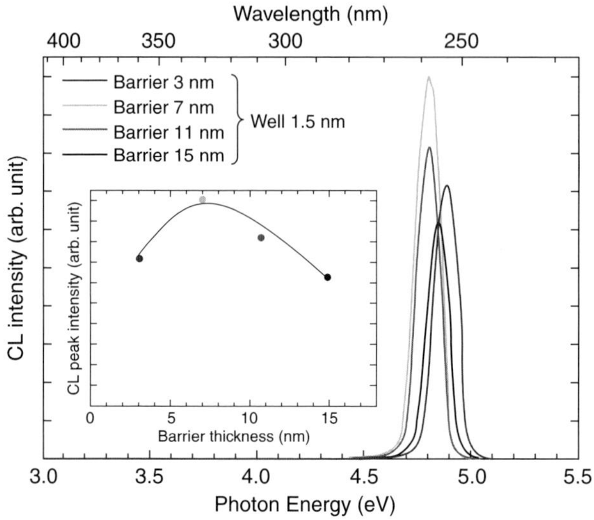

FIG. 7.11. Barrier thickness dependence of CL intensity. The barrier thicknesses were 3, 7, 11, and 15 nm, the well thickness was fixed at 1.5 nm, and the total MQW thickness was 600 nm. The MQW thickness was controlled by varying the number of MQW pairs.

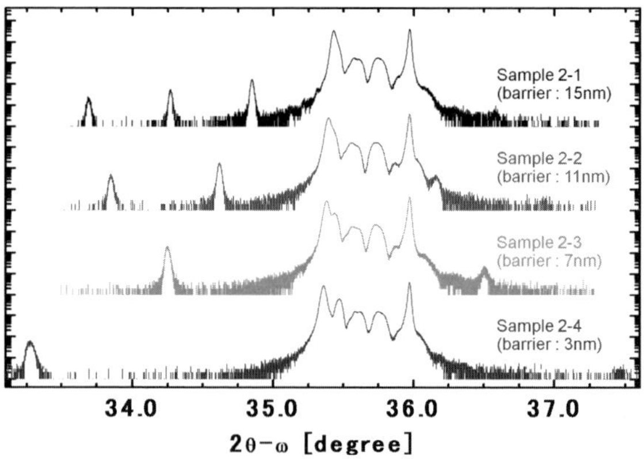

FIG. 7.12. Results of 2θ-ω scan of AlGaN MQWs for different barrier layer thicknesses.

centers before reaching the wells. Thus, we confirmed that the optimum barrier thickness is approximately 7 nm. Figure 7.12 shows the results of an XRD (0002) 2θ-ω scan for the AlGaN MQW samples. A clear satellite peak was observed for all four samples, confirming that the samples had an abrupt superlattice interface. Figure 7.13 shows a reciprocal lattice map of the AlGaN

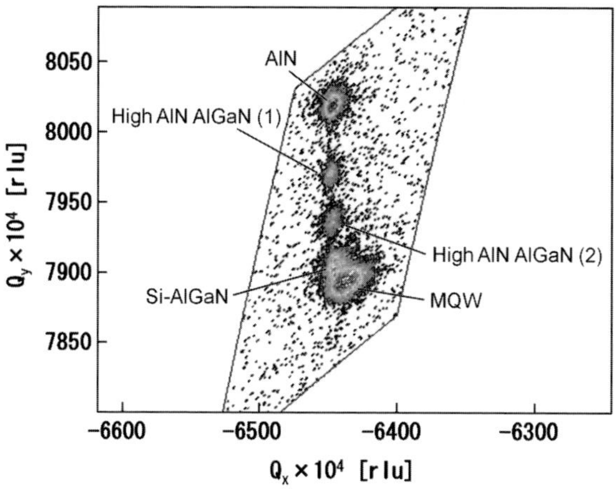

FIG. 7.13. Reciprocal lattice maps of AlGaN MQWs (−1−124) for samples with well and barrier thicknesses of 1.5 and 7 nm, respectively.

MQWs (−1−124) for the samples with well and barrier thicknesses of 1.5 and 7 nm, respectively. Clear reciprocal lattice points corresponding to each layer were observed, indicating that the two high-AlN-mole-fraction AlGaN layers and the Si-doped AlGaN layer, which were inserted as buffer layers, were almost coherently grown on the AlN substrate. Moderate lattice relaxation was observed for the MQWs, meaning that the MQW structure contributed to stress relaxation.

It is essential to examine the thickness of emission layers in electron beam excitation. Here, we fabricated three MQW samples with different total thicknesses. The thicknesses of the well and barrier layers were fixed at 1.5 and 7 nm, respectively. The number of periods of the MQW structure was set to 50, 75, or 100 to obtain MQW structures with total thicknesses of 425, 637, or 850 nm, respectively. Figure 7.14 shows the change in CL peak intensity with acceleration voltage for the AlGaN MQWs with different total thicknesses of the MQW structure. For all the samples, the CL peak intensity increased with increasing acceleration voltage of the electron beam and then saturated. The acceleration voltage at which the CL peak intensity started to saturate decreased with decreasing total thickness of the MQW structure: the CL peak intensity for the sample with fifty periods (approximately 0.4 μm) was about half that for the sample with 100 periods (approximately 0.8 μm) at 10 kV. This was because the penetration depth of the electron beam was increased by increasing the acceleration voltage and the electron beam transmitted through the MQW structure with a smaller total thickness. Therefore, the thickness of the emission layers should be set considering the penetration depth of the electron beam in electron-beam-excited light emission. In this study, a high-emission output was obtained using

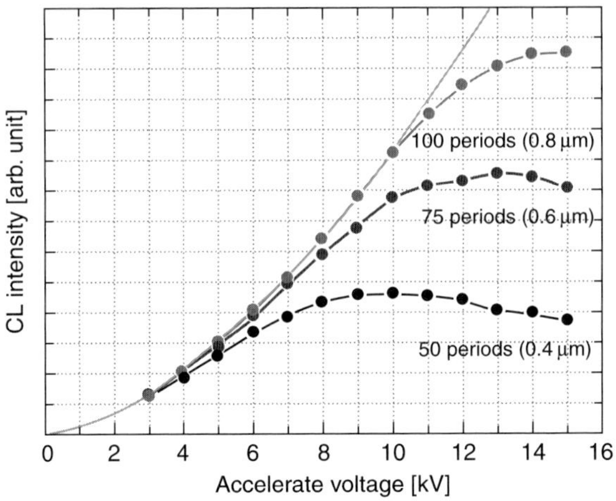

FIG. 7.14. Change in CL peak intensity with acceleration voltage for AlGaN MQWs with different total thicknesses of the MQW structure.

a 10 kV electron beam when the total thickness of the MQW structure was 0.8 µm.

The doping of impurities into quantum wells has been proposed as a means of improving electric and optical characteristics. Modulation doping into quantum wells has also been proposed as a technique applicable to high-output laser diodes (LDs), LEDs, and high-electron-mobility transistors. These methods use the change of the band structure caused by impurity doping, and impurities are generally doped into barrier layers. Nevertheless, the effect of impurity doping on electron beam excitation has not yet been clarified.

Here, we fabricated four samples and compared their characteristics to examine the effect of doping Si into quantum wells on electron beam excitation: (1) no Si was doped, (2) Si was doped only into the barrier layers, (3) Si was doped only into the well layers, and (4) Si was doped into both the well and barrier layers. For all four samples, the thicknesses of the well and barrier layers were 1.5 and 7 nm, respectively, the number of periods was 75, and the total thickness of the MQW structure was 637 nm. The AlN mole fractions of the well layers (Si-doped $Al_yGa_{1\text{-}y}N$ layers) and the barrier layers (Si-doped $Al_xGa_{1\text{-}x}N$ layers) were $y = 0.59 \pm 0.01$ and $x = 0.73 \pm 0.01$, respectively. The flow rate of CH_3SiH_3 was 0.45 nmol/min during the fabrication of the Si-doped samples.

Figure 7.15 shows the results of CL evaluation for the four samples. The acceleration voltage of the electron beam was 5 kV and the irradiation current was 3×10^{-10} A. A deep-UV CL peak was observed at a wavelength of approximately 255 nm for all four samples. The CL intensities for the sample with the Si-doped well layers and the sample with the Si-doped well and barrier layers were much higher than those for the undoped sample and the sample with the

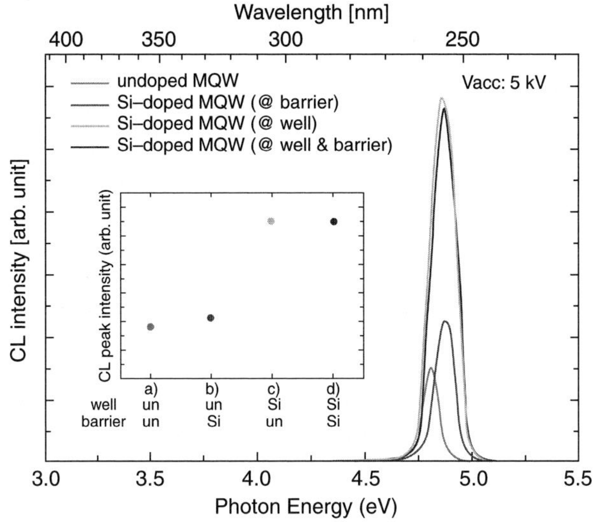

FIG. 7.15. CL intensities for AlGaN MQW samples with and without Si-doped well and barrier layers.

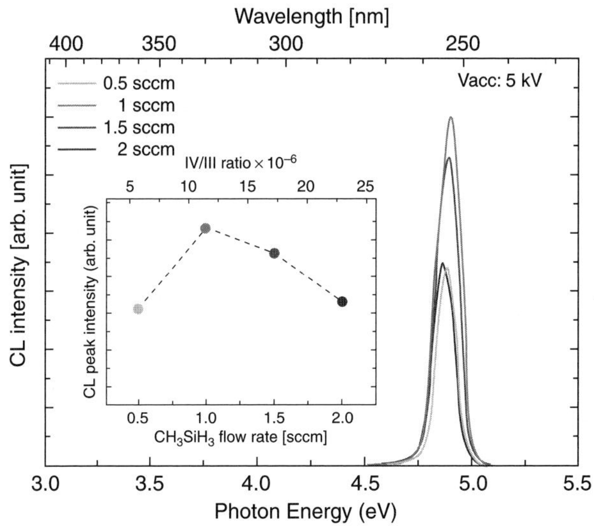

FIG. 7.16. CL intensities for AlGaN MQW samples with Si-doped well layers for different CH_3SiH_3 flow rates.

Si-doped barrier layers. This may be because the defect concentration in the well layers, where carriers were concentrated, was decreased by doping Si, resulting in efficient radiative recombination.

Finally, the doping of Si into well layers was examined for different CH_3SiH_3 flow rates. For the samples used in this examination, Si was doped only into the well layers at a CH_3SiH_3 flow rate of 0.5, 1, 1.5, or 2 sccm. Figure 7.16 shows the results of CL evaluation for the above samples. It was found that there was a peak of the CL intensity for the different amounts of doped Si. In this examination, the highest CL intensity was observed at a CH_3SiH_3 flow rate of 0.45 nmol/min. This marked difference in CL intensity for the different amounts of doped Si is considered to support the effectiveness of Si doping into the well layers of MQW structures.

7.4 Fabrication of AlGaN MQWs for electron-beam target for deep-ultraviolet light sources

In an attempt to satisfy the technical requirements of deep-UV LEDs, we focus on a UV light source using AlGaN as a target for electron-beam (EB) pumping, because the target does not require p-type AlGaN. Such UV light sources with a simple structure are also suitable for use as large-scale light sources. Furthermore, the emission wavelength can be designed without using a filter by changing the AlN mole fraction in the AlGaN. Recently, Watanabe *et al.* have fabricated a device using BN powder with an emission wavelength of 225 nm, a maximum output power of 1 mW, and a power efficiency of 0.6% [25]. However, it was difficult to control the emission wavelength. Oto *et al.* obtained 240-nm emission with an output power of 100 mW by EB pumping using 128-nm-thick $Al_xGa_{1-x}N/AlN$ quantum wells, and a power efficiency of 40% was achieved with no metal back layer on the EB input surface [26], though a detailed investigation of the characteristics has not been reported yet. In this study we have grown Si-doped AlGaN MQWs on AlN/sapphire templates by LP-MOVPE, and investigated their characteristics in detail for the purpose of developing AlGaN MQW targets for EB-pumped deep-UV light sources for industrial applications. With the optimized well and barrier thicknesses, we next evaluated the deep-UV light output power. Figure 7.17 shows the deep-UV light output power and conversion efficiency of an AlGaN MQW target, with optimized well and barrier thicknesses, pumped by a 10-kV, 5-mm-diameter EB. The AlGaN MQW with an Al metal back layer (thickness 20 nm) was pumped by the EB, and the deep-UV light emitted from the target that passed through the sapphire substrate to the outside was measured using a calibrated silicon photodetector. The spectral sensitivity of the detector was checked before and after the evaluation, and it was confirmed that there was no change in sensitivity. As shown in Fig. 7.16, the deep-UV light output power gradually increases with EB input power. The AlGaN MQW target exhibits a deep-UV light output power of more than 15 mW at a wavelength of 256 nm, and the estimated conversion efficiency is more than

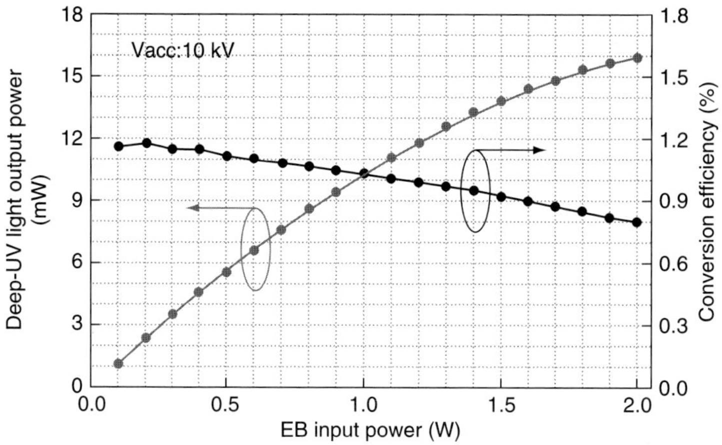

FIG. 7.17. Deep-UV light output power and conversion efficiency of optimized AlGaN MQW target pumped by a 10-kV, 5-mm-diameter EB.

0.75% at an EB input power of 2.0 W. In the CL analysis for the structure optimization described previously, the peak intensity of the optimized structure was about ten times higher than that of an AlGaN single-layer structure [23]. However, in the deep-UV light output power evaluation here, the light output power of the optimized structure for an EB input power of 1.0 W was only about five times higher than that of an AlGaN single-layer structure. It seems to be caused by the fact that the target temperature depends on EB input power, and light extraction efficiency depends on the buffer layer structure.

7.5 Conclusions

The growth of Si-doped AlGaN on an AlN/sapphire substrate as an underlying layer was performed using LP-MOVPE. The AlGaN films were grown with different AlN mole fractions by changing the growth temperature. GaN and AlGaN with an AlN mole fraction of 0.28 were free from stress because of the 3D growth. In the case of AlGaN with an AlN mole fraction of nearly 0.4, the lattice mismatch between the AlGaN and underlying AlN had an adverse effect on the crystal quality. AlGaN with an AlN mole fraction of more than 0.6 was coherently grown on AlN, and in the samples with an AlN mole fraction of more than 0.6 the crystal quality was high because of the coherent growth on the underlying AlN. Moreover, AlGaN was subjected to Si doping. The carrier concentration increased linearly with the Si concentration up to a Si concentration of 2×10^{18} cm^{-3}, indicating that the activation rate was approximately 1.

The structure of Si-doped AlGaN MQW targets was optimized for application to EB-pumped deep-UV light sources by the CL analysis. It was found that the optimum well and barrier layer thicknesses were 1.5 and 7 nm, respectively. We evaluated the deep-UV light emission from Si-doped AlGaN MQW targets that were pumped by a 10-kV, 5-mm-diameter EB. A deep-UV light output power of more than 15 mW with a conversion efficiency of more than 0.75% was achieved at a wavelength of 256 nm and an EB input power of 2.0 W. These results indicate the potential of using a p-type-AlGaN-free simple structure for AlGaN MQW targets for application to EB-pumped deep-UV light sources.

Acknowledgments

The author would like to thank Professor K. Hiramatsu, Mr Y. Shimahara, and Mr S. Ochia of Mie University, Mr F. Fukuyo and Mr Y. Kobayashi of Hamamatsu Photonics K.K., and Professor N. Kuwano of Kyushu University for their useful discussions. This work was partially supported by the Akasaki Research Center of Nagoya University, Grants-in-Aid for Scientific Research (Nos. 21360007 and 21560014), and Scientific Research on Priority Areas (No. 18069006) from the Ministry of Education, Culture, Sports, Science, and Technology (MEXT), and the Adaptable and Seamless Technology Transfer Program Through Target-Driven R&D (A-STEP) of the Japan Science and Technology Agency (JST).

References

[1] H. Hirayama, S. Fujikawa, N. Noguchi, J. Norimatsu, T. Takano, K. Tsubaki, and N. Kamata: *Phys. Status Solidi A* **206** (2009) 1176.

[2] C. Pernot, M. Kim, S. Fukahori, T. Inazu, T. Fujita, Y. Nagasawa, A. Hirano, M. Ippommatsu, M. Iwaya, S. Kamiyama, I. Akasaki, and H. Amano: *Appl. Phys. Express* **3** (2010) 061004.

[3] S. J. Hearne, J. Han, S. R. Lee, J. A. Floro, D. M. Follstaedt, E. Chason, and I. S. T. Tsog: *Appl. Phys. Lett.* **76** (2000) 1534.

[4] S. Einfeldt, V. Kirchner, H. Heinke, M. Diebelberg, S. Figge, K. Vogeler, and D. Hommel: *J. Appl. Phys.* **88** (2000) 7029.

[5] D. G. Zhao, D. S. Jiang, J. J. Zhu, Z. S. Liu, S. M. Zhang, H. Yang, U. Jahn, and K. U. Ploog: *J. Cryst. Growth* **310** (2008) 5366.

[6] Y. Koide, N. Itoh, N. Sawaki, and I. Akasaki: *Jpn. J. Appl. Phys.* **27** (1988) 1156.

[7] K. Ito, K. Hiramatsu, H. Amano, and I. Akasaki: *J. Cryst. Growth* **104** (1990) 533.

[8] M. Z. Peng, L. W. Guo, J. Zhang, X. L. Zhu, N. S. Yu, J. F. Yan, H. H. Liu, H. Q. Jia, H. Chen, and J. M. Zhou: *J. Cryst. Growth* **310** (2008) 1088.

[9] Y. Kida, T. Shibata, H. Naoi, H. Miyake, K. Hiramatsu, and M. Tanaka: *Phys. Status Solidi A* **194** (2002) 498.

[10] L. J. Schowalter, Y. Shusterman, R. Wang, I. Bhat, G. Arunmozhi, and G. A. Slack: *Appl. Phys. Lett.* **76** (2000) 985.

[11] Y. Kida, T. Shibata, H. Miyake, and K. Hiramatsu: *Jpn. J. Appl. Phys.* **42** (2003) 572.

[12] R. Q. Jin, J. P. Liu, J. C. Zhang, and H. Yang: *J. Cryst. Growth* **268** (2004) 35.

[13] N. Kato, S. Sato, T. Sumii, N. Fujimoto, N. Okada, M. Imura, K. Balakrishnan, M. Iwaya, S. Kamiyama, H. Amano, I. Akasaki, H. Maruyama, T. Noro, T. Takagi, and A. Bandoh: *J. Cryst. Growth* **298** (2007) 215.

[14] M. Z. Peng, L. W. Guo, J. Zhang, N. S. Yu, X. L. Zhu, J. F. Yan, Y. Wang, H. Q. Jia, H. Chen, and J. M. Zhou: *J. Cryst. Growth* **307** (2007) 289.

[15] D. B. Li, M. Aoki, T. Katsuno, H. Miyake, K. Hiramatsu, and T. Shibata: *J. Cryst. Growth* **298** (2007) 372.

[16] K. X. Chen, Q. Dai, W. Lee, J. K. Kim, E. F. Schubert, J. Grandusky, M. Mendrick, X. Li, and J. A. Smart: *Appl. Phys. Lett.* **93** (2008) 192108.

[17] T. Asai, K. Nagata, T. Mori, K. Nagamatsu, M. Iwaya, S. Kamiyama, H. Amano, and I. Akasaki: *J. Cryst. Growth* **311** (2009) 2850.

[18] H. K. Cho, J. Y. Lee, S. R. Jeon, and G. M. Yang: *J. Cryst. Growth* **233** (2001) 667.

[19] C. R. Lee: *J. Cryst. Growth* **246** (2002) 25.

[20] A. F. Wright: *J. Appl. Phys.* **82** (1997) 2833.

[21] T. Takeuchi: Doctoral thesis, Faculty of Science and Technology, Meijo University, Nagoya (1998).

[22] Y. Shimahara, H. Miyake, K. Hiramatsu, F. Fukuyo, T. Okada, H. Takaoka, and H. Yoshida: *Jpn. J. Appl. Phys.* **50** (2011) 095502.

[23] Y. Shimahara, H. Miyake, K. Hiramatsu, F. Fukuyo, T. Okada, H. Takaoka, and H. Yoshida: *Appl. Phys. Express* **4** (2011) 042103.

[24] R. G. Banal, M. Funato, and Y. Kawakami: *Phys. Status Solidi C* **7** (2010) 2111.

[25] K. Watanabe, T. Taniguchi, T. Niiyama, K. Miya, and M. Taniguchi: *Nat. Photonics* **3** (2009) 591.

[26] T. Oto, R. G. Banal, K. Kataoka, M. Funato, and Y. Kawakami: *Nat. Photonics* **4** (2010) 767.

8 Optical and structural properties of InGaN light-emitters on non-polar and semipolar GaN

Michael Kneissl and Tim Wernicke

Today group III-nitride-based visible-light-emitting diodes (LEDs) exhibiting record external quantum efficiencies and InGaN quantum well laser diodes in the near-UV to the green spectral range have been demonstrated. Nevertheless, a large number of technological problems and scientific questions remain. One of the most critical issues for InGaN-based LEDs and laser diodes is the presence of large polarization fields along the (0001) direction. These can be significantly reduced or completely eliminated by growing on non- and semipolar GaN surfaces. Controlling the polarization fields in III-nitride heterostructures is particularly critical for light emitters in the blue and green spectral regions. Triggered by the first demonstration non-polar GaN quantum wells grown on $LiAlO_2$ by Waltereit and colleagues in 2000 (Waltereit *et al.*, 2000), impressive advances in the area of non-polar and semipolar nitride semiconductors and devices have been achieved. Today, a large variety of group III-nitride heterostructures free of polarization fields exhibit exceptional performance and the fundamental understanding of polar, semipolar, non-polar nitrides has made significant leaps forward. In the following chapter we will review the different approaches for heteroepitaxial growth of non- and semipolar GaN films on sapphire, SiC, silicon, and $LiAlO_2$ substrates, and discuss challenges related to surface morphologies, structural defects, and optical properties. Different strategies for defect reduction for heteroepitaxially grown GaN layers will also be discussed. In addition, the structural and optical properties of InGaN quantum wells (QWs) grown on m-plane and different semipolar surfaces, i.e., $(10\bar{1}2)$, $(10\bar{1}1)$, $(20\bar{2}1)$, $(11\bar{2}2)$, as well as c-plane GaN substrates, will be reviewed. The dependence of the indium incorporation efficiency on the growth surface orientations as well as the emission properties, in particular the polarization of the light emission, for InGaN QWs grown on semi- and non-polar surfaces will be discussed. We will also present studies on the effects of the surface orientation and the excitation stripe orientation on the optical polarization, gain, and the threshold for amplified stimulated emission (ASE) in non- and semipolar laser diodes. Finally, the gain characteristics of InGaN single and multiple quantum-well lasers on m-plane GaN, as well as different semipolar surfaces, will be discussed.

8.1 Spontaneous and piezoelectric polarization in InGaN/GaN quantum wells on c-plane, semipolar, and non-polar crystal orientations

The initial driving force for the development of non-polar and semipolar nitrides was and still is the reduction of polarization fields in conventional (0001) c-plane InGaN/GaN QWs, which lead to long radiative recombination times and poor internal quantum efficiencies (IQE) due to the quantum-confined Stark effect (QCSE) (Miller *et al.*, 1984; Takeuchi *et al.*, 1997; Berkowicz *et al.*, 2000). The origin of the spontaneous polarization is the lack of inversion symmetry in wurtzite crystal structures, the polar nature of the Ga-N bond (represented by different electro-negativity values), and the unequal bonding lengths between Ga and N along c- and a-directions (Piprek, 2007). Therefore, in InGaN/GaN heterostructures interface charges are generated due to the non-vanishing spontaneous and piezoelectric polarization. For InGaN QWs grown on GaN, piezoelectric fields are generated by the compressive strain that the InGaN layer experiences. As the covalent bonds within the c-plane are compressed, the bond length along the c-direction increases (Piprek, 2007). This leads to a further deviation of the c/a ratio which in turn leads to a stronger polarization, denoted as piezoelectric polarization. Due to the symmetry of the wurtzite unit cell, the field is always oriented along the [0001] c-axis. Considering a slab of III-N bulk crystal, the dielectric displacement is constant within and outside the crystal. As a result, an electric field and its associated sheet charges at the crystal surface are generated. In the picture of band-structure diagrams, the electric field causes a tilt of the conduction and valence bands. However, if the bands cross the Fermi level, free carriers are generated and the fields are screened. Typically, surface sheet charges are screened at a distance of ~40 nm in GaN (Netzel *et al.*, 2009). Considering a quantum well within bulk GaN (or another III-nitride alloy), only the change of the total polarization ΔP, i.e., the sum of spontaneous and piezoelectric polarization, at the quantum well interfaces is important for the calculation of the polarization field. The spontaneous polarization field along growth direction can be calculated by $\Delta\, P^{sp}\ (\vartheta) = (P^{sp}_{barrier} - P^{sp}_{well}) \cos\vartheta$, were $P^{sp}_{barrier}$ and P^{sp}_{well} are the spontaneous polarizations of the barrier and well, interpolated by Vegard's law from the binary constants ($P^{sp}_{AlN} = -0.081$, $P^{sp}_{GaN} = -0.029$, $P^{sp}_{InN} = -0.032$, (Bernardini *et al.*, 1997)) and ϑ is the inclination angle of the growth plane to the c-plane, i.e., $\vartheta = 0$ for polar and $\vartheta = 90°$ for non-polar nitrides. Due to the symmetry of the strain tensor and the piezoelectric tensor, there is no difference between m-plane-type $(10\bar{1}l)$ and a-plane-type $(11\bar{2}l)$ semipolar planes. To obtain the piezoelectric polarization the full strain of a structure has to be calculated. Due to the reduced in-plane symmetry in semipolar GaN, this leads to anisotropic and shear strain, making it necessary to consider all components of the piezoelectric tensor (Romanov *et al.*, 2006). This leads (for a negative value of the piezoelectric tensor element e_{15}) to a zero crossing of the piezoelectric polarization for a certain semipolar orientation. This zero crossing depends on the exact values of the material constants (Feneberg *et al.*, 2006), which are still being debated. However, a number of experimental results (Raß, 2012; Feneberg *et al.*, 2010) suggest that the parameters proposed by Romanov *et al.* (2006) are a very good fit. A detailed calculation of the total and piezoelectric polarization in pseudomorphically strained InGaN on GaN and strained AlGaN on GaN heterostructures can be found in Romanov *et al.* (2006).

For InGaN/GaN heterostructures, the difference in spontaneous polarization is rather small. Large piezoelectric constants in comparison to classical III/V compounds (Bernardini *et al.*, 1997) and a large lattice mismatch between InN and GaN of 11% makes piezoelectric fields dominating in the InGaN/GaN system causing 98% of the polarization field (for c-plane: $\Delta P^{sp}_{In_{0.15}Ga_{0.85}N/GaN} = 0.0005$ C/m, $\Delta P^{sp}_{In_{0.15}Ga_{0.85}N/GaN} = 0.022$). The internal electric field for an $In_{0.15}Ga_{0.88}N$/GaN QW in dependence of ϑ is depicted in Fig. 8.1. For the c-plane InGaN QW the field F has a magnitude of nearly 2 MV/cm. The zero crossing angle ϑ_s, can be found at $\sim 45°$ independent of the indium content. Very close to this point is the $(10\bar{1}2)$ orientation with $\vartheta = 43.19°$ (F = 0.2 MV/cm). More commonly used orientations are $(11\bar{2}2)$ at $\vartheta = 58.4°$ (F = –0.35 MV/cm), $(10\bar{1}1)$ at $\vartheta = 61.96°$ (F = –0.4 MV/cm), $(20\bar{2}1)$ at $\vartheta = 75.09°$ (F = –0.35 MV/cm) and the $(10\bar{1}0)$ at $\vartheta = 90°$ (F = 0 MV/cm) orientations (all angles given for GaN). In fact, Schrödinger–Poisson calculations at a carrier density of 10^{18} cm^{-3} show that these fields have a much smaller impact on the transition energy than the different strain states of the QWs (Wernicke *et al.*, 2012) (see chapter 4b), i.e., such small fields are easily screened

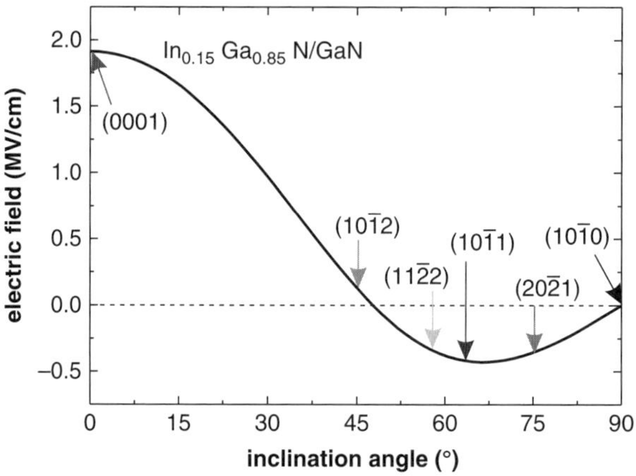

FIG. 8.1. Electric field in dependence of the inclination angle ϑ for strained InGaN/GaN QWs. Reprinted with permission from Wernicke *et al.* (2012), © 2012 IOP Publishing Ltd.

in devices and other properties, such as surface morphology (Wernicke *et al.*, 2011; Ploch *et al.*, 2012*a*; Ploch *et al.*, 2012*c*), indium uptake (Wernicke *et al.*, 2012; Browne *et al.*, 2012) (see also Section 8.4.1), critical layer thickness (Ploch *et al.*, 2012*b*; Hardy *et al.*, 2012), etc. become more important.

AlGaN/GaN heterostructures exhibit a much larger difference of the spontaneous polarization and a smaller lattice mismatch. Therefore the spontaneous polarization for c-plane AlGaN/GaN heterostructures is nearly by a factor of two larger than the piezoelectric polarization. For semipolar AlGaN this is no longer true due to the complex interdependence of the polarization field on the inclination angle with respect to the c-plane. For strained AlGaN layers on GaN (relevant for electron blocking layers), at $\sim 70°$ piezoelectric and spontaneous polarization cancel each other out, leading to a zero crossing of the polarization field. However, calculations for AlGaN strained to AlN and AlGaN, as these are important for UV-LEDs, are still missing. Also, more conclusive experimental data are needed to confirm the calculations for the InGaN/GaN and the AlGaN/GaN material systems. There are indications that applying Vegard's law to spontaneous and piezoelectric material parameters imposes non-negligible errors to the calculations (Miller *et al.*, 2002; Winzer *et al.*, 2005).

The polarization fields in InGaN/GaN quantum wells result in the QCSE, which has significant effects on the carrier recombination (Miller *et al.*, 1984). The field-induced spatial separation of electron and hole wavefunctions leads to a reduced wavefunction overlap and consequently to increased radiative recombination lifetimes (Berkowicz *et al.*, 2000). Since non-radiative and radiative recombination processes are competing, longer radiative recombination lifetimes lead to a decrease of the internal quantum efficiency. It should be noted, however, that the non-radiative recombination lifetime also depends on the orientation (Langer *et al.*, 2011). Nevertheless, smaller fields allow for much larger quantum well thicknesses in LED heterostructures that in turn reduces LED droop, i.e., drop in external quantum efficiency at higher carrier densities (Gardner *et al.*, 2007; Pan *et al.*, 2012). In addition, the modal gain in semipolar lasers increases with thicker QWs due to the increased wavefunction overlap (Scheibenzuber *et al.*, 2009) (see also chapter 6a). Another benefit of semipolar LEDs is the reduced blue-shift of the emission wavelength with increasing injection current due to the screening of polarization fields by free carriers (Kuokstis *et al.*, 2002). This is a big problem for polar c-plane LED; which experience large emission wavelength shifts and changes in color temperature with the drive current. One drawback, however, is that semipolar and non-polar InGaN quantum-well LEDs require a substantially higher indium mole fraction in order to obtain the same emission wavelength as polar LEDs. Calculations as well as experimental data show that the polarization fields in c-plane QWs lasers are not fully screened even for high carrier densities at lasing threshold. Actually the red-shifted emission in polar lasers due to the QCSE helps to reach green laser emission at lower indium mole fractions (Strauß *et al.*, 2011).

Also it is important to note that the direction of the polarization field in InGaN heterostructures changes for substrate inclination angles above 45°. This does not affect the recombination in quantum wells but affects the carrier injection for structures typically grown from the n-side to the p-side (Kawaguchi *et al.*, 2012). Therefore it can be advantageous to grow devices on orientations with flipped polarity, e.g. $(20\bar{2}\bar{1})$ instead of $(20\bar{2}1)$.

8.2 Performance characteristics of violet, blue, and green (0001) c-plane InGaN quantum well LEDs and laser diodes

(0001) c-plane InGaN light-emitting diodes (LEDs) already exhibit outstanding levels of performance. This, of course, can also be attributed to two decades of intense research worldwide, since the first demonstration of blue III-nitride-based LEDs on (0001) GaN/sapphire by Nakamura *et al.* (1991, 1994, 1995). Figure 8.2 shows the measured external quantum efficiencies (EQE) of InGaN single and multiple quantum well LEDs grown on c-plane (0001) sapphire substrates emitting in the violet-to-green spectral range. In addition, the EQEs of InGaP-based LEDs emitting in the red, orange, and yellow wavelength spectrum are plotted. As can be seen from Fig. 8.2 the EQE of InGaN-based LED peaks near 450 nm and rapidly decreases for longer wavelengths. Although yellow and even red-emitting c-plane InGaN QW LEDs have been demonstrated (Mukai *et al.*, 1999), their efficiencies cannot compete with the performance of InGaP-based red and yellow LEDs. However, the peak efficiencies of InGaP LEDs appear also to be limited to a narrow spectral range near 650 nm, and is rapidly decreasing for shorter wavelengths. This situation creates a spectral

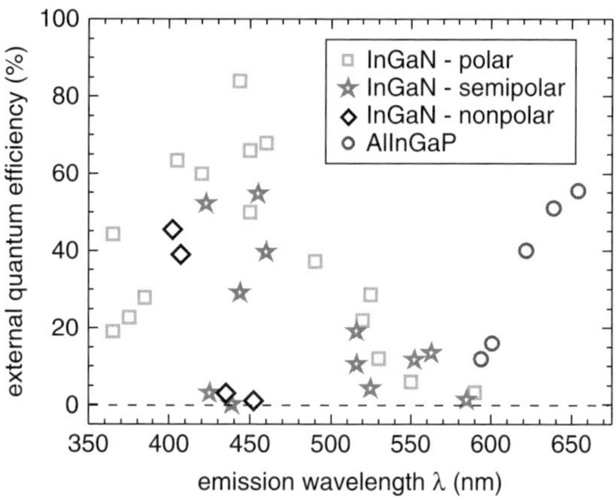

FIG. 8.2. Reported external quantum efficiencies (EQEs) for c-plane, non-polar, and semipolar InGaN quantum-well LEDs as well InGaP-based LEDs at different emission wavelengths (Morita *et al.*, 2004; Nichia Corporation, 2012*a*).

region in the green–yellow wavelength range between about 530 nm and 600 nm, where all semiconductor LEDs exhibit relatively poor external quantum efficiencies—a circumstance that is often addressed as the so-called "green gap". For InGaN quantum-well LEDs the drop in EQE originates from multiple causes: (1) at longer wavelengths the increasing indium mole fraction in the InGaN QWs results in an increase in the spontaneous and piezoelectric polarization charges at the InGaN/GaN interfaces (Schmidt *et al.*, 2007). The resultant polarization fields result in a spatial separation of the electron and hole envelope wave-functions and a red-shift of the emission wavelength due to the quantum-confined Stark effect (QCSE) (Miller *et al.*, 1984). Simulations of the electron-hole wavefunction overlap for (0001) InGaN/GaN quantum well show that the envelope electron-hole wavefunction overlap is reduced by more than 50 % when going from a blue emitter to a green or yellow LED (Arif *et al.*, 2008). Consequently, the radiative recombination lifetimes in the InGaN QWs increase significantly at longer emission wavelengths (Chichibu *et al.*, 1999). (2) In addition, at higher indium mole fraction the formation of extended and point defects in the InGaN layers is enhanced, leading to shorter non-radiative recombination lifetimes (Lu *et al.*, 2010; Chichibu *et al.*, 2006). Since the internal quantum efficiency η_{IQE} depends on the radiative lifetime τ_{rad} as well as non-radiative lifetime τ_{nr} according to following relationship

$$\eta_{IQE} = \frac{R_{rad}}{R_{rad} + R_{nr}} = \frac{1/\tau_{rad}}{1/\tau_{rad} + 1/\tau_{nr}}$$

both of these detrimental effects will lead to a reduction in the internal and consequently also external quantum efficiency of the light-emitting devices. Obviously, the effects also depend on additional factors, such as the threading dislocation density and temperatures (Karpov and Makarov, 2002; Chichibu *et al.*, 2006) and may be partially suppressed by potential barriers forming threading dislocations (Hangleiter *et al.*, 2005).

Furthermore, the EQE also changes with the injection current and current densities—a phenomenon that is also often described as "droop" or "efficiency droop". As can be seen in Fig. 8.3, droop is observed for GaN-based LEDs at all emission wavelengths starting at different current densities. Droop is a significant impediment for high-power LEDs, e.g. for application in solid-state lighting, where high light output power levels at low chip cost are required, which can be achieved only by operating these devices at high current densities. For example, the first generation of high-power InGaN MQW LEDs were operated at a dc current of 350 mA with chip sizes in the order of 1 mm^2. Currently, high-power LEDs are operating at currents up to 2 A, which corresponds to current densities of 200 A/cm^2 for the same chip size, and future generations of LEDs will certainly be driven at even higher current levels. The physical mechanisms underlying the droop effect are still not completely understood, and are intensively discussed among a number of research groups. Proposed mechanisms for the droop effect include Auger recombination (Shen *et al.*, 2007; David and

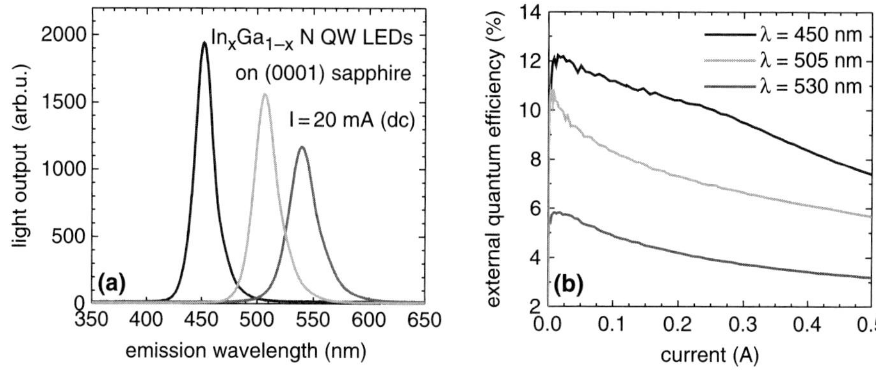

FIG. 8.3. (a) Emission spectra for blue, cyan, and green (0001) c-plane InGaN MQW LEDs measured at a forward current of 20 mA. (b) Measured external quantum efficiencies for blue, cyan, and green (0001) c-plane InGaN MQW LEDs vs. dc forward current. Courtesy of C. Reich (Technical University, Berlin).

Grundmann, 2010; Laubsch *et al.*, 2009; Kioupakis *et al.*, 2011), carrier leakage (Kim *et al.*, 2007*b*; Piprek, 2010), density-activated defect recombination (DADR) (Hader *et al.*, 2010), and changing radiative recombination rate and rate equation analysis (Ryu *et al.*, 2009). It is not the intention here to discuss the pros and cons of the different possible mechanisms. However, almost regardless of the specific cause, the droop effect is certainly influenced by the strong polarization fields in c-plane light-emitting heterostructures. For example, the large polarization fields in InGaN/GaN QWs limit the thickness of the QWs, since thicker QWs will result in a stronger separation of electron and hole envelope wavefunctions, due to the QCSE. Thick InGaN QWs, however, are preferred for reducing the carrier density in the QWs, in order to reduce Auger recombination and carrier leakage-related droop effects. Therefore, reducing or eliminating the polarization fields in InGaN QWs should also lead to a reduced droop in III-nitride-based LEDs.

Another challenge for (0001) c-plane light-emitters is the change in peak emission wavelength with increasing current density. The high carrier densities at high current densities lead to a partial screening of the polarization field in the InGaN QWs, which in turn reduces the quantum-confined Stark effect. As a consequence, the peak emission is shifted to shorter wavelengths at higher current densities (Mukai *et al.*, 1999). This effect is more pronounced for longer-wavelength LEDs, where the higher indium mole fraction in the InGaN QWs leads to stronger polarization fields. Figure 8.3 shows the emission spectra and external quantum efficiencies for blue and green c-plane InGaN MQW LEDs measured at different current densities, and Fig. 8.4 shows the change in emission

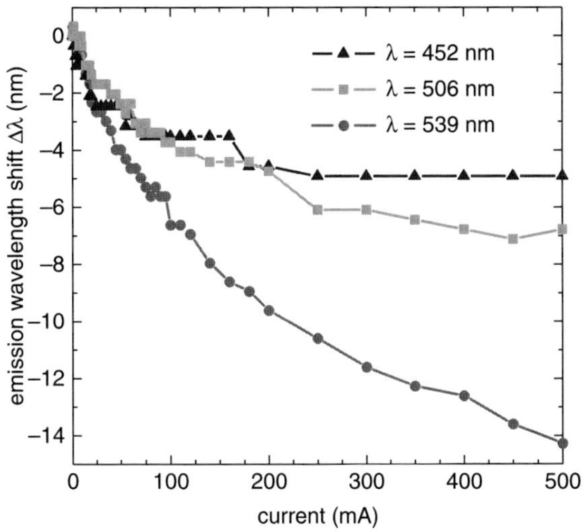

FIG. 8.4. Change in peak emission wavelength vs. dc forward current for blue, cyan, and green (0001) c-plane InGaN MQW LEDs. Courtesy of C. Reich (Technical University, Berlin).

wavelength for the same set of LED vs. the drive current. As can be seen, in all cases the emission shifts to shorter wavelengths at higher currents due to the screening of the polarization fields with higher carrier densities. It should be noted that part of the wavelength shift can be attributed to the filling of potential fluctuations that originate from compositional inhomogeneities in the InGaN QWs. However, this alone cannot explain the observed behavior.

Similarly, c-plane InGaN QW laser diodes are affected by the polarization fields. Although the first violet laser diodes were demonstrated in late 1995 (Nakamura *et al.*, 1996*a*; Nakamura *et al.*, 1996*b*; Nakamura and Fasol, 1997) and continuous-wave operation with low threshold current densities were shown just a few years later (Nakamura *et al.*, 1998*a*; Nakamura *et al.*, 1998*b*), the development of blue- and especially green-wavelength InGaN laser diodes lagged significantly compared to the rapid development of high-efficiency blue and green LEDs. For many years the realization of green InGaN-based laser diodes was considered extremely difficult, if not impossible. Although many factors contribute to these challenges, the large polarization fields in c-plane InGaN QWs were considered to be one of the main causes for the great challenges to realize green laser diodes. Therefore, the development of laser diodes on non-polar and semi-polar GaN surfaces is considered one of the most promising approaches for the realization of high-efficiency and high-power true green laser diodes.

Nevertheless, by advancing the waveguide heterostructure and reducing nano-scopic and microscopic defects in the InGaN materials, several groups have

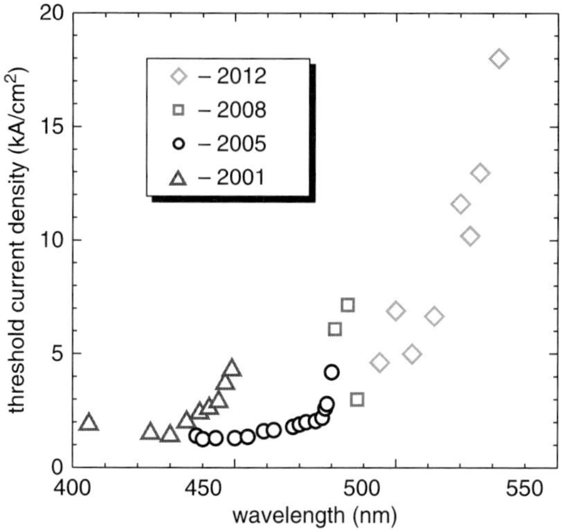

FIG. 8.5. Reported threshold current densities of (0001) c-plane InGaN QW laser diodes for different emission wavelengths. The different sets of data also represent the different time-periods in which the data have been obtained (Kozaki *et al.*, 2007; Kim *et al.*, 2008; Queren *et al.*, 2009; Avramescu *et al.*, 2009; Miyoshi *et al.*, 2009; Nichia Corporation *et al.*, 2012*b*; Avramescu *et al.*, 2010; Lutgen *et al.*, 2010; Lermer *et al.*, 2010; Schwarz and Scheibenzuber, 2011; Müller *et al.*, 2011).

been able to push the emission wavelength of c-plane InGaN QW lasers well into the green spectral range. Figure 8.5 shows the reported threshold current densities of c-plane InGaN QW laser diodes for different emission wavelengths reported for different time-periods. Currently c-plane green laser diodes emitting at wavelengths as long as 532 nm have been demonstrated (Schwarz and Scheibenzuber, 2011). At slightly shorter wavelengths, cw operation of ridge-waveguide laser diodes with output power levels of more than 100 mW have been shown (Schwarz and Scheibenzuber, 2011). Almost simultaneously, high-power green laser diodes on different semipolar GaN surfaces have been demonstrated by several groups (Kubota *et al.*, 2008; Okamoto *et al.*, 2009; Yoshizumi *et al.*, 2009; Enya *et al.*, 2009; Kelchner *et al.*, 2009; Tyagi *et al.*, 2010; Raring *et al.*, 2010; Yanashima *et al.*, 2012) emitting at wavelengths as long as 535 nm, with a light output of more than 100 mW. At the moment the performance characteristics of green lasers on polar and semipolar GaN are very close and which of these approaches will succeed will be determined sometime in the future. Parameters such as threshold current densities, wallplug efficiency, and output power will play an important role in that decision, but other important criteria, such as laser diode lifetimes, manufacturability, and chip cost will certainly play an important role as well.

8.3 Growth of non-polar and semipolar GaN buffer layers for device applications

Most of the excellent non- and semipolar device results were achieved on GaN bulk substrates (see Sections 8.5 and 8.6). Therefore we will review GaN growth on bulk substrates. However, for mass production, low-cost large-area substrates are desirable. These must be heteroepitaxial substrates, since low-cost large-area semipolar bulk GaN substrates are currently not available.

8.3.1 *Growth of GaN on low-defect bulk GaN substrates*

The growth of GaN layers on low defect density GaN bulk substrates is governed by surface diffusion processes. However, interaction with threading dislocations can lead to the formation of macroscopic defects (Wernicke *et al.*, 2011), even at the low density of $\sim 10^5$–1×10^7 cm^{-2} present in bulk substrates (Fujito *et al.*, 2008). These can be avoided using substrate miscut or proper growth conditions that limit the adatom diffusion length.

In Fig. 8.6 the macroscopic surface morphologies (Normarski contrast images) of on-axis $(10\bar{1}2)$, $(10\bar{1}1)$, $(11\bar{2}2)$, $(20\bar{2}1)$ and $(10\bar{1}0)$ GaN layers grown with different temperatures and pressures are shown. The $(10\bar{1}0)$ surfaces exhibit pyramidal structures for all growth parameters. These form at screw dislocations when a

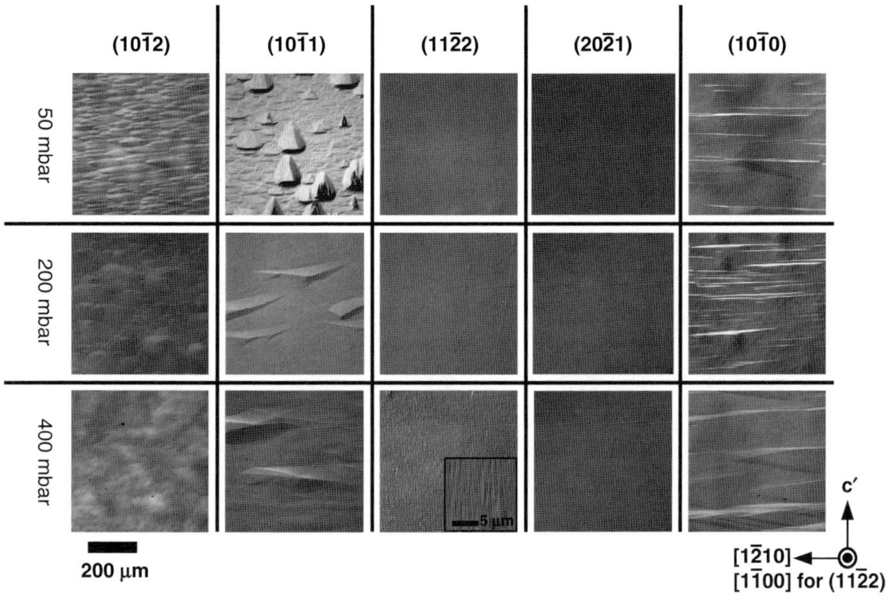

FIG. 8.6. Normarski contrast micrographs of $(10\bar{1}2)$, $(10\bar{1}1)$, $(11\bar{2}2)$, $(20\bar{2}1)$, and $(10\bar{1}0)$ GaN surfaces grown at 980°C and varied reactor pressure.

pinned surface step leads to a local enhancement of the growth rate (Farrell *et al.*, 2010). On m-plane GaN this pyramidal surface structure can be avoided using miscut substrates (Hirai *et al.*, 2007*b*). The miscut leads to a defined step orientation and a sufficiently high step density to completely hinder the formation of growth spirals. Also, for layers grown on semipolar GaN macroscopic features can be observed: triangular pyramids on $(10\bar{1}1)$, shallow round pyramids on $(10\bar{1}2)$, and elongated hillocks on $(11\bar{2}2)$. For $(10\bar{1}1)$ triangular pyramids, a dark spot was found in the center which indicates the presence of a dislocation (not shown). These are strong indications that dislocations cause macroscopic features also in semipolar GaN. However, by changing the growth parameters these features can be reduced or even eliminated. Two different classes of plane can be distinguished. The low-indexed $(10\bar{1}1)$ and $(10\bar{1}2)$ smooth with increasing reactor pressure, whereas the higher-indexed $(11\bar{2}2)$ and $(20\bar{2}1)$ planes are smoothest for low reactor pressures. A similar behavior is known from the stability a-plane vs. m-plane facet (Sun *et al.*, 2008). The m-plane facet is more stable, and smooth layers can be grown at various growth conditions. The a-plane is rather unstable, and for a full coalescence low reactor pressure and high temperatures are required (Chen *et al.*, 2004).

Looking at the microscopic surface morphology of $(11\bar{2}2)$ and $(20\bar{2}1)$ layers, distinct stripe patterns are visible (Fig. 8.7). For the $(11\bar{2}2)$ surfaces, undulations with short period are found along $[1\bar{1}00]$, and with a longer period along $[11\bar{2}\bar{3}]$ (Ploch *et al.*, 2012*a*). On $(20\bar{2}1)$ surfaces, undulations with short period are found along $[10\bar{1}4]$, which is the projection of the c-plane onto the growth plane, also denoted as c′ (Ploch *et al.*, 2012*c*). No undulations were found along $[11\bar{2}0]$. These stripe patterns are the result of anisotropic surface diffusion and provide a less stable surface (in comparison to the $(10\bar{1}1)$ and $(10\bar{1}0)$ surfaces). Surface diffusion activation energies were calculated for $(10\bar{1}0)$ and $(11\bar{2}0)$ GaN (Lymperakis

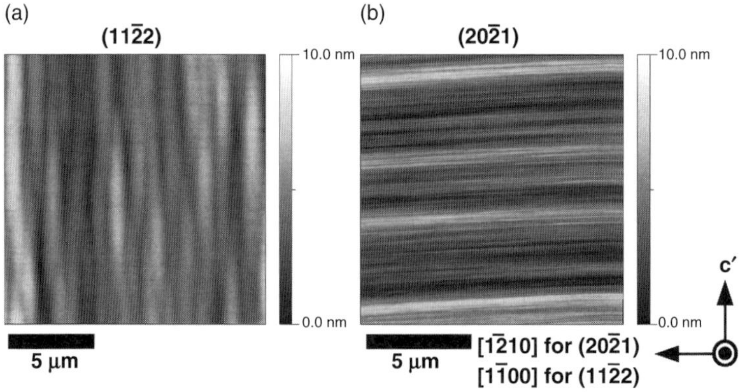

FIG. 8.7. AFM image of a) $(11\bar{2}2)$, and b) $(20\bar{2}1)$ surfaces. Images are courtesy of S. Ploch (Technical University, Berlin).

and Neugebauer, 2009), showing that adatoms can diffuse easily along close-spaced nitrogen-atom rows. For $(10\bar{1}0)$ these are oriented along $[1\bar{2}10]$, and the activation energy for diffusion is 4.5 times lower than along the perpendicular [0001]. For a-plane the closely spaced nitrogen-atom rows are aligned parallel [0001], and the activation energy for diffusion along the perpendicular $[1\bar{1}00]$ direction is higher by a factor of two. Geometrical considerations suggest the same for the $(10\bar{1}l)$-type tilted m-plane surfaces (like $(10\bar{1}1)$, $(10\bar{1}2)$, and also $(20\bar{2}1)$) and $(11\bar{2}l)$-type tilted a-plane surfaces (like $(11\bar{2}2)$) (Dinh, 2012). In fact, anisotropic diffusion can explain the formation of undulations on both $(11\bar{2}2)$ and $(20\bar{2}1)$ surfaces. In this case the undulation is formed by adatom diffusion, and its period is proportional to the diffusion length. For $(11\bar{2}2)$ surfaces the undulation with high spatial frequency is oriented along $[1\bar{1}00]$ which exhibits a high diffusion barrier on $(11\bar{2}l)$-type surfaces. Longer diffusion occurs along $[11\bar{2}\bar{3}]$. In fact, from the temperature dependence of the undulation periods, activation energies of 0.8 eV and 1.3 eV along $[11\bar{2}\bar{3}]$ and $[1\bar{1}00]$, respectiviely, were derived (Ploch *et al.*, 2012*a*). For $(20\bar{2}1)$ the undulations are oriented along the c′ direction, which exhibits a high diffusion barrier on $(10\bar{1}l)$-type surfaces. However, in order to explain the occurrence of undulations at all, the facet stability has to be taken into account (Ploch *et al.*, 2012*c*). As mentioned above, facets that exhibit a very high stability are the $(10\bar{1}1)$ and $(10\bar{1}0)$ facets. Therefore it is energetically favorable to form areas with $(10\bar{1}1)$ and $(10\bar{1}0)$-like surfaces. For $(11\bar{2}2)$ surfaces, the $10\bar{1}1$ plane are titled by $\pm 26°$ towards $[1\bar{1}00]$. Therefore undulations along $[1\bar{1}00]$ are stabilized by the presence $(10\bar{1}1)$ micro-facets. For the $(20\bar{2}1)$ surface orientation the $(10\bar{1}1)$ plane is titled by 13° towards $[10\bar{1}4]$, and the $(10\bar{1}0)$ plane is tilted by 15° towards $[\bar{1}01\bar{4}]$. This stabilizes undulations along $[10\bar{1}4]$.

For heteroepitaxial $(11\bar{2}2)$ GaN the same dependencies were found. However, the presence of stacking faults deteriorates the surface (Ploch *et al.*, 2012*a*), as was also reported for m-plane GaN (Hirai *et al.*, 2007*a*).

8.3.2 *Growth of GaN on planar heteroepitaxial substrates*

In order to reduce the LED device fabrication costs, large-area low-cost substrates are in demand. Polar c-plane LEDs have been demonstrated on (0001) sapphire and (111) silicon with up to a 6-inch diameter (Dadgar *et al.*, 2006; Lu *et al.*, 2011). Therefore the easiest approach is to grow heteroepitaxial GaN on foreign substrates, e.g. sapphire and silicon. The resultant growth plane depends on the local atomic arrangement and lattice mismatch, i.e., for every substrate material and orientation an epitaxial relationship is defined. An overview of different orientations that have been realized is given in Fig. 8.8. Non-polar $(11\bar{2}0)$ a-plane GaN was first grown on $(10\bar{1}2)$ r-plane sapphire (Sano and Aoki, 1976), but can also be grown on $(11\bar{2}0)$ a-plane SiC (Craven *et al.*, 2004*b*). $(10\bar{1}0)$ m-plane GaN was realized on (100) γ-$LiAlO_2$ (Hellman *et al.*, 1997), $(10\bar{1}0)$ m-plane SiC (Gardner *et al.*, 2005), and $(10\bar{1}0)$ m-plane sapphire. Also, semipolar

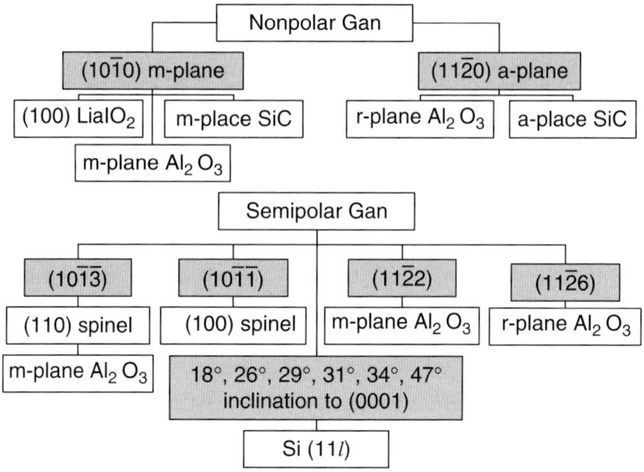

FIG. 8.8. Overview of realized surface orientations for planar heteroepitaxial growth.

GaN can be grown heteroepitaxially. (10$\bar{1}\bar{1}$) GaN was realized on (100) spinel (Baker *et al.*, 2005), (10$\bar{1}\bar{3}$) GaN on (110) spinel (Baker *et al.*, 2005) and (10$\bar{1}$0) m-plane sapphire (Matsuoka and Hagiwara, 2001; Baker *et al.*, 2006), and (11$\bar{2}$2) GaN on (10$\bar{1}$0) m-plane sapphire as well. Semipolar (11$\bar{2}$6) GaN was grown on (10$\bar{1}$2) r-plane sapphire by using an AlInN nucleation layer (Bastek *et al.*, 2010). Semipolar GaN of (10$\bar{1}$l) type with relatively small inclination angles of 18°, 26°, 29°, 31°, 34°, and 47° can be grown on high-indexed Si substrates by nucleation of (0001) GaN on Si (111) microfacets (Ravash *et al.*, 2010). Obviously, the epitaxial relationship for all these substrates is not unique. For example, on m-plane sapphire, (10$\bar{1}$0), (10$\bar{1}$3), and (11$\bar{2}$2), GaN was grown. Actually, many more orientations can occur (Ploch *et al.*, 2010), but none of the others were ever observed to be dominant. A selection of the orientation can be reached by *ex situ* pretreatment to achieve (10$\bar{1}$0) orientation (Armitage and Hirayama, 2008), nitridation of the surface to achieve (11$\bar{2}$2) orientation (Baker *et al.*, 2006; Ploch *et al.*, 2010), and *vice versa*, to achieve (10$\bar{1}$3) orientation without nitridation. Also, composition and growth conditions of the nucleation layer, e.g. AlN (Bougrioua *et al.*, 2007) or high-temperature GaN nucleation layers to achieve (11$\bar{2}$2) GaN on m-plane sapphire (Lee *et al.*, 2010; Zhu *et al.*, 2010), as well as the subsequent growth process to select a single domain (Ploch *et al.*, 2011) can lead to a layer with a single surface orientation. Another issue with heteroepitaxial growth is twinning (formation of domains with different in-plane relationship) due to an ambiguous epitaxial relationship, as observed for (10$\bar{1}$1) GaN on spinel (Kaeding *et al.*, 2006) or (10$\bar{1}$3) GaN on sapphire (Frentrup *et al.*, 2011). This can be solved by using the proper substrate miscut (Kaeding *et al.*, 2006).

But even for phase-pure semipolar or non-polar GaN, basal-plane stacking faults (BSF) and associated defects are typically observed for all heteroepitaxial layers grown on planar substrates with a very high density (BSF 10^5–10^6 cm^{-1}, threading dislocation density $\sim 10^{10}$ cm^{-2} (Craven *et al.*, 2002*a*; Baker *et al.*, 2005; Venngus *et al.*, 2007)). BSF are faults in the wurtzite ABAB stacking order, e.g. ABABCBCB, and represent a two-dimensional cubic inclusion within a hexagonal matrix. Different types of stacking faults are possible (Stampfl and Van de Walle, 1998), classified by the number of cubically coordinated layers and the displacement vector. Associated defects are partial dislocations as well as prismatic stacking faults (Li *et al.*, 2005). BSF can be identified by transmission electron microscopy as well as by their luminescence (Li *et al.*, 2005), especially the I_1 BSF at 3.42 eV. In terms of defect density, m-plane GaN is exceptional. Here the BSF density and threading dislocation density are surprisingly low (10^4 cm^{-1} and 10^9 cm^{-2}), possibly due to a very small lattice mismatch (Neumann *et al.*, 2009). However, due to an ambiguity in the nucleation site, stacking mismatch domains occur (Waltereit *et al.*, 2000; Wernicke, 2010) which might cause the formation of BSF.

Stacking faults do not contribute to non-radiative recombination, but they deteriorate the surface morphology and thereby heterointerfaces. Also, the associated defects contribute to non-radiative recombination.

Several reasons for stacking-fault formation were observed: (1) growth of $(000\bar{1})$ at the early coalescence (Wu *et al.*, 2003), (2) ambiguous epitaxial relationship relating to the stacking sequence (Vennéguès *et al.*, 2008), (3) formation at monoatomic steps in the substrate (Vanfleet *et al.*, 2003), and (4) relaxation of strain (Cho *et al.*, 2008; Wu *et al.*, 2010; Wu *et al.*, 2011).

8.3.3 *Strategies for defect reduction for heteroepitaxially grown GaN*

As described in the previous section, heteroepitaxial layers grown on planar foreign substrates exhibit very-high-defect densities. Many different attempts were undertaken to reduce the density of BSFs and dislocations. A BSF density of less than $\sim 10^4$ and dislocation densities below 10^8 cm^{-2} would be desirable, since for such values optically pumped green lasing was observed (Strittmatter *et al.*, 2010). The observation in c-plane GaN that the threading dislocation density decreases with thickness by dislocation annihilation applies to stacking faults and partial dislocations in non-polar GaN (Tavernier *et al.*, 2004). In principle, the partial dislocations could bend and annihilate together with the stacking fault. However, climb or glide on the c-plane of these dislocations seems to be hindered by the presence of the stacking faults, making it impossible to use the growth of thick layers for defect reduction. In semipolar GaN a dislocation and BSF reduction was observed, leading to sufficiently low-defect densities (Usikov *et al.*, 2010). The mechanism is unresolved as yet, but it could be caused by glide of partial dislocations in the c-plane due to shear stress that is not present in non-polar GaN (Romanov *et al.*, 2006). Another defect-reduction approach successfully employed for c-plane GaN is three-dimensional growth. This idea was applied to non- and semipolar GaN as well (Hollander *et al.*, 2008; Sun *et al.*, 2009*a*; Sun *et al.*, 2009*b*). A reduction of the BSF and dislocation density was achieved, but the defect densities are not yet at the desired levels.

Also, *in situ* and *ex situ* nanomasking with SiN (Chakraborty *et al.*, 2006*b*; Johnston *et al.*, 2009) and ScN (Johnston *et al.*, 2009) or TiN (Tavernier *et al.*, 2004) has been examined, and typically a strong reduction of threading dislocation density could be observed, but no or very little reduction of the BSF density.

The only successful approach to reduce the BSF density to a level at or below $\sim 10^4\,\text{cm}^{-1}$ based on a maskless *in situ* technique was to exploit the strain relaxation for BSF reduction. This can be accomplished by introducing low-temperature AlN layers (Dadgar *et al.*, 2011). The SF terminate at the (Al,Ga)N interface by the formation of partial misfit dislocations. However, this mechanism works only for $(10\bar{1}1)$-type planes (Schulz *et al.*, 2013). After a strong BSF reduction at the interface, the dislocation density can be reduced further by dislocation annihilation after growth of thick layers down to a level near $10^8\,\text{cm}^{-2}$ (Usikov *et al.*, 2010).

Another approach also known from c-plane GaN is lateral epitaxial overgrowth (ELOG) (Beaumont *et al.*, 2001). For this technique, GaN is partly covered by a lithographically patterned dielectric mask (e.g., SiO_2 or SiN_x). During overgrowth, GaN does not grow on the mask, but adatoms diffuse to the GaN surface and are incorporated in the overgrown GaN. This way, GaN can grow vertically and laterally to form a coalesced layer (see Fig. 8.9). Defects from the underlying GaN layer are terminated at the mask, and the laterally grown region above the mask is mostly defect-free. This technique was successfully applied to a-plane (Wu *et al.*, 2003), m-plane (Haskell *et al.*, 2005) and $(11\bar{2}2)$ GaN (Bougrioua *et al.*, 2007). Usually, stripe masks are employed which fit well to the surface symmetry of semipolar and non-polar GaN. Due to the low symmetry of the surface, the stripe orientation is very crucial (Craven *et al.*, 2002*b*) as it strongly influences the defect reduction as well as the formation of facets, the lateral growth rate, and the coalescence of the layers (Netzel *et al.*, 2008). Here we will discuss briefly ELOG of a-plane GaN with the low-index stripe orientations of $[1\bar{1}00]$, $[1\bar{1}01]$, and [0001] stripe orientations (90°, 45° and 0° to the c-direction). As depicted in Fig. 8.10 for [0001] stripes, the overgrown crystal develops $10\bar{1}0$ m-plane facets that lead to some dislocation reduction (as can be seen from CL), but due to the equivalency of the $10\bar{1}0$ surfaces the lateral growth rate is limited. Also, the inclined top facets impede coalescence. The monochromatic CL image at 3.42 eV also show that BSF can be found in the coherently grown region (window) as well as in the laterally grown region (wing). Since the BSF are perpendicular to the stripe orientation they can enter the laterally overgrown region. The $[1\bar{1}01]$ stripes develop $(10\bar{1}\bar{1})$, $(11\bar{2}0)$, $(01\bar{1}1)$, and $(\bar{1}102)$ facets which exhibit very different growth rates. The fast lateral growth of the $(01\bar{1}1)$ and $(\bar{1}102)$ facets causes a strong dislocation reduction. The inclined

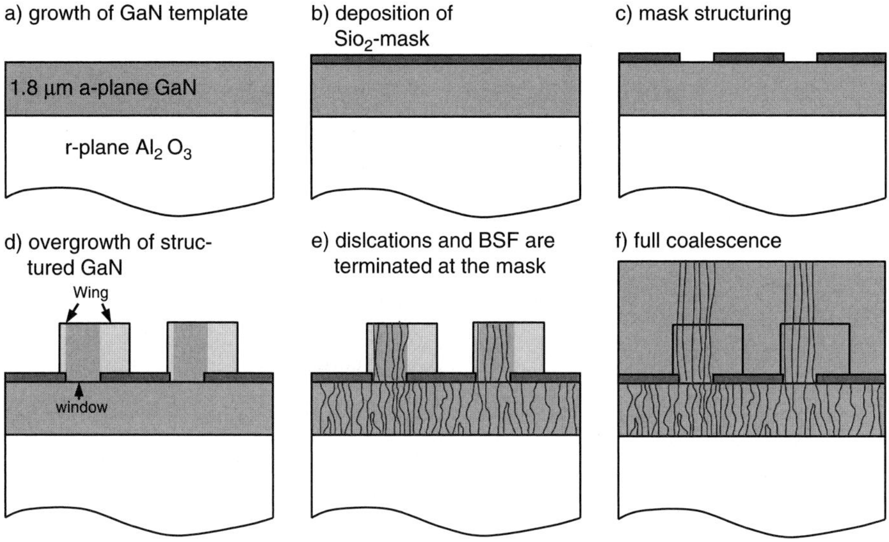

FIG. 8.9. Working priciple of epitaxial lateral overgrowth.

$(10\bar{1}\bar{1})$ facets are overgrown during the coalescence, leading to a reduction of threading dislocation density even in the wing regions (Wernicke *et al.*, 2009). The fast lateral growth and the $(11\bar{2}0)$ facet also support coalescence of the layers. But also for this stripe orientation, BSF can enter the wing region, since they are inclined to the stripe direction. $[1\bar{1}00]$ stripes exhibit a rectangular profile with $(000\bar{1})$, $(11\bar{2}0)$, and (0001) facets, again with very anisotropic growth rates. The CL mapping shows a very strong reduction of threading dislocation and BSF density in the [0001]-grown wing (Wu *et al.*, 2008; Bastek *et al.*, 2008). This shows that BSF can be reduced by lateral overgrowth only if the growth direction is perpendicular to the basal plane. By side-wall epitaxy (Imer *et al.*, 2006) a further reduction of the defect densities could be realized. These studies show why many attempts to reduce the BSF density do not produce sufficiently low BSF densities if the crystallographic nature of this defect is not taken into account. In all cases the BSF easily penetrate overgrown regions, as long as the growth is not in the [0001] direction.

Although the defect densities obtained by the ELOG approach are very good, growth on stripe-patterned substrates can achieve even lower defect densities and can be employed for arbitrary surface orientation. The basic idea is to nucleate (0001) oriented GaN on a tilted side facet of a foreign substrate and to achieve a semipolar surface by coalescence. Growth on Si <111> etched into (001) Si by KOH was reported first (Honda *et al.*, 2001). Using differently oriented Si substrates $(10\bar{1}1)$ (Honda *et al.*, 2002), $(11\bar{2}2)$ (Tanikawa *et al.*, 2008*a*), $(11\bar{2}0)$ (Tanikawa *et al.*, 2008*b*) and $(10\bar{1}0)$ (Ni *et al.*, 2010), oriented GaN was grown.

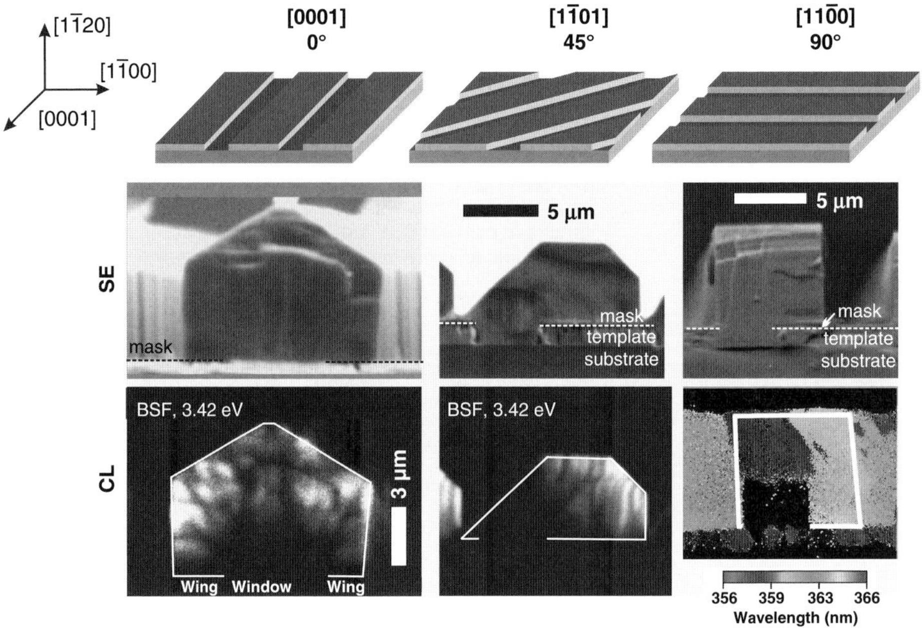

FIG. 8.10. SEM and CL images of a-plane ELOG with [0001], [1$\bar{1}$01], and [1$\bar{1}$00] stripe orientation (Netzel *et al.*, 2008; Bastek *et al.*, 2010). Images are courtesy of U. Zeimer (FBH) and B. Bastek.

Also, sapphire can be structured so that (0001) c-plane GaN nucleates on either (0001) c-plane or (11$\bar{2}$0) a-plane-like surfaces. With this technique, (10$\bar{1}$0) (Okada *et al.*, 2008), (11$\bar{2}$0) (Okada *et al.*, 2011*a*), (10$\bar{1}$1) (Schwaiger *et al.*, 2010), (11$\bar{2}$2) (Schwaiger *et al.*, 2011) and (20$\bar{2}$1) (Okada *et al.*, 2011*b*) GaN layers have been fabricated. The key for such layers is to control nucleation sites by masking with SiO_2 and by the nucleation conditions. In addition, achieving coalescence is a major issue requiring optimization of substrate miscut and growth conditions. For structured substrates the c-plane nucleation allows rather low threading dislocation densities which are further reduced by the subsequent growth, and dislocation densities as low as $1 \times 10^5\,\mathrm{cm}^{-2}$ have been achieved (Murase *et al.*, 2011). Also, the formation of stacking faults during nucleation can be prevented in this way. However, the layers still exhibit a low density of stacking faults (the density of BSF bundles is typically $2 \times 10^3\,\mathrm{cm}^{-1}$ (Schwaiger *et al.*, 2011)) due to an exposed (000$\bar{1}$) facet during coalescence. But approaches for eliminating these residual stacking faults were proposed (Schwaiger *et al.*, 2011). Heteroepitaxial growth of low-defect-density non- and semipolar GaN is much more challenging than c-plane growth, especially due to the formation of basal-plane stacking faults and associated defects. So far, the most promising approach appears to be the use of c-plane nucleation on side facets of structured sapphire or silicon substrates.

8.4 Growth of InGaN layers and quantum wells on m-plane and different semipolar surfaces, i.e. $(10\bar{1}2)$, $(10\bar{1}1)$, $(20\bar{2}1)$, $(11\bar{2}2)$

InGaN quantum wells are the most important part of visible emitting LEDs and lasers. Recombination dynamics in non- and semipolar InGaN are significantly different, not only due to the reduced polarization fields (see also Section 8.1) but also due to the reduced symmetry of the surfaces leading to anisotropic and shear strain. Also, there are major differences in the growth of InGaN due to different surfaces and reconstructions leading to different adatom bonding strengths, adatom incorporation, and adatom diffusion barriers (see Section 8.3.1). This leads to a different indium incorporation for the different non- and semipolar surfaces.

8.4.1 *Indium incorporation efficiency for different surface orientations*

The indium incorporation depends strongly on the surface orientation (Wernicke *et al.*, 2012). However, a conclusive picture cannot yet be drawn, since growth parameters have an impact on the relative indium incorporation on different planes (Browne *et al.*, 2012). Thus, the reported relative indium incorporation of the $(11\bar{2}2)$ varies from higher (Tanikawa *et al.*, 2011; Wernicke *et al.*, 2012) to similar (Jönen *et al.*, 2012) to lower incorporation (Browne *et al.*, 2012) in comparison to the (0001) c-plane. Because the presence of morphological defects, e.g. caused by dislocations or BSF, typically increases the indium incorporation (Chakraborty *et al.*, 2005*b*), we will focus only on QW grown on high-quality bulk substrates. Firstly, the indium content of a semipolar InGaN layer is not easily accessible by X-ray diffraction (XRD) due to a triclinic distortion of the unit cell caused by anisotropic strain and shear strain (Romanov *et al.*, 2006). Different methods of analyzing the XRD data were proposed for pseudomorphically strained InGaN layers on GaN (Young *et al.*, 2011) as well as for layers with an arbitrary strain state (Jönen *et al.*, 2012). Additionally, measurements on quantum wells are more difficult, since the material volume is usually small and evaluation requires the knowledge of the exact barrier and quantum-well thickness. The comparison of the emission wavelength, e.g. from PL or EL measurements, is much easier and more reliable than the comparison of the indium content. However, extracting the relative indium incorporation requires complex calculations that imposes rather large uncertainties. Finally, it must be kept in mind that indium content from luminescence experiments relate to regions with the highest indium content (Kuokstis *et al.*, 2002), whereas XRD provides the average indium content. Larger localization energies therefore cause a deviation of local PL indium content and average XRD indium contents. The emission energy for a set of QWs grown on polar, semipolar, and non-polar GaN with different growth temperatures at a gas phase composition x_{gas} of 25% is shown in Fig. 8.11 (Wernicke *et al.*, 2012). The following order-of-emission energies can

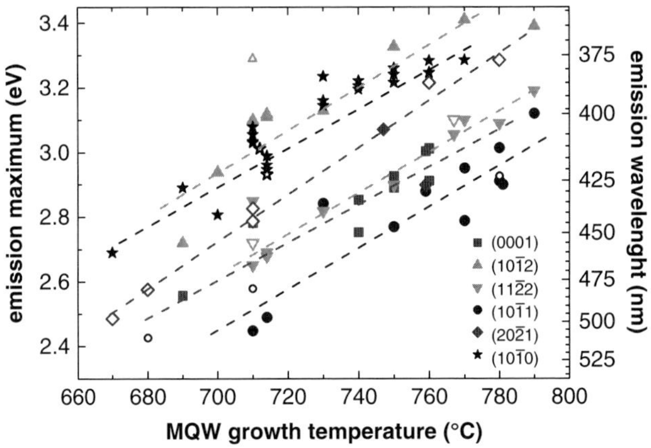

FIG. 8.11. Emission wavelength of polar, semipolar and non-polar QWs. Reprinted with permission from Wernicke *et al.* (2012), © 2012 IOP Publishing Ltd.

be extracted: (10$\bar{1}$1) < (11$\bar{2}$2) = (0001) < (20$\bar{2}$1) < (10$\bar{1}$0) = (10$\bar{1}$2). This ratio can be found for the whole temperature range. This means that long-wavelength emitters on (10$\bar{1}$1) can be grown at substantially higher temperatures than on the c-plane, and even more so on (10$\bar{1}$0) GaN. From this set of data the corresponding indium content was calculated using k.p theory and Schrdinger–Poisson calculations (Schade *et al.*, 2011*b*). Again, a hierarchy for the indium incorporation efficency can be found throughout the entire temperature range: (10$\bar{1}$1) > (11$\bar{2}$2) > (0001) = (20$\bar{2}$1) = (10$\bar{1}$0) = (10$\bar{1}$2) (Wernicke *et al.*, 2012). Thus the (10$\bar{1}$1) surface shows the highest indium incorporation. Also, the (11$\bar{2}$2) exhibits increased indium incorporation, whereas incorporation of indium on the other planes is relatively similar. We also compared the indium incorporation into thick (>30 nm) InGaN layers on (0001) and (20$\bar{2}$1) GaN, which can be accurately measured by XRD. The indium incorporation for (20$\bar{2}$1) InGaN was lower by a factor of ~2 at 725°C, and x_{gas} up to 41% was measured (Wernicke *et al.*, 2012). This is in contradiction to the results of the QWs, and indicates a much higher localization energy in (20$\bar{2}$1) InGaN, a strong (vertical) segregation a c-plane or a higher polarization field in (20$\bar{2}$1). Overall, final conclusions cannot be drawn yet, and additional optical and structural investigations are necessary. Although the indium incorporation is highest on (10$\bar{1}$1), the best device results are obtained on other planes (see Sections 8.4 and 8.5). Other factors such as point-defect incorporation and critical layer thickness must be considered as well.

8.4.2 *Optical properties of non-polar and semipolar InGaN QWs*

The reduced impact of the QCSE in semipolar InGaN QWs can be seen in the smaller shift of the emission wavelength with increasing well-width (Craven *et al.*, 2004*a*) and increasing injection-current (Sharma *et al.*, 2005). In addition, as the wavefunction overlap in non-and semipolar QWs increases, the recombination lifetime decreases (Netzel *et al.*, 2010*a*). This could of course be caused by a short non-radiative life time—even at 10 K—which we could exclude by comparison to c-plane samples (Netzel *et al.*, 2011*b*). In addition, the exciton binding energy in non- and semipolar QWs is enhanced (Netzel *et al.*, 2010*b*). This allows the observation of excitonic recombination rather than bimolecular recombination at room temperature for $(10\bar{1}0)$ and $(11\bar{2}2)$ InGaN quantum wells. PL measurements also revealed a delayed S-shape of the quantum-well luminescence emission energy (Netzel *et al.*, 2011*a*; Wernicke *et al.*, 2012). This is a sign for an increased localization energy (Li *et al.*, 2005). InGaN QWs on (0001) c-plane typically exhibit an S-shape turning point of ~150 K, whereas semi- and non-polar InGaN QWs can exhibit turning points at temperatures exceeding 300 K (Netzel *et al.*, 2011*b*). This is accompanied by an increased full-width-half-maximum (FWHM) of the emission spectra. In order to differentiate between long-range fluctuations and microscopic localization, micro-PL measurements were performed. These scans show very small fluctuations of the emission energy in $5 \times 5\,\mu\text{m}^2$ areas as shown in Fig. 8.12. The emission energy of the $(11\bar{2}2)$ $In_{0.24}Ga_{0.76}N$ varies by only 12 meV and the minimum intensity is 78 % of the local maximum, which is small in comparison to c-plane QWs. Single spectra (spatial resolution is 400 nm) reveal a local FWHM that increases

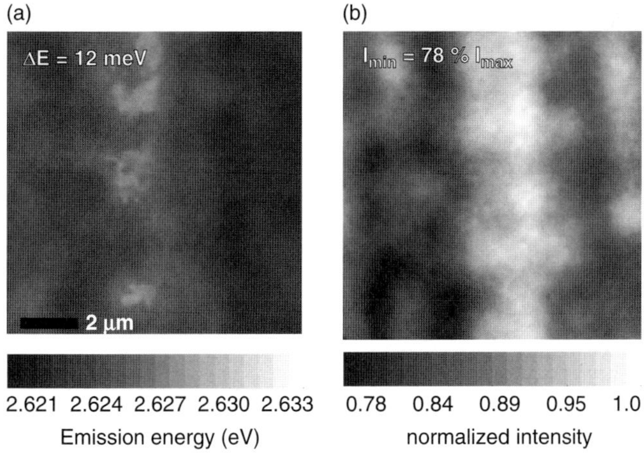

FIG. 8.12. $5 \times 5\,\mu\text{m}^2$ micro-PL image of an $(11\bar{2}2)$ InGaN QW. Courtesy of Lukas Schade (University of Freiburg).

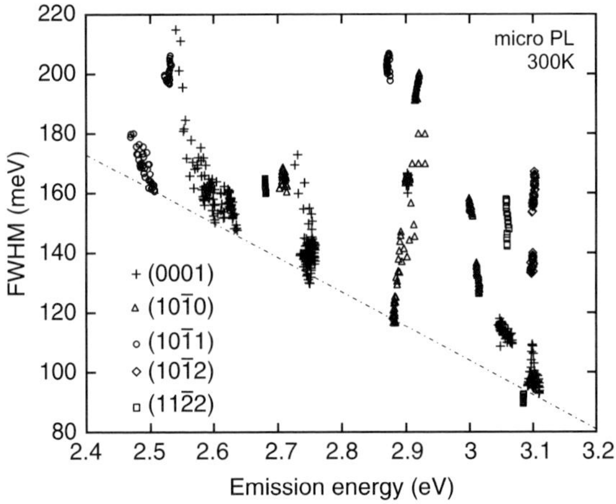

FIG. 8.13. Local FWHM from micro-PL of a variety of non- and semipolar QWs. A reduced transition energy leads a strong increase of the local FWHM. Courtesy of Lukas Schade (University of Freiburg).

strongly with the emission wavelength (Fig. 8.13). This is another indication to highly localized states in non- and semipolar QWs.

The reduced quantum confined Stark effect is not the only difference for the recombination in semipolar and c-plane QWs. The anisotropic and shear strain that causes the zero crossing of the piezoelectric field (see Section 8.1) also affects the band structure (Schade *et al.*, 2011*b*), causing polarized emission and a shift of the transition energy (Wernicke *et al.*, 2012). The reduction of surface symmetry lifts the isotropic symmetry of the hole wavefunction within the basal plane. The three typical valance subbands of (0001) InGaN (light-hole LH, heavy-hole HH, and split-off hole SH band) mix into new subbands. The symmetry of the wavefunction defines the selection rules and thus the polarization of the emitted light. In c-plane GaN the light emitted from the LH and HH band is polarized perpendicular to the c-axis isotropically in the basal plane, and the SH is polarized parallel to the c-axis. For non-polar InGaN the topmost valence band (called A1) is formed from a mix of LH and HH, and is polarized mostly perpendicular to the c-direction. For semipolar QWs, the valence bands are formed again from a mix of LH and HH bands, and with increasing shear strain (maximum at 45°) a portion of the SH band. This leads to a reduced degree of polarization of the single subbands. The calculations also predict an orthogonal polarization of the second-highest (B1) subband (Fig. 8.14), at least for small k-values (Schade *et al.*, 2011*a*). The model also describes the polarization switching (change of the subband polarization from perpendicular to c towards parallel to c) observed for (11$\bar{2}$2) InGaN quantum wells by increasing the indium content. The switching point is present for all indium content, and dominant c'

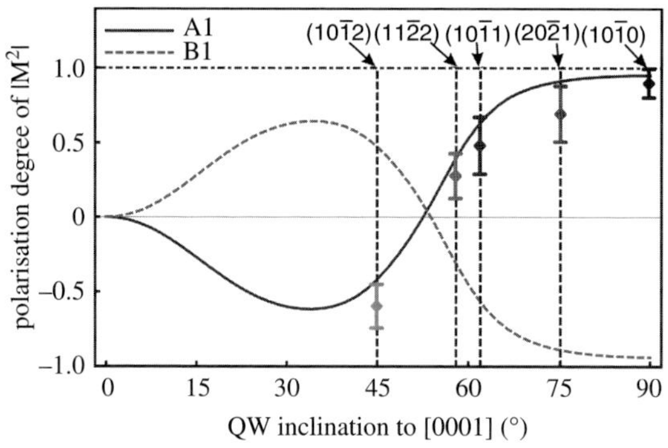

FIG. 8.14. Optical polarization degree of the top-most subbandlevel A1 and the second top-most subbandlevel B1 of an $In_{0.15}Ga_{0.85}N$ QW strained to GaN in dependence of the tilt angle to the c-plane. A polarization degree of 1 represents total polarization with E perpendicular to the c axis. A polarization degree of -1 corresponds to a polarization of the subband parallel to c′ (the projection of the c-axis). Measurement points represent the polarization at 5 K of violet-emitting InGaN QWs (∼15 % indium content). At this temperature the B1 subband is not populated, and the A1 polarization can be measured directly (Schwarz and Scheibenzuber, 2011; Schade *et al.*, 2012*b*). Reprinted with permission from Schade *et al.* (2011*b*), © 2011 Wiley-VCH Verlag GmbH & Co. KGaA, Weinheim.

polarization was experimentally verified for (10$\bar{1}$2) InGaN QWs. The switching point increases to higher angles with increasing indium content to a maximum angle of 69° (calculated for InN on GaN). The (20$\bar{2}$1) and (10$\bar{1}$0) cannot be affected by polarization switching. Since subband polarization and distance depend strongly on strain, strain relaxation influences the polarization (Koslow *et al.*, 2012). The polarization degree of an InGaN QW at room temperature is not only given by the subband polarization, but is further reduced by occupation of the orthogonally polarized B1 subband. Therefore, a high degree of polarization can be achieved only for inclination angles >65° for a wide range of indium content, due to high subband polarization and subband spacing (Schade *et al.*, 2011*a*; Schade *et al.*, 2011*b*). Also, inclination angles between 15° and 55° might allow for a high degree of polarization for sufficiently high indium content if the subband distance is large enough. Only the (10$\bar{1}$1) exhibits a rather low degree of polarization for all indium content. Experimentally observed polarization values agree mostly with the calculated ones (Schade *et al.*, 2011*a*). The trend is even observed in localization centers with different indium content causing the inhomogeneous broadening (Schade *et al.*, 2012*a*). Surprisingly, the polarization observed in (20$\bar{2}$1) QWs is much lower than expected (Schade *et al.*, 2011*a*).

Although the subband polarization fits the calculated values very well, the measured subband spacings are much larger, leading to a higher degree of polarization at room temperature than expected from the calculation (Schade *et al.*, 2011*b*).

This effect of split subbands can be used to produce LEDs with strongly polarized emission, and also increases the optical gain in lasers (see Section 8.6). The effect can also be used to extract the quasi-Fermi level and thereby the carrier density in the quantum well (Schade *et al.*, 2012*b*).

8.5 Performance characteristics of non-polar and semipolar InGaN QW LEDs

8.5.1 *External quantum efficiencies and emission wavelength*

Within the past decade the efficiency and spectral range of non- and semipolar LEDs has increased rapidly. First, non- and semipolar InGaN QW LEDs were demonstrated on heteroepitaxially-grown defect-reduced $(11\bar{2}0)$ GaN on a-plane sapphire substrates (Chakraborty *et al.*, 2004) as well as on $(10\bar{1}\bar{1})$ and $(10\bar{1}\bar{3})$ oriented GaN grown on (100) and (110) spinel ($MgAl_2O_4$) substrates (Chakraborty *et al.*, 2005*a*). However, due to the relatively large threading dislocation densities and basal-plane stacking faults in these heteroepitaxially grown layers, the light output of the LEDs just barely exceeded one mW, and the external quantum efficiencies were a fraction of 1% (Chakraborty *et al.*, 2005*a*). The performance of non- and semipolar LED improved significantly with the development of low-defect-density bulk GaN substrates cut along non- and semipolar orientations. First, blue- and violet-emitting InGaN QW LEDs were demonstrated on low-defect-density $(10\bar{1}0)$ m-plane GaN single crystals by Okamoto *et al.* as well as Chakraborty *et al.* with EQE of 3.1 % and 1.1 %, respectively (Okamoto *et al.*, 2006; Chakraborty *et al.*, 2006*a*). Simultaneously, first blue, green, and even amber InGaN quantum-well LEDs were realized on $(11\bar{2}2)$ semipolar bulk GaN substrates with EQEs of 4.0 %, 4.9 %, and 1.6 %, respectively (Funato *et al.*, 2006). After these initial breakthroughs, the efficiency of non- and semipolar LEDs increased significantly within a short period of time. Schmidt *et al.* (Schwarz and Kneissl, 2007) reported on high-efficiency non-polar InGaN QW violet LEDs on low-defect-density bulk m-plane GaN substrates with EQEs of 38.9 %, and high-efficiency semipolar blue LEDs on free-standing $(10\bar{1}\bar{1})$ bulk GaN substrates have been demonstrated by Zhong *et al.* (2007) with EQEs of 29%. Shortly thereafter, Sato *et al.* (2007) reported high-power green InGaN multiple-quantum-well (MQW) light-emitting diodes grown on low extended-defect-density semipolar $(11\bar{2}2)$ bulk GaN substrate with peak external quantum efficiencies of more than 12 %. More recently, high-power and high-efficiency blue InGaN LEDs have also been demonstrated by Koslow *et al.* on semipolar $(30\bar{3}3)$ GaN substrates with peak EQEs of 26.5 % (Koslow *et al.*, 2010), and Pan *et al.* on semipolar $(20\bar{2}\bar{1})$ GaN with maximum external quantum efficiencies of 50.1 % (Koslow *et al.*, 2010).

Figure 8.15 shows examples of InGaN MQW and SQW LED heterostructure on semipolar $(10\bar{1}\bar{1})$ and $(20\bar{2}\bar{1})$ GaN substrate, respectively. The active region of the LED on $(10\bar{1}\bar{1})$ GaN substrate is comprised of six periods of 3-nm thick InGaN quantum wells separated by 20-nm wide undoped GaN barriers. The MQW stack is covered by a 10-nm thick nominally undoped $Al_{0.15}Ga_{0.85}N$ electron blocking layer followed by a 200-nm thick Mg-dopded GaN contact layer. Although the design of the device heterostructure appears quite straightforward and is very similar to c-plane LEDs, there are some remarkable differences. Kim *et al.* (2007*a*), as well as Zhong *et al.* (2007), demonstrate high-efficiency non- and semipolar blue LEDs on m-plane and $(10\bar{1}\bar{1})$ GaN with a nominally undoped $Al_{0.15}Ga_{0.85}N$ electron blocking layers (EBL). This design detail is very different from c-plane LED, where the functionality of the EBL within an LED heterostructure depends on the ability to incorporate p-type AlGaN:Mg electron blocking layers. Another interesting detail is the utilization of 12-nm thick InGaN single-quantum-well active region in semipolar LEDs, as described e.g. by Pan *et al.* (2012), and also shown in Fig. 8.15b). This design is a direct result of the reduced polarization fields in semipolar InGaN/GaN heterostructures, which, due to a significantly reduced piezoelectric field, enables the use of thick InGaN QWs without detrimental effect on the electron-hole envelope wavefunction overlap. As a consequence of the larger QW thickness and the shorter radiative recombination lifetimes, the carrier densities for a fixed current density are also reduced, which could reduce higher-order carrier-density effects such as Auger-type recombination. This should result in much smaller efficiency droop at increasing

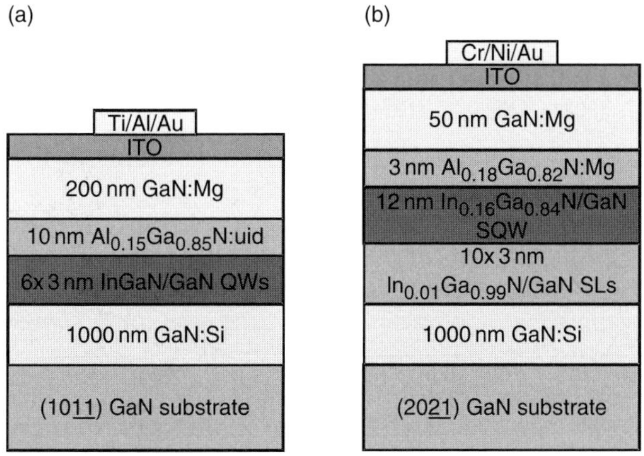

FIG. 8.15. Schematic of a typical InGaN multiple-quantum-well (MQW) LED heterostructure grown on (a) semipolar $(10\bar{1}\bar{1})$ GaN substrate (Zhong *et al.*, 2007), and (b) a InGaN single-quantum-well (SQW) LEDs on $(20\bar{2}\bar{1})$ GaN (Pan *et al.*, 2012).

current densities, as was also by observed (Pan *et al.*, 2012), where the EQE dropped from 50.1 % at 100 A/cm^2 to 41.2 % at a current density of 400 A/cm^2.

8.5.2 *Polarization effects and efficiency droop*

As described in the previous section, the indium incorporation efficiency during growth of InGaN quantum wells and layers depends strongly on the surface orientation of the layers. Whereas non-polar GaN surfaces seem to exhibit reduced indium incorporation efficiencies, some of the semipolar surface orientations appear to have a high indium incorporation compared to InGaN on polar (0001) surfaces (Wernicke *et al.*, 2012). Therefore, some of the semipolar orientations, in particular ($10\bar{1}1$), ($11\bar{2}2$) and ($20\bar{2}1$), seem ideally suited for realizing long-wavelength light emitters, as can be also seen in Fig. 8.16.

A number of groups have investigated the magnitude and direction of the polarization field in polar and semipolar ($11\bar{2}2$) InGaN/GaN quantum wells (QWs) by bias-dependent photoluminescence (Funato *et al.*, 2010) and electroreflectance spectroscopy (Renner *et al.*, 2002; Shen *et al.*, 2009; Shen *et al.*, 2010). These investigations reveal significantly reduced polarization field strength in semipolar InGaN/GaN quantum wells compared to c-plane QWs. Whereas the polarization field in InGaN/GaN QWs is in the range of 1–1.4 MV/cm, even at moderate indium mole fractions as low as 9 %, the polarization field in ($11\bar{2}2$) QWs was found to be −0.68 MV/cm and 0.26 MV/cm, even at high indium contents of 38 % and 49 %, respectively. These studies seem to also reveal polarization switching, e.g. a piezoelectric polarization field that is oriented in the opposite direction

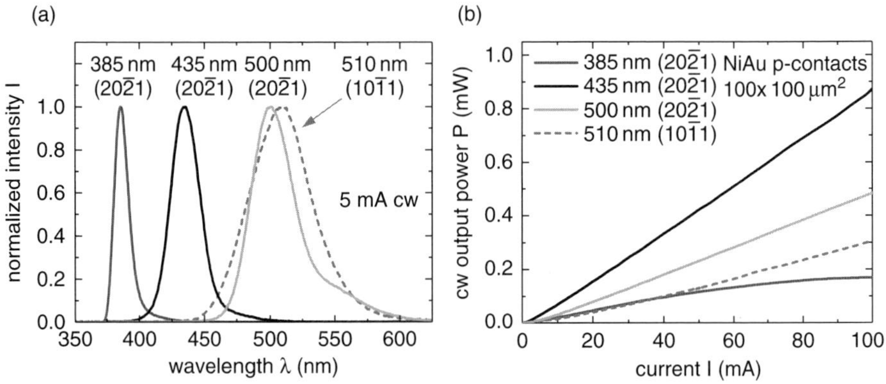

FIG. 8.16. (a) Emission spectra of violet, blue, and green InGaN QW LEDs on different semipolar GaN surface orientations (Raß, 2012). (b) Light output vs. current (LI) and current–voltage (IV) characteristics of violet, blue, and green InGaN SQW LEDs on different semipolar GaN surface orientations (Raß, 2012). Courtesy of J. Rass (Technical University, Berlin).

compared to the piezoelectric field in c-plane InGaN/GaN QWs. The observed polarization switching behaviour seems to depend on the indium mole fraction in the InGaN QW and the resultant pseudomorphic strain in the heterostructure. Modelling results by Yan *et al.* (2010) show that the ordering of the two top-most valence bands of $(11\bar{2}2)$ semipolar InGaN on GaN is highly dependent on the shear strain and the anisotropic strain in the c-plane, which might explain the observed change of the electric field orientation for $(11\bar{2}2)$ InGaN/GaN QWs.

8.6 Performance characteristics of non-polar and semipolar InGaN QW lasers

8.6.1 *Gain characteristics of InGaN quantum-well lasers on non-polar and semipolar GaN and effects of the excitation stripe orientation*

Due to the broken symmetry in semi- and non-polar III-nitride laser heterostructures, the optical properties will become directional-dependent. This anisotropy affects in particular the optical gain in non- and semipolar InGaN laser diodes. The anisotropic strain in InGaN quantum wells pseudomorphically grown on non- and semipolar GaN substrates is further adding to the anisotropy in the optical gain. Recently, Scheibenzuber *et al.* (Schwarz and Scheibenzuber, 2011) as well as Rass *et al.* (2010*b*, 2012) have theoretically and experimentally investigated the optical gain and polarization for different directions of the laser resonators as well as surface orientations. Furthermore, due to the birefringence of wurtzite III-nitride materials, other optical properties also become directional-dependent, e.g. the real part of the refractive index. As a consequence, the laser threshold current densities as well as the optical polarization of the laser light depend strongly on the surface orientation, the direction of the laser resonator, and the strain in the InGaN/GaN quantum wells. For example, investigation of InGaN quantum-well laser diodes grown on $(11\bar{2}2)$ GaN substrate show very different threshold power densities as well as polarization characteristics, depending on whether the laser resonator is oriented along the $[10\bar{1}0]$ m-direction or the $[11\bar{2}3]$ c′-direction. The threshold power densities for amplified stimulated emission for resonators oriented along the $[10\bar{1}0]$ m-direction are almost twice as large as threshold power densities for the resonators along the $[11\bar{2}3]$ c′-direction. This behaviour can be explained by anisotropic optical gain for the two different resonator orientations. Figure 8.17 shows the measured and calculated gain spectra for $(11\bar{2}2)$ InGaN SQW laser heterostructures with resonator orientations along the $[10\bar{1}0]$ m-direction or the $[11\bar{2}3]$ c′-direction. For similar excitation power densities the highest peak gain is obtained for laser resonators oriented along the $[11\bar{2}3]$ c′-direction and the TE polarized optical modes. As a consequence, the lowest threshold power densities are also obtained for the same resonator orientation, and the laser emission is predominantly TE polarized. This behaviour is confirmed by theoretical calculation of the optical gain in $(11\bar{2}2)$ laser diodes (Scheibenzuber *et al.*, 2009; Scheibenzuber, 2012) that is shown in Fig. 8.17 b)

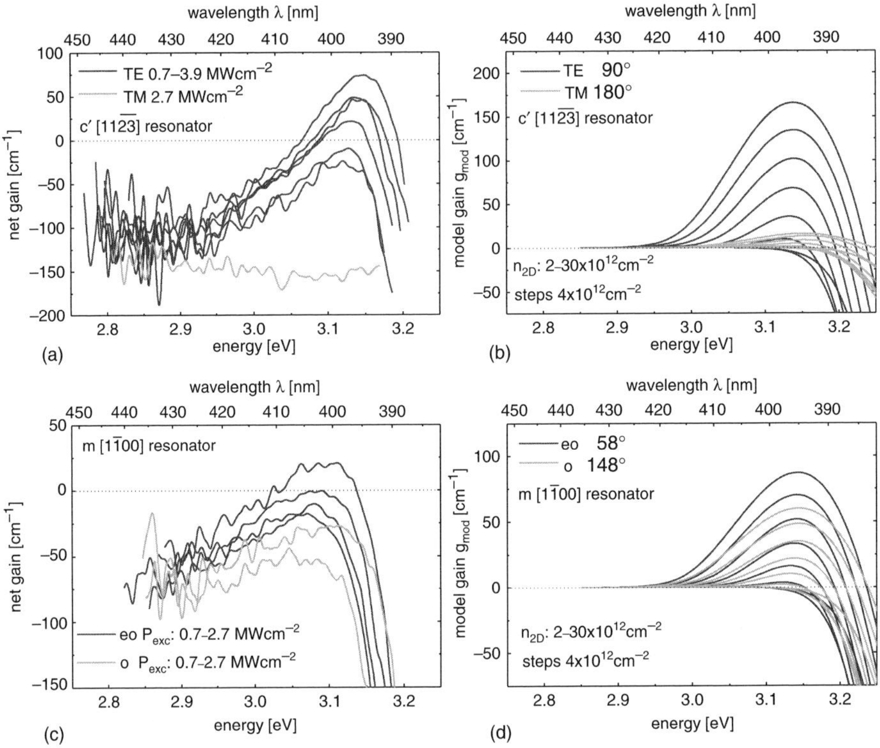

FIG. 8.17. Measured (left) and calculated (right) gain spectra for InGaN SQW laser heterostructures on (11$\bar{2}$2) GaN substrates. On top are the gain spectra measured for c′ [11$\bar{2}$3] resonator orientation, and on the bottom the gain spectra for the m [1$\bar{1}$00] resonator orientation (Rass *et al.*, 2012). Reprinted with permission from Rass *et al.* (2011), © 2011 American Institute of Physics.

and d). The calculation are based on 6x6 k.p theory, assuming pseudomorphically strained InGaN quantum wells and taking into account the effects of the polarization fields and homogeneous as well as inhomogeneous broadening of the gain spectra (Scheibenzuber *et al.*, 2009; Scheibenzuber, 2012).

Figure 8.18 shows the valence band alignments at the gamma point for pseudomorphically strained InGaN quantum wells on (0001) c-plane, semipolar (11$\bar{2}$2), and (10$\bar{1}$0) m-plane GaN. For (0001) InGaN QW laser diodes the strain within the c-plane is isotropic and therefore also the valence band structure is isotropic. Heavy- and light-hole valence bands are basically degenerate at the gamma point and contribute evenly to the optical gain. Consequently, (0001) InGaN QW lasers exhibit predominantly TE polarized gain and show lasing in the TE mode regardless of the resonator orientation. For non-c-plane InGaN

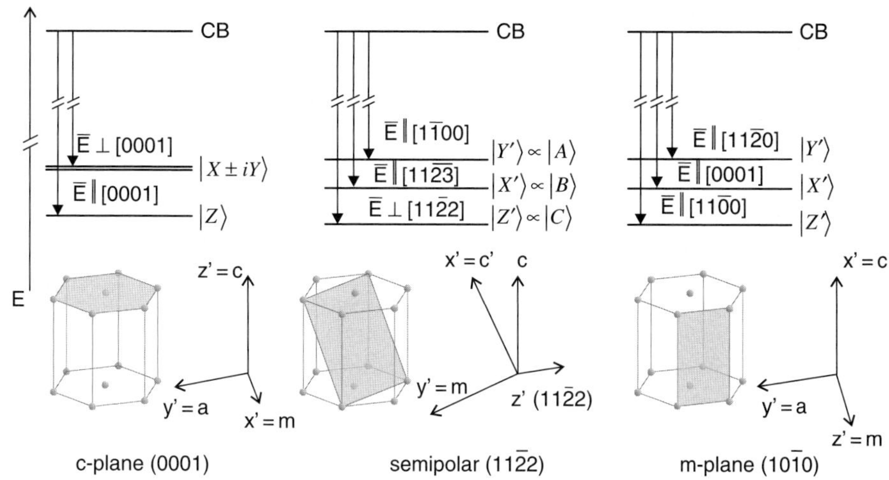

FIG. 8.18. Valence-band alignments at the gamma point for pseudomorphically strained InGaN quantum wells on (0001) c-plane, semipolar (11$\bar{2}$2), and (10$\bar{1}$0) m-plane GaN. The different optical transitions and their corresponding polarization are also depicted. Note that the spacing of the energy levels is just illustrative and not quantitatively exact. The x′, y′, z′ coordinates refer to a coordinate system that is tilted by the inclination angle α, compared to the x, y, z coordinate system of the (0001) crystal (Raß, 2012). The z′ axis corresponds to the growth direction. Courtesy of J. Rass (Technical University, Berlin).

QWs the strain is anisotropic, reducing the symmetry of the band structure. Therefore, it is expected that the optical gain on non- and semipolar InGaN QWs shows a directional dependence, and the degeneracy of the heavy and light hole valence bands is lifted. For non-polar (10$\bar{1}$0) m-plane InGaN quantum wells the X′-band is the energetically highest valence band, and hence optical transitions and gain exhibit TE polarized emission with the electric field vector E perpendicular to the (0001) c-axis. The 6 x 6 k.p calculations of Scheibenzuber *et al.* (2009) also show that the optical gain in non-polar InGaN QWs should be strongly TE, and that the peak gain for the same charge carrier density should be significantly larger than for polar (0001) QWs, which should result in a significant reduction in the threshold current density for non-polar InGaN quantum-well laser diodes. Due to the anisotropic strain, the valence bands in semipolar InGaN quantum wells are a superposition of different quantum levels. As a result, the optical transitions include contributions from different polarization states which depend on the indium mole fraction and strain in the QWs, as well as the specific orientation of the growth surface. For (11$\bar{2}$2) InGaN QWs the A1 valence band is the top-most valence band followed by the B1 valence band. Modelling of the gain characteristics of $(11\bar{2}2)$ InGaN QWs by Scheibenzuber *et al.* (2009) shows that the material gain for TE mode for semipolar quantum wells with a c′ $[11\bar{2}\bar{3}]$ resonator orientation is significantly higher than the material gain for the TM. This behaviour was also confirmed in gain measurement by Rass *et al.* (2012) that are plotted in Fig. 8.17. These experiments also demonstrated that the net gain for the c′ $[11\bar{2}\bar{3}]$ resonator orientation is higher than for the resonator aligned along the m $[1\bar{1}00]$ direction. Simulations by Scheibenzuber *et al.* (2009) also showed that for similar charge-carrier densities, semipolar InGaN QWs exhibit much higher TE optical gain than polar (0001) QWs, which should translate in a significant reduction in the threshold current density for semipolar laser diodes. Figure 8.19 (a) exhibits the light-output vs. pump-power characteristic for optically excited InGaN QW lasers grown on $(11\bar{2}\bar{2})$ GaN substrates. As can be seen, the threshold power density P_{th} of lasers with the resonators oriented along the c′ $[11\bar{2}3]$ is nearly a factor of two smaller than for laser resonators oriented along the m $[1\bar{1}00]$ direction, confirming the anticipated behaviour. The situation in semipolar laser diodes is further complicated due to the birefringence of wurtzite crystal materials. This leads to different confinement factors for TE, TM polarized modes and to the formation of ordinary and extraordinary modes for resonators along the m $[1\bar{1}00]$ direction (Rass *et al.*, 2012). In Fig. 8.19 (b) the polarization of eigenmodes for lasers in $(11\bar{2}2)$ GaN substrates is plotted for excitation power densities above the amplified stimulated emission (ASE) threshold and was normalized to the peak power. As can be seen, the polarization angle of the c′ $[11\bar{2}3]$ has its maximum at 90° corresponding to the TE mode, whereas the m $[1\bar{1}00]$ exhibits a maximum at a polarization angle near 55°, corresponding to the extraordinary mode. A systematic study of blue and violet lasers on different semipolar and m-plane GaN substrates, plotted in Fig. 8.19 (c), shows that for all resonators oriented along the c′ direction the ASE is always TE-polarized. In contrast, lasers with the resonator oriented along the m or a direction always exhibit ASE polarization angles Θ, close to the growth plane angle α. In the latter case the optical polarization is governed by the birefringence of the III-nitride materials. Nevertheless, the preferred resonator alignment remains the c′ $[11\bar{2}3]$ direction, since lasers oriented in the m $[1\bar{1}00]$ direction always exhibit higher threshold-power densities (Raß, 2012).

8.6.2 *Fabrication of laser cavities (e.g., etched, cleaved facets)*

The fabrication of smooth mirror facets for laser heterostructures grown on non- and semipolar GaN is quite challenging. The best approach depends on the surface orientation as well as on the orientation of the laser resonator. On (0001) c-plane laser diodes grown on bulk GaN substrates, fabrication of smooth cleaved facets is quite straightforward. The laser resonators grown on the polar c-plane surface are typically oriented in the m-direction. Since the m-plane also is a low-energy surface it provides a natural cleaving plane for the formation of smooth cleaved facets that are perpendicular to the c-plane. Therefore, the

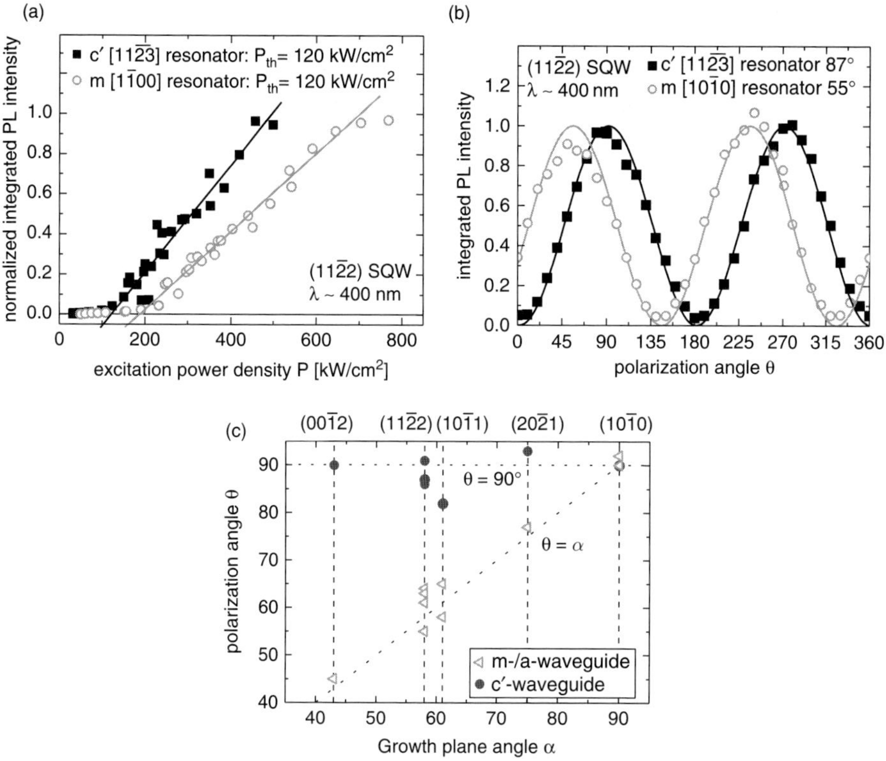

FIG. 8.19. (a) Light-output vs. pump-power density (L-P) characteristic for semipolar InGaN QW lasers grown on (11$\bar{2}\bar{2}$) GaN substrates with resonators oriented along the c′ [11$\bar{2}$3] and m [1$\bar{1}$00] direction (Rass *et al.*, 2010*b*). (b) Polarization of eigenmodes for semipolar InGaN QW lasers grown on (11$\bar{2}\bar{2}$) GaN substrates. The integrated ASE intensities are normalized to the maximum intensity. (c) Polarization angle of the amplified stimulated emission for a variety of non- and semipolar laser heterostructures vs. the growth-plane angle with respect to the c-plane. Reprinted with permission from Rass *et al.* (2011), © 2011 American Institute of Physics.

fabrication of high-quality mirrors on c-plane GaN is quite similar to the well-established approaches of GaAs and InP-based laser diodes. In the case of laser diodes on non-polar surfaces, e.g. lasers on m-plane GaN, two different resonator orientations are possible, either aligned along the c-direction or aligned along the a-direction. The laser-scribing and cleaving process utilizes a focused UV laser beam from a frequency-tripled 355-nm Nd:YAG laser that scribes narrow grooves into the backside of the GaN substrate. The grooves are typically 10–20 μm wide and reach about half-way into the GaN substrate. The laser-scribed groove determines the cleavage direction for the laser facet, which is obtained in a second step by manually cleaving along the scribed grooves.

The quality of the facets depends on the substrate orientation and on the orientation of the resonator, i.e. the crystal plane that is being cleaved. For laser diodes on bulk m-plane GaN, both resonator orientation, i.e. in the [0001] c-direction and [11$\bar{2}$0] a-direction, the laser-scribing and cleaving process yields smooth facets that are aligned perpendicular to the laser resonator. Figure 8.20 (a) shows a scanning electron microscope image of a (11$\bar{2}$0) facet of a laser heterostructure grown on a bulk m-plane GaN substrate obtained by laser-scribing and cleaving. As can be seen from the SEM images, very smooth mirror faces are obtained with an rms roughness of less than 1 nm measured by atom-force microscopy (AFM). Cleaving the facet of a laser diode grown

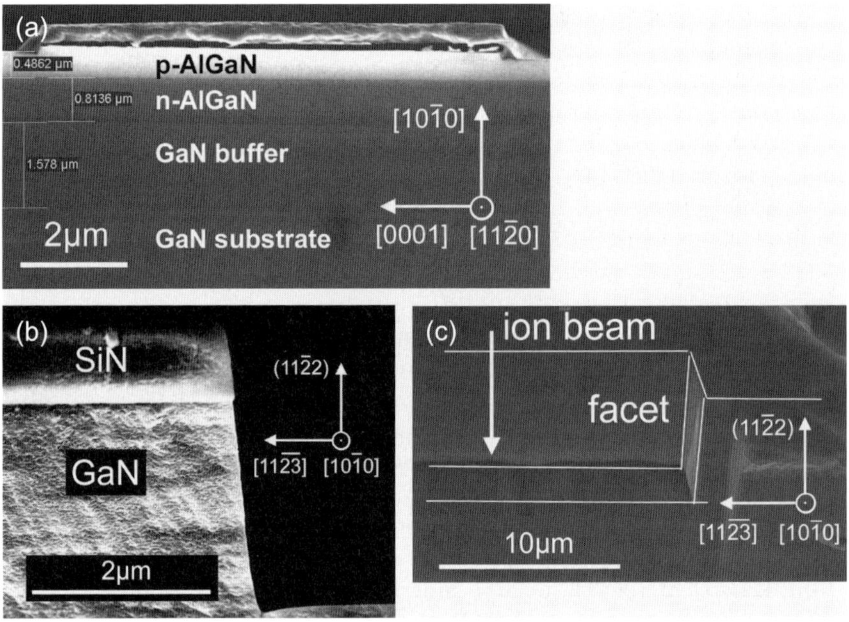

FIG. 8.20. (a) Scanning electron microscope (SEM) image of a (11$\bar{2}$0) facet of a laser heterostructure on a bulk m-plane GaN substrate obtained by laser-scribing and cleaving. (b) SEM image of a (11$\bar{2}\bar{3}$) facet of a laser heterostructure on semipolar (11$\bar{2}$2) GaN obtained by inductively coupled plasma (IPC) etching. (c) SEM image of a (11$\bar{2}\bar{3}$) facet after focused ion-beam etching (FIB). Reprinted with permission from Rass *et al.* (2010*a*), © 2010 Wiley-VCH Verlag GmbH & Co. KGaA, Weinheim.

on semipolar bulk GaN is much more difficult, and the facet quality depends strongly on the resonator direction. For example, facets for InGaN quantum-well laser diodes grown on $(11\bar{2}2)$ bulk GaN can be cleaved either along the $[10\bar{1}0]$ m-direction or $[11\bar{2}\bar{3}]$ c′-direction. Whereas the $[10\bar{1}0]$ laser facets show relatively smooth surfaces, no low-energy cleavage plane exist for the resonator cleaved in the $[11\bar{2}\bar{3}]$ c′-direction. When facets are cleaved along the $[11\bar{2}\bar{3}]$ c′-direction the cleaved surfaces exhibit large steps and terraces that are not perpendicular to the resonator direction. The rough surfaces are probably generated by cleaves along competing lower-energy surfaces, resulting in large steps and terraces. Jens Raß and colleagues have systematically studied the formation of facets by laser-scribing and cleaving on laser heterostructures grown on different non- and semipolar bulk GaN substrates, as well as with different resonator orientations (Rass *et al.*, 2010*a*; Raß, 2012). An overview of the resultant facet quality on the different growth planes and resonator orientations is summarized in Table 8.1.

Another approach for realizing laser facets is by dry etching. The advantage of this technique is that the facet quality depends only slightly on the resonator and substrate orientation. Therefore, this approach is suitable for almost all non- and semipolar laser diodes. The patterning process consists of three steps. First a 500-nm thick SiNx layer is deposited by plasma-assisted chemical vapor deposition (PECVD), followed by a layer of photoresist, which is photolithographically patterned in order to define openings in the SiNx mask for the subsequent dry-etching process. After the SiNx mask is formed, the group-III-nitride laser heterostructure is etched by a chlorine-based inductively coupled plasma (ICP) etching or reactive ion etching step. Figure 8.20 (b) shows an SEM image of a $(11\bar{2}\bar{3})$ facet of a laser heterostructure grown on a semipolar $(11\bar{2}2)$ bulk GaN substrate obtained by inductively coupled plasma (IPC) etching. By optimizing the dry-etching conditions, i.e. accelerating voltages and substrate temperature, nearly vertical mesa sidewalls with a facet angle of 83° relative to

Table 8.1 Quality of cleaved facets for lasers on c-plane and different non- and semipolar GaN surface orientations and for different resonator orientations (Rass *et al.*, 2010*a*).

Laser resonator orientations

Growth planes		polar c	c′ facets	m- or a-plane facets
	$(10\bar{1}0)$	smooth	N.A.	smooth
	$(10\bar{1}1)$	N.A.	smooth, tilted	smooth, terraces
	$(10\bar{1}2)$	N.A.	smooth, tilted	smooth, terraces
	$(11\bar{2}2)$	N.A.	steps, terraces	smooth

the laser waveguide plane can be obtained. To further improve the facet quality, additional wet chemical treatment steps can be used, e.g. by etching the exposed mesa sidewalls in hot KOH dissolved in ethylene glycol (Rass *et al.*, 2010*a*). Alternatively, the facet can be smoothed as well as realigned close to the ideal 90° angle by subsequent focused ion-beam etching (FIB) step. Figure 8.20 (c) shows an SEM image of an extremely smooth and perpendicular $(11\bar{2}\bar{3})$ facet after focused ion-beam etching (FIB). Although this technique yields excellent result it is not suitable for mass production of laser facets, due to the slow etching process. Overall, the dry-etching approach results in good mirror facets, though with the disadvantage that the laser facet extends only a few microns from the top surface, which can cause some interference of the laser beam with the remaining substrate.

8.6.3 *State-of-the-art of non-polar and semipolar InGaN laser diodes*

Since the demonstration of the first non-polar laser diode on m-plane GaN in 2007 by Schmidt *et al.* (Schwarz and Kneissl, 2007), and more or less simultaneously by Okamoto *et al.* (2007), progress in the performance characteristics of InGaN laser diodes on non- and semipolar GaN has been tremendous. At present, current-injection InGaN quantum-well laser diodes have been realized on a number of different semipolar surface orientations, including laser diodes on $(11\bar{2}2)$, $(10\bar{1}\bar{1})$, $(20\bar{2}1)$, $(20\bar{2}\bar{1})$, and $(30\bar{3}1)$ GaN substrates (Asamizu *et al.*, 2008; Tyagi *et al.*, 2007; Enya *et al.*, 2009; Huang *et al.*, 2011; Hsu *et al.*, 2010). The laser emission wavelengths have been pushed towards the true green spectral range (Takagi *et al.*, 2012; Schwarz and Scheibenzuber, 2011) and high-power blue laser diodes (Farrell *et al.*, 2011; Schwarz and Scheibenzuber, 2011) have been demonstrated.

Figure 8.21 summarizes the reported performance characteristics of InGaN QW laser diodes grown on different non- and semipolar substrates in the violet, blue, and green spectral regions. As can be seen, laser diodes with low threshold current densities can be achieved both for non- and semipolar surface orientations, and lasers operating continuous-wave (cw) have been realized in the entire spectral range. Nevertheless, which of these differently oriented non- and semipolar orientations will be best suited to meet the requirements put forward by applications such as laser projection and micro-displays is still to be decided. The ideal orientation will certainly depend on the specific emission wavelength that is being targeted. Obviously, achieving the lowest threshold current densities, lowest operating voltages, highest differential quantum efficiencies, highest output power levels, and long lifetimes will determine the choice of the ideal substrate orientation. All these parameters are governed by a number of different factors that themselves depend on the surface orientation. Critical factors such as indium incorporation efficiency, low-resistance Mg- and Si-doped layers, optical gain, transverse mode confinement, strain relaxation, and formation of defect heterostructure interfaces, homogeneity of the InGaN, InAlN,

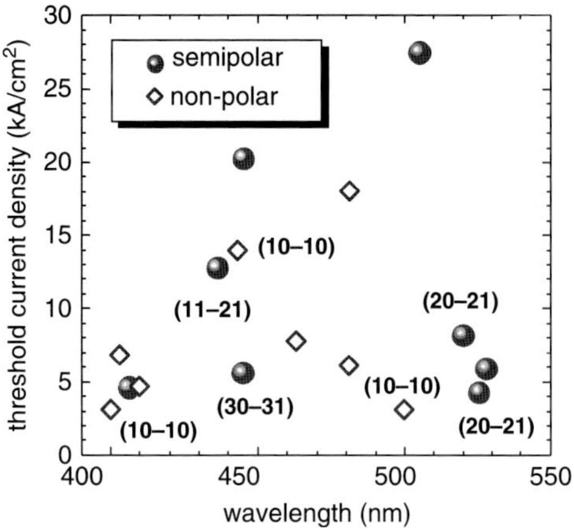

FIG. 8.21. Threshold current densities for violet, blue, and green InGaN QW laser diodes grown on different non- and semipolar substrates (Hsu *et al.*, 2010; Takagi *et al.*, 2012; Yoshizumi *et al.*, 2009; Farrell *et al.*, 2011; Huang *et al.*, 2011; Enya *et al.*, 2009; Raring *et al.*, 2011; Kelchner *et al.*, 2009; Tyagi *et al.*, 2010; Asamizu *et al.*, 2008; Okamoto *et al.*, 2009; Okamoto *et al.*, 2007; Hsu *et al.*, 2012; Sizov *et al.*, 2011; Melo *et al.*, 2012; Adachi *et al.*, 2010).

and AlGaN layers, and ease of fabrication, e.g. formation of facets and laser resonators, will have to be taken into account. For example, high-efficiency and low-threshold blue laser diodes have been realized on m-plane GaN substrate showing light-output power-levels of several Watts (Raring *et al.*, 2011). On the other hand, recent experimental data suggest that the m-plane surface is not well suited for obtaining true green laser diodes, which has been attributed mainly to the low incorporation efficiency on $(10\bar{1}0)$ surfaces (Wernicke *et al.*, 2012; Northrup, 2009). In order to obtain true green laser diodes emitting in the 520–535 nm spectral range, currently the most promising approach appears to be growth on GaN substrates with $(20\bar{2}1)$, $(20\bar{2}\bar{1})$, and $(30\bar{3}1)$ orientation (Huang *et al.*, 2011; Schwarz and Scheibenzuber, 2011; Takagi *et al.*, 2012). High-power green laser diodes on semipolar $(20\bar{2}1)$ GaN substrates operating at wavelengths beyond 530 nm with cw output power levels of more than 100 mW and wallplug efficiencies of nearly 8 % (Takagi *et al.*, 2012). Laser diodes on semipolar $(20\bar{2}1)$ GaN substrates hold the current record for the longest emission wavelength ever achieved with a group-III-nitride-based device with laser emission at 535.7 nm (Takagi *et al.*, 2012). Nevertheless, other semipolar orientations also show promising results, e.g. recently AlGaN cladding-free InGaN quantum-well laser diodes grown $(30\bar{3}1)$ GaN have been demonstrated with threshold current densities as low as 5.6 kA/cm^2 and emission wavelength near 445 nm

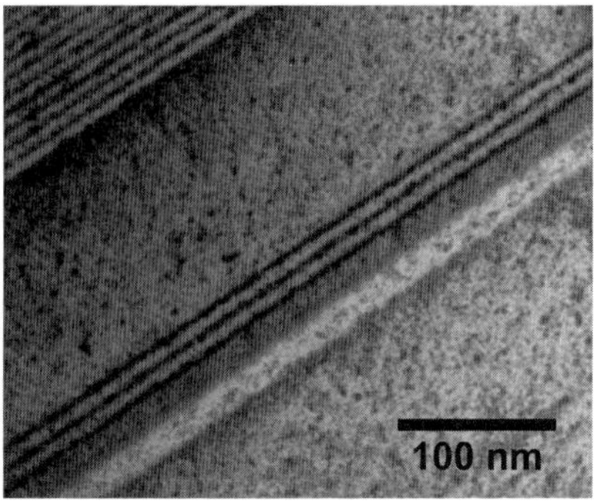

FIG. 8.22. Cross-sectional transmission electron microscope image of a triple InGaN quantum-well laser heterostructure grown on $(10\bar{1}0)$ GaN. Courtesy of P. Veit (Otto von Guericke University, Magdeburg).

(Hsu *et al.*, 2010). Remarkably, an extremely small wavelength shift of only 4 nm between the peak electroluminescence wavelength and the laser emission wavelength has been observed. This is similar to blue laser diodes on m-plane GaN, and considerably lower than the wavelength shift for polar blue laser diodes on (0001) GaN, indicating very small piezoelectric fields in the InGaN quantum wells (QWs). Another peculiarity of non- and semipolar lasers is the use of AlGaN-cladding-free heterostructure designs. Whereas (0001) laser diodes almost always incorporate AlGaN n- and p-cladding layers in their separate confinement heterostructure in order to increase the transverse mode confinement, this is often not the case for non- and semipolar lasers (see Fig. 8.22). One reason is the anisotropic strain in lattice-mismatched AlGaN cladding layers grown on non- and semipolar GaN substrates, resulting in stress relaxation and the formation of misfit dislocation at the GaN/AlGaN heterostructure interface (Young *et al.*, 2010). Thus the critical layer thickness for AlGaN/GaN heterostructure is significantly reduced, and consequently AlGaN-cladding-free laser heterostructures employing thick InGaN waveguide layers have been developed, and low threshold lasers have been successfully demonstrated (Rass *et al.*, 2012; Farrell *et al.*, 2011).

Just recently, the first non-polar vertical-cavity surface-emitting laser diode (VCSEL) on m-plane GaN substrates has been realized by Holder *et al.* (2012), emitting near 412 nm. The laser cavities were formed by selective photoelectrical chemical etching (PEC) in order to remove the GaN substrate and subsequent deposition of SiO_2/Ta_2O_5 distributed Bragg reflectors on the top and bottom to form the VCSEL cavity. Although the threshold current densities of these first non-polar VCSELs were very high, the VCSEL devices exhibited very interesting optical properties, such as polarization locking of the laser emission along the $[1\bar{2}10]$ a-direction, which is a consequence of the band alignment of the wurtzite crystal structure.

8.7 Summary and outlook

As the examples described in this chapter have shown, the device characteristics of non- and semipolar laser diodes as well as LEDs have already reached outstanding levels of performance. High-power non- and semipolar blue laser diodes have been realized, and the emission wavelength of semipolar laser diodes has been extended well into the green spectral range. Despite the short time since their first demonstration, non- and semipolar lasers are already competing seriously with conventional group-III-nitride laser diodes on (0001) c-plane GaN. In addition, high-efficiency blue, green, yellow, and even red LEDs have been demonstrated on non- and semipolar GaN. Besides high output power and external quantum efficiencies, these devices show unusual optical properties, e.g. strong optical polarization of the spontaneous emission, which may be interesting for applications in LCD displays. Some of these LEDs also show very-low-efficiency droops due to the reduced polarization field in non- and semipolar nitrides. This may enable high-power and high-current-density operation of blue, green, and white LEDs, which could be a significant competitive advantage in the high-power LED lighting market. Whether non- and semipolar laser diodes and LEDs will eventually emerge as the winning approach, and if so, which surface orientations will be the best choice, is still open. The decision will depend on many factors, some of which are not related to the superior physical properties but are more mundane, such as fabrication cost and device yields. Nevertheless, this discussion has shown that InGaN light emitters on non- and semipolar surfaces provide a wide parameter space for advancing the performance of nitride LEDs and lasers—an opportunity that should not be underestimated nor missed.

Acknowledgments

The authors thank J. Rass, S. Ploch, M. Frentrup, L. Schade, W. Scheibenzuber, U.T. Schwarz, C. Netzel, U. Zeimer, B. Basktek, F. Bertram, P. Veit, Z. Wu, J. Christen, F. Ponce, V. Hoffmann, A. Knauer M. Herms, and M. Weyers for performing measurements, data evaluation, and fruitful discussions. We also thank T. Tessaro, T. Petzke, and H. Lawrenz for technical support. We thank J. Stellmach for help with the formatting. We grateful acknowledge partial funding by the DFG within the Project PolarCon and SFB787.

References

Adachi, M., Yoshizumi, Y., Enya, Y., Kyono, T., Sumitomo, T., Tokuyama, S., Takagi, S., Sumiyoshi, K., Saga, N., Ikegami, T. *et al.* (2010). *Appl. Phys. Express*, **3**(12), 1001.

Arif, R. A., Zhao, H., Ee, Y. K., and Tansu, N. (2008). *IEEE Journal of Quantum Electronics*, **44**(6), 573–580.

Armitage, R. and Hirayama, H. (2008). *Appl. Phys. Lett.*, **92**(9), 092121–092121.

Asamizu, H., Saito, M., Fujito, K., Speck, J. S., DenBaars, S. P., and Nakamura, S. (2008). *Appl. Phys. Express*, **1**(9), 091102.

Avramescu, A., Lermer, T., Müller, J., Eichler, C., Bruederl, G., Sabathil, M., Lutgen, S., and Strauß, U. (2010). *Appl. Phys. Express*, **3**(6), 061003.

Avramescu, A., Lermer, T., Müller, J., Tautz, S., Queren, D., Lutgen, S., and Strauß, U. (2009). *Appl. Phys. Lett.*, **95**(7), 071103–071103.

Baker, T. J., Haskell, B. A., Wu, F., Fini, P. T., Speck, J. S., and Nakamura, S. (2005). *Jpn. J. Appl. Phys.*, **44**, L920.

Baker, T. J., Haskell, B. A., Wu, F., Speck, J. S., and Nakamura, S. (2006). *Jpn. J. Appl. Phys.*, **45**, L154–L157.

Bastek, B., August, O., Hempel, T., Christen, J., Wieneke, M., Bläsing, J., Dadgar, A., Krost, A., and Wendt, U. (2010). *Appl. Phys. Lett.*, **96**, 172102.

Bastek, B., Bertram, F., Christen, J., Wernicke, T., Weyers, M., and Kneissl, M. (2008). *Appl. Phys. Lett.*, **92**, 212111.

Beaumont, B., Vennégués, P., and Gibart, P. (2001). *Phys. Status Solidi (b)*, **227**(1), 1–43.

Berkowicz, E., Gershoni, D., Bahir, G., Lakin, E., Shilo, D., Zolotoyabko, E., Abare, A. C., Denbaars, S. P., and Coldren, L. A. (2000, Apr). *Phys. Rev. B*, **61**(16), 10994–11008.

Bernardini, F., Fiorentini, V., and Vanderbilt, D. (1997, Oct.). *Phys. Rev. B*, **56**(16), R10024–R10027.

Bougrioua, Z., Laügt, M., Vennégués, P., Cestier, I., Gühne, T., Frayssinet, E., Gibart, P., and Leroux, M. (2007). *physica status solidi (a)*, **204**(1), 282–289.

Browne, D. A., Young, E. C., Lang, J. R., Hurni, C. A., and Speck, J. S. (2012). *Journal of Vacuum Science & Technology A: Vacuum, Surfaces, and Films*, **30**(4), 041513.

Chakraborty, A., Baker, T. J., Haskell, B. A., Wu, F., Speck, J. S., Denbaars, S. P., Nakamura, S., and Mishra, U. K. (2005*a*). *Jpn. J. Appl. Phys.*, **44**, L945–L947.

Chakraborty, A., Haskell, B. A., Keller, S., Speck, J. S., DenBaars, S. P., Nakamura, S., and Mishra, U. K. (2004). *Appl. Phys. Lett.*, **85**, 5143.

Chakraborty, A., Haskell, B. A., Masui, H., Keller, S., Speck, J. S., DenBaars, S. P., Nakamura, S., and Mishra, U. K. (2006*a*). *Jpn. J. Appl. Phys.*, **45**, 739.

Chakraborty, A., Keller, S., Meier, C., Haskell, B. A., Keller, S., Waltereit, P., DenBaars, S. P., Nakamura, S., Speck, J. S., and Mishra, U. K. (2005*b*). *Appl. Phys. Lett.*, **86**(3), 031901.

Chakraborty, A., Kim, K. C., Wu, F., Speck, J. S., DenBaars, S. P., and Mishra, U. K. (2006*b*). *Appl. Phys. Lett.*, **89**(4), 041903.

Chen, C., Shatalov, M., Kuokstis, E., Adivarahan, V., Gaevski, M., Rai, S., and Asif Khan, M. (2004). *Jpn. J. Appl. Phys*, **43**, L1099–L1102.

Chichibu, S. F., Abare, A. C., Mack, M. P., Minsky, M. S., Deguchi, T., Cohen, D., Kozodoy, P., Fleischer, S. B., Keller, S., Speck, J. S., Bowers, J.E, Hu, E., Mishra, U. K., Coldren, L. A., DenBaars, S. P., Wada, K., Sota, T., and Nakamura, S. (1999). *Materials Science and Engineering: B*, **59**(1), 298–306.

Chichibu, S. F., Uedono, A., Onuma, T., Haskell, B. A., Chakraborty, A., Koyama, T., Fini, P. T., Keller, S., DenBaars, S. P., Speck, J. S., Mishra, U. K., Nakamura, S., Yamaguchi, S., Kamiyama, S., Amano, H., Akasaki, I., Han, J., and Sota, T. (2006). *Nature Materials*, **5**(10), 810–816.

Cho, Y. S., Sun, Q., Lee, I. H., Ko, T. S., Yerino, C. D., Han, J., Kong, B. H., Cho, H. K., and Wang, S. (2008). *Appl. Phys. Lett.*, **93**(11), 111904.

Craven, M. D., Lim, S. H., Wu, F., Speck, J. S., and DenBaars, S. P. (2002*a*). *Appl. Phys. Lett.*, **81**(3), 469–471.

Craven, M. D., Lim, S. H., Wu, F., Speck, J. S., and DenBaars, S. P. (2002*b*). *Appl. Phys. Lett.*, **81**(7), 1201–1203.

Craven, M. D., Waltereit, P., Speck, J. S., and DenBaars, S. P. (2004*a*). *Appl. Phys. Lett.*, **84**, 496.

Craven, M. D., Wu, F., Chakraborty, A., Imer, B., Mishra, U. K., DenBaars, S. P., and Speck, J. S. (2004*b*). *Appl. Phys. Lett.*, **84**, 1281.

Dadgar, A., Hums, C., Diez, A., Bläsing, J., and Krost, A. (2006). *J. Crys. Growth*, **297**(2), 279–282.

Dadgar, A., Ravash, R., Veit, P., Schmidt, G., Muller, M., Dempewolf, A., Bertram, F., Wieneke, M., Christen, J., and Krost, A. (2011). *Appl. Phys. Lett.*, **99**(2), 021905.

David, A. and Grundmann, M. J. (2010). *Appl. Phys. Lett.*, **96**(10), 103504.

Dinh, D. (2012). Ph.D. thesis, TU Berlin.

Enya, Y., Yoshizumi, Y., Kyono, T., Akita, K., Ueno, M., Adachi, M., Sumitomo, T., Tokuyama, S., Ikegami, T., Katayama, K., and Nakamura, T. (2009). *Appl. Phys. Express*, **2**, 082101.

Farrell, R. M., Haeger, D. A., Hsu, P. S., Schmidt, M. C., Fujito, K., Feezell, D. F., DenBaars, S. P., Speck, J. S., and Nakamura, S. (2011). *Appl. Phys. Lett.*, **99**(17), 171113.

Farrell, R. M., Haeger, D. A., Chen, X., Gallinat, C. S., Davis, R. W., Cornish, M., Fujito, K., Keller, S., DenBaars, S. P., Nakamura, S., and Speck, J. S. (2010). *Appl. Phys. Lett.*, **96**, 231907.

Feneberg, M., Lipski, F., Sauer, R., Thonke, K., Wunderer, T., Neubert, B., Brückner, P., and Scholz, F. (2006). *Appl. Phys. Lett.*, **89**, 242112.

Feneberg, M., Thonke, K., Wunderer, T., Lipski, F., and Scholz, F. (2010). *J. Appl. Phys.*, **107**(10), 103517.

Frentrup, M., Ploch, S., Pristovsek, M., and Kneissl, M. (2011). *physica status solidi (b)*, **248**(3), 583–587.

Fujito, K., Kiyomi, K., Mochizuki, T., Oota, H., Namita, H., Nagao, S., and Fujimura, I. (2008). *physica status solidi (a)*, **205**(5), 1056–1059.

Funato, M., Ueda, M., Inoue, D., Kawakami, Y., Narukawa, Y., and Mukai, T. (2010). *Appl. Phys. Express*, **3**(7), 1001.

Funato, M., Ueda, M., Kawakami, Y., Narukawa, Y., Kosugi, T., Takahashi, M., and Mukai, T. (2006). *Jpn. J. Appl. Phys.*, **45**, L659–L662.

Gardner, N. F., Kim, J. C., Wierer, J. J., Shen, Y. C., and Krames, M. R. (2005). *Appl. Phys. Lett.*, **86**(11), 111101.

Gardner, N. F., Muller, G. O., Shen, Y. C., Chen, G., Watanabe, S., Gotz, W., and Krames, M. R. (2007). *Appl. Phys. Lett.*, **91**(24), 243506.

Hader, J., Moloney, J. V., and Koch, S. W. (2010). *Appl. Phys. Lett.*, **96**, 221106.

Hangleiter, A., Hitzel, F., Netzel, C., Fuhrmann, D., Rossow, U., Ade, G., and Hinze, P. (2005). *Phys. Rev. Lett.*, **95**(12), 127402.

Hardy, M. T., Hsu, P. S., Wu, F., Koslow, I. L., Young, E. C., Nakamura, S., Romanov, A. E., DenBaars, S. P., and Speck, J. S. (2012). *Appl. Phys. Lett.*, **100**(20), 202103.

Haskell, B. A., Baker, T. J., McLaurin, M. B., Wu, F., Fini, P. T., DenBaars, S. P., Speck, J. S., and Nakamura, S. (2005). *Appl. Phys. Lett.*, **86**, 111917.

Hellman, E. S., Liliental-Weber, Z., and Buchanan, D. (1997). *MRS Internet J. Nitride Semicond. Res.*, **2**, 30.

Hirai, A., Haskell, B. A., McLaurin, M. B., Wu, F., Schmidt, M. C., Kim, K. C., Baker, T. J., DenBaars, S. P., Nakamura, S., and Speck, J. S. (2007*a*). *Appl. Phys. Lett.*, **90**(12), 121119.

Hirai, A., Jia, Z., Schmidt, M. C., Farrell, R. M., DenBaars, S. P., Nakamura, S., Speck, J. S., and Fujito, K. (2007*b*). *Appl. Phys. Lett.*, **91**, 191906.

Holder, C., Speck, J. S., DenBaars, S. P., Nakamura, S., and Feezell, D. (2012). *Appl. Phys. Express*, **5**(9), 092104.

Hollander, J. L., Kappers, M. J., McAleese, C., and Humphreys, C. J. (2008). *Appl. Phys. Lett.*, **92**, 101104.

Honda, Y., Kameshiro, N., Yamaguchi, M., and Sawaki, N. (2002). *J. Crys. Growth*, **242**(1), 82–86.

Honda, Y., Kawaguchi, Y., Ohtake, Y., Tanaka, S., Yamaguchi, M., and Sawaki, N. (2001). *J. Crys. Growth*, **230**(3), 346–350.

Hsu, P., Hardy, M. T., Wu, F., Koslow, I., Young, E. C., Romanov, A. E., Fujito, K., Feezell, D. F., DenBaars, S. P., Speck, J. S., and Nakamura, S. (2012). *Appl. Phys. Lett.*, **100**(2), 021104.

Hsu, P. S., Kelchner, K. M., Tyagi, A., Farrell, R. M., Haeger, D. A., Fujito, K., Ohta, H., DenBaars, S. P., Speck, J. S., and Nakamura, S. (2010). *Appl. Phys. Express*, **3**(5), 2702.

Huang, C. Y., Hardy, M. T., Fujito, K., Feezell, D. F., Speck, J. S., DenBaars, S. P., and Nakamura, S. (2011). *Appl. Phys. Lett.*, **99**(24), 241115.

Imer, B. M., Wu, F., DenBaars, S. P., and Speck, J. S. (2006). *Appl. Phys. Lett.*, **88**, 061908.

Johnston, C. F., Kappers, M. J., Moram, M. A., Hollander, J. L., and Humphreys, C. J. (2009). *J. Crys. Growth*, **311**(12), 3295–3299.

Jönen, H., Bremers, H., Rossow, U., Langer, T., Kruse, A., Hoffmann, L., Thalmair, J., Zweck, J., Schwaiger, S., Scholz, F. *et al.* (2012). *Semiconductor Science and Technology*, **27**(2), 024013.

Kaeding, J. F., Iza, M., Sato, H., DenBaars, S. P., Speck, J. S., and Nakamura, S. (2006). *Jpn. J. Appl. Phys.*, **45**, L536–L538.

Karpov, S. Y. and Makarov, Y. N. (2002). *Appl. Phys. Lett.*, **81**(25), 4721–4723.

Kawaguchi, Y., Huang, C. Y., Wu, Y. R., Yan, Q., Pan, C. C., Zhao, Y., Tanaka, S., Fujito, K., Feezell, D., Van de Walle, C. G. *et al.* (2012). *Appl. Phys. Lett.*, **100**, 231110.

Kelchner, K. M., Lin, Y. D., Hardy, M. T., Huang, C. Y., Hsu, P. S., Farrell, R. M., Haeger, D. A., Kuo, H. C., Wu, F., Fujito, K. *et al.* (2009). *Appl. Phys. Express*, **2**(7), 1003.

Kim, K. C., Schmidt, M. C., Sato, H., Wu, F., Fellows, N., Saito, M., Fujito, K., Speck, J. S., Nakamura, S., and DenBaars, S. P. (2007*a*). *physica status solidi (RRL) Rapid Research Letters*, **1**(3), 125–127.

Kim, K. S., Son, J. K., Lee, S. N., Sung, Y. J., Paek, H. S., Kim, H. K., Kim, M. Y., Ha, K. H., Ryu, H. Y., Nam, O. H. *et al.* (2008). *Appl. Phys. Lett.*, **92**(10), 101103.

Kim, M. H., Schubert, M. F., Dai, Q., Kim, J. K., Schubert, E. F., Piprek, J., and Park, Y. (2007*b*). *Appl. Phys. Lett.*, **91**, 183507.

Kioupakis, E., Rinke, P., Delaney, K. T., and Van de Walle, C. G. (2011). *Appl. Phys. Lett.*, **98**(16), 161107.

Koslow, I. L., Hardy, M. T., Shan Hsu, P., Dang, P. Y., Wu, F., Romanov, A., Wu, Y. R., Young, E. C., Nakamura, S., Speck, J. S. *et al.* (2012). *Appl. Phys. Lett.*, **101**(12), 121106.

Koslow, I. L., Sonoda, J., Chung, R. B., Pan, C. C., Brinkley, S., Ohta, H., Nakamura, S., and DenBaars, S. P. (2010). *Jpn. J. Appl. Phys.*, **49**(8), 0203.

Kozaki, T., Nagahama, S., and Mukai, T. (2007). In *Proc. SPIE*, Volume 6485, pp. 648503–648508.

Kubota, M., Okamoto, K., Tanaka, T., and Ohta, H. (2008). *Appl. Phys. Express*, **1**(1), 072201.

Kuokstis, E., Yang, J. W., Simin, G., Kahn, M. Asif, Gaska, R., and Shur, M. S. (2002). *Appl. Phys. Lett.*, **80**, 977–979.

Langer, T., Kruse, A., Ketzer, F. A., Schwiegel, A., Hoffmann, L., Jönen, H., Bremers, H., Rossow, U., and Hangleiter, A. (2011). *physica status solidi (c)*, **8**(7–8), 2170–2172.

Laubsch, A., Sabathil, M., Bergbauer, W., Strassburg, M., Lugauer, H., Peter, M., Lutgen, S., Linder, N., Streubel, K., Hader, J. *et al.* (2009). *physica status solidi (c)*, **6**(S2), S913–S916.

Lee, S. N., Kim, K. K., Nam, O. H., Kim, J. H., and Kim, H. (2010). *physica status solidi (c)*, **7**(7–8), 2043–2045.

Lermer, T., Schillgalies, M., Breidenassel, A., Queren, D., Eichler, C., Avramescu, A., Müller, J., Scheibenzuber, W., Schwarz, U., Lutgen, S. *et al.* (2010). *physica status solidi (a)*, **207**(6), 1328–1331.

Li, Q., Xu, S. J., Xie, M. H., and Tong, S. Y. (2005). *Europhys. Lett.*, **71**(6), 994.

Lu, F., Lee, D., Byrnes, D., Armour, E., and Quinn, W. (2011). *Science China Technological Sciences*, **54**(1), 33–37.

Lu, I. L., Wu, Y. R., and Singh, J. (2010). *J. Appl. Phys.*, **108**(12), 124508.

Lutgen, S., Avramescu, A., Lermer, T., Queren, D., Müller, J., Bruederl, G., and Strauss, U. (2010). *physica status solidi (a)*, **207**(6), 1318–1322.

Lymperakis, L. and Neugebauer, J. (2009). *Phys. Rev. B*, **79**(24), 241308.

Matsuoka, T. and Hagiwara, E. (2001). *physica status solidi (a)*, **188**(2), 485–489.

Melo, T., Hu, Y. L., Weisbuch, C., Schmidt, M. C., David, A., Ellis, B., Poblenz, C., Lin, Y. D., Krames, M. R., and Raring, J. W. (2012). *Semiconductor Science and Technology*, **27**(2), 024015.

Miller, D. A. B., Chemla, D. S., Damen, T. C., Gossard, A. C., Wiegmann, W., Wood, T. H., and Burrus, C. A. (1984, Nov.). *Phys. Rev. Lett.*, **53**(22), 2173–2176.

Miller, E. J., Yu, E. T., Poblenz, C., Elsass, C., and Speck, J. S. (2002). *Appl. Phys. Lett.*, **80**(19), 3551–3553.

Miyoshi, T., Masui, S., Okada, T., Yanamoto, T., Kozaki, T., Nagahama, S., and Mukai, T. (2009). *Appl. Phys. Express*, **2**(6), 2201.

Morita, D., Yamamoto, M., Akaishi, K., Matoba, K., Yasutomo, K., Kasai, Y., Sano, M., Nagahama, S., and Mukai, T. (2004). *Jpn. J. Appl. Phys.*, **43**, 5945.

Mukai, T., Yamada, M., Nakamura, S. *et al.* (1999). *Jpn. J. Appl. Phys.*, **38**(part 1), 3976–3981.

Müller, J., Strauß, U., Lermer, T., Brüderl, G., Eichler, C., Avramescu, A., and Lutgen, S. (2011). *physica status solidi (a)*, **208**(7), 1590–1592.

Murase, T., Tanikawa, T., Honda, Y., Yamaguchi, M., Amano, H., and Sawaki, N. (2011). *Jpn. J. Appl. Phys.*, **50**(1), 01AD04–01AD04–3.

Nakamura, S. and Fasol, G. (1997). *The Blue Laser Diode: GaN Based Light Emitters and Lasers.* Springer, Berlin.

Nakamura, S., Mukai, T., and Senoh, M. (1991). *Jpn. J. Appl. Phys.*, **30**(12A), L1998–L2001.

Nakamura, S., Mukai, T., and Senoh, M. (1994). *Appl. Phys. Lett.*, **64**(13), 1687–1689.

Nakamura, S., Senoh, M., Iwasa, N., and Nagahama, S. (1995). *Jpn. J. Appl. Phys.*, **34**(7), L797–L799.

Nakamura, S., Senoh, M., Nagahama, S., Iwasa, N., Yamada, T., Matsushita, T., Kiyoku, H., and Sugimoto, Y. (1996*a*). *Jpn. J. Appl. Phys. Part 2 Letters*, **35**, L74–L76.

Nakamura, S., Senoh, M., Nagahama, S., Iwasa, N., Yamada, T., Matsushita, T., Kiyoku, H., Sugimoto, Y., Kozaki, T., Umemoto, H. *et al.* (1998*a*). *Jpn. J. Appl. Phys.*, **37**, L309–L312.

Nakamura, S., Senoh, M., Nagahama, S., Iwasa, N., Yamada, T., Matsushita, T., Kiyoku, H., Sugimoto, Y., Kozaki, T., Umemoto, H. *et al.* (1998*b*). *Jpn. J. Appl. Phys.*, **37**, L627–L629.

Nakamura, S., Senoh, M., Nagahama, S., Iwasa, N., Yamada, T., Matsushita, T., Sugimoto, Y., and Kiyoku, H. (1996*b*). *Appl. Phys. Lett.*, **69**(26), 4056–4058.

Netzel, C., Hatami, S., Hoffmann, V., Wernicke, T., Knauer, A., Kneissl, M., and Weyers, M. (2011*a*). *physica status solidi (c)*, **8**(7–8), 2151–2153.

Netzel, C., Hoffmann, V., Wernicke, T., Kanuer, A., and Weyers, M. (2010*a*). *J. Appl. Phys.*, **107**, 033510.

Netzel, C., Hoffmann, V., Wernicke, T., Knauer, A., Weyers, M., and Kneissl, M. (2010*b*). *physica status solidi (c)*, **7**(7–8), 1872–1874.

Netzel, C., Mauder, C., Wernicke, T., Reuters, B., Kalisch, H., Heuken, M., Vescan, A., Weyers, M., and Kneissl, M. (2011*b*). *Semiconductor Science and Technology*, **26**(10), 105017.

Netzel, C., Skriniarova, J., Herms, M., Wenzel, H., Hoffmann, V., Knauer, A., and Weyers, M. (2009). *physica status solidi (c)*, **6**(S2), S691–S694.

Netzel, C., Wernicke, T., Zeimer, U., Brunner, F., Weyers, M., and Kneissl, M. (2008). *J. Crys. Growth*, **310**, 8.

Neumann, W., Mogilatenko, A., Wernicke, T., Richter, E., Weyers, M., and Kneissl, M. (2009). *J. Microsc.*, **237**, 308–313.

Ni, X., Wu, M., Lee, J., Li, X., Baski, A., Özgür, Ü., and Morkoç, H. (2010). In *OPTO*, pp. 760220. International Society for Optics and Photonics.

Nichia Corporation (2012*a*). 2012 Nichia led catalogue, website, 2012, available online at http://www.nichia.co.jp.

Nichia Corporation (2012*b*). Details of the commercial blue and green (product number ndg1116) laser diodes can be found at http://www.nichia.co.jp.

Northrup, J. E. (2009). *Appl. Phys. Lett.*, **95**(13), 133107.

Okada, N., Kawashima, Y., and Tadatomo, K. (2008). *Appl. Phys. Express*, **1**(11), 111101.

Okada, N., Oshita, H., Kurisu, A., and Tadatomo, K. (2011*a*). *Jpn. J. Appl. Phys.*, **50**(3), 5602.

Okada, N., Oshita, H., Yamane, K., and Tadatomo, K. (2011*b*). *Appl. Phys. Lett.*, **99**(24), 242103.

Okamoto, K., Hiroaki, O., Chichibu, S. F., Ichihara, J., and Takasu, H. (2007). *Jpn. J. Appl. Phys.*, **46**(9), L187–L189.

Okamoto, K., Kashiwagi, J., Tanaka, T., and Kubota, M. (2009). *Appl. Phys. Lett.*, **94**(7), 071105.

Okamoto, K., Ohta, H., Nakagawa, D., Sonobe, M., Ichihara, J., and Takasu, H. (2006). *Jpn. J. Appl. Phys.*, **45**, L1197–L1199.

Pan, C. C., Tanaka, S., Wu, F., Zhao, Y., Speck, J. S., Nakamura, S., DenBaars, S. P., and Feezell, D. (2012). *Appl. Phys. Express*, **5**(6), 2103.

Piprek, J. (2007). *Nitride Semiconductor Devices: Principles and Simulation.* Volume 590. Wiley Online Library.

Piprek, J. (2010). *physica status solidi (a)*, **207**(10), 2217–2225.

Ploch, S., Bum Park, J., Stellmach, J., Schwaner, T., Frentrup, M., Niermann, T., Wernicke, T., Pristovsek, M., Lehmann, M., and Kneissl, M. (2011). *J. Crys. Growth*, **331**(1), 25–28.

Ploch, S., Frentrup, M., Wernicke, T., Pristovsek, M., Weyers, M., and Kneissl, M. (2010). *J. Crys. Growth*, **312**, 2171–2174.

Ploch, S., Wernicke, T., Dinh, D. V., Pristovsek, M., and Kneissl, M. (2012*a*). *J. Appl. Phys.*, **111**(3), 033526.

Ploch, S., Wernicke, T., Frentrup, M., Pristovsek, M., Weyers, M., and Kneissl, M. (2012*b*). *Appl. Phys. Lett.*, **101**(20), 202102.

Ploch, S., Wernicke, T., Thalmair, J., Lohr, M., Pristovsek, M., Zweck, J., Weyers, M., and Kneissl, M. (2012*c*). *J. Crys. Growth*, **356**, 70.

Queren, D., Avramescu, A., Bruderl, G., Breidenassel, A., Schillgalies, M., Lutgen, S., and Strauß, U. (2009). *Appl. Phys. Lett.*, **94**(8), 081119.

Raring, J. W., Hall, E. M., Schmidt, M. C., Poblenz, C., Li, B., Pfister, N., Feezell, D. F., Craig, R., Speck, J. S., DenBaars, S. P. *et al.* (2010). In *OPTO*, p. 760218. International Society for Optics and Photonics.

Raring, J. W., Schmidt, M. C., Poblenz, C., Li, B., Chang, Y. C., Mondry, M. J., Lin, Y. D., Krames, M. R., Craig, R., Speck, J. S., DenBaars, S. P., and Nakamura, S. (2011). *Proc. SPIE*, **7939**, 79390Y–1.

Raß, J. (2012). Ph.D. thesis, TU Berlin.

Rass, J., Wernicke, T., John, W., Krüger, O., Vogt, P., Einfeldt, S., Weyers, M., and Kneissl, M. (2010*a*). *physica status solidi (a)*, **207**, 1361.

Rass, J., Wernicke, T., Ploch, S., Brendel, M., Kruse A., Hangleiter, A., Scheibenzuber, W., Schwarz, U. T., Weyers, M., and Kneissl, M. (2011). *Appl. Phys. Lett.*, **99**(17), 171105.

Rass, J., Wernicke, T., Ploch, S., Brendel, M., Kruse, A., Hangleiter, A., Scheibenzuber, W., Schwarz, U. T., Weyers, M., and Kneissl, M. (2012). *Gallium Nitride Materials and Devices*, **7**(8262), 1.

Rass, J., Wernicke, T., Scheibenzuber, W. G., Schwarz, U. T., Kupec, J., Witzigmann, B., Vogt, P., Einfeldt, S., Weyers, M., and Kneissl, M. (2010*b*). *physica status solidi (RRL)*, **4**(1–2), 1–3.

Ravash, R., Blaesing, J., Dadgar, A., and Krost, A. (2010). *Appl. Phys. Lett.*, **97**, 142102.

Renner, F., Kiesel, P., Dohler, G. H., Kneissl, M., Van de Walle, C. G., and Johnson, N. M. (2002). *Appl. Phys. Lett.*, **81**(3), 490–492.

Romanov, A. E., Baker, T. J., Nakamura, S., and Speck, J. S. (2006). *J. Appl. Phys.*, **100**(2), 023522.

Ryu, H. Y., Kim, H. S., and Shim, J. I. (2009). *Appl. Phys. Lett.*, **95**(8), 081114.

Sano, M. and Aoki, M. (1976). *Jpn. J. Appl. Phys.*, **15**(10), 1943–1950.

Sato, H., Tyagi, A., Zhong, H., Fellows, N., Chung, R. B., Saito, M., Fujito, K., Speck, J. S., DenBaars, S. P., and Nakamura, S. (2007). *physica status solidi (RRL) Rapid Research Letters*, **1**(4), 162–164.

Schade, L., Schwarz, U. T., Wernicke, T., Ploch, S., Weyers, M., and Kneissl, M. (2012*a*). *phys. status solidi (c)*, **9**, 700.

Schade, L., Schwarz, U. T., Wernicke, T., Rass, J., Ploch, S., Weyers, M., and Kneissl, M. (2011*a*). *Appl. Phys. Lett.*, **99**(5), 051103.

Schade, L., Schwarz, U. T., Wernicke, T., Rass, J., Ploch, S., Weyers, M., and Kneissl, M. (2012*b*). *Proc. SPIE*, **8262**, 9.

Schade, L., Schwarz, U. T., Wernicke, T., Weyers, M., and Kneissl, M. (2011*b*). *physica status solidi (b)*, **248**(3), 638–646.

Scheibenzuber, W. G. (2012). Ph.D. thesis, University of Freiburg.

Scheibenzuber, W. G., Schwarz, U. T., Veprek, R. G., Witzigmann, B., and Hangleiter, A. (2009). *Phys. Rev. B*, **80**(11), 115320.

Schmidt, M. C., Kim, K. C., Farrell, R. M., Feezell, D. F., Cohen, D. A., Saito, M., Fujito, K., Speck, J. S., DenBaars, S. P., and Nakamura, S. (2007). *Jpn. J. Appl. Phys.*, **46**, L190–L191.

Schulz, T., Markurt, T., Albrecht, M., Dadgar, A., Ravash, R., M. Wienecke, P. Veit, and Krost, A. (2013), in press.

Schwaiger, S., Argut, I., Wunderer, T., Rosch, R., Lipski, F., Biskupek, J., Kaiser, U., and Scholz, F. (2010). *Appl. Phys. Lett.*, **96**(23), 231905.

Schwaiger, S., Metzner, S., Wunderer, T., Argut, I., Thalmair, J., Lipski, F., Wieneke, M., Bläsing, J., Bertram, F., Zweck, J. *et al.* (2011). *physica status solidi (b)*, **248**(3), 588–593.

Schwarz, U. T. and Kneissl, M. (2007). *physica status solidi (RRL) Rapid Research Letters*, **1**(3), A44–A46.

Schwarz, U. T. and Scheibenzuber, W. G. (2011). *Optics and Photonics News*, **22**(9), 38–44.

Sharma, R., Pattison, P. M., Masui, H., Farrell, R. M., Baker, T. J., Haskell, B. A., Wu, F., DenBaars, S. P., Speck, J. S., and Nakamura, S. (2005). *Appl. Phys. Lett.*, **87**, 231110.

Shen, H., Garrett, G. A., Wraback, M., Zhong, H., Tyagi, A., DenBaars, S. P., Nakamura, S., and Speck, J. S. (2010). *physica status solidi (c)*, **7**(10), 2378–2381.

Shen, H., Wraback, M., Zhong, H., Tyagi, A., DenBaars, S. P., Nakamura, S., and Speck, J. S. (2009). *Appl. Phys. Lett.*, **94**(24), 241906.

Shen, Y. C., Mueller, G. O., Watanabe, S., Gardner, N. F., Munkholm, A., and Krames, M. R. (2007). *Appl. Phys. Lett.*, **91**, 141101.

Sizov, D., Bhat, R., Song, K., Allen, D., Paddock, B., Coleman, S., Hughes, L. C., and Zah, C. (2011). *Appl. Phys. Express*, **4**(10), 2103.

Stampfl, C. and Van de Walle, C. G. (1998). *Phys. Rev. B*, **57**(24), 15052–15055.

Strauß, U., Avramescu, A., Lermer, T., Queren, D., Gomez-Iglesias, A., Eichler, C., Müller, J., Brüderl, G., and Lutgen, S. (2011). *physica status solidi (b)*, **248**(3), 652–657.

Strittmatter, A., Teepe, M., Knollenberg, C., Yang, Z., Chua, C., Johnson, N. M., Spiberg, P., Ivantsov, V., Syrkin, A., Shapovalov, L. *et al.* (2010). *Proc. SPIE*, **7616**, 76160H.

Sun, Q., Kong, B. H., Yerino, C. D., Ko, T. S., Leung, B., Cho, H. K., and Han, J. (2009*a*). *J. Appl. Phys.*, **106**, 123519.

Sun, Q., Leung, B., Yerino, C. D., Zhang, Y., and Han, J. (2009*b*). *Appl. Phys. Lett.*, **95**, 231904.

Sun, Q., Yerino, C. D., Ko, T. S., Cho, Y. S., Lee, I. H., Han, J., and Coltrin, M. E. (2008). *J. Appl. Phys.*, **104**, 093523.

Takagi, S., Enya, Y., Kyono, T., Adachi, M., Yoshizumi, Y., Sumitomo, T., Yamanaka, Y., Kumano, T., Tokuyama, S., Sumiyoshi, K. *et al.* (2012). *Appl. Phys. Express*, **5**(8), 2102.

Takeuchi, T., Sota, S., Katsuragawa, M., Komori, M., Takeuchi, H., Amano, H., and Akasaki, I. (1997). *Jpn. J. Appl. Phys.*, **1**(4A Part 2), L382–L385.

Tanikawa, T., Hikosaka, T., Honda, Y., Yamaguchi, M., and Sawaki, N. (2008*a*). *physica status solidi (c)*, **5**(9), 2966–2968.

Tanikawa, T., Honda, Y., Yamaguchi, M., and Amano, H. (2011). *physica status solidi (a)*, **208**(5), 1175–1178.

Tanikawa, T., Rudolph, D., Hikosaka, T., Honda, Y., Yamaguchi, M., and Sawaki, N. (2008*b*). *J. Crys. Growth*, **310**(23), 4999–5002.

Tavernier, P. R., Imer, B., DenBaars, S. P., and Clarke, D. R. (2004). *Appl. Phys. Lett.*, **85**, 4630–4632.

Tyagi, A., Farrell, R., Kelchner, K. M., Huang, C., Hsu, P. S., Haeger, D. A., Hardy, M. T., Holder, C., Fujito, K., Cohen, D. A., Ohta, H., Speck, J. S., DenBaars, S. P., and Nakamura, S. (2010). *Appl. Phys. Express*, **3**, 011002.

Tyagi, A., Zhong, H., Chung, R. B., Feezell, D. F., Saito, M., Fujito, K., Speck, J. S., DenBaars, S. P., and Nakamura, S. (2007). *Jpn. J. Appl. Phys.*, **46**, L444–L445.

Usikov, A., Soukhoveev, V., Shapovalov, L., Syrkin, A., Ivantsov, V., Scanlan, B., Nikiforov, A., Strittmatter, A., Johnson, N., Zheng, J. G. *et al.* (2010). *physica status solidi (a)*, **207**(6), 1295–1298.

Vanfleet, R. R., Simmons, J. A., Maruska, H. P., Hill, D. W., Chou, M. C. C., and Chai, B. H. (2003). *Appl. Phys. Lett.*, **83**, 1139–1141.

Vennéguès, P., Chauveau, J. M., Korytov, M., Deparis, C., Zuniga-Perez, J., and Morhain, C. (2008). *J. Appl. Phys.*, **103**(8), 083525.

Vennéguès, P., Bougrioua, Z., and Guehne, T. (2007). *Jpn. J. Appl. Phys.*, **46**, 4089–4095.

Waltereit, P., Brandt, O., Trampert, A., Grahn, H. T., Menniger, J., Ramsteiner, M., Reiche, M., and Ploog, K. H. (2000). *Nature*, **406**(6798), 865–868.

Wernicke, T. (2010). Ph.D. thesis, TU Berlin.

Wernicke, T., Ploch, S., Hoffmann, V., Knauer, A., Weyers, M., and Kneissl, M. (2011). *physica status solidi (b)*, **248**(3), 574–577.

Wernicke, T., Schade, L., Netzel, C., Rass, J., Hoffmann, V., Ploch, S., Knauer, A., Weyers, M., Schwarz, U., and Kneissl, M. (2012). *Semiconductor Science and Technology*, **27**(2), 024014.

Wernicke, T., Zeimer, U., Netzel, C., Brunner, F., Knauer, A., Weyers, M., and Kneissl, M. (2009). *J. Crys. Growth*, **311**, 2895.

Winzer, A. T., Goldhahn, R., Gobsch, G., Link, A., Eickhoff, M., Rossow, U., and Hangleiter, A. (2005). *Appl. Phys. Lett.*, **86**, 181912.

Wu, F., Craven, M. D., Lim, S. H., and Speck, J. S. (2003). *J. Appl. Phys.*, **94**, 942.

Wu, F., Lin, Y. D., Chakraborty, A., Ohta, H., DenBaars, S. P., Nakamura, S., and Speck, J. S. (2010). *Appl. Phys. Lett.*, **96**(23), 231912.

Wu, Z., Fischer, A. M., Ponce, F. A., Bastek, B., Christen, J., Wernicke, T., Weyers, M., and Kneissl, M. (2008). *Appl. Phys. Lett.*, **92**, 171904.

Wu, Z. H., Tanikawa, T., Murase, T., Fang, Y. Y., Chen, C. Q., Honda, Y., Yamaguchi, M., Amano, H., and Sawaki, N. (2011). *Appl. Phys. Lett.*, **98**(5), 051902.

Yan, Q., Rinke, P., Scheffler, M., and Van de Walle, C. G. (2010). *Appl. Phys. Lett.*, **97**(18), 181102.

Yanashima, K., Nakajima, H., Tasai, K., Naganuma, K., Fuutagawa, N., Takiguchi, Y., Hamaguchi, T., Ikeda, M., Enya, Y., Takagi, S. *et al.* (2012). *Appl. Phys. Express*, **5**(8), 2103.

Yoshizumi, Y., Adachi, M., Enya, Y., Kyono, T., Tokuyama, S., Sumitomo, T., Akita, K., Ikegami, T., Ueno, M., Katayama, K. *et al.* (2009). *Appl. Phys. Express*, **2**(9), 2101.

Young, E. C., Gallinat, C. S., Romanov, A. E., Tyagi, A., Wu, F., and Speck, J. S. (2010). *Appl. Phys. Express*, **3**, 111002.

Young, E. C., Romanov, A. E., and Speck, J. S. (2011). *Appl. Phys. Express*, **4**(6), 061001.

Zhong, H., Tyagi, A., Fellows, N. N., Wu, F., Chung, R. B., Saito, M., Fujito, K., Speck, J. S., DenBaars, S. P., and Nakamura, S. (2007). *Appl. Phys. Lett.*, **90**, 233504.

Zhu, T., Johnston, C. F., Kappers, M. J., and Oliver, R. A. (2010). *J. Appl. Phys.*, **108**, 083521.

9 GaN-based single-nanowire devices

Rudeesun Songmuang and Eva Monroy

9.1 Introduction

The downscaling of semiconductor devices is a continuous trend in the microelectronics industry. Following this trend, semiconductor nanowires (NWs) become a powerful and diverse kind of functional nanomaterials for electronics, optoelectronics, and biotechnology. An attractive feature of these one-dimensional (1D) nanostructures is the fact that their chemical and physical characteristics (chemical composition, size, electronic or optical properties) can be rationally controlled during their synthesis in a predictable manner. Furthermore, NWs provide an opportunity to combine different materials, since their large surface-to-volume ratio allows misfit strain to be elastically released without dislocation formation, which extends the range of size and band-gap engineering beyond the limits of planar systems.

This chapter presents a summary of new device concepts incorporating GaN NWs as active media. The text is structured as follows:

- *Nanowire synthesis*: We introduce the growth processes to achieve controlled-by-design GaN-based NW heterostructures and discuss the resultant structural and optical properties.
- *Energy conversion*: We first present a comprehensive study of single GaN NWs operated as photoconductors—devices which are characterized by giant photocurrent gain and by a visible rejection of several orders of magnitude. The potential application of In-containing core-shell structures to photovoltaics is also discussed. Finally, we introduce the operation of NWs as piezoelectric transducers for energy harvesting.
- *Nanoelectronics*: Progress in the performance of single-NW field-effect transistors (FETs) is discussed. Then, the potential of band-gap engineering via NW heterostructures for nanoelectronic applications is demonstrated via their electrical characteristics. For instance, the two-dimensional electron gas (2DEG) formed in GaN/AlN/AlGaN NW radial heterostructures can be operated as a nanoscale high-mobility transistor. On the other hand, AlN axial-double-barriers inserted in GaN NW can show different transport regimes, opening the possibility to realize single electron transistors (SETs) or resonant tunneling diodes (RTDs), depending on the heterostructure design.
- *Sensorics*: Due to their chemical robustness and its large surface-to-volume ratio, the application of GaN NWs as chemical sensors has been proposed.

Functionalization by metal or oxide nanoparticles is applied to enhance the selectivity and sensitivity of their optical/electrical properties to the environment.

9.2 Nanowire synthesis

A number of growth techniques can be applied to spontaneously synthesize crystalline GaN NWs; for example, the reaction of Ga/Ga_2O_3 mixtures with flowing NH_3 in anodic alumina membranes (Cheng *et al.* 1999), the arc-discharge in N_2 atmosphere (Han *et al.* 2000) or the direct reaction of Ga and NH_3 in the presence of In power (Chen and Yeh 2000), etc. We can classify the most common procedures to synthesize GaN NWs into two standard approaches: catalytic-based vapor-liquid-solid (VLS) mechanisms, and catalyst-free growth.

In this section an overview of both approaches for synthesizing GaN NWs as well as their heterostructures is briefly presented. The electronic properties of GaN/Al(Ga)N nanostructures in NWs probed by optical characterization is also discussed.

9.2.1 *Catalyst-induced NW growth*

Catalyst-induced growth of semiconductor NWs is generally attributed to a VLS mechanism, which was first demonstrated by Wagner and Ellis (1964) in the Si material system. In this process, metal nanoclusters are heated up above the eutectic temperature, resulting in the formation of alloy liquid droplets. The metal/semiconductor phase diagram and/or knowledge of their chemical reactivity are key criteria for selecting the catalyst. While being fed to the growth system, the vapor atoms of the growing material become adsorbed at the droplet surface and are then transferred into the crystal phase via the supersaturation of the alloy liquid droplets. The NW formation occurs at the interface between the supersaturated droplets and the substrate because of the faster chemical reaction and nucleation of the crystal phase at that area. The NW diameter is determined by the size of the droplet, which is in turn controlled by the original size of the metallic clusters. Randomly produced catalytic nanoparticles lead to randomly positioned NWs with a broad range of diameters. The catalytic synthesis of crystalline GaN NWs can involve metals such as In (Chen *et al.* 2001), Co (Chen *et al.* 2001), Fe (Kuykendall *et al.* 2003; Duan 2000), Ni (Chen *et al.* 2001; Johnson *et al.* 2002; Lan *et al.* 2004), Ni/Co (Chen *et al.* 2001), Fe/Co (Chen *et al.* 2001), Au (Kuykendall *et al.* 2003; Liu *et al.* 2005; Tang *et al.* 2008), Au/Pd (Kuykendall *et al.* 2003; Nam *et al.* 2005) or Ta (Shi *et al.* 2010).

Kuykendall *et al.* (2003) applied metal-initiated metal–organic chemical vapor deposition (MOCVD) to synthesize VLS-GaN NWs. Tri-methylgallium (TMG) and NH_3 sources were used as Ga and N precursors. A thin film of Ni, Fe, or Au was thermally evaporated onto silicon or *c*-plane or *a*-plane sapphire. The obtained NWs were single crystals for all the used substrates and catalysts (scanning electron microscopy (SEM) and transmission electron microscopy (TEM)

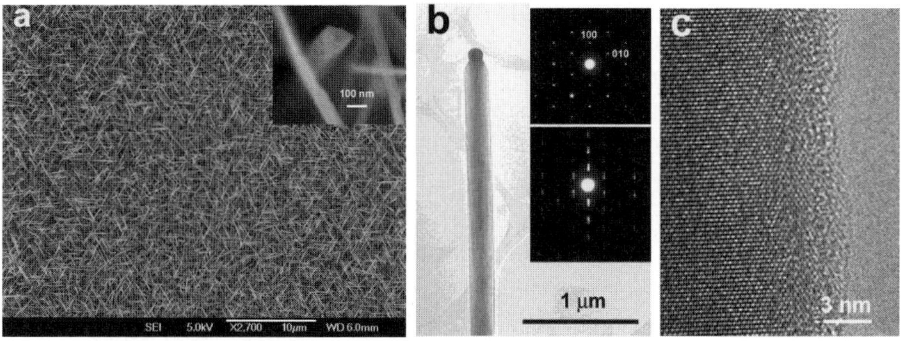

FIG. 9.1. (a) SEM image of the GaN NWs grown on a gold-coated c-plane sapphire substrate. The inset shows an NW with its triangular cross-section. (b) TEM image of a GaN NW with a gold metal-alloy droplet on its tip. The insets are electron diffraction patterns taken along the [001] zone axis. The lower inset is the same electron diffraction pattern, but purposely defocused to reveal the wire growth direction. (c) Lattice-resolved TEM image of the NW. (Reprinted, with permission, from Kuykendall *et al.* (2003), © 2003 by the American Chemical Society.)

images in Fig. 9.1(a–c)). For NWs grown on Fe-coated sapphire, the predominant NW growth direction was [110], while Ni-coated or Au-coated sapphire led to NW growth along the [210] direction, with a triangular cross-section enclosed by (112), (1–12), and (001) side planes (inset of Fig. 9.1(a)). GaN NWs growing along the [001] direction were rarely observed.

Although chemical vapor deposition (CVD) is a widely selected technique for the catalytic growth of III-N NWs, there were some efforts to synthesize VLS-GaN NWs by other growth techniques. There was an attempt to use an Ni catalyst to induce the formation of GaN NWs on Si substrates by plasma-assisted molecular beam epitaxy (PAMBE) (Chèze *et al.* 2010, 2010b). However, the metal catalyst strongly deteriorates the structural and optical properties in comparison to catalyst-free NWs grown by the same technique. GaN NWs can also be grown by hydride vapor phase epitaxy (HVPE), but the dissociation of Ni–Au catalyst during the synthesizing process can terminate the NW growth (Seryogin *et al.* 2005).

A key feature of semiconductor NWs is the possibility to adjust electronic properties by heterostructuring (Lauhon *et al.* 2002), and to modulate the incorporation of n-type and/or p-type dopants (Cui *et al.* 2000; Duan *et al.* 2001). In the case of GaN NWs, electrical studies show that unintentionally-doped NWs are intrinsically n-type (Huang *et al.* 2002; Kim *et al.* 2002). P-type doping has been demonstrated by introduction of Mg impurities during the CVD process of Ni-induced GaN NWs on c-plane sapphire (Zhong *et al.* 2003), using NH_3, Ga, and Mg_3N_2 as precursors. Such Mg-doped GaN NWs have wurtzite crystal structure and grow along the ⟨0001⟩ direction with an epitaxial relationship with c-plane sapphire. Evidence of active Mg dopants was obtained from electrical transport measurements.

Concerning bandstructure engineering in the NWs, heterostructures can be created either along (axial) or perpendicular (radial or core-shell) to the growth axis. Based on the nature of CVD or MOCVD growth, most of the heterostructures synthesized by these techniques are radial heterostructures. Figure 9.2 presents the schematic and scanning transmission electron

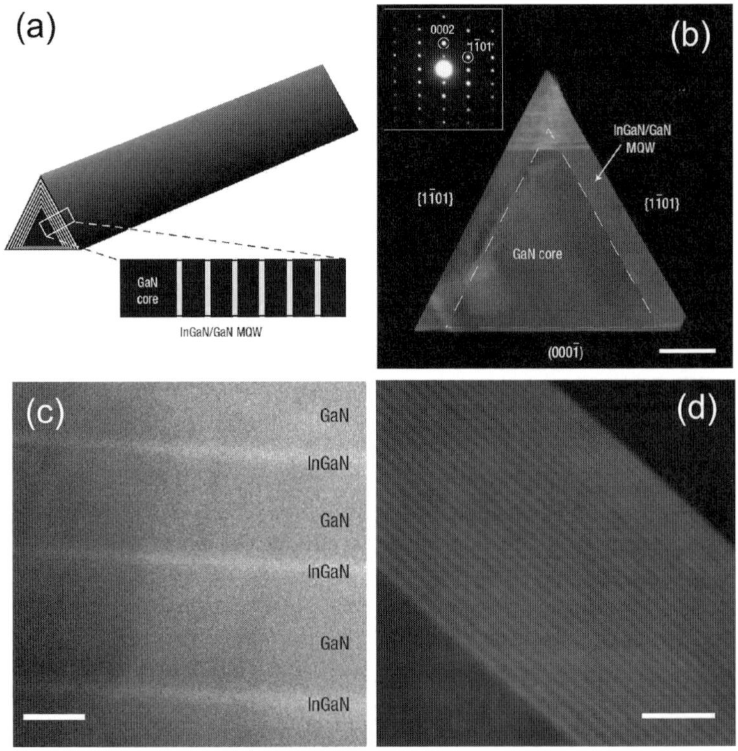

FIG. 9.2. InGaN MQW NWs: (a) Schematic diagram of an MQW NW and magnified cross-sectional view of a NW facet highlighting the InGaN/GaN MQWs. InGaN layers are indicated in yellow. (b) Dark-field cross-section STEM image recorded along the [11-20] zone axis of a 26 MQW NW structure. Dashed lines indicate core/shell interface. The scale bar is 100 nm. Inset: Corresponding electron diffraction pattern indexed for the [11-20] zone axis. Dark-field cross-section STEM images recorded along the [11-20] direction at the {1-101} facet of (c) 3 MQW and (d) 26 MQW NWs; scale bars are 20 nm. (Reprinted with permission from Macmillan Publishers Ltd, *Nature Materials*, Qian *et al.* (2008).)

microscopy (STEM) images of core-shell InGaN multiple-quantum-well (MQW) NWs grown by MOCVD on *r*-plane sapphire (Qian *et al.* 2004, 2008). The synthesis principle is the following. First, the NW axial growth is activated by the VLS mechanism using Ni catalyst, leading to GaN NWs growing along the [11-20] direction with a triangular cross-section. Then, the growth conditions must be altered to favor the radial deposition of InGaN MQW on the NW sidewall surface. The In composition is adjusted by varying the growth temperature. This growth concept has been applied to implement core/multi-shell NWs with an *n*-GaN core and $In_xGa_{1-x}N$/GaN/*p*-AlGaN/*p*-GaN shells, to be operated as nanoscale light-emitting diodes (Qian *et al.* 2005).

Instead of using external catalysts, one of the constituents of the NW material can serve as the VLS catalyst. Self-catalytic behavior was investigated *in situ* by real-time TEM while heating a GaN thin-film in a vacuum (Stach *et al.* 2003). GaN decomposition leads to the formation of isolated liquid Ga nanoparticles which behave as catalysts for NW growth. Self-catalytic NWs are also observed by direct reaction of Ga with NH_3 (He *et al.* 2001), direct evaporation of GaN powders (Zhou *et al.* 2003), or heating a mixture of GaN and In powders (Kim *et al.* 2010). An advantage of a self-catalytic process is that undesired contamination from foreign metal catalysts can be avoided.

9.2.2 *Catalyst-free NW growth*

Several growth techniques can spontaneously provide GaN NWs without a metal catalyst, which allows a straightforward adaptation of conventional doping and heterostructure growth approaches. Kim *et al.* (2002b) reported catalyst-free GaN NWs deposited on sapphire by low-temperature HVPE. The NW morphology depends on growth conditions such as growth temperature and time, as shown in Fig. 9.3. The potential of this technique is supported by the demonstration of *p*- and *n*-type doping, the growth of dislocation-free InGaN/GaN MQWs in GaN NWs, and the growth of InGaN alloyed NWs (Kim *et al.* 2003, 2004, 2004b). Later on, Kuykendall *et al.* (2007) applied low-temperature HVPE to tune the In composition of single-crystalline $In_xGa_{1-x}N$ NWs over the entire compositional range.

Alternatively, catalyst-free GaN NWs can also be generated via selective-growth MOCVD (Deb *et al.* 2005; Hersee *et al.* 2006; Bergbauer *et al.* 2010), as illustrated in Fig. 9.4(a). Those NWs grow along the $\langle 0001 \rangle$ wurtzite axis are enclosed by *m*-plane sidewalls. To prevent the NW coalescence, Hersee *et al.* (2006) applied alternate pulsed flow of TMG and NH_3 once the NWs emerged out of the SiN_x hole-mask. In contrast, Bergbauer *et al.* (2010) showed that it is possible to keep continuous flow conditions to realize GaN NWs on SiO_2-patterned *c*-plane sapphire. An important parameter is a well-adjusted H_2/N_2 carrier-gas mixture. Increasing the H_2 fraction promotes the NW geometry rather than the formation of pyramid-like structures (Bergbauer *et al.* 2010, 2011; Li *et al.* 2011). As an alternative to lithographically defined

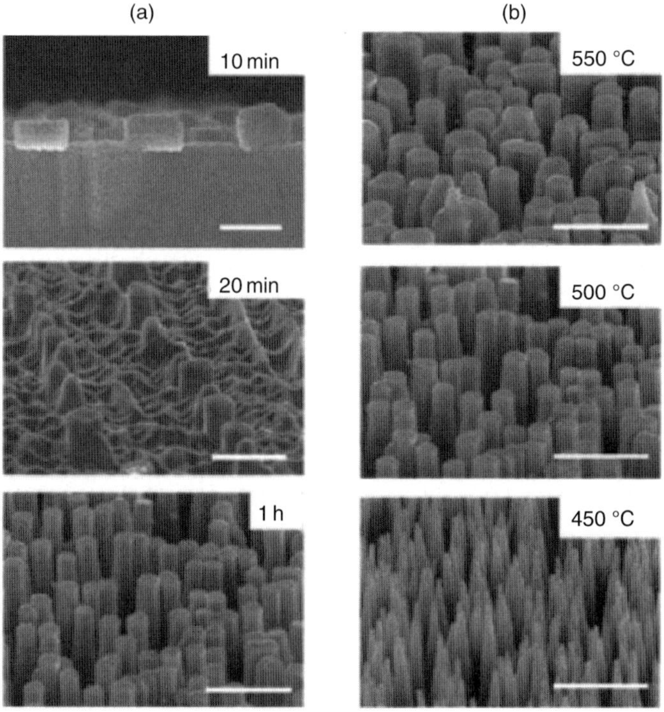

FIG. 9.3. SEM images of vertically aligned GaN NWs grown on *c*-sapphire: (a) As a function of time. Scale bars are 100 nm, 200 nm, and 500 nm for growth times 10 min, 20 min, and 60 min. (b) As a function of the substrate temperature. Scale bars are 500 nm. (Reprinted, with permission, from Kim *et al.* (2003), © 2003, Wiley-VHC Verlag GmBH & Co, KGaA, Weinheim.)

templates, the *in situ* deposition of a thin SiN_x layer on *c*-sapphire was found to promote the formation of GaN NWs (Koester *et al.* 2010). The NW morphology and structural quality are strongly affected by the duration of the SiN_x deposition and the NW nucleation time, while their vertical growth is enhanced by an additional SiH_4 flow. Finally, the crystal polarity of the substrate also plays a significant role on the morphology of the grown nanostructures: GaN NWs with vertical sidewalls present N polarity (Fig. 9.4(b)) whereas pyramidal-shaped structures are Ga-polar (Fig. 9.4(c)) (Chen *et al.* 2010; Li *et al.* 2011). The change of shape is attributed to the stability of the different crystallographic planes in the MOCVD atmosphere (Li *et al.* 2011).

Concerning band-gap engineering of catalyst-free MOCVD-grown *c*-GaN NWs, radial InGaN/GaN heterostructures have been fabricated on the *m*-plane sidewalls (Koester *et al.* 2011). The emission wavelength can be controlled by

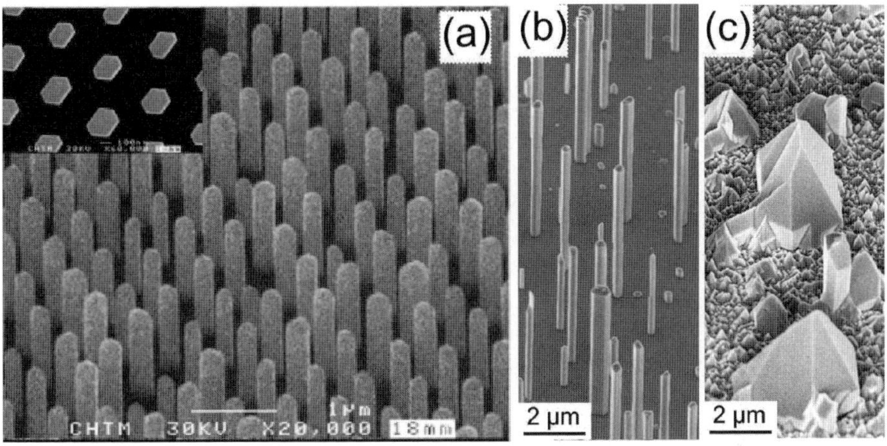

FIG. 9.4. Tilted SEM view of (a) a GaN NW array (inset shows plan view and reveals the hexagonal symmetry of the NWs). (Adapted with permission from Hersee *et al.* (2006), © 2006 by the American Chemical Society.) (b) self-assembled GaN NWs on N-polar GaN (c) GaN pyramids on Ga-polar GaN. (Reprinted with permission from Chen *et al.* (2010), © 2010 by the American Institute of Physics.)

varying the In content in the MQWs and the quantum-well (QW) thickness. The absence of piezoelectric field due to the non-polar heterostructure nature has been verified. However, a limitation of GaN catalyst-free NWs grown by MOCVD is their relatively large diameter, usually several hundreds of nanometers, which hinders efficient elastic strain relaxation in lattice-mismatched heterostructures. Recently, Chen *et al.* (2011) suggested a catalyst-free method to obtain by MOCVD nanoscale GaN wires exhibiting high optical quality. The approach is to grow GaN NWs on bare *c*-sapphire using very low precursor flow while keeping similarly low V/III ratio compared to the standard GaN NW growth, which favors a reaction-limited growth regime.

On the other hand, PAMBE is known as a standard way to achieve nanometer-size catalyst-free GaN NWs. The NWs are generally defect-free and grow along the *c*-axis of the wurtzite structure. The spontaneous growth of NWs was initially demonstrated on sapphire (Yoshizawa *et al.* 1997, 1998) and Si substrates (Sánchez-García *et al.* 1998). The GaN NW morphology can be controlled by technological parameters such as N/III ratio and substrate temperature. N-rich atmosphere was found to favor the NW geometry, while Ga excess promotes two-dimensional (2D) growth (Calleja *et al.* 2000; Ristic *et al.* 2008; Fernández-Garrido *et al.* 2009). Additionally, high growth temperature is found to be mandatory, since low-temperature growth activates the formation of

GaN 2D compact layers even under N-rich conditions (Songmuang *et al.* 2007; Fernández-Garrido *et al.* 2009).

The NW shape is very sensitive to the nucleation process. GaN NWs grown directly on Si(111) substrates are usually not well oriented perpendicular to the substrate surface, possibly due to the formation of thin SiN_x at the early stage of the deposition (Grandal *et al.* 2007). The vertical alignment can be drastically improved by inserting a thin AlN layer prior to the NW growth (Songmuang *et al.* 2007). The thickness and the morphology of AlN buffer layers significantly influence GaN NW morphology (Bertness *et al.* 2007; Sekiguchi *et al.* 2007; Ristic *et al.* 2008; Brubaker *et al.* 2011).

During the NW nucleation stage, it was suggested that the deposited GaN initially forms three-dimensional (3D) GaN islands, driven by the strain accommodated inside the film (Songmuang *et al.* 2007; Ristic *et al.* 2008; Landré *et al.* 2009; Consonni *et al.* 2010), and later transform into the wire-like structure. Consonni *et al.* (2010) attributed the shape evolution to a drastic change in the total free energy, mainly caused by the anisotropy of surface energy. After this first stage, GaN NW growth proceeds by two contributions: namely, the direct material deposition on the NW top and the Ga diffusion along the wire sidewalls up to their apex (Debnath *et al.* 2007; Calarco *et al.* 2007; Songmuang *et al.* 2007; Ristic *et al.* 2008). The marker technique based on NW heterostructures showed that the NW axial growth rate is limited by the supplied N flux, revealing that the contribution of N adatom diffusion is negligible. In addition, a significant contribution of Ga diffusion shifts the Ga-limited and N-limited regimes of GaN NWs growth from those in 2D GaN layers (Songmuang *et al.* 2010). In contrast, the contribution of Al atom diffusion to the NW growth was found to be insignificant due to the small diffusion length of Al at the NW growth temperature.

Similar to the case of GaN NWs grown by MOCVD, the polarity of PAMBE-grown GaN NWs is under debate. Cherns *et al.* (2008) and Foxon *et al.* (2009) reported that the growth of GaN on 5-nm thick AlN on *c*-sapphire under N-rich conditions gave defect-free Ga-polar NWs emanating from a compact, highly defective N-polar layer, indicating the polarity-dependent growth rate of the NWs. They claimed that the preferential growth of Ga-polar NWs is consistent with the faster growth of Ga-polar inversion domains in a surrounding N-polar matrix during the growth of continuous GaN layers by MBE under N-rich (Romano and Myers 1997). Consistently, Brubaker *et al.* (2011) reported that although the AlN buffer layer shows a mixed crystallographic polarity indicated by piezoresponse force microscopy and polarity-sensitive etching, convergent beam electron diffraction (CBED) indicates that GaN NWs are Ga-polar. In contrast, Hestroffer *et al.* (2011) used resonant X-ray diffraction and CBED to show that GaN NWs grown on Si substrate were N-polar, independent of the presence or otherwise of a thin AlN buffer layer. This observed N-polarity agrees with findings using aberration-corrected STEM with HAADF imaging and the newly developed annular bright-field imaging (De la Mata *et al.* 2012; den Hertog *et al.* 2012).

In order to realize NW-based devices, the control of NW doping is essential. It was found that unintentionally doped GaN NWs are n-type, with a carrier concentration that depends on the background doping in the growth chamber. Adding Si impurities during the NWs growth can further increase the electron concentration; however, it also modifies the NW morphology. Furtmayr *et al.* (2008) suggested that the presence of Si enhances the nucleation on the wire sidewall, resulting in an increased NW diameter. The presence of Mg impurities also leads to a continuous increase of the radial growth rate with a simultaneous reduced axial growth. Mg incorporation to attain p-type conductivity requires a decrease of the growth temperature, which drastically increases the NW lateral growth. Despite the morphology modification induced by adding impurities during the growth, light-emitting devices based on p-n junctions in NWs were demonstrated (Kikuchi *et al.* 2004).

Unlike MOCVD, using PAMBE it is more difficult to synthesize core-shell heterostructures than axial heterostructures. This fact is attributed to the high NW density in the range of $10^{10}\,\mathrm{cm}^{-3}$ and to the directionality of the impinging molecular beams, which results in shadowing effects and non-uniformity of the shell. Nevertheless, Carnevale *et al.* (2011) showed that axial or radial GaN/AlN heterostructures can be achieved by adjusting separately the NW nucleation and the adatom mobility during the NW growth. Once the nucleation process is concluded, high growth temperature activates the axial heterostructure formation (Fig. 9.5(a)), while low growth temperature reduces adatom mobility and thus promotes the formation of coaxial GaN/AlN superlattices (Fig. 9.5(b)).

Pioneer work on axial AlGaN/GaN heterostructures in GaN NWs were demonstrated by Ristic *et al.* (2003, 2005, 2005b). However, Al(Ga)N sections in GaN NWs present a significant lateral growth and induce a shell around the base of the GaN NWs, because of the low Al adatom diffusion length (Carlaco *et al.* 2007; Songmuang *et al.* 2010). Furthermore, the different kinetics of the atomic species leads to a non-uniform chemical distribution in ternary insertions, whether they are AlGaN (Allah *et al.* 2012) or InGaN (Tourbot *et al.* 2010; Chang *et al.* 2010). In the case of AlGaN barriers, it was proposed that the lateral growth-rate depends on the material composition (Furtmayr *et al.* 2011). Besides, non-uniform material distribution along the radius of the AlGaN insertions was observed, suggesting a spontaneous formation of Al-rich shell around Al-poor core in AlGaN barriers (Songmuang *et al.* 2011; Allah *et al.* 2012), which creates a complex strain distribution in the nanoinsertions.

The electronic properties of axial GaN/Al(Ga)N nanodisks embedded in GaN NWs have been deeply investigated by optical studies. The photoluminescence (PL) of polar GaN/Al(Ga)N nanodisks in NWs showed signatures of both quantum confinement and polarization-induced Stark effect (Ristic *et al.* 2003, 2005b; Renard *et al.* 2009; Furtmayr *et al.* 2011; Songmuang *et al.* 2011). However, the Stark effect in NW heterostructures is less pronounced than in QWs due to the particular strain distribution in the nanodisk structures, which

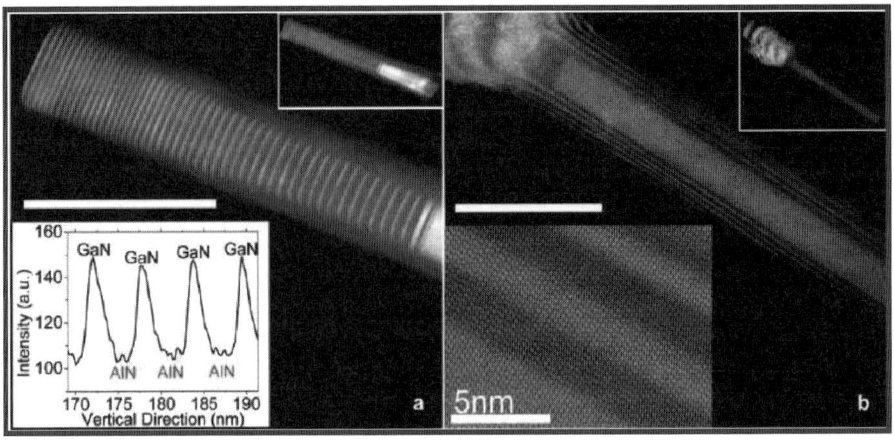

FIG. 9.5. STEM images of 3D GaN/AlN NW heterostructures. (a) Fifty-period vertically-aligned AlN/GaN NW superlattice. Top inset: Zoomed-out image of the NW. Bottom inset: Example of a profile intensity scan measuring compositional modulation along the NW *c*-axis. (b) Five-period coaxially aligned AlN/GaN NW superlattice. Top inset: Zoomed-out image of the NW. Bottom inset: Section of coaxial layers imaged with atomic resolution. Dark (light) areas correspond to AlN (GaN) due to Z-contrast. All scale-bars correspond to 100 nm unless labeled otherwise. (Reprinted with permission from Carnevale *et al.* (2011), © 2011 by the American Philosophical Society.)

can involve elastic relaxation by the sidewalls and compression by the Al-rich NW shell.

NWs are also an attractive medium to access the inherent properties of single nano-objects. Renard *et al.* (2008) used micro-PL to probe single GaN/AlN nanodisks embedded in single GaN NWs, showing a PL line width in the range of 2 meV which is attributed to excitonic transition, as illustrated in Fig. 9.6(a–b). Photon-correlation measurements provided a proof that the transition stems from a single discrete level, confirming the quantum dot-like behavior of the disk (Renard *et al.* 2009). Rigutti *et al.* (2010) compared the photoluminescence spectrum from individual GaN NWs containing single and multiple $Al_xGa_{1-x}N$/GaN quantum discs. A single-line spectrum was found from single nanodisks, while the multiple nanodisks showed multiple peaks with total spectral broadening that depends on the Al content in the barrier. Correlating with TEM images of the same NWs, the spectral dispersion was attributed to the strain relaxation in the nanodisk, which is strongly affected by the Al-rich shell, and to monolayer fluctuations of the nanodisk thickness. The same conclusion was found in the spatially resolved cathodoluminescence of AlN/GaN NW heterostructures (Zagonel *et al.* 2011).

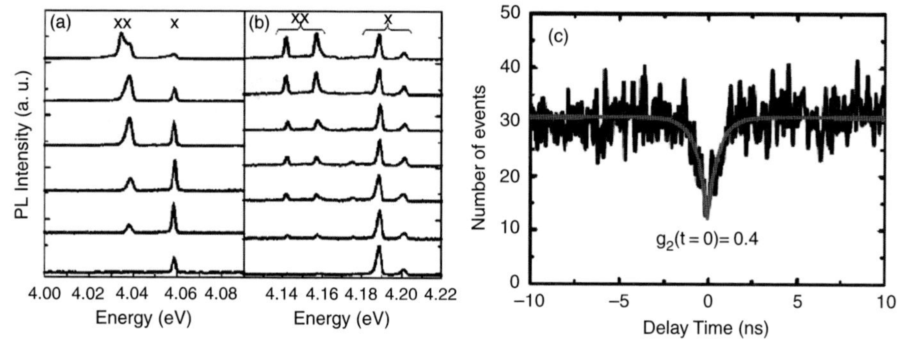

FIG. 9.6. Power-dependent spectra of nano-disk emission at 5 K. Each spectrum has been normalized to its maximum intensity. Excitation power (increasing from the bottom) = 10, 50, 100, 200, 300, and 500 μW for (a) and 25, 100, 120, 170, 200, 250, and 300 μ*W* for (b). (Adapted, with permission, from Renard *et al.* (2008), © 2008 by the American Chemical Society.) (c) Hanbury-Brown and Twiss coincidence histogram for the PL at 4 K of a single GaN/AlN nanodisk in a NW. The continuous line is a fit of the experimental data that allows extracting $g_2(t = 0)$. (Reprinted, with permission, from Renard *et al.* (2009), © 2009 by the American Physical Society.)

9.3 Energy conversion

In this section, we first present a comprehensive study of single GaN NWs operated as photoconductors. Then, the application of In-containing NWs structures to photovoltaics is discussed; finally, we analyze the performance of GaN and InN NWs as piezoelectric converters for energy harvesting.

9.3.1 *Photoconductive detection*

The photodetector capabilities of single NWs are under study for a variety of materials, such as Si (Zhang *et al.* 2008; Alvarez *et al.* 2011), Ge (Kim 2010b), GaAs (Schricker *et al.* 2006); ZnO (Soci *et al.* 2007; Chen *et al.* 2010b; Peng *et al.* 2011), ZnTe (Cao *et al.* 2011), CdS (Gu and Lauhon 2006), or SnO_2 (Hu *et al.* 2011). Focusing on III-nitrides, a number of publications address the behavior of single GaN NWs as photoconductors (Calarco *et al.* 2005; Polenta *et al.* 2008; Richter *et al.* 2008; Chen *et al.* 2007, 2008; Sanford *et al.* 2010; Rigutti *et al.* 2010b; De Luna Bugallo *et al.* 2011; González-Posada *et al.* 2012, 2012b).

For a correct interpretation of the NW performance, structural parameters such as growth axis, crystallographic orientation of the sidewalls, NW diameter, and doping level are particularly relevant. Indeed, the angle between the polarization vector and the NW growth axis determines the internal electric field in heterostructured NWs, and the distribution of surface states and Fermi-level pinning at the sidewall facets has an impact on the free carrier distribution in the NWs, due to their large surface-to-volume ratio. In the case of GaN NWs, surface states induce an upwards band bending at the surface that can result in a total depletion of the NW. Applying the abrupt depletion approximation to a cylindrical structure (Luscombe and Frenzen 2002; Dobrokhotov *et al.* 2006; Sanford *et al.* 2010), total depletion occurs for an NW radius

$$R_c = \sqrt{\frac{4\epsilon\epsilon_0\Phi}{e^2 N_d}}$$

where $\epsilon\epsilon_0$ are the dielectric constant and the permitivity of vacuum, Φ is the surface potential (conduction band edge—Fermi level), e is the electron charge, and N_d is the doping level.

In this section, and unless indicated, we focus on (000-1)-oriented NWs with {10-10} m-plane sidewall facets, which is the most common crystallographic configuration for self-assembled NWs grown by PAMBE or MOCVD. For the fabrication of single-NW devices, NWs are generally detached from their original substrate and dispersed on SiO_2/Si or SiN_x plates, and contacted by e-beam lithography.

Dark current. Figure 9.7 compares typical current–voltage (I–V) curves of single n-i-n and undoped (000-1)-oriented NWs with a diameter of ≈ 50 nm,

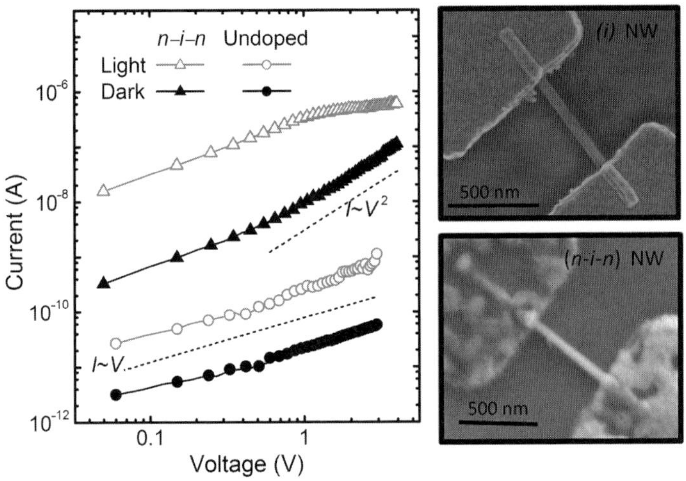

FIG. 9.7. Left: $I-V$ characteristics at room temperature of single n-i-n and undoped NWs measured at room temperature in the dark and under UV illumination. Right: Top-view SEM image of an n-i-n (top) and undoped (bottom) contacted single NW. Black symbols correspond to the dark current; grey symbols correspond to illumination with $\approx 0.2\ \mathrm{W/cm^2}$ of UV light ($\lambda = 244$ nm). (Reprinted from González-Posada *et al.* (2012b).)

measured in the dark (in the air and in vacuum) and under ultraviolet (UV) illumination (in the air). *N-i-n* structures consist of two Si-doped *n*-type edges and an undoped middle section with a nominal length of 400 nm. Focusing on the characteristics in the dark, we observe a strong dependence of the dark current on the measurement atmosphere. Measurements in the air render current levels one order of magnitude lower than in vacuum (Bertness *et al.* 2011; González-Posada *et al.* 2012b), in agreement with results in CdS NWs (Gu and Lauhon 2006). In the dark, *n-i-n* NWs present the $I \propto V^{\alpha}$ ($\alpha \geq 2$) behavior characteristic of space-charge-limited transport, as expected due to the surface-induced depletion of the non-intentionally-doped sections.

Space-charge-limited current was first studied in insulators, where it is the dominant mechanism when the charge injected by the electrodes exceeds the free carrier density in the intrinsic material. In ideal insulators the $I - V$ curve presents a characteristic $I \propto V^2$ dependence (Mott–Gurney law) (Rose 1955; Lampert 1956). However, in low-dimensional materials such as NWs, the presence of surface states leads to $I \propto V^{\alpha}$ with $\alpha > 2$ (Gu and Lauhon 2006). In the literature, space-charge-limited current is observed for undoped GaN NWs with diameters below 100 nm (Calarco *et al.* 2005, 2011; Sanford *et al.* 2010; González-Posada *et al.* 2012). Space-charge-limited current has also been reported as the dominant transport mechanism in undoped CdS (Gu and Lauhon 2006), GaAs (Schricker *et al.* 2006), or ZnO (Soci *et al.* 2010) NWs.

On the other hand, undoped NWs in Fig. 9.7 present current densities at least two orders of magnitude lower than *n-i-n* devices. The approximately linear $I{-}V$ characteristic is explained by the trap-filling process characteristic of insulators prior to the application of the voltage (Lampert 1956). Under illumination, undoped NWs display larger on/off current contrast in comparison to *n-i-n* NWs. However, the overall photocurrent level in undoped NWs is smaller by at least one order of magnitude.

Light absorption and polarization. In general, the coupling of the light into the NW can be confirmed by analyzing the photoresponse as a function of the polarization angle, θ, when exciting with linearly polarized light. The absorption properties of semiconducting NWs are strongly dependent on the polarization of the incident radiation. The two major mechanisms responsible for this phenomenon are (i) the modification of energy distribution by size quantization (Peter and Kerry 1990; Xinyuan *et al.* 2004), and (ii) the dielectric confinement of the optical electric field due to the difference in the dielectric constants of the NW, ϵ_{NW}, and the environment, ϵ_{env} (Maslow *et al.* 2006; Ruda and Shik 2005; González-Posada *et al.* 2012). While the former mechanism is significant only in very thin NWs (diameter < 10 nm), the relevance of the latter is dictated by the $\epsilon_{NW}/\epsilon_{env}$ ratio, though it has to be treated differently if the NW diameter is much smaller than the light wavelength. In the case of thin NWs (diameter < 100 nm), due to the suppression of the perpendicular component of the electric field vector inside the NW, the ratio of the absorption coefficient for light polarization parallel and perpendicular to the NW axis is given by (Ruda and Shik 2005)

$$\frac{k_{//}}{k_{\perp}} = \left| \frac{\epsilon_{NW} + \epsilon_{env}}{2\,\epsilon_{env}} \right|^2$$

This theoretical ratio is about 30 for GaN. For thicker NWs the non-uniform distribution of the field inside the NW must also be taken into account, and the $k_{//}/k_{\perp}$ ratio becomes strongly dependent on the excitation wavelength.

The inset in Fig. 9.8 presents the photocurrent from a single NW exposed to linearly polarized UV light (González-Posada *et al.* 2012). The anisotropy of the optical absorption results in the photocurrent varying as $I = I_0\cos(2\theta)$, where θ is the light polarization angle with respect to the NW axis (Maslow *et al.* 2006).

Linearity, gain, and spectral response. The quantitative analysis of the photoresponse requires an assessment of the linearity of the device with the impinging optical power, P. Figure 9.8 presents the results of the characterization performed under continuous-wave (CW) illumination, as well as at different light-chopping frequencies. The photocurrent scales sublinearly with P following approximately a power law $I \propto P^{\beta}$ with $\beta < 1$, though both the photocurrent magnitude and the value of β depend on the frequency. A sublinear response with the optical power has been reported in NWs based on other materials, such as Si (Zhang

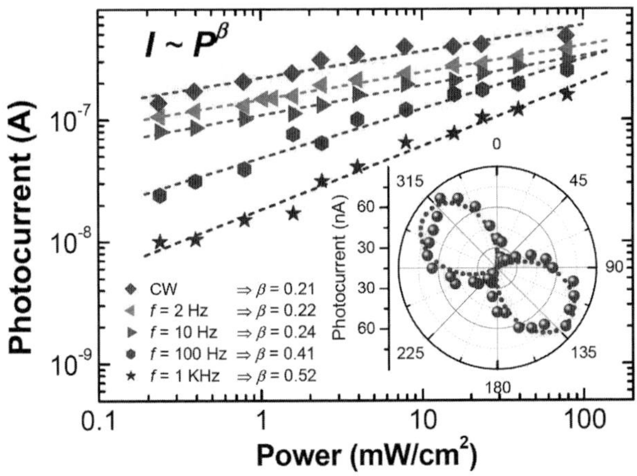

FIG. 9.8. Photocurrent variation from a single NW as a function of the excitation power ($\lambda = 244$ nm) and frequency. Inset: Photocurrent dependence with the polarization angle of a linearly polarized white-light source and simulated angle dependence for a NW, following the $I \propto \cos(2\theta)$ proportionality. (Reprinted with permission from González-Posada *et al.* (2012), © 2012 American Chemical Society.)

et al. 2008), Ge (Kim *et al.* 2010b), ZnO (Soci *et al.* 2007; Chen *et al.* 2010b), SnO_2 (Hu *et al.* 2011), or ZnTe (Cao *et al.* 2011). This strong non-linearity, also comparable to observations in GaN 2D photoconductors (Monroy *et al.* 2003), is generally associated with high gain, persistent photoconductivity effects, and a strong photoresponse to excitation below the GaN band gap. The photodetector gain, G, defined as the number of electrons detected per absorbed photon, can be estimated from the photocurrent, I, via the equation:

$$G = \frac{I}{P}\frac{h\,c}{\lambda\,e}$$

where h is Planck's constant, c is the speed of light, and λ is the excitation wavelength. Note that this equation provides an underestimation of G, since it assumes the detector internal quantum efficiency to be unity, i.e. the light reflected or transmitted through the NW is neglected. Taking the NW size into account, G reaches values of $10^6 - 10^7$ for an irradiance of 0.2 mW/cm^2.

The magnitude of G can vary as a function of the measuring atmosphere and the doping profile in the NW. Whereas for *n-i-n* NWs gain measurements in vacuum and in the air are in the same range ($\pm 5\%$), for undoped NWs the gain measured in vacuum can be more than one order of magnitude higher than in the air. Despite the enhancement of the photocurrent in vacuum, its sublinear behavior with the optical power remains unchanged, i.e. β remains constant within an error bar $\pm 5\%$. A similar enhancement of the photoresponse in vacuum has been reported in ZnO NWs (Soci *et al.* 2007).

Figure 9.9 presents the spectral photoresponse of a single GaN NW (*n-i-n* or undoped, samples grown by PAMBE) measured by the lock-in technique at various frequencies. The photocurrent spectra present a flat spectral response for wavelengths above the GaN band gap ($\lambda < 350$ nm), while the response to $\lambda > 450$ nm is below the resolution limit of the system (González-Posada *et al.* 2012, 2012b). Similar results are obtained in the case of MOCVD-grown GaN NWs (De Luna Bugallo *et al.* 2011). In order to verify the NW blindness to visible light, they were exposed to 0.5 W/cm^2 of the 488 nm line of an Ar laser. The NWs show no sensitivity to this illumination, neither under CW illumination nor at different chopping frequencies (2–100 Hz), which confirms a UV (350-nm)/visible (488 nm) contrast of more than six orders of magnitude. This result sets a critical difference in performance with GaN 2D photoconductors, whose visible rejection ratio decreases markedly when decreasing the measuring frequency, to the point of being lower than one decade for CW measurements (Monroy *et al.* 2003). This poor visible rejection of GaN 2D photoconductors, orders-of-magnitude worse than expected from the spectral variation of the GaN absorption coefficient, is explained by the fact that the photocurrent gain is associated with charge separation at extended defects. In the case of GaN NWs, the huge UV/visible contrast is understood as a result of the absence of extended defects, except for the surface dangling bonds. Furthermore, the NW sidewalls are {10-10} *m*-planes, whose dangling bonds do not create occupied states within

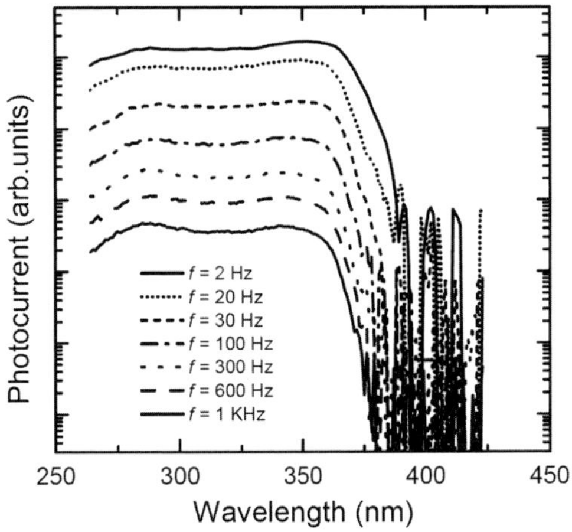

FIG. 9.9. Spectral response of a single NW under 3 V bias at a measurement frequency of 2–1000 Hz. Data are corrected by the lamp emission spectrum, taking the sublinear power dependence of the NW into account. (Reprinted, with permission, from González-Posada *et al.* (2012), © 2012, American Chemical Society.)

the band gap (Van de Walle and Segev 2007). The cutoff wavelength can be shifted into the visible spectral range by incorporating InGaN insertions in the NW structure (De Luna Bugallo *et al.* 2011).

Time response. *N-i-n* NWs present non-exponential dynamics with an initial decay constant around 4–10 ms independent of the measuring environment (González-Posada *et al.* 2012). In the case of undoped NWs, their behavior is similar to *n-i-n* devices when operated in the air, but persistent photoconductivity effects in the range of tens to hundreds of seconds are activated under vacuum (González-Posada *et al.* 2012b).

Photodetection mechanism. Under illumination, we observe a variation of the NW conductivity $\Delta\sigma$. Keeping in mind that the conductivity is given by $\sigma = en\mu$, where n is the carrier density and μ is the carrier mobility, a change in conductivity, $\Delta\sigma = e(\mu\Delta n + n\Delta\mu)$, can occur either due to change in the carrier concentration, Δn, or to a change in the carrier mobility, $\Delta\mu$. The carrier concentration should scale linearly with the excitation, in contrast with the observed non-linear behavior of the photocurrent. Moreover, the carrier lifetime measured by PL in the *ns* range (Pfüller *et al.* 2010) is in contradiction with the photocurrent decay times described above (millisecond times for *n-i-n* NWs and significantly longer for undoped NWs). We can hence conclude that the photodetector response is dominated by $\Delta\mu$, which is affected by the band bending induced by the surface and by the scattering associated with surface states (González-Posada *et al.* 2012b). Thus, the non-linear photocurrent gain in NWs is assigned to hole-trapping at surface states (Calarco *et al.* 2005; Soci *et al.* 2010), which results in a spatial separation of the electron and hole. When light is switched off, photogenerated carriers recombine at a non-exponential rate, due to the time-dependent potential barrier associated with the surface band bending. The current recovery involves also a rearrangement of the surface charge, which is at the origin of the slow photocurrent components. The recombination process is accelerated in presence of adsorbed oxygen, which decreases the carrier lifetime (Reshchikov *et al.* 2009; Foussekis *et al.* 2009; Pfüller *et al.* 2010).

The different response of n-i-n NWs and undoped NWs is assigned to the different location and behavior of the Fermi level at the surface. Surface states play a major role on the undoped NW photocurrent dynamics due to the unpinned Fermi level at the clean m-plane surfaces (Carterm and Stampflm 2009; Bertelli *et al.* 2009). In the case of n-i-n NWs, the pinning of the Fermi level close to the conduction band at the n-regions and the reduction of the surface potential due to residual silicon on the sidewalls reduce the photoinduced sweep of the Fermi level, preventing persistent effects and reducing the environment sensitivity. Therefore, an important projection of the present work is that the doping profile of the NW is a parameter critical for determining not only its performance as a photodetector, but also its functionalization capabilities for use as a chemical sensor.

9.3.2 *Photovoltaics*

Strategies on photovoltaic research today pursue two major targets: reduction of production costs and increase of conversion efficiency (Luque 2011). The former aim motivates efforts on low-cost materials such as amorphous Si, polycrystalline Si, CuInSe, CuInGaSe, etc. However, the maximum efficiency achieved with these materials is below 15% because of their high impurity/defect densities, which reduce the minority carrier diffusion length. On the other hand, efforts to improve the efficiency are oriented towards multi-junction architectures using III–V compounds (GaInP/GaInAs/Ge or GaInP/$Ga_xIn_{1-x}As/Ga_yIn_{1-y}As$). These devices attain 30% conversion efficiency for commercial modules and hold the record efficiency of 44% for an GaInNAs solar cell under concentrated light (announced by Solar Junction in October 2012). However, III–V tandem solar cells are based on Ge or GaAs substrates, both of which are expensive and limited in size.

The consideration of InGaN as a potential alternative for multi-junction devices stems from its direct band gap spanning from 0.65 eV to 3.42 eV. An additional interest of the InGaN alloy is the naturally obtained alignment between the conduction band of n-type $In_{0.46}Ga_{0.54}N$ and the valence band of p-type Si, which might be an advantage for the implementation of InGaN-Si tandem devices (Hsu and Walukiewicz 2008). Current results on 2D InGaN solar cells render values of open circuit voltage, V_{OC}, in the 2–2.4 eV range with a fill factor, FF, up to 80%, though the solar conversion efficiency remains poor (record at 3.4% under AM1.5G at 200 suns and room temperature, Yamamoto *et al.* 2012). The low efficiency is linked to the growth challenges of In-rich InGaN layers, which have a tendency to phase separation into In-rich and Ga-rich clusters. In addition, *p*-doping of In-rich InGaN must overcome the strong propensity of InGaN to be unintentionally *n*-type doped. Currently, this problem is circumvented either using *p*-type GaN (Farrell *et al.* 2011; Yamamoto *et al.* 2012) or an In-graded InGaN *p*-layer (Yamamoto *et al.* 2012b)—approaches that are confronted with the generation of defects associated with the large lattice mismatch between InN and GaN ($\approx 11\%$).

The NW geometry appears as a promising substitute or complement to the 2D technology thanks to the efficient strain relaxation that is provided by the free lateral surface. Furthermore, semiconductor NWs allow the independent optimization of light absorption (light trapping in the NW layer) and carrier collection, while reducing the active material volume (Tian *et al.* 2007). First realizations of NW solar cells based on dye-sensitized ZnO or TiO_2 NWs appeared in 2002, and they were followed by intense research on Si NWs (Tian *et al.* 2007; Kempa *et al.* 2008; Garnett and Yang 2008) and the first demonstrations of III–V NW solar cells (Colombo *et al.* 2009; Czaban *et al.* 2009; Goto *et al.* 2009). Regarding III-nitride materials, a study of single III-N NWs operated as photovoltaic devices has been reported by Dong *et al.* (2009) using InGaN/GaN core/shell NWs with triangular cross-section (Ni catalyst, MOCVD growth). Single-NW devices yielded an open-circuit voltage $V_{OC} = 1{-}2\,\mathrm{V}$, a short-circuit current density $J_{SC} = 390 - 59\,\mu\mathrm{A/cm}^2$, and a best fill factor FF $= 56\%$. Their conversion efficiency remains $\approx 0.19\%$ under AM1.5G conditions due to the relatively low In content. Devices incorporating GaN NW ensembles on silicon substrates have been demonstrated following two approaches: GaN *p-i-n* junction within the NW using silicon as a support (catalyst-free PAMBE growth) (De Luna Bugallo *et al.* 2010), and hybrid *n*-silicon/*p*-GaN NW device (Au catalyst, CVD growth) (Tang *et al.* 2008b). The former device can be considered as a validation of the technological approach, although the devices are selective UV detectors with a peak zero-bias responsivity of 90 mA/W at 356 nm wavelength. In the case of hybrid structures, the devices profit from the broadband photoresponse of the silicon substrate. Cells with $V_{OC} \approx 0.95\,V$, $J_{SC} \approx 7.6\,\mathrm{mA/cm}^2$ and FF $= 38\%$ were demonstrated, reaching a conversion efficiency of $\approx 2.73\%$ under AM1.5G conditions.

9.3.3 *Energy harvesting via piezoelectric effects*

The capability of piezoelectric materials to convert mechanical energy into electrical energy can be exploited for the fabrication of nanogenerators or self-powering nano-devices (Zhang *et al.* 2011). Piezoelectric NWs are particularly suitable as electromechanical transducers because of their high length-to-diameter ratio. The low maturity of the field results in numerous technological choices being investigated in terms of material (PZT, ZnO, CdS, GaN, or InN), NW shape (cylindrical, hexagonal, conical), and NW dimensions (length between 1 μm and a few tens of microns, diameters between a few tens of nm and a few microns). Recently, rising attention focuses on wurtzite materials, namely ZnO,

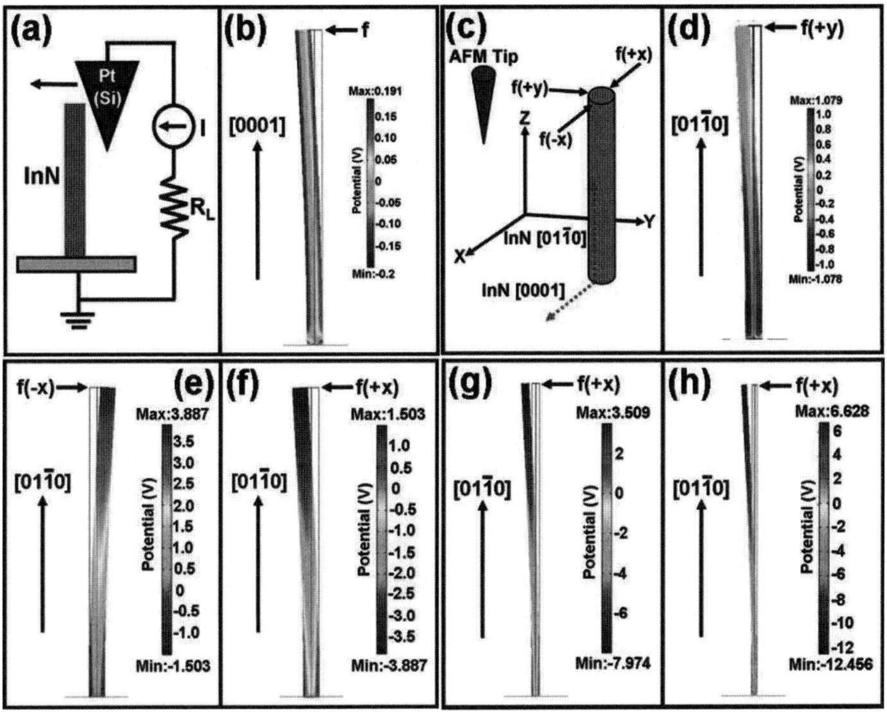

FIG. 9.10. (a) Schematic of AFM measurement system. (b) Calculated piezopotential distribution at the cross-section of [0001] growth InN NW with length = 1 μm and diameter = 50 nm. (c) Coordinate system used to define the direction of lateral force applied from the AFM tip. (d–f) Calculated piezopotential distributions at the cross section of a [01-10]-oriented InN NW under lateral force of 80 nN from three different directions. The dimensions of InN NW are length = 1 μm and diameter = 50 nm. (g,h) Calculated piezopotential distributions at the cross-section of [01-10]-oriented InN NW with different length under a lateral force of 80 nN. (g) length = 2 μm and diameter = 50 nm. (h) length = 3 μm and diameter = 50 nm. (Reprinted with permission from Huang *et al.* (2010), © 2010 Wiley-VHC Verlag GmBH & Co. KGaA, Weinheim.)

GaN, InN or CdS (Wang and Song 2006; Wang *et al.* 2007; Su *et al.* 2007; Lin *et al.* 2008; Lu *et al.* 2009; Huang *et al.* 2010, 2010b; Xu *et al.* 2011; Zhang *et al.* 2011), due to their coupling of semiconducting and piezoelectric properties, which opens prospects for on-chip integration of energy-conversion structure and signal-processing electronics or additional sensing capabilities.

A number of piezoelectric measurements on III-nitride NWs have been made. Huang *et al.* (2010) bent GaN NWs with an atomic-force microscope (AFM), using a conductive tip to measure the induced voltages. The samples under study consisted of three-fold symmetrically distributed GaN NW arrays which were epitaxially grown on GaN-on-sapphire templates by MOCVD using Ni as catalyst. These GaN NWs present a triangular cross-section enclosed by (000-1), (2-1-12) and (-2112) planes, and the angle between the NWs and the substrate surface is about 62°. Scanning the tip across the GaN NW array produces an average piezoelectric output voltage of -20 mV on a 500-$M\Omega$ load resistance.

Similar experiments have been performed on ≈ 1 μm long individual (000-1)-oriented GaN NWs synthesized catalyst-free by PAMBE on Si(111) wafers. Su *et al.* (2007) recorded current peaks of ≈ 0.03 nA when bending a single NW, whereas Xu *et al.* (2011) measured an electric potential of 150 mV for intrinsic GaN NWs, and as high as 200 mV for n-doped GaN/AlN (8 nm)/GaN heterostructured NWs.

InN has also been considered as a potential material for nanogenerators. With this purpose, Huang *et al.* (2010b) characterized the piezoelectric response of InN NWs synthesized by a VLS method using Au nanoparticles as catalyst and ammonia and pure In as precursors, leading to a high density of (1-100)-oriented NWs. Measurements on 1–3 μm long InN NWs with a diameter of 25–100 nm render an average output voltage of -120 mV, with some peak output voltages reaching -1000 mV. The large signal is explained by the non-polar growth axis, which should theoretically provide a response twenty times larger than the one expected in NWs with identical geometry grown along [0001] (see Fig. 9.10).

9.4 Nanoelectronics

9.4.1 *GaN NW field-effect transistors*

Semiconductor NWs have the potential to become key building blocks for a variety of electronic devices, including FETs. NW-FETs are interesting both as a vehicle for investigating basic carrier-transport behavior and as a promising element for high-performance electronics. NW-FETs have been fabricated from various semiconductors (Cui *et al.* 2003; Wang *et al.* 2003; Zheng *et al.* 2004; Gretak *et al.* 2004; Xiang *et al.* 2006; Dayeh *et al.* 2007) and with different gate geometries, such as back-gate (Fig. 9.11(a)) or top-gate. Due to their freestanding cylindrical shape, all-around gate devices (Bryllert *et al.* 2006) can be implemented in NWs, enabling lower off-state (passive) power consumption in comparison to planar devices (Auth *et al.* 1997).

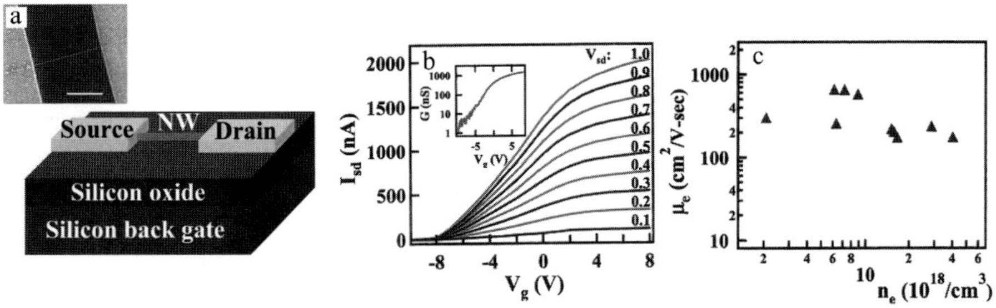

FIG. 9.11. (a) Schematic of a NW-FET, and (inset) FE-SEM image of a GaN NW-FET. Scale bar is 2 μm. (b) $I_{sd} - V_g$ data recorded for values of $V_s d = 0.1 - 1\,V$. (Inset) Conductance (G) vs. gate voltage, and (c) electron mobility of ten different devices as a function of the carrier concentration. (Adapted with permission from Huang *et al.* (2002), © 2002 by the American Chemical Society.)

Generally, GaN-based NW-FETs are realized by transferring GaN NWs onto an oxide-covered heavily-doped Si wafer. At least two metallic contacts are defined on the NWs in order to create ohmic source and drain electrodes. The gate voltage can be applied by a fabricated top-gate or using the Si substrate as a back-gate. Both the carrier nature and the carrier concentration in the NW channel can be extracted from the FET characteristics (Cui *et al.* 2000). In a linear operating regime, the electron density is given by

$$n = \frac{Q_{tot}}{e\pi r^2 L}$$

where Q_{tot} is the total charge in the NW, r is the NW radius, and L is the NW channel length. The total charge is estimated by $Q_{tot} = CV_{th}$, where C is the NW capacitance and V_{th} is the threshold voltage required to deplete the wire. The capacitance per unit length of the cylindrical-plane geometry is given by

$$C = \frac{2\pi\epsilon\epsilon_0 L}{\ln(2h/r)}$$

where h is the thickness of the oxide substrate. Furthermore, the carrier mobility, μ, can be derived from the transconductance of the NW-FET by the expression

$$G(V_{ds}) = \frac{\partial I_{sd}}{\partial V_g} = \frac{\mu C V_{sd}}{L^2}$$

where I_{sd}, V_{sd} and V_g are the source-to-drain current, the source-to-drain voltage, and the gate voltage, respectively

Huang *et al.* (2002) has demonstrated FETs based on individual GaN NWs synthesized via a laser-assisted catalytic growth method. The studied NWs are single crystals with wurtzite structure. The NW diameter and length are in the order of 10 nm and 10 μm, respectively (inset of Fig. 9.11(a)). The gate-dependent $I_{sd} - V_{sd}$ characteristics at room temperature shown in Fig. 9.11(b) confirms that the NWs are n-type, since the device is turned off by applying a certain negative V_g. Interestingly, these devices exhibit a good switching behavior with a transconductance $G > 1$ μS (inset of Fig. 9.11(b)). The electron mobility in these GaN NW-FETs is in the range of 150–650 $\mathrm{cm}^2/\mathrm{Vs}$, with an electron concentration around 10^{18}–10^{19} cm^{-3}. This mobility value is substantially larger than those in NW-FETs based on single GaN NWs grown by CVD (Kim *et al.* 2002; Stern *et al.* 2005; Chang *et al.* 2006), though it is comparable to values measured in NWs grown by direct reaction of NH_3 with flowing gallium vapor (Motayed *et al.* 2007, 2008). However, it must be kept in mind that the measured transconductance is greatly influenced by the source and drain contact resistance, and hence the extracted mobility may not reflect the true intrinsic value (Lu and Lieber 2006).

An important issue concerning NW-FET characteristics is the extremely large NW surface-to-volume ratio, which renders the total channel conductivity especially sensitive to the surface and hence to the environment. Indeed, attempts to correlate the electrical performance of GaN NW-FETs with their structural properties concluded that NWs with smaller diameters exhibit significantly lower mobility than the larger ones (Motayed *et al.* 2007), which is explained by the effect of the sidewall scattering on the overall mobility (Sundaram and Mizel 2004). These surface effects make NW-FET ultrasensitive nanosensors for detecting a wide range of gases, chemicals, and biomedical species (Patolsky *et al.* 2007).

9.4.2 *GaN NW single-electron transistors*

In nanoscale systems, electrical transport is governed mainly by the electron charging and the quantized energy in a confined structure (Kastner 1992). De Franceschi *et al.* (2003) demonstrated that single InP NWs can behave as a single electron transistor (SET) at a low enough temperature, when the charging energy becomes greater than the thermal energy. However, so far, electrical investigations of GaN NW-FETs focus mostly on their room-temperature operation and their transistor characteristics, while the quantum transport phenomena and their potential for quantum devices remain relatively unexplored.

The signature of single-electron tunneling transport was observed in single GaN NWs by Kim *et al.* (2004c). Electrical transport measurements were performed on back-gated NW-FETs based on either lightly or heavily n-type doped GaN NWs synthesized by a thermal vapor deposition method (Lyu *et al.* 2003). The $I_{sd} - V_g$ characteristics at low temperature reveal distinct oscillations in both types of NWs. This observation is ascribed to a single-electron tunneling effect, in which the electron charging energy (E_c) is associated with the addition of one electron to the NW channel. This energy can be obtained from the relation $E_c = e^2/C$, where C is the capacitance of the NWs described previously. The irregularity of the observed $I_{sd} - V_g$ oscillations suggests that multiple confined islands form randomly along the NWs, which might originate from various defects and/or an inhomogeneous dopant distribution.

Later on, low-temperature electron transport was studied in NW-FETs based on *n-i-n* GaN NWs grown by PAMBE (Songmuang *et al.* 2010b). This growth technique offers high-purity wurtzite GaN NWs grown along the *c* axis. In addition, it is straightforward to control the doping along the wire axis, making the transport through single high-quality GaN islands possible. This *n-i-n* GaN NW-FETs are completely turned off by a certain negative V_g, revealing the *n*-type character of the NW channel.

The black trace in Fig. 9.12(a) is $I_{sd} - V_{sd}$ of a single *n-i-n* GaN NW-FETs with a linear characteristic at room temperature, while the grey trace reveals a non-linear behavior at 4.2 K. The suppression of I_{sd} around zero bias is a signature of the Coulomb-blockade effect, which becomes dominant at low temperature. This effect is also visible in the variation of the differential conductance $(\partial I_{sd}/\partial V_{sd})$ vs. V_g and V_{sd}, shown in Fig. 9.12(b). The sequence of diamond-shape dark regions correspond to the values of $V_{sd} - V_g$ where the transport is entirely suppressed. This trace of single-electron tunneling implies that the undoped region of the NW confines a single electronic island, at least for some gate voltage ranges. Within each Coulomb diamond, the island contains a well-defined, integer number of electrons. Similar phenomena were reported in modulation-doped silicon NWs (Hofheinz *et al.* 2008).

The height of the Coulomb diamonds can provide an estimation value of E_c, which is around a few meV in this case. The diamond size measured along the horizontal axis, ΔV_g, provides the coupling capacitance to the back gate, $C_g = e/\Delta V_g \approx 3\,aF$, corresponding to an island length of $\approx$ 100 nm. The estimated size is comparable to the extension of the undoped region of the *n-i-n* GaN wire. This suggests a relatively low amount of crystal disorder, which leads to the formation of multiple electronic islands resulting in irregular and overlapping Coulomb diamonds.

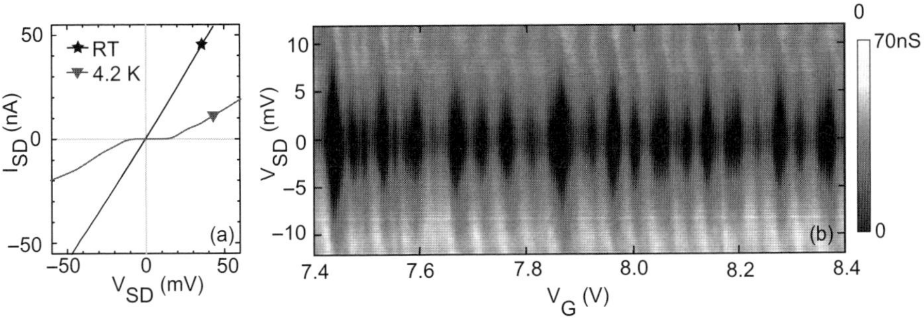

FIG. 9.12. (a) Current–voltage characteristic of a single *n-i-n* NW at room temperature (black trace) and at 4.2 K (grey trace). (b) Color scale plot of the differential conductance $\partial I_{sd}/\partial V_{sd}$ versus (V_g, V_{sd}) of a single *n-i-n* NW. The measurement was taken at 4.2 K. (Adapted with permission from Songmuang *et al.* (2010b), © 2010 by the American Chemical Society.)

9.4.3 *GaN/AlN/AlGaN core-shell NW high-electron-mobility transistors*

Semiconductor heterostructures confining a 2DEG have served as a key platform for high-speed electronic applications and fundamental physics studies. In such heterostructures, the impurity scattering in the conduction channel is suppressed, which significantly improves the carrier mobility and the device performance. NW radial heterostructures are a promising option to further miniaturize 2DEG devices. Moreover, the efficient strain relaxation in the NW geometry inhibits the formation of dislocations, which behave as electron-trapping and scattering media. The reduced dimensionality, together with the band-gap engineering potential in NW radial heterostructures, open the possibility to further enhance the carrier mobility via the quantum confinement, as shown in InAs/InP (Jiang *et al.* 2007) and $Si_{1-x}Ge_x$/Ge core shell NWs (Xiang *et al.* 2006; Nah *et al.* 2012).

GaN/AlN/AlGaN NW radial heterostructures synthesized by MOCVD via VLS mechanism using Ni as catalyst (Fig. 9.13(a)) can be operated as high-performance nanoscale transistors (Li *et al.* 2006). These NWs grow along the ⟨11-20⟩ direction and are enclosed by (0001) and {-110-1} facets. The large internal electric field induced by spontaneous and piezoelectric polarization across the GaN/AlN/AlGaN radial heterojunction leads to the formation of an electron gas inside the GaN core. The thin epitaxial AlN interlayer between the GaN core and AlGaN outer shell is used for reducing alloy scattering from the AlGaN shell and for improving the electron confinement.

To evaluate their electrical characteristic, NW-FETs were fabricated in a standard back-gate geometry. The measurement of $I_{sd} - V_{sd}$ as a function of

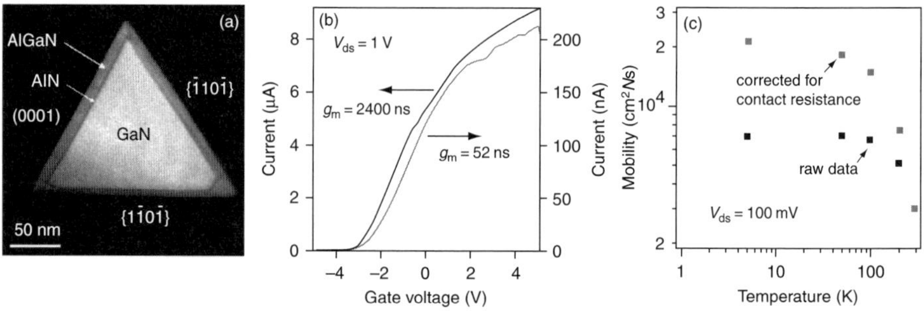

FIG. 9.13. (a) HAADF-STEM image of a GaN/AlN/AlGaN NW cross-section. (b) Transfer characteristics of GaN/AlN/AlGaN (grey) and GaN (black) NWs for $V_{sd} = 1\,V$. (c) Measured (black symbols) and intrinsic (grey symbols) electron mobility of a single GaN/AlN/AlGaN NW at different temperatures. The intrinsic values were obtained after the correction for contact resistance. (Adapted with permission from Li *et al.* (2006), © 2006 by the American Chemical Society.)

V_g reveals that the GaN/AlN/AlGaN NW-FET operates in an n-type depletion mode with a large peak transconductance of 2.4 μS at $V_{sd} = 1\,V$. This value is higher than that of a reference undoped GaN NW-FET with the same diameter (Fig. 9.13(b)). Temperature-dependent measurements show that the on-state conduction and the transconductance of the GaN/AlGaN/AlN NW-FETs increase when decreasing the measurement temperature, in contrast to the behavior observed in Si-doped GaN NW-FETs (Yu *et al.* 2005) and 2D doped semiconductors, which show a decrease in the device on-current at low temperature as the charge carriers from the dopants are frozen. These results confirm the formation of the confined electron gas in radial NW heterostructures.

The intrinsic electron mobility extracted from transconductance measurements with a correction to account for the contact resistance are 2,300 and 21,000 cm^2/Vs at room temperature and 5 K, respectively. The mobility at room temperature is several times higher than that reported in n-type GaN NW-FET (650 cm^2/Vs) (Huang *et al.* 2002). In addition, the electron mobility was found to increase drastically as the temperature decreases from room temperature to 100 K, which is attributed to the reduction of phonon scattering (Fig. 9.13(c)). Below 100 K, the mobility saturates due to the suppression of phonon scattering, as observed in conventional planar HEMT structures (Skierbiszewski *et al.* 2005; Henriksen *et al.* 2005).

The device performance was significantly improved by incorporating a 6-nm-thick high-k ZrO_2 gate dielectric and metal top-gate electrodes, resulting in on-current and transconductance values of 500 mA/mm and 420 mS/mm, respectively. An almost-ideal sub-threshold slope of 68 mV per decade and an on/off current ratio of 10^7 were obtained for the 1-μm-long device.

9.4.4 *GaN/AlN axial-heterostructure resonant tunneling devices*

Resonant tunneling diodes (RTD) are unipolar quantum devices which are attractive for terahertz-frequency applications. Oscillators based on RTDs diodes have recently shown operation frequencies up to 1.1 THz. (Suzuki *et al.* 2010; Feiginov *et al.* 2011). The large and adjustable band offsets of III-N heterostructures are a useful feature for the design of devices relying on quantum confinement and tunneling transport. This unique property is expected to provide a larger peak-to-valley ratio in resonant tunneling current compared to other materials. However, the negative differential resistance (NDR) observed in GaN/AlN RTDs grown on the polar (0001) face exhibits instabilities and reproducibility issues (Kikuchi *et al.* 2002; Golka *et al.* 2006; Bayram *et al.* 2010) stemming from the high dislocation density and the strong polarization-induced electric field. III-Nitride NW heterostructures offer an alternative approach with an additional degree of freedom for strain relaxation, which not only provides dislocation-free material, but also attenuates the piezoelectric effects. In fact, the potential of NW heterostructures for nanoscale RTDs was firstly demonstrated in InP/InAs NW heterostructure (Björk *et al.* 2002).

Recently, electron-resonant-tunneling transport was presented in single defect-free *n-i-n* GaN NWs containing closely-spaced (4 nm) AlN double barriers (Songmuang *et al.* 2010b) grown catalyst-free by PAMBE (Fig. 9.14(a)). The devices were fabricated with a typical back-gate NW-FET geometry, placing the double-barrier structure between the source and drain contacts. Their low-temperature $I_{sd} - V_{sd}$ characteristics show much stronger non-linearity than that of a reference single *n-i-n* GaN NW, with I_{sd} being asymmetrically suppressed over a larger range of V_{sd} (Fig. 9.14(b)). This pronounced asymmetry is ascribed to the polarization discontinuities at the GaN/AlN interfaces, which cause a depletion and an accumulation region at the two sides of the AlN double barrier.

NDR features appear at both negative and positive bias, and are attributed to the onset of tunneling via the confined levels between the AlN barriers. As in conventional RTDs, NDR occurs when a quantized level lines up with the conduction band of the injecting contact (Sze 1981). The value of V_{sd} displaying NDR

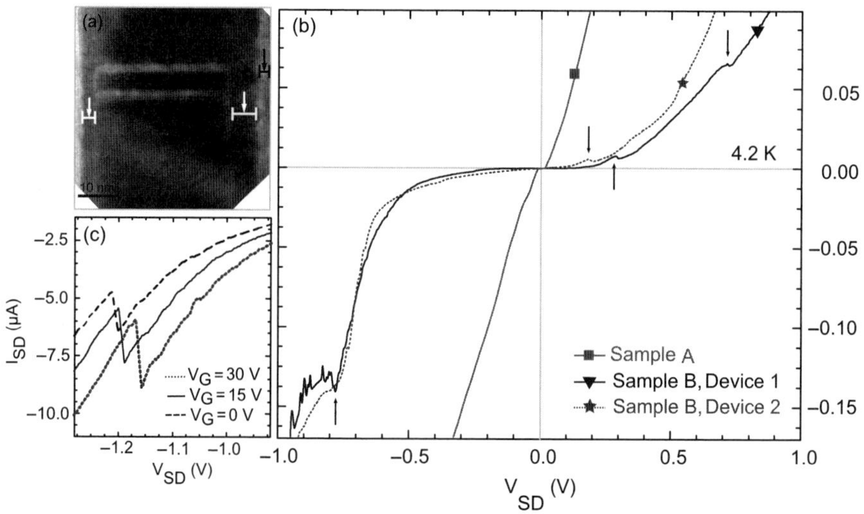

FIG. 9.14. (a) HRTEM image a NW section containing a double-barrier heterostructure, i.e. a pair 2-nm-thick AlN barriers (bright contrast) separated by a 6-nm-high GaN section. The GaN lateral growth around the AlN barriers is resolved in the image (white arrows). The sharp contrast observed at the right facet of the NW (black arrow) corresponds to native gallium oxide. (b). $I_{sd} - V_{sd}$ characteristics at 4.2 K of GaN NWs with two closely-spaced AlN tunnel barriers in comparison to that of *n-i-n* GaN NWs. (c) $I_{sd} - V_{sd}$ characteristic showing the evolution of the NDR appearing at negative V_{sd} for different V_g. (Adapted, with permission, from Songmuang *et al.* (2010b), © 2010 by the American Chemical Society.)

features can be tuned by adjusting the electrostatic potential via the applied V_g, in order to shift the resonance energy level (Fig. 9.14(c)). Differing from AlN/GaN RTDs based on 2D layers, the NDR features in NWs are reproducible for both bias sweeping directions, and no degradation of the device characteristic is found after repeated measurements.

Rigutti *et al.* (2010c) studied the electrical transport through GaN NWs containing multiple GaN/AlN nanodisks. They found that the GaN shell surrounding GaN/AlN nanodisks significantly influences the device characteristics. When the GaN shell does not exist, the current flows mainly though the multiple nanodisks, and a reproducible NDR is observed and attributed to the electron tunneling through the electronic states available in those nanodisks. On the contrary, the NDR feature disappears when a GaN shell is present, as the current flows mainly through the shell close to the lateral surface.

There has been an attempt to take advantage from the suppressed internal electric field of AlN/GaN heterostructures grown on the non-polar orientation of III-N material by growing the AlN/GaN double barriers on the *m*-plane sidewall of a GaN NW by using MBE technique (Carnevale *et al.* 2012). The electron transport was measured by using an SEM equipped with nanomanipulators, capable of contacting individual NWs. The NDR found in the ensemble and individual AlN/GaN core-shell NWs is explained by the resonant tunneling through the AlN/GaN double barrier. Individual NW measurements show a negative differential resistance at room temperature with a peak current density of 5×10^5 A/cm^2.

9.4.5 *GaN/AlN axial-heterostructure single-electron transistors*

Mastering the electron tunneling transport opens the way towards single-electron transistors (SETs). These devices have been intensively studied in III-As and Si (Tarucha *et al.* 1996; Kouwenhoven 2001; Hanson *et al.* 2007). In contrast, there are very few reports on the potential of such devices in III-N materials (Chou *et al.* 2005, 2006; Nakaoka *et al.* 2007). It was firstly shown that InP/InAs NW heterostructure can be operated as single-electron transistors (Thelander *et al.* 2003), which offers a way to study few-electrons InP/InAs quantum dots (Björk *et al.* 2004, 2005).

The potential of GaN/AlN NW heterostructures for single-electron tunneling devices was recently demonstrated (Songmuang *et al.* 2010b). Devices were fabricated with back-gate geometry using defect-free GaN/AlN double-barrier NW axial heterostructures grown by PAMBE. An essential requirement for single-electron transport is a sizeable conduction at small bias voltages, attained by lowering quantized energy levels at or below the Fermi level, E_F (Tarucha *et al.* 1996). In III-N nanostructures grown along the polar direction, the polarization-induced triangular shape of the potential profile along the NW growth axis enables the lowest electron confined level E_1 to be below E_F simply by increasing the GaN insertion height (Fig. 9.15(a)). Following this principle,

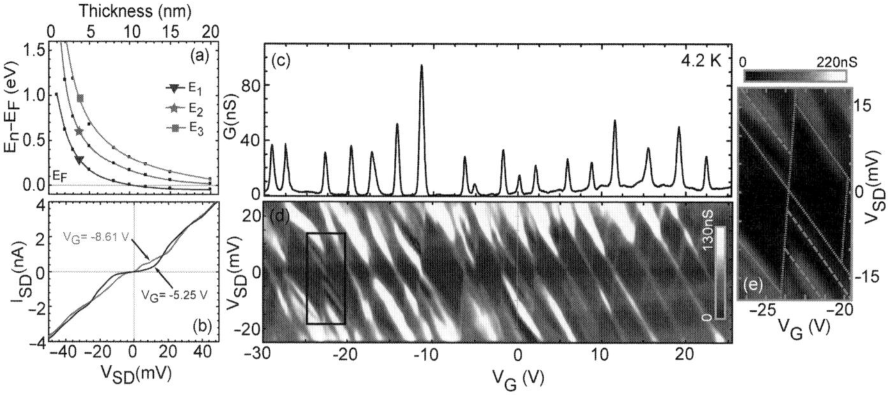

FIG. 9.15. (a) Plot showing the 1D calculation of the three lowest quantized levels for a GaN quantum dot between two 2-nm thick AlN barriers as a function of the dot thickness. (b) $I_{sd} - V_{SD}$ characteristic of a GaN NW with two AlN tunnel barriers at different V_g (c) Differential conductance versus V_g, revealing Coulomb blockade peaks. The measurement was made by using the lock-in technique with an alternative frequency of 13.305 Hz and an excitation amplitude of 500 μV. (d) Color scale plot of $\partial I_{sd}/\partial V_{sd}$ vs. V_g and V_{sd}. All measurements were taken at 4.2 K. (e) Zoom-in on (d) at the region delimited by a black square. Peaks in $\partial I_{sd}/\partial V_{sd}$ denoting the onset of tunneling via ground and excited states have been highlighted by dotted and dashed lines, respectively. (Adapted with permission from Songmuang *et al.* (2010b), © 2010 by the American Chemical Society.)

the distance between AlN double-barrier were designed to be around 6 nm, larger than the one required for the fabrication of GaN/AlN NW-RTDs.

Figure 9.15(b) shows two $I_{sd} - V_{sd}$ characteristics at 4.2 K and different gate voltages. The gate-dependent suppression of I_{sd} on a 10-mV range around zero bias is due to the Coulomb-blockade effect. This is confirmed by the conductance oscillations and by the characteristic diamond-shape features in the color plot of $\partial I_{sd}/\partial V_{sd}(V_g, V_{sd})$ (Fig. 9.15(c)–(d)). The asymmetric shape of the Coulomb diamonds denotes a different capacitive coupling of the GaN island to the source and the drain leads, ascribed to the polarization-induced asymmetry of the conduction-band profile. The characteristic size of the Coulomb diamonds gives a charging energy $E_c \approx 10$ meV and a gate capacitance $C_g \approx 0.1\,aF$, corresponding to an island size of a few nm, which matches well with the height of the GaN insertion between the AlN barriers.

The magnified view in Fig. 9.15(e) shows additional structures appearing as lines (i.e. $\partial I_{sd}/\partial V_{sd}$ peaks) parallel to the diamond edges, ascribed to the onset of single-electron tunneling through some excited states in the GaN island. Their presence constitutes direct evidence of a discrete energy spectrum: the motion of electrons within the dot is quantized not only along the NW axis but also in the transverse plane due to the finite NW diameter. From the separation between the observed excited-state lines and the corresponding diamond edges, the level separation is in the range of 1−10 meV. Because the NW diameter is an order of magnitude larger than the GaN quantum-dot height, this energy distance should be associated to the size quantization in the transverse plane.

9.5 Sensorics

In a chemical sensor the active surface area is one of the important factors determining the detection limits or sensitivity. Due to their large surface-to-volume ratio, NWs are well suited for direct measurement of changes in their optical or electrical properties (e.g. PL, conductance/resistance, impedance) when exposed to various analytes. Furthermore, in the case of GaN, its chemical inertness and capability of operating in extreme environments (high temperatures, presence of radiation, extreme pH levels) is highly desirable for sensor design.

Chen *et al.* (2009) and González-Posada *et al.* (2012b) have reported the dependence of the conductivity, photocurrent, and photocurrent decay time of single GaN NWs on the ambient conditions, notably on the presence of oxygen (Chen *et al.* 2009; González-Posada *et al.* 2012b) and hydrogen (Chen *et al.* 2009). The results are comparable even if the works describe the performance of m-oriented CVD-deposited NWs (Chen *et al.* 2009) and (000-1)-oriented PAMBE-grown NWs (González-Posada *et al.* 2012b). An enhancement of the chemical sensitivity of GaN NWs by insertion of an AlN barrier has been proposed by den Hertog *et al.* (2012). The photocurrent response of single GaN NWs to UV illumination in vacuum, air, and pure oxygen at room temperature are illustrated in Fig. 9.16. As a function of the measuring environment, two major differences are identified: (i) the average steady-state photocurrent measured in vacuum is higher than the one in air/oxygen; (ii) the transient photocurrent shows that the photoresponse in air or oxygen is faster than in vacuum. These observations are assigned to adsorbate-induced variations of the surface band bending and carrier lifetime.

Although these effects can find applications in the domain of chemical sensors, the specificity of the surface/adsorbate interactions is limited. The idea of functionalizing or decorating the NW surface with metal or metal-oxide nano-particles or nano-clusters aims at solving this deficiency.

Teubert *et al.* (2011) and Wright *et al.* (2009) have studied the Pt-functionalization of GaN NWs for detection of H_2, measuring chemically induced variations of the NW PL or the NW resistance, respectively. With this purpose, NW samples were coated with a Pt layer of a nominal thickness of 5–7 nm, which spontaneously arranges forming Pt islands. Figure 9.17 plots the PL response of GaN NWs with embedded GaN/AlGaN nanodisks to H_2 and O_2, comparing bare surfaces and NWs functionalized by deposition of Pt. Hydrogen-induced dipole fields can suppress the surface recombination in Pt-coated NWs leading

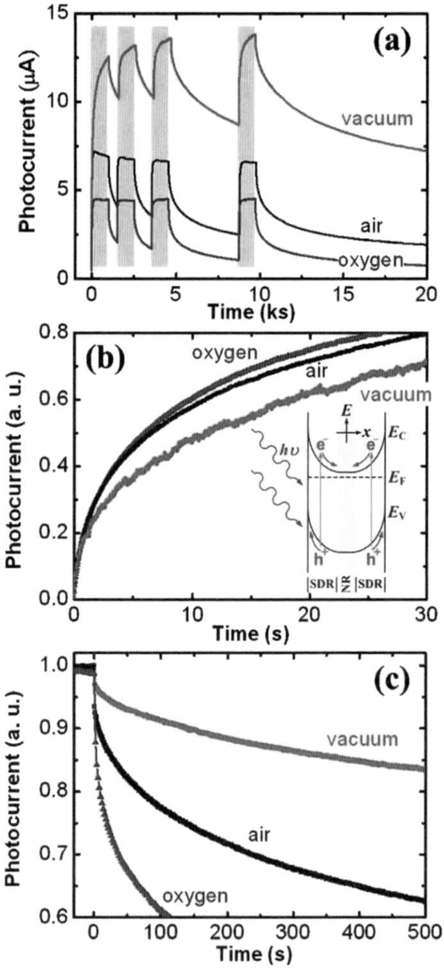

FIG. 9.16. (a) Different photocurrent responses of GaN NWs to 325-nm UV excitation measured in vacuum, air, and pure oxygen. The shadow shows the duration with photoexcitation. (b) Normalized photocurrent rise curves in vacuum, air, and pure oxygen. Inset: A schematic of the electron-hole spatial separation mechanism induced by surface band bending in a GaN NW. SDR denotes surface depletion region and NR denotes neutral region. (c) Normalized photocurrent decay curves in vacuum, air, and pure oxygen. (Reprinted, with permission from Chen *et al.* (2009), © 2009 American Institute of Physics.)

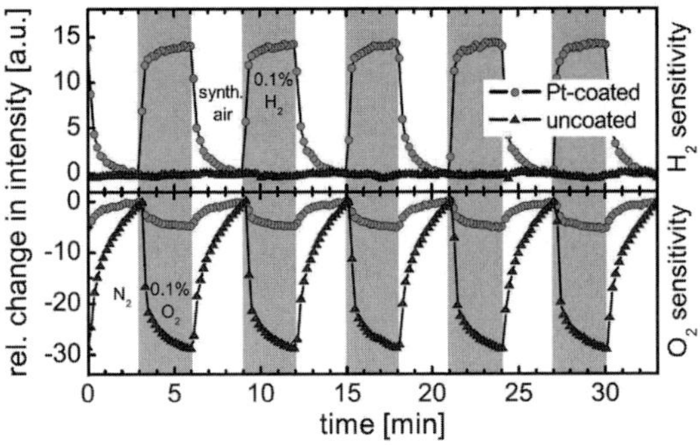

FIG. 9.17. Transient response of the relative PL intensity from AlGaN nanodisks inserted in GaN NWs to variations of the gaseous environment ($T = 150°C$). In the presence of hydrogen (H_2 concentration of 0.1% in synthetic air) an increase of the PL intensity for Pt-coated NWs is observed. In contrast, a decrease of the PL signal is detected in the presence of oxygen (O_2 concentration of 0.1% in nitrogen) which can be suppressed by Pt coverage. (Reprinted, with permission, from Teubert *et al.* (2011), © 2011 IOP Publishing Ltd.)

to an enhanced PL intensity in hydrogen-containing atmosphere. For uncoated NWs, the oxygen-induced enhancement of the surface recombination induces a reduction in PL intensity.

Regarding the effect on the NW resistance, non-linear relative variations of ≈1.7% and ≈1.9% are obtained when exposing the Pt-coated NWs to 200 ppm and 2000 ppm of H_2 in N_2, respectively. Better results are obtained by replacing Pt with Pd, which results in relative responses of ≈7.4% at 200 ppm and ≈9.1% at 1500 ppm of H_2 in N_2 (Lim *et al.* 2008). More recently, relative responses of 34.1% at 100 ppm H_2 were achieved by anealing the Pd-coated GaN to form Ga_2Pd_5 nanodots (Kim *et al.* 2011).

Aluri *et al.* (2011) demonstrated highly selective and sensitive sensors using GaN NWs decorated with TiO_2 nanoclusters. Hybrid sensor devices were developed by fabricating two-terminal devices using single GaN NWs followed by the deposition of TiO_2 nanoclusters using RF magnetron and annealing at 700°C. The catalytic properties of TiO_2 enables selective sensing of aromatic compounds with additional selectivity for methyl group substitution, i.e. these sensors can distinguish toluene from other aromatic compounds. In the proposed device the reaction of the analyte is measured as a variation of the photocurrent under UV excitation. Thus, a change of photocurrent was observed when the sensors are exposed to the vapors of aromatic compounds (toluene, benzene, ethylbenzene, and xylene mixed with air), while they had no response to non-aromatic organic compounds such as methanol, ethanol, isopropanol, chloroform, acetone, and 1,3-hexadiene. The sensitivity range for the noted aromatic compounds was from 1% down to 50 ppb at room temperature. The authors speculate that photoinduced oxygen desorption and subsequent capture of holes by organic adsorbate molecules on the surface of TiO_2 clusters produces a local gating effect, which is responsible for the sensing action.

By addition of sputtered Pt, single GaN/(TiO_2-Pt) NWs become only sensitive to methanol, ethanol, and hydrogen, whereas higher carbon-containing alcohols (n-propanol, iso-propanol, n-butanol) did not produce any sensor response (Aluri *et al.* 2012). Reference single GaN/Pt NWs were only sensitive to hydrogen, and not to methanol or ethanol. Figure 9.18) shows the comparative summary of the sensing behavior of GaN/TiO_2 NWs, GaN/(TiO_2-Pt) NWs and GaN/Pt NWs to 1000 ppm of various analytes in air. The sensitivity is defined as the relative variation of the single NW resistance when exposed to the analyte, i.e. $R_{gas}-R_{air}/R_{air}$, where R_{gas} and R_{air} are the resistances of the sensor in the presence of the analyte-air mixture and in the presence of air only, respectively (R_{air} is replaced by $R_{nitrogen}$ for H_2 sensing experiments).

On the other hand, semiconductor NWs, having a diameter comparable to biologically-relevant molecules, are interesting nanostructures for biosensing

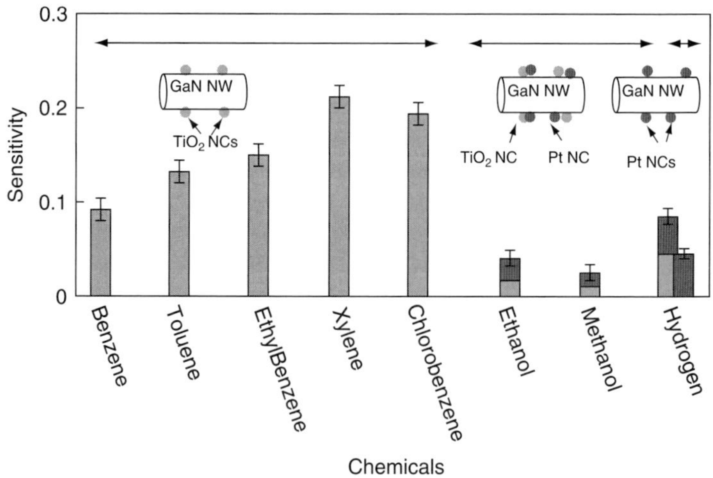

FIG. 9.18. Comparative sensing behavior of functionalized GaN NWs to 1000 ppm of various analytes in air: light grey bar graphs represent single GaN/TiO_2 NWs, light and dark grey bars represent single GaN/(TiO_2-Pt) NWs, and the dark grey plot represents single GaN/Pt NWs. Other chemicals which did not produce any response in any one of the sensors are not included in the plot. The zero line is the baseline response to 20 sccm of air and N_2. The error bars represent the standard deviation of the mean sensitivity values for every chemical computed for five different devices with diameters in the range of 200–300 nm. (Reprinted with permission from Aluri *et al.* (2012), © 2012 IOP Publishing Ltd.)

devices (Cui *et al.* 2001). Recent studies have demonstrated successful application of GaN NWs as transducers in highly sensitive label-free DNA-sensing, using cyclic voltammetry, electrochemical impedance spectroscopy, and PL techniques (Chen *et al.* 2009b; Ganguly *et al.* 2009). The DNA-immobilized GaN NWs are found to possess distinct Faradaic characteristics compared to the unmodified NWs. Based on these results, Chen *et al.* (2011b) have demonstrated label-free identification of specific DNA sequences by connecting a sample of GaN NWs synthesized on Si by Au-catalyzed CVD to the gate of a commercial n-MOSFET, reaching a detection limit in the level of 10^{-18} molar, about six orders of magnitude lower than that of GaN 2D layers in the same configuration.

9.6 Conclusions

In summary, this chapter is an introduction to new device concepts incorporating GaN NWs as active elements. The NW geometry set the bases for ultimate device miniaturization, and its large surface-to-volume ratio extends the range of size and band-gap engineering beyond the limits of planar systems. The development of new methods to synthesize defect-free III-N NWs with controlled-by-design chemical and physical characteristics (geometrical dimensions, doping profile, radial and axial heterostructuring...) has enabled the introduction of these nano-objects in application domains such as energy conversion, nanoelectronics, optoelectronics, and sensorics. First prototypes of single-NW photoconductors, In-containing core-shell-NW photovoltaic devices, piezo-electric nanogenerators, single-NW FETs, SETs, RTDs, and opto-chemical and electro-chemical transducers, have laid the foundations of a cutting-edge technology with the potential to revolutionize the semiconductor field.

Acknowledgments

The authors acknowledge support from the French National Research Agency via "UVLamp" (ANR-2011-NANO-027) and "COSMOS" (ANR JCJC 302286E) projects, and from the EU ERC-StG "TeraGaN" (#278428) project.

References

Allah, R. F.; Ben, T.; Songmuang, R.; and González, D. (2012). *Appl. Phys. Express*, **5**, 045002.

Aluri, G. S.; Motayed, A.; Davydov, A. V.; Oleshko, V. P.; Bertness, K. A.; Sanford N. A.; and Mulpuri, R. V. (2011). *Nanotechnology*, **22**, 295503.

Aluri, G. S.; Motayed, A.; Davydov, A. V.; Oleshko, V. P.; Bertness, K. A.; Sanford N. A.; and Mulpuri, R. V. (2012). *Nanotechnology*, **23**, 175501.

Auth, C. P. and Plummer, J. D. (1997). *IEEE Electron Dev. Lett.*, **18**, 011914.

Alvarez, J.; Ngo, I.; Gueunier-Farret, M.-E.; Kleider, J.-P.; Linwei, Y.; Cabarrocas, P. R.; Perraud, S.; Rouvire, E.; Celle, C.; Mouchet, C.; and Simonato, J.-P. (2011). *Nanoscale Res. Lett.*, **6**, 110.

Bayram, C.; Vashaei, Z.; and Razeghi, M. (2010). *Appl. Phys. Lett.*, **97**, 181109.

Bergbauer, W.; Strassburg, M.; Kölper, Ch.; Linder, N.; Roder, C.; Lähnemann, J.; Trampert, A.; Fündling, S.; Li, S. F.; Wehmann, H.-H.; and Waag, A. (2010). *Nanotechnology*, **21**, 305201.

Bergbauer, W.; Strassburg, M.; Kölper, C.; Linder, N.; Roder, C.; Lähnemann, J.; Trampert, A.; Fündling, S.; Li, S. F.; Wehmann, H. H.; and Waag, A. (2011). *J. Cryst. Growth*, **315**, 164.

Bertelli, M.; Löptien, P.; Wenderoth, M.; Rizzi, A.; Ulbrich, R. G.; Righi, M. C.; Ferretti, A.; Martin-Samos, L.; Bertoni, C. M.; and Catellani, A. (2009). *Phys. Rev. B*, **80**, 1115324.

Bertness, K. A.; Roshko, A.; Mansfield, L. M.; Harvey, T. E.; and Sanford, N. A. (2007). *J. Cryst. Growth*, **300**, 94.

Bertness, K. A.; Sanford, N. A.; and Davydov, A. V. (2011). *IEEE J. Quant. Electr.*, **17**, 847.

Björk, M. T.; Ohlsson, B. J.; Thelander, C.; Persson, A. I.; Deppert, K.; Wallenberg, L. R.; and Samuelson, L. (2002). *Appl. Phys. Lett.*, **81**, 4458.

Björk, M. T.; Thelander, C.; Hansen, A. E.; Jensen, L. E.; Larsson, M. W.; Wallenberg, L. R.; and Samuelson, L. (2004). *Nano Lett.*, **4**, 1621.

Björk, M.; Fuhrer, A.; Hansen, A.; Larsson, M.; Fröberg, L.; and Samuelson, L. (2005). *Phys. Rev. B* , **72**, 201307(R).

Brubaker, M. D.; Levin, I.; Davydov, A. V.; Rourke, D. M.; Sanford, N. A.; Bright, V. M.; and Bertness, K. A. (2011). *J. Appl. Phys.*, **110**, 053506.

Bryllert, T.; Wernersson, L. E.; Fröberg, L. E.; and Samuelson, L. (2006). *IEEE Electron Dev. Lett.*, **27**, 323.

Calarco, R.; Marso, M.; Richter, T.; Aykanat, A. I.; Meijers, R.; Hart, A.v.d.; Stoica, T.; and Lüth, H. (2005). *Nano Lett.*, **5**, 981.

Calarco, R.; Meijers, R. J.; Debnath, R. K.; Stoica, T.; Sutter, E.; and Lu, H. (2007). *Nano Lett.*, **7**, 2248.

Calarco, R.; Stoica, T.; Brandt, O.; and Geelhaar, L. (2011). *J. Mater. Res.*, **7**, 2248.

Calleja, E; Sánchez-García, M.; Sánchez, F.; Calle, F.; Naranjo, F.; Muñoz, E.; Jahn, U.; and Ploog, K. (2000). *Phys. Rev. B*, **14**, 2157.

Cao, Y. L.; Liu, Z. T.; Chen, L. M.; Tang, Y. B.; Luo, L. B.; Jie, J. S.; Zhang, W. J.; Lee, S. T.; and Lee, C. S. (2011). *Optics Express*, **19**, 6100.

Carter, D. J. and Stampfl, C. (2009). *Phys. Rev. B*, **79**, 195302.

Carnevale, S. D.; Yang, J.; Phillips, P. J.; Mills, M. J.; and Myers, R. C. (2011). *Nano Lett.*, **11**, 866.

Carnevale, S. D.; Marginean, C.; Phillips, P. J.; Kent, T. F.; Sarwar, a. T. M. G.; Mills, M. J.; and Myers, R. C. (2012). *Appl. Phys. Lett.*, **100**, 142115.

Chang, C.; Chi, G.; Wang, W.; Chen, L.; Chen, K.; and Ren, F. (2006). *J. Electron Mater.* **35**, 738.

Chang, Y.-L.; Wang, J. L.; Li, F.; and Mi, Z. (2010). *Appl. Phys. Lett.*, **96**, 013106.

Chen, C. C. and Yeh, C. C. (2000). *Adv. Mater.*, **12**,738.

Chen, C. C.; Yeh, C. C.; Chen, C. H.; Yu, M. Y.; Liu, H. S.; Wu, J. J.; Chen, K. H.; Chen, L. C.; Pong, J. Y.; Chen, Y. F. (2001). *J. Am. Chem. Soc.*, **123**, 2791.

Chen, R.-S.; Chen, H.-Y.; Lu, C.-Y.; Chen, K.-H.; Chen, C.-P.; Chen L.-C.; and Yang, Y.-J. (2007). *Appl. Phys. Lett.*, **91**, 223106.

Chen, R.-S.; Wang, S.-W.; Lan, Z.-H.; Tsai, J. T.-H.; Wu, C.-T.; Chen, L.-C.; Chen, K.-H.; Huang, Y.-S.; and Chen, C.-C. (2008). *Small*, **4**, 925.

Chen, R.-S.; Lu, C.-Y.; Chen, K.-H.; and Chen, L.-C. (2009). *Appl. Phys. Lett.*, **95**, 233119.

Chen, C. P.; Ganguly, A.; Wang, C. H.; Hsu, C. W.; Chattopadhyay, S.; Hsu, Y. K.; Chang, Y. C.; Chen, K. H.; and Chen, L. C. (2009b). *Anal. Chem.*, **81**, 36.

Chen, X. J.; Perillat-Merceroz, G.; Sam-Giao, D.; Durand, C.; and Eymery, J. (2010). *Appl. Phys. Lett.*, **97**, 151909.

Chen, M.-W.; Chen, C.-Y.; Lien, D.-H.; Ding, Y.; and He, J.-H. (2010b). *Optics Express*, **18**, 14836.

Chen, X. J.; Gayral, B.; Sam-Giao, D.; Bougerol, C.; Durand, C.; and Eymery, J. (2011). *Appl. Phys. Lett.*, **99**, 251910.

Chen, C. P.; Ganguly, A.; Lu, C. Y.; Chen, T. Y.; Kuo, C.-C; Chen, R. S.; Tu, W. H.; Fisher, W. B.; Chen, K. H.; and Chen, L. C. (2011b). *Anal. Chem.*, **83**, 1938.

Cheng, G. S.; Zhang, L. D.; Zhu, Y.; Fei, G. T.; Li, L.; Mo, C. M.; and Mao, Y. Q. (1999). *Appl. Phys. Lett.*, **75**, 2455.

Cherns, D.; Meshi, L.; Griffiths, I.; Khongphetsak, S.; Novikov, S. V.; Farley, N.; Campion, R. P.; and Foxon, C. T. (2008). *Appl. Phys. Lett.*, **92**, 121902.

Chèze, C.; Geelhaar, L.; Jenichen, B.; and Riechert, H. (2010). *Appl. Phys. Lett.*, **97**, 153105.

Chèze, C.; Geelhaar, L.; Brandt, O.; Weber, W. M.; Riechert, H.; Münch, S.; Rothemund, R.; Reitzenstein, S.; Forchel, A.; Kehagias, T.; Komninou, P.; Dimitrakopulos, G. P.; and Karakostas, T. (2010b). *Nano Research*, **3**, 528.

Chou, H. T.; Lscher, S.; Goldhaber-Gordon, D.; Manfra, M. J.; Sergent, A. M.; West, K. W.; and Molnar, R. J. (2005). *Appl. Phys. Lett.* **86**, 073108.

Chou, H. T.; Goldhaber-Gordon, D.; Schmult, S.; Manfra, M. J.; Sergent, A. M.; and Molnar; R. J. (2006). *Appl. Phys. Lett.* **89**, 033104.

Colombo, C.; Heiβ, M.; Graetzel, M.; and Fontcuberta i Morral, A. (2009). *Appl. Phys. Lett.*, **94**, 173108.

Consonni, V.; Knelangen, M.; Geelhaar, L.; Trampert, A.; and Riechert, H. (2010). *Phys. Rev. B*, **81**, 085310.

Cui, Y. and Lieber, C. M. (2000). *Science*, **291**, 891.

Cui, Y.; Wei, Q.; Park, H.; and Lieber, C. M. (2001). *Science*, **293**, 1289.

Cui, Y.; Zhong, Z. H.; Wang, D. L.; Wang, W. U.; and Lieber, C. M. (2003). *Nano Lett.*, **3**, 149.

Czaban, J. A.; Thompson, D. A.; and LaPierre R. R. (2009). *Nano Lett.*, **9**, 148.

Dayeh, S. A.; Aplin, D. P. R.; Zhou, Z. T.; Yu, P. K. L.; Yu, E. T.; and Wang, D. L. (2007). *Small*, **3**, 326.

De Franceschi, S.; van Dam, J. A.; Bakkers, E. P. A. M.; Feiner, L. F.; Gurevich, L.; and Kouwenhoven, L. P. (2003). *Appl. Phys. Lett.*, **83**, 344.

De Luna Bugallo, A.; Tchernycheva, M.; Jacopin, G.; Rigutti, L.; Julien, F. H.; Chou, S.-T.; Lin, Y.-T.; Tseng, P.-H.; and Tu, L.-W. (2010). *Nanotechnol.*, **21**, 315201.

De la Mata, M.; Magen, C.; Gazquez, J.; Utama, M. I. B.; Heiss, M.; Lopatin, S.; Furtmayr, F.; Fernández-Rojas, C. J.; Peng, B.; Morante, J. R.; Rurali, R.; Eickhoff, M.; Fontcuberta I Morral, A.; Xiong, Q.; and Arbiol, J. (2012). *Nano Lett.*, **12**, 2579.

De Luna Bugallo, A.; Rigutti, L.; Jacopin, G.; Julien, F. H.; Durand, C.; Chen, X. J.; Salomon, D.; Eymery, J.; and Tchernycheva, M. (2011). *Appl. Phys. Lett.*, **98**, 233107.

Deb, P.; Kim, H.; Rawat, V.; Oliver, M.; Kim, S.; Marshall, M.; Stach, E.; Sands, T. (2005). *Nano Lett.*, **5**, 1847.

Debnath, R. K.; Meijers, R.; Richter, T.; Stoica, T.; Calarco, R.; and Lüth, H. (2007). *Appl. Phys. Lett.*, **90**, 123117.

den Hertog, M. I.; González-Posada, F.; Songmuang, R.; Rouviere, J.-L.; Fournier, T.; Fernandez, B.; and Monroy, E. (2012). *Nano Lett.*, **12**, 5691.

Dobrokhotov, V.; McIlroym, D. N.; Grant Norton, M.; Abuzir, A.; Yeh, W. J.; Stevenson, I.; Pouy, R.; Bochenek, J.; Cartwright, M.; Wang, L.; Dawson, J.; Beaux, M.; and Bervena, C. (2006). *J. Appl. Phys.*, **99**, 104302.

Dong, Y.; Tian, B.; Kempa, T. J.; and Lieber, C. M. (2009). *Nano Lett.*, **9**, 2183.

Duan, X. and Lieber, C. M. (2000). *J. Am. Chem. Soc.*, **122**, 188.

Duan, X.; Huang, Y.; Cui, Y.; Wang, J.; Lieber, C. M. (2001). *Nature*, **409**, 66.

Farrell, R. M.; Neufeld, C. J.; Cruz, S. C.; Lang, J. R.; Iza, M.; Keller, S.; Nakamura, S.; DenBaars, S. P.; Mishra, U. K.; and Speck, J. S. (2011). *Appl. Phys. Lett.*, **98**, 201107.

Feiginov, M.; Sydlo, C.; Cojocari, O.; and Meissner, P. (2011). *Appl. Phys. Lett.*, **99**, 233506.

Fernández-Garrido, S.; Grandal, J.; Calleja, E.; Sánchez-Garca, M. A.; López-Romero, D. (2009). *J. Appl. Phys.*, **106**, 126102.

Foxon, C. T.; Novikov, S. V.; Hall, J. L.; Campion, R. P.; Cherns, D.; Griffiths, I.; and Khongphetsak, S. (2009). *J. Cryst. Growth*, **311**, 3423.

Foussekis, M.; Baski, A. A.; and Reshchikov, M. A. (2009). *Appl. Phys. Lett.*, **94**, 162116.

Furtmayr, F.; Vielemeyer, M.; Stutzmann, M.; Arbiol, J.; Estrad, S.; Peir, F.; Morante, J. R.; and Eickhoff M. (2008). *J. Appl. Phys.*, **104**, 034309.

Furtmayr, F.; Teubert, J.; Becker, P.; Conesa-Boj, S.; Morante, J. R.; Chernikov, A.; Schäfer, S.; Chatterjee, S.; Arbiol, J.; and Eickhoff, M. (2011). *Phys. Rev. B.*, **84**, 205303.

Ganguly, A.; Chen, C. P.; Lai, Y. T.; Kuo, C. C.; Hsu, C. W.; Chen, K. H.; and Chen, L. C. (2009). *J. Mater. Chem.*, **19**, 928.

Garnett, E. C. and Yang, P. (2008). *JACS*, **130**, 9224.

Glas, F. (2006). *Phys. Rev. B.*, **74**, 121302(R).

Golka, S.; Pflügl, C.; Schrenk, W.; Strasser, G.; Skierbiszewski, C.; Siekacz, M.; Grzegory, I.; and Porowski, S. (2006). *Appl. Phys. Lett.*, **88**, 172106.

González-Posada, F.; Songmuang, R.; Den Hertog, M.; and Monroy, E. (2012). *Nano Lett.*, **12**, 172.

González-Posada, F.; Songmuang, R.; den Hertog, M. I.; and Monroy, E. (2012b). arXiv:1212.1591.

Goto, H.; Nosaki, K.; Tomioka, K.; Hara, S.; Hiruma, K.; Motohisa, J.; and Fukui, T. (2009). *Appl. Phys. Express*, **2**, 035004.

Grandal, J.; Sánchez-García, M. A.; Calleja, E.; Luna, E.; and Trampert, A. (2007). *Appl. Phys. Lett.*, **91**, 021902.

Greytak, A. B.; Lauhon, L. J.; Gudiksen, M. S.; and Lieber, C. M. (2004). *Appl. Phys. Lett.*, **84**, 4176.

Gu, Y. and Lauhon, L. J. (2006). *Appl. Phys. Lett.*, **89**, 143102.

Han, W.; Redlich, P.; Ernst, F.; and Rühle, M. (2000). *Appl. Phys. Lett.*, **76**, 652.

Hanson, R.; Petta, J. R.; Tarucha, S.; and Vandersypen, L. M. K. (2007). *Rev. Mod. Phys.*, **79**, 1217.

He, M.; Zhou, P.; Mohammad, S. N.; Harris, G. L.; Halpern, B.; Jacobs, R.; Sarney, W. L.; and Salamanca-Riba, L. (2001). *J. Cryst. Growth*, **231**, 357.

Henriksen, E. A.; Syed, S.; Ahmadian, Y.; Manfra, M. J.; Baldwin, K. W.; Sergent, A. M.; Molnar, R. J.; and Stormer, H. L. (2005). *Appl. Phys. Lett.* **86**, 252108.

Hersee, S. D.; Sun, X.; and Wang, X. (2006). *Nano Lett.*, **6**, 1808.

Hestroffer, K.; Mata, R.; Camacho, D.; Leclere, C.; Tourbot, G.; Niquet, Y. M.; Cros, A.; Bougerol, C.; Renevier, H.; and Daudin, B. (2010). *Nanotechnology*, **21**, 415702.

Hofheinz, M.; Jehl, X.; Sanquer, M.; Molas, G.; Vinet, M.; and Deleonibus, S. (2006). *Appl. Phys. Lett.*, **89**, 143504.

Hu, L.; Yan, J.; Liao, M.; Wu, L.; Fang, X. (2011). *Small*, **8**, 1012.

Huang, Y.; Duan, X.; Cui, Y.; and Lieber, C. M. (2002). *Nano Lett.*, **2**, 101.

Huang, C.-T.; Song, J.; Lee, W.-F.; Ding, Y.; Gao, Z.; Hao, Y.; Chen, L.-J.; and Wang, Z. L. (2010). *J. Am. Chem. Soc.*, **132**, 4766.

Huang, C.-T.; Song, J.; Tsai, C.-M; Lee, W.-F.; Lien, C.-H; Gao, Z.; Hao, Y.; Chen, L.-J.; and Wang, Z. L. (2010b). *Adv. Mater.*, **22**, 4008.

Hsu, L. and Walukiewicz, W. (2008). *J. Appl. Phys.*, **104**, 024507.

Jahn, U.; Ristic, J.; and Calleja, E. (2007). *Appl. Phys. Lett.*, **90**, 161117.

Jiang, X.; Xiong, Q.; Nam, S.; Qian, F.; Li, Y.; and Lieber, C. M. (2007). *Nano Lett.*, **7**, 3214.

Johnson, J. C.; Choi, H. J.; Knutsen, K. P.; Schaller, R. D.; Yang, P.; and Saykally, R. J. (2002). *Nature. Mat.*, **1**, 106.

Kastner, M. A. (1992). *Rev. Mod. Phys.*, **64**, 849.

Kempa, T. J.; Tian, B.; Kim, D. R.; Hu, J.; Zheng, X.; and Lieber, C. M. (2008). *Nano Lett.*, **8**, 3456.

Kikuchi, A.; Bannai, R.; Kishino, K.; Lee, C.-M.; and Chyi, J. I. (2002). *Appl. Phys. Lett.*, **81**, 1729.

Kikuchi, A.; Kawai, M.; Tada, M.; and Kishino, K. (2004). *Jpn. J. Appl. Phys.*, **43**, L 1524.

Kim, J. R.; So, H. M.; Park, J. W.; Kim, J. J.; Kim, J.; Lee, C. J.; and Lyu, S. C. (2002). *Appl. Phys. Lett.*, **80**, 3548.

Kim, H.-M.; Kim, D. S.; Park, Y. S.; Kim, D.Y.; Kang, T. W.; and Chung, K. S. (2002b). *Adv. Mater.*, **14**, 991.

Kim, H. M.; Cho, Y. H.; and Kang, T. W. (2003). *Adv. Mater.*, **15**, 232.

Kim, H. M.; Kang, T. W.; and Chung, K. S. (2003b). *Adv. Mater.*, **15**, 567.

Kim, H.; Cho, Y.; Lee, H.; Kim, S.; Ryu, S.; Kim, D.; Kang, T.; and Chung, K. (2004). *Nano Lett.*, **4**, 1059.

Kim, H.-M.; Lee, H.; Kim, S. I.; Ryu, S. R.; Kang, T. W.; and Chung, K. S. (2004b). *Phys. Stat. Sol. B*, **241**, 2802.

Kim, J. R.; Kim, B. K.; Lee, I.; Kim, J.-J.; Kim, J.; Lyu, S.; and Lee, C. (2004c). *Phys. Rev. B*, **69**, 233303.

Kim, H. W.; Kim, H. S.; Na, H. G.; Yang, J. C.; Kim, S. S.; and Lee, C. M. (2010). *Chem. Eng. J.*, **165**, 720.

Kim, C.-J.; Lee, H.-S.; Cho, Y.-J.; Kang, K.; and Jo, M.-H. (2010b). *Nano Lett.*, **10**, 2043.

Kim, S. S.; Park, J. Y.; Choi, S. W.; Kim, H. S.; Na, H. G.; Yang, J. C.; Lee, C.; and Kim, H. W. (2011). *Int. J. Hydrogen Energy*, **36**, 2313.

Koester, R.; Hwang, J. S.; Durand, C.; Dang, L. S.; and Eymery, J. (2010). *Nanotechnology*, **21**, 015602.

Koester, R.; Hwang, J.-S.; Salomon, D.; Chen, X.; Bougerol, C.; Barnes, J.-P.; Le Si Dang, D.; Rigutti, L.; de Luna Bugallo, A.; Jacopin, G.; Tchernycheva, M.; Durand, C.; and Eymery, J. (2011). *Nano Lett.*, **11**, 4839.

Kouwenhoven, L. E. O. P.; Marcus, C. M.; McEuen, P. L.; Tarucha, S.; and Robert, M. (2001). *Rep. Prog. Phys.*, **64**, 901.

Kuykendall, T.; Pauzauskie, P.; Lee, S.; Zhang, Y.; Goldberger, J.; and Yang, P. (2003). *Nano Lett.*, **3**, 1063.

Kuykendall, T.; Ulrich, P.; Aloni, S.; and Yang, P. (2007). *Nature Mater.*, **6**, 951.

Lampert, M. A. (1956). *Phys. Rev.*, **103**, 1648.

Lan, Z. H.; Liang, C. H.; Hsu, C. W.; Wu, C. T.; Lin, H. M.; Dhara, S.; Chen, K. H.; Chen, L. C.; and Chen, C. C. (2004). *Adv. Funct. Mater.*, **14**, 233.

Landré, O.; Bougerol, C.; Renevier, H.; and Daudin, B. (2009). *Nanotechnology*, **20**, 415602.

Lauhon, L. J.; Gudiksen, M. S.; Wang, D.; and Lieber, C. M. (2002). *Nature*, **420**, 57.

Li, Y.; Xiang, J.; Qian, F.; Gradečak, S.; Wu, Y.; Yan, H.; Blom, D. A.; and Lieber, C. M. (2006). *Nano Lett.*, **6**, 1468.

Li, S. F.; Fuendling, S.; Wang, X.; Merzsch, S.; Al-Suleiman, M. A. M.; Wei, J. D.; Wehmann, H.-H.; Waag, A.; Bergbauer, W.; and Strassburg, M. (2011). *Cryst. Growth Des.*, **11**, 1573.

Lim, W.; Wright, J. S.; Gila, B. P.; Johnson, J.; Ural, A.; Anderson, T.; Ren, F.; and Pearton, S. J. (2008). *Appl. Phys. Lett.*, **93**, 072109.

Lin, Y. F.; Song, J.; Ding, Y; Lu, S. Y.; and Wang, Z. L. (2008). *Appl. Phys. Lett.*, **92**, 022105.

Liu, B.; Bando, Y.; Tang, C.; Xu, F.; and Golberg, D. (2005). *Appl. Phys. Lett.*, **87**, 073106.

Lu, W. and Lieber, C. M. (2006). *J. Phy. D: Appl. Phys.*, **39**, R387.

Lu, W.; Xie, P.; and Lieber, C. M. (2008). *IEEE Trans. Electron Dev.*, **55**, 2859.

Lu, M. Y.; Song, J.; Lu, M. P. Lee, C. Y.; Chen, L. J.; and Wang, Z. L. (2009). *ACS Nano*, **3**, 357.

Luque, A. (2011). *J. Appl. Phys.*, **110**, 031301.

Luscombe, J. H. and Frenzen, C. L. (2002). *Solid-State Electronics*, **46**, 885.

Lyu, S. C.; Cha, O. H.; Suh, E. K.; Ruh, H.; Lee, H. J.; and Lee, C. J. (2003). *Chem. Phys. Lett.*, **367**, 136.

Maslow, A. V.; Bakunov, M. I.; and Ningb, C. Z. (2006). *J. Appl. Phys.*, **99**, 024314.

Monroy, E.; Omnès, F.; and Calle, F. (2003). *Semicond. Sci. Technol.*, **18**, R33.

Motayed, A.; Vaudin, M.; Davydov, A.V.; Melngailis, J.; He, M.; and Mohammad, S. N. (2007). *Appl. Phys. Lett.*, **90**, 043104.

Motayed, A.; Davydov, A. V.; Mohammad, S. N.; and Melngailis, J. (2008). *J. Appl. Phys.*, **104**, 024302.

Nah, J.; Dillen, D. C.; Varahramyan, K. M.; Banerjee, S. K.; and Tutuc, E. (2012). *Nano Lett.*, **12**, 108.

Nakaoka, T.; Kako, S.; Arakawa, Y.; and Tarucha, S. (2007). *Appl. Phys. Lett.*, **90**, 162109.

Nam, C. Y.; Kim, J. Y.; and Fischer, J. E. (2005). *Appl. Phys. Lett.*, **86**, 193112.

Patolsky, F.; Timko, B. P.; Zheng, G.; and Lieber, C. M. (2007). *MRS Bulletin*, **32**, 142.

Peng, S.-M.; Su, Y.-K.; Ji, L.-W.; Young, S.-J.; Tsai, C.-N.; Wu, C.-Z.; Chao, W.-C.; Cheng, W.B.; and Huang, C.-J. (2011). *Electrochem. and Solid-State Lett.*, **14**, J13.

Peter, C. S. and Kerry, J. V. (1990). *Appl. Phys. Lett.*, **57**, 545.

Pfüller, C.; Brandt, O.; Grosse, F.; Flissikowski, T.; Chze, C.; Consonni, V.; Geelhaar, L.; Grahn, H. T.; and Riechert, H. (2010). *Phys. Rev. B*, **82**, 045320.

Polenta, L.; Rossi, M.; Cavallini, A.; Calarco, R.; Marso, M.; Meijers, R.; Richter, T.; Stoica, T.; and Lüth, H. (2008). *AcsNano*, **2**, 287.

Qian, F.; Li, Y.; Gradečak, S.; Wang, D.; Barrelet, C. J.; and Lieber, C. M. (2004). *Nano Lett.*, **4**, 1975.

Qian, F.; Gradečak, S.; Li, Y.; Wen, C. Y.; and Lieber, C. M. (2005). *Nano Lett.*, **5**, 2287.

Qian, F.; Li, Y.; Gradečak, S.; Park, H.-G.; Dong, Y.; Ding, Y.; Wang, Z. L.; and Lieber, C. M. (2008). *Nature Mat.*, **7**, 701.

Renard, J.; Songmuang, R.; Bougerol, C.; Daudin, B.; and Gayral, B. (2008). *Nano Lett.*, **8**, 2092.

Renard, J.; Songmuang, R.; Tourbot, G.; Bougerol, C.; Daudin, B.; and Gayral, B. (2009). *Phys. Rev. B*, **80**, 12130.

Reschikov, M. A.; Behrends, A.; Bakin, A.; and Waag, A. J. (2009). *Vac. Sci. Technol B*, **27**, 1688.

Richter, T.; Lüth, H.; Meijers, R.; Calarco, R.; and Marso, M. (2008). *Nano Lett.*, **8**, 3056.

Rigutti, L.; Teubert, J.; Jacopin, G.; Fortuna, F.; Tchernycheva, M.; De Luna Bugallo, A.; Julien, F.; Furtmayr, F.; Stutzmann, M.; and Eickhoff, M. (2010). *Phys. Rev. B*, **82**, 235308.

Rigutti, L.; Tchernycheva, M.; De Luna Bugallo, A.; Jacopin, G.; Julien, F. H.; Zagonel, L. F.; March, K.; Stephan, O.; Kociak, M.; and Songmuang, R. (2010b). *Nano Lett.*, **10**, 2939.

Rigutti, L.; Jacopin, G.; De Luna Bugallo, A.; Tchernycheva, M.; Warde, E.; Julien, F. H.; Songmuang, R.; Galopin, E.; Largeau, L.; and Harmand, J. C. (2010c). *Nanotechnology*, **21**, 425206.

Ristic, J.; Calleja, E.; Sánchez-García, M. A.; Ulloa, J. M.; Sánchez-Páramo, J.; Calleja, J. M.; Jahn, U.; and Trampert, A. (2003). *Phys. Rev. B*, **68**, 125305.

Ristic, J.; Calleja, E.; Trampert, A.; Fernández-Garrido, S.; Rivera, C.; Jahn, U.; and Ploog, K. H. (2005). *Phys. Rev. Lett.*, **94**, 146102.

Ristic, J.; Rivera, C.; Calleja, E.; Fernández-Garrido, S.; Povoloskyi, M.; and Di Carlo, A. (2005b). *Phys. Rev. B*, **72**, 85330.

Ristic, J.; Calleja, E.; Fernández-Garrido, S.; Cerutti, L.; Trampert, A.; Jahn, U.; and Ploog, K. H. (2008). *J. Cryst. Growth*, **310**, 4035.

Romano, L. T. and Myers, T. H. (1997). *Appl. Phys. Lett.*, **71**, 3486.

Rose, A. (1955). *Phys. Rev.*, **97**, 1538.

Ruda, H. E. and Shik, A. (2005). *Phys. Rev. B*, **72**, 115308.

Sánchez-García, M. A.; Calleja, E.; Monroy, E.; Sanchez, F. J.; Calle, F.; Muñoz, E.; and Beresford, R. (1998). *J. Cryst. Growth*, **183**, 23.

Sanford, N. A.; Blanchard, P. T.; Bertness, K. A.; Mansfield, L.; Schlager, J. B.; Sanders, A. W.; Roshko, A.; Burton, B. B.; and George, S. M. (2010). *J. Appl. Phys.*, **107**, 034318.

Schricker, A. D.; Davidson III, F. M.; Wiacek, R. J.; and Korgel, B. A. (2006). *Nanotechnology*, **17**, 2681.

Sekiguchi, H.; Nakazato, T.; Kikuchi, A.; and Kishino, K. (2007). *J. Cryst. Growth*, **300**, 259.

Seryogin, G.; Shalish, I.; Moberlychan, W.; and Narayanamurti, V. (2005). *Nanotechnology*, **16**, 2342.

Shi, F.; Li, H.; and Xue, C. (2010). *J. Mater Sci. Mater. Electron.*, **21**, 1249.

Skierbiszewski, C.; Dybko, K.; Knap, W.; Siekacz, M.; Krupczynski,W.; Nowak, G.; Boaekowski, M.; Lusakowski, J.; Wasilewski, Z. R.; Maude, M.; Suski, T.; and Porowski, S. (2005). *Appl. Phys. Lett.*, **86**, 102106.

Soci, C.; Zhang, W. J.; Xiang, B.; Dayeh, S. A.; Aplin, D. P. R.; Park, J.; Bao, X. Y.; Lo, Y. H.; and Wang, D. (2007). *Nano Lett.*, **7**, 1003.

Soci, C.; Zhang, A.; Bao, X.-Y.; Kim, H.; Lo, Y.; and Wang, D. J. (2010). *Nanosci. Nanotechnol.*, **10**, 1430.

Songmuang, R.; Landré, O.; and Daudin, B. (2007). *Appl. Phys. Lett.*, **91**, 251902.

Songmuang, R.; Ben, T.; Daudin, B.; González, D.; and Monroy, E. (2010). *Nanotechnology*, **21**, 295605.

Songmuang, R.; Katsaros, G.; Monroy, E.; Spathis, P.; Bougerol, C.; Mongillo, M.; and De Franceschi, S. (2010b). *Nano Lett.*, **10**, 3545.

Songmuang, R.; Kalita, D.; Sinha, P.; den Hertog, M.; André, R.; Ben, T.; González, D.; Mariette, H.; and Monroy, E. (2011). *Appl. Phys. Lett.*, **99**, 141914.

Stach, E.; Pauzauskie, P. J.; Kuykendall, T.; Goldberger, J.; He, R.; and Yang, P. (2003), *Nano Lett.*, **3**, 867.

Stern, E.; Cheng, G.; Cimpoiasu, E.; Klie, R.; Guthrie, S.; Klemic, J.; Kretzschmar, I.; Steinlauf, E.; Turner-Evans, D.; Broomfield, E.; Hyland, J.; Koudelka, R.; Boone, T.; Young, M.; Sanders, A.; Munden, R.; Lee, T.; Routenberg, D.; and Reed, M. A. (2005). *Nanotechnology*, **16**, 2941.

Su, W. S.; Chen, Y. F.; Hsiao, C. L.; and Tu, L. W. (2007). *Appl. Phys. Lett.*, **90**, 063110.

Sundaram, V. S. and Mizel, A. (2004). *J. Phys. Con. Matt.*, **16**, 4697.

Suzuki, S.; Asada, M.; Teranishi, A.; Sugiyama, H.; and Yokoyama, H. (2010). *Appl. Phys. Lett.* **97**, 242102.

Sze, S. M. (1981). *Physics of Semiconductor Devices* (2nd edn.). Wiley, New York.

Tang, Y. B.; Bo, X. H.; Lee, C. S.; Cong, H. T.; Cheng, H. M.; Chen, Z. H.; Zhang, W. J.; Bello, I.; and Lee, S. T. (2008). *Adv. Funct. Mat.*, **18**, 3515.

Tang, Y. B.; Chen, Z. H.; Song, H. S.; Lee, C. S.; Cong, H. T.; Cheng, H. M.; Zhang, W. J.; Bello, I.; and Lee, S. T. (2008b). *Nano Lett.*, **8**, 4191.

Tarucha, S.; Austing, D.; Honda, T.; Van der Hage, R. J.; and Kouwenhoven, L. (1996). *Phys. Rev. Lett.*, **77**, 3613.

Teubert, J.; Becker, P.; Furtmayr, F.; and Eickhoff, M. (2011). *Nanotechnology*, **22**, 275505.

Thelander, C.; Martensson, T.; Björk, M. T.; Ohlsson, B. J.; Larsson, M. W.; Wallenberg, L. R.; and Samuelson, L. (2003). *Appl. Phys. Lett.*, **83**, 2052.

Tian, B.; Zheng, X.; Kempa, T. J.; Fang, Y.; Yu, N.; Yu, G.; Huang, J.; and Lieber, C. M. (2007). *Nature*, **449**, 885.

Tourbot, G.; Bougerol, C.; Grenier, A.; Den Hertog, M.; Sam-Giao, D.; Cooper, D.; Gilet, P.; Gayral, B.; and Daudin, B. (2011). *Nanotechnology*, **22**, 075601.

Van de Walle, C. G. and Segev, D. (2007). *J. Appl. Phys.*, **101**, 081704.

Wagner, R. S. and Ellis, W. C. (1964). *Appl. Phys. Lett.*, **4**, 89.

Wang, D. W.; Wang, Q.; Jaey, A.; Tu, R.; Dai, H. J.; Kim, H.; McIntyre, P. C.; Krishnamohan, T.; and Saraswat, K. C. (2003). *Appl. Phys. Lett.*, **83**, 2432.

Wang, Z. L. and Song, J. (2006). *Science*, **312**, 248.

Wang, X.; Song, J.; Liu, J.; and Wang, Z. L. (2007). *Science*, **316**, 102.

Wright, J. S.; Lim, W.; Gila, B. P.; Pearton, S. J.; Johnson, J. L.; Ural, A.; and Ren F. (2009). *Sensors and Actuators B*, **140**, 196.

Xiang, J.; Lu, W.; Hu, Y.; Wu, Y.; Yan, H.; and Lieber, C. M. (2006). *Nature*, **441**, 489.

Xinyuan, Z.; Wei, C. M.; Yang, L.; and Chou, M. Y. (2004). *Phys. Rev. Lett.*, **92**, 236805.

Xu, X.; Potié, A.; Songmuang, R.; Lee, J.; Bercu, B.; Baron, T.; Salem, B.; and Montès, L. (2011). *Nanotechnol.*, **22**, 105704.

Yamamoto, S.; Mori, M.; Kuwahara, Y.; Fujii, T.; Nakao, T.; Kondo, S.; Iwaya, M.; Takeuchi, T.; Kamiyama, S.; Akasaki, I.; and Amano, H. (2012). *phys. stat. sol. (RRL)*, **6**, 145.

Yamamoto, S.; Mori, M.; Nakao, T.; Kondo, S.; Kuwahara, Y.; Morita, Y.; Fujii, T.; Sugiyama, T.; Iwaya, M.; Takeuchi, T.; Kamiyama, S.; Akasaki, I.; and Amano, A. (2012b). *SPIE Proc.*, **8262**, 37.

Yoshizawa, M.; Kikuchi, A.; Mori, M.; Fujita, N.; and Kishino, K. (1997). *Jpn. J. Appl. Phys.*, **36**, L459.

Yoshizawa, M.; Kikuchi, A.; Fujita, N.; Kushi, K.; Sasamoto, H.; and Kishino, K. (1998). *J. Cryst. Growth*, **189/190**, 138.

Yu, H. Y.; Kang, B. H.; Park, C. W.; Pi, U. H.; Lee, C. J.; and Choi, S. Y. (2005). *Appl. Phys. A.*, **81**, 245.

Zagonel, L. F.; Mazzucco, S.; Tencé, M.; March, K.;. Bernard, R.; Laslier, B.; Jacopin, G.; Tchernycheva, M.; Rigutti, L.; Julien, F. H.; Songmuang, R.; and Kociak, M. (2011). *Nano Lett.*, **11**, 568.

Zhang, A.; You, S.; Soci, C.; Liu, Y.; Wang, D.; and Lo, Y.-H. (2008). *Appl. Phys. Lett.*, **93**, 121110.

Zhang, Y.; Liu, Y.; and Wang, Z. L. (2011). *Adv. Mater.*, **23**, 3004.

Zheng, G.F.; Lu, W.; Jin, S.; and Lieber, C. M. (2004). *Adv. Mater.*, **16**, 1890.

Zhong, Z.; Qian, F.; Wang, D.; and Lieber, C. M. (2003). *Nano Lett.*, **3**, 343.

Zhou, S. M.; Feng, Y. S.; and Zhang, L. D. (2003). *Chem. Phys. Lett.*, **369**, 610.

10 Advanced photonic and nanophotonic devices

Jean-Yves Duboz

GaN materials are currently reaching a maturity which allows them to be used for many exploratory fundamental thematics and applications. Among them, advanced photonic devices are rapidly emerging. This fast progress arises mainly from the fact that the concepts behind these devices have been explored and demonstrated with other materials—in particular, GaAs-based heterostructures. The motivation for copying or extending this research in GaN-based heterostructures lies in the specific properties of nitrides (band-gap energies and offsets, excitonic parameters) which permit demonstration of the same concepts at much higher temperatures for electrical/optical injection. Hence, these novel devices have a chance to move from research laboratories to real-life applications. The operating wavelength also is a major difference, changing from the near-infrared for arsenides to the visible/near-UV for nitrides. While many novel devices based on GaAs are targeting fiber-optics communications, novel devices based on GaN are aiming at other applications, some of them being identified, for instance in displays, optical storage, and processing, and some of them still remaining to be defined. Among these novel devices, I will describe the progress in the field of polariton lasers and Bose–Einstein condensation in planar microcavities. Although nitride cavities still have much smaller Q factors than arsenide or CdTe cavities, quite comparable results have been obtained thanks to the intrinsic excitonic features of GaN. I will also describe photonic crystal nanoresonators, where the results are improving very rapidly, and microdisk resonators. Although quality factors in these cavities are decades below the state of the art reached in Si or GaAs, interesting results have already been obtained. Micro- and nanowires, with axial or whispering gallery modes, will finally be described. Applications to devices such as nanowire emitters will be briefly mentioned.

10.1 Planar microcavities

10.1.1 *Linear regime: basics*

A planar microcavity is an active medium with a refractive index n_r and a thickness L in between two planar mirrors, forming a cavity of length L_{cav}. Note that L_{cav} can be larger than L for non metallic mirrors. The optical transmission spectrum shows resonances for wavelengths equal to $2n_rL_{cav}/p$, where p is the order of the mode. Microcavities are designed to operate with small p values so that the free spectral range (FSR) between the p mode and the next one is large, and only one longitudinal mode (with energy E_c) will be considered. The width of the cavity mode Γ_c is related to the cavity Q factor ($Q = E_c/\Gamma_c$). Q describes the energy stored in the cavity or the reinforcement (compared to free space) of the electromagnetic field. As Q increases, the interaction of the field with the active medium increases and leads to peculiar effects. This is obtained for large mirror reflectivities R. The cavity finesse defined by

$$F = \frac{FSR}{\Gamma_c} = \frac{\pi}{2arcsin\left(\frac{1-R}{2\sqrt{R}}\right)} \tag{10.1}$$

takes a simple expression $F = \pi/(1-R)$ when R approaches unity. The active medium, a semiconductor, has excited states called excitons corresponding to correlated electron-hole pairs. The system is out of equilibrium and needs to be pumped by an external (optical or electrical) excitation. Excitons can relax and emit a photon at the energy E_X. In a cavity the photon will be reflected by mirrors and stay in the cavity for a time equal to $\hbar/\Gamma_c = Q\hbar/E_c$, assuming that the photon energy is close to the cavity mode energy. As Q increases, this time increases and can be long enough so that the photon is reabsorbed by the medium before escaping the cavity. In this case, the excitation in the system is partly a photon (when bouncing back and forth in the cavity) and partly an electronic excitation (when absorbed in the medium). It is described in terms of a polariton—a particle that mixes a photon and an exciton. The system is said to be in the strong coupling regime. The condition for reaching the strong coupling is that the interaction time (time to emit or to absorb a photon) is shorter than the photon lifetime (limited by the escape from the cavity through the mirror) and the exciton lifetime (limited by non-radiative decay). This condition can also be expressed in terms of energy. If the homogeneous spectral broadening of the exciton is Γ_X, then one can calculate the energy of the two new eigenmodes resulting from the coupling (of energy Ω) of the photon with the exciton :

$$E_{\pm} = \frac{E_c + E_X}{2} \pm \sqrt{\left(\frac{E_c - E_X}{2}\right)^2 + \Omega^2 - \left(\frac{\Gamma_X - \Gamma_c}{2}\right)^2} \tag{10.2}$$

When the photon and exciton mode energies are similar, this expression becomes

$$E_{\pm} = \frac{E_c + E_X}{2} \pm \sqrt{\Omega^2 - \left(\frac{\Gamma_X - \Gamma_c}{2}\right)^2} \tag{10.3}$$

A splitting between the two modes appears if $\Omega > 1/2(\Gamma_X - \Gamma_c)$. Slightly different expressions of the spliting can be found (Kavokin 2003), and the condition that is generally used is $\Omega > \Gamma_X$ and Γ_c, or $\Omega > \sqrt{\Gamma_X \times \Gamma_c}$. This is the strong coupling regime, characterized by the Rabi energy 2Ω which is the mode-splitting energy when the photon and exciton mode broadening can be neglected.

Ω describes the intensity of the coupling between the photon and the exciton. It is given by

$$\Omega^2 = \frac{\left(1+\sqrt{R}\right)^2}{2\sqrt{R}} \frac{e^2}{\epsilon_0 n_r^2 m_0} \frac{f_{osc}}{L_{cav}} \tag{10.4}$$

For a high-cavity finesse, Ω^2 can be expressed as $F\alpha_0 L\Gamma_X\Gamma_c/\pi$, where α_0 is the excitonic absorption coefficient. The strong coupling condition $\Omega > \sqrt{\Gamma_X \times \Gamma_c}$ becomes $\alpha_0 L > \pi/F = 1 - R = T$. This means that the photon absorption in the cavity must be larger than the losses due to the mirror transparency, which is the condition intuitively expressed earlier. As the excitonic absorption coefficient in nitrides in very high, the strong coupling condition is actually not so difficult to meet in terms of cavity finesse. With $\alpha_0 = 5 \times 10^4 cm^{-1}$ in bulk GaN, L = 40 nm (this length of interaction takes the field distribution into account), which gives a 0.8 reflection coefficient. Let us take R = 0.85. This corresponds to a finesse of 20. In a typical microcavity with Bragg mirrors, the cavity length L_{cav} is much larger (by a factor 5 to 10) than the active medium thickness L due to a non-negligible penetration depth of light in the Bragg mirrors. Hence, the typical mode order p is ranging from 5 to 10. For p = 8, the free spectral range is of the order of 400 meV. The F = 20 cavity thus has an optical mode broadening of 20 meV and a Q factor of 160. At 300 K the exciton broadening is in the range of 15–20 meV (Siozade 2000). Hence, the strong coupling regime can be achieved at room temperature with a modest Q-factor cavity.

10.1.2 *Linear regime: results*

The first observation of strong coupling in GaN was achieved in a modest Q factor (Q = 60) cavity (Antoine-Vincent 2003) at low temperature and then at room temperature (Semond 2005). Compared to GaAs, one peculiarity of nitrides is that the spin–orbit coupling is small and the three excitons A, B, and C related to the valence band maxima are close to each other and can couple to the cavity mode. At low temperature, this leads to a fine polariton structure where A and B polaritons can be identified (Sellers 2006*a*). At room temperature, A, B, and C excitons are not clearly resolved but still contribute to the coupling, with an oscillator strength depending on polarization (Butte 2006). It was pointed out that the oscillator strength is one order of magnitude larger in bulk GaN than in bulk GaAs. This explains why the strong coupling could be observed in modest Q-factor cavities in GaN, while the same observation in II-VIs or in arsenides required larger Q factors (Q = 350 in GaAs QWs (Weisbuch 1992), Q = 650 in ZnCdSe QWs (Kelkar 1995), Q = 415 in CdTe QWs (LeSiDang 1998)). In addition, the latter achievements in GaAs and II-VIs were based on QWs, i.e. less material than with a bulk active region. The strong coupling was also achieved with GaN QWs in a cavity with larger Q values (200–800) (Feltin 2006; Christmann 2006*a*). Again, it was pointed out that the oscillator strength in nitride QWs is ten times larger than in arsenide QWs, explaining that the strong coupling regime can be obtained in spite of broader exciton and photon modes. The impact of the inhomogeneous excitonic broadening on the strong coupling was analysed. The strong coupling condition is related to the exciton coherence time, and is thus related to the homogeneous broadening. However, when the inhomogeneous broadening increases, the number of excitons coupled to the mode decreases (or their effective oscillator strength decreases) and the strong coupling may disappear. The transition value for the inhomogeneous broadening was estimated to be 45 meV, i.e. larger than the Rabi splitting of 30 meV (Christmann 2006*a*). However, when GaN QWs are replaced by InGaN QWs, the much larger inhomogeneous broadening prevents any strong coupling observation, in contradiction to a claim for a strong coupling observation published by another group (Tawara 2004). The inhomogeneous broadening of the cavity mode due in plane fluctuations of the cavity thickness is also a key parameter. By reducing the area of the optical analysis the Q factor of the cavity is significantly increased from 200 to 2800 (Christmann 2006*b*). The nearly optimized cavity was finally obtained by increasing the number of QWs, leading to a Rabi splitting of 50 meV at 300 K (Christmann 2008).

Another specificity of nitride is the fact that they can be grown on various substrates, still keeping good optical and electrical properties. This has been demonstrated for LEDs or HEMTs, where high-performance devices have been fabricated on GaN, sapphire, SiC, and Si substrates. This is in contrast with arsenides, where homoepitaxy has been the only approach so far for obtaining device-quality material. The first observation of strong coupling was made in GaN grown on Si (Antoine-Vincent 2003), where the Si acts as a mirror. Latter, a bottom nitride Bragg mirror with larger reflectivity was added (Sellers 2006*a*). Cavities were also grown on sapphire. In this case, the bottom Bragg mirror was AlInN/GaN lattice-matched to GaN (Feltin 2006*a*), thus yielding both a high Q factor (a few thousands) and a good active region crystallographic quality (narrow exciton linewidth) (Feltin 2006*b*; Christopoulos 2007). Today, the highest reported Q factor in nitrides is 6400 (Butte 2009*a*). Cavities have also been grown on 6H-SiC (Tawara 2004) and on GaN substrates (Lu 2011). While the top mirror is almost always a dielectric Bragg mirror, various approaches have been used for the bottom mirror. In addition to epitaxial AlGaN/AlGaN or AlInN/AlGaN Bragg mirrors already mentioned, dielectric stacks are also used as bottom Bragg mirrors after substrate removal. This was done first for LEDs on Si in the weak coupling regime (Duboz 2003), and was extended to cavities in the strong coupling regime (Bejtka 2008). The technological process takes advantage of the easy removal of the Si substrate. Removing substrates others than Si is more difficult but has nevertheless been achieved with SiC by reactive ion etching, leading to a $Q = 400$ cavity after bottom-mirror deposition (Tawara 2004), and with sapphire by laser lift-off leading to a $Q = 500$–1000 after bottom-mirror deposition (Pattison 2007; Song 1999).

Interestingly enough, the technological development achieved in nitride cavities has allowed development of ZnO-based cavities which include an epitaxial

Table 10.1 Typical exciton energy (in bulk and in QWs) and observed Rabi splitting energies in various material systems.

Material	$E_X^{3D}(meV)$	$E_X^{2D}(meV)$	Rabi splitting (meV)
GaAs	4.8	$8-14$	$6-20$
CdTe	10	25	$20-30$
GaN	26	50	$40-90$
ZnO	60	>80	100

nitride bottom mirror (Medard 2009; Guillet 2011*a*; Guillet 2011*b*) and also development of perovskite microcavities (Lanty 2011). Table 10.1 shows that the larger exciton binding energy, associated to a larger oscillator strength, in GaN (also in ZnO) leads to larger Rabi splitting energies than in smaller band-gap materials. This chapter does not cover cavities in the weak coupling regime. We just recall that very early cavities were fabricated in the 1990s for LEDs and optically pumped VCSELs (Redwing 1996; Krestnikov 1999; Someya 1998). While the main differences lie in the active region (thickness, homogeneous broadening), the cavity fabrication is similar for weak and strong coupling regimes, and high Q cavities have been fabricated by hybrid approaches with a bottom epitaxial mirror (Kako 2002) or by a fully dielectric approach involving substrate removal (Song 2000*a*; Song 2000*b*; Martin 2001).

10.1.3 *Non-linear regime: basics*

In the strong coupling regime, the system can be described in terms of polaritons which are composite bosons. While this has no consequence in the linear regime, it becomes a key factor in the non-linear regime. Cavity polaritons have a dispersion (energy versus in-plane wave vector) which depends on the detuning between the bare cavity mode and the exciton mode. In the so-called positive detuning (the cavity mode energy is larger than the exciton energy, so that uncoupled modes do not overlap) the anticrossing of coupled modes leads to a lower polariton branch which is dominated by the exciton dispersion with a shallow minimum at k = 0. In the negative detuning the lower polariton branch has a deep minimum at k = 0 dominated by its photon part. In the k = 0 region where the polariton is dominated by its photon part (in particular in the negative detuning situation), the polariton mass can be deduced from the optical mode dispersion to be equal to

$$m_{pol} = \frac{p\pi\hbar n_r}{cL_{cav}} = \frac{n_r^2 h}{c\lambda} = \frac{n_r^2}{c^2}E_{phot} \tag{10.5}$$

where E_{phot} is the photon energy. For a GaN cavity, this is about $2 \times 10^{-5}m_0$. Compared to excitons, which have a mass of the order of m_0, polaritons have a much lower mass. It is interesting to compare this with GaAs and CdTe, where the photon energy is smaller but the refractive index n_r is larger, so that the polariton mass is about the same. Bosons distribute in energy following the Bose–Einstein statistics:

$$f_{BE}(E, T, \mu) = \frac{1}{exp\left(\frac{(E-\mu)}{kT}\right) - 1} \tag{10.6}$$

where the particle energy E is positive or nul (we take the ground-state energy equal to zero as a reference) and μ is negative or nul. The chemical potential μ is given by the fact that the sum of all occupied states should give the number of particles. The chemical potential increases with the particle density. In 1D or 2D one can show that the sum over all states, except the E = 0 state, diverges when the chemical potential tends to zero. Hence, one can always find a non-zero value for μ which leads to an arbitrary large density of bosons. However, in 3D, the sum over all states, except the E = 0 state, is convergent, and even for $\mu = 0$, cannot be larger than a critical density n_c. Hence, if the density exceeds n_c, the extra particles $n - n_c$ must condense in the E = 0 ground state which has an infinite occupation for $\mu = 0$. This is the Bose–Einstein condensation in 3D proposed by Bose and Einstein in 1925, and experimentally demonstrated in 1995 in trapped alkali atoms leading to the Nobel Prize in 2001 (Nobel 2001). In 1D and 2D, the system is not forced to fully condense in the ground state but will locally condense in droplets, and the system experiences a phase transition towards a superfluid state, refered to as the Kosterlitz–Thouless transition (Kavokin 2003). It is interesting to note that the size of a droplet can be comparable to the size of the real system under study (less than 100 μm), so that locally the BE condensation and the KT transition are equivalent to each other (Kavokin 2003). This transition occurs at a finite temperature, below which a normal phase and a superfluid phase coexist. Although not strictly equivalent, the Bose–Einstein condensation and the KT transition are similar to each other, and can be qualitatively described by the same rough and simple consideration. On the one hand, in the KT transition in 2D at the temperature T_{KT}, the density of particles in the superfluid phase is given by $n_s = \frac{2mkT_{KT}}{\pi\hbar^2}$. On the other hand, the criterion for condensation is that the de Broglie wavelength becomes comparable to the distance between particles:

$$\lambda_{DB} = \sqrt{\frac{2\pi\hbar^2}{mkT}} = (1/n_{2D})^{1/2} \tag{10.7}$$

Both criteria turn out to be similar, within a factor of 2. Let us emphasize that this description is very basic. In the high-temperature-high density region, the interactions between bosons, which are an essential ingredient of the KT transition, play an important role, and a more refined approach is needed (Kavokin 2003). The important point arising from our basic description is the following. By using the Bose–Einstein condensation criterion, one immediately observes that the

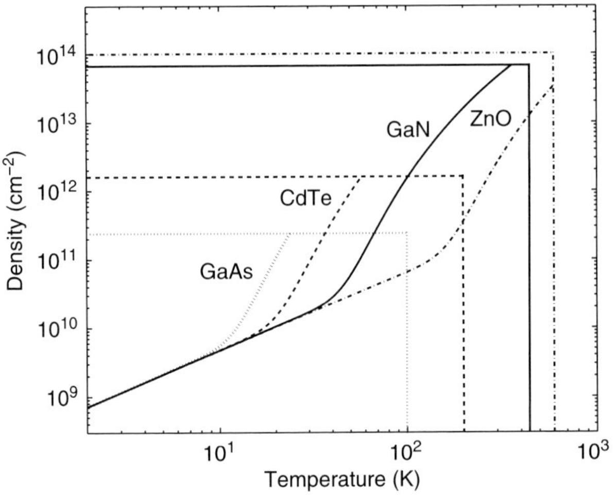

FIG. 10.1. Phase diagram for GaAs, CdTe, GaN and ZnO based microcavities at zero detuning. At high temperature, the system is in the weak coupling regime due to thermal exciton dissociation. Also, at high densities, Coulombic interactions and Mott transition break excitons and lead to a weak coupling regime. (From Kavokin (2003).)

transition temperature increases as 1/m. Compared to atoms, or even to excitons, polaritons are much lighter and offer a perspective of condensation at much higher temperatures.

Figure 10.1 shows the phase diagram for GaAs, CdTe, GaN, and ZnO. One observes that condensation can occur at much higher temperatures in GaN and ZnO than in GaAs or CdTe. The diagram also shows that the polariton condensate can make a transition at high density towards a VCSEL. This happens because of the composite nature of the polariton, which is not a pure boson, which induces polariton–polariton interactions. When the distance between polaritons or excitons becomes comparable to the exciton radius, then the Coulombic interactions can no longer be neglected, and induce a transition from the exciton gas to an electron-hole plasma (polariton Mott transition). The criterion for the excitonic Mott transition in 2D can be roughly written as

$$(n_{Mott})^{-1/2} = 2 \times r_{exc} = \frac{e^2}{4\pi\epsilon_0\epsilon E_{exc}} \tag{10.8}$$

where ϵ is the static dielectric constant of the material. Equation 10.8 shows that the transition density increases as $E_{exc}^2\epsilon^2$. Table 10.2 shows the exciton Bohr radius and exciton Mott density for 2D and 3D systems for various semiconductors.

The phase diagram in Fig. 10.1 shows this transistion as an horizontal line. Above this transition, excitons and polaritons are broken and the system is no longer in the strong coupling regime. At high temperature, excitons and

Table 10.2 Exciton Bohr radius and Mott transition exciton density (in bulk and in QWs) in various material systems.

Material	3D Bohr rad. nm	3D Mott dens. cm^{-3}	2D Bohr rad. nm	2D Mott dens. cm^{-2}
GaAs	11.6	7×10^{16}	$8.5(QW7nm)$	2×10^{11}
CdTe	6.8	4×10^{17}	$5.0(QW4nm)$	1×10^{12}
GaN	2.8	5×10^{18}	$2.5(QW2nm)$	4×10^{12}

polaritons are broken by the thermal energy when $kT > E_{exc}$. The transition temperature increases as the exciton energy. This phase diagram shows that GaN, together with ZnO, are ideal candidates for condensation at high temperature. In addition to thermodynamic consideration, the condensation also depends on kinetics. Contrary to cold atoms, polaritons are out of equilibrium and must be created continuously. Before condensing in the ground state, polaritons have to be created in excited states and then have to relax their energy along the dispersion curve. The injection is achieved either electrically or, more frequently, optically. Excitons are created at high energy, relax in the lower polariton branch (LPB) at large wavevectors by optical phonon emission, and then relax along the LPB by acoustic phonon emission and other inelastic scattering phenomena. The energy and wavevector are conserved at each phonon emission. For negative detuning (and even for small positive detuning), the LPB shows a clear minimum at k = 0 (trap), a flat region at large k, and an inflexion region between. Relaxation along the LPB in the flat region is mediated by acoustic phonons. In the inflexion region the slope ($>10^5$ m/s) becomes much larger than the sound velocity (<8000 m/s in GaN (Deger 1998)) and the energy–wavevector conservation can no longer be verified. Relaxation by acoustic phonon emission becomes inefficient, and polaritons accumulate in this region (relaxation bottleneck) (Kavokin 2003). Other processes such as polariton–polariton and electron–polariton scattering become the most effective relaxation mechanisms.

While the inflexion point in the LPB tends to prevent condensation, it can become beneficial in other cases. The middle of the inflexion region can be seen as a symetry point between the k = 0 region and the flat region at larger k. Then, a polariton–polariton scattering between two polaritons sitting in the inflexion-point region can lead to one polariton scattered to k = 0 (ground state) and the other one scattered to a large k value, with energy and wavevector conservation. This scattering can be stimulated by the polariton occupancy of the ground state (boson effect), which can be achieved by sending a weak optical beam normal to the cavity. As a result, the ground-state occupancy will be increased and the beam at normal incidence amplified. This stimulated scattering or parametric amplification first became evident in 2000 in GaAs cavities (Savvidis 2000).

10.1.4 *Non-linear regime: results*

Although not within the scope of this chapter, we recall some results obtained in the weak coupling regime. Optically pumped VCSELs have been obtained in high Q cavities (Someya 1998; Someya 1999; Song 2000*a*; Kako 2002). The threshold is clearly observed, and can be expressed either in terms of absorbed power in the active region (a few kW/cm^2) or in terms of electron-hole pair density (a few 10^{19} cm^{-3}). The non-linearity of the emission around the threshold allows measurement of the β factor in the range of 2×10^{-2}. Later, current-injected VCSELs were demonstrated at low temperature (Lu 2008*a*) and room temperature (Higuchi 2008; Omae 2009; Lu 2010; Kasahara 2011). The β factor in an electrical device is reduced to 5×10^{-3}, due to electrical contacts, and the Q factor is in the range of 200 with a bottom epitaxial mirror (Lu 2010). With a more complex process and a full dielectric cavity, Q can be as high as 3700 (Higuchi 2008). These devices are in the weak coupling regime due to the reduced number of QWs, and/or the inhomogeneous broadening of the InGaN QWs, which reduces their effective oscillator strength.

In the strong coupling regime, polariton lasers were devised a long time ago (Immamoglu 1996). The theory predicted that polariton lasers should be obtained in GaN at room temperature (Malpuech 2002). It took a few years, however, before a real breakthrough was achieved. Shortly after the observation of condensation effects in CdTe at low temperature in 2006 (Kasprzak 2006), polariton lasing was observed at room temperature in GaN (Christopoulos 2007). The cavity was made of bulk GaN with a bottom AlInN/AlGaN Bragg mirror and a top dielectric mirror. Figure 10.2 shows the non-linearity of the emission at threshold, suggesting a β factor of 1.5×10^{-4}, which is rather low. The emission remains at the lower polariton branch energy with a small energy shift due only to polariton–polariton interactions. The emission becomes very narrow ($\leq$1.7 meV) at threshold, consistent with a sudden increase of the temporal

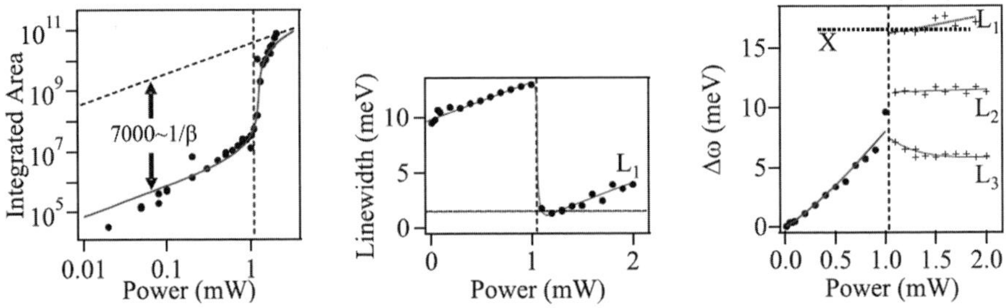

FIG. 10.2. Luminescence properties from a GaN-based cavity as a function of pump power at room temperature. (Left) Integrated intensity; (middle) luminescence linewidth; (right) energy shift. (Reprinted with permission from Christopoulos (2007), © 2007 by the American Physical Society.)

coherence (a temporal coherence of 700 fs has been directly measured in agreement with the emission linewidth) and the massive occupancy of the ground state. Authors estimate that the ground-state occupancy reaches 10 above threshold.

The optical pumping power at threshold is 1 mW (or 30 W/cm^2), corresponding to an estimated polariton density of 2×10^{18} cm^{-3}. This is one order of magnitude lower than in VCSELs operating in the weak coupling regime, thus realizing the hopes for a low threshold (ideally thresholdless) laser. This density remains smaller than the Mott density for excitons measured by Binet *et al.* (Binet 1999), in the range of 3×10^{18} cm^{-3}. As interactions between polaritons are twice smaller than between excitons, a polariton density of 2×10^{18} cm^{-3} can clearly be considered to be below the polariton Mott transition. Interestingly, no relaxation bottleneck was observed, suggesting that the dynamics might be more favorable at room temperature than at low temperature, which is an additional advantage of nitrides operating at room temperature. One year after this breakthrough, the same team observed polariton lasing in a similar cavity but with GaN quantum wells as the active region (Christmann 2008). The threshold pumping power density was about 20 W/cm^2, comparable to the value with bulk GaN, and the exciton density less than 10^{10} cm^{-2} in each quantum well, which is two orders of magnitude below the Mott density in a QW (Table 10.2).

Detailed study of the threshold power density as a function of temperature and detuning emphasised kinetic (dynamics of relaxation along the LPB at low temperature) or thermodynamic (thermal escape of polaritons from the ground state at high temperature) dominating effects (Butte 2009*b*). Optimum detuning between cavity and exciton modes, i.e. leading to the lowest pumping-power threshold, moves towards more negative values when the temperature increases (Levrat 2010). The polarization of the light emitted by the condensate is an interesting property. In GaAs (Ballili 2007) and CdTe (Kasprzak 2006) quantum-well cavities, the light is linearly polarized along some crystallographic direction, due to some residual anisotropy. In the case of a cavity based on bulk GaN, light emission is found to be polarized (linearly or circularly) within each pulse, but with a random change from pulse to pulse (Baumberg 2008). Hence, the ground state appears to be degenerate and a spontaneous symmetry breaking appears in the condensate, which is considered to be evidence of a BEC. An important aspect of the condensation is the spatial coherence. In the condensed phase or condensed droplet, the system should exhibit lateral spatial coherence. In other words, one should be able to construct interferences between two separate regions of the emission spot. This has been observed in CdTe (Kasprzak 2006) and GaAs-based condensates (Wertz 2010; Deng 2007; Ballili 2007), but not yet in GaN condensates.

Polariton lasing has also been observed in a hybrid system where the ZnO active region is grown on top of nitride Bragg mirrors (Guillet 2011*a*; Guillet 2011*b*). Although not a pure GaN result, this achievement clearly follows from the developments made in the field of nitride cavities, and deserves to be mentioned here.

It has been observed that depending on temperature and detuning, the device can exhibit a laser action either in the weak coupling (VCSEL) (Guillet 2011*a*) or in the strong coupling regime (polariton laser) (Guillet 2011*b*). Polariton lasing was claimed at room temperature also in a similar nitride/ZnO cavity by another group (Lu 2012). But the clearest results in the domain remain within the French consortium which has obtained the full condensation phase diagram at room temperature in a ZnO cavity (private 2012). Still in the field of hybrid systems, a quite surprising result has been obtained in a full dielectric cavity where a GaN nanowire has been inserted in the central region (Das 2011). The nanowire (60-nm diameter, 750 nm long) was grown by MBE, detached and deposited on an SiO_2 layer on top of an SiO_2/TiO_2 Bragg mirror. Then SiO_2 and a second SiO_2/TiO_2 Bragg mirror were deposited, leading to a symmetric and full dielectric cavity. The advantage of this structure is to combine a very-high-quality material (nanowires are almost defect free) and a cavity with very-high-reflectivity DBRs and a small effective length, hence leading to a large Rabi splitting. As a matter of fact, a Rabi splitting of 48 meV was measured. The detuning was chosen negative (−20 meV), which is the most favorable case for condensation at room temperature. When increasing the pump power, a first threshold was observed, and was attributed to polariton lasing, as shown in Fig. 10.3. The exciton density at threshold is estimated to be in the range of $2 \times 10^{16} - 1 \times 10^{17}$ cm^{-3}, which is surprisingly low. A second threshold is observed at much higher (by a factor 2700) power than the first one, attributed to a weak coupling transition and laser operation. Similar observations

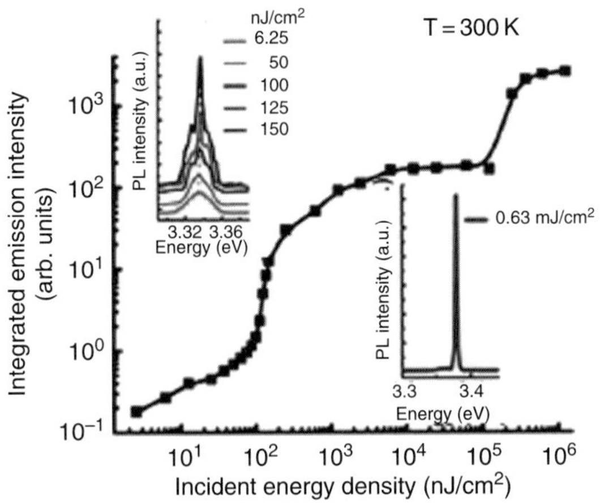

FIG. 10.3. Integrated luminescence from a GaN nanowire inserted in a cavity as a function of pump energy at room temperature. The insets show the corresponding emission spectra. (Reprinted with permission from Das (2011), © 2011 by the American Physical Society.)

(i.e. two consecutive thresholds) had been made previously in GaAs cavities (Nelsen 2009; Bajoni 2008).

As can be seen from the aforementioned GaN polariton results, progress in nitrides is following that in GaAs with a couple of years delay. In 2008 a few groups reported polariton emission under current injection in GaAs cavities at temperatures from 10 K to 300 K (Khalifa 2008; Bajoni 2008*b*; Tsintzos 2008; Tsintzos 2009). From a theoretical point of view, a current-injected polariton laser should be feasible in nitrides (Solnyshkov 2009). In nitride cavities, however, electrical injection is more difficult, in particular on the p side. Injection

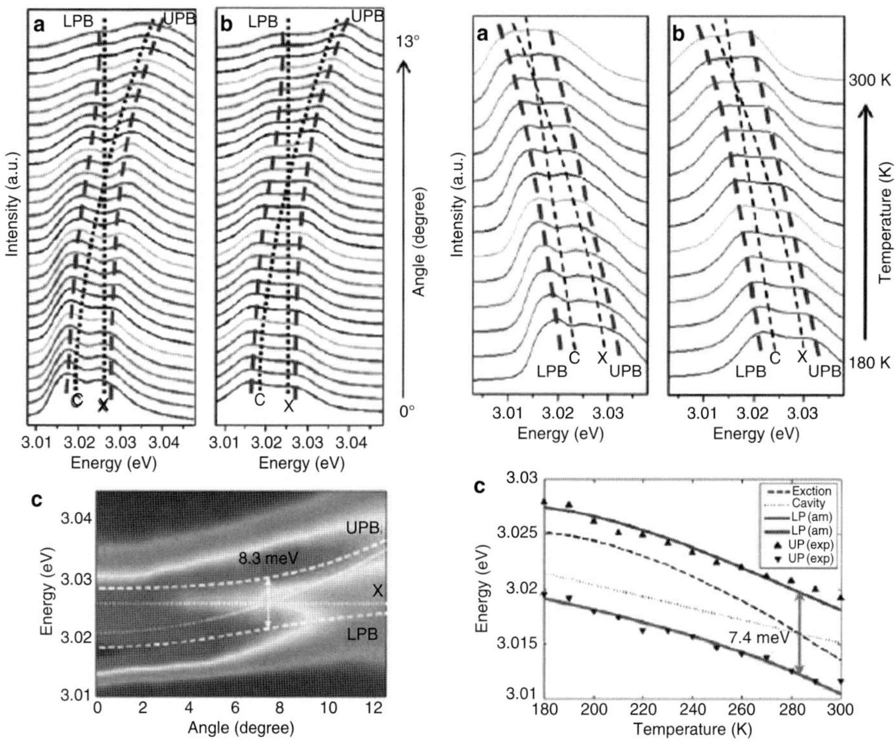

FIG. 10.4. Left: Experimental angle-resolved electroluminescence spectra (a) at 180 K and calculated angle-resolved transmission spectra (b) from 0° to 13°. (c) Measured polariton angular dispersion. The horizontal dotted lines indicate the bare exciton mode, and the curve dotted line the cavity mode. Right: Experimental temperature-dependent electroluminescence spectra (a) and calculated temperature-dependent transmission spectra (b) from 180 to 300 K. (c) Extracted experimental peaks, and calculated uncoupled exciton and photon modes and polariton modes. (Reprinted with permission from Lu (2011), © 2011 by the American Chemical Society.)

through epitaxial DBRs is highly challenging, and is impossible for dielectric DBRs. In these conditions, intracavity contacts are to be preferred, and lead to good performance VCSELs (Kasahara 2011). A polariton LED has been reported at room temperature (Lu 2011). The cavity was grown on a GaN substrate, with an epitaxial nitride DBR, an InGaN/GaN QW-based active region, and a top dielectric mirror. Intracavity contacts were used in order to inject the current. Figure 10.4 shows the electroluminescence as a function of temperature, showing the anticrossing at a temperature close to 300 K. The anticrossing was also evidenced by studying the luminescence as a function of angle. An original approach combining an electrically driven LED and MQW structure inside a cavity has been described by the EPFL group (Butte 2011). According to a rough estimation, an LED with a realistic efficiency is powerful enough to optically pump the QWs to a level sufficient to reach condensation. The experimental demonstration remains to be done. The current situation is that there has not been any report of a polariton laser with electrical injection so far. However, given the fast learning curve in this field, it is quite realistic to state that an electrical polariton laser based on nitrides should be achieved in the coming years. Similarly, one should be able, in the near future, to create and manipulate polaritons in nitrides as is done in arsenides (Ferrier 2011; Tosi 2012), with the difference that it might be possible to do it at room temperature and under higher densities, opening new avenues in optics.

10.2 Photonic crystal

Photonic crystals (PCs) were proposed in 1987 by E. Yablonovitch (Yablonovitch 1987) and S. John (John 1987). These pioneers had to face some criticism (Maddox 1990) for three years before the concept was demonstrated (Ho 1990; Yablonovitch 1991). The photonic band-gap material is a generalization to 3D of the grating or of the Bragg mirror. The destructive interference between all diffracted waves can completely forbid the propagation of waves in any direction, any polarization, and for a given spectral window. Then, the concept has been enlarged to structures where the propagation is severely affected in many directions. This is particularly the case for semiconductors where the fabrication of 3D PCs is very challenging, and where most fabricated structures are 2D+1D: in one direction, the light is confined in a waveguide and the propagation along the guide is controlled by a 2D grating. In this case, the 2D grating may forbid the propagation of guided modes but does not forbid the propagation of non-guided modes. PC can be used mainly in two different ways. First, optical modes close to the band-gap edges present peculiar properties (slow propagation, negative effective index) which can be used for increasing the interaction with the material and for special optic effects (super prism, negative lens, invisibility cap...). They can also be used as multidirectional feedback devices to obtain surface emitting sources (high-power lasers or LEDs). Second, a cavity can be introduced in the PC leading to a high-Q and small-volume optical resonator. This is the world of nanophotonics with low-power quantum devices. In this case, the spontaneous emission rate of an emitter ideally located inside such a cavity can be enhanced by the so-called Purcell factor, which is given by

$$F_P = \frac{3}{4\pi^2}\frac{Q}{V_{eff}}\left(\frac{\lambda}{n_r}\right)^3 \tag{10.9}$$

where Q is the cavity quality factor, n_r is the refractive index, and V_{eff} is the effective mode volume. The interest in photonic crystal cavities can be immediately seen from this expression, as the modal volume can be made very small, in the range of the wavelength in the material, λ/n_r. This can be understood easily in the simplest cavity obtained by altering one period, which is equal to λ/n_r, thus localizing the field on such a scale. Many fascinating results were achieved in such cavities fabricated in a membrane based on silicon or arsenides. The Purcell effect (observation of a Purcell factor significantly larger or smaller than unity) in the weak coupling regime was observed with a single GaAs quantum dot in a photonic crystal membrane nanocavity (Badolato 2005). The strong coupling was observed with an arsenide single quantum dot (Yoshie 2004; Reithmaier 2004). Low-threshold nanolasers based on PCs were demonstrated with InAs quantum dots (Bordas 2009; Nomura 2009). In terms of pure cavity performance, Q values in excess of 1 million and volumes close to λ^3 have been reported, mostly in Si-based PC cavities. A good review of best results obtained in photonic crystals can be found in Sciencewatch (2008). In the nitrides, the first PC was fabricated by Coquillat in 2001 (Coquillat 2001). The main objective was to exploit resonant (propagating) modes and phase-matching possibilities in order to increase the generation of second harmonics (Coquillat 2004). As a result, the PC (1D grating in this case) allowed an increase in the second harmonic generation at 395 nm by a factor of 5000 (Torres 2004). Resonant PC propagating modes in a thin GaN/sapphire slab were observed, in spite of the mode coupling with the substrate, showing that PC effects can be observed even in non-ideal (but easier-to-fabricate) configurations (Rosenberg 2005). The effect of substrate coupling on the band gap and band edges was also investigated. While the band gap is still visible in a transmission experiment in a GaN/sapphire slab, the coupling with the substrate reduces the transmission contrast (Caro 2008). Still in the GaN/sapphire system, the most sudied problem has been the extraction of light by a PC in a light-emitting diode. The starting problem is the internal reflection, which traps the light inside the material and reduces the LED efficiency. Among various solutions, PCs have been proposed to extract the light (Oder 2004; Wierer 2004; Kim 2005). The extraction improvement remained, however, limited to a factor of 1.5 to 2. The reasons for this quite disappointing result were accurately analysed by David (David 2005; David 2006*a*), who undoubtedly reported the most sophisticated study of light extraction by PCs in LEDs. The first problem is the distance between the emitting QWs and the PC. It has to be short to maximize the coupling. However, electrical injection prevents excessive reduction of this distance. Hence, results obtained by photoluminescence cannot be directly transfered to LEDs. Another point to mention about photoluminescence is that it is often measured in conditions where the internal quantum efficiency is high, which increases the extraction. Actually, any photon-recycling scheme gives an extraction efficiency of unity in a material where the internal quantum efficiency is unity. In an LED, some internal losses due to contacts, doped layers, and non-radiative recombination cannot be supressed. Second, the PC extraction must be omnidirectional, as the light emission is omnidirectional in the QWs. Square PC and even triangular PC are not ommidirectional. Hence, David has proposed (David 2006*b*) the use of Archimedean tilings which combine triangular and square lattices in the PC. The principle is to increase the number of equivalent points in the reciprocal lattice which extract the light (i.e. lead after diffraction to a wavevector in the extraction air cone). The Archimedean PC allows an increase of this number to 12, compared to 6 in the triangular lattice, and is more efficient than the triangular PC in GaN/air slabs. Unfortunately, for air/GaN/sapphire slabs, this no longer holds. This is the reason why LEDs with a removed substrate and a PC etched in the n:GaN layer were proposed and fabricated (David 2006*c*; Cho 2008), with, however, a performance strongly limited by light absorption on metal contacts (David 2006*c*). Another alternative is to embed the PC in the structure with a regrowth process (David 2008; Matioli 2009). This yields a better extraction of all guided modes (Matioli 2010), leading to an extraction efficiency in the range of 0.5–0.6.

Although impressive, similar or even larger extraction efficiencies are obtained by less sophisticated approaches. A detailed analysis was made by Wiesmann (Wiesmann 2009). The extraction efficiency of an optimized PC can reach 0.7, which is the same value as the one reached with a rough surface. A simple explanation is that an optimized PC should provide extraction omnidirectionality, which the rough surface does also. In the Fourier domain the rough surface has a larger number of components than the PC, each of them having a smaller amplitude, finally resulting in a comparable efficiency (Wiesmann 2009). Directionality might actually be a clearer bonus of PC LEDs for applications where the brilliance is an issue (Wiesmann 2009). A half-intensity emission angle of 17° is obtained with a PC and a very thin GaN LED structure (Lai 2010), while the corresponding angle in a Lambertian source would be 60°. Such PCs can be used to fabricate surface-emitting lasers if the material is pumped at high intensity and has some gain. The stimulated emission can then be observed (Chang 2006). However, in the absence of vertical confinement the stimulated emission hardly leads to laser operation. The vertical feedback was provided by a Bragg reflector grown below the active region, while the PC was etched on the top surface (Lu 2008*b*). The feedback can be undestood as follows. In the plane of the layer the light is diffracted for some resonant modes of the PC. For these modes, the field intensity is reinforced and the modal gain is larger. As the diffraction k vector in the plane is comparable to the light k vector, the light is

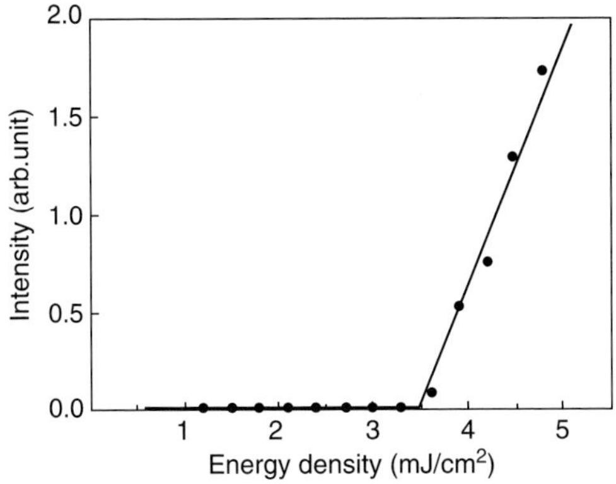

FIG. 10.5. Photoluminescence intensity as a function of pump power in a PC laser at room temperature. (Reprinted with permission from Lu (2008*b*), © 2008 American Institute of Physics.)

diffracted perpendicular to the surface, upwards or downwards. The bottom mirror redirects all the light towards the top surface. A clear laser action has been demonstrated, as shown in Fig. 10.5, with a linewidth of 0.1 nm above threshold (Lu 2008*a*). By varying the PC period, and keeping the same hole/period ratio, one observes that the lasing wavelength does not vary as the period, as one could naively expect. This is due to the fact that the gain in InGaN QWs is not spectrally broad enough. Hence, the system chooses the PC resonant mode that has the best spectral overlap with the gain spectrum. As shown in Fig. 10.6, the lasing wavelengths for various PC periods correspond to different PC modes, Γ, M, and K.

Another interesting feature in this laser is the polarization. As shown in Fig. 10.7, the polarization of each laser operating at a given PC mode is given by the symmetry of the mode. Due to the six-fold symmetry of the PC, one could expect that the resultant polarization is nul. However, imperfections in the PC are sufficient to break the symmetry, and the system chooses one of the six possible directions and the laser emission is thus polarized (Lu 2008*c*). Such a structure obviously does not provide rigorous control of spontaneous emission (many modes in the plane, and even more out of the plane), and thus leads to a modest, but not so small, β factor of 5×10^{-3}, smaller by a factor of 2 than in a VCSEL, but larger than in a ridge laser by a factor of 100.

While the PC etched on the top surface can be used for optically pumped lasers, it induces injection problems for electrical lasers. An alternative, as for LEDs (David 2008; Matioli 2009; Matioli 2010), is to embed the PC in the structure (Matsubara 2008), as shown in Fig. 10.8. Below threshold, the mode dispersion can be measured by angle-resolved luminescence. The resonant edge modes are observed as shown in Fig. 10.9. In the Γ region, the detail of the band

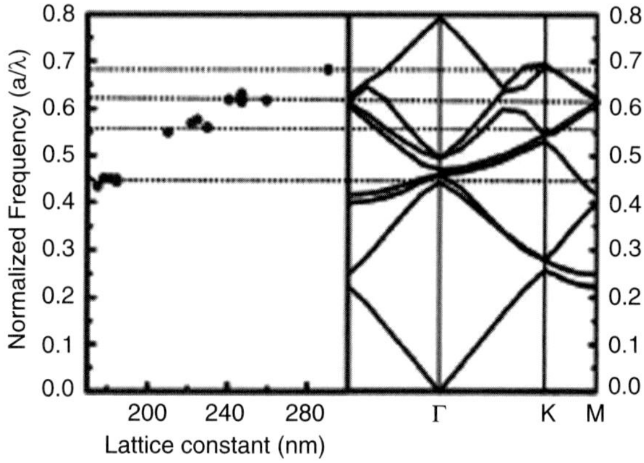

FIG. 10.6. Lasing frequency (normalized to the PC period) as a function of PC period in a PC laser. The calculated band diagram allows comparison of the lasing frequencies with the calculated PC resonant modes. (Reprinted with permission from Lu (2008*b*), © 2008 American Institute of Physics.)

structure is nicely resolved. Above threshold, only one mode is lasing, at almost normal angle and with a small divergence of 1°.

As already explained, the second thematic area for PCs is the world of nanophotonics. Nanocavities have been fabricated in GaN with 2D PCs in a membrane. In the first report by UCSB (Choi 2005; Meier 2006), the membrane was fabricated by a selective photoelectrochemical etching of a sacrificial InGaN layer underneath the GaN layer. The resultant Q factors were a modest 300 in an H2 circular cavity (7 missing holes) and 800 for an L7 linear cavity (seven missing holes in the Γ-K direction), at wavelengths in the 460–480 nm range. Random structural defects and surface roughness were believed to be limiting factors. The main difficulty in this process was the membrane fabrication. Alternative approaches are based on substrate removal. While sapphire is difficult to remove and requires a laser lift-off process, SiC can be etched chemically, or more precisely, photoelectrochemically. Such a nanocavity has been fabricated (Arita 2007). A Q factor of 2400 has been achieved in an L7 cavity at room temperature and at a wavelength of 382 nm. More recently, the same group was able to reach Q values above 5000 at 380 nm in a nanobeam PC (tapered array of holes in a narrow ridge) (Sergent 2012*a*), and even above 6300 (current record value) with a layer transfer process (Sergent 2012*b*).

Silicon being by far the easiest substrate to process and to remove, a quite obvious way to produce GaN membranes is to grow nitrides on Si substrates. This approach has been pioneered by CRHEA-CNRS since 2007. The first problem to solve was to produce efficient emitters in a thin GaN layer. Indeed, the membrane needs to be thin in order to support a small number of guided modes (ideally

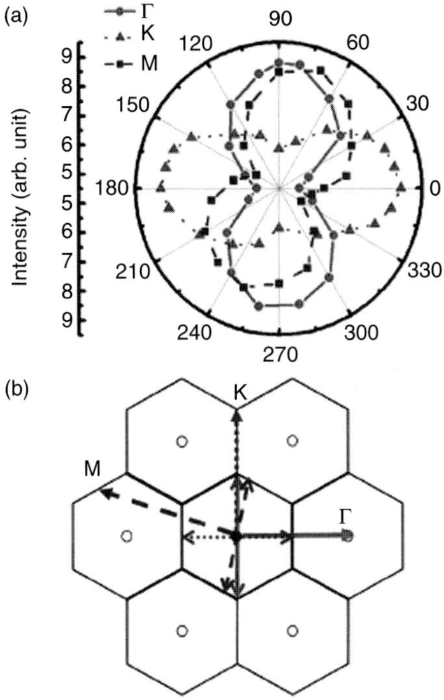

FIG. 10.7. Top: Measured polarization for various lasers operating in the Γ, M, and K modes of the PC. Bottom: The reciprocal space of the PC allows the visualize the three main beam directions and polarizations. (Reprinted with permission from Lu (2008*c*), © American Institute of Physics.)

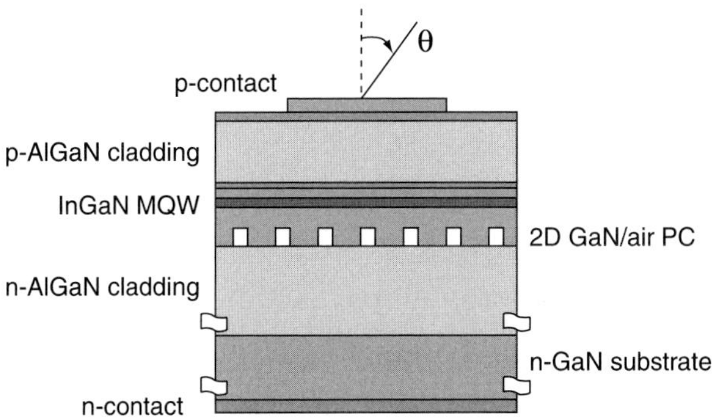

FIG. 10.8. Embedded PC in a laser structure. (From Matsubara (2008), reprinted with permission from AAAS.)

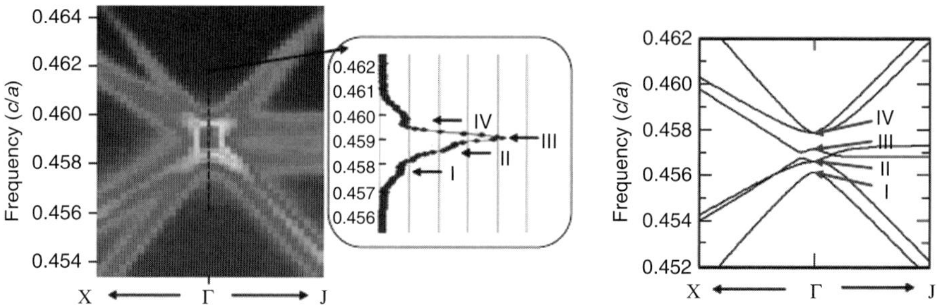

FIG. 10.9. Left: PC modes observed by angle resolved luminescence. Right: calculated photonic band structure. (From Matsubara (2008), reprinted with permission from AAAS.)

the membrane must be $\lambda/2n_r$ thick to be single mode, i.e. typically 83 nm for a wavelength of 400 nm and a refractive index of 2.4). Then the difficulty is to achieve a high crystallographic quality in such a thin layer grown by heteroepitaxy with a large lattice-mismatch and a resultant large dislocation density. QWs grown in such conditions have a poor radiative efficiency, in particular at room temperature. Using quantum dots can be a solution, as the 3D confinement leads to a smaller sensitivity of recombination to the defects, as was demonstrated with efficient GaN/AlGaN QDs at room temperature (Sergent 2009). The second issue, which was initially thought to be a problem, was the difficulty in etching GaN with a process that could respect the hole diameter and give smooth and vertical etched surfaces so as to maximize the Q value. For this reason, an original approach was followed. The PC was first etched in Si, benefiting from the mature Si process, and the GaN layer was grown on the Si PC by conformal epitaxy. The Si was finally underetched through the holes or through additional larger apertures, leading to a thin AlN membrane with GaN QDs. The photoluminescence at room temperature revealed a Q factor of 1800 at 425 nm in an L3 cavity (Neel 2011). Although appealing, this approach turned out to suffer from problems such as parasitic epitaxy in the holes, and could not reach larger Q values. A more classical approach was then used. The GaN was grown first, and then the PC was fabricated. Progress in ICP etching has led to the production of high-quality structure, with smooth sidewalls, and Q values as high as 4800 were recently achieved in Notomi cavities (SamGiao 2012). Other groups such as EPFL have also followed this approach, obtaining a Q value of 5200 at 420 nm in an L7 cavity (Vico Trivino 2012). This remains quite far from record values obtained in GaAs or in Si. However, such cavities proved to be good enough for showing laser effects. A first report was based on cavities (H1,H2,H3) in a triangular PC etched in bulk GaN (Lai 2007). The vertical emission showed a single line for the small H1 cavity but many lines for larger H2 and H3 cavities, suggesting the presence of different transverse modes in the cavities. The

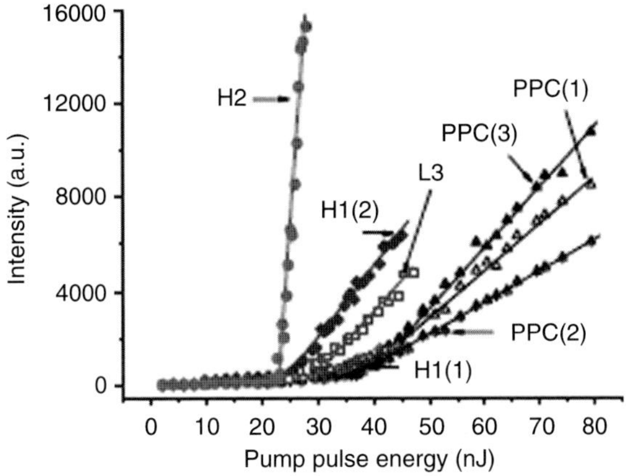

FIG. 10.10. Photoluminescence intensity as a function of pump power in PC structures at room temperature. (Reprinted with permission from Lin (2011).)

second report related to a PC membrane laser. The structure (Lin 2011) was copied from the first UCSB structure. The GaN membrane was fabricated by photoelectrochemical etching of a sacrificial InGaN layer. Various cavities were fabricated (H1, H2, L3), and also resonant band-edge modes have been used. The laser effect was demonstrated at room temperature by optical pumping, as shown in Fig. 10.10. From the position of the lasing peak, it remains difficult to clearly assign the emission to cavity modes or to band-edge modes. The emission is clearly polarized even for H1 and H2 cavities, where a net polarization is not expected due to the six-fold sysmmetry. Obviously, the symmetry is broken by some imperfections and the light is emitted with a preferential polarization, and the polarization ratio can reach 25.

10.3 Microdisks

Microdisks are optical resonators, with lateral dimensions in the micron range (or a few microns) and a vertical light confinement in the range of the wavelength in the material. They were first studied in the early 1990s in the context of semiconductor lasers (McCall 1992). The light confinement is due to total internal reflection at the semiconductor/air interface. Optical modes are travelling around the periphery of the disk, and the resonance condition defining the modes can be simply expressed as $2\pi R n_{eff} = m\lambda$, where n_{eff} is the effective optical index taking into account the vertical confinement of the light. This formula based on geometrical optics assumes that the light is following the disk periphery very closely. In reality, the photon modes have a non-zero lateral extension and are not restricted to the disk periphery. In addition, some modes can be reflected further inside the disk. This can be described accurately by the Maxwell equations and the Bessel functions in cylindrical coordinates. The resultant whispering gallery modes can be described by two numbers: the azimuthal m number describing the number of periods of the electromagnetic field amplitude along the periphery, as in the simplified formula (the field intensity having 2m maxima or minima), and the radial number n giving the number of intensity maxima along the radius. Modes with n = 1 correspond to modes close to the periphery, while modes with larger n numbers are located closer to the disk center. Taking into account the light polarization TM ($\vec{B}$ perpendicular to the disk) or TE, modes are described by $TM_{m,n}$ and $TE_{m,n}$. A common approximation is to take a field equal to zero at the disk periphery (similar to the infinite-well approximation for an electron in a quantum well). However, the field is not nul at the interface, and there is an evanescent field extending outside the disk. In a cylindrical geometry one can show that there is some energy leaking outside the disk, which limits the Q value. A rough estimate of Q can be taken as (McCall 1992; Slusher 1993):

$$Q = \frac{1}{7}e^{2mJ} \tag{10.10}$$

with $J = \frac{1}{2}Log(\frac{1+x}{1-x}) - x$ and $x = \sqrt{1 - \frac{1}{n_{eff}^2}}$. The vertical light confinement in nitride microdisks gives typical values of 1.75 for n_{eff}, leading to x = 0.82 and J = 0.33. Hence, the radiation-limited Q value varies in nitride disks as

$$Q_{rad} \approx \frac{1}{7}e^{0.66m} \approx \frac{1}{7}e^{0.66\times 2\pi R n_{eff}/\lambda} \tag{10.11}$$

At a wavelength of 350 nm, it varies as $\frac{1}{7}e^{20\times R}$, with the radius R expressed in microns. We immediately observe that the radiation-limited Q value is very large, showing that radiation is not the limiting factor. Unfortunately, it also shows that perfect microdisks are very poor emitters. In addition, the energy is emitted in a continuum around the disk and not easy to collect. The residual absorption also introduces a limit for Q. One can relate the energy broadening δE ($Q_\alpha = E/\delta E$) to the photon lifetime ($\delta E = \hbar/\tau$), and relate the photon lifetime to the absorption α by ($c/n_{eff} \times \tau \times \alpha = 1$), and one easily finds Q_α:

$$Q_\alpha = \frac{2\pi n_{eff}}{\alpha\lambda} \tag{10.12}$$

Assuming a residual absorption coefficient 1000 times lower than the band-edge value, which is quite easy to achieve in a good-quality material ($\alpha = 50$ cm^{-1}), one obtains $Q_\alpha = 6200$ at 350 nm. This shows that residual absorption can be a limiting factor in particular in large disks where other limiting mechanisms are negligible. The scattering by surface roughness is the third limitation for the Q value. The Rayleigh scattered energy varies as λ^{-4}, so that Q_{ss} varies as λ^3. A detailed calculation (Borselli 2005) leads to the following expression:

$$Q_{ss} = \frac{3\lambda^3}{8\pi^{5/2}} \frac{(n_{\mathit{eff}}^2 - 1)n_r^2}{(n_r^2 - 1)^2 n_{\mathit{eff}}^2} \frac{R}{L_c h \sigma^2} \tag{10.13}$$

where we have assumed that the surrounding medium is air, and n_r is the material index, h is the disk thickness, L_c is the correlation distance of the roughness, and σ is the roughness amplitude. Q_{ss} decreases as the square of the roughness, and decreases when the index contrast increases. In this respect, nitrides can take advantage of a smaller index contrast than smaller band-gap materials.

The modal effective volume can be made quite small. This is not as intuitive as in the case of photonic crystal cavitites, but whispering-gallery modes (modes with n = 1) have effective volumes calculated by

$$V_{\mathit{eff}} = \frac{\int \epsilon E^2 d^3 r}{max(\epsilon E^2)} \tag{10.14}$$

and found to be in the range of a few $(\lambda/n_r)^3$ (Srinivasan 2006). ϵ is the dielectric function at optical frequencies, equal to the square of the optical index. The modal volume is found to decrease almost linearly with the disk diameter. However, for small disks, Q becomes radiation-limited and the Q/V_{eff} ratio strongly decreases with the decreasing diameters. For large disks, Q is limited by other mechanisms such as residual absorption, and Q/V_{eff} decreases with increasing diameters. As a result, the largest ratio is found for diameters of a few microns, where Q can reach 10^5 and the effective volume is about $2(\lambda/n_r)^3$.

One severe drawback of microdisks is the radiation pattern, which is isotropic in the plane of the disk. Hence, efficiently collecting the emitted light is challenging. A solution can be to couple the disk with a linear waveguide or a fiber taper. Another solution can be to deform the disk slightly from the circular geometry and introduce preferred radiation directions. The typical localization of the light in the disk and exit angle of the light from the disk may depend strongly on the disk deformation in these so-called chaotic resonators (Gmachl 1998). A defect can also be introduced in the disk, leading to preferential localizations and exit angles for the light. All these approaches must trade off the Q value for the directionality.

The first results in microdisks were obtained in GaAs during the 1995–2005 period with InAs QD emitters, and included lasers, then the observation of the Purcell effect, and then of the strong coupling regime (see references in Srinivasan (2006)). In nitrides, first whispering-gallery modes were observed in 1998 (Mair 1998) in thin layers grown on sapphire and large disks (9 μm diameter). Whispering-gallery modes were observed also in 2 μm thick GaN layers (not single mode in the vertical direction) and very large disks (25−750 μm diameters) (Chang 1999). Observed modes were high-order radial modes (not circulating at the disk periphery) which could be brought to lasing at 365 nm by optical pumping. The directionality issue was adressed in a spiral-shaped disk or pillar of large size (500 μm diameter) (Chern 2003). The diameter increases slightly counter-clockwise (CCW) and abruptly returns to its initial value at the notch position. This reduces the Q factor in particular for mode-propagating CCW, but leads to an increased emission at the notch position and along a certain direction. Results on more sophisticated microdisks (small diameter, vertical single mode) were obtained in 2004 at UCSB, with an optically pumped microdisk laser emitting at 415 nm (Haberer 2004). The disk was obtained by selective photo-electro-chemical etching of an InGaN sacrificial layer, and the emitting region was made of InGaN QWs. Lasing was observed at 10 K with a large threshold power density (a few MW/cm^2) on 4 μm diameter disks. The device was largely improved later, and CW lasing was observed in 2006 at room temperature with a threshold power density as low as 300 W/cm^2 (Tamboli 2007). One additional reason for such a low threshold is the small diameter (1.2 μm) and the reduced number of modes overlapping with the gain region. A similar approch based on selective etching of nitrides was demonstrated with AlInN as a sacrificial layer (Simeonov 2008). Room-temperature lasing was achieved from 409 to 475 nm with a threshold power density of 166 kW/cm^2 in microdisks with a diameter of 3–5 μm. These microdisks have Q factors of up to 3500, limited by the QW absorption. The absorption due to the emitting medium is a key issue. While the residual absorption can be made negligible in a cold cavity (without emitter), it is inherent in cavities with emitters and can limit the Q factor. A good solution to minimize it is to use an active medium with a large Stokes shift, i.e. absorbing at higher energy than the emission energy: in other words, reducing the density of states at the emission energy. The use of quantum dots is obviously a good solution, as will be shown later in this section. In InGaN quantum wells, localization due to lateral variations of In content and/or of thickness (in particular, in the presence of an internal field) leads to a situation quite similar to the QD case. This has been beautifully shown by the UCSB team. While the microdisks fabricated on polar material (leading to the low-threshold laser mentioned above) have typical Q factors of 4000, similar disks fabricated on non-polar material have Q factors of 1500–2000, limited by absorption in InGaN QWs, where the Stark effect is absent and the Stokes shift is small (Tamboli 2009). As for photonic crystals, the under-etching of Si can be exploited to fabricate microdisks. The first report concerned a simple GaN on Si (Choi 2006), and lasing was observed at 365 nm at 4 K on 20 μm-diameter disks. The main problem when growing the active region directly on Si is the crystal quality. Indeed, the layer has to be thin (about $\lambda/2$), and it is very challenging to reduce the dislocation density in such a thin layer grown on a highly mismatched substrate. This is a first motivation to use quantum-dot emitters, as the 3D confinement makes them less sensitive to defects. The second motivation is to reduce the density of states, as already explained. This approach has been pionnered by CRHEA, and intense emission has been observed in QDs inserted in very thin AlN and AlGaN layers grown on

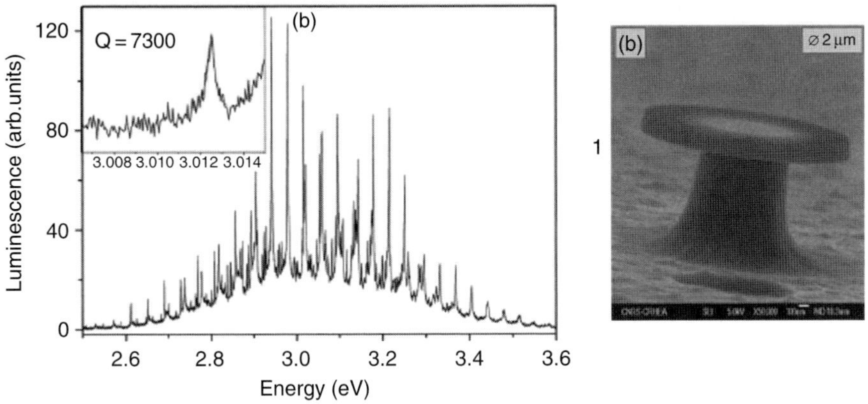

FIG. 10.11. Left: Room-temperature PL spectrum (excitation power density $60 W cm^{-2}$) of a 5 µm diameter disk with embedded GaN/AlN dots. Right: SEM picture of a 2 µm diameter disk. (Reprinted with permission from Mexis (2011).)

Si up to room temperature (Sergent 2009). The emission from these QDs has been used to measure Q factors as large as 7300 at a wavelength of about 400 nm for n = 1 modes (propagating at the periphery of the disk) in 5 µm disks, as shown in Fig. 10.11 (Mexis 2011). The Q value is smaller for n = 2 modes (propagating closer to the center) due to the absorption in the central post. Q is found to be smaller when the AlN layer is replaced by AlGaN layer due to residual absorption. Finally, Q values are smaller in 2 µm diameter disks than in 5 µm disks, as the modes overlap more with the post, both for n = 1 and n = 2, as shown in Fig. 10.12 (Mexis 2011). The mode volumes, in units of

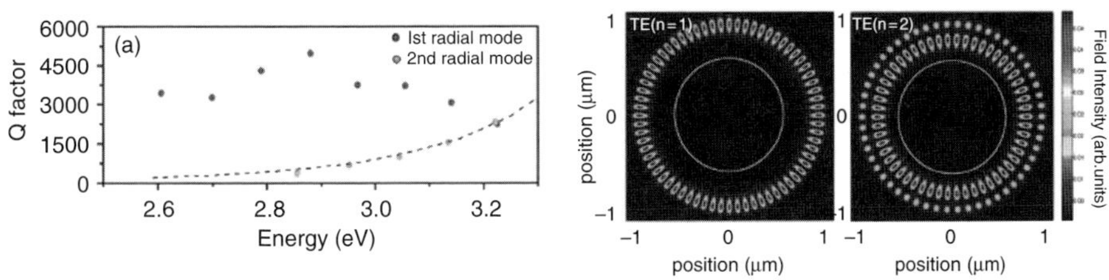

FIG. 10.12. (a) Left: Q value of the first and the second radial-order modes in a 2-µm disk; the dashed line is proportional to the inverse of the overlap of the second radial mode with the post that limits their Q factor. Right: Simulated TE field distribution for first (n = 1) and second (n = 2) radial-order modes in a 2.06-µm disk (the grey open circle represents the post periphery). (Reprinted, with permission, from Mexis (2011).)

$(\lambda/n_r)^3$ are about 6 and 20, leading to Purcell factors ($F_P = \frac{3}{4\pi^2}\frac{Q}{V_{eff}}$) of about 60 and 30 for the 2-μm and 5-μm disks respectively. The effect is thus large, and should be observable, but has not been observed so far experimentally. Reasons for that are the homogeneous broadening at 300 K that reduces the effective Purcell effect, and the dominating non-radiative recombination which masks the effect in time-resolved PL measurements.

10.4 Nanowires

Nanowires for cavity quantum electrodynamics have been developed in arsenides first, where a Purcell factor of 5 has been measured in a 1 μm diameter pillar (Gerard 1998), corresponding to a coupling coefficient β in the vertical mode of the pillar of 0.9. Compared to planar cavities where β is in the range of 10^{-2}, micropillars emit photons more efficiently in a given (vertical) mode, which is a key issue for single photon sources, for the generation of squeezed states of light, and in particular for quantum cryptography. A few years later a clear photon antibunching ($g^{(2)}(0) = 0.18$) was measured in a micropillar coupled (in the strong coupling regime) to a single InGaAs quantum dot (Press 2007). This single-photon source has a Purcell factor of 61 and a β factor of 0.97 (we note that the QD linewidth is much narrower at low temperature than the cavity mode, so that the spectral overlap is unity). The Q factor reaches 20 000, but the pillar diameter is quite large (1.8 μm), so that the mode volume is not that small (in the range of 18 in reduced units). GaAs/AlGaAs nanowires also allowed observation of 0D polariton lasing (Bajoni 2008).

Nanowires have also been developed in nitrides for many applications and with diameters ranging from a few tens of nm to above 1 μm (microwires). Given the nanometric size and the general problem of crystallographic quality in GaN, the bottom-up approach (direct growth of wires) has been the preferred option (instead of processing planar layers as in arsenides). Indeed, this approach allows the eradication of strain problems and related dislocations which arise in heteroepitaxy of nitrides on foreign substrates. Plasma molecular beam epitaxy turned out to be the ideal tool for growing nanowires. The first nanowires were obtained in 1997 on sapphire at Sophia University in Tokyo (Yoshizawa 1997), with diameters in the range of 50–100 nm. Similar nanowires were obtained by the ISOM group in Madrid on sapphire and Si (111) (Calleja 2000). AlGaN nanowires were also grown (Al content of 0.20–0.30, diameter of 30 nm), with inserted GaN disks (Ristic 2003), with properties only slightly different from quantum-well properties due to surface potential and lateral variation of strain (Ristic 2005*a*). We will not discuss the details of growth of these nanowires, but focus on optical resonators. Nanocavities were obtained by the Madrid group (Ristic 2005*b*) with GaN nanodisks inserted between GaN/AlGaN Bragg mirrors. The optical properties, however, did not exhibit any particular feature, showing that these structures remain far from being optimized. In particular, in these small-diameter columns (20–60 nm), the optical modes largely extend laterally in air, so that a more accurate calculation would be needed for all layer thicknesses in order to match the cavity and exciton energies.

The Sophia University group has fabricated nanoLEDs based on nanowires grown by plasma MBE (Kikuchi 2004; Kishino 2012). The CEA-CNRS group in Grenoble also developed plasma MBE nanowires (Landre 2009). Their interest in terms of extraction efficiency in LEDs has been discussed (Henneghien 2011). As LEDs and lighting are not the scope of this chapter, these issues will not be discussed here. A review can be found elsewhere (Li 2012).

Accurately positioning the nanocolumns turns out to be crucial for many applications, and many approaches have been sucessfully proposed (Hersee 2006, Sekiguchi 2008), but have not led to improved optical nanoresonators.

Other methods, such as the VLS growth with a Ni catalyst, have been used to produce GaN nanowires with a diameter of 30–150 nm and a length of several hundreds of micrometers (Johnson 2002). This very large length/diameter ratio (typically 100) favors the stimulated emission of axial modes, while whispering-gallery modes suffer from high radiation losses. Hence, stimulated emission was observed in a $L = 16$ μm long wire, with a mode spacing of 1.7 nm consistent with the expected value $\frac{\lambda^2}{2n_r L}$. The cavity finesse (1.4) is low, as the facet reflectivity is low, and the optical mode linewidth is comparable to the free spectral range, leading to barely visible modes. However, the Q value reaches 100–1000 as the cavity is very long, and lasing emission could be observed (Johnson 2002). The threshold pump power is, however, quite high, with an electron-hole pair density above the exciton Mott transition (Binet 1999). Nanowires were also fabricated by MOCVD in the group of C. Lieber (Zhong 2003; Qian 2004; Qian 2005). Non-polar nanowires with a triangular section have been obtained (Gradečak 2005), with typical dimensions similar to those grown by VLS. Again, axial Fabry–Perot modes were observed, and laser action was observed under optical pumping. The improvement in overall quality (material, roughness) led to a reduced threshold power of 22 kW/cm^2. By inserting InGaN QWs in a core shell geometry in these triangular wires, emission and laser action could be achieved over a large spectral range (400–480nm) (Qian 2008). The overlap between the optical mode in the center of the triangular guide and the wells grown on the triangle edges was 0.24 for 26 QWs (0.12 for 13 QWs), which gives an overlap of about 0.009 per QW. This is actually larger than the usual values (0.007 per QW) in edge-emitting lasers with AlGaN or even AlInN cladding layers. This results comes as a surprise, as QWs in edge-emitting lasers are grown in the center of the guide where the mode amplitude is maximum, while they are grown offset from the mode maximum in these nanowires. However, the overall mode-confinement is higher in the nanowire, as the GaN/air interface has a much higher index contrast than the AlGaN/GaN interface, finally leading to a better overlap in nanowires. Apart from these details there is no major difference between these long nanowire lasers and a conventional edge-emitting laser, which would be narrow and short. The fabrication of the nanowire-based optically injected laser is easier, as there is no processing step. However, for electrical devices the nanowire approach is much more complicated, and has not yet allowed fabrication of electrical injection lasers. Given their aspect ratio, the core-shell geometry is by far the most favorable configuration for fabricating lasers in these long wires.

In shorter and broader nanowires (or microwires), radial (meaning non-axial) modes are in competition with axial modes. Two types of radial mode can propagate. First, Fabry–Perot modes can resonate between the facets of hexagonal wires when the diameter (in hexagons, "diameter" should be understood as the distance between facets opposite each other) is a multiple of $\lambda/2n_r$ (about 75 nm at a wavelength of 365 nm). Second, whispering-gallery modes can propagate as in microdisks (see the previous section) when the diameter is a multiple of $\lambda/n_r\pi$ (for n = 1 modes circulating close to the periphery). Hence, both modes can coexist in some nano–micro wires. This has been observed in GaN wires with a diameter of about 2 μm (Coulon 2012). A significant difference between whispering-gallery modes in wires and disks is the dispersion along the wire axis, which reduces the apparent mode Q factor if the measurement is not angle-resolved (Q of up to 100 was measured). When the measurement is angle-resolved, the Q factor is as high as 700 in 5-μm diameter wires (Trichet

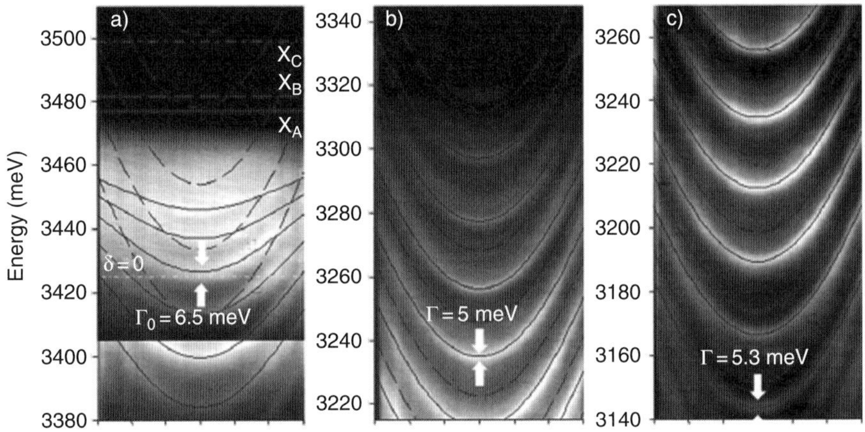

FIG. 10.13. Measured polarized photoluminescence intensity (grey scale with contrast sometimes adjusted to maximize the visibility of the dispersion branches) of a GaN microwire at T = 10 K with angular resolution in the wire-axis direction. The free exciton energies labeled XA, XB, and XC are indicated by straight horizontal lines (solid, dashed, and dot–dashed, respectively). The solid lines are the calculated dispersion branches in the strong coupling regime. The calculated uncoupled modes are shown by dashed lines. The energy corresponding to polaritons with zero detuning between excitons and uncoupled gallery modes is indicated by the horizontal ($\delta = 0$) dashed line. A similar result, though more noisy, was achieved at 300 K (not shown here). (Reprinted with permission from Trichet (2012), © 2012 IOP Publishing Ltd.)

2012). In doped GaN, the exciton broadening and the residual absorption lead to a weak coupling between the exciton and the optical mode. In undoped GaN, a strong coupling is observed with a large Rabi splitting energy of 115 meV at temperatures from 10 to 300 K, while the transition linewidth is as small as 7.5 meV, leading to a very large splitting/linewidth ratio of 15 (Trichet 2012) and a good splitting visibility, as can be seen in Fig. 10.13. Attempts were carried out to produce lasing from these polaritons, but without success. We recall, to finish, that polariton lasing was achieved in a nanowire, but the nanowire was simply used as the emitting region inserted in a dielectric planar cavity, as has been described in the planar cavity section (Das 2011).

10.5 Conclusion

Nanophotonics in nitrides is following the avenues opened by other materials such as arsenides. In many areas the gap in performance between arsenides and nitrides has been reduced, and nitride even caught up in the domain of room-temperature polariton lasing. As in other materials, nanophotonics in nitrides has remained so far very academic. In terms of pure science, the specificities of nitrides (the same could be written for ZnO) are the exciton stability at high temperature and density, and the strong exciton–phonon interaction. This has led to the demonstration of some aspects which were not observable in other materials. However, one must admit that these specificities and possible advantages of nitrides for pure science in nanophotonics remain quite limited. Then, a legitimate question could be: Why should we keep exploring nanophotonics with nitrides? The first answer is that many breakthroughs in science were made in areas where they were not expected, and exploring and pushing the limits of current knowledge remains essential. There are still specificities of nitrides which have not been fully explored, such as the small spin–orbit coupling and the expected long spin coherence lifetime, the large phonon energy, the electronic band structure, and the possibility of having very energetic carriers (under electric or optical excitation), and many others. Hence, domains such as spin optronics or non-linear photonics in materials very far from equilibrium may have a better chance of emerging in nitrides rather than in any other material. The second answer is related to applications. So far, no application has been found for nanophotonics in nitrides. However, this remark also holds for other materials, and simply follows from the current immaturity of nanophotonics. Applications are not to be expected in the short term. In the longer term, nitrides will benefit from higher operation temperature and higher power level than many other materials. They will also benefit from a wider spread (in terms of industrial importance, and the number of industrial players) than any other optically active semiconductors, so that integrating nanophotonics with other nitride devices or developing nanophotonic devices starting from the existing nitride technology will be easier than with any other material. Compared to arsenides, there will be no real competition, as the spectral regions covered by both materials do not overlap. While arsenide nanophotonics is largely dominated by the optical emission, detection, and processing of photons in the near-infrared (1–1.6 μm), with applications in optical computing, quantum calculation, cryptography, and so on, nitrides will probably be used for visible photons with applications such as optical storage, displays, and biophotonics (most biomolecules are sensitive in the near UV and visible range). Nanophotonic devices should allow the emission of light with low optical power but high brightness and directionality, and under low electrical power. This last point is essential. Most applications require electrical devices, and this is clearly an area where GaN nanophotonics should develop and improve, if it is to aim at moving from research laboratories to commercial devices.

References

Adele, C., *et al.* (2006), Room-temperature continuous-wave lasing in GaN/InGaN microdisks, *Nature Photonics* 1, 52, 61.

Antoine-Vincent, N., *et al.* (2003), Observation of Rabi splitting in a bulk GaN microcavity grown on silicon, *Phys. Rev. B* 68(15), 153313.

Arita, M., *et al.* (2007), AlN air-bridge photonic crystal nanocavities demonstrating high quality factor, *Appl. Phys. Lett.* 91, 051106.

Badolato, A., *et al.* (2005), Deterministic coupling of single quantum dots to single nanocavity modes, *Science* 308, no. 5725, 1161.

Bajoni, D., *et al.* (2008), Polariton light-emitting diode in a GaAs-based microcavity, *Phys. Rev. B* 77, 113303.

Bajoni, D., *et al.* (2008), Polariton laser using single micropillar GaAs-GaAlAs semiconductor cavities, *Phys. Rev. Lett.* 100, 047401.

Ballili, R., *et al.* (2007), Bose–Einstein condensation of microcavity polaritons in a trap, *Science* 316, 1007.

Baumberg, J., *et al.* (2008), Spontaneous polarization buildup in a room temperature polariton laser, *Phys. Rev. Lett.* 101, 136409.

Bejtka, K., *et al.* (2008), Strong light-matter coupling in ultrathin double dielectric mirror GaN microcavities, *Appl. Phys. Lett.* 92, 241105.

Bejtka, K., *et al.* (2008), Fabrication and characterization of ultrathin double dielectric mirror GaN microcavitie, *Semicond. Sci. Tech.* 8, 045008.

Binet, F., *et al.* (1999), High-excitation photoluminescence in GaN: hot carrier effects and the Mott transition, *Phys. Rev. B* 60, 4715.

Borselli, M., *et al.* (2005), Beyond the Rayleigh scattering limit in high Q silicon microdisks; theory and experiment, *Optics Express* 13, 1515.

Butte, R., *et al.* (2006), Room-temperature polariton luminescence from a bulk GaN microcavity, *Phys. Rev. B* 73, 033315.

Butte, R., *et al.* (2009a), Room temperature polariton lasing in III-nitride microcavities: a comparison with blue GaN-based vertical cavity surface emitting lasers, in: Proceedings of SPIE, San Jose, CA, USA, pp. 721616-721619-16.

Butte, R., *et al.* (2009b), Phase diagram of a polariton laser from cryogenic to room temperature, *Phys. Rev. B* 80, 233301.

Butte, R. and Grandjean, N. (2011), A novel class of coherent light emiters: polariton lasers, *Semicond. Sci. Technol.* 26, 014030.

Calleja, E., *et al.* (2000), Luminescence properties and defects in GaN nanocolumns grown by molecular beam epitaxy, *Phys. Rev. B* 62, 16826.

Caro, J., *et al.* (2008), Transmission measurements of the photonic band gap of GaN photonic crystal slab, *Appl. Phys. Lett.* 93, 051117.

Chang, L.-M., *et al.* (2006), Laser emission from GaN photonic crystals, *Appl. Phys. Lett.* 89, 071116.

Chang, S., *et al.* (1999), Stimulated emission and lasing in whispering-gallery modes of GaN microdisk cavities, *Appl. Phys. Lett.* 75, 166.

Chern, G., *et al.* (2003), Unidirectional lasing from InGaN multiple-quantum-well spiral-shaped micropillars, *Appl. Phys. Lett.* 83, 1710.

Cho, H.K., *et al.* (2008), Laser lift off GaN thin film photonic crystal GaN based light emitting diode, *IEEE Photonics Techn. Lett.* 20, 2096.

Choi, Y.-S., *et al.* (2005), GaN blue photonic crystal membrane nanocavities, *Appl. Phys. Lett.* 87, 243101.

Choi, H.-W., *et al.* (2006), Lasing in GaN microdisks pivoted on Si, *Appl. Phys. Lett.* 89, 211101.

Coquillat, D., *et al.* (2001), First Observations of 2D Photonic Crystal Band Structure in GaNSapphire Epitaxial Material, *physica status solidi (a)*, 183: 135. Observations of band structure and reduced group velocity in epitaxial GaN-sapphire 2D photonic crystals, *Appl. Phys. B* 73, 591.

Coquillat, D. (2004), Equifrequency surfaces in a two-dimensional GaN-based photonic crystal, *Optics Express* 12, Issue 6, 1097.

Coulon, P.-M. (2012), GaN microwires as optical microcavities: whispering gallery modes Vs Fabry-Perot modes, *Optics Express* 20, 18707.

Christmann, G., *et al.* (2006a), Impact of inhomogeneous excitonic broadening on the strong exciton photon coupling in QW nitride microcavities, *Phys. Rev. B* 73, 153305.

Christmann, G., *et al.* (2006b), Impact of disorder on high quality factor III-V nitride microcavities, *Appl. Phys. Lett.* 89, 261101.

Christmann, G., *et al.* (2008), Large vacuum Rabi splitting in MQW GaN based microcavity in the strong coupling regime, *Phys. Rev. B* 77, 085310.

Christmann, G., *et al.* (2008), Room temperature polariton lasing in a GaN/AlGaN multiple quantum well microcavity, *Appl. Phys. Lett.* 93, 051102.

Christopoulos, S., *et al.* (2007), Room-Temperature Polariton Lasing in Semiconductor Microcavities, *Phys. Rev. Lett.* 98, 126405.

Das, A., *et al.* (2011), Room Temperature Ultralow Threshold GaN Nanowire Polariton Laser, *Phys. Rev. Lett.* 107, 066405.

David, A., *et al.* (2005), Photonic bands in two-dimensionally patterned multimode GaN waveguides for light extraction, *Appl. Phys. Lett.* 87, 101107.

David, A., *et al.* (2006a), Photonic-crystal GaN light-emitting diodes with tailored guided modes distribution, *Appl. Phys. Lett.* 88, 061124.

David, A., *et al.* (2006b), GaN light-emitting diodes with Archimedean lattice photonic crystals, *Appl. Phys. Lett.* 88, 073510.

David, A., *et al.* (2006c), Photonic crystal laser lift-off GaN light-emitting diodes, *Appl. Phys. Lett.* 88, 133514.

David, A., *et al.* (2008), GaN/InGaN light emitting diodes with embedded photonic crystal obtained by lateral epitaxial overgrowth, *Appl. Phys. Lett.* 92, 113514.

Deger, C., *et al.* (1998), Sound velocity of AlGaN thin films obtained by surface acoustic wave measurements, *Appl. Phys. Lett.* 72, 2400.

Deng, H., *et al.* (2007), Spatial coherence of a polariton condensate, *Phys. Rev. Lett.* 99, 126403.

Duboz, J.-Y., *et al.* (2003), Microcavity Light Emitting Diodes Based on GaN Membranes Grown by Molecular Beam Epitaxy on Silicon, *Jpn. J. of Appl. Phys.* 42, part 1, No 1, 118.

Feltin, E., *et al.* (2006a), Crack-free highly reflective AlInN/AlGaN Bragg mirrors for UV applications, *Appl. Phys. Lett.* 88, 051108.

Feltin, E., *et al.* (2006b), Room temperature polariton luminescence from a GaN/AlGaN quantum well microcavity, *Appl. Phys. Lett.* 89, 071107.

Ferrier, L., *et al.* (2011), interactions in confined polariton condensates, *Phys. Rev. Lett.* 106, 126401.

Fujii, T., *et al.* (2004), Microcavity effect in GaN based LEDs formed by laser lift off and Etch-back technique, *Jpn. J. of Appl. Phys.* 43, n3B, L411.

Gerard, J.-M., *et al.* (1998), Enhanced Spontaneous Emission by Quantum Boxes in a Monolithic Optical Microcavity, *Phys. Rev. Lett.* 81, 1110.

Gmachl, C., *et al.* (1998), High-Power Directional Emission from Microlasers with Chaotic Resonators, *Science*, Vol. 280, no. 5369, 1556.

Gradečak, S., *et al.* (2005), GaN nanowire lasers with low lasing thresholds, *Appl. Phys. Lett.* 87, 173111.

Guillet, T., *et al.* (2011a), Laser emission with excitonic gain in a ZnO planar microcavity, *Appl. Phys. Lett.* 98, 211105.

Guillet, T., *et al.* (2011b), Polariton lasing in a hybrid bulk ZnO microcavity, *Appl. Phys. Lett.* 99, 161104.

Guillet, T., *et al.* (2012), High quality factor photonic resonators for nitride quantum dots, *physica status solidi (b)*, 249, 449.

Haberer, E., *et al.* (2004), Free-standing, optically pumped, GaN/InGaN microdisk lasers fabricated by photoelectrochemical etching, *Appl. Phys. Lett.* 85, 5179.

Henneghien, A.-L., *et al.* (2011), Optical anisotropy and light extraction efficiency of MBE grown GaN nanowires epilayers, *Optics Express*, Vol. 19 Issue 2, 527.

Hersee, S.D., *et al.* (2006), The Controlled Growth of GaN Nanowires, *Nanoletters* 6, 8, 1808.

Higuchi, Y., *et al.* (2008), Room temperature CW lasing of a GaN based vertical cavity surface emitting laser by current injection, *Appl. Phys. Express* 1, 121102.

Ho, K., *et al.* (1990), Existence of a photonic gap in periodic dielectric structures, *Phys. Rev. Lett.* 65, 3152.

Imamoglu, A., *et al.* (1996), Nonequilibrium condensates and lasers without inversion: Exciton-polariton lasers, *Phys. Rev. A* 53, 4250.

John, S. (1987), Strong localization of photons in certain disordered dielectric superlattices, *Phys. Rev. Lett.* 58, 2486.

Johnson, J. (2002), Single gallium nitride nanowire lasers, *Nature Materials* 1, 106.

Kako, S., *et al.* (2002), Observation of enhanced spontaneous emission coupling factor in nitride based vertical cavity surface emitting laser, *Appl. Phys. Lett.* 80, 722.

Kasahara, D., *et al.* (2011), Demonstration of blue and green GaN based vertical cavity surface emitting lasers by current injection at room temperature, *Appl. Phys. Express* 4, 072103.

Kasprzak, J., *et al.* (2006), Bose–Einstein condensation of exciton polaritons, *Nature* 443, 409.

Kasprzak, J., *et al.* (2008), Formation of an exciton polariton condensate: thermodynamic versus kinetic regimes, *Phys. Rev. Lett.* 101, 146404.

Kavokin, A. and Malpuech, G. (2003), Cavity Polaritons, Thin Films and Nanostructures, Vol. 32, Elsevier, The Netherlands. Kavokin, A., Baumberg, J., Malpuech, G., and Laussy, F., (2007), *Microcavities*, Oxford University Press, New York, USA.

Kelkar, P., *et al.* (1995), Excitons in a II-VI semiconductor microcavity in the strong-coupling regime, *Phys. Rev. B* 52, R5491.

Khalifa, A., *et al.* (2008), Electroluminescence emission from polariton states in GaAs-based semiconductor microcavities, *Appl. Phys. Lett.* 92, 061107.

Kikuchi, A., *et al.* (2004), InGaN/GaN Multiple Quantum Disk Nanocolumn Light-Emitting Diodes Grown on (111) Si Substrate, *Jpn. J. Appl. Phys.* Vol. 43 Part 2, No. 12A, L1524.

Kim, D.-H., *et al.* (2005), Enhanced light extraction from GaN-based light-emitting diodes with holographically generated two-dimensional photonic crystal patterns, *Appl. Phys. Lett.* 87, 203508.

Kishino, K., *et al.* (2012), Near-Infrared InGaN Nanocolumn Light-Emitting Diodes Operated at 1.46 μm, *Appl. Phys. Express*, 5, 031001.

Krestnikov, I., *et al.* (1999), Room-temperature photopumped InGaN/GaN/AlGaN vertical-cavity surface-emitting laser, *Appl. Phys. Lett.* 75, 1192.

Lai, C.-F., *et al.* (2010), Optimized microcavity and photonic crystal parameters of GaN-based ultrathin-film light-emitting diodes for highly directional beam profiles, *IEEE Photonics Technol. Lett.* 22, 1547.

Lai, C.-F., *et al.* (2007), Lasing characteristics of a GaN photonic crystal nanocavity light source, *Appl. Phys. Lett.* 91, 041101.

Landre, O., *et al.* (2009), Nucleation mechanism of GaN nanocolumns grown on (111) Si by molecular beam epitaxy, *Nanotechnology* 20, 415602.

Lanty, G., *et al.* (2011), Hybrid cavity polaritons in a ZnO-perovskite microcavity, *Phys. Rev. B* 84, 195449.

Le Si Dang, *et al.* (1998), Stimulation of polariton luminescence in semiconductor microcavity, *Phys. Rev. Lett.* 81, 3920.

Levrat, J., *et al.* (2010), Condensation phase diagram of cavity polaritons in GaN based microcavities: Experiment and theory, *Phys. Rev. B* 81, 125305.

Li, S., *et al.* (2012), GaN based nanorods for solid state lighting, *J. Appl. Phys.* 111, 071101.

Lin, C.-H., *et al.* (2011), A GaN photonic crystal membrane laser, *Nanotechnology* 22, 025201.

Lu, T.-C., *et al.* (2008a), CW lasing of current injection blue GaN based vertical cavity surface emitting laser, *Appl. Phys. Lett.* 92, 141102.

Lu, T.-C., *et al.* (2010), Continuous wave operation of current injected GaN vertical cavity surface emitting lasers at room temperature, *Appl. Phys. Lett.* 97, 071114.

Lu, T.-C., *et al.* (2011), Room temperature current injection polariton light emitting diode with a hybrid microcavity, *NanoLetters* 11, 2791.

Lu, T.-C., *et al.* (2012), Room temperature polariton lasing versus photon lasing in a ZnO based hybrid microcavity, *Optics Letters* 20, 5530.

Lu, T.-C., *et al.* (2008b), GaN-based two-dimensional surface-emitting photonic crystal lasers with AlN/GaN distributed Bragg reflector, *Appl. Phys. Lett.* 92, 011129.

Lu, T.-C., *et al.* (2008c), Characteristics of GaN-based photonic crystal surface emitting lasers, *Appl. Phys. Lett.* 93, 111111.

McCall, *et al.* (1992), Whispering gallery mode microdisk lasers, *Appl. Phys. Lett.* 60, 289.

Maddox, J. (1990), Photonic band gap bite the dust, *Nature* 348, 481.

Malpuech, G., *et al.* (2002), Room temperature polariton lasers based on GaN micrcavities, *Appl. Phys. Lett.* 81, 412.

Mair, R., *et al.* (1998), Optical modes within III-nitride multiple quantum well microdisk cavities, *Appl. Phys. Lett.* 72, 1530.

Martin, R., *et al.* (2001), Optical spectroscopy of GaN microcavities with thickness controlled using plasma etchback, *Appl. Phys. Lett.* 79, 3029.

Matioli, E., *et al.* (2009), Growth of embedded photonic crystals for GaN-based optoelectronic devices, *J. Appl. Phys.* 106, 024309.

Matioli, E., *et al.* (2010), High extraction efficiency GaN-based photonic-crystal light-emitting diodes: Comparison of extraction lengths between surface and embedded photonic crystals, *Appl. Phys. Express* 3, 032103. High extraction efficiency light-emitting diodes based on embedded air-gap photonic-crystals, *Appl. Phys. Lett.* 96, 031108.

Matsubara, H., *et al.* (2008), GaN photonic-crystal surface-emitting laser at blue-violet wavelengths, *Science* 319, 5862, 445.

Medard, F., *et al.* (2009), Experimental observation of strong light-matter coupling in ZnO microcavities: Influence of large excitonic absorption, *Phys. Rev. B* 79, 125302.

Meier, C., *et al.* (2006), Visible resonant modes in GaN-based photonic crystal membrane cavities, *Appl. Phys. Lett.* 88, 031111.

Mexis, M., *et al.* (2011), High quality factor nitride-based optical cavities: microdisks with embedded GaN/Al(Ga)N quantum dots, *Optics Letters* 36, 12, 2203.

Neel, D., *et al.* (2011), AlN photonic crystal nanocavities realized by epitaxial conformal growth on nanopatterned silicon substrate, *Appl. Phys. Lett.* 98, 261106.

Nelsen, B., *et al.* (2009), Lasing and polariton condensation: Two distinct transitions in GaAs microcavities with stress traps, *J. Appl. Phys.* 105, 122414.

Nobel Prize in Physics (2001): Eric Cornell, Wolfgang Ketterle, Carl Wieman.

Nomura, M., *et al.* (2009), Photonic crystal nanocavity laser with a single quantum dot gain, *Optics Express* 17, 18,15975.

Oder, T., *et al.* (2004), III-nitride blue and ultraviolet photonic crystal light emitting diodes, *Appl. Phys. Lett.* 84, 466.

Omae, K., *et al.* (2009), Improvement in lasing characteristics of GaN based vertical cavity surface emitting lasers fabricated using a GaN substrate, *Appl. Phys. Express* 2, 052101.

Pattison, P., *et al.* (2007), Gallium nitride based microcavity light emitting diodes with 2λ effective cavity thickness, *Appl. Phys. Lett.* 90, 031111.

Peter, E., *et al.* (2005), Exciton photon strong coupling regime for a single quantum dot embedded in a microcavity, *Phys. Rev. Lett.* 95, 067401.

Press, D., *et al.* (2007), Photon antibunching from a single quantum-dot-microcavity system in the strong coupling regime, *Phys. Rev. Lett.* 98, 117402.

Private (2012), Private communication from a consortium of French laboratories (CRHEA, L2C, LPN and LASMEA).

Qian, F., *et al.* (2004), Gallium nitride-based nanowire radial heterostructures for nanophotonics, *Nano Lett.*, 4 (10), 1975.

Qian, F., *et al.* (2005), Core/multishell nanowire heterostructures as multicolor, high-efficiency light emitting diodes, *Nano Lett.*, 5 (11), 2287.

Qian, F., *et al.* (2008), Multi quantum well nanowire heterostructures for wavelength controlled lasers, *Nature Materials* 7, 701.

Redwing, J., *et al.* (1996), An optically pumped $GaN/AlGaN$ vertical cavity surface emitting laser, *Appl. Phys. Lett.* 69, 1.

Reithmaier, J.P., *et al.* (2004), Strong coupling in a single quantum dot semiconductor microcavity system, *Nature* 432, 7014, 197.

Ristic, J., *et al.* (2003), Characterization of GaN quantum discs embedded in AlxGa1-xN nanocolumns grown by molecular beam epitaxy growth, *Phys. Rev. B* 68, 125305.

Ristic, J., *et al.* (2005a), Carrier-confinement effects in nanocolumnar GaN/AlxGa1-xN quantum disks grown by molecular-beam epitaxy, *Phys. Rev. B* 72, 085330.

Ristic, J., *et al.* (2005b), Columnar AlGaN/GaN nanocavities with AlN/GaN Bragg reflectors grown by molecular beam epitaxy on Si(111), *Phys. Rev. Lett.* 94, 146102.

Rosenberg, A., *et al.* (2005), Guided resonances in asymmetrical GaN photonic crystal slabs observed in the visible spectrum, *Optics Express* 13, Issue 17, 6564.

SamGiao, D., *et al.* (2012), High quality factor AlN nanocavities embedded in a photonic crystal waveguide, *Appl. Phys. Lett.* 100, 191104.

Savvidis, P., *et al.* (2000), Angle resonant stimulated polariton amplifier, *Phys. Rev. Lett.* 84, 1547.

Sciencewatch (2008), High Q photonic crystal nanocavities, http://sciencewatch.com/ana/st/photo-cry/rfmap2/.

Sekiguchi, H., *et al.* (2008), Ti-mask selective-area growth of GaN by RF-plasma assisted molecular beam epitaxy for fabricating regularly arranged $InGaN/GaN$ Nanocolumns, *Appl. Phys. Express* 1, 124002.

Sellers, I., *et al.* (2006a), Strong coupling of light with A and B excitons in GaN microcavities grown on silicon, *Phys. Rev. B* 73, 033304.

Sellers, I., *et al.* (2006b), Polariton emission and reflectivity in GaN microcavities as a function of angle and temperature, *Phys. Rev. B* 74, 193308.

Semond, F., *et al.* (2005), Strong light matter coupling at room temperature in simple geometry GaN microcavities grown on silicon, *Appl. Phys. Lett.* 87, 021102.

Sergent, S., *et al.* (2009), GaN quantum dots grown on silicon for free standing membrane photonic structures, *Appl. Phys. Express* 2, 051003.

Sergent, S., *et al.* (2012a), High Q (>5000) AlN nanobeam photonic crystal cavity embedding GaN quantum dots, *Appl. Phys. Lett.* 100, 121103.

Sergent, S., *et al.* (2012b), High Q photonic crystal nanobeam cavities fabricated by layer transfer, *Appl. Phys. Lett.* 101, 101106.

Simeonov, D., *et al.* (2008), High quality nitride based microdisks obtained via selective wet etching of AlInN sacrificial layers, *Appl. Phys. Lett.* 92, 171102.

Slusher, R., *et al.* (1993), Threshold characteristics of semiconductor microdisk lasers, *Appl. Phys. Lett.* 63, 1310.

Siozade, L., *et al.* (2000), Temperature dependence of optical properties of h GaN films studied by reflectivity and ellipsometry, *Jpn. J. Appl. Phys.* Part 1, 39, 20.

Solnyshkov, D., *et al.* (2009), Theory of an electrically injected bulk polariton laser, *Appl. Phys. Lett.* 94, 011110.

Someya, T., *et al.* (1998), Lasing emission from an InGaN vertical cavity surface emitting laser, *Jpn. J. Appl. Phys.* 37, L1424.

Someya, T., *et al.* (1999), Room temperature lasing at blue wavelengths in gallium nitride microcavities, *Science*, 285, 5435, 1905.

Song, Y.-K., *et al.* (1999), A vertical cavity light emitting InGaN quantum well heterostructure, *Appl. Phys. Lett.* 74, 3441.

Song, Y.-K., *et al.* (2000a), A quasicontinuous wave, optically pumped violet vertical cavity surface emitting laser, *Appl. Phys. Lett.* 76, 1662.

Song, Y.-K., *et al.* (2000b), Resonant-cavity InGaN quantum-well blue light emitting diodes, *Appl. Phys. Lett.* 77, 1744.

Srinivasan, K., *et al.* (2006), Cavity Q, mode volume, and lasing threshold in small diameter AlGaAs microdisks with embedded quantum dots, *Optics Express* 14, N3, 1094.

Tamboli, T., *et al.* (2007), Room temperature continuous-wave lasing in GaN/InGaN microdisks, *Nature Photonics* 1, 61.

Tamboli, T., *et al.* (2009), Observation of whispering gallery modes in nonpolar m plane GaN microdisks, *Appl. Phys. Lett.* 94, 251116.

Tawara, T., *et al.* (2004), Cavity polaritons in InGaN microcavities at room temperature, *Phys. Rev. Lett.* 92, 256402.

Torres, J., *et al.* (2004), Giant second-harmonic generation in a one-dimensional GaN photonic crystal, *Phys. Rev. B* 69, 85105-1.

Tosi, G., *et al.* (2012), Sculpting oscillators with light within a non-linear quantum fluid, *Nature Physics* 8, 190.

Trichet, A., *et al.* (2012), From strong to weak coupling regime in a single GaN microwire up to room temperature, *New J. Phys.* 14, 073004.

Tsintzos, S., *et al.* (2008), A GaAs polariton light emitting diode operating near room temperature, *Nature* 453, 372–375.

Tsintzos, S., *et al.* (2009), Room temperature GaAs exciton polariton light emitting diode, *Appl. Phys. Lett.* 94, 071109.

Vico Trivino, N., *et al.* (2012), High quality factor two dimensional GaN photonic crystal cavity membranes grown on silicon substrate, *Appl. Phys. Lett.* 100, 071103.

Weisbuch, C., *et al.* (1992), Observation of the coupled exciton photon mode spliting in a semiconductor quantum cavity, *Phys. Rev. Lett.* 69, 3314.

Wierer, J., *et al.* (2004), InGaN/GaN quantum well heterostructure light-emitting diodes employing photonic crystal structures, *Appl. Phys. Lett.* 84, 3885.

Wiesmann, C., *et al.* (2009), Photonic crystal LEDs - designing light extraction, *Laser and Photonics Reviews* 3 (3), 262.

Wertz, E., *et al.* (2010), Spontaneous formation and optical manipulation of extended polariton condensates, *Nature Physics* 6, 860.

Yablonovitch, E. (1987), Inhibited Spontaneous Emission in Solid-State Physics and Electronics, *Phys. Rev. Lett.* 58, 2059.

Yablonovitch, E., *et al.* (1991), Photonic band structure: The face centered cubic case employing nonspherical atoms, *Phys. Rev. Lett.* 67, 2295.

Yoshie, T., *et al.* (2004), Vacuum Rabi splitting with a single quantum dot in a photonic crystal nanocavity, *Nature* 432 (7014), 200.

Yoshizawa, M., *et al.* (1997), Growth of self-organized GaN nanostructures on Al_2O_3(0001) by RF radical source molecular beam epitaxy, *Jpn. J. Appl. Phys.* 36, L459.

Zhong, Z., *et al.* (2003), Synthesis of p type gallium nitride nanowires for electronic and photonic nanodevices, *Nano Lett.* 3 (3), 343.

11 Nitride-based electron devices for high-power/high-frequency applications

Yvon Cordier, Tatsuya Fujishima, Bin Lu, Elison Matioli, and Tomás Palacios

Progress in the development of III-nitride materials has made possible the fabrication of many different types of electron devices. This chapter describes the main advantages of GaN-based electronics as well as some of the main electronic devices developed with this material system. Section 11.1 introduces the key material properties and figures of merit that make III-nitride materials ideal for many high-power/high-frequency applications. Section 11.2 focuses on the different GaN diode structures developed so far, while Section 11.3 discusses GaN bipolar transistors. Finally, section 11.4 studies field-effect transistors, both lateral and vertical.

11.1 Relevant GaN properties for electron devices

The unique and broad spectrum of properties of wurtzite III-nitride materials offers an unprecended level of flexibility for the design of advanced electron devices. Table 11.1 shows some of these material properties and the main advantages of these materials for electronics. First, their band gap is direct and tunable from 0.7 eV in the case of InN to 6.2 eV in AlN. The large band-gap energies enable large conduction and valence band discontinuities in many heterostructures, significantly helping with carrier confinement. These materials are also highly polar, which allows the use of polarization doping to induce two- and three-dimensional charge layers without the need of doping [Yu97, Ambacher99, Rajan04]. Both the electron mobility and velocity (ν_s) in GaN are high, especially in the two-dimensional electron gas (2DEG) induced at the interface of $Al_xIn_yGa_{1\text{-x-y}}N/GaN$ heterostructures. This enables the fabrication of devices with high cut-off frequencies (for instance, high electron mobility transistors, HEMTs), but also for low-loss power switching devices that need low on-state resistances. In addition, the large band gap induces a very low intrinsic carrier density (n_i) that ensures good charge control even at high temperatures. Furthermore, the combination of these properties with a high critical electric field (E_c), up to ten-fold higher in GaN than in Si, makes these materials ideal for power applications. Finally, the thermal conductivity (Θ_K) of GaN is similar to that of silicon, while that of AlN is 30% higher, which helps to efficiently remove the generated heat in these devices. The only significant drawback of III-nitrides

Table 11.1 Main material properties of some of the most commonly used semiconductors.

300K parameter	Si	GaAs	6H-SiC	4H-SiC	Diamond	GaN	AlN	InN
E_{gap} (eV)	1.1 Indirect	1.4 Direct	2.9 Indirect	3.3 Indirect	5.5 Indirect	3.4 Direct	6.1 Direct	0.7 Direct
n_i (cm^{-3})	$1.5 \times 10^{+10}$	$1.8 \times 10^{+6}$	1.6×10^{-6}	5×10^{-9}	1×10^{-27}	2×10^{-10}	9×10^{-34}	$1 \times 10^{+3}$
Electron mobility μ_e ($cm^2/V.s$)	1400	7000–8500**	600	800	1900	1000–2000*	1100	3600
Hole mobility μ_p ($cm^2/V.s$)	600	400	40	40	1600	30	14	30
Saturation velocity $\nu_s(\times 10^7$ cm/s)	1	1–1.2	1.5	1.8	1.9	1.5–2	1.5	1
Maximum velocity $\nu(\times 10^7$ cm/s)	1	2.2	1.9	2.2	2.7	2.8	1.6	4
Critical electric field $E_c(\times 10^6$ V/cm)	0.3	0.4	4	3	10	3.3	6–15	1
Thermal conductivity Θ_K (W/cm.K)	1.5	0.5	5	4.9	6–20	1.5	2	1
Relative dielectric constant ε_r	12	13	9.8	9.8	5.7	9.5–10.4	9.1	15

* Measured on (AlGa)N/GaN HEMT structures.
** Measured on (AlGa)As/GaAs HEMT structures.

relates to the low hole mobility and the poor p-type doping activation, which has prevented the fabrication of high-performance hole-based electronic devices until now.

Beyond the individual material properties, many figures of merit (FOM) have been proposed to benchmark the merit of different material systems for a given application. One of the most widely used at one time is the Johnson Figure of Merit (JFOM) [Johnson65], which gives an idea of material suitability for high-power applications at high frequencies:

$$JFOM = \left(\frac{E_c \times \nu_s}{2\pi}\right)^2$$

Table 11.2 shows a comparison of the JFOM in the main semiconductor families. It is clear that wide band-gap materials like GaN, AlN, diamond, and silicon carbide are excellent candidates for high-frequency power applications. Other FOMs have been proposed for more specific cases [Baliga89]. Among these, the Keyes FOM considers thermal limitations due to transistor switching:

$$KFOM = \Theta_K \times \left(\frac{c\nu_s}{4\pi\varepsilon}\right)^{0.5}$$

where c is the speed of light. The Baliga FOM was developed to compare the ultimate performance of field-effect transistors for low-frequency power-switching applications, where the conduction losses dominate:

$$BFOM = \varepsilon \times \mu_e \times E_{gap}{}^3$$

while the Baliga high-frequency FOM (BHFFOM) benchmarks these devices at high frequency, where switching losses dominate:

$$BHFFOM = \mu_e \times E_c^2 \times \ V_G^{1/2}/2V_{Br}^{3/2}$$

where V_G is the gate drive voltage and V_{Br} the breakdown voltage. The latter can be related to an experimental figure of merit involving the on-state specific resistance R_{on} of the device and its critical electric field:

$$FOM = V_{Br}^2/R_{on} \equiv \varepsilon \times \mu_e \times E_c^3/4$$

Table 11.2 Comparison of different FOM calculated with data from Table 11.1, and normalized to Si.

	Si	GaAs	6H-SiC	4H-SiC	Diamond	GaN	AlN	InN
JFOM	1	2	400	324	4000	270–480	900–5600	11
KFOM	1	0.4	4.5	4.8	8.0–27.0	1.4	1.9	0.6
BFOM	1	13	6	12	80	17–34	100	0,8
BHFFOM	1	10	76	57	1500	86–172	314–1960	28
CFOM	1	4	310	275	5400–18000	108–290	480–2980	24

Finally, a Combined Figure of Merit (CFOM) was developed to simultaneously account for the high-frequency, high-power and high-temperature performance [Shur04]:

$$CFOM = \Theta_K \times \varepsilon \times \mu \times \nu_s \times E_c^2$$

No matter what FOM is used, wide-band-gap materials dominate high-frequency and high-power applications. Among wide-band-gap semiconductors, GaN-based materials are arguably the most mature ones in terms of wafer-size availability, technology, and market prospects.

11.2 GaN two-terminal devices

There is a wide variety of GaN two-terminal devices. In this section we describe those of main interest for electronic applications, starting with Schottky and bipolar junction diodes. Other devices, such as Gunn-effect diodes, tunnel junctions, and finally resonant tunneling diodes, are also presented.

11.2.1 *High-breakdown Schottky diodes*

The Schottky diode is probably the simplest, but still relevant, electron device based on GaN. The high breakdown voltage achievable in GaN and its combination with a high carrier concentration in the channel to lower on-state resistance make it ideal for high-breakdown rectifiers. Such power rectifiers are mainly developed for replacing 600 V silicon-based diodes in switching applications such as power-factor correction (PFC) [Alquier12]. The extremely short switching time of Schottky devices (ns) and the lack of reverse recovery thanks to its unipolar character are important advantages of these diodes [Zhou06]. SiC devices have been the main option for these applications for many years, though it has been reported recently that GaN rectifiers present equal or even better performance at a fraction of the cost [Cohen05].

GaN-based Schottky rectifiers are generally vertical devices based on several micron-thick n-type GaN films, and a highly doped contact layer (Fig. 11.1). Depending on the thickness and the doping level of the top layer (drift layer), the breakdown is limited by punch-through or by GaN avalanche [Bandic98] (Fig. 11.2). Various device architectures have been proposed for high-breakdown-voltage (HBV) Schottky diodes, depending on the substrate used for GaN growth (Fig. 11.1). The growth of GaN is typically achieved on insulating sapphire substrates, or on conductive silicon carbide or silicon substrates. In all these cases it is necessary to remove the substrate or use through-wafer vias to make vertical devices, due to either the insulating character of the substrate or the highly resistive AlN buffer layers used for the growth. This problem does not exist when the diodes are grown directly on highly doped GaN substrates, though the substrate cost in this case dominates the availability of these devices.

Another approach for fabricating Schottky diodes is to develop a pseudo-vertical device for which a low-doped thick GaN layer is grown on top of a

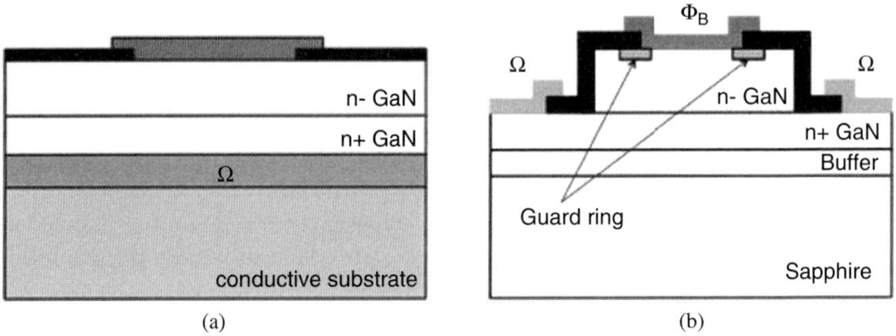

FIG. 11.1. Schematic cross-section of a vertical Schottky diode (a), and a pseudo-vertical Schottky diode (b). (After [Alquier12], Copyright 2012 The Japan Society of Applied Physics.)

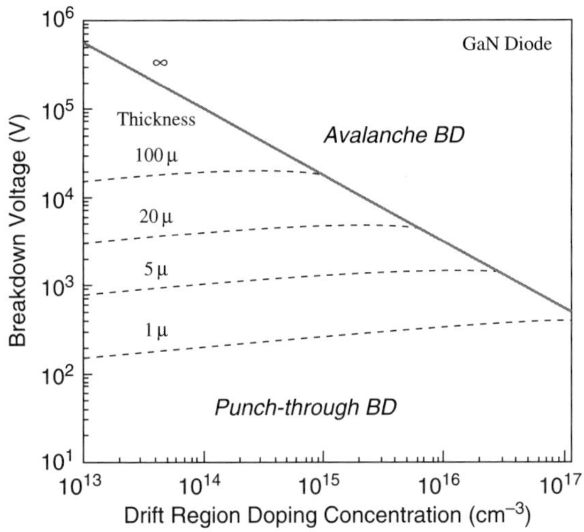

FIG. 11.2. Calculated breakdown voltage as a function of doping concentration and thickness of the drift region in GaN diodes. Reprinted from [Pearton00], Copyright (2000) with permission from Elsevier.

highly doped one. In this case it is necessary to etch several micrometers of GaN to be able to deposit the ohmic contact on the highly doped bottom layer. The critical step here is the control of the etching depth by reactive ion etching (RIE) or inductively coupled plasma etching (ICP).

The ohmic contact formation on highly doped GaN layer is easily obtained after deposition and annealing of metal stacks with Ti that benefit from the formation of conductive TiN [Lin94] or $TiAl_3$ [Ménard10]; this provides contact resistances in the 10^{-6}–10^{-5} Ωcm^2 range. The Schottky contact on GaN can be obtained by deposition of metals with a large work function (e.g., Au,

Pd, Ni, and Pt). The Schottky barrier height (Φ_B) is typically of 1 eV or less, indicating the probable presence of Fermi-level pinning at the metal/GaN interface [Roccaforte10]. This is certainly due to electron tunneling via surface defects such as threading dislocations, as confirmed by the dependence of the reverse leakage current with the crystal quality [Alquier12]. The diode current obeys the thermionic emission theory at low bias, while more complex transport involving thermionic field emission and tunneling occurs at large reverse bias. The reverse leakage currents of simple diodes are generally larger than expected, and associated breakdown voltages of the pseudo-vertical devices are often below 100 V. This is partly due to the electric field crowding at the edges of the Schottky contact. In order to lower the electric field in this region, the combination of a p-type doped guard ring and a field plate is suitable. Unfortunately, Mg implantation and activation annealing steps have been reported as very difficult, and the use of resistive guard rings with Mg, In, Ar, or He is preferred [Liu09]. Using this technology, diodes with a forward capability of 100 A/cm^2 at 1.2 V and with more than 600 V reverse breakdown voltage have been demonstrated on GaN on sapphire substrates, in spite of the large dislocation density typical of GaN heteroepitaxy [Alquier12]. The use of low-dislocation-density GaN substrates helps for the reduction of the reverse leakage current and the enhancement of the breakdown voltage. 5-μm thick GaN layers grown on GaN substrates with threading dislocation densities lower than 10^6 cm^{-2} led to the fabrication of diodes with a breakdown voltage over 1100 V, a forward current capability of 500 A/cm^2 at 1.46 V, a specific on-state resistance of 0.71 mΩ.cm^2, and a resultant figure of merit ($V_{Br}{}^2/R_{on}$) of 1.7 GW/cm^2 [Saitoh10]. The achievement of such a defect reduction is, however, still very difficult for GaN grown on large-area silicon wafers, even when noticeable improvements have been made with 4-μm thick crack-free continuous GaN films on 4- and 6-in. Si substrates [Frayssinet11] and with 5-μm thick GaN on 3C-SiC/Si(111) substrates [Cordier10].

Finally, an important challenge for applications is that GaN Schottky diodes experience performance degradation while increasing the temperature. As shown in Fig. 11.3, the reverse leakage current increases as well as the series resistances at forward bias larger than 1V.

The main difficulty for the realization of HBV Schottky diodes is the sensitivity of the leakage current to the crystal quality and to surface treatments such as plasma etching, implantation, and annealing [Pearton99, Pearton00]. An alternative approach to vertical or pseudo-vertical devices is the fabrication of planar diodes. A planar GaN diode can be obtained with both the ohmic contact and the Schottky contact deposited on top of a lightly doped GaN layer. In that case, however, the device is severely limited by a high on-state resistance (in spite of the achievement of low-resistance ohmic contacts via the implantation of low-doped GaN layers [Burm97]), and in terms of the trade-off between the on-state resistance and the breakdown voltage. In a similar manner, replacing GaN with an alloy such as AlGaN helps in achieving higher Schottky barrier and larger breakdown voltage [Zhang00], but the crystal and surface quality as well

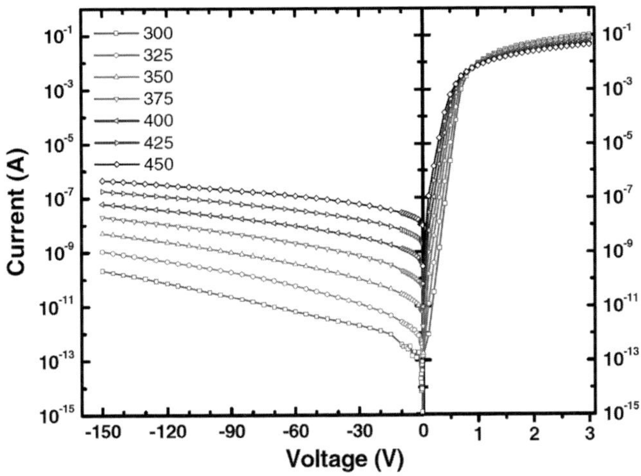

FIG. 11.3. I–V characteristics of 300-μm-diameter Schottky diodes at temperatures between 300 and 450 K. Reprinted from [Wang11], Copyright (2011) with permission from IOP.

as the lower electron mobility are generally important drawbacks with respect to GaN diodes. An attractive alternative consists in combining the advantage of a wide band-gap barrier material such as AlGaN or AlInN with the low-access resistance of the two-dimensional electron gas formed at the interface with GaN [Chen08]: namely, a high electron mobility transistor (HEMT) structure (see Section 11.4.1).

11.2.2 *Bipolar GaN diodes (p-n or p-i-n)*

p-i-n diodes are expected to have larger reverse blocking voltages than Schottky diodes due to the lower leakage current of the reverse-biased junction [Zhang01], [Mochizuki11]. However, the junction barrier leads to a higher forward turn-on voltage. Furthermore, when switching from forward to reverse bias, the minority carriers stored in quasi-neutral regions must be removed. This is responsible for longer switching times (in relation to minority carrier lifetime), and their diffusion in the direction opposite to the initially applied electric field leads to reverse current spikes and switching losses. In addition, these vertical devices share some common challenges with the Schottky-based devices, such as the sensitivity of leakage current to crystal defects such as dislocations [Kozodoy98], and the necessity to implement surface passivations and field plates for high breakdown voltage. The device structure of a GaN bipolar diode generally consists of an n+/n bottom layer capped with a p GaN layer doped with Mg to form the junction, and terminated with a p+ contact layer. The doping level is generally below 10^{16} cm^{-3} to avoid premature breakdown at the junction. The on-state resistance of these devices depends critically on the p-doping efficiency and the surface preparation for ohmic contact formation. However, the low Mg activation in III-N semiconductors makes necessary the incorporation of a very large density of Mg atoms, in the range of 10^{17} to 10^{19} per cm^3. In spite of higher contact resistivity than for n-type contacts, the forward-biased p-n diode can benefit from photon recycling effects so that relatively low on-state resistances can be achieved [Hatakeyama11].

Like Schottky rectifiers, the fabrication of devices on GaN free-standing substrates has quickly improved device performance [Limb06, Hatakeyama11]. p-n diodes with field-plates demonstrated breakdown voltages over 1100 V, a knee voltage of about 2 V (Fig. 11.4), and specific on-resistance of 0.4 mΩ·cm^2, with a resultant figure of merit ($V_{Br}{}^2/R_{on}$) of 3.0 GW/cm^2.

As in the case of Schottky diodes, high temperatures enhance the reverse leakage current but with a lower magnitude for large reverse bias [Kozodoy98], [Uesugi11]. At the same time, the increase in temperature lowers the forward threshold voltage of the diode.

11.2.3 *Gunn diodes*

The bulk negative differential resistance (NDR) effect is a very popular phenomenon in III–V compounds such as GaAs and InP. In these materials it is due to the transfer of electrons between the high-mobility central valley and the

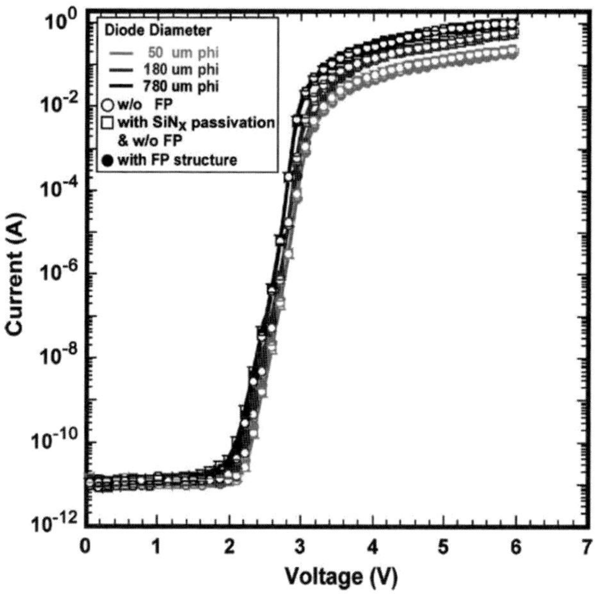

FIG. 11.4. Forward I–V characteristics of GaN p-n junction diodes on free-standing GaN substrates. (After [Hatakeyama11], copyright © 2011 IEEE.)

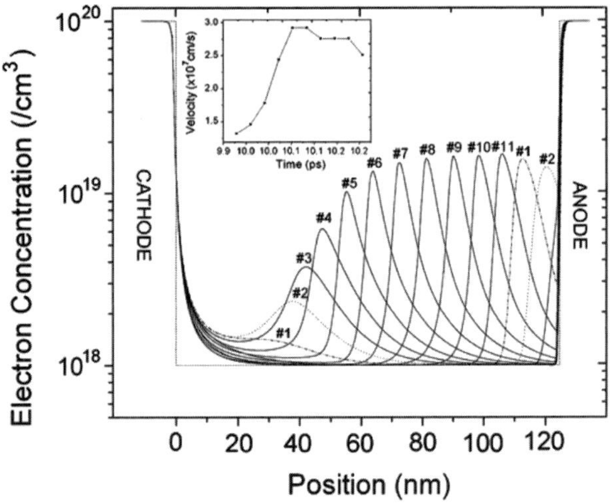

FIG. 11.5. Electron concentration vs. position at successive instances of time (each time interval is 30 fs, apart from the previous one) showing an electron accumulation layer in transit towards the anode. The insert shows the average electron accumulation layer velocity as a function of time during transit. Reprinted from [Aslan11], Copyright (2011) with permission from Elsevier.

low-mobility satellite valley, and leads to a decrease and saturation of the electron velocity beyond a critical electric field. As shown in Fig. 11.5, this leads to a charge inhomogeneity (dipole), traveling from the cathode to the anode of a diode (named Gunn diode), and finally gives rise to transit-time current oscillations that can be exploited for the microwave and millimeter-wave signal generation. The significantly higher electrical strength of wide-band-gap semiconductors allows operation with higher doping levels and at higher bias, which should lead to oscillators with much better power performance than their III–V counterparts.

Various possibilities are suggested for the nature of the NDR in nitrides. The first one, as in GaAs, is the electron intervalley transfer [Kolnik95], but the large energy barrier separating neighbor valleys in GaN (2.1 eV) may push the effect to very large electrical fields. Another possibility is the inflexion of the central valley [Krishnamurthy97]. Wraback measured the electron velocity along the c axis of wurtzite GaN with femtosecond optical pulse-probe experiments using 80-fs pump and probe pulses [Wraback03]. Transient velocity overshoot was observed along the c axis at electric fields as low as 130 kV/cm. According to Monte Carlo calculations, at electric fields below 300 kV/cm this velocity overshoot is not associated with intervalley transfer, but with band non-parabolicity in the central valley. Calculations of transport in the basal plane suggest that a similar velocity overshoot would need much higher electric fields ($\sim$175 kV/cm) in this plane.

Previously, a critical field of 180 kV/cm was measured on thick GaN samples [Barker05] with 200-ns-long pulse measurements. At this field, a maximum velocity of 2.5×10^7 cm/s was measured in the basal plane. When the pulse width is reduced to 10 ns, AlGaN/GaN heterostructures with a two-dimensional electron gas (2DEG) showed an apparent saturation velocity near 3.1×10^7 cm/s, at a field of 140 kV/cm. The latter indicates that sample heating (lattice heating and hot phonons) may play an important role in such measurement. It is also worth noting that the NDR effect is also expected for zincblende GaN [Kolnik95].

In a classical model the main criterion of the active layers for Gunn instabilities is the (doping $\times$ length) product. This criterion is based on the fact that the accumulated electron domain growth rate should be higher than the transit frequency through the layer [Alekseev00]; furthermore, the threshold electric field for NDR determines the maximum doping level. The frequency capability of GaN diodes is, however, still controversial. Indeed, the ultimate frequency is highly dependent on the involved relaxation times. Alekseev reported that the frequency capability of GaN-based NDR devices was of 700 GHz in the case of the intervalley-transfer-based NDR and 4 THz in the case of the injection-based NDR (relaxation time of 0.15 ps). However, Aslan has indicated that collision-dominated Monte Carlo simulations of the velocity-field relation typically do not take into account the relatively long (0.35 ps) non-equilibrium longitudinal optical (LO) phonon lifetime in GaN, and more important, the Monte Carlo method was often used even when the diode was too short for assuming a transport dominated by collisions (i.e., 100 nm) [Aslan11]. Alternatively, based on the velocity–field relation derived from the pulse-probe experiments reported by Wraback, the use of higher electric fields in an ultra-short GaN channel (i.e., 250 nm) was proposed to ensure operation in the quasi-ballistic transport regime. As shown in Figure 11.6, compared with Monte Carlo results [Wraback03], such a regime can lead to noticeably high-velocity overshoots (as high as 7×10^5 m/s). Device simulations in such submicron channel devices project power-added efficiency (PAE) up to 3.4% at 1.5 THz frequency [Aslan11]. At the same time, simulations for GaN diodes with thermal effects predicted a PAE of 1–2% at frequencies slightly below 200 GHz for 1–2-μm-channel devices on submicron-GaN-thickness films [Tang09].

As the NDR is expected in both vertical and basal plane directions, in bulk GaN as well as for 2DEGs a variety of device configurations has been investigated. The simplest one is the N+/N/N+ diode shown in Fig. 11.7. An oscillator can be built with such a diode when a high enough DC bias is applied inside an RLC circuit [Alekseev00], [Aslan11]. The fact that a relatively high electric field is necessary to induce NDR has an important impact on the device design and technology, as electromigration and other degradation effects (e.g. avalanche breakdown) have to be avoided [Mutamba06]. An alternative approach to accelerate electrons without having to increase the external bias is to insert graded-composition AlGaN barriers in a manner similar to the way by which graded gap AlGaAs emitters were used in GaAs devices [Montanari05].

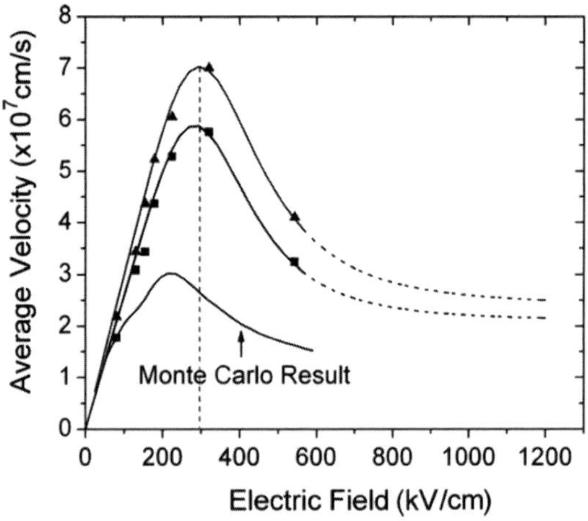

FIG. 11.6. Velocity-field characteristics derived for 125- and 250-nm-long devices. Reprinted from [Aslan11], Copyright (2011) with permission from Elsevier.

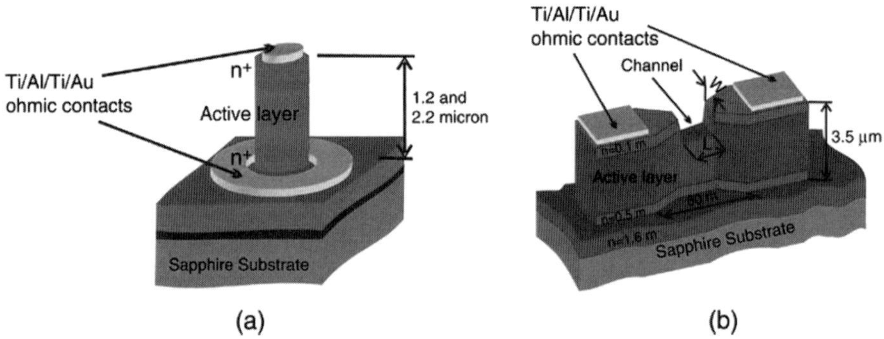

FIG. 11.7. Schematics of the N+/N/N+ diode structures: (a) vertical (b) lateral designs. Reprinted from [Mutamba06], Copyright (2006) with permission from Elsevier.

Some of the previous issues (including air breakdown) have been partly solved by Yilmazoglu, who has shown bias oscillations in a pulse reflected from a vertical GaN Gunn diodes (5μm, $n = 1 \times 10^{17}$ cm^{-3}) connected to a series inductance and subjected to 70-ns bias pulses. The voltage across the GaN Gunn diode terminals was about 200 V and the DC peak power 190 W, which emphasizes that thermal management is a major issue for GaN Gunn diodes [Yilmazoglu08].

11.2.4 *Tunnel diodes*

Electron tunnel junctions and devices are used in many applications, such as oscillators, amplifiers, frequency converters [Sze81], and detectors, as well as electrical connections inside light-emitting devices [Grundmann07] or multi-junction solar cells. The GaN tunnel junction is, however, quite peculiar due to the large polarization-induced electric fields present in c-oriented wurtzite heterostructures, and the large band-gap of these materials. If one examines the case of the interband electron tunneling, it is clear that GaN is not a good candidate compared to heavily doped germanium p-n junction diodes [Esaki58]. This is due to the large band-gap of GaN (3.4 eV), coupled with the difficulty in achieving highly doped p-type GaN which result in p-n junctions with wide depletion regions (~25 nm). However, the polarization-induced electric field present at the AlGaN/GaN or the InGaN/GaN interfaces can be sufficient to enable the correct alignment of the conduction and valence bands of a properly designed structure.

As shown in Figure 11.8, the polarization field of 12 MV/cm at the AlN/GaN interface is sufficient to align the bands for an AlN thickness below 5 nm [Leconte07]. If the GaN layer located on the top of the AlN barrier is p-type doped, it is then easy for holes to tunnel through the thin barrier into the n-type doped layer, even for 0 V bias. Such a Zener diode is more conductive under reverse bias due to the asymmetry of the band diagram, and can be used for the rectification of small AC signals [Simon09].

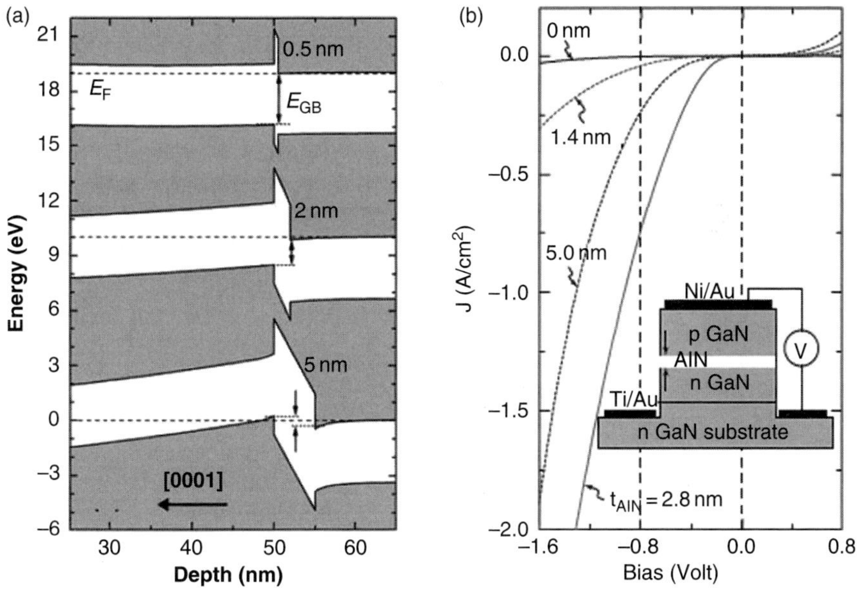

FIG. 11.8. Left: Band diagram of GaN/AlN/GaN single-barrier structures with different AlN barrier thickness and a GaN cap layer of 50 nm. Reprinted from [Leconte07], Copyright (2007) with permission from IOP. Right: Current–voltage characteristics of a Zener diode under small applied biases. Reprinted figure with permission from [Simon09]. Copyright (2009) by the American Physical Society.

The use of narrower-bandgap nitride films would improve the performance of tunnel junctions. For instance, an InGaN/GaN/InGaN structure is expected to have better tunneling than the previously discussed GaN/AlN/GaN, though the growth of thick indium-rich InGaN layers is still a challenge. An alternative method is based on using GaN/InGaN/GaN junctions [Krishnamoorthy10]. The band diagram of such a junction is shown in Fig. 11.9. One can see that the inserted 6.4-nm thick $In_{0.33}Ga_{0.67}N$ layer is able to align the p and n GaN regions at 0 V. It should be noted that the structure has to be grown with N-polarity in order to keep the p-type doped layer on the top. In spite of the thickness and the presence of spikes due to the band discontinuities, the lower band-gap of InGaN results in larger current densities compared to the AlN-based diode or a reference GaN p+/n+ junction. The reverse current density is 100 A/cm^2 at 1 V, which is high enough to enable the use of tunnel junctions in III-nitride devices such as LEDs, lasers, and solar cells.

The resonant tunneling effect can be used for building high-speed digital circuits [Haddad97] or for injecting electrons with a precisely controlled energy in vertical unipolar devices such as quantum cascade lasers, resonant hot electron transistors, or long-wavelength intersubband detectors (see Chapter 12). Resonant tunneling diodes (RTDs) are the basic building-blocks of all these devices. In these diodes, the tunnel current flow reaches a maximum when the energy of electrons from the emitter is resonant whith one of the quantum-well energy states. For further increase of the bias, the tunnel current flow decreases sharply and the minimum of the total current is determined by leakage, scattering, and thermionic current flow.

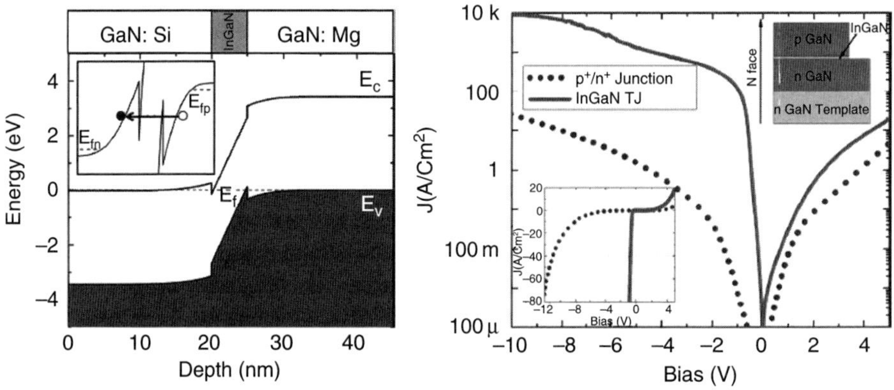

FIG. 11.9. Left: Equilibrium energy-band diagram of GaN/$In_{0.33}Ga_{0.67}N$/GaN zero bias interband tunnel junction; inset: band diagram at reverse bias showing interband tunneling. Right: Characteristics of the GaN/$In_{0.33}Ga_{0.67}N$/GaN tunnel junction (solid line) and standard GaN p+/n+ junction (dotted line). Reprinted with permission from [Krishnamoorthy10]. Copyright 2010, American Institute of Physics.

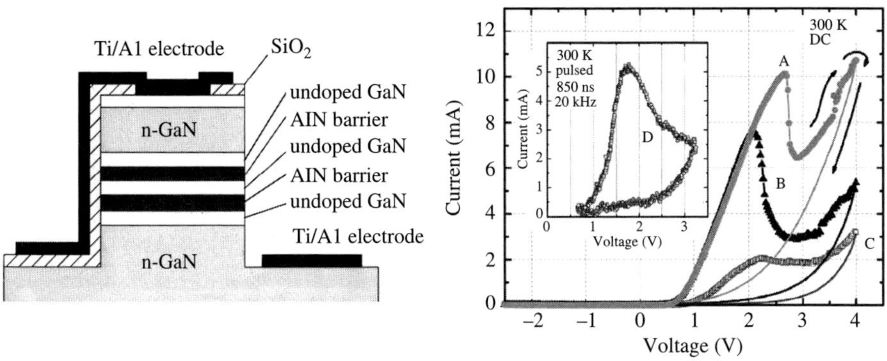

FIG. 11.10. Left: Cross-sectional view of a GaN/AlN/GaN RTD device. Reprinted with permission from [Kikuchi02]. Copyright 2002, American Institute of Physics. Right: Room-temperature DC *IV* characteristics of AlN/GaN/AlN RTD devices on a GaN substrate. Reprinted with permission from [Golka06]. Copyright 2006, American Institute of Physics.

A RTD device basically consists of a 0.5–2 nm thick GaN quantum well located between 1–3 nm AlGaN barriers and N-type doped GaN contact layers (Fig. 11.10). The occurrence of a resonant tunneling effect induces a sharp decrease of the electrical current passing through the double-barrier structure when electrons from the emitter region no longer tunnel due to the misalignment of the energy of electrons on the emitter side with the quantized state within the central quantum well (negative differential resistance). In principle, the AlGaN/GaN double-barrier diodes offer advantages such as material robustness, high breakdown field, and large conduction-band discontinuity ($\Delta E_c \sim 1.7$–1.9 eV for AlN/GaN). However, the asymmetric band diagram induced by the presence of the polarization fields at the interfaces of c-oriented wurtzite structures is a challenge for the design of RTDs in nitride semiconductors (Fig. 11.11). Preliminary results show asymmetric I–V characteristics (Fig. 11.10). Furthermore, the sensitivity of this device to point defects and dislocations acting as traps or leakage path makes the realization of resonant tunneling difficult and highly non-uniform (Fig. 11.10). Many authors noticed the instability of the I–V characteristics [Foxon03, Golka06] independently of dislocation density. As these structures were grown at relatively low temperatures by plasma-source molecular beam epitaxy (MBE), traps in AlN barriers are probably involved [Hermann04, Sakr11]. This pushed Bayram to grow AlN/GaN/AlN RTD structures by metal–organic chemical vapor deposition (MOCVD) on low-dislocation-density GaN on sapphire templates [Bayram10a]. This produced I–V characteristics with current peak-to-valley ratio and hysteresis comparable to MBE-grown RTDs on GaN substrates by Golka. The growth of low Al content (Al $\sim$0.2) AlGaN/GaN/AlGaN structures on low-dislocation-density GaN has been shown to improve the stability of I–V characteristics (by using MOCVD [Bayram10b] and MBE [Li12]).

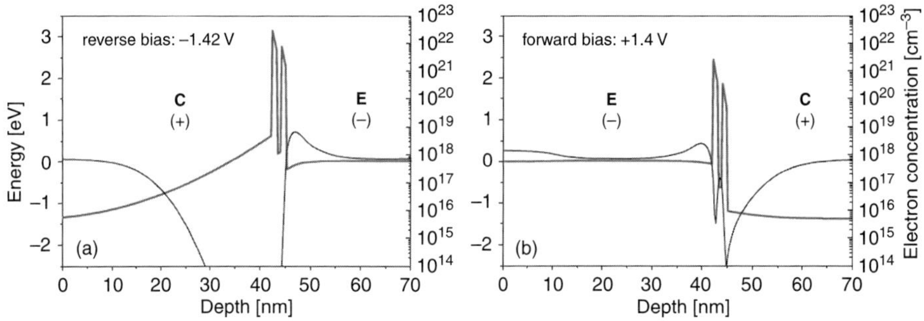

FIG. 11.11. Potential profile and electron concentration for an applied voltage of (left) −1.42 V (reverse direction, T = 300 K), and (right) +1.40 V (forward direction, T = 300 K). Reprinted from [Indlekofer02], Copyright (2002) with permission from Wiley.

The asymmetric conduction-band profile due to the polarization-induced electric fields is responsible for asymmetric current–voltage behavior, which is undesirable for most of the high-frequency applications. As the asymmetry is due mainly to the GaN band bending at the top and at the bottom of the double-barrier structure (Fig. 11.11), the insertion of a graded AlGaN emitter or collector has been proposed to make the structure more symmetric [Berland11]. Another proposed solution consists in growing the structure on non-polar GaN (e.g. m-plane GaN susbtrate) [Bayram10b], or on cubic GaN [Zainal10].

11.3 Bipolar junction transistors and heterojunction bipolar transistors (BJT and HBT)

Bipolar junction transistors (BJT) are made of two back-to-back p-n diodes (Fig. 11.12). Under normal operating conditions, the first junction (emitter-base) is forward-biased so that minority carriers are injected into the base while the other junction (base-collector) is reverse-biased, and minority carriers transiting trough the base are collected in the collector layer. A small change of the emitter–base junction-barrier energy has a very important effect on the number of minority carriers injected into the base. Most of these carriers reach the collector and, as a consequence, the collector current is controlled by small changes in the base bias. As the voltage at the base–collector junction can be much higher than the one at the emitter–base junction, power amplification arises in the transistor.

The current gain defined as the ratio of the collector current over the base current is an important figure of merit for bipolar transistors. It represents the modulation efficiency of the minority carrier current in the collector with respect to the base current. The base current is due to the recombination of carriers in the base, minority carrier injection from the base into the emitter, and leakage current at the base–collector junction. A good bipolar transistor necessitates that

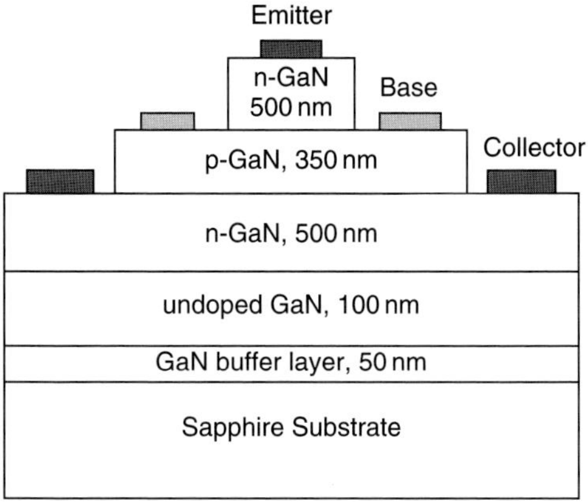

FIG. 11.12. Schematic drawing of the structure of an n–p–n bipolar junction transistor. Reprinted with permission from [Yoshida99]. Copyright 1999, American Institute of Physics.

the minority carriers injected into the base reach the collector region without recombining with majority carriers or defects. For this to happen, a thin base layer with high carrier mobility and velocity is required. At the same time, a thin base requires a high doping level to limit the increase of the base resistance as well as the reduction of the effective base width under high collector bias (Early effect). As the doping level in the base should be lower than the one in the emitter to limit the injection of carriers from the base into the emitter, it is clear that the design of a bipolar junction transistor is made of multiple trade-offs. The use of an emitter with a wider band-gap energy than the base (emitter–base heterojunction) drastically reduces the undesirable injection of carriers from the base in this heterojunction bipolar transistor (HBT). Furthermore, this allows reduction of the doping level of the emitter, which reduces the junction capacitance and is beneficial for high-frequency applications. In addition, a wide-band-gap collector is often used for enhancing the breakdown voltage in these transistors.

The high critical electric field and electron velocity of GaN makes this material system very attractive for high-voltage HBT devices [Xing01]. In addition, GaN HBTs offer two important advantages over field-effect devices for microwave power amplification: higher current densities, and linearity. Furthermore, for high-frequency applications, the shrinking of the transistor gate length in field-effect devices is more technologically challenging than to control the base thickness through heteroepitaxy. However, GaN is a direct band-gap material and the radiative recombination in the base cannot be completely avoided, which limits the efficiency of the collector current modulation. But the main difficulty arises from the difficulty of achieving high p-type doping in GaN or AlGaN (i.e. above 10^{18} cm^{-3}) as well as sufficient hole mobility. For these reasons, the n-p-n transistors are preferred to p-n-p ones, and some authors proposed the n-GaN/p-SiC/n-SiC hybrid heterojunction as an alternative [Pankove94]. Although such a device demonstrated a current density of 1.8 kA/cm^2 and a power density of 30 kW/cm^2, it suffered from low breakdown voltage (below 10 V) and noticeable leakage of current, preventing common-emitter operation. The low breakdown voltage is believed to be due to the combination of high doping levels and crystal defects at the GaN/SiC heterojunction.

Recent progress made in the doping of GaN with Mg has led to the demonstration of pure GaN n-p-n BJTs. Even with the limited performance of these early devices, these transistors exhibited excellent thermal stability, with no degradation at 300°C [Yoshida99]. The device-processing technology is, however, difficult due to the required dry etching of the emitter region, which damages the base layer. In order to overcome this difficulty, the GaN emitter can be selectively regrown onto the base after patterning a dielectric mask deposited after the growth of the base–collector junction [Limb00]. Limb *et al.* reported high-voltage operation (>80V) but with a limited current gain of 3, due to the high lateral resistance of the base layer.

As discussed previously, the use of AlGaN/GaN, GaN/InGaN and AlGaN/InGaN heterojunctions allow significant improvement in the performance of HBTs. In addition, thanks to the large polarization fields present in wurtzite materials, graded composition alloys can be grown for lowering the base resistance [Asbeck00]. n-p-n AlGaN/GaN HBTs with a current gain of 35 have been obtained [Xing01]. Pearton reported current gains of 15–20 in n-p-n structures [Pearton00]. More recently, p-n-p AlGaN/GaN HBTs grown on GaN substrates have demonstrated a maximum current gain of 85 and a maximum collector current density of 7.3 kA/cm^2 at a collector–emitter voltage of 30 V, which corresponds to a maximum power-dissipation density of 219 kW/cm^2 [Kumakara08]. The same devices grown on sapphire exhibited noticeably lower current gain, attributed to the larger dislocation density and self-heating.

p-type doped InGaN base layers have a resistivity of only one tenth of that in p-GaN layers, which has led to the development of AlGaN/InGaN and GaN/InGaN HBTs [Makimoto02, 03, 04]. In addition, the chemical stability of InGaN layers simplifies the emitter etching process and allows the regrowth of an external base layer for an even lower access resistance. This is crucial for low-off-set voltages. In these InGaN-base devices, the presence of band-offsets and polarization charges can significantly increase the breakdown voltage (up to 100 V), but at the expense of the current gain, so that graded composition layers at the base–collector junctions are inserted with resultant differential current gain (dI_c/dI_b) of several thousands and breakdown voltages up to 50 V. Moreover, the insertion of a graded composition InGaN layer at the emitter–base interface has been proposed to provide a larger energy barrier for holes, which allows higher emitter-injection efficiency, higher current gain, and possibly the achievement of an emitter–base junction with higher crystal quality [Chung06, Lee10, Lochner11]. In [Lee10], a GaN/$In_{0.03}Ga_{0.97}N$ n-p-n DHBT with

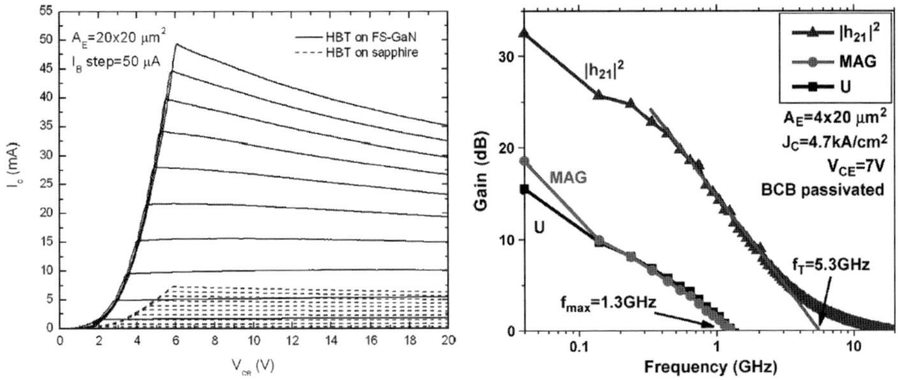

FIG. 11.13. Left: Common-emitter I–V characteristics of HBTs with $20 \times 20\ \mu m^2$ grown on free-standing GaN and sapphire substrates. Reprinted with permission from [Lochner11]. Copyright 2011, American Institute of Physics. Right: Measured $|h_{21}|^2$, maximum available gain (MAG), and unilateral power gain (U) of a GaN/InGaN DHBT with emitter area AE $= 4 \times 20\ \mu m^2$. (After [Shen11], copyright © 2011 IEEE.)

a peak current gain of 105 and a collector current density >6.5 kA/cm^2 has been demonstrated on a sapphire substrate. In [Lochner11] a device made on free-standing GaN substrate produced a maximum collector current of 12.3 kA/cm^2 and a current gain of 90. In this device (Fig. 11.13, left), the reduction in localized states (traps), as well as threading dislocation density, may have contributed to significant improvement of the diffusion length of minority carriers.

Continuous improvement in material quality and device technology has also led to a significant increase in the high-frequency capability of GaN HBTs. As shown in Fig. 11.13 (right) [Shen11], a device grown on a high-quality GaN-on-sapphire template with a current density above 16 kA/cm^2 has demonstrated a maximum current-gain cut-off frequency of 5.3 GHz at 4.7 kA/cm^2 and a power-gain cut-off frequency of 1.3 GHz. This is still far from the theoretical projections of more than 200 GHz for a 50-nm base width [McCarthy01a], but there is significant room for improvement by optimizing the device topology. Unfortunately, like SiC, GaN and related compounds still suffer from a low hole mobility, which severely limits the performance of HBTs. This limitation has fostered the development of alternative unipolar vertical devices, such as the hot-electron transistor [Shur00, Dasgupta11].

11.4 Field-effect transistors

Field-effect transistors have been the most successful GaN electron devices until now. They are widely used in RF power amplification and high-voltage power-switching. In this section, we will review some of the most common field-effect transistor structures, device design, and state-of-the-art performance for these two applications. Section 11.4.1 will discuss lateral high electron mobility transistors (HEMT) for high-frequency power amplifiers, while Section 11.4.2 will introduce the main differences in the design of these devices when used in high-voltage switching applications. Finally, Section 11.4.3 focuses on vertical field-effect transistors used mainly in power applications.

11.4.1 *High-frequency GaN-based HEMTs for RF and mixed-signal applications*

AlGaN/GaN high-electron-mobility transistors (HEMTs) were the first nitride-based HEMT structures to be developed [Khan94]. The AlGaN/GaN benefits from high electron mobility (~2000 $cm^2/V{\cdot}s$) and high two-dimensional electron gas (2DEG) density (~10^{13} cm^{-2}). Contrary to other material systems, such as AlGaAs/GaAs or Si/SiGe, the high 2DEG density is not induced by doping, but by donor-like surface states at the AlGaN surface, which accumulate at the interface of AlGaN/GaN due to spontaneous and piezoelectric polarization electric fields inside the AlGaN layer [Ibbetson00]. The 2DEG density in this case is determined mainly by the thickness and aluminum content in the AlGaN barrier layer (Fig. 11.14).

To use these devices in power amplifiers, it is important to ensure that the power gain cut-off frequency f_{MAX} is several times larger than the intended operating frequency, as this will determine the gain and, to some degree, the

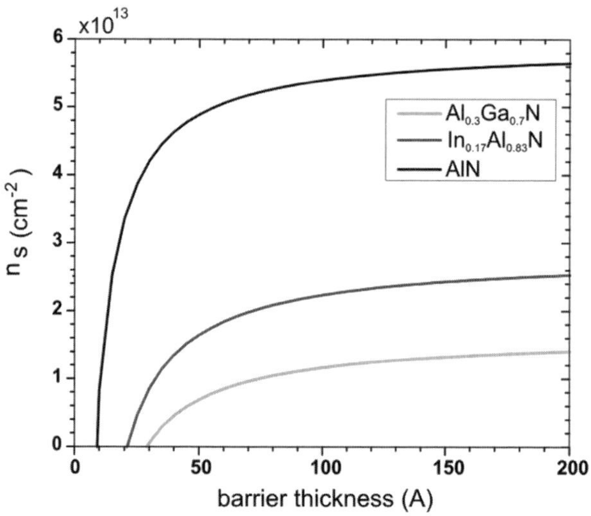

FIG. 11.14. 2D electron gas density in GaN HEMTs as a function of the barrier thickness and composition.

efficiency of the amplifier [Chung10]. f_{MAX} can be calculated with the following expression:

$$f_{MAX} \cong \frac{f_T}{2\sqrt{(R_i + R_s + R_g)g_0 + 2\pi f_T R_g C_{gd}}}$$

where C_{gd} is the gate-drain capacitance, g_0 is the output conductance, and R_i, R_s, and R_g are the input, source, and gate resistances, respectively. The optimization of f_{MAX} relies closely on the maximization of f_T as well as on the reduction of parasitic elements such as R_i, R_s, R_g, C_{gd}, and g_0 [Tasker89, Nidhi06].

The reduction of gate length is an important factor to improve f_T and therefore f_{MAX}, but as the gate length is reduced, the horizontal electric field increases and becomes comparable to the vertical field, which confines the carriers on the 2D channel. This two-dimensional distribution of the electric field under the gate results in short-channel effects which degrade the device performance by shifting the threshold voltage, increasing the output conductance (g_0), reducing the transconductance and resulting in poor pinch-off and limited frequency performance [Sze81]. These degrading effects can be significantly reduced by maintaining a large aspect ratio (L_g/t_b, where L_g is the transistor gate length and $\mathrm{t_b}$ the barrier thickness). Along with ultra-short gate lengths to maximize f_T, the search for new barrier materials and back-barrier structures to better confine the electrons in the channel and provide a suitable aspect ratio (L_g/t_b) are key technologies for increasing the frequency performance of these devices.

Minimizing the parasitic elements is just as important. The reduction of contact resistance has been investigated through the use of recessed ohmic contacts [Buttari02] and regrown contacts [Dasgupta10, Guo12]. Ultra-low contact resistance of 27 $\Omega \cdot \mu\mathrm{m}$ has been achieved in N-polar GaN transistors by regrowing InN in the ohmic contacts, which allows the contact to be made to a narrow-bandgap material [Dasgupta10]. In parallel, the access resistances from the contacts to the gate need to be minimized by reducing the source-to-drain distance or, ultimately, by using a self-aligned process [Shinohara11a, Nidhi09]. The optimization of T-shaped gates is also essential to reduce gate resistance R_g and gate capacitance C_{gd} [Chung10, Denninghoff12]. The following sections will discuss these and other technologies yielding state-of-the-art frequency performance in nitride-based HEMTs.

Scaling transistor gate length. The geometrical shape of the gate directly determines its resistance and capacitance. The gate resistance R_g is proportional to $\rho W/A$, where ρ is the gate-metal resistivity, W is the gate width, and A is the cross-sectional area of the gate. As the gate length is reduced to maximize frequency performance, the gate resistance becomes prohibitively high due to the reduction of the area A. The implementation of T-shaped gates efficiently reduces gate resistance through a larger metal (hat) on top of the short gate (foot). To reduce additional parasitic capacitances, the distance between the hat and the channel, or the height of the gate foot, needs to be as large as possible, though it is limited by mechanical stability of the T-gate. The gate-to-drain capacitance in the saturation regime, given approximately by $C_{gd} = 2\epsilon W/(1 + 2X_{dep}/L_g)$, where X_{dep} is the extension of the depletion region [Ladbrooke89], can be reduced by shrinking the gate foot L_g.

The shape of the T-gate has been improved, and high-aspect-ratio T-gates have been demonstrated on both gallium polar (Ga-polar) and nitrogen polar (N-polar) HEMTs. For example, a N-polar GaN/AlGaN MIS-HEMT has shown an f_{MAX} of 351 GHz [Denninghoff12], where an L_g of 80 nm and 1.1-μm-tall T-gates with a 370-nm-tall foot were used to simultaneously minimize gate resistance (R_g) and parasitic gate–drain capacitance (C_{gd}) [Denninghoff12]. A more aggressive scaling of the gate length combined with the optimization of contact and access resistances pushed the limits of both f_T and f_{MAX} in nitride-based HEMTs. A 40-nm-gate implemented on a AlN/GaN/AlGaN double heterojunction (DH) HEMT structure resulted in a peak f_T of 220 GHz and a peak f_{MAX} of 400 GHz [Shinohara10]. Later, in 2011, Shinohara *et al.* set the state-of-the-art DC and RF performance through a deeply scaled self-aligned gate operating in both depletion mode (D-mode) and enhancement mode (E-mode). A 20-nm-gate AlN/GaN/AlGaN double-heterojunction HEMT exhibited a simultaneous $f_T/f_{MAX} = 310/364$ GHz (Fig. 11.15) [Shinohara11b].

Although the transistor gate length plays an important role in determining the transistor operating frequency, as the frequency increases, other parameters also become important and ultimately dominant. For example, in 2006 the use of a 30-nm gate length in an $Al_{0.4}Ga_{0.6}N/GaN$ HFET on sapphire substrate was demonstrated [Higashiwaki06]. This HEMT structure showed a maximum drain current density of 1.49 A/mm, and a peak extrinsic transconductance of 402 mS/mm, yielding an f_T/f_{MAX} of 181/186 GHz. A high aspect ratio was maintained by the 8-nm thick AlGaN barrier. More recently, a similar gate length of 30 nm was applied to a lattice-matched $In_{0.17}Al_{0.83}N/GaN$ HEMT

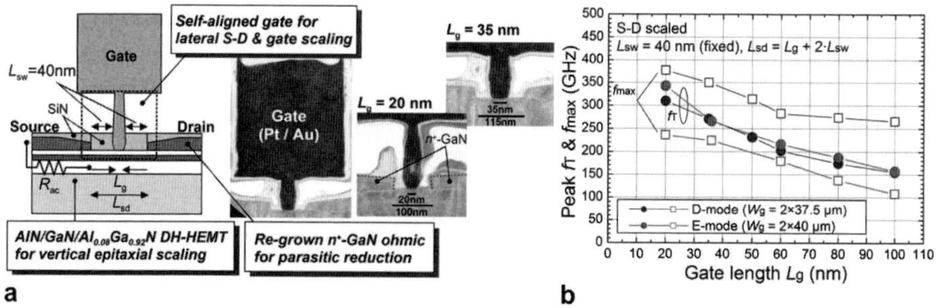

FIG. 11.15. (a) Deeply-scaled self-aligned gates of 35 nm and 20 nm were fabricated on an AlN/GaN/AlGaN double-heterojunction HEMT. (b) Peak f_T/f_{MAX} versus gate length showing its scalability with gate length. (After [Shinohara11b].)

on a SiC substrate with optimized access resistances and carrier confinement, yielding a much higher f_T of 300 GHz due to the significant reduction in parasitic delays and short-channel effects [Lee11a]. These results show the importance of scaling the gate length to improve f_T, but also highlight the need of combining other technologies to effectively reduce the parasitic resistances and capacitances. The following sections address these technologies as well as the search for new nitride-based barrier materials.

Barrier thickness scaling. As the gate length is scaled, a suitable aspect ratio (L_g/t_b) needs to be maintained to avoid short-channel effects and minimize the output conductance. The recess of the AlGaN barrier under the gate region can be used to improve the aspect ratio and minimize short-channel effects [Ando03, Chini04]. However, dry etching the AlGaN barrier can induce damage on the channel under the gate and degrade the mobility of the carriers. Also, dry etching often yields a non-uniform recess across the wafer and therefore large inhomogeneity on the device performance. New barrier materials with larger polarization discontinuities than the traditional $Al_xGa_{1-x}N$ ($x < 0.35$) allow reduced barrier thicknesses, improved gate control over the electrons in the channel, and a subsequent improvement in short-channel effects. Some of these new barrier materials include $In_{1\text{-x}}Al_xN_y$, AlN, $Al_xIn_yGa_{1\text{-x-y}}N$, and N-polar GaN. In the following we discuss specific technologies applied to these materials yielding state-of-the art performance.

InAlN/GaN. The use of a GaN lattice-matched $In_{0.17}Al_{0.83}N$ layer as an alternative barrier for AlGaN/GaN-based HEMTs was first proposed by Kuzmík [Kuzmík01]. Its polarization-charge discontinuity to the underlying GaN ($\Delta P_0 = 2.73 \times 10^{13}\ e \cdot cm^{-2}$), solely due to spontaneous polarization, is almost three times larger than in $Al_{0.2}Ga_{0.8}N$/GaN ($\Delta P_0 = 1.18 \times 10^{13}\ e \cdot cm^{-2}$), which accounts for both spontaneous and piezoelectric polarization fields. This allows large 2DEG density from thinner barriers [Butte07, Lecourt11], which enables excellent gate-to-channel aspect ratios down to very short gate-lengths. The feasibility of ultrathin-barrier InAlN/GaN HEMTs was demonstrated in a 3-nm-thick InAlN barrier [Medjdoub08], with an outstanding stability at high temperatures of up to 1000°C [Medjdoub06]. This lattice-matched barrier also potentially avoids reliability concerns associated with strain in the conventional lattice-mismatched AlGaN/GaN heterostructures [Kuzmík09]. The effect of barrier strain on device reliability in transistors based on the $In_xAl_{1-x}N$/AlN/GaN heterostructure has been investigated experimentally for both tensile and compressive strain fields. While devices under tensile and compressive strain showed large degradation and loss of drain current after electrical stress, the nearly-lattice-matched $In_{0.17}Al_{0.83}N$ barrier exhibited very low degradation along with relatively high current density, suggesting better reliability of the nearly-lattice-matched InAlN as barrier materials [Kayis12].

The frequency performance of InAlN/GaN HEMTs has improved significantly over recent years. From the initial f_T/f_{MAX} of 26/40 GHz [Medjdoub07],

Wang *et al.* reported lattice-matched InAlN/GaN HEMTs with an f_T/f_{MAX} of 82/161 GHz with a 100-nm gate-length and 25 nm Al_2O_3 passivation [Wang10]. By improving material quality, shrinking the gate length to 55 nm, and selectively removing the passivation, the performance was increased to f_T/f_{MAX} of 205/191 GHz [Sun10a, Sun10b]. However, despite these results the device performance is still below the theoretical predictions. A significant boost in frequency performance was achieved by applying an O_2 plasma treatment under the gate [Lee11b]. These devices showed a more than two orders of magnitude reduction in the gate leakage and a 30%–50% increase in f_T, resulting in $f_T = 245$ GHz for a gate length of 30 nm. The use of the O_2 treatment in the device access regions also improves the frequency performance by reducing the gate-length extension [Wang11a]. This resulted in an increase of f_T from 125 to 210 GHz for a 60-nm rectangular gate with a high-DC drain-current density of 2.1 A/mm and a peak extrinsic transconductance of 487 mS/mm.

The use of a thin layer of InGaN as back-barrier to the electrons in the channel increases the carrier confinement, since the polarization field in the InGaN layer raises the conduction band in the GaN buffer with respect to the GaN channel [Palacios06]. This improvement in confinement and short-channel effects was demonstrated in devices with an $In_{0.15}Ga_{0.85}N$ back-barrier and a 10-nm-thick InAlN barrier on a 26-nm thick GaN channel. A peak f_T of 300 GHz with an on-resistance of 1.2 $\Omega \cdot$mm and an extrinsic transconductance $g_{\mathrm{m.ext}}$ of 530 mS/mm [Lee11a] was obtained in devices with a 30-nm gate-length. While, in principle, further scaling down the GaN channel thickness would also improve carrier confinement in the channel, in practice it degrades the carrier mobility in the channel due to the proximity of the interface and alloy scattering caused by the InGaN back barrier layer. This higher interface scattering caused a reduction of the effective electron velocity from 1.45×10^7 cm/s to 1.16×10^7 cm/s when thinning down the channel from 26 nm to 4 nm, degrading its frequency performance [Lee11c].

AlN/GaN. The AlN/GaN system allows for further reduction of the barrier thickness while maintaining very high carrier densities at the heterostructure interface. There has been great progress on the material quality [Binari97, Alekseev99], resulting in electron sheet densities up to 3.65×10^{13} cm^{-2} along with relatively high electron mobilities of 1460 cm^2/V.s utilizing extremely thin (5 nm) AlN layers [Smorchkova00]. The use of advanced fabrication technologies in AlN/GaN HEMTs has led to the demonstration of record frequency performances in these devices. First, Al_2O_3 gate dielectrics deposited by atomic-layer deposition (ALD) have been used to reduce gate leakage in highly scaled AlN/GaN MOS-HEMTs. An ultrathin 2-nm-thick Al_2O_3 dielectric and a gate length of 40 nm reduced the forward-bias gate current by two orders of magnitude without degrading the DC and RF performance [Corrion11]. In addition, the use of AlGaN back-barrier and n^+-GaN ohmic contacts regrown by MBE resulted in better carrier confinement and lower parasitic resistances, demonstrating a peak

f_T/f_{MAX} of 220/400 GHz in 40-nm-gate devices [Shinohara10]. The relatively low f_T compared to f_{MAX} was limited by an excess drain delay associated with the extended drain depletion at higher V_{ds}. The reduction of the source–drain distance down to 170 nm significantly reduced the electron transit time at higher V_{ds} and enhanced the average electron velocity. This technology combined with n^+-GaN ohmic regrowth and $Al_{0.08}Ga_{0.92}N$ back-barrier exhibited a small on-resistance of 0.44 Ωmm, maximum drain current density of 2.3 A/mm, peak extrinsic transconductance of 905 mS/mm, and a f_T/f_{MAX} of 260/394 GHz with a 45-nm gate length [Shinohara11a]. Finally, a self-aligned gate technology was developed to further enhance both DC and RF performance by simultaneously reducing the gate length to 20 nm and the source-drain distance to 100 nm, which further reduced drain delay and enhanced the electron velocity to 1.5×10^7 cm/s. This resulted in $R_{on} = 0.29\ \Omega \cdot$ mm, $I_{dmax} = 2.7$ A/mm and a peak extrinsic $g_m = 1.04$ S/mm. This technology was applied for both depletion and enhancement mode devices exhibiting record frequency performances in both operation modes. The depletion-mode device demonstrated a f_T/f_{MAX} of 310/364 GHz, and the enhancement-mode device showed a f_T/f_{MAX} of 343/236 GHz (Fig. 11.16) [Shinohara11b].

InAlGaN/GaN. Nitride-based quaternary barriers have also attracted interest recently, due to the challenge to obtain carrier mobilities in InAlN/GaN heterostructures at the level of those of other nitride-based heterostructures, such as AlGaN/GaN or AlN/GaN. This limitation is believed to be due to interface roughness scattering [Cao07], which is higher than in other GaN HEMTs due to the immiscibility between AlN and InN [Takayama01]. InAlGaN quaternaries are expected to have a narrower immiscibility gap [Takayama01], which is reflected in the high mobilities (>1700 cm^2/Vs at $n_s = 1.8 \times 10^{13}$ cm^{-2}) observed in lattice-matched $In_{0.16}Al_{0.74}Ga_{0.10}N$/AlN/GaN heterostructures [Ketteniss10].

Excellent performance has been demonstrated on $In_{0.13}Al_{0.83}Ga_{0.04}N$/AlN/GaN transistors on SiC substrates. These devices show a drain-current density of 2.1 A/mm, a peak extrinsic transconductance of 548 mS/mm, and a high f_T of 220 GHz with a 66-nm gate-length. This high-frequency performance is the result of high electron mobility (1790 cm^2/Vs) at a carrier density of 1.8×10^{13} cm^{-2} [Wang11b].

N-polar GaN. While several methods have been investigated to reduce the ohmic contact resistance for Ga-face AlGaN/GaN HEMTs, it has been difficult to achieve values lower than 200 Ωμm.[1] N-polar GaN offers the possibility of achieving very low contact resistance values, since the contacts are not made

[1]Recently, very-low-contact resistances have been obtained in Ga-face devices by etching away the AlGaN barrier and regrowing n^+-GaN contacts, yielding contact resistances down to 50 Ωμm· [Guo12].

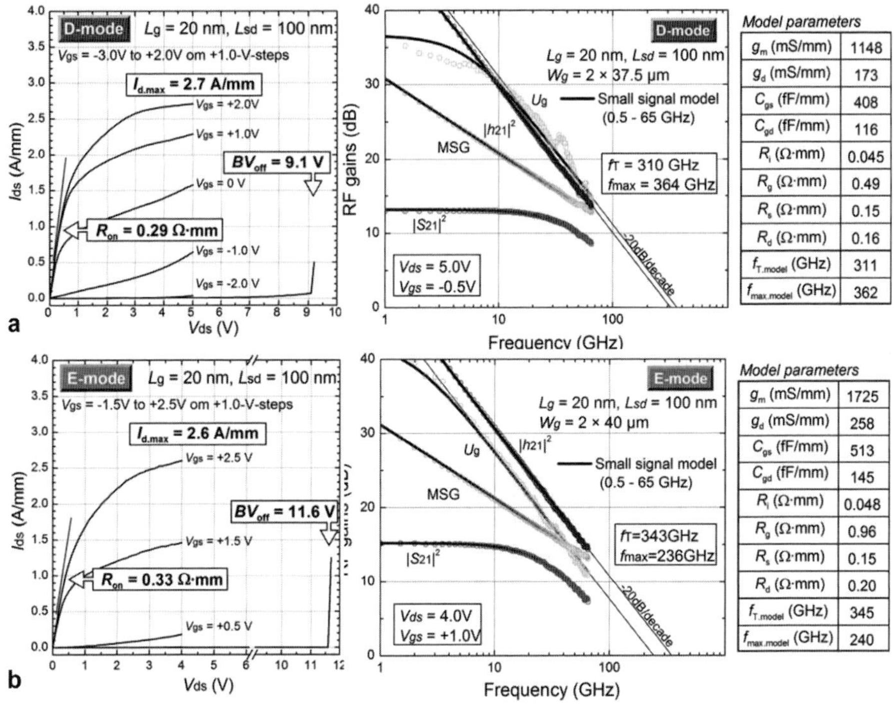

FIG. 11.16. DC and RF performances of the deeply scaled self-aligned gate AlN/GaN/AlGaN double-heterojunction for D-mode (a) and E-mode (b) devices. (After [Shinohara11b].)

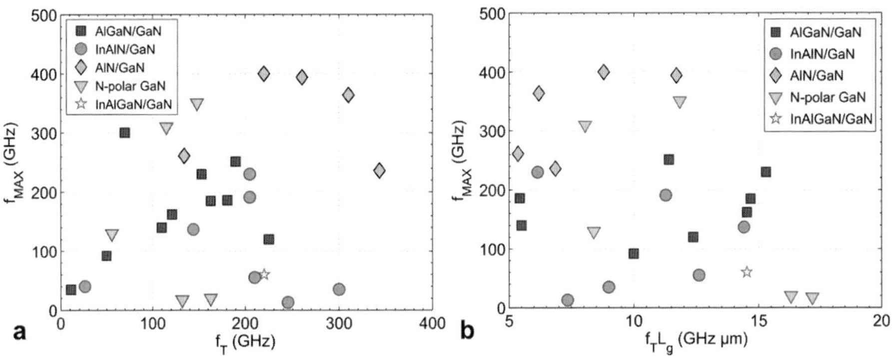

FIG. 11.17. Benchmark of state-of-the-art f_{MAX} versus f_T (a) and versus the product $f_T L_g$ (b) for different nitride-based systems.

through a wide-band-gap barrier material such as AlGaN [Rajan07]. The high electron densities (1×10^{13} cm^{-2}) of metal–InN ohmic combined with a graded $In_xGa_{1-x}N$ layer between the GaN channel and surface InN resulted in an ultra-low ohmic contact resistance of 27 Ω·μm [Dasgupta10]. Moreover, the reversed polarization fields of N-polar GaN/Al(In,Ga)N offers other advantages over Ga-polar Al(In,Ga)N/GaN HEMTs, such as improved electron confinement [Park11] and flexibility to scale the gate-to-2DEG distance without changing the charge density in the 2DEG, as the Al(In,Ga)N barrier in N-polar devices is on the backside of the 2DEG. The frequency performance of N-polar GaN HEMTs has rapidly improved [Nidhi09, Nidhi10] with a demonstrated f_{MAX} of 351 GHz [Denninghoff12] and f_T of 163 GHz [Nidhi11]. These results were achieved through the combination of gate-length scaling, reduction of vertical gate-to-channel distance and the lateral access region, and minimization of parasitic access resistance and capacitance (see Fig. 11.17).

Power amplifiers. Large breakdown voltage, high carrier concentration and mobility, and excellent thermal and chemical stability make nitride transistors ideal for high-performance power amplifiers. Nitride-based HEMTs on SiC substrates offer more than one order of magnitude higher power densities than any other semiconductor technology. 41.4 W/mm of output power density has been demonstrated with a power-added efficiency (PAE) of 60% at 2 GHz [Wu06]. In addition, the smaller size of GaN RF transistors, compared to Si or GaAs devices, significantly simplifies circuit design and impedance matching.

The availability of resistive silicon susbtrates (resistivity above 1kΩ·cm) makes possible the development of III-nitride RF devices on this substrate [Cordier02]. GaN HEMTs on silicon offer a lower-cost solution to high-efficiency power amplification below the X band. As in the case of GaN HEMTs on SiC substrates, there has been a steady improvement in device performance. By 2002, saturated power densities of less than 2 W/mm at 2 GHz had been obtained. In 2003, 6.6 W/mm [Betash03], and in 2004, 12 W/mm [Johnson04], were demonstrated at 2 GHz, while in 2006, 5.1 W/mm were reported at 18 GHz [Ducatteau06].

Power generation at mm and sub-mm wave frequencies represent the next challenge for GaN power amplifiers. The combination of high-frequency performance, described in the previous sections, with the high breakdown and current capability of GaN devices, has allowed the recent demonstration of impresive results at these frequencies. For example, at Ka-band, 10 W/mm at 40 GHz have been measured with PAE of 33% [Palacios05]. W-band GaN monolithic microwave integrated circuits (MMIC) with output an power of an 842 mW at 88 GHz in continuous wave operation [Micovic10] have been recently demonstrated, which represents 96% more power than the highest power reported for the best competing solid-state technology based on InP [Chen98, Samoska11]. More recently, a state-of-the-art 1.7 W, W-band power amplifier MMICs with power added efficiencies in excess of than 20%, and a small signal gain of 21 dB has been demonstrated. This was achieved through the use of a 150-nm T-gate GaN HEMT with an output power exceeding 300 mW and a peak-power added efficiency (PAE) of 37% [Brown11]. By combining twelve GaN MMICs in a low-loss radialline combiner network, a W-band solid-state power amplifier with an output power of 5.2W at 95 GHz has been achieved with an overall combined efficiency of 8.75% [Schellenberg10].

11.4.2 *Lateral GaN high-voltage power transistors*

In addition to RF applications, the unique properties of GaN-based HEMTs have also attracted great interest in high-voltage power-switching applications, where the transistors are switched between their off and on states, as shown in Fig. 11.18. These devices are the centerpiece of power electronic circuits.

Two important parameters of a power-switching transistor are the off-state blocking voltage (BV), also known as breakdown voltage (V_{Br}), and on-state resistance (R_{on}), many times given as the specific on-resistance, $R_{on} \cdot A$, to take into account the total area of the transistor (A).

The breakdown voltage of AlGaN/GaN high-voltage HEMTs has been steadily improved over the years. In 2001 the first 1 kV AlGaN/GaN HEMT grown on semi-insulating SiC substrate was reported [N.-Q. Zhang01a] with $R_{on} \cdot A$ of only 3.4 mΩ·cm^2. In 2006, 1.6 kV breakdown voltage was reported in an AlGaN/GaN HEMT grown on sapphire substrate [Tipirneni06]. In the same year, the breakdown voltage of the AlGaN/GaN HEMT grown on SiC reached up to 1.9 kV, with $R_{on} \cdot A$ as low as 2.2 mΩ·cm^2 [Dora06a]. In 2012, Lee *et al.* demonstrated InAlN/GaN HEMTs grown on SiC with a breakdown voltage of 3 kV and a specific on-resistance of 4.3 mΩ·cm^2 [Lee12], while Sun *et al.* fabricated AlGaN/GaN HEMTs with a breakdown of 1.8 kV and $R_{on} \cdot A$ equal to 1.9 mΩ·cm^2 [Sun12]. These encouraging results, many times better than conventional Si devices, have demonstrated the great potential of AlGaN/GaN HEMTs as the next generation of power switches.

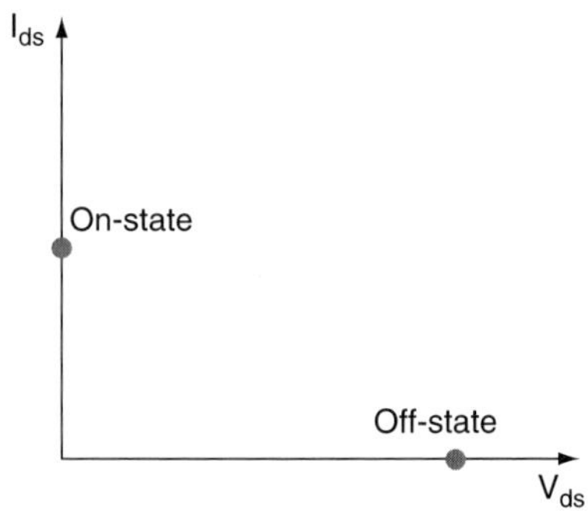

FIG. 11.18. On–off states of an ideal power-switching transistor.

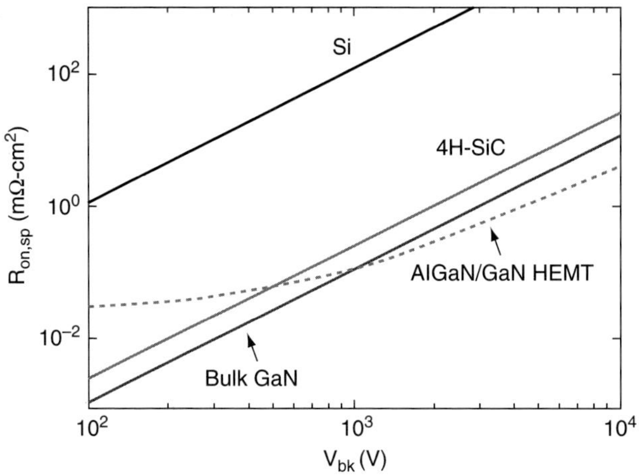

FIG. 11.19. Specific on-resistance as a function of breakdown voltage for Si, 4H-SiC bulk GaN, and AlGaN/GaN HEMT. The performance advantage in the specific resistance of the AlGaN/GaN HEMT structure decreases at low voltages due to the impact of the contact resistances, which has not been taken into account in the vertical structures (Si, 4H-SiC, and Bulk GaN).

The theoretical specific on-resistance $R_{on} \cdot A$ as a function of breakdown voltage V_{Br} (FOM in Section 11.1) of Si, SiC, and GaN power transistors is shown in Fig. 11.19 (solid lines). In this plot, the limit lines are calculated for vertical power transistors with a uniformly doped drift region [Baliga96]:

$$\mathrm{R}_{on} \cdot \mathrm{A} = 4\mathrm{V}_{Br}{}^2/\varepsilon_\mathrm{s}\mu\mathrm{E}_\mathrm{c}^3$$

In this expression, ε_s is the semiconductor dielectric constant, μ is the carrier mobility, and E_c is the critical electric field for breakdown.

For lateral AlGaN/GaN HEMTs, the limit of specific on-resistance can be calculated by assuming uniform lateral electric field distribution in the drift region between the gate and drain [Tipirneni06, N.-Q. Zhang01a], and the on-resistance is dominated by drift-region resistance and contact resistance [Tipirneni06]:

$$R_{on} \cdot A = \left(2R_c + R_{sh}\frac{V_{Br}}{E_c}\right)\left(\frac{V_{Br}}{E_c} + \frac{2R_c}{R_{sh}}\right)$$

where R_c is the contact resistance, R_{sh} is the 2DEG sheet resistance, E_c is the GaN critical electric field, V_{Br}/E_c is the drift-region length, and R_c/R_{sh} is the contact transfer length. As shown in Fig. 11.19 (dashed line), typical AlGaN/GaN HEMTs with sheet resistance of 400 Ω/square and contact resistance of 0.5 Ω·mm offer a 2–3 orders-of-magnitude reduction of specific on-resistance compared to Si, and better performance than SiC. At lower voltages the contact resistance of lateral AlGaN/GaN HEMTs starts to play a bigger role in the on-resistance than in the Si and SiC vertical devices.

Due to the low cost and large size of Si wafers, AlGaN/GaN HEMTs grown on Si(111) substrates promise to reduce the cost of GaN devices. Since the demonstration of AlGaN/GaN HEMTs on 100-mm-diameter Si(111) substrate in 2002 [Brown02], substantial progress has been achieved. 6-inch GaN-on-Si wafers have now been made available commercially and recently AlGaN/GaN HEMT structures have also been demonstrated on 200-mm-diameter Si(111) substrate [Cheng12].

In addition to the buffer and gate leakage currents, the AlGaN/GaN HEMTs on Si substrates have an additional leakage path through the conductive Si substrate when compared with AlGaN/GaN HEMTs grown on insulating sapphire [Fontserè13] and semi-insulating SiC substrates, as illustrated in Fig. 11.20. This drain-to-substrate leakage causes the breakdown voltage to saturate at large gate-to-drain separations, at which the AlGaN/GaN HEMTs reach the maximum breakdown voltage [Visalli09, Lu10a]. The limitation of the Si substrate on the maximum device breakdown voltage can be overcome by transferring the AlGaN/GaN HEMTs from the Si substrate onto an insulating substrate [Lu10b, Srivastava10], after which a 1.5 kV breakdown voltage can be achieved from devices with an original breakdown voltage of less than 600 V [Lu10b]. However, this removal of the Si substrate can potentially increase fabrication cost and device thermal resistance.

The breakdown voltage of AlGaN/GaN HEMTs Si substrates is closely related to the wafer growth conditions. For example, screw dislocations have been shown to increase the leakage of GaN devices [McCarthy01b, Hsu02], and nitrogen vacancies and impurity donors have also been proposed for the origin of n-type background doping in GaN [Neugebauer94, Look03]. Several wafer-growth technologies have been developed to improve the breakdown voltage of AlGaN/GaN HEMTs. Deep-level, such as iron and carbon, doping have been used to compensate for the unintentional n-type background doping in GaN to reduce the buffer leakage current [Choi06, Kato07]. The growth of thick buffer layers on Si substrates has also proven to be an effective way to improve

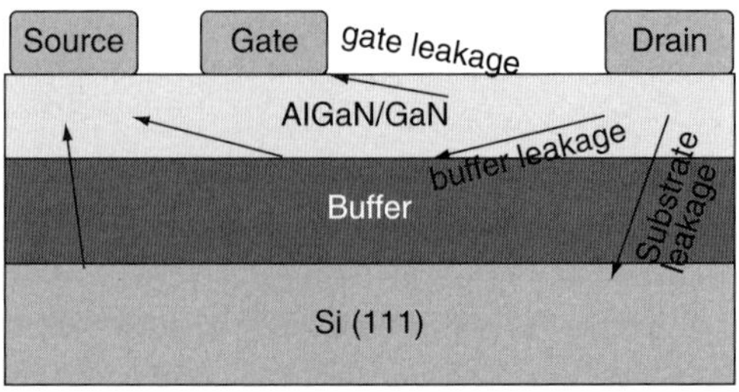

FIG. 11.20. AlGaN/GaN off-state leakage currents with floating Si substrate.

device breakdown voltage [Iwakami07, Ikeda08, Selvaraj09, Visalli09, Rowena11]. The thick buffer layer reduces the screw dislocations and increases the GaN epi-layer vertical resistance, thus reducing the substrate leakage current and increasing breakdown voltage [Rowena11]. Combining the C-doping and a 6-μm-thick buffer structure, AlGaN/GaN HEMTs on 4-inch Si substrate have reached a high breakdown voltage of 1.8 kV [Ikeda08]. However, the growth of a thick III-n buffer on Si substrate can cause large wafer bow and cracking in the GaN film, due to the large lattice and thermal mismatch, while the incorporation of carbon and iron in the buffer can cause carrier-trapping in AlGaN/GaN HEMTs [Klein01, Rudzinski06, Ramdani12], increasing the device on-resistance. Due to these issues, achieving high-breakdown-voltage AlGaN/GaN HEMTs on Si substrate is still one of the most challenging problems in GaN power devices.

Trapping and dynamic on-resistance. Device instability related to current collapse, DC-RF dispersion, or poor dynamic on-resistance has limited the performance of GaN devices over the last few years. As shown in Fig. 11.21, the current collapse or dynamic on-resistance characterizes the degradation of transistor performance under high-stress (high drain and/or gate bias) switching (pulsing or high-frequency continuous wave operation) conditions compared to low-stress DC conditions.

The two most widely used methods for characterizing the trapping effects in AlGaN/GaN HEMTs are gate-lag and drain-lag measurements [Green00, Binari02, Verzellesi05]. Short-duration pulses (<1 μs) are used to avoid the self-heating (determined mainly by the substrate thermal conductivity) that can contribute to the current collapse [Cordier11]. Automatic pulsed I–V measurement systems capable of pulsing gate and drain independently have become increasingly valuable due to their flexibility in biasing the transistors at arbitrary bias points [McGovern05].

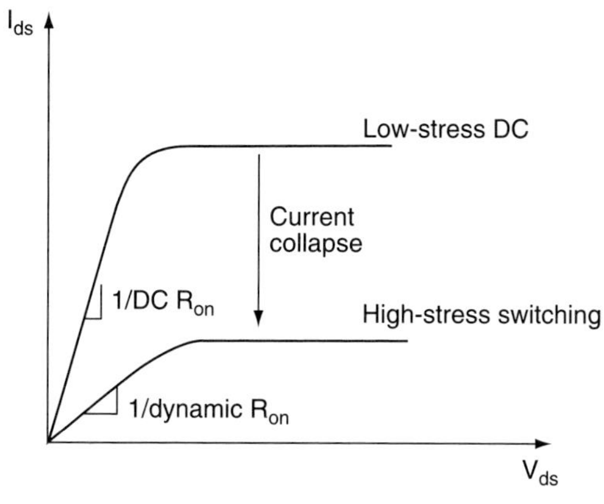

FIG. 11.21. Current collapse and dynamic R_{on}.

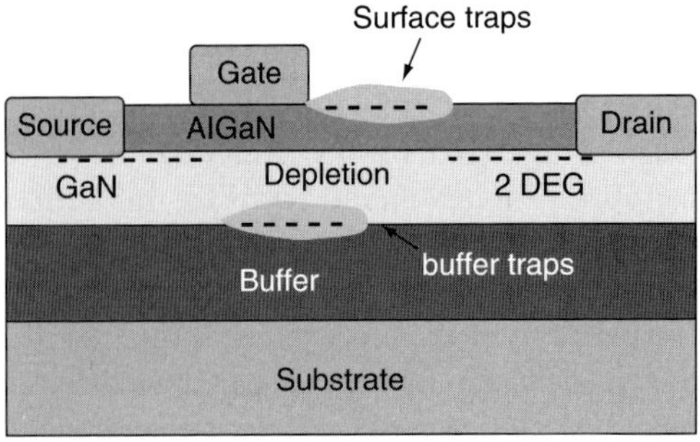

FIG. 11.22. Charges trapped at the surface and buffer of an AlGaN/GaN HEMT.

However, these systems cannot reproduce the complex waveforms used in power electronics. In order to extract device dynamic on-resistance in the large drain-voltage swing, a voltage-clamping sensing circuit can be used [Lu11].

Charge-trapping in deep levels in the buffer and in surface states of the nitride semiconductor (Fig. 11.22) has been the most widely accepted explanation for the origin of current collapse in AlGaN/GaN HEMTs [Klein01, Rudzinski06, Binari01, Vetury01, Koley03, Verzellesi05]. For GaN buffer with deep traps, when the device is biased at high drain-source voltage, the buffer traps are filled by high-energy channel electrons. The resultant negatively charged traps therefore deplete the 2DEG in the channel, causing current collapse and DC-RF dispersion [Klein00, 01]. For the surface-trapping, the most accepted model is the theory of a "virtual gate" [Vetury01] formed between the gate and drain of AlGaN/GaN HEMTs by electrons tunneling from the gate to the surface at high drain-to-gate bias conditions [Green00, Vetury01, Koley03]. Similarly, the electrons can also be trapped inside the AlGaN barrier layer [Mitrofanov04, Joh11]. Several deep trap levels corresponding to the slow transient of device turn-on/off current have been extracted [Vertiatchikh02, Mizutani03a-b, Mitrofanov04, Tapajna10, Joh11]. However, the exact origin of these trap levels is not completely understood yet. A recent study has shown the improvement of pulsed I–V performance of AlGaN/GaN HEMTs when measured in vacuum [Pei09], suggesting the influence of ambient air on device-trapping. Indeed, Gao *et al.* showed that water vapor in the air can cause significant current collapse in AlGaN/GaN [Gao12] HEMTs, which may lead to reconsideration of the conventional surface electron-trapping model.

Surface passivation technology has been developed to prevent the formation of a "virtual gate" on the surface and reduce the trapping effects in AlGaN/GaN HEMTs. While several dielectric materials have been studied [Green00, Lu09, Arulkumaran04], Si_xN_y has been the most widely used passivation layer. The actual mechanism by which the surface Si_xN_y passivation works has not been completely understood yet. For the surface passivation, some of the current hypotheses include the prevention of electrons from reaching the semiconductor surface [Vetury01, Koley03], reduction of trap-state density [Vertiatchikh02, Hashizume01], and change of surface donor states [Vetury01]. A recent study has also shown that the hydrophobic property of Si_xN_y plays an important role in the reduction of current collapse [Gao12]. Alternative passivation methods have also been developed, such as using a p-GaN cap layer or a very thick GaN cap layer to screen and reduce the surface potential fluctuation [Coffie02, Shen04].

There is a trade-off between passivation and breakdown voltage in AlGaN/GaN HEMTs. Negative charge trapped in surface states helps to compensate for the positive charge in the depletion region between the gate and drain in the AlGaN/GaN HEMTs in off-state, reducing the peak electric field and increasing the breakdown voltage [N.-Q. Zhang01b, Tan02]. When the surface of AlGaN/GaN HEMTs is passivated with dielectrics such as Si_xN_y, the electric field at the gate edge increases, which also enhances the gate leakage and reduces the breakdown voltage of the transistors. In order to reduce the high electric field at the gate edge, various gate and source field-plate structures have been used [N.-Q. Zhang00, Saito03, Xing04, Saito07], as shown in Fig. 11.23. Combining the surface passivation and field-plates, high-voltage AlGaN/GaN HEMTs with more than 600 V breakdown and low trapping effects can be achieved [Saito07].

Normally-off GaN devices. Due to the high charge density induced by the polarization difference at the AlGaN/GaN heterostructure, these HEMTs are typically depletion-mode (D-mode or normally-on) devices. However, normally-off enhancement-mode (E-mode) transistors are easier to drive by external circuits, and are highly desirable in power electronics applications due to safety considerations.

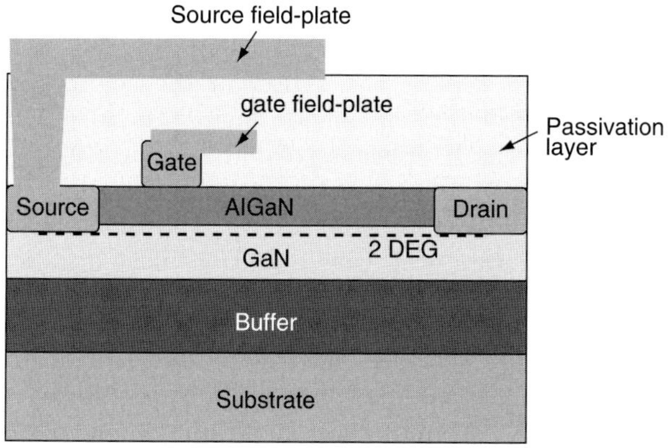

FIG. 11.23. Schematic of an AlGaN/GaN HEMT with gate and source field-plates.

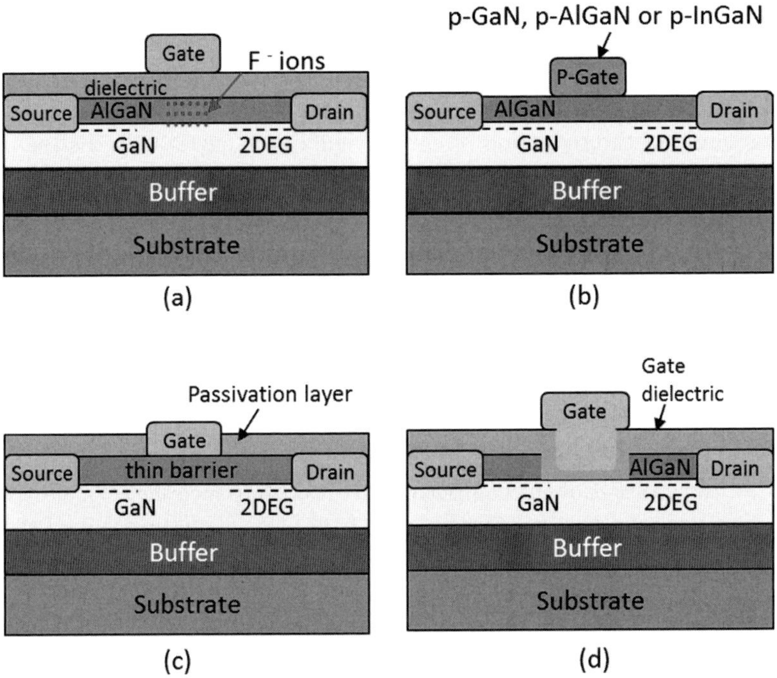

FIG. 11.24. Example of E-mode AlGaN/GaN HEMTs fabricated by: (a) fluorine plasma treatment, (b) p-type gate, (c) surface potential engineering, and (d) gate recess.

Fluorine-based plasma treatment underneath the gate region has been used to make E-mode AlGaN/GaN HEMTs [Cai05, Chu11], as shown in Fig. 11.24(a). The negatively-charged fluorine ions F^- implanted into the AlGaN layer from the fluorine-based plasma deplete the 2DEG in the channel, converting the D-mode AlGaN/GaN HEMT into an E-mode device [Yuan08]. Although Mizuno *et al.* [Mizuno07] showed the F^- plasma treated E-mode AlGaN/GaN HEMTs are stable at 200°C, other reports have observed negative shifts in the threshold voltage with increasing temperature [Wang09]. Threshold voltage instability after high field stress and gate over-drive conditions have also been reported in these devices [Yi07, Ma10]. Other than the fluorine plasma, E-mode AlGaN/GaN HEMTs by hydrogen-plasma treatment have also been demonstrated, though they show similar instability issues [Lu09].

Another type of normally-off E-mode AlGaN/GaN HEMT is the p-GaN, p-AlGaN or p-InGaN gate AlGaN/GaN transistor [Hu00, Suh06, Uemoto07, Shimizu08], shown in Fig. 11.24(b). Due to the built-in potential between the p-type gate and the channel, the 2DEG underneath the gate is depleted at zero gate bias, forming the E-mode transistor. In these devices the trade-off between the 2DEG density in the device access region and the threshold voltage needs to be considered. High 2DEG density with low access resistance outside the gate results in low threshold voltage, while high threshold voltage results in high access resistance. Due to the large forward gate leakage current when the gate p/n junction is turned-on, these devices also have limited gate voltage swing. As a result, threshold voltage above 1 V and high current density are difficult to achieve in this structure.

A third approach for fabricating E-mode devices is based on engineering the surface potential of the transistor structure [Ohmaki06, Higashiwaki07, Medjdoub10], as shown in Fig. 11.24(c). It has been observed that the deposition of a thin dielectric on the surface of AlGaN/GaN HEMT increases the 2DEG density [Green00, Arulkumaran04]. The E-mode devices using this method have a very thin AlN or AlGaN barrier layer where the 2DEG is formed only after the deposition of such a surface passivation dielectric layer. Since no passivation dielectric is underneath the gate, the channel remains depleted. The limitation of this technology is its marginal positive threshold voltage and very low voltage swing.

Since the 2DEG in the AlGaN/GaN HEMTs is induced by the net polarization field in the AlGaN barrier [Ibbetson00], recess-etching of the AlGaN barrier in the gate region depletes the 2DEG in the channel and shifts the device threshold voltage into positive values [Lanford05, Saito06], as shown in Fig. 11.24(d). There has been great progress in recessed-gate E-mode AlGaN/GaN HEMTs. Combing the gate recess with gate dielectric, devices with threshold voltage larger than 2 V have been demonstrated [Oka08, Kanamura10, Lu10c]. The recessed-gate metal–insulator–semiconductor HEMTs (MIS-HEMTs) have high drain current density [Kanamura10, Im10, Lu10c], due to their large forward gate-bias capability. One of the major concerns of this approach is, however, the impact of the timed dry-etching step on the threshold voltage uniformity. A new AlGaN/GaN HEMT structure has been proposed to improve this uniformity [Ota09].

Another challenge of recessed-gate E-mode AlGaN/GaN MIS-HEMTs is the degradation of channel electron mobility, which makes the channel resistance the major contribution of the total device resistance. In order to reduce the channel resistance and maintain high breakdown voltage, a dual-gate structure has been proposed where a short-channel recessed-gate E-mode AlGaN/GaN HEMT is connected to a long-channel D-mode AlGaN/GaN HEMT [Lu10c] (Fig. 11.25). Since the E-mode channel length is short, its impact on the total resistance of the combined structure is reduced. The depletion of the channel underneath the long-channel D-mode gate helps to shield the E-mode short channel from high drain voltage, improving the breakdown voltage. However, the short-channel E-mode region is very sensitive to the electrostatic integrity of the entire device design. The off-state leakage can be large if the short-channel gate is not well shielded from the drain voltage. To solve this problem, a 3D tri-gate structure has been recently proposed in E-mode AlGaN/GaN HEMTs [Lu12] (Fig. 11.26). Due

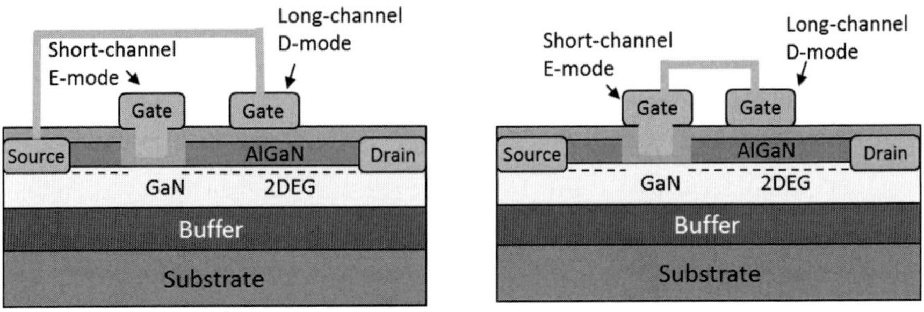

FIG. 11.25. Dual-gate cascade normally-off E-mode AlGaN/GaN HEMTs.

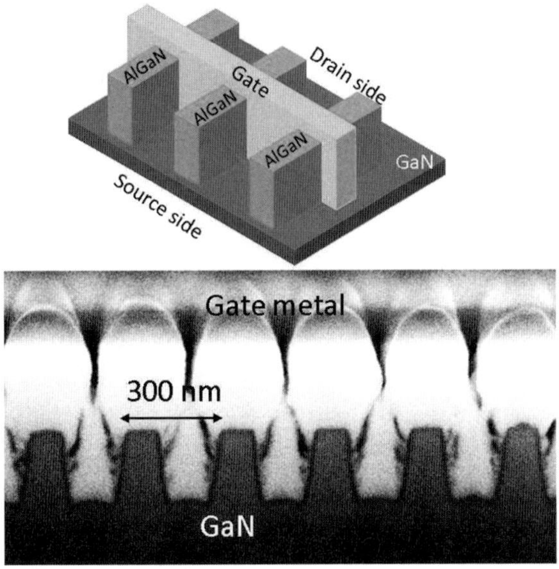

FIG. 11.26. Tri-gate AlGaN/GaN HEMT. (Adapted from [Lu12], copyright © 2012 IEEE.)

to excellent channel control, very-low-leakage current (<1μA/mm) high-voltage normally-off AlGaN/GaN HEMT can be realized.

11.4.3 *Vertical GaN power transistors*

In spite of the excellent performance demonstrated by GaN HEMT transistors in power applications, their current density, reliability, and cost is limited by the poor scalability of lateral structures. This has motivated several groups to develop vertical GaN transistors, and this work has leveraged the rapid development of GaN bulk substrates used for GaN laser diodes and GaN high-power LEDs.

Different device structures have been studied, such as GaN trench-gate MOSFETs [Otake07, 08, Kodama08], GaN vertical HEMTs [Yaegashi11], and GaN-CAVETs (current aperture vertical electron transistors) [Ben-Yaacov04, Kanechika07]. This section will discuss in detail the device structures and characteristics of GaN-CAVETs and GaN trench-gate MOSFETs.

CAVET. GaN-CAVETs are similar in structure to conventional Si-DMOSFETs (double-diffused metal-oxide semiconductors) (Fig. 11.27). The high-electric-field region of a GaN-CAVET is below the gate electrode instead of at the surface (as in a GaN HEMT), so a higher breakdown voltage is expected compared with conventional GaN-HEMTs (Fig. 11.27b, right). Also, as the high electric field peaks far from the surface where the surface states are, dispersion-free devices are expected [Chowdhury12].

The standard device fabrication process flow and schematic device structure of a GaN-CAVET are shown in Fig. 11.28. First, an n^+-GaN layer with $\sim 10^{18}$ cm^{-3} doping is grown on a c-plane sapphire substrate by MOCVD. This layer will become the drain contact of the device. For this, an n^+-GaN substrate can also be used [Chowdhury12]. An n-GaN drift layer is then grown with a doping density around $\sim 10^{16}$ cm^{-3}, followed by an n-GaN current spreading layer with a doping in the $\sim 10^{17}$ cm^{-3} range. Then, a Ti/Ni metal layer on PECVD SiO_2 is deposited as a hard mask to define the aperture size, and Mg ions are implanted to form the current blocking layer (CBL). Typical conditions for this ion implantation are an energy of 50 keV and a dose of 1×10^{15} cm^{-2}. The size and doping density of the aperture region strongly affects the total resistance of the GaN-CAVET. The Ti/Ni and SiO_2 hard masks are then

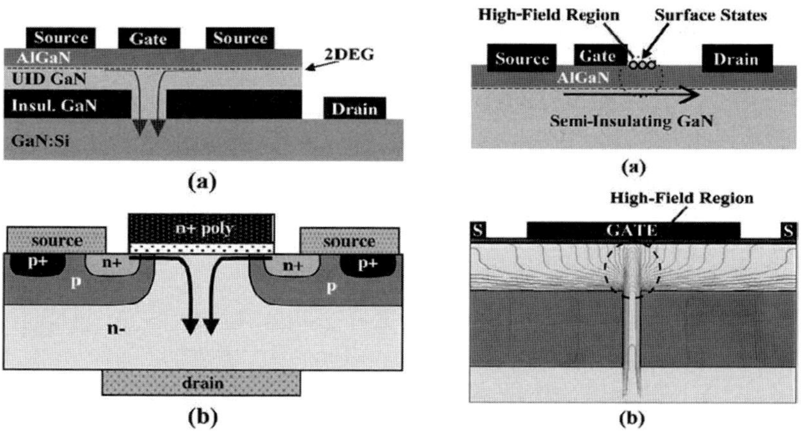

FIG. 11.27. Left: Schematic current flow of (a) GaN-CAVET and (b) DMOSFET. Reprinted with permission from [Ben-Yaacov04]. Copyright 2004, American Institute of Physics. Right: Schematic electric distribution of (a) GaN-HEMT and (b) GaN-CAVET.

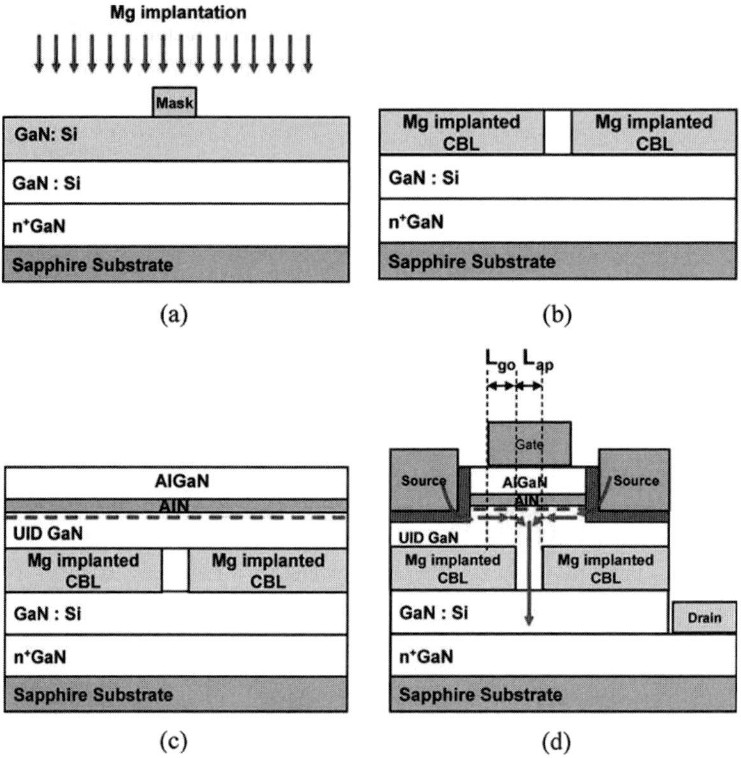

FIG. 11.28. Main fabrication steps of a GaN-CAVET. (a) Implantation of Mg ions to form the CBL. An implantation mask is used to cover the aperture. (b) Removal of implantation mask by wet etching. (c) Regrowth of AlGaN/AlN/GaN, (d) Completed CAVET structure with implanted source contacts and n^+ layer as a drain contact. (After [Chowdhury08], copyright © 2008 IEEE.)

removed by wet etching, and the unintentionally-doped (uid)-GaN access region and (AlN) and AlGaN barriers are then regrown on this implanted wafer by MOCVD. During the regrowth, Mg ions have been reported to diffuse into the regrown AlGaN/(AlN)/GaN layer, which degrades the transport properties of the channel and shifting the threshold voltage. To prevent this diffusion the AlGaN/GaN structure can be grown at low temperature by molecular beam epitaxy [Chowdhury12]. A thin layer of carbon-doped GaN is also grown at the regrowth interface in order to compensate for the residual Si-doping typically seen at this interface and which causes leakage current.

The ohmic contacts in these devices are typically formed by Si ion implantation to prevent the metal spikes seen in alloyed contacts [Dora06b], which could cause a short through the current-blocking layer. When a foreign substrate is used, an etching step is required to form the drain contact on the drain layer. Usually, Cl_2-based reactive ion etching (RIE) is used to etch III-nitrides. The mesa isolation also uses the same gas system. Ti/Al/Ni contacts are deposited as the source and drain contacts, while a Ni/Au/Ni metal stack has been used for the gate electrode. To demonstrate an E-mode CAVET, the wafer is exposed to CF_4 plasma before gate-metal deposition [Chowdhury08]. Also, to reduce the gate leakage current, a CAVET with SiN gate insulator deposited by MOCVD has been reported [Chowdhury12].

The DC and pulsed I_{ds}–V_{ds} characteristics of a D-mode GaN-CAVET are shown in Fig. 11.29(a). The maximum drain current density (I_{max}) and the extrinsic transconductance (g_m) were 5.6×10^3 A/cm^2 and 54 mS/mm, respectively, while the pinch-off voltage (V_p) of the device was -2 V. The DC-RF dispersion of this device was relatively small, because the AlGaN surface was not exposed to a high electric field. The transfer characteristics (I_{ds}–V_{gs}) of the GaN-CAVET with different CF_4 plasma exposure times are shown in Fig. 11.29(b). The corresponding threshold voltages (V_{th}) were -1.5, -0.5, and 0.6 V, for no plasma, and 5 and 10 min plasma exposure, respectively.

GaN vertical MOSFET. There are at least two types of GaN trench-gate MOSFETs: V-shaped [Otake08] and U-shaped [Kodama08]. The schematic cross-sectional structure of the V- and U-shape GaN vertical MOSFET is shown in Fig. 11.30. The most important advantage of these GaN vertical MOSFETs as compared to the CAVET is their intrinsic E-mode operation (threshold voltage >3 V). On the other hand, it is very challenging to form a damage-free MOS interface by using Cl_2-based dry chemistry to form the gate trench.

For the fabrication of GaN vertical MOSFETs, the n-GaN drift layer and the p-GaN channel layer are grown on an n^+-GaN substrate by MOCVD. There are

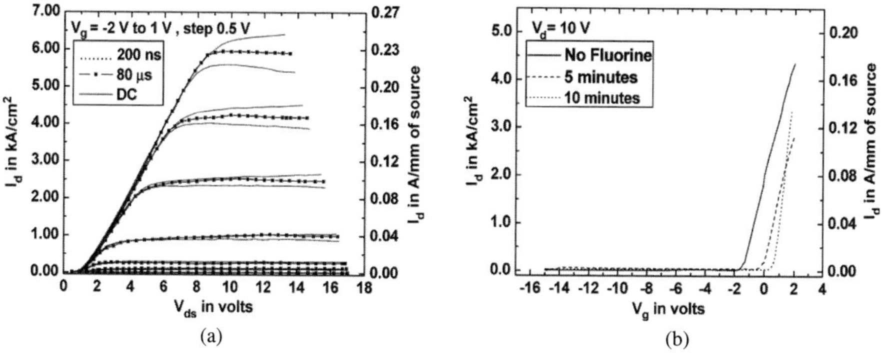

(a) (b)

FIG. 11.29. (a) DC and pulsed I_d–V_{ds} measurement of a depletion-mode CAVET. (After [Chowdhury08].). (b) Comparison of I_d as a function of V_g at $V_{ds} = 10$ V as a function of V_g for 5 min, 10 min, and no CF_4 plasma exposure. (After [Chowdhury08], copyright © 2008 IEEE.)

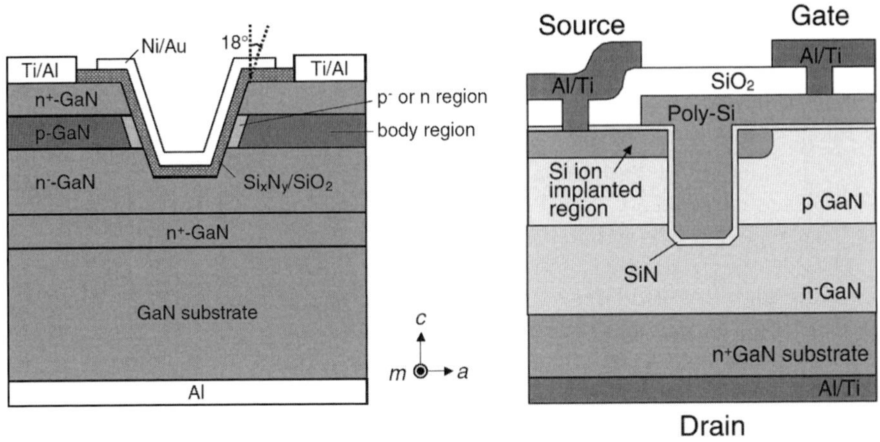

FIG. 11.30. Device structures of (left) V-shaped [Otake08] and (right) U-shaped [Kodama08] GaN trench-gate MOSFETs. Copyright 2008 The Japan Society of Applied Physics.

two different ways to form the n-GaN source layer: a) direct growth, and b) Si ion implantation. In the case of Si ion implantation, Si ions are implanted into the p-GaN layer to form n^+-GaN source region. Then an activation annealing process is performed at around 950°C in N_2 atmosphere. The trench structure is formed by inductively coupled plasma (ICP) etching with Cl-based gases ($Cl_2/SiCl_4$, Cl_2/BCl_3). The slope of the trench walls can be changed with the etching conditions. The source ohmic metal deposition and annealing processes are typically done before the trench fabrication in order to protect the GaN sidewalls from the high-temperature process. To form the U-shape vertical GaN-MOSFETs, a TMAH (tetramethylammonium hydride)-based wet etching at 85°C is used after the dry etching of a V-shape trench to obtain the smooth m-plane surface. Two different gate dielectrics have been reported for these structures: a) a 100-nm-thick Si_xN_y film deposited by low-pressure chemical vapor deposition (LP-CVD); and b) a double-layered structure consisting of 1-nm Si_xN_y/100-nm SiO_2 deposited by electron-cyclotron-resonance (ECR) plasma deposition. The role of the 1-nm Si_xN_y layer is to stabilize the trench surface. Al-based metals deposited by electron-beam evaporator or plasma-assisted sputtering deposition are widely used for source and drain electrodes. As the gate electrode, Ni/Au film deposited by e-beam evaporator and phosphorous-doped polysilicon film deposited by LP-CVD have been reported.

The transfer characteristics (I_{ds}–V_{gs}) of V-shaped and U-shaped GaN vertical MOSFETs are shown in Fig. 11.31. Both structures show enhancement-mode operation. The measured threshold voltages of V-shaped and U-shaped GaN vertical MOSFET are 3.7 V and 10 V. However, the calculated threshold voltages for the V-shaped and U-shaped GaN vertical MOSFET are around 20 V and 40 V, respectively, from the acceptor density in the p-GaN channel layer. This

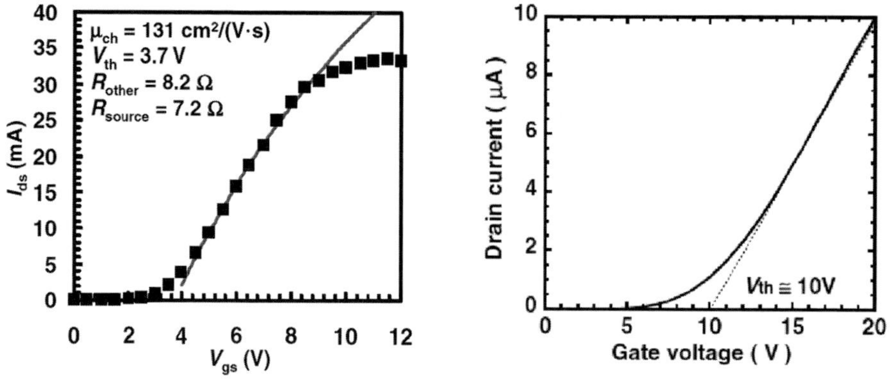

FIG. 11.31. Left: I_{ds}–V_{gs} characteristics of V-shaped GaN trench MOSFET at $V_{ds} = 1$ V. I_{ds}–V_{gs} curve calculated with fitted parameters and estimated resistances also shown. (After [Otake08].) Right: drain current–gate voltage characteristics of U-shaped GaN trench MOSFET at drain voltage 0.1V. (After [Kodama08], Copyright 2008 The Japan Society of Applied Physics.)

difference between experimental and theoretical values is due to the existence of a parasitic n-type layer at the sidewalls of the gate trench due to the generation of N-vacancies and/or etch damage. The net doping density in that region is therefore very different from that in bulk p-GaN. Figure 11.32 shows the DC characteristics for the V-shaped devices. A channel mobility of 131 cm^2/(V·s)

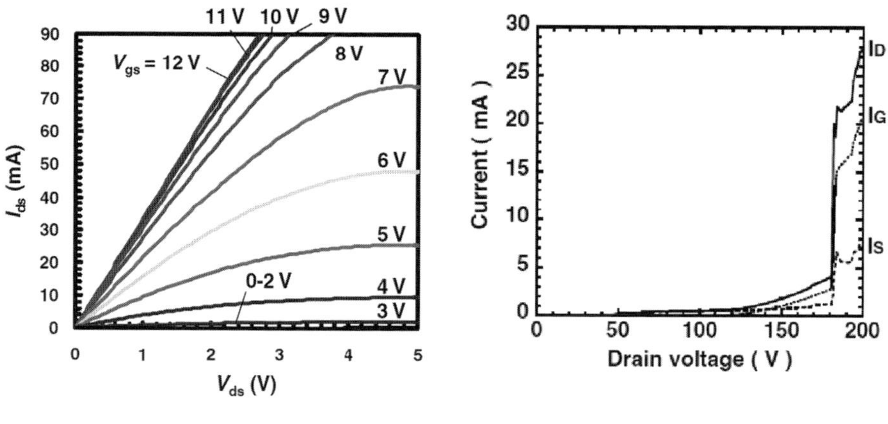

FIG. 11.32. Left: Transfer characteristics of V-shaped GaN trench MOSFET for gate voltages from 0 to 12 V with 1 V steps. (After [Otake08].) Right: drain breakdown characteristics of U-shaped GaN trench MOSFET at the source and gate voltage 0 V. The I_D, I_G, and I_S labels indicate the drain current, the gate current, and the source current, respectively. (After [Kodama08], Copyright 2008 The Japan Society of Applied Physics.)

can be extracted from these measurements. This low value significantly increases the total on-resistance. The drain current and on-resistance at $V_{ds} = 1$ V and $V_{gs} = 12$ V are 33.2 mA and 9.3 mΩ·cm^2. The breakdown voltage of a vertical-trench GaN-MOSFET is determined by the acceptor density in the p-GaN channel layer, the thickness of p-GaN channel layer, the donor density in the n-GaN drift layer, the thickness of the n-GaN drift layer, and the gate oxide thickness. The drain breakdown voltage of U-shaped GaN vertical MOSFETs at a gate voltage 0 V is 180 V (Fig. 11.32, right). The strong increase in the gate current at the breakdown voltage indicates that the breakdown is occurring at the gate region. The on-resistance in this device is below 10 mΩ·cm^2.

11.5 Conclusion

During the past few years, GaN electronics has undergone very important developments in terms of frequency and power capability. Starting with simple structures made of GaN, p-i-n and Schottky diodes confirmed the great potentional of III-nitrides for power applications. The good control of the growth of InGaN alloys has made possible the fabrication of heterojunction bipolar transistors with improved performance. At the same time, high-electron-mobility transistors based on AlGaN/GaN heterostructures, ultrathin-barrier AlN/GaN and AlInN/GaN heterostructures show impressive performance in terms of breakdown voltages and operation frequencies. Even tunnel transport devices such as resonant tunneling diodes and devices based on the Gunn effect are quickly improving in performance due to advances in the growth, device structure, and fabrication technology of all these GaN-based electronic devices. However, in spite of all these promising results, the material quality is still severely limiting the maximum performance of GaN electronics. The use of AlN and GaN bulk substrates, which are quickly being developed for the optoelectronics market, offer new opportunities for achieving the ultimate performance of GaN electronics.

References

[Alekseev99] E. Alekseev, *et al.*, *Electron. Lett.* 35, p. 2145 (1999).

[Alekseev00] E. Alekseev, *et al.*, *Solid-State Electronics* 44, p. 941 (2000).

[Alquier12] D. Alquier, *et al.*, *Jpn. J. Appl. Phys.* 51, 01AG08 (2012).

[Ambacher99] O. Ambacher, *et al.*, *J. Appl. Phys.* 85, no. 6, p. 3222 (1999).

[Ando03] Y. Ando, *et al.*, *IEDM Tech. Dig.*, p. 2311–2314, (2003).

[Arulkumaran04] S. Arulkumaran, *et al.*, *Appl. Phys. Lett.*, vol. 85, no. 4, p. 613 (2004).

[Asbeck00] P. M. Asbeck, *et al.*, *Solid-State Electronics* no. 44, p. 211 (2000).

[Aslan11] B. Aslan and L. F. Eastman, *Solid-State Electronics* 64, p. 57 (2011).

[Baliga89] B. J. Baliga, *IEEE Electron Device Lett.*, vol. 10, no. 10, p. 455 (1989).

[Baliga96] B. J. Baliga, *IEEE Trans. Electron Devices*, vol. 43, no. 10, p. 1717 (1996).
[Bandic98] Z. Z. Bandic, *et al.*, *Appl. Phys. Lett.* 74, p. 1266 (1998).
[Barker05] J. M. Barker, *et al.*, *J. Appl. Phys.* 97, p. 063705 (2005).
[Bayram10a] C. Bayram, *et al.*, *Appl. Phys. Lett.* 96, p. 042103 (2010).
[Bayram10b] C. Bayram, *et al.*, *Appl. Phys. Lett.* 97, p. 181109 (2010).
[Behtash03] R. Behtash, *et al.*, *Electronics Letters*, 39 (7), p. 626 (2003).
[Ben-Yaacov04] I. Ben-Yaacov, *et al.*, *J. Appl. Phys.*, 95, 2073, (2004).
[Berland11] K. Berland, *et al.*, *Phys. Rev. B* 84, p. 245313 (2011).
[Binari97] S. C. Binari, *et al.*, *Solid-State Electron.* 41, p. 177 (1997).
[Binari01] S. C. Binari, *et al.*, *IEEE Trans. Electron Devices*, vol. 48, no. 3, p. 465 (2001).
[Binari02] S. C. Binari, *et al.*, *Proceedings of the IEEE*, vol. 90, no. 6, p. 1048 (2002).
[Brown02] J. D. Brown, *et al.*, *Solid-State Electronics*, vol. 46, p. 1535 (2002).
[Brown11] A. Brown, *et al.*, *Microwave Symposium Digest (MTT)*, IEEE MTT-S International (2011).
[Burm97] J. Burm, *et al.*, *Appl. Phys. Lett.*, vol. 70, p. 464 (1997).
[Buttari02] D. Buttari, *et al.*, *IEEE Electron Device Lett.*, vol. 23, no. 2, p. 76 (2002).
[Butte07] R. Butte, *et al.*, *J. Phys. D: Appl. Phys.*, vol. 40 p. 6328 (2007).
[Cai05] Y. Cai, *et al.*, *IEEE Electron Device Lett.*, vol. 26, no. 7, p. 435 (2005).
[Cao07] Y. Cao and D. Jena, *Appl. Phys. Lett.*, vol. 90, no. 18, p. 182112 (2007).
[Chen98] Y. C. Chen, *et al.*, *IEEE Microwave and Guided Wave Letters*, vol. 8, p. 399 (1998).
[Chen08] W. Chen, *et al.*, *Appl. Phys. Lett.*, vol. 92, no. 25, p. 253501 (2008).
[Cheng12] K. Cheng, *et al.*, *Appl. Phys. Express*, vol. 5, p. 011002 (2012).
[Chini04] A. Chini, *et al.*, *IEEE Electron Device Lett.*, vol. 25, no. 4, p. 229, (2004).
[Choi06] Y. C. Choi, *et al.*, *IEEE Trans. Electron Devices*, vol. 53, no. 12, p. 2926 (2006).
[Chowdhury08] S. Chowdhury, *et al.*, *IEEE Electron Device Lett.*, 29, 543 (2008).
[Chowdhury12] S. Chowdhury, *et al.*, *IEEE Electron Device Lett.*, 33, 41 (2012).
[Chu11] R. Chu, *et al.*, *IEEE Electron Device Lett.* vol. 32, p. 632 (2011).
[Chung06] T. Chung, *et al.*, *Appl. Phys. Lett.* 88, p. 183501 (2006).
[Chung10] J. W. Chung, *et al.*, *IEEE Electron Device Lett.* vol. 31, no. 3, p. 195 (2010).
[Coffie02] R. Coffie, *et al.*, *IEEE Electron Device Lett.* vol. 23, no. 10, p. 588 (2002).
[Cohen05] I. Cohen, *et al.*, *Proc. IEEE Applied Power Electronics Conf.* (APEC), p. 311 (2005).
[Cordier02] Y. Cordier, *et al.*, *Electronics Letters*, vol. 38, no. 2, p. 91 (2002).

[Cordier10] Y. Cordier, *et al.*, 3rd International Symposium on Growth of III-Nitrides, Montpellier, p. MoP-21 (2010).

[Cordier11] Y. Cordier, *et al.*, 35th Workshop On Compound Semiconductor Devices and Integrated Cirduits (WOCSDICE) Catania, Italy, p. 89 (2011).

[Corrion11] A. Corrion, *et al.*, *IEEE Electron Device Lett.* vol. 32, no. 8 (2011).

[Dasgupta10] S. Dasgupta, *et al.*, *Appl. Phys. Lett.*, vol. 96, no. 14, p. 143, 504-1-3 (2010).

[Dasgupta11] S. Dasgupta, *et al.*, *IEEE Electron. Device. Lett.*, vol. 32, no. 9, p. 1212 (2011).

[Denninghoff12] D. J. Denninghoff, *et al.*, *IEEE Electron Device Lett.* vol. 33, no. 6 (2012).

[Dora06a] Y. Dora, *et al.*, *IEEE Electron Device Lett.* vol. 27, no. 9, p. 713 (2006).

[Dora06b] Y. Dora, *et al.*, *IEEE Electron Device Lett.* 27, 529 (2006).

[Ducatteau06] D. Ducatteau, *et al.*, *IEEE Electron. Device Lett.* vol. 27, no. 1, p. 7 (2006).

[Esaki58] L. Esaki, *Phys. Rev.* 109, p. 603 (1958).

[Fontserè13] A. Fontserè, *et al.*, *Appl. Phys. Lett.* 102, 093503 (2013).

[Foxon03] C. T. Foxon, *et al.*, *Phys. Status Solidi C 0*, p. 2389 (2003).

[Frayssinet11] E. Frayssinet, *et al.*, *Physica Status Solidi C 8*, no. 5, p. 1479 (2011).

[Gao12] F. Gao, *et al.*, *IEEE Electron Device Letters*, in press.

[Golka06] S. Golka, *et al.*, *Appl. Phys. Lett.* 88, p. 172106 (2006).

[Green00] B. M. Green, *et al.*, *IEEE Electron Device Letters*, vol. 21, no. 6, p. 268 (2000).

[Grundmann07] M. J. Grundmann, *et al.*, *Phys. Status Solidi C* 4, no. 7, p. 2830 (2007).

[Guo12] J. Guo, *et al.*, *IEEE Electron Device Lett.* vol. 33, no. 4 (2012).

[Haddad97] G. I. Haddad and P. Mazumder, *Solid-State Electronics* vol. 41, no. 10, p. 1515 (1997).

[Hashizume01] T. Hashizume, *et al.*, *J. Vac. Sci. Technol. B.* vol. 19, no. 4, p. 1675 (2001).

[Hatakeyama11] Y. Hatakeyama, *et al.*, *Electron Device Letters*, IEEE, vol. 32, no. 12, p. 1674 (2011).

[Hermann04] M. Hermann, *et al.*, *phys. stat. sol. (c)* 1, no. 8, p. 2210 (2004).

[Higashiwaki06] M. Higashiwaki, *et al.*, *Jpn. J. Appl. Phys.* vol. 45, no. 42, p. L1111 (2006).

[Higashiwaki07] M. Higashiwaki, *et al.*, *IEEE Trans. Elec. Dev.* vol. 54, no. 6, p. 1566 (2007)

[Hsu02] J. W. P. Hsu, *et al.*, *Appl. Phys. Lett.* vol. 81, no. 1, p. 79 (2002).

[Hu00] X. Hu, *et al.*, *Electronics Letters*, vol. 36, no. 8, p. 753 (2000).

[Ibbetson00] J. P. Ibbetson, *et al.*, *Appl. Phys. Lett.* vol. 77, no. 2, p. 250 (2000).

[Ikeda08] N. Ikeda, *et al.*, Proc. 20th International Symposium on Power Semiconductor Devices & ICs (ISPSD), p. 287 (2008).

[Im10] K.-S. Im, *et al.*, *IEEE Electron Device Lett.* vol. 31, no. 3, p. 192 (2010).

[Indlekofer02] K. M. Indlekofer, *et al.*, *Phys. Status Solidi B* 234, p. 769 (2002).

[Iwakami07] S. Iwakami, *et al.*, *Jpn. J. Appl. Phys.* vol. 46, no. 24, p. L587 (2007).

[Joh11] J. Joh and J. A. del Alamo, *IEEE Trans. Electron Devices*, vol. 58, no. 1, p. 132 (2011).

[Johnson65] E. O. Johnson, *RCA rev.* vol. 26, p. 163 (1965).

[Johnson04] J. W. Johnson, *et al.*, *IEEE Electron. Device. Lett.* vol. 25, no. 7, p. 459 (2004).

[Kanamura10] M. Kanamura, *et al.*, *IEEE Electron Device Lett.* vol. 31, p. 189 (2010).

[Kanechika07] M. Kanechika, *et al.*, *Jpn. J. Appl. Phys.* 46, L503 (2007).

[Kato07] S. Kato, *et al.*, *J. Crystal Growth*, vol. 298, p. 831 (2007).

[Kayis12] C. Kayis, *et al.*, *Phys. Status Solidi RRL* 6, no. 4, p. 163 (2012).

[Ketteniss10] N. Ketteniss, *et al.*, *Semicond. Sci. Technol.* vol. 25, no. 7, p. 075013 (2010).

[Khan94] M. A. Khan, *et al.*, *Appl. Phys. Lett.* vol. 65, p. 1121 (1994).

[Kikuchi02] A. Kikuchi, *et al.*, *Appl. Phys. Lett.* 81, p. 1729 (2002).

[Klein00] P. B. Klein, *et al.*, *J. Appl. Phys.* vol. 88, no. 5, p. 2843 (2000).

[Klein01] P. B. Klein, *et al.*, *Appl. Phys. Lett.* vol. 78, no. 21, p. 3527 (2001).

[Kodama08] M. Kodama, *et al.*, *Appl. Phys. Express* 1, 021104 (2008).

[Koley03] G. Koley, *et al.*, *IEEE Trans. Electron Devices*, vol. 50, no. 4, p886 (2003).

[Kolnik95] J. Kolnik, *et al.*, *J. Appl. Phys.* 78 (2), p. 1033 (1995).

[Kozodoy98] P. Kozodoy, *et al.*, *Appl. Phys. Lett.* 73, p. 975 (1998).

[Krishnamoorthy10] S. Krishnamoorthy, *et al.*, *Appl. Phys. Lett.* 97, p. 203502 (2010).

[Krishnamurthy97] S. Krishnamurthy, *et al.*, *Appl. Phys. Lett.* 71(14), p. 1999 (1997).

[Kumakara08] K. Kumakura and T. Makimoto, *Appl. Phys. Lett.* 92, p. 153509 (2008).

[Kuzmík01] J. Kuzmík, *IEEE Electron Device Lett.* vol. 22, no. 11, p. 510 (2001).

[Kuzmík09] J. Kuzmík, *et al.*, *J. Appl. Phys.* vol. 106, no. 12, p. 124503 (2009).

[Ladbrooke89] P. H. Ladbrooke, *MMIC Design: GaAs FETs and HEMTs*, Artech House (1989).

[Lanford05] W. B. Lanford, *et al.*, *Elect. Lett.* vol. 41, no. 7, p. 449 (2005).

[Leconte07] S. Leconte, *et al.*, *Semicond. Sci. Technol.* 22, p. 107 (2007).

[Lecourt11] F. Lecourt, *et al.*, *IEEE Electron Device Lett.* vol. 32, no. 11, p. 1538 (2011).

[Lee10] Y.-C. Lee, *et al.*, *IEEE Transactions on Electron. Devices*, vol. 57, no. 11, p. 2964 (2010).

[Lee11a] D. S. Lee, *et al.*, *IEEE Electron Device Lett.* vol. 32, no. 11 (2011).

[Lee11b] D. S. Lee, *et al.*, *IEEE Electron Device Lett.* vol. 32, no. 6 (2011).

[Lee11c] D. S. Lee, *et al.*, *IEDM Tech. Dig.* p. 457 (2011).

[Lee12] H.-S. Lee, *et al.*, *IEEE Electron Device Lett.* vol. 33, no. 7, p. 982 (2012).
[Li12] D. Li, *et al.*, *Appl. Phys. Lett.* 100, p. 252105 (2012).
[Limb00] J. B. Limb, *et al.*, *Appl. Phys. Lett.* vol. 76, no. 17, p. 2457 (2000).
[Limb06] J. B. Limb, *et al.*, *Electronics Letters*, vol. 42, no. 22, p. 1313 (2006).
[Lin94] M. E. Lin, *et al.*, *Appl. Phys. Lett.* 64, p. 1003 (1994).
[Liu09] L. Liu, *et al.*, *Mater. Sci. Forum* 600, p. 1251 (2009).
[Lochner11] Z. Lochner, *et al.*, *Appl. Phys. Lett.* 99, p. 193501 (2011).
[Look03] D. C. Look, *et al.*, *Appl. Phys. Lett.* vol. 83, no. 17, p. 3525 (2003).
[Lu09] B. Lu, *et al.*, 67th Device Research Conference, p. 59 (2009).
[Lu10a] B. Lu, *et al.*, *IEEE Electron Device Letters*, Vol. 31, no. 4, p. 302, (2010).
[Lu10b] B. Lu and T. Palacios, *IEEE Electron Device Lett.* vol. 31, no. 9, p. 951 (2010).
[Lu10c] B. Lu, *et al.*, *IEEE Electron Device Lett.* vol. 31, p. 990 (2010).
[Lu11] B. Lu, *et al.*, Compound Semiconductor Integrated Circuit Symposium (CSICS), IEEE, Waikoloa, HI, USA (2011).
[Lu12] B. Lu, *et al.*, *IEEE Electron Device Letters*, vol. 33, no. 3, p. 360 (2012).
[Luo02] B. Luo, *et al.*, *Appl. Phys. Lett.* vol. 80, no. 9, p. 1661 (2002).
[Ma10] C. Ma, *et al.*, *IEEE IEDM Tech. Dig.* (2010).
[Makimoto02] T. Makimoto, *et al.*, *Phys. stat. sol.* (c) 0, no. 1, p. 95 (2002).
[Makimoto03] T. Makimoto, *et al.*, *Appl. Phys. Lett.* 83, p. 1035 (2003).
[Makimoto04] T. Makimoto, *et al.*, *Appl. Phys. Lett.* 84, p. 1964 (2004).
[McCarthy01a] L. McCarthy, *et al.*, *IEEE Trans. Electron. Devices*, vol. 48, no. 3, p. 543 (2001).
[McCarthy01b] L. McCarthy, *et al.*, *Appl. Phys. Lett.* vol. 78, no. 15, p. 2235 (2001).
[McGovern05] P. McGovern, *et al.*, IEEE MTT-S International Microwave Symposium Digest (2005).
[Medjdoub06] F. Medjdoub, *et al.*, International Electron Devices Meeting, 35.7 (2006).
[Medjdoub07] F. Medjdoub, *et al.*, *Electron. Lett.*, 43, p. 309 (2007).
[Medjdoub08] F. Medjdoub, *et al.*, *IEEE Electron Device Lett.* vol. 29, no. 5, p. 422 (2008).
[Medjdoub10] F. Medjdoub, *et al.*, *IEEE Electron Device Letters*, vol. 31, no. 2, p. 111 (2010).
[Ménard10] O. Ménard, *et al.*, *Phys. Status Solidi* C, 7, p. 112 (2010).
[Micovic10] M. Micovic, *et al.*, *IEEE MTT-S Int. Microw. Symp. Dig.*, p. 237 (2010).
[Mitrofanov04] O. Mitrofanov and M. Manfra, *J. Appl. Phys.*, vol. 95, no. 11, p. 6414 (2004).
[Mizutani03a] T. Mizutani, *et al.*, *Phys. Stat. Sol. (a)*, vol. 200, no. 1, p. 195 (2003).

[Mizutani03b] T. Mizutani, *et al.*, *Jpn. J. Appl. Phys.* vol. 42, p. 424 (2003).
[Mizuno07] H. Mizuno, *et al.*, *Phys. Stat. Sol. (c)*, vol. 4, no. 7, p. 2732 (2007).
[Mochizuki11] K. Mochizuki, *et al.*, *IEEE Trans. Electron Devices*, vol. 58, no. 7, p. 1979 (2011).
[Montanari05] S. Montanari, *et al.*, *Solid-State Electronics* 49, p. 245 (2005).
[Mutamba06] K. Mutamba, *et al.*, *Superlattices and Microstructures* 40, p. 363 (2006).
[Neugebauer94] J. Neugebauer and C. G. Van de Walle, *Phys. Review B*, vol. 50, no. 11, p. 8067 (1994).
[Nidhi06] Nidhi, *et al.*, *IEEE Electron Device Lett.* vol. 27, no. 11, p. 877 (2006).
[Nidhi09] Nidhi, *et al.*, *IEDM Tech. Dig.* p. 955 (2009).
[Nidhi10] Nidhi, *et al.*, *IEEE Electron Device Letters*, vol. 31, no. 12, p. 1437 (2010).
[Nidhi11] Nidhi, *et al.*, Device Research Conference (DRC), 69th Annual, p. 141 (2011).
[Oka08] T. Oka and T. Nozawa, *IEEE Electron Device Lett.* vol. 29, p. 668 (2008).
[Ota09] K. Ota, *et al.*, *IEDM Tech. Dig.*, p. 153 (2009).
[Otake07] H. Otake, *et al.*, *Jpn. J. Appl. Phys.* 46, L599 (2007).
[Otake08] H. Otake, *et al.*, *Appl. Phys. Express*, 1, 011105 (2008).
[Ohmaki06] Y. Ohmaki, *et al.*, *Jpn. J. Appl. Phys.*, vol. 45, no. 44, p. L1168 (2006).
[Palacios05] T. Palacios, *et al.*, *IEEE Elect. Dev. Lett.*, vol. 26, p. 78 (2005).
[Palacios06] T. Palacios, *et al.*, *IEEE Electron Device Lett.*, 27, no. 1, (2006).
[Pankove94] J. Pankove, *et al.*, *Tech. Dig. Int. Electron Devices Meet.*, p. 389 (1994).
[Park11] P. S. Park and S. Rajan, *IEEE Trans. Electron Devices*, vol. 58, no. 3, p. 704 (2011).
[Pearton99] S. J. Pearton, *et al.*, *J. Appl. Phys.*, vol. 86, p. 1 (1999).
[Pearton00] S. J. Pearton, *et al.*, *Materials Science and Engineering R: Reports*, vol. 30, Issues 3–6, 1, p. 55 (2000).
[Pei09] Y. Pei, *et al.*, *IEEE Electron Device Letters*, vol. 30, no. 4 (2009).
[Rajan04] S. Rajan, *et al.*, *Appl. Phys. Lett.* vol. 84, no. 9, pp. 1591–1593 (2004).
[Rajan07] S. Rajan, *et al.*, *J. Appl. Phys.* 102, p. 044501 (2007).
[Ramdani12] M. Ramdani *et al.*, *Solid-State Electronics*, no. 75, p. 86 (2012).
[Roccaforte10] F. Roccaforte, *et al.*, *Appl. Surf. Sci.*, 256, p. 5727 (2010).
[Rowena11] I. B. Rowena, *et al.*, *IEEE Electron Device Letters*, vol. 32, no. 11, p. 1534 (2011).
[Rudzinski06] M. Rudzinski, *et al.*, *IEEE Trans. Electron Devices*, vol. 53, no. 9, p. 2413 (2006).
[Saito03] W. Saito, *et al.*, *IEEE Trans. Electron Devices*, vol. 50, no. 12, p. 2528 (2003).

[Saito06] W. Saito, *et al.*, *IEEE Trans. Elec. Devices*, vol. 53, no. 2, p. 356 (2006).

[Saito07] W. Saito, *et al.*, *IEEE Trans. Electron Devices*, vol. 54, no. 8, p. 1825 (2007).

[Saitoh10] Y. Saitoh, *et al.*, *Applied Physics Express*, 3, p. 081001 (2010).

[Sakr11] S. Sakr, *et al.*, *Appl. Phys. Lett.* 99, p. 142103 (2011).

[Samoska11] L. A. Samoska, *IEEE Transactions on terahertz science and technology*, vol. 1, no. 1 (2011).

[Schellenberg10] J. Schellenberg, *et al.*, *IEEE IMS Symp. Dig.* (2010).

[Selvaraj09] S. L. Selvaraj, *et al.*, *IEEE Electron Device Letters*, vol. 30, no. 6, p. 587 (2009).

[Shen04] L. Shen, *et al.*, *IEEE Electron Device Letters*, vol. 25, no. 1, p. 7 (2004).

[Shen11] S.-C. Shen, *et al.*, *IEEE Electron Device Letters*, vol. 32, 8, p. 1065 (2011).

[Shimizu08] M. Shimizu, *et al.*, *Jpn. J. Appl. Phys.*, vol. 47, no. 4, p. 2817 (2008).

[Shinohara10] K. Shinohara, *et al.*, IEDM 10-675 (2010).

[Shinohara11a] K. Shinohara, *et al.*, *IEEE Electron Device Lett.* vol. 32, no. 8 (2011).

[Shinohara11b] K. Shinohara, *et al.*, IEDM 11-456 (2011).

[Shur00] M. S. Shur, *et al.*, *Appl. Phys. Lett.*, vol. 76, no. 22, p. 3298 (2000).

[Shur04] M. S. Shur and R. F. Davis, *GaN-base Materials and Devices, Growth, Fabrication, Characterization and Performance*, Chap. 3, p. 179 (2004).

[Simon09] J. Simon, *et al.*, *Phys. Rev. Lett.* 103, p. 026801 (2009).

[Smorchkova00] I. P. Smorchkova, *et al.*, *Applied Phys. Lett.* vol. 77, no. 24, p. 3998, (2000).

[Srivastava10] P. Srivastava, *et al.*, *IEEE Electron Device Letters*, vol. 31, no. 8, p. 851 (2010).

[Suh06] C. S. Suh, *et al.*, 64th Device Research Conference, p. 163 (2006).

[Sun10a] H. Sun, *et al.*, *IEEE Electron Device Lett.*, vol. 31, no. 4, p. 293 (2010).

[Sun10b] H. Sun, *et al.*, *IEEE Electron Device Lett.*, vol. 31, no. 9, p. 957 (2010).

[Sun12] M. Sun, et al. *Appl. Phys. Express*, vol. 5, no. 7, p. 074202 (2012).

[Sze81] S. M. Sze, *Physics of Semiconductor Devices*, Wiley-Interscience (1981).

[Tan02] W. S. Tan, *et al.*, *Appl. Phys. Lett.*, vol. 80, no. 17, p. 3207 (2002).

[Tang09] X. Tang, *et al.*, *Appl. Phys. Lett.*, 95, p. 142102 (2009).

[Tasker89] P. J. Tasker and B. Hughes, *IEEE Electron Device Lett.*, vol. 10, no. 7, p. 291 (1989).

[Takayama01] T. Takayama, *et al.*, *J. Cryst. Growth*, vol. 222, no. 1/2, p. 29 (2001).

[Tapajna10] M. Tapajna, *et al.*, *IEEE Electron Device Lett.*, vol. 31, no. 7 (2010).

[Tipirneni06] N. Tipirneni, *et al.*, *IEEE Electron Device Lett.*, vol. 27, no. 9, p. 716 (2006).

[Uemoto07] Y. Uemoto, *et al.*, *IEEE Trans. Elec. Dev.*, vol. 54, no. 12, p. 3393 (2007).

[Uesugi11] T. Uesugi and T.Kachi, *Jpn. J. Appl. Phys.*, 50, p. 031005 (2011).

[Vertiatchikh02] A. V. Vertiatchikh, *et al.*, *Electronics Letters*, vol. 38, no. 8, p. 388 (2002).

[Verzellesi05] G. Verzellesi, *et al.*, *Microelectronics Reliability*, vol. 45, p. 1585 (2005).

[Vetury01] R. Vetury, *et al.*, *IEEE Trans. Electron Devices*, vol. 48, no. 3, p. 560 (2001).

[Visalli09] D. Visalli, *et al.*, *Jpn. J. Appl. Phys.* vol. 48, p. 04C101 (2009).

[Wang09] R. Wang, *et al.*, *Solid-State Electronics*, vol. 53, p. 1 (2009).

[Wang10] H. Wang, *et al.*, *physica status solidi (c)*, vol. 7, p. 2440 (2010).

[Wang11] Y. Wang, *et al.*, *Semicond. Sci. Technol.* 26, p. 022002 (2011).

[Wang11a] R. Wang, *et al.*, *IEEE Electron Device Lett.* vol. 32, no. 7, p. 892 (2011).

[Wang11b] R. Wang, *et al.*, *IEEE Electron Device Lett.* vol. 32, no. 9, p. 1215 (2011).

[Wraback03] M. Wraback, *et al.*, *Appl. Phys. Lett.* vol. 82, no. 21, p. 2674 (2003).

[Wu06] Y.-F. Wu, *et al.*, Proceedings of the 64th Device Research Conference, New York, USA. p. 151 (2006).

[Xing01] H. Xing, *et al.*, Proc. of 28th International Symposium on Compound Semiconductors (ISCS 2001) p. ThM-7.

[Xing04] H. Xing, *et al.*, *IEEE Electron Device Lett.* vol. 25, no. 4, p. 161 (2004).

[Yaegashi11] S. Yaegashi, *et al.*, *Phys. Status Solidi C*, 8, 45 (2011).

[Yi07] C. Yi, *et al.*, *IEEE IEDM Tech. Dig.*, p. 389 (2007).

[Yilmazoglu08] O. Yilmazoglu, *IEEE Trans. Electron. Devices*, vol. 55, no. 6, p. 1563 (2008).

[Yoshida99] S. Yoshida and J. Suzuki, *J. Appl. Phys.*, 85, p. 7931 (1999).

[Yu97] E. T. Yu, *et al.*, *Appl. Phys. Lett.*, 71, no. 19, pp. 2794–2796 (1997).

[Yuan08] L. Yuan, *et al.*, *Appl. Phys. Lett.*, vol. 92, p. 102109 (2008).

[Zainal10] N. Zainal, *et al.*, *Appl. Phys. Lett.*, 97, 112102 (2010).

[Zhang00] A. P. Zhang, *et al.*, *Appl. Phys. Lett.*, 76, p. 3816 (2000).

[Zhang01] A. P. Zhang, *et al.*, *IEEE Trans. Electron. Devices*, vol. 48, no. 3, p. 407 (2001).

[N.-Q. Zhang00] N.-Q. Zhang, *et al.*, *IEEE Electron. Device Lett.*, vol. 21, no. 9, p. 421 (2000).

[N.-Q. Zhang01a] N.-Q. Zhang, *et al.*, *Phys. stat. sol. (a)*, vol. 188, no. 1, p. 213 (2001).

[N.-Q. Zhang01b] N.-Q. Zhang, *et al.*, *IEDM Technical Digest* (2001).

[Zhou06] Y. Zhou, *et al.*, *Appl. Phys. Lett.*, vol. 88, p. 113509 (2006).

12 Intersubband transitions in low-dimensional nitrides

Maria Tchernycheva and François H. Julien

12.1 Introduction

Intersubband transitions (ISB) are resonant optical transitions between confined states of semiconductor heterostructures either in the conduction band or in the valence band. They were first observed in the 1970s at far-infrared wavelengths in Si accumulation layers [Kamg 74]. In 1985, West and Eglash reported the observation of 10-µm wavelength ISB absorption in $GaAs/Al_{0.3}Ga_{0.7}As$ quantum wells (QW) [West 85]. This was made possible by the progress of epitaxial growth techniques allowing atomic-layer control of layer thicknesses. In the pioneering work of West and Eglash, it was shown that ISB transitions are characterized by a large optical dipole moment, whose length scales like the QW thickness, and that the resonant wavelength can be tailored in a wide spectral range through a proper choice of the layer thickness. During the ensuing years, extensive investigations led to the demonstration of ISB transitions in various material systems, such as InGaAs/AlInAs, InAs/AlSb, or SiGe/Si in the form of QWs [Paie 06], but also quantum dots (QD) [Sauv 97, Webe 99]. The interest was motivated by fundamental aspects but even more by the development of a new class of control-by-design optoelectronic devices such as, for example, focal-plane arrays of quantum-well infrared photodetectors (QWIP) [Levi 87] or quantum cascade lasers (QCL) [Fais 94].

Using conventional III–V semiconductors, the operation wavelength of ISB devices such as QWIPs or QCLs can be tuned from the mid-infrared to the THz spectral range. Operation at short wavelengths is limited by the material transparency and by the available conduction-band offset (~0.35 eV for direct gap GaAs/AlGaAs and ~0.52 eV for lattice-matched InGaAs/InAlAs). To access the near-infrared spectral domain and in particular the 1.3–1.55-µm wavelength range used for fiber-optics telecommunications, materials with a band offset above 1.5 eV are required. There are only few semiconductor systems capable of producing ISB transitions at telecommunication wavelengths: namely, InGaAs/AlAs(Sb) [Neog 99, Hira 94], ZnSe/BeTe [Akim 01, Akim 02], CdS/ZnSe/BeTe [Li 06a, Cong 07], GaInAsN/AlAs [Ma 07] and GaN/Al(Ga)N [Suzu 99, Gmac 00, Kish 02b, Helm 03b]. Among these materials, nitride semiconductors have shown the highest potential for device applications. This is due to their remarkable properties, such as a high conduction-band offset (~1.75 eV for AlN/GaN) [Tche 06b] and their remote lateral valleys, as well as an ultra-short ISB absorption recovery time of the order of 140–400 fs [Iizu 00, Hebe 02, Hama 04, Hama 05, Wang 05, Wang 06], which is ten times faster than in InGaAs QWs [Iizu 00, Hira 94]. However, in order to reach short infrared wavelengths, ultrathin layers are required because of the rather large electron effective mass in nitrides. This explains why the development of nitride ISB devices has been crucially dependent on the progress in the epitaxial growth techniques of these materials with atomic-layer thickness control; namely, molecular beam epitaxy (MBE) or metal organic chemical vapor deposition (MOCVD). Finally, in addition to short-infrared-wavelength ISB devices, III-nitrides are finding appealing applications in the THz frequency domain because of their large optical phonon energy.

This chapter presents an overview of the physics and device applications of ISB transitions in III-nitrides. It is organized along the following lines. Section 12.2 addresses the theoretical aspects of nitride ISB transitions. Section 12.3 reviews the spectroscopic investigations of ISB transitions in QWs and QDs as well as polar, semipolar, and cubic nitrides. Sections 12.4–12.6 recast the present state-of-the-art of nitride ISB devices for optical modulation, detection, and light-emission.

12.2 Intersubband transitions in nitride heterostructures: theoretical aspects

First, theoretical study of ISB transitions in GaN/Al(Ga)N QWs was performed by Suzuki *et al.* [Suzu 97]. The authors have predicted the possibility of tuning the ISB transition energy to the telecommunication wavelengths and to achieve 1.55 μm for 1.3–1.5-nm-thick GaN QWs with barriers containing at least 80% Al. Electron scattering rates have been calculated [Suzu 98] showing that for a 1.5-nm-thick GaN/$Al_{0.8}Ga_{0.2}N$ QW the dominant mechanism for ISB relaxation is the interaction with longitudinal-optical phonons with a characteristic time of $\sim$100 fs. The carrier thermalization to the subband minimum (intrasubband relaxation) is due to electron–phonon, electron–electron, and electron-impurity scattering, and is as fast as $\sim$10 fs. The quantum confinement in GaN/Al(Ga)N heterostructures have been further simulated by different authors [Kish 02b, Kish 02a, Suzu 03, Iizu 02, Hosh 02]. Kishino *et al.* [Kish 02b, Kish 02a] have calculated the ISB energy e_{12} in a GaN/AlN QW as a function of its thickness using the effective-mass approximation. This study allowed evidence for a strong impact of the internal electric field on the transition energy in large QWs, and to calculate a lower limit for the ISB wavelength equal to 1.07 μm. Similar calculations by Iizuka *et al.* [Iizu 02], taking the electric field in the well as a fitting parameter, allowed estimation of its value as 5–6 MV/cm. Hoshino *et al.* [Hosh 02] showed that the ISB energy increases when increasing the internal field. Finally, Suzuki *et al.* [Suzu 03] have performed self-consistent calculations by solving Schrödinger–Poisson equations accounting for the non-parabolicity of the conduction band of GaN.

In the following, we expose the theoretical framework used for the description of ISB transitions in nitride heterostructures.

12.2.1 *Effective-mass approximation*

The basic optical properties of ISB transitions can be derived using the envelope function formalism [Bast 81, Whit 81, Bast 82]. This formalism is less general than *ab initio* methods, and its validity is limited to the vicinity of high-symmetry points of the Brillouin zone. However, due to the flexibility and the relative computational simplicity of this model it is widely used to describe the quantum confinement in semiconductor heterostructures. In the envelope-function approximation, the wavefunction is developed over the basis formed by the periodic components of the Bloch functions of the bulk material, and then only a small number of terms corresponding to the nearest energy bands is retained. In this way, the wavefunction component varying slowly on the crystal-cell scale–the envelope function–is separated from the rapidly varying component:

$$\varphi(\mathbf{r}) = \sum_i u_{i,\mathbf{k}=0}(\mathbf{r})\psi_i(\mathbf{r}), \tag{12.1}$$

where $u_{i,\mathbf{k}=0}(\mathbf{r})$ is the periodic part of the Bloch function at the center of an energy band i, and $\psi_i(\mathbf{r})$ is the corresponding envelope function.

To describe the ISB transitions, the sum can be often limited to only one energy band—the so-called effective-mass approximation [BenD 66].

The envelope function satisfies the Schrödinger equation:

$$\left(-\frac{\hbar^2}{2}\boldsymbol{\nabla}\left[\frac{1}{M^*(\mathbf{r})}\right]\boldsymbol{\nabla} + U_H(\mathbf{r})\right)\psi(\mathbf{r}) = E\psi(\mathbf{r}),$$

where $U_H(\mathbf{r})$ is the heterostructure potential and $\left[\frac{1}{M^*(\mathbf{r})}\right]$ is the inverse effective mass tensor which can be diagonalized:

$$\left[\frac{1}{M^*(\mathbf{r})}\right] = \begin{pmatrix} \frac{1}{m_x^*(\mathbf{r})} & 0 & 0 \\ 0 & \frac{1}{m_y^*(\mathbf{r})} & 0 \\ 0 & 0 & \frac{1}{m_z^*(\mathbf{r})} \end{pmatrix}.$$

For the QW case, the translational invariance in the (x, y) layer plane allows factorization of the envelope function into plane waves in the x and y direction and z-dependent envelope function $\psi_l(\mathbf{r}) = \frac{1}{\sqrt{S}}e^{i\mathbf{k}_{||}\cdot\mathbf{r}_{||}}\Psi_l(z)$, where $\mathbf{k}_{||}$ and $\mathbf{r}_{||}$ are the wavevector and the coordinate in the layer plane, and S is the considered crystal surface. The function $\Psi(z)$ satisfies the one-dimensional Schrödinger equation:

$$\left(-\frac{\hbar^2}{2m^*}\frac{d^2}{dz^2} + U_H(z)\right)\Psi(z) = E\Psi(z) \tag{12.2}$$

with continuity of the $\Psi(z)$ and $\frac{1}{m^*(z)}\Psi'(z)$ at the QW interfaces [BenD 66].

12.2.2 *Band non-parabolicity*

The influence of other energy bands induces a non-parabolicity of the energy dependence on the wavevector **k** [Nels 87, Bast 81, Bast 82, Leav 91, Sirt 94]. Despite the high band-gap of GaN, implying a weak coupling between the conduction and the valence bands, the non-parabolicity effect in GaN/AlN QWs is important for near-infrared ISB transitions [Tche 06b]. The non-parabolicity mainly impacts the excited states in the heterostructure, and reduces the ISB transition energy with respect to the parabolic case.

To obtain a correct prediction of the energy levels, the confinement in nitride heterostructures can be described using the 8-band **k·p** model [Kand 08]. However, this approach is computationally heavy. Alternatively, the non-parabolicity effect can be accounted for by using a simplified model with the energy-dependent effective mass [Tche 06b]. For GaN/AlN heterostructures with the layer plane perpendicular to the c-axis, the following expression for the energy-dependent effective mass in GaN was proposed [Vard 09]:

$$m^*(E) = 0.22m_0(1 + 0.068E + 0.003E^2), \tag{12.3}$$

where E is an energy in eV, measured from the conduction band edge.

12.2.3 *Hartree approximation*

To induce ISB absorption in a heterostructure, the fundamental energy state must be populated with carriers. This is achieved by doping the active region. For doped structures, the model of one electron in an empty conduction band is no longer valid. In particular, one should take into account the Coulomb interaction of electrons with ionized donors as well as between free carriers. These phenomena can be introduced via the Hartree approximation, in which the Coulomb interactions between individual carriers are replaced by an average electrostatic potential. The energy levels and the corresponding envelope functions are calculated by solving self-consistently the Schrödinger and Poisson equations:

$$\begin{cases} \left(-\frac{\hbar^2}{2m^*}\frac{d^2}{dz^2} + U_H(z) + U_{SC}(z)\right)\Psi_i(z) = E_i\Psi_i(z) \\ \frac{d^2}{dz^2}U_{SC}(z) = \frac{e^2}{\varepsilon}\left(N_D^+(z) - \sum_i n_i^S|\Psi_i(z)|^2\right) \end{cases} \tag{12.4}$$

where $U_{SC}(z)$ is the self-consistent Coulomb potential, which represents the average potential of the other free carriers and of the ionized impurities, e is the elementary charge, $N_D^+(z)$ is the volumic density of ionized doping impurities, and n_i^S is the surface carrier density in th i^{th} subband.

The surface carrier density at equilibrium equals

$$n^S = \int \rho_{2\mathrm{D}}(E)f(E)dE,$$

where $\rho_{2D}(E)$ is the density of states in the QW: $\rho_{2D}(E) = \frac{m^*}{\pi\hbar^2}\sum_i Y(E - E_i)$ (where $Y(x)$ is a Heaviside step function), and $f(E)$ is the Fermi–Dirac occupation statistics $f(E) = \frac{1}{1+\exp\frac{E-E_F}{kT}}$.

For the ith subband the surface carrier density is given by

$$n_i^S = \frac{m^* kT}{\pi\hbar^2}\ln\left(1 + \exp\left(\frac{E_F - E_i}{kT}\right)\right), \tag{12.5}$$

and the Fermi energy E_F can be determined from the electrical neutrality condition:

$$\sum_i n_i^S - \int_{-\infty}^{+\infty} N_D^+(z)dz = 0. \tag{12.6}$$

12.2.4 *Internal electric field*

GaN/AlN heterostructures grown in the polar direction sustain large internal electric fields in the layers induced by the difference of the spontaneous and piezoelectric polarization between the well and barrier materials [Bern 97]. The polarization discontinuity gives rise to bound charges of opposite sign at each interface and consequently to an internal electric field of opposite sign in the well and barrier layers. For a thin layer of GaN within an AlN matrix, the internal field in the GaN well can be as large as 10 MV/cm.

As illustrated in Fig. 12.1, the heterostructure potential $U_H(z)$ in nitride QWs has a saw tooth shape given by:

$$U_H(z) = \begin{cases} eF_w(z - z_1) \\ eF_b(z - z_2) + \Delta E^c \end{cases}$$

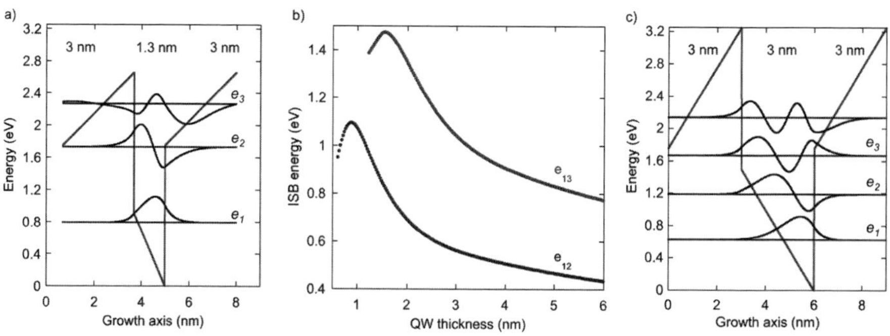

FIG. 12.1. a) Conduction-band profile and energy levels for a 1.3-nm-thick GaN/AlN QW. b) Calculated energy for e_{12} and e_{13} ISB transitions for a GaN/AlN QW with 3-nm-thick barriers. c) The same as a), but for a 3-nm-thick QW.

where F_w (F_b) is the electric field in the QW (barrier), ΔE^c is the conduction band offset, and the coordinates z_i are defined by the heterostructure geometry.

In the case of nitride superlattices, the internal electric field is often calculated assuming zero potential drop at each period[1] [Bern 98, Lero 98]:

$$-e(L_b F_b + L_w F_w) = 0,$$

where L_w (L_b) is the QW (barrier) thickness. Taking into account the continuity of the perpendicular component of the electric displacement vector **D** at the heterointerfaces, we obtain:

$$\varepsilon_0 \varepsilon_w F_w + P_w = \varepsilon_0 \varepsilon_b F_b + P_b,$$

where ε_0 is the vacuum permittivity, ε_w (ε_b) is the relative permittivity of the QW (barrier) material, and P_w (P_b) is the polarization in the QW (barrier). The internal field can be expressed as:

$$F_w = \frac{P_b - P_w}{\varepsilon_0} \frac{L_b}{L_b \varepsilon_w + L_w \varepsilon_b}$$

$$F_b = -\frac{P_b - P_w}{\varepsilon_0} \frac{L_w}{L_b \varepsilon_w + L_w \varepsilon_b}. \tag{12.7}$$

These expressions show that the electric field in the QW increases with the polarization discontinuity and with the barrier thickness, and decreases with the well thickness.

The solution of the Schrödinger equation (12.2) for an undoped nitride QW can be expressed analytically in terms of Airy functions. A numerical solution is required for doped heterostructures to account for the subband population.

Figure 12.1b shows the calculated dependence of the ISB transition energy on the well thickness for an undoped GaN/AlN QW with 3-nm-thick AlN barriers. The calculations were performed using effective mass approximation assuming a conduction-band offset of 1.75 eV and a non-parabolicity in GaN given by eq. (12.3). Two regimes can be distinguished for the ISB energy dependence as illustrated in Figs. 12.1a and b on the example of a thin (1.3-nm) and a thick (3-nm) GaN/AlN QW. For small well thicknesses, the e_2 state is confined by the QW interfaces, which means that the transition energy is governed mostly by the well thickness, i.e. by the quantum size effect. When reducing the well thickness, the calculated e_{12} transition energy increases, goes through a maximum for ~0.9 nm, and then decreases. Indeed, when the QW thickness is reduced below 0.9 nm, the e_2 state becomes delocalized in the triangular potential in the right barrier, and is almost insensitive to the QW thickness, whereas the energy of the e_1 state continues to increase. Experimentally, the ISB absorption vanishes when the well thickness is reduced because of the delocalization of the e_2 state in the

[1]The validity of periodic conditions will be discussed in Section 12.2.5.

barrier layers and the consequent reduction of the associated oscillator strength. For well thicknesses above 2 nm, the ground and the first excited states are confined by the V-shaped potential in the GaN well. The transition energy is therefore ruled by the magnitude of the internal field.

The electric field imposes a lowest limit on the ISB energy that can be attained. Indeed, in a wide well both e_1 and e_2 energy levels are confined by the triangular potential, and the effective QW size is reduced. The lowest ISB energy for GaN/AlN heterostructure is estimated to be ~0.25 eV (λ ~5 μm). To reach longer wavelengths, the internal electric field should be reduced, for example, by reducing the Al content in the barriers, as discussed in Section 12.3.4.

The external electric field can be used to efficiently tune the ISB transition energy via the Stark effect. For a 2.5-nm-thick QW a variation of the field by ~0.5 MV/cm shifts the transition energy by more than 50 meV. This effect can be exploited to design electro-optical modulators [Holm 06].

For thick enough QWs, an ISB absorption from the ground state e_1 to the second excited state e_3 can be observed [Hosh 02]. This transition is allowed by the built-in electric field that breaks the potential symmetry. As shown in Fig. 12.1b, the calculated e_{13} transition energy is maximal for QW thickness of ~1.5 nm.

12.2.5 *Band bending*

The polarization discontinuity between heterostructure layers discussed previously not only modifies the ISB transition energy but also leads to strong band-bending effects, as well as the formation of a depletion layer on one side of the active region and of an accumulation layer with a 2D electron gas on the other side of the active region. In addition, it has major consequences on the distribution of carriers within the active layers.

The band-bending in the active region depends on the cladding layer composition [Mach 09]. This effect is illustrated in Fig. 12.2, which presents the conduction-band profile and electron concentration in a ten-period GaN/AlN QW superlattice sandwiched between either GaN or $Al_{0.5}Ga_{0.5}N$ claddings. The QW and barrier thickness is 2 nm. The QWs are n-doped at 2×10^{19} cm^{-3}, while the claddings are n-doped at 5×10^{18} cm^{-3}. The calculations were performed using a Schrödinger–Poisson solver (C-band) with material parameters from [Vurg 03]. It can be seen that for GaN claddings (Fig. 12.2a) the band profile is distorted. A large depletion zone appears at the upper heterostructure/cladding interface, and a 2D electron gas is formed at the lower cladding/heterostructure interface. In spite of the homogeneous doping in the QWs, the electron concentration is strongly inhomogeneous: only four QWs over ten are populated. This effect is of major importance for the design of ISB devices, since unpopulated QWs do not contribute to the ISB absorption.

The band-bending can be avoided by a proper choice of the cladding layer composition. AlGaN claddings with an Al percentage equal to the average Al

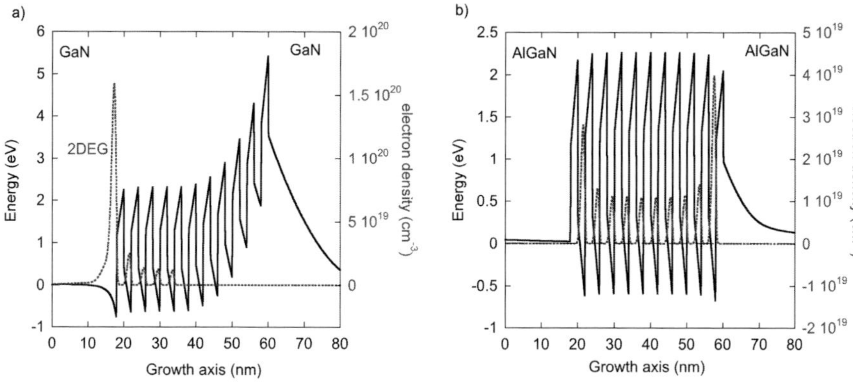

FIG. 12.2. Conduction-band profile and electron concentration for a ten-period GaN/AlN superlattice sandwiched between either two GaN (a) or two $Al_{0.5}Ga_{0.5}N$ (b) claddings. The energy origin is set at the Fermi energy. (After [Tche 10].)

content in the active region should be used to obtain a flat potential profile and an homogeneous electron concentration. This situation is illustrated in Fig. 12.2b. In this case, the periodic conditions (12.7) can be applied for all QWs of the superlattice. In the case of GaN claddings, the periodic conditions are valid only for the first four QWs.

12.2.6 *Many-body effects*

The previously described Hartree approximation (12.4) for a doped QW predicts a screening of the internal electric field by free carriers. Because the screened potential profile in the well tends to flatten, a reduction of the ISB transition energy when increasing the carrier concentration is predicted by the Hartree model. However, experimental measurements show the opposite behavior: the ISB energy for doped samples is systematically blueshifted with respect to the non-intentionally doped samples [Helm 03b, Kand 10]. This apparent contradiction can be clarified by taking into consideration the many-body interactions, which become important at large carrier concentrations. They are, namely, the depolarization shift, the excitonic correction, and the electron–electron exchange interaction.

The depolarization shift α, which reflects the resonant screening of the external infrared field by the collective oscillation of the electron plasma, results in an increase of the transition energy. The excitonic correction β stems from the Coulomb interaction between the excited electron and the quasi-hole formed in the ground subband. In contrast to the previous effect, it leads to a reduction of the ISB energy. The effective absorption energy of the electron gas, e_{12}^{coll}, is given by

$$e_{12}^{\text{coll}} = e_{12}\sqrt{1+\alpha-\beta},$$

where e_{12} is ISB transition energy for a single electron transition [Alle 76, Ando 77]. Assuming only two subbands, the depolarization correction can be expressed as [Alle 76]:

$$\alpha = \frac{2e^2 n^S}{\varepsilon_0 \varepsilon_r e_{12}} \int_{-\infty}^{\infty} dz \left(\int_{-\infty}^{z} dz' \Psi_1(z')\Psi_2(z') \right)^2,$$

where ε_r is the high-frequency dielectric constant of GaN, n^S is the surface carrier density, and Ψ_1 and Ψ_2 are the envelope functions of the ground and first excited subbands, respectively.

The excitonic correction can be estimated as [Blos 89]:

$$\beta = -\frac{2n^S}{e_{12}} \int_{-\infty}^{\infty} dz \left|\Psi_1(z)\right|^2 \left|\Psi_2(z)\right|^2 \frac{\partial V_{xc}[n(z)]}{\partial n(z)},$$

where the exchange-correlation energy $V_{xc}(n(z))$ is expressed as the following functional of the 3D carrier density $n(z)$ [Gunn 76, Helm 99]:

$$V_{xc}(n(z)) = -\frac{2}{r_s}\left(\frac{9}{4\pi^2}\right)^{\frac{1}{3}} \frac{e^2}{8\pi\varepsilon_0\varepsilon_r a^*}\left(1 + 0.0545 r_s \ln\left(1+\frac{11.4}{r_s}\right)\right), \quad (12.8)$$

with dimensionless parameter $r_s = \sqrt[3]{\frac{3}{4\pi(a^*)^3 n(z)}}$ and effective Bohr radius $a^* = \frac{4\pi\varepsilon_0\hbar^2}{m_0 e^2}\frac{\varepsilon_r}{m^*/m_0}$.

For the electron–electron exchange interaction, different approaches have been proposed. In the Hartree–Fock approximation the exchange interaction appears as an additional term in the Schrödinger equation [Fock 30]. This term has a non-local nature, which renders the exact solution very computationally intensive. To avoid computational difficulties, one approach is to estimate the contribution of the exchange interaction using perturbation theory [Band 88, Band 89], where Hartree envelope functions are used as a basis and the first perturbative correction of the energy is calculated. Bandara *et al.* showed that the exchange interaction mainly affects the fundamental subband e_1 by lowering its energy, which results in an increase of the ISB transition energy. If one neglects the thermal excitation, the modification of the e_1 subband energy is given by

$$\Delta e_1^{\text{exch}} = -\frac{e^2}{2\varepsilon_0\varepsilon_r}\int_{-\infty}^{\infty} dz \int_{-\infty}^{\infty} dz' \int_0^{k_F} \frac{k'dk'}{2\pi}\frac{e^{-k'|z-z'|}}{k'}|\Psi_1(z')|^2|\Psi_1(z)|^2,$$

where ε_r is the dielectric constant of the QW material, the integration is carried out over the modulus of the in-plane wavevector k', k_F is the Fermi wavevector related to the surface carrier density n^S by $k_F = \sqrt{2\pi n^S}$, and Ψ_1 is the envelope function for the fundamental subband calculated within Hartree approximation.

The exchange correction predicts a rapid increase of the ISB transition energy with carrier concentration [Tche 06b, Helm 03b]. However, since the approach of Bandara *et al.* is based on a perturbation theory, its predictions are valid only for moderate carrier concentrations, for which $\Delta e_1^{\text{exch}} << e_{12}$.

Another approach to account for the exchange interaction is to follow the local density approximation [Hedi 71] and to replace the non-local term in the Schrödinger equation by a local effective exchange-correlation potential $V_{xc}(n(z))$ defined in (12.8). The $V_{xc}(n(z))$ is added to the Schrödinger equation and the system (12.4) becomes:

$$\begin{cases} \left(-\dfrac{\hbar^2}{2m^*}\dfrac{d^2}{dz^2} + U_H(z) + U_{SC}(z) + V_{xc}(n(z))\right)\Psi_i(z) = E_i\Psi_i(z) \\ \dfrac{d^2}{dz^2}U_{SC}(z) = \dfrac{e^2}{\varepsilon}(N_D^+(z) - \sum_i n_i^S|\Psi_i(z)|^2) \end{cases} \tag{12.9}$$

The effect of many-body corrections is illustrated in Fig. 12.3, showing the e_{12} transition energy calculated for a 7-nm-thick GaN QW with 4-nm-thick $Al_{0.1}Ga_{0.9}N$ barriers as a function of the surface carrier density. The effect of each many-body correction is shown separately in order to compare their relative importance. The open circles represent the e_{12} energy obtained from self-consistent

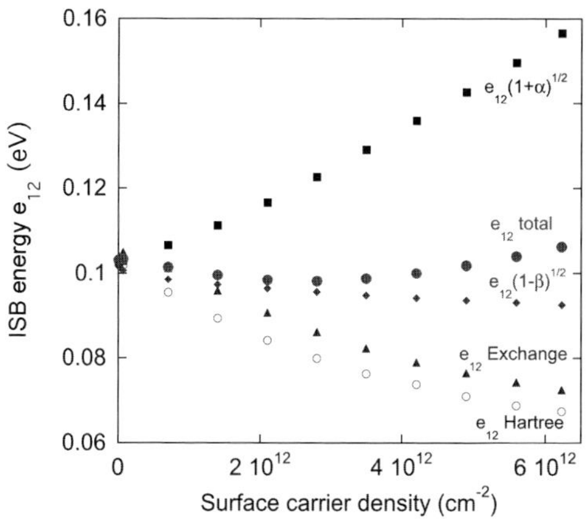

FIG. 12.3. ISB transition energy e_{12} as a function of surface electron density calculated in the Hartree approximation (open circles), including exchange-correlation potential (full triangles), depolarization shift (full squares) and excitonic shift (full diamonds) and taking into account all many-body corrections (full circles).

solution of Schrödinger and Poisson equations (Hartree approximation (12.4)). As expected, when accounting only for screening of the internal field, the ISB energy decreases with carrier concentration. The triangles show the solution of Schrödinger and Poisson equations (12.9), where the exchange-correlation potential has been added. The exchange interaction tends to increase the ISB transition energy compared to Hartree approximation, though this interaction alone cannot invert the global trend. The squares represent the depolarization shift $e_{12}\sqrt{1+\alpha}$, which translates into a rapid increase of the ISB energy with carrier density. The diamonds show the excitonic shift $e_{12}\sqrt{1-\beta}$, which competes with depolarization shift and reduces the ISB energy. When all many-body corrections are included (the circles), the resultant ISB energy increases for carrier concentrations above $3 \cdot 10^{12}$ cm^{-2}. At large carrier concentrations, the ISB transition energy blueshift is dominated by the depolarization shift.

12.2.7 *Optical properties of intersubband transitions*

The transition rate between subbands $|i>$ and $|f>$ induced by a plane wave of amplitude F and frequency ω is given by the Fermi golden rule:

$$| < f|W|i > |^2 = \frac{2\pi}{\hbar}\frac{e^2F^2}{4{m^*}^2\omega^2}| < f|\boldsymbol{\varepsilon}\cdot\mathbf{p}|i > |^2\delta(E_f - E_i - \hbar\omega),$$

where $\mathbf{p}$ is the impulse operator, $\boldsymbol{\varepsilon}$ is the polarization vector of the electromagnetic wave, m^* is the effective mass, E_f (E_i) is the energy of the final (initial) subband, and $\delta(E)$ is the delta function reflecting the energy conservation.

The matrix element $< f|\boldsymbol{\varepsilon}\cdot\mathbf{p}|i >$ can also be expressed in terms of the dipole operator:

$$< f|\boldsymbol{\varepsilon}\cdot\mathbf{p}|i >= \frac{im^*(E_f - E_i)}{e\hbar}\boldsymbol{\varepsilon}\cdot\boldsymbol{\mu}_{if},$$

where the dipole momentum of the i–f transition is defined as $\boldsymbol{\mu}_{if} =< f|e\mathbf{r}|i >$. For a GaN/AlN QW absorbing at telecommunication wavelengths, the dipole momentum length μ_{if}/e is about 25% of the QW thickness.

Using the expression of the wavefunction as a product of a rapidly varying function and a slowly varying envelope function (12.1), the matrix element can be approximated by [Bast 88]:

$$\begin{aligned}< f|\boldsymbol{\varepsilon}\cdot e\mathbf{r}|i > \simeq\ & < u_f(\mathbf{r})|\boldsymbol{\varepsilon}\cdot e\mathbf{r}|u_i(\mathbf{r}) >< \psi_f(\mathbf{r})|\psi_i(\mathbf{r}) > \\ & + < u_f(\mathbf{r})|u_i(\mathbf{r}) >< \psi_f(\mathbf{r})|\boldsymbol{\varepsilon}\cdot e\mathbf{r}|\psi_i(\mathbf{r}) > .\end{aligned} \tag{12.10}$$

The first term in (12.10) is non-zero only when $u_i(\mathbf{r})$ and $u_f(\mathbf{r})$ are of opposite parity. It corresponds to the interband transitions. Since the $u_n(\mathbf{r})$ functions are orthogonal, the second term couples only states from the same band and gives rise to the ISB transitions.

Taking into account the envelope function expression in a QW $\psi_n(\mathbf{r}) = \frac{1}{\sqrt{S}} e^{i\mathbf{k}_{||}\cdot\mathbf{r}_{||}} \Psi_n(z)$, and using that $E_f - E_i \sim \hbar\omega$ due to the energy conservation, the ISB transition rate can be rewritten as

$$|<f|W|i>|^2 \sim \frac{2\pi}{\hbar}\frac{e^2F^2}{4}\varepsilon_z^2|<\Psi_f(z)|z|\Psi_i(z)>|^2\delta(\mathbf{k}_{||}^i - \mathbf{k}_{||}^f)\delta(E_f - E_i - \hbar\omega),$$

where ε_z is the z projection of the light-polarization vector.

From this expression, important properties of the ISB transitions can be derived. Firstly, the ISB transitions are possible only between states having the same in-plane wavevector $\mathbf{k}_{||}^i = \mathbf{k}_{||}^f$. For parabolic bands this implies that the transition energy $E_i - E_f$ and the transition rate does not depend on $\mathbf{k}_{||}$. Secondly, the absorption is forbidden between states having the same parity of the envelope function. Thirdly, the ISB transitions can be induced only by an electromagnetic wave with a non-zero component along the quantization axis z. The ISB absorptions are therefore TM-polarized, and they can be easily distinguished from any other absorption in the material. Finally, the transition rate depends only on the envelope functions and can be tailored by the QW design.

The ISB absorption coefficient is defined as the energy loss by unit volume by unit time, divided by the intensity of incident light:

$$\alpha(\omega) = \frac{\hbar\omega}{\Omega nc\varepsilon_0\frac{F^2}{2}}\sum_{i,f}2|<f|W|i>|^2(f(E_i) - f(E_f))\delta(E_f - E_i - \hbar\omega),$$

where Ω is the active volume, n is the refractive index, and the summation is taken over all possible initial and final states. For the absorption between two subbands $|i>$ and $|f>$ of a QW, the sum can be expressed as a difference of populations of the two subbands:

$$\sum_{i,f}2\left(f(E_i) - f(E_f)\right) = S\left(n_i^S - n_f^S\right),$$

where n_l^S is the carrier suface density in the corresponding subband, and S is the crystal surface. The expression of the ISB absorption coefficient between subbands i and f is then given by

$$\alpha(\omega) = \frac{\pi(E_f - E_i)}{nc\varepsilon_0\hbar}\frac{S}{\Omega}(n_i^S - n_f^S)\varepsilon_z^2\mu_{if}^2\delta(E_f - E_i - \hbar\omega)$$

where $\mu_{if} =<\Psi_f(z)|ez|\Psi_i(z)>$.

It is also convenient to introduce the absorption cross-section σ such that $\alpha(\omega)L = \sigma(\omega)(n_i^S - n_f^S)$:

$$\sigma(\omega) = \frac{\pi(E_f - E_i)}{nc\varepsilon_0\hbar}\frac{\sin^2\theta}{\cos\theta}\mu_{if}^2\delta(E_f - E_i - \hbar\omega)$$

(θ being the angle between the light-propagation direction inside the QW and the direction perpendicular to the layers, and L is the active layer thickness) and the oscillator strength

$$\mathrm{f}_{if} = \frac{2m_0(E_f - E_i)}{e^2\hbar^2}\mu_{if}^2,$$

which defines the coupling efficiency between two subbands.

To account for a finite width of the ISB absorption, the delta function for energy conservation is replaced by a phenomenological function reflecting the transition broadening $g(E)$. For example, the intersublevel absorption in GaN/AlN quantum dots is well reproduced with a Gaussian function $g(E) = \frac{1}{\sqrt{\pi}\hbar\Gamma}\exp(-\frac{(E_f-E_i-\hbar\omega)^2}{(\hbar\Gamma)^2})$ [Guil 06], whereas the ISB absorption in GaN/AlN QWs has a Lorentzian shape $g(E) = \frac{\hbar\Gamma}{\pi}\frac{1}{(E_f-E_i-\hbar\omega)^2+(\hbar\Gamma)^2}$ [Tche 06b] ($\hbar\Gamma$ denotes the half-width at half-maximum of the absorption resonance).

The integrated absorbance I_{if} between the subbands $|i>$ and $|f>$ can be written as

$$I_{if} = \int \alpha(\omega)Ld\omega = \frac{e^2\pi\hbar\left(n_i^S - n_f^S\right)}{2m_0nc\varepsilon_0}\frac{\sin^2\theta}{\cos\theta}\mathrm{f}_{if}.$$

12.3 Intersubband spectroscopy of nitride quantum wells and quantum dots

12.3.1 *Near-infrared intersubband absorption in polar GaN/AlN quantum wells*

Interest in ISB transitions in nitride heterostructures arose in the late 1990s after theoretical work by Suzuki *et al.*, showing that ISB transitions in GaN/Al(Ga)N can cover the telecommunication domain [Suzu 97]. First experimental studies carried out on MOCVD grown GaN/$Al_{0.65}Ga_{0.35}N$ QWs have demonstrated ISB absorptions at 3–4 μm [Suzu 99]. At the beginning 2000, shorter wavelengths (1.41–4.15 μm) were achieved in MBE-grown GaN/$Al_xGa_{1-x}N$ QWs with $x = 0.65$–0.85 [Gmac 00, Gmac 01a, Hebe 02, Zhou 03]. MBE-grown GaN QWs with pure AlN barriers were studied for the first time by Iizuka *et al.* [Iizu 02], showing ISB absorption at 1.3–2.2 μm. The shortest ISB wavelength of $\lambda = 1.08$ μm was reported by Kishino *et al.* [Kish 02b] for a three-monolayer (ML) (~0.8 nm) thick GaN/AlN QW. Helman *et al.* [Helm 03b] studied the dependence of the ISB energy in GaN/AlN QWs on the doping, and showed that the ISB wavelength experiences a blueshift when increasing the carrier concentration.

Figure 12.4a shows a typical example of transmission spectra of MBE-grown GaN/AlN QWs measured at room temperature in a multi-pass waveguide configuration [Tche 06b]. The samples contain 20 periods of GaN QWs with 3 nm thick AlN barriers. The QWs are n-doped with Si at 5×10^{19} cm^{-3}. The ISB absorption is observed only for TM-polarized light, in agreement with polarization selection rules. The absorption shifts towards lower energies, with increasing QW width from 4 MLs (1 nm) to 7 MLs (1.8 nm).

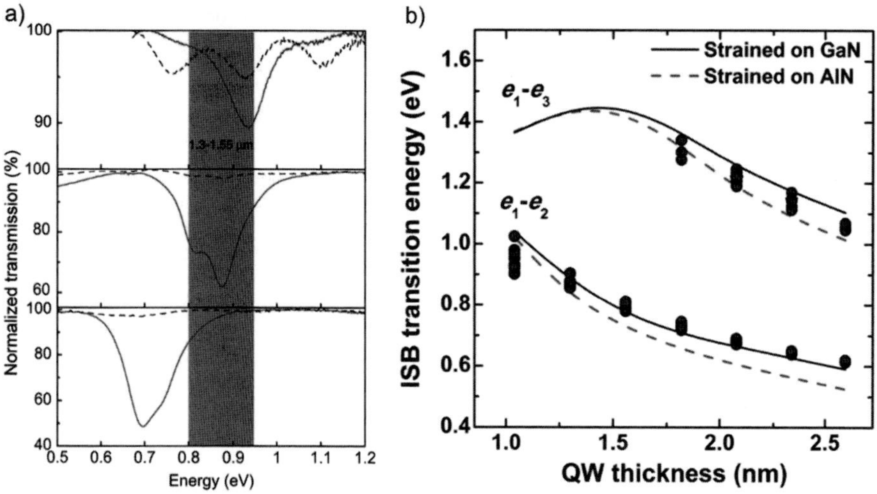

FIG. 12.4. a) Room-temperature transmission spectra of GaN/AlN QWs for TM- (solid lines) and TE-polarized (dashed lines) light. The QW thickness from top to bottom is equal to 1, 1.3, and 1.8 nm. b) Calculated energy for e_{12} and e_{13} ISB transitions in a GaN/AlN QW with 3-nm barriers. Circles show the experimental data, and solid (dashed) curves correspond to calculations in the hypothesis of a structure strained on GaN (AlN). (After [Kand 08].)

Comparison of the measured and calculated ISB transition energy is reported in Fig. 12.4b. Calculations were performed using the Nextnano^3 8-band $\mathbf{k}\cdot\mathbf{p}$ Schrödinger–Poisson solver for a GaN/AlN superlattice within two different hypotheses of a heterostructure fully strained on GaN and on AlN [Kand 08]. For a GaN well thickness of between 4 and 6 ML (1–1.5 nm), the e_{12} transition energy covers the spectral range used for fiber-optic telecommunication. For thick QWs the ISB transition energy is sensitive to the strain state, which has an impact on the piezoelectric component of the internal electric field. The internal field is enhanced for a tensile-strained structure (strained on GaN), which results in an increase of the ISB transition energy compared to the case of a compressively strained structure (strained on AlN).

The typical full width at half maximum (FWHM) of the ISB absorption in GaN/AlN QWs is between 60 and 120 meV. A record low broadening of 40 meV at $\lambda = 1.43$ µm ($\Delta E/E \sim 4.6\%$) has been reported by Nevou *et al.* [Nevo 06b] in non-intentionally doped GaN/AlN QWs studied by a photoinduced absorption technique. The absorbance resonance either exhibits a Lorentzian lineshape or is structured with two or three well-defined peaks, each of them having a Lorentzian shape. The structuring of ISB absorption of GaN/AlN QWs is interpreted as originating from absorbance in well regions with different thicknesses equal to an integer number of monolayers [Tche 06b]. Indeed, thickness fluctuations of

±1 ML (±0.26 nm), which occur during the growth, create potential minima deep enough to localize carriers even at room temperature. For 4–6-ML-thick GaN QWs, a 1-ML increase of the thickness translates into a strong ISB energy shift of about 60 meV. This value is comparable to the broadening factor, and therefore results in structuring of the absorbance spectrum instead of inhomogeneous broadening. For comparison, in material systems such as GaAs/AlGaAs or InGaAs/InAlAs, the minimal thickness necessary to have two bound levels in the well is much larger than in the GaN/AlN system, of the order of several nanometers. For such large thicknesses a fluctuation of 1 ML results only in a small shift of the ISB absorption energy contributing to the broadening [Helm 99].

The MOCVD technique, because of its large growth rate, is less suited for the fabrication of QWs absorbing at telecommunication wavelengths with thickness control down to 1 ML. An additional difficulty is related to the Ga/Al interdiffusion favored by the high growth temperature. However, ISB absorption wavelength close to telecommunication range ($\lambda \sim$1.6 μm) has been reported in MOCVD grown GaN/AlN QWs [Nico 06, Nico 07]. It has been shown that the interdiffusion depends on the strain state in the heterostructure, and can be strongly reduced in heterostructures under compressive strain grown on AlN on sapphire templates [Nico 07].

12.3.2 *Intersubband transitions in In-containing heterostructures*

In order to reduce the strain accumulation and the defect formation in superlattices, In-containing alloys have been proposed as a solution. The substitution of AlGaN in the barriers by $Al_{1-x}In_xN$ allows the reduction of the lattice mismatch (the exact lattice matching to GaN is achieved for $x = 0.18$) [Butt 07]. It was shown that thick AlInN layers can be grown on GaN templates without crack formation to form low-loss waveguides at 1.55 μm [Lupu 08]. Nicolay *et al.* have reported the observation of ISB absorption at $\lambda \sim$3 μm in MOCVD-grown lattice-matched GaN/$Al_{0.82}In_{0.18}N$ QWs [Nico 05]. ISB absorption between 2.3 and 2.9 μm has been also observed in MBE-grown GaN/AlInN QWs [Mali 09]. It should be noted that the conduction-band offset is reduced by adding In in the barriers. Therefore, the near-infrared spectral range cannot be attained for lattice-matched AlInN/GaN structures. This material combination remains well-suited for ISB devices operating at $\lambda >$ 2.5 μm.

An alternative strategy for strain management consists in designing strain-compensated heterostructures. This can be achieved using GaInN/AlInN superlattice on GaN substrates. The advantage of this solution with respect to GaN/AlInN heterostructures is that it preserves a high conduction-band offset compatible with ISB transitions in the near-infrared range. ISB absorptions at $\lambda \sim$ 1.5 μm have been demonstrated in MBE-grown strain-compensated GaInN/AlInN QWs [Cywi 06, Skie 06, Moty 06, Kudr 08, Cywi 08].

12.3.3 *Coupled GaN/AlN quantum wells*

To design the active region of many ISB devices, coupled QWs are required. Because of the high conduction-band offset between GaN and AlN, the design of coupled QWs can be based either on an ultra-thin AlN barrier or on a ternary alloy in the barrier to achieve efficient coupling.

Coupling between GaN QWs was first investigated by C. Gmachl *et al.* [Gmac 01b, Hebe 02]. The spectroscopic studies of GaN/$Al_{0.65}Ga_{0.35}N$-coupled QWs with the coupling barrier thickness varied from 0.7 nm to 6 nm have revealed broad ISB absorptions (FWHM ~300 meV) presenting multiple peaks in $\lambda = 1.35-2.9$ μm spectral range. The QWs were highly doped (10^{20} cm^{-3}) in order to populate the ground state and to establish a common reference energy at the Fermi level. The structuring of the ISB absorptions in symmetric double QWs was attributed to transitions towards excited states exhibiting anticrossing.

Quantum confinement in coupled GaN/AlN QWs has been modeled by Suzuki *et al.* [Suzu 03] in order to design structures absorbing at 1.3–1.55 μm for ultrafast switching applications. These calculations, performed within the envelope function formalism, proposed the use of ultra-thin (0.25 nm) AlN layers to achieve coupling. The ISB absorption with e_{12} and e_{13} ISB absorption at telecommunication wavelengths in double GaN/AlGaN QWs was also modeled by Lei *et al.* [Lei 08b, Lei 08a] and by Cen *et al.* [Cen 09b] showing strong dependence of the transition energy on the internal field.

Coupled GaN/AlN QWs with an ultra-thin coupling barrier have been studied experimentally by Tchernycheva *et al.* [Tche 06a]. Figure 12.5 (middle) shows a TEM image of coupled QWs with a 0.5-nm-thick AlN barrier grown by plasma-assisted MBE. Excellent control of the coupling-barrier thickness from period to period was achieved. The absorbance spectra of two coupled QW samples with different QW thickness (Fig. 12.5, left) exhibit a pronounced ISB absorption revealing two peaks. These peaks are attributed to the e_{12} transition between the ground states of both QWs and the e_{13} transition between the ground state of the second QW in growth order, and the excited state delocalized in both QWs. Comparison of spectroscopic results with simulations showed that the description of the ultra-thin barrier using an abrupt potential profile and the internal field calculated following the same formula as for thick layers [Suzu 03] produce wrong predictions of the transition energies. Agreement with experiment is achieved when the ultra-thin AlN barrier is approximated by a triangular potential introduced in a thick GaN QW, as shown in Fig. 12.5 (right). This can be understood by taking into account the theoretical prediction by Bernardini *et al.* [Bern 98] that the potential drop at the GaN/AlN heterointerfaces is not abrupt. Based on *ab initio* calculations, it has been shown that the potential at the GaN/AlN interface varies almost linearly over 1 ML between the GaN and the AlN values. This effect can be disregarded when dealing with heterostructures with layer thicknesses larger than 1 nm. However, when the layer thickness is as small as 1–2 ML, the offset spreading should be taken into account to obtain reliable predictions of ISB transition energies.

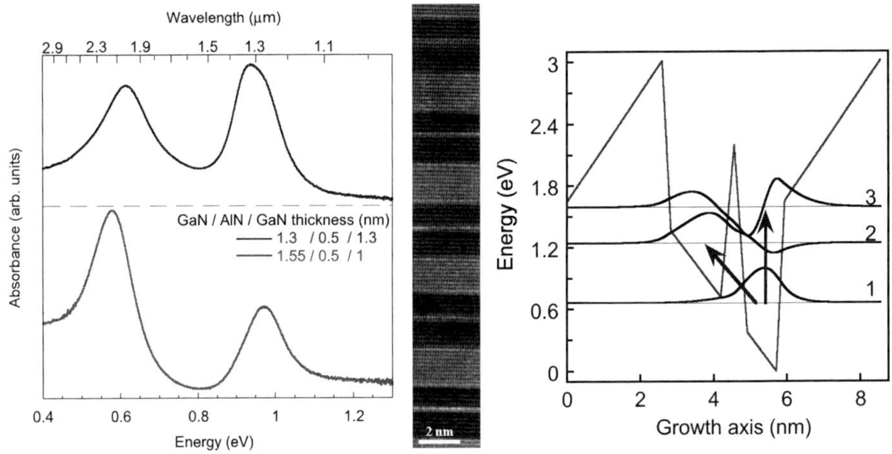

FIG. 12.5. Left – Room-temperature absorbance spectra for TM-polarized light of two GaN/AlN coupled QW samples with well and barrier thicknesses of 1.3 nm/0.5 nm/1.3 nm/2.9 nm and 1.55 nm/0.5 nm/1 nm/2.9 nm. Middle – Transmission electron microscopy image of the second sample. The growth axis is directed upwards. Dark (bright) regions correspond to GaN (AlN). Right – Calculated conduction band profile and the corresponding electronic levels of the second sample. (After [Tche 06a].)

An alternative approach for coupled QWs using a ternary alloy coupling barrier has been studied by Driscoll *et al.* [Dris 07]. The authors fabricated double QWs with an AlGaN coupling barrier with 39–53% Al content, whereas the separating barrier was composed of pure AlN. With this strategy, strong interwell coupling was observed for a coupling barrier thickness of 1.3 nm. These structures have been further used by the authors for development of all optical switches, where an optical control pulse produces a large Stark shift of the ISB absorption via a redistribution of electrons in coupled QWs [Li 06b], as discussed in Section 12.4.1.

12.3.4 *Mid-infrared intersubband absorption in AlGaN/GaN quantum wells*

Until recently, most research on GaN-based ISB transitions has been focused on near-infrared applications, with a challenge to reach the 1.3–1.55 μm spectral range. However, nitrides can also produce interesting applications in the mid- and far-infrared domain. Due to the extended material transparency, nitride ISB devices can cover the entire range of 1.3 μm to 13 μm and 19 μm to THz. The energy of the longitudinal-optical (LO) phonon is much higher than in other III–V materials (92 meV for GaN to be compared with 36 meV for GaAs), so that nitride ISB devices can access the spectral range forbiden to other III–V materials because of their Reststrahlen absorption.

To push the ISB transitions in nitride nanostructures to mid-infrared wavelengths, the internal field must be reduced. This can be achieved by changing the growth plane to non-polar, such as m-plane {1-100} or a-plane {11-20}, or semipolar, such as {11-22} [Laho 08a]. Alternatively, cubic GaN can also be used [DeCu 07, Mach 11]. However, the growth in non-polar and semipolar directions and the growth of cubic nitrides is much less mature than the polar growth.

For [0001]-oriented GaN/AlGaN heterostructures, polarization effects can be attenuated by lowering the Al content in the barriers and increasing the QW width. Figure 12.6(left) shows the calculated ISB energy as a function of the QW width for different compositions of 2-nm-thick AlGaN barriers. This calculation performed for a nominally undoped QW shows that in order to tune the transition to 8–12 μm wavelength range, the Al content in the barriers should be chosen close to 10–20% and the well width should be 4–7 nm. The calculation also predicts an ISB absorption below the Reststrahlen band of GaN for 10% Al composition and a QW width above 9 nm.

ISB absorption in GaN/AlGaN QWs up to ~4 μm has been reported in early works [Suzu 99, Ng 02] and later extended to 5.3 μm by Péré-Laperne *et al.* [Pere 09]. Kandaswamy *et al.* have further shown in a systematic investigation that the absorption in GaN/AlGaN QWs can be shifted almost to the Reststrahlen band of GaN [Kand 09]. Figure 12.6(right) displays the ISB absorption spectra of Si-doped GaN/AlGaN superlattices grown by plasma-assisted MBE on sapphire and on semi-insulating GaN-on-Si(111) templates. The ISB absorption peak is tuned in a large spectral range and attains 9 μm for 7-nm-thick $GaN/Al_{0.1}Ga_{0.9}N$ QWs.

In practice, the ISB absorption energy of AlGaN/GaN QWs is larger than that predicted by the theoretical calculations of Fig. 12.6(left). This is due to the modification of the ISB transition energy by the QW doping. A systematic

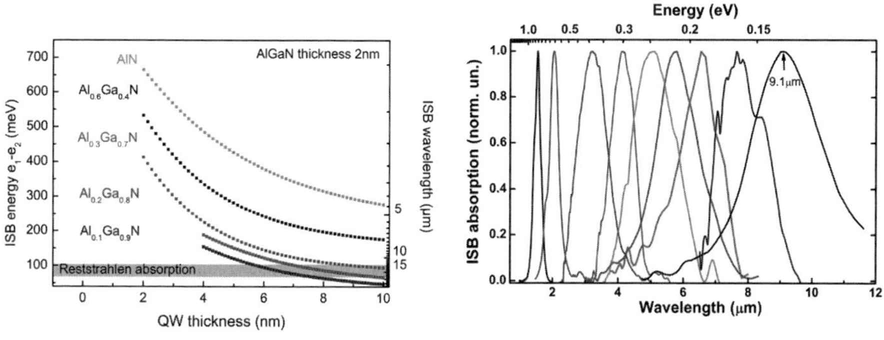

FIG. 12.6. Left: Calculated e_{12} ISB absorption in $GaN/Al_{1-x}Ga_xN$ QWs as a function of QW thickness for different x. Right: ISB absorption spectra of GaN/AlGaN QWs. (After [Kand 09].)

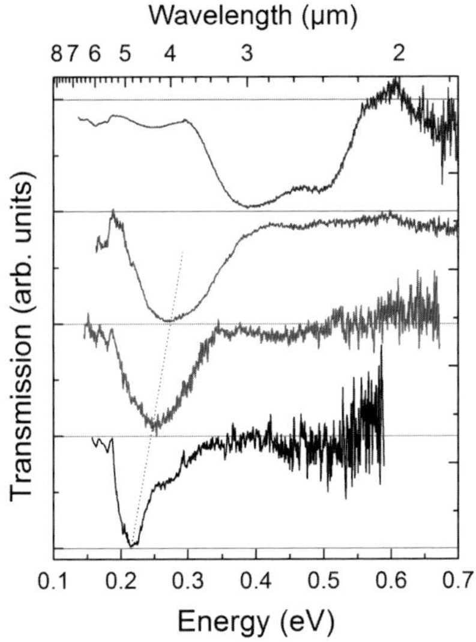

FIG. 12.7. Transmission spectra of Si-doped 7-nm-thick GaN/$Al_{0.1}Ga_{0.9}N$ QWs. Si doping increases from 1×10^{19} cm^{-3} (bottom curve) to 2×10^{20} cm^{-3} (top curve).

investigation [Kand 10] of the influence of the surfacic electron concentration on the ISB absorption of GaN/AlGaN QWs has shown that increasing the carrier concentration results in a strong blueshift of the ISB absorption, which is accompanied by a large enhancement of the broadening. This effect is illustrated in Fig. 12.7, showing the transmission spectra for 7-nm-thick GaN/$Al_{0.1}Ga_{0.9}N$ QWs doped with Si at a nominal concentration ranging from $n = 1 \times 10^{19}$ cm^{-3} to 2×10^{20} cm^{-3} (equivalent surfacic concentration $n^S = 7 \times 10^{12}$ cm^{-2} to 1.4×10^{14} cm^{-2}). The ISB absorption is progressively blueshifted, and the shift attains 50% of the e_{12} transition energy for the maximal doping. The blueshift is attributed to many-body effects dominated by the depolarization shift, as discussed in Section 12.2.6.

12.3.5 *Tuning ISB transitions to the terahertz frequency domain in polar GaN/AlGaN quantum wells*

The terahertz (THz) spectral region is of particular importance for applications such as medical diagnostics, security screening, trace molecule sensing, astronomical detection, or non-invasive quality control. In terms of sources, this region is covered with arsenide-based quantum cascade lasers. However, the main limitation for these devices is their cryogenic operating temperature. Nitride materials are predicted to provide a solution for THz lasers operating at temperatures extending beyond 300 K [Bell 09].

The first step in view of the development of GaN THz QCLs is the tuning of the ISB transition to the far-infrared domain. As presented in Section 12.3.7, ISB absorption at THz frequencies has been observed in cubic GaN/AlGaN QWs [Mach 11], but this is very challenging in the case of polar wurtzite nitrides because of the internal electric field. The strategy followed for tuning the ISB absorption to the mid-infrared region may potentially provide a solution, though it requires a very low Al content in the barriers as well as a thick QW. In this case the transition energy is difficult to control, since it is extremely sensitive to the carrier concentration via screening of the internal field and many-body effects.

A more robust design for achieving ISB absorption in the THz has been proposed by Machhadani *et al.* [Mach 10]. The main idea is to use a step QW to approach a flat potential profile. The design is illustrated in Fig. 12.8(left), showing the conduction-band profile and the square of the envelope functions for a structure consisting of an Si-doped 3-nm-thick GaN QW, 15-nm-thick $Al_{0.05}Ga_{0.95}N$ step barrier, and a 3-nm-thick $Al_{0.1}Ga_{0.9}N$ spacer barrier. The composition in the step AlGaN barrier is chosen equal to the average Al content in the structure, therefore the potential in the step barrier is almost flat except for a slight band-bending due to the Coulomb interaction between electrons and ionized donors. The ground and excited electronic states are confined in the GaN well and $Al_{0.05}Ga_{0.95}N$ step barrier. The predicted e_{12} energy is 10 meV (2.4 THz).

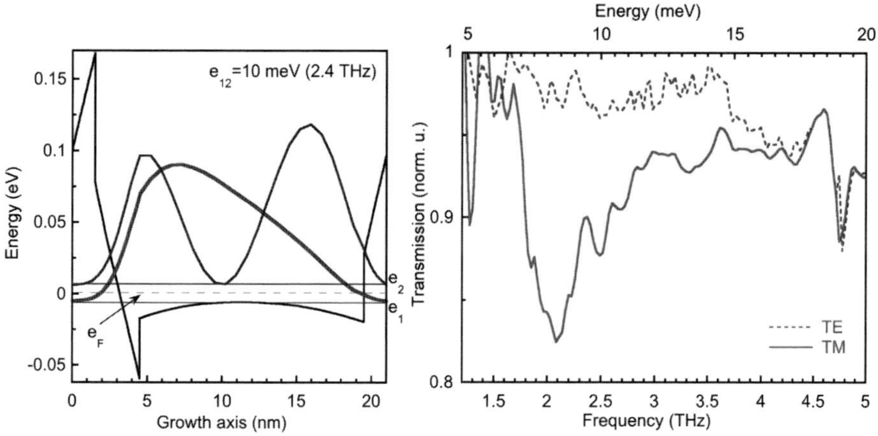

FIG. 12.8. Left: Calculated conduction-band profile and confined states in a step $GaN/Al_{0.05}Ga_{0.95}N/Al_{0.1}Ga_{0.9}N$ QW. Right: Far-infrared transmission spectra for TM- and TE-polarized light, revealing an ISB absorption at 2.1 THz. (After [Mach 10].)

A sample consisting of forty periods of step QWs was grown by plasma-assisted MBE on GaN-on-Si template. The transmission measurement shown in Fig. 12.8(right) reveals an ISB absorption peaked at 2.1 THz (8.7 meV), in good agreement with theoretical predictions. The rather large FWHM of the absorption equal to 0.72 THz (that is, $\Delta\lambda/\lambda = 34\%$) is attributed to the electron impurity and electron–electron scattering. By reducing the step barrier thickness to 10 nm, the ISB absorption was shifted to 4.2 THz [Mach 10].

12.3.6 *Intersubband transitions in semipolar GaN/AlN quantum wells*

As mentioned in Sections 12.3.4 and 12.3.5, minimizing the internal electric field constitutes a major challenge for tuning the ISB absorption to the mid- and far-infrared spectral range. For near-infrared applications, the device design in some particular cases may take advantage of the internal field (as, for example, in quantum cascade detectors described in Section 12.5.3). However, for most applications in the near-infrared the internal field presents a number of drawbacks. First, it reduces the oscillator strength of the ISB transition with respect to a square QW of the same thickness, leading to a lower ISB absorption or emission efficiency. In addition, it induces a band-bending, creates depleted regions in heterostructures, and modifies the alignment of coupled QWs. This in general complicates the design of ISB devices, in which a high degree of control over the QW population and over the energy-level alignment is required. It is therefore desirable to reduce the internal electric field while keeping a high conduction-band offset.

This can be achieved by changing the growth direction in order to bring the electrical polarization vector in the layer plane. One possibility is to use a non-polar growth plane, such as m-plane {1-100} or a-plane {11-20}. However, growth of non-polar III-nitrides is challenging due to the strong anisotropy of the surface properties, resulting in layers with a high density of crystalline defects. An alternative approach is the growth on semipolar planes, such as {11-22}. The advantage of the semipolar orientation is that it allows to considerably reduce the internal electric field [Chak 05, Roma 06] while presenting a lower in-plane anisotropy than non-polar surfaces [Laho 07, Laho 08b].

The observation of ISB absorptions in GaN/AlN semipolar QWs has been reported by L. Lahourcade *et al.* [Laho 08a]. The non-intentionally doped GaN/AlN multiple QWs were studied by photoinduced absorption infrared spectroscopy, which revealed ISB resonances in the 2.4–3.8 μm range. The measured ISB energies together with the results of interband photoluminescence spectroscopy confirmed the strong reduction of the internal electric field in the QWs to ~0.6 MV/cm [Laho 08a].

Figure 12.9 shows the comparison of the ISB absorption in polar and semipolar QWs. Semipolar and polar Si-doped GaN/AlN superlattices with a QW width from 1.2 to 3 nm and AlN thickness of 3 nm were simultaneously

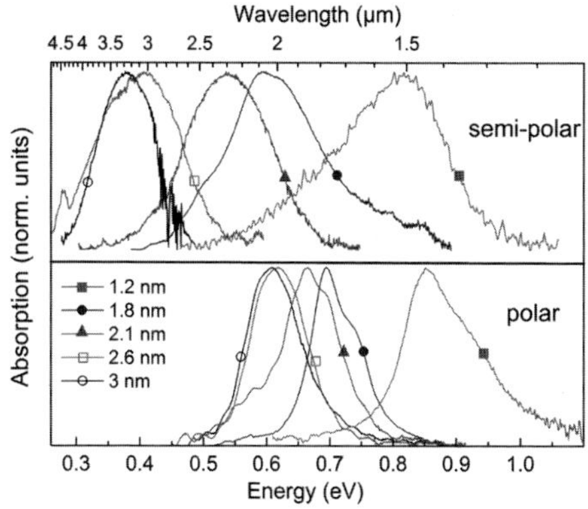

FIG. 12.9. ISB absorption spectra for semipolar (top) and polar (bottom) GaN/AlN QWs with different well thickness (barrier thickness equals 3 nm). (After [Mach 13].)

grown on m-plane and c-plane sapphire substrates. As seen, the ISB absorption in semipolar QWs is redshifted. It covers the 1.5–3.3 μm range for semipolar structures compared to 1.45–2 μm in polar structures. This red shift of the transition energy in large semipolar QWs with respect to the polar case confirms the electric field reduction. We also note that the ISB absorption has a different spectral shape: for polar QWs the ISB spectrum is the sum of Lorentzian resonances, whereas a Gaussian shape is observed in the semipolar case. This dissimilarity can be ascribed to a different carrier localization in these two cases.

12.3.7 *Intersubband transitions in cubic GaN/Al(Ga)N quantum wells*

The issue of the internal electric field inherent to the polar nature of wurtzite crystals is removed in zincblende nitride materials. Although the stable phase for nitride crystals is the hexagonal phase, cubic material can be synthesized under specific growth conditions using substrates of cubic symmetry [As 09]. Cubic GaN exhibits a slightly smaller direct band-gap (3.299 eV [Vurg 03]) than its hexagonal counterpart. Cubic AlN has an indirect band-gap, which value is still not well known (Vurgaftman *et al.* [Vurg 03] recommend a value of 4.9 eV for the X-valley gap). Despite this uncertainty, a high conduction-band offset of 1.4 ± 0.1 eV at Γ point has been reported for cubic GaN/AlN heterostructures [Mach 11, Miet 11]. The electron effective mass in cubic GaN is in the range $0.11m_0$ [Pugh 99]–$0.17m_0$ [Suzu 96] (to be compared to $0.2m_0$ in wurtzite GaN [Vurg 03]). These material parameters allow tuning of the ISB transition energy in cubic GaN/AlN QWs to the near-infrared domain.

The near-infrared ISB absorptions in cubic GaN/AlN QWs have been observed initially by DeCuir *et al.* [DeCu 07]. Infrared quantum-well photodetector

(QWIP) using cubic material has been demonstrated [DeCu 08]. Resonant tunneling in cubic GaN/AlGaN double-barrier structures has also been reported [Zain 10].

Machhadani *et al.* [Mach 11] have shown further that by changing the QW thickness and Al content in GaN/AlGaN QWs it is possible to tune the ISB absorption wavelength over a huge spectral interval from 1.4 μm (214 THz) to 63 μm (4.76 THz). Figure 12.10 shows the low-temperature far-infrared transmission spectrum of a stack of 12-nm-thick $GaN/Al_{0.05}Ga_{0.95}N$ QWs. The ISB absorption is observed at 19.7 meV (4.76 THz). This result demonstrates the interest in cubic nitrides for applications in the THz spectral region. It should be noted that the LO-phonon energy in cubic GaN is almost the same as in wurtzite GaN (92.7 meV [Braz 06]), whereas the effective mass is much smaller. A lower effective mass would result in a higher gain and lower threshold current density in QCLs based on this material system.

12.3.8 *Intersublevel absorption in GaN/AlN quantum dots*

Quantum dots are semiconductor nanostructures presenting quantum confinement in three spatial directions. Contrary to QWs, QDs can exhibit TE-polarized absorption for intersublevel (or intraband) transitions between states confined in

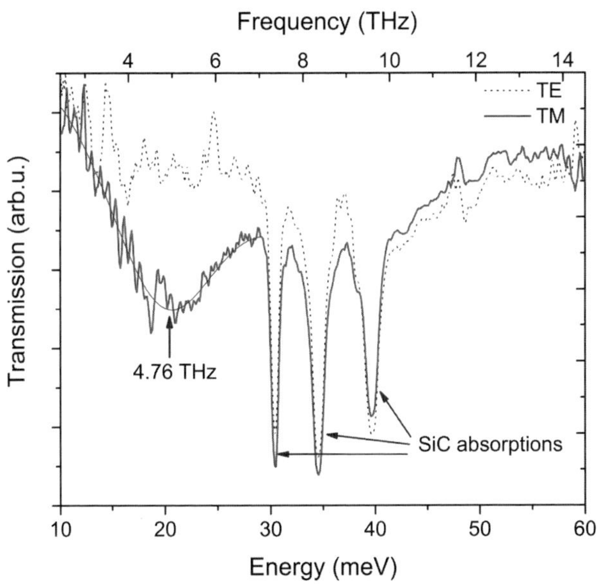

FIG. 12.10. Far-infrared transmission spectra of cubic $GaN/Al_{0.05}Ga_{0.95}N$ QWs for TM- and TE-polarized light at 4 K. ISB absorption is observed at 4.76 THz. (After [Mach 11].)

the layer plane. The three-dimensional confinement also modifies the electron–phonon interaction in this system, resulting in a longer dephasing time than in the QW case [Nevo 08].

GaN/AlN QDs can be formed by MBE using the so-called Stranski–Krastanov mode, in which the strain relaxation during the growth leads to island formation [Daud 97]. The QDs have the shape of a truncated hexagonal pyramid with {10-13} facets standing on 2-ML-thick wetting layer [Cham 04]. The resultant dot surfacic density is large ($10^{11}-10^{12}$ cm^{-2}), which is a key point for obtaining efficient intraband absorption.

The energy states of GaN/AlN QDs were calculated using an 8-band $\mathbf{k}\cdot\mathbf{p}$ model [Andr 00, Andr 01, Vard 09] and using a tight-binding method accounting for the electric-field screening by carriers [Ranj 03]. The calculations show that the electrons are localized at the pyramid apex due to the internal electric field. It was also demonstrated that the built-in field in GaN/AlN dots can provide a strong additional lateral confinement for carriers localized in the dot [Will 05].

The observation of intraband absorption was first reported for large GaN/AlN QDs grown on Si (111) and sapphire substrates [Moum 03]. The intraband absorption of nominally undoped QDs with a height of 4–4.5 nm and a diameter of ~20 nm grown by ammonia MBE was probed by photoinduced absorption spectroscopy. The QDs exhibit a TM-polarized intraband absorption at 0.5–0.6 eV attributed to the transition between the ground state s and the first excited stated, with one node of the envelope function along the growth axis (denoted as p_z state) [Moum 03]. The intraband absorption for TE light polarization in these QDs was observed at 0.15 eV attributed to the s-p_x transition, where p_x is the excited state in the layer plane [Helm 03a].

GaN/AlN QDs of a smaller height (1–1.5 nm) were shown to present intraband absorption at telecommunication wavelength [Tche 05, Guil 06]. Si-doped GaN/AlN QDs were grown by plasma-assisted MBE on AlN-on-sapphire templates. The structure consists of twenty periods of QDs with 3-nm-thick AlN barriers and one additional uncapped QD layer on the surface to assess the dot morphology by AFM (Fig. 12.11, left). Strong TM-polarized intraband absorptions peaking at wavelengths ranging from 1.38 to 1.67 μm were observed at room temperature. The transition energy is governed by the QD height and is almost diameter-independent. The effect of various parameters (the amount of GaN deposited, the substrate temperature, and the growth interruption time after each GaN layer) was analyzed. In particular, the increase of either the substrate temperature or the growth interruption time leads to the formation of bigger QDs and shifts the intraband transition to lower energy. This is illustrated in Fig. 12.11(right), showing the intraband absorption spectra of QDs grown at different substrate temperatures. The s-p_x TE-polarized absorption in similar samples was evidenced by photocurrent spectroscopy. The transition energy was found to be equal to 0.15 eV (0.23 eV) for a QD diameter of 16 nm (11 nm) respectively [Vard 09].

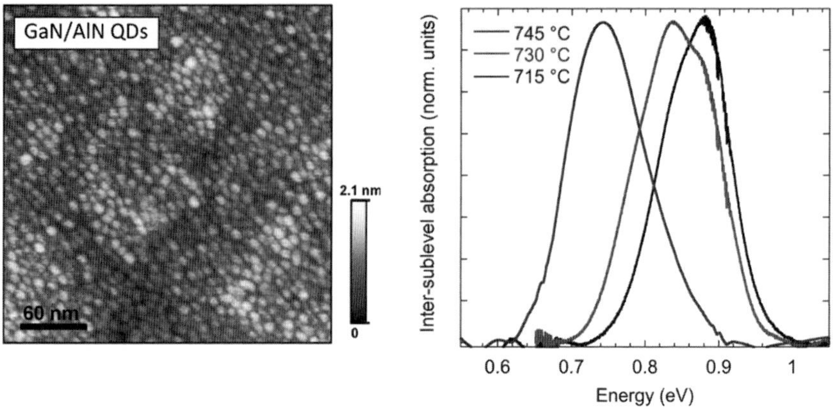

FIG. 12.11. Left: AFM image of GaN/AlN QDs grown on the sample surface. Right: Intersublevel absorption of twenty planes of Si-doped GaN/AlN QDs grown at different substrate temperatures. (After [Tche 05].)

In the following sections we review recent progress in the field of ISB devices based on nitride QWs and QDs.

12.4 GaN-based intersubband light modulators

12.4.1 *All-optical switches*

One remarkable feature of nitride-based materials is the extremely short recovery time of the ISB absorption. As pointed out by Suzuki and Iizuka [Suzu 98], this is a consequence of very efficient scattering processes via electron–LO-phonon interaction in these highly polar materials. The ISB absorption recovery time has been measured using time-resolved pump-probe techniques to be in the range of 140–400 fs in GaN/AlGaN single and coupled QWs at room temperature [Iizu 00, Gmac 01a, Hebe 02, Rapa 03, Hama 04, Iizu 05]. This offers prospects for ultrafast ISB devices operating at multi-Tbit/s data-rates.

One ISB device exploiting the ultrafast absorption recovery time is the all-optical switch, which was originally demonstrated at long infrared wavelengths using the GaAs/AlGaAs material system [Noda 93]. Suzuki *et al.* proposed making use of GaN/ AlGaN QWs for achieving all-optical switching at telecommunication wavelengths [Suzu 97, Suzu 00]. The principle of operation illustrated in Fig. 12.12 is based on the saturation of the ISB absorption by an intense control pulse. In the absence of the control pulse, the signal at the wavelength of the ISB transition is completely absorbed. In the presence of the control pulse, the signal experiences no attenuation because the ISB absorption is saturated. These ultrafast all-optical devices may become building blocks for future ultra-broadband all-optical networks with enhanced capacity and functionality,

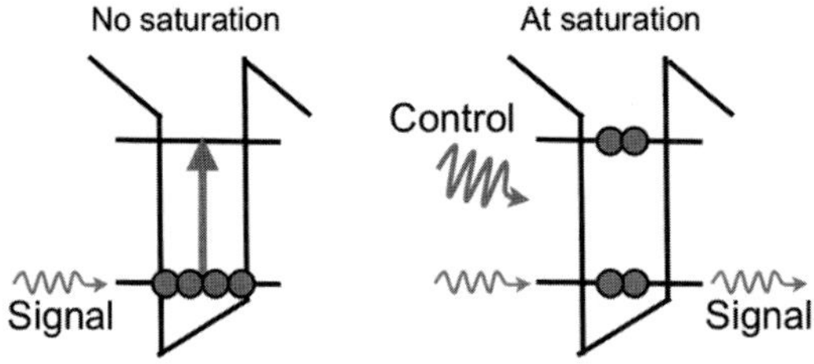

FIG. 12.12. Principle of operation of an all-optical switch based on ISB absorption saturation in GaN/AlGaN QWs.

in which the information is transmitted and processed in the optical domain [Wada 00].

Optical modulation due to the ISB absorption saturation in GaN/AlN multiple QWs embedded in a ridge waveguide structure grown by MBE was first demonstrated in 2004 by Iizuka *et al.* [Iizu 04]. Using a 120-pJ control pulse at 1.7 μm wavelength, the extinction ratio for the 1.55-μm signal pulse was measured to be 2.4 dB, and the device exhibited a response time of 0.36 ps. It was found that the extinction ratio could be further enhanced by improving the crystalline quality of the material in order to minimize the polarization-dependent losses (PDL) induced by the edge dislocations, which lead to attenuation of the control pulse propagating in the waveguide [Iizu 06b]. By combining MOCVD growth of the GaN lower cladding layer and MBE growth of the GaN/AlN QWs and upper cladding layer, a significant reduction of the PDL was observed. Under 150 (25) pJ control pulses, an extinction ratio of 11.5 (3) dB and a response time of 110 fs were demonstrated for 400-μm-long tapered waveguides [Iizu 05, Iizu 06a]. All-optical switching at a wavelength as short as 1.43 μm was reported for a waveguide with AlN cladding layers [Kumt 07, Shim 07].

In order to reduce the control pulse energy for switching, Iizuka *et al.* proposed to decrease the cross-section of the waveguide and to make use of spot-size conversion using tapered waveguides [Iizu 09]. An extinction ratio of 5 dB was demonstrated with only 25 pJ control pulse energy. Another strategy for reducing the control pulse energy was investigated by Li *et al.* [Li 07a], aimed at minimizing the optical losses induced by the high density of defects at the interface between the nitride buffer and sapphire substrate, as well as increasing the optical mode overlap within the active region. By using a waveguide with an AlN lower cladding, thirty periods of GaN/AlN QWs, and a GaN upper cladding, a 10-dB modulation depth at 1.55 μm wavelength could be achieved with only thirty-eight pJ control pulses [Li 07b]. Theoretical predictions show that the control pulse energy for switching could be further reduced by a factor over 30 using properly designed AlN/GaN/AlGaN coupled QWs [Li 06b].

Another approach for reducing the switching energy is to make use of GaN/AlN QDs. Indeed, it was pointed out that the electron dephasing time is significantly longer in GaN-based QDs than in QWs [Nevo 08], which should translate into a smaller saturation intensity. In addition, the large density of GaN/AlN QDs, which is typically achieved through MBE growth, ensures efficient intraband absorption. The ultrafast relaxation and absorption saturation has been investigated by Nevou *et al.* in a 200-period GaN/AlN QD superlattice [Nevo 09]. The recovery time of the s-p_z intraband absorption at $\lambda = 1.55$ μm was assessed using degenerate pump-probe experiments at room temperature. As shown in Fig. 12.13a, the measurements reveal a multi-exponential decay of the differential transmission with an ultrafast (~160 fs) and a slower (~1.5 ps) time constant attributed respectively to the excited state lifetime and to the thermalization time in the ground state.

Figure 12.13b shows the normalized transmittance through the QD sample as a function of the incident power density and intensity. The saturation intensity is estimated to be in the range 0.15–1.4 W/μm^2, which is significantly smaller than the value of 9.46 W/μm^2 reported for GaN/AlN QWs [Li 07b]. A detailed comparison of third-order non-linearities leading to absorption saturation between GaN/AlN QWs and QDs has been investigated in [Vald 08].

12.4.2 *Electro-optical modulators*

Electro-optical modulators are key components for the fiber-optics telecommunication systems at $\lambda = 1.55$ μm. The present driving technology relies on the quantum-confined Stark effect of interband absorption in InGaAsP-based QWs. These interband modulators allow high modulation bandwidths in excess of 100 GHz. It was predicted that ISB modulators would provide new technological solutions for even higher-bandwidth and low-driving-voltage devices

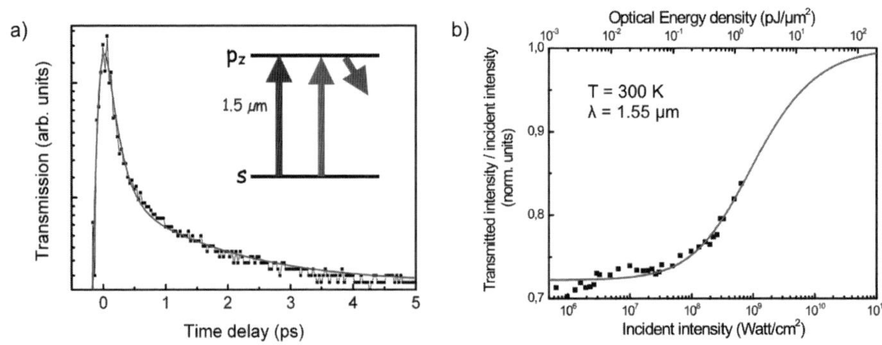

FIG. 12.13. a) – Differential transmission at 1.55 μm through a 200 period GaN/AlN QD superlattice as a function of the pump-probe delay. The solid line is a multi-exponential fit. b) – Normalized transmittance at 1.55 μm through a 200 period GaN/AlN QD superlattice as a function of the incident power density and intensity. (After [Nevo 09].)

almost insensitive to the input power and temperature [Holm 07]. In addition, they would provide better handling of chirp issues during the commutation process with respect to current technologies based on interband electro-absorption [Holm 06]. Frequency chirp broadens the spectral bandwidth of the modulated optical signal and introduces additional performance degradation when the signal propagates [Hui 09].

Electro-absorption ISB modulators have been investigated at long infrared wavelengths in the GaAs/AlGaAs material system based either on Stark shifting of ISB absorption [Harw 87], photoinduced ISB absorption [Juli 91], or charge-transfer between coupled QWs [Vodj 91]. In view of an electro-optical modulator at telecommunication wavelengths, Holmström proposed the use of the Stark shift of ISB absorption of GaN/AlGaN/AlN step QWs under an applied bias [Holm 06]. He predicted, for a 13-μm-long ridge-waveguide device containing four-step QWs, an *RC*-limited speed of f_{-3dB}=100 GHz with a 50 Ω shunt resistance, and a DC peak-to-peak voltage of 2.8 V for a 10-dB extinction ratio.

Nevou *et al.* reported on an ISB electro-absorption modulation device operating at room temperature, which relies on electron tunneling in GaN/AlN-coupled quantum wells (CQW) [Nevo 07a]. The active structure grown by RF plasma assisted MBE consists of a 3-nm-thick well, which acts as an electron reservoir, of a 1-nm-thick AlN coupling barrier, and of a narrow 1-nm-thick well designed to exhibit ISB absorption at 1.3 μm. The TEM image of three periods of the CQWs is displayed in Fig. 12.14b, showing the good reproducibility of the layer thickness. Similar structure has also been reproduced by using ammonia-MBE [Duss 10]. As illustrated in Fig. 12.14a, by applying a positive bias on the CQW structure, electrons are transferred from the reservoir well to the active well, which gives rise to electro-absorption modulation at $\lambda = 1.3$–1.6 μm. Under reverse bias, electrons tunnel back to the reservoir well, and the device becomes transparent at telecommunication wavelengths. Figure 12.15a shows room-temperature differential transmission spectra for positive and negative applied biases. For positive

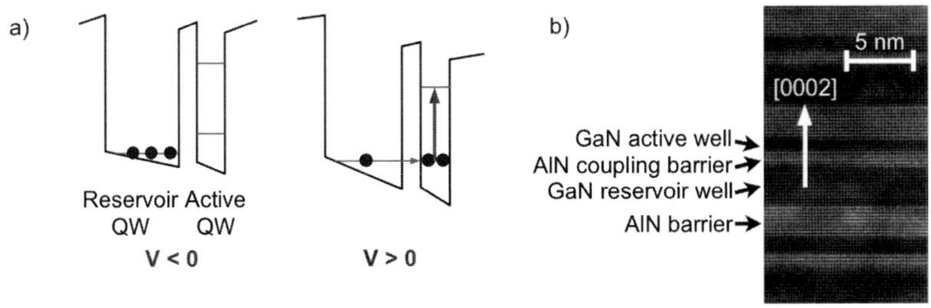

FIG. 12.14. a) Principle of operation of an electro-absorption modulator based on electron tunneling between GaN/AlN CQWs. b) High-resolution transmission electronic microscopic image of the CQW modulator structure. The dark (bright) regions correspond to GaN (AlN) layers. (After [Nevo 07a].)

applied biases, the ISB absorption at $\lambda = 2.3$ μm corresponding to the reservoir well decreases, while the absorption at $\lambda = 1.2$–1.67 μm, corresponding to the active-well increases. This reflects the charge transfer from the reservoir to the active QW. An opposite behavior is observed for the negative applied bias. Both modulation peaks at $\lambda = 2.3$ μm and $\lambda = 1.4$ μm exhibit a blueshift of ∼30 meV when the bias is increased from −30 V to +30 V. This blueshift is a consequence of the Stark shift of the ISB transition in the reservoir and active wells.

The response time of the CQW modulator was investigated by processing mesa devices with mesa sizes down to 15×15 μm^2 [Khei 08]. As shown in Fig. 12.15b, the optical modulation bandwidth at 1.3 μm wavelength increases when the mesa size is reduced, and is *RC*-limited at 3 GHz for the smaller mesas. The intrinsic speed of the CQW modulator is expected to be much faster, since the limiting mechanism is the electron tunneling time through the thin AlN coupling barrier, which is of the order of a few ps.

Another interesting approach for ISB light modulation was investigated by Baumann *et al.* [Baum 06b]. The structure resembles that of a high-electron-mobility transistor. A GaN/AlN superlattice with 1.5-nm-thick well and barrier layers was grown on top of a GaN buffer layer. A metallic gate contact was fabricated on top of a 100-nm-thick SiN dielectric layer deposited on the surface. Under negative applied bias, the superlattice is depleted and therefore transparent at

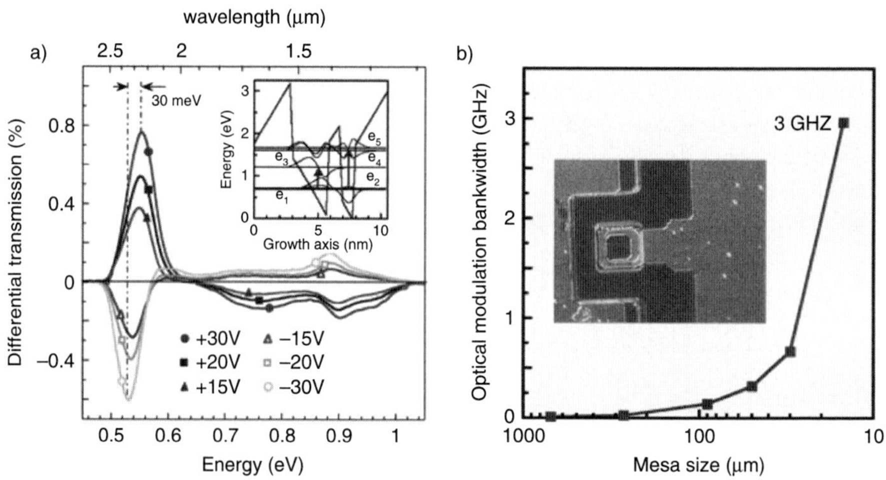

FIG. 12.15. a) Differential transmission spectra at 300 K of a 270×270 μm^2 mesa CQW modulator for different applied biases; inset: conduction-band potential profile showing the ISB absorption at 2.3 μm in the reservoir well and 1.38 μm in the active well. b) Optical modulation bandwidth as a function of mesa size. The inset shows a top-view microscope image of the mesa device. (After [Khei 08].)

1.5 μm. Only an ISB absorption at 2.2 μm was observed arising from the two-dimensional electron gas (2DEG) formed at the interface between the GaN buffer layer and the superlattice. By applying a positive gate bias, electrons are transferred from the 2DEG to the GaN/AlN QWs, giving rise to an additional ISB absorption at 1.5 μm.

For compatibility with fiber-optics technology it is desirable to integrate the electro-absorption modulator into an optical waveguide with a strong optical mode overlap within the active region to enhance the modulation depth. Figure 12.16 shows a sketch of a GaN-based ridge-waveguide ISB modulator relying on an electron population/depletion scheme [Mach 09]. The active region consists of only three periods of 1.3-nm-thick GaN wells with 3-nm-thick AlN barriers, and is inserted in the core of a 1-μm-thick $Al_{0.5}Ga_{0.5}N$ waveguiding layer. The operation of the device is illustrated in Fig. 12.16a. Under negative applied bias, the 3 QWs are depleted and the modulator is transparent. In turn, under positive applied bias, the ground state of each well is below the Fermi level, and the electron population of the wells gives rise to ISB absorption at 1.5 μm wavelength. The device was grown by RF plasma MBE on an AlN on *c*-sapphire template and processed into 2.6-mm-long and 50-μm-wide ridge waveguides with top and bottom metallic contacts. Figure 12.17a shows the waveguide transmission at room temperature under various applied biases measured with an injection-lensed fiber and a tunable semiconductor laser diode. For TE-polarized light, the transmission is practically flat and does not depend on applied bias. In contrast, for TM-polarized light the spectra reveal an ISB absorption increasing with applied bias as a result of the population of the ground state of the QWs. One can also notice a 32-meV blueshift of the absorption

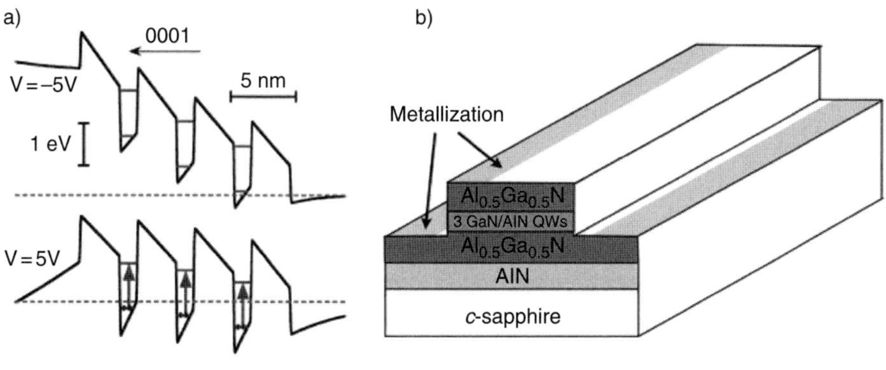

FIG. 12.16. a) Conduction-band profile of a 3 QW depletion modulator under −5 and 5 V applied bias as a function of the distance from the surface. The dashed line shows the Fermi energy. b) Sketch of the ridge-waveguide modulator device. Only part of the top contact layer is metallized to minimize plasmon losses at 1.5 μm. (After [Mach 09].)

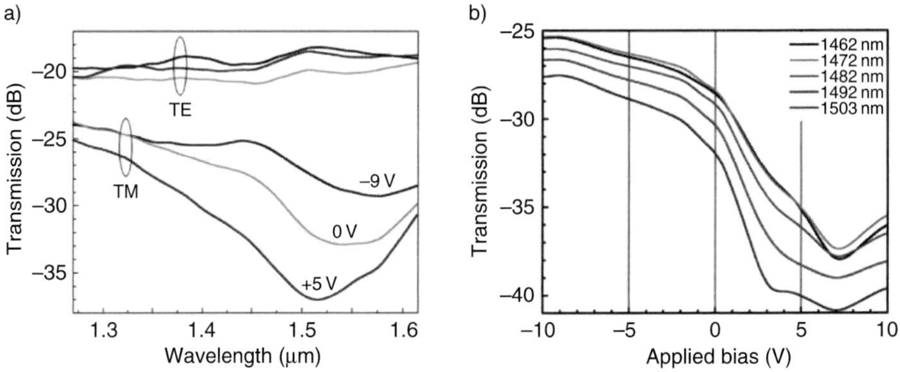

FIG. 12.17. a) Waveguide transmission spectra of the depletion modulator for TE- and TM-polarized light under an applied bias of −9 V, 0 V, and 5 V, respectively. b) TM-polarized transmission of the depletion modulator waveguide device versus static applied bias for various wavelengths. The transmission is not corrected for the coupling losses.

peak with increasing bias, which is due to the Stark shift of the ISB transition. Figure 12.17b shows the TM-polarized transmission of the modulator versus applied bias for different wavelengths. The modulation depth is of the order of 13.5 dB for −9V/+7V applied bias and 10 dB for 5 V voltage swing. Note that a value of 12 dB is required for optical modulators in order to achieve 10^{-12} bit error rates in current-technology fiber-optics data-transmission systems.

All modulation devices described so far rely on amplitude modulation of the ISB absorption. Based on Kramers–Kronig relations, the ISB absorption should also translate into a variation of the refractive index at wavelengths in the vicinity of the ISB resonance. Such an effect opens prospects for the realization of ISB phase modulators. Refractive index modulation by ISB transitions has been studied at long infrared wavelengths in GaAs/AlGaAs step QWs [Dupo 93]. Fast electrical control of ISB phase modulation would be desirable at 1.3–1.55 μm wavelengths in view of optical networks, switching, and routing devices. A few theoretical works have addressed the potential of GaN-based QWs for ISB refractive index modulation [Cen 09a]. The refractive index nonlinearities in GaN/AlN waveguides have been observed by Li *et al.*, presenting prospects for all-optical cross-phase modulators [Li 08].

ISB phase modulation at telecommunication wavelengths has been investigated using the same structure as the GaN-based waveguide depletion modulator shown in Fig. 12.16 [Lupu 12]. By analyzing the oscillations of the waveguide transmission spectrum due to the beating interference between the fundamental and higher-order modes, the refractive index variation associated with the ISB absorption of the 3 GaN/AlN QWs could be determined. Figure 12.18 shows

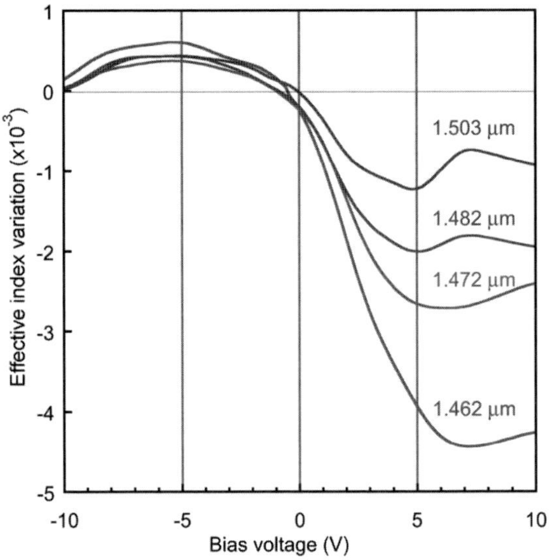

FIG. 12.18. Refractive index shift versus bias of the depletion modulator for different wavelengths. (After [Lupu 12].)

the refractive index shift versus bias at room temperature for wavelengths in the 1.462–1.503 μm range. The index variation with bias from complete depletion to full population of the QWs is around 5×10^{-3}. This value is similar to the typical index variation achieved in InP, and is an order of magnitude higher than the index variation obtained in silicon. It can be enhanced further by increasing the electron concentration in the GaN/AlN QWs or the number of QWs.

12.5 GaN-based intersubband photodetectors

12.5.1 *Quantum well infrared photodetectors*

Quantum well infrared photodetectors (QWIP) rely on ISB absorption and photocurrent generation from the transfer of electrons photoexcited from the fundamental state of the QWs into the continuum under an applied bias [Levi 87, Levi 93]. QWIPs have been studied thoroughly in standard III–V semiconductors for long-infrared-wavelength focal-plane arrays and imagers. The first GaN-based QWIP was demonstrated in 2003 by Hofstetter *et al.* [Hofs 03]. The device grown by MBE on a *c*-sapphire substrate relies on twenty n-doped GaN/AlN QWs with well and barrier thickness of 2 nm sandwiched between two $Al_{0.5}Ga_{0.5}N$ contact layers. The photoresponse of the device peaked at 1.76 μm and allowed detection at 1.55 μm. The peak responsivity was estimated to be 0.1 mA/W at 10 K and the photocurrent could be observed up to T = 170 K. In 2005, Ushida *et al.* reported a GaN/AlN QWIP operating at room temperature at 1.47 μm with a top Schottky contact [Uchi 05]. The device contains 50 QWs with 1.42-nm-thick GaN wells and 3.14-nm-thick Si-doped AlN barriers. The maximum responsivity was estimated to be 0.11 mA/W for a DC bias of 15 V at room temperature. Photoconductive devices were also demonstrated at a peak wavelength of 1.6 μm using strain-compensated GaN/AlN superlattices [Baum 05a]. The large dark current associated with these photoconductive QWIPs probably arising from defect-mediated leakage paths prevented operation beyond 120 K. Later, Hofstetter *et al.* demonstrated photovoltaic QWIPs at telecommunication wavelengths operating up to room temperature, using GaN/AlN superlattices grown by MBE [Hofs 06] or MOCVD [Baum 06a]. These photovoltaic devices were shown able to detect a modulated laser beam at 1.55 μm at frequencies up to 2.37 GHz [Gior 07]. Although a combination of ISB absorption and tunneling effects within the superlattice were originally evoked to explain the photovoltaic mechanism of detection [Baum 05b], it was shown through a systematic study of short-period GaN/AlN superlattices with different well and barrier thicknesses that the photovoltaic detection mechanism arises from optical rectification [Hofs 07]. Such a second-order optical non-linear process is allowed in GaN/AlN QWs because of the asymmetric potential induced by the spontaneous and piezoelectric polarization discontinuity [Nevo 06c]. It was shown that by replacing the GaN/AlN QWs by QDs, a factor of 60 could be gained in terms of maximum responsivity, and that the temperature for peak responsivity could be as large as 160 K instead of 80 K for QWs [Hofs 10].

QWIP detectors have also been investigated at far-infrared wavelengths by Sudradjat *et al.* [Sudr 12]. The device relies on a double-step QW design, where two different (Al)GaN compositions are used both in the wells and in the barriers in order to create a virtually flat multiple-quantum-well potential-energy profile. The device exhibits photocurrent at a peak wavelength of 23 μm (13 THz frequency) up to 50 K, with a responsivity around 7 mA/W.

12.5.2 *Quantum dot infrared photodetectors*

The first quantum dot infrared photodetectors (QDIP) in the GaN/AlN material system were reported at liquid-nitrogen temperature by Doyennette *et al.* [Doye 05], and at room temperature by Vardi *et al.* [Vard 06]. Samples containing twenty-one periods of 1-nm-high GaN QDs with 3-nm-thick AlN barriers were grown by MBE on AlN-on-sapphire templates. The electron population of the QDs was provided by silicon doping of the GaN layer at a nominal concentration of 1×10^{20} cm^{-3}. The QD density was estimated to be in the range $5{-}7 \times 10^{11}$ cm^{-2}. All samples exhibit TM-polarized absorption in the 1.45–1.67 μm range ascribed to the intraband transition from the ground s state to the p_z excited state [Doye 06]. In contrast to standard QDIPs based on vertical transport, the operation of these QD detectors relies on in-plane conductivity of photoexcited electrons in the 2-ML-thick GaN wetting layer. The current is collected by two Ti/Al/Ti/Au metal contacts deposited on the

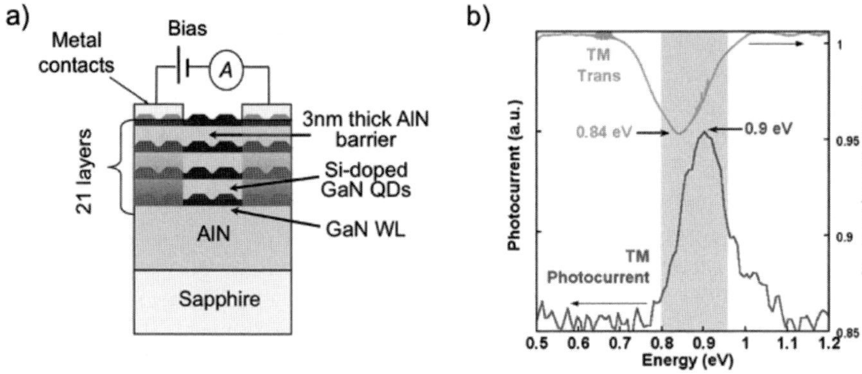

FIG. 12.19. a) Sketch of the GaN/AlN QDIP structure. b) TM-polarized transmission and photocurrent spectra at room temperature of a QDIP at 1.37 μm.

surface and annealed to enhance metal diffusion into the semiconductor layers, as illustrated in Fig. 12.19a.

Figure 12.19b shows transmission and photocurrent spectra at room temperature of a GaN/AlN QDIP at telecommunication wavelengths. The device exhibits photocurrent only for TM-polarized light, following the intraband s-p_z selection rules. Because the p_z state is at a lower energy than the WL groundstate, it was pointed out that the transfer of photoexcited electrons to the WL could involve the absorption of optical phonons [Doye 06]. This mechanism explains the blueshift of the photocurrent spectra with respect to the QD absorption. This interpretation was further supported by the observation of an exponential increase of the photocurrent with temperature. The peak responsivity of the GaN/AlN QDIPs was measured to be in the range 8–12 mA/W at 300 K under 10 V applied bias. Further studies have shown that deep levels in the AlN barrier layers may also contribute to the photocurrent, giving rise to negative photoconductivity effects [Vard 10].

Photocurrent under TE-polarized light attributed to bound-to-bound s-p_{xy} intraband transitions in GaN/AlN QDs has also been reported in the 5–8-μm mid-infrared spectral range [Vard 09]. The photocurrent mechanism was explained in terms of lateral electron-hopping, giving rise to in-plane conductivity.

12.5.3 *Quantum cascade photodetectors*

Quantum cascade detectors (QCD) are ISB photovoltaic devices, which can be operated at zero bias [Gend 04]. The unit cell of a QCD device consists of an active QW where ISB absorption takes place, and of an extractor stage used to transfer photoexcited electrons from one period to the next one, giving rise to a photovoltage in an open-circuit configuration, or to a photocurrent if the device is loaded on a resistor. The absence of dark current stemming from the operation at zero bias is particularly favorable to enhance the signal-to-noise ratio and to improve the dynamic range and integration time of infrared imagers. Another appealing feature of QCDs is their intrinsically low capacitance, which enables high-frequency response. QCDs have been demonstrated in several material systems: namely, GaAs/AlGaAs, InGaAs/InAlAs, and InGaAs/AlAsSb, operating at wavelengths in the range 2.1–84 μm [Gior 09]. For mid-infrared QCDs, the design of the extractor stage relies on multiple QWs whose thickness is sequentially increased to achieve a ladder of ground states separated by the energy of one LO phonon. This design allows for fast electron transfer through the extractor via enhanced electron–LO-phonon interaction.

GaN-based quantum cascade photodetectors The first GaN-based QCD was demonstrated in 2008 by Vardi *et al.* [Vard 08a]. In contrast to GaAs- or InP-based QCDs, the design of the extractor region makes use of the polarization discontinuity between GaN, AlN, and AlGaN. Figure 12.20a shows the conduction-band profile of this GaN-based QCD. The active QW consists of an n-doped 6 ML-(1.56-nm)-thick GaN well with a 4-ML-(1-nm)-thick AlN barrier. The extractor is formed by a superlattice containing 4-ML-thick $Al_{0.25}Ga_{0.75}N$ wells and 4-ML-thick AlN barriers. The composition of the AlGaN layers has been chosen to engineer the band-bending in the extractor region in order to achieve an energy ladder of the ground states of the extractor QWs with an energy separation close to the LO-phonon energy. The device grown by MBE contains forty periods of active regions between two $Al_{0.25}Ga_{0.75}N$ contact layers n-doped with Si at 1×10^{19} cm^{-3}.

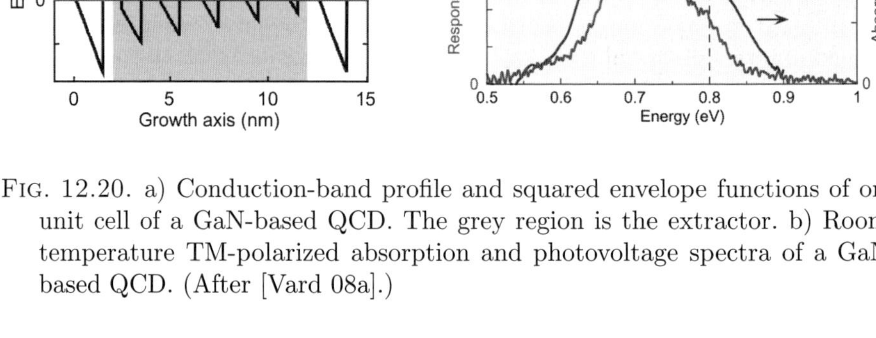

FIG. 12.20. a) Conduction-band profile and squared envelope functions of one unit cell of a GaN-based QCD. The grey region is the extractor. b) Room-temperature TM-polarized absorption and photovoltage spectra of a GaN-based QCD. (After [Vard 08a].)

Figure 12.20b shows the ISB absorption and photovoltage spectrum measured at room temperature under TM-polarized irradiation. The absorption spectrum is structured into three peaks corresponding to the absorption of active QWs with a thickness of 5, 6, and 7 MLs, respectively. The photovoltage is maximum at a wavelength of 1.7 μm, which corresponds to an active QW thickness of 6 MLs. The fact that the responsivity is smaller for a QW thickness of 5 or 7 MLs is explained by the misalignment of the active QW excited state with respect to the highest energy state in the extractor. The calibrated peak responsivity was measured to be ~10 mA/W (~1000 V/W) at $\lambda = 1.7$ μm. The responsivity decreases from its maximum value by a factor of ~3 at $\lambda = 1.5$ μm.

In order to tune the peak QCD responsivity at 1.3–1.55 μm, Sakr *et al.* have reported an optimized design of the extractor and active QWs [Sakr 10]. The active region is composed of a 1.5-nm-thick GaN QW doped with silicon at 5×10^{19} cm^{-3} and 1-nm-thick AlN barriers. The extractor consists of four periods of 1-nm-thick $Al_{0.3}Ga_{0.7}N$ QWs with 1.5-nm-thick AlN barriers. The QCD structure with forty periods sandwiched between Si-doped $Al_{0.3}Ga_{0.7}N$ contact layers was grown by MBE. Figure 12.21a shows a TEM image, which reveals abrupt interfaces as well as good reproducibility of the layer thickness. As shown in Fig. 12.21b, the photovoltage spectrum at room temperature is peaked at 1.42 and 1.50 μm, with a full width at half maximum of 120 meV ($\lambda = 1.37$–1.57 μm) in line with the absorption spectrum.

The presence of the internal field offers an additional degree of freedom for the design of GaN-based QCDs, making the band engineering quite flexible. It also provides a symmetry-breaking of the potential in the active QW, which permits ISB transitions not only between the ground e_1 and first excited state e_2 but also

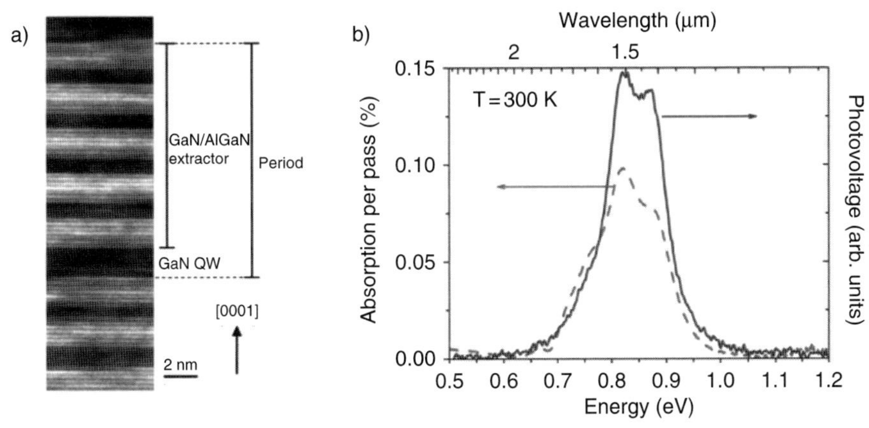

FIG. 12.21. a) Transmission electron microscopy (TEM) image of one unit cell of a 1.5-μm QCD. b) Photovoltage (solid curve) and absorption (dashed curve) spectra of the 1.5-μm QCD for TM-polarized light. (After [Sakr 10].)

between e_1 and the second excited state e_3, the latter transition being forbidden in a symmetric potential. This feature has been exploited for demonstrating a two-color GaN/AlGaN QCD [Sakr 12a]. The design relies on an active GaN QW supporting three bound states and on a multi-QW stage with two state ladders for efficient extraction of electrons from the e_2 and e_3 states of the active QW (see Fig. 12.22a). A sample containing forty periods sandwiched between two Si-doped $Al_{0.3}Ga_{0.7}N$ contact layers was grown by ammonia-MBE on an AlN-on-c-sapphire template. Figure 12.22b presents the ISB photovoltage spectrum of the QCD device at room temperature. Two photovoltage peaks are observed at 1.7 and 1 μm under TM-polarized illumination. These two peaks correspond to the photovoltage generated by the e_1e_2 and e_1e_3 transitions, respectively, in agreement with the ISB absorption measurements and with the simulation of the band diagram. It should be noted that the 1-μm detection wavelength is the shortest value ever reported for an ISB detector, whatever the material system.

Another illustration of the flexibility of the design of GaN-based QCDs is the demonstration of simplified QCD device with an alloy extractor [Sakr 12b]. As shown in Fig. 12.23a, the sophisticated multiple QW extractor region is replaced by an $Al_{0.58}Ga_{0.42}N$ thick layer, whose composition is chosen to produce an internal field of −0.4 MV/cm and achieve a graded potential. A forty-period structure was grown by ammonia MBE on an AlN-on-c-sapphire template. The active QW consists of an 8-ML-thick GaN layer n-doped with Si and two AlN barriers with a thickness of 2 nm and 1 nm, in growth order respectively. The thinner AlN barrier is designed to force photoexcited electrons to predominantly tunnel to the right extractor in growth order. As seen in Fig. 12.23a, the active QW supports two bound states while the thick alloy layer supports more than ten

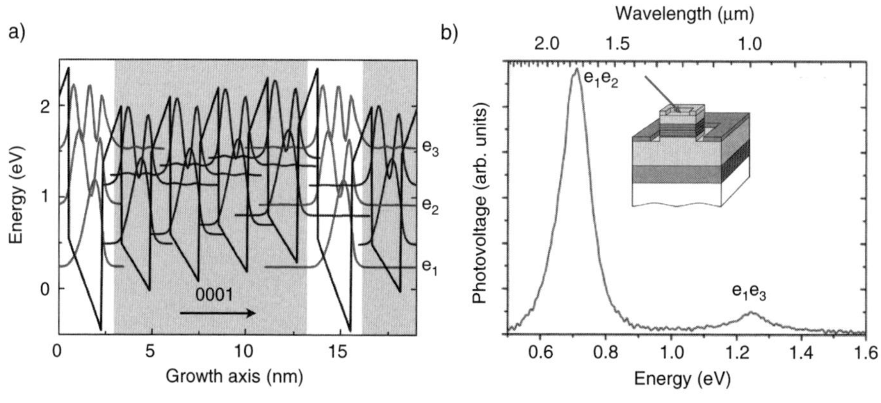

FIG. 12.22. a) Conduction-band diagram of one period of a two-color QCD. b) Photovoltage spectrum of the two-color QCD for TM-polarized light. (After [Sakr 12a].)

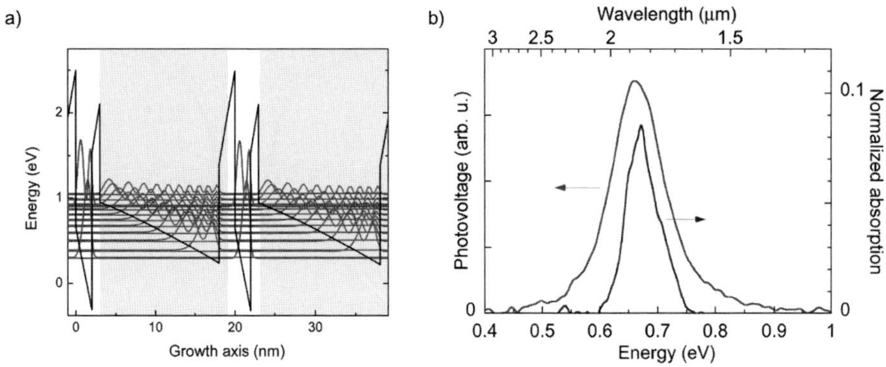

FIG. 12.23. a) Conduction-band diagram of two periods of an alloy extractor QCD. b) Photovoltage and absorption spectrum of the QCD for TM-polarized light. (After [Sakr 12b].)

bound states, with some of the excited states in close resonance with the active QW excited state. The thickness and composition of the alloy extractor have been chosen in order to achieve an energy separation between the ground state of the extractor and that of the next-period active QW close to the LO-phonon energy. As illustrated in Fig. 12.23b, the device exhibits at room temperature both TM-polarized ISB absorption and photovoltage at a peak wavelength of 1.9 μm. This simplified design offers many advantages in terms of robustness and growth simplification. It is also easily scalable in terms of target wavelength.

Ultrafast operation of quantum cascade photodetectors GaN-based QCDs are intrinsically very fast devices. The intrinsic speed limitation is determined by the carrier transit time from one period to the next one across the extractor stage. The intrinsic speed of GaN/AlGaN QCDs has been assessed by time-resolved degenerate pump-probe measurements at 1.55 μm wavelength of the device photocurrent [Vard 11].

Figure 12.24 shows the photocurrent as a function of the pump-probe delay. The dip at zero pump-probe delay is due to the saturation of the ISB absorption in the active QWs. The dynamics of the photocurrent is governed by competing mechanisms, i.e. the ISB relaxation of photoexcited electrons to the ground state of the active QW and the tunneling of electrons into the extractor followed by the transit through the extractor. Based on a rate equation model [Vard 11], the best fit to the experimental measurements, shown by the solid line in Fig. 12.24, corresponds to an ISB scattering time in the active QW of 0.1 ps and a transit time through the extractor of 1 ps. The intrinsic frequency bandwidth of the GaN-based QCD is then expected to be above 160 GHz.

In practical QCD devices, the frequency bandwidth is limited by the extrinsic RC time constant, where R is the access resistance and C the active region

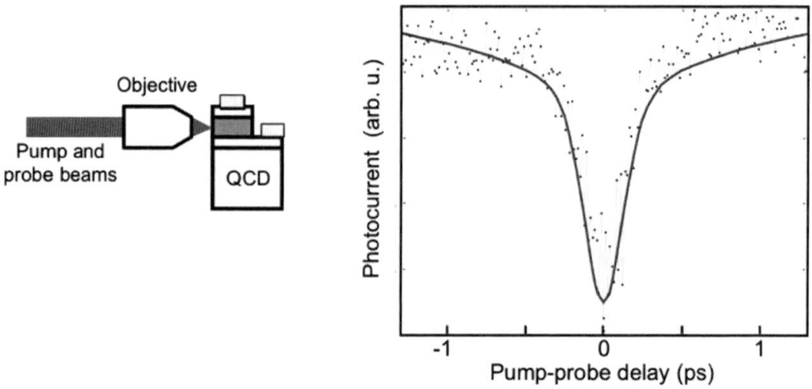

FIG. 12.24. Pump-probe measurements of the QCD photocurrent with 200 fs pulses at 1.55 μm wavelength (left). QCD photocurrent at room temperature versus the pump-probe delay (right). The peak intensity of the pump and probe beams is around 270 MW/cm^2. The dots are experimental data. The solid curve is a theoretical fit. (After [Vard 11].)

capacitance. The electro-optical frequency response of GaN/AlGaN QCDs was first investigated by Vardi *et al.* [Vard 08b]. The devices based on the design described in [Vard 08a] were processed in the form of square mesas with 50 Ω coplanar access lines for the top and bottom metallic contacts. The size of the mesas was 25 × 25 and 17 × 17 μm^2. The central part of the top surface of the mesas was kept free from metallization in order to allow illumination at a 45-angle of incidence onto the surface. The electro-optical response was investigated by measuring the photocurrent at $\lambda = 1.55$ μm under illumination by an RF-modulated laser diode. As shown in Fig. 12.25, the frequency response is that of a first-order *RC* filter with a cut-off frequency at −3 dB of 11.4 GHz and 19.7 GHz for the 25 × 25 μm^2 and 17 × 17 μm^2 mesa detectors, respectively. The value of the access resistance and of the active region capacitance was estimated from separate S-parameter measurements. It was found that the capacitance scales with the mesa surface, and is as low as 45 fF for the 17 × 17 μm^2 mesas. The access resistance was deduced to be 210 (180) Ω for the 17 × 17 μm^2 (25 × 25 μm^2) mesas.

In order to push the cut-off frequency even further, it is necessary to decrease both the capacitance and the contact resistance. The capacitance can be reduced by decreasing the device surface or by increasing the number of periods of active regions. In order to improve the top contact resistance, one solution is to metallize the whole top surface of the mesa QCDs. This imposes illumination of the mesa QCD from the side facet. Figure 12.26a shows a top view of a QCD processed along these guidelines [Sakr 13]. The sample structure is the same as shown in Fig. 12.21 (described in Ref [Sakr 10]). The QCDs with a peak detection wavelength of 1.5 μm are processed in the form of square mesas with a size of

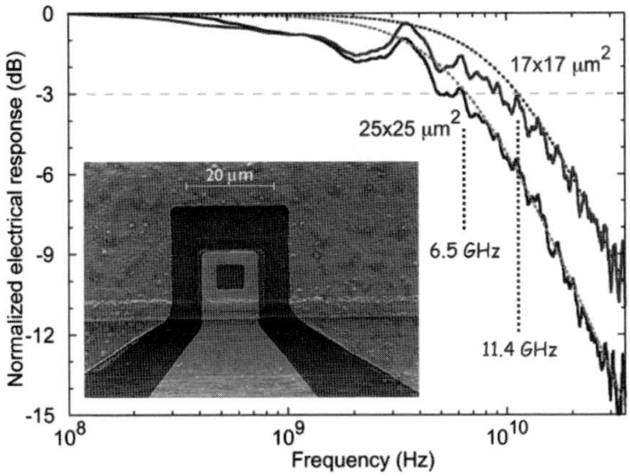

FIG. 12.25. Frequency response of the 25 × 25 and 17 × 17 μm^2 mesa QCDs at room temperature under 45° angle surface illumination. The dotted lines show the simulated frequency response using the access resistance and capacitance values deduced from S-parameter measurements. The inset shows a top-view scanning electron microscope image of the 17 × 17 μm^2 mesa QCD. (After [Vard 08b].)

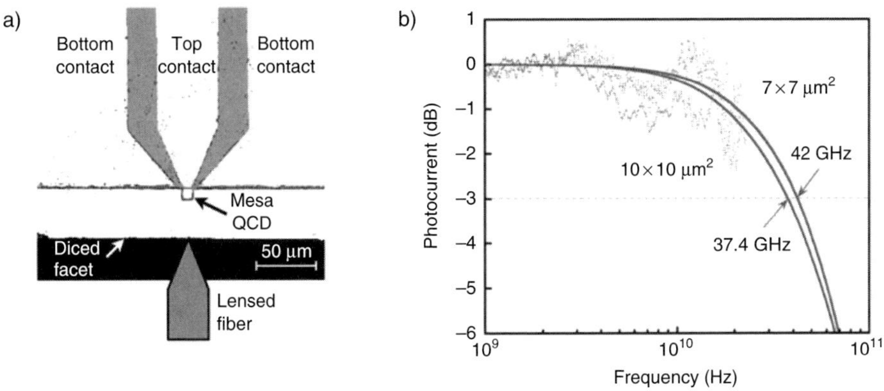

FIG. 12.26. a) Top-view microscopic image of the ultrafast mesa QCD. b) Frequency response of the 10 × 10 μm^2 and 7 × 7 μm^2 mesa QCDs at room temperature under side-facet illumination at 1.55 μm wavelength by a lensed fiber. The dots are measurements. The solid lines are the fits based on the impedance values deduced from S-parameter measurements. (After [Sakr 13].)

$10 \times 10\ \mu m^2$ and $7 \times 7\ \mu m^2$ along with RF impedance-matched access lines. The samples are then diced and polished as close as possible to the mesa detector facet in order to allow illumination by a lensed fiber at normal incidence on the side facet. Figure 12.26b shows the frequency response of the photocurrent. As seen, the −3 dB frequency bandwidth has been increased with respect to previous work to 37.4 and 42 GHz for the $10 \times 10\ \mu m^2$ and $7 \times 7\ \mu m^2$ mesa detectors, respectively. These values set a new record for ISB detectors, whatever the material system.

12.6 GaN-based intersubband light emitters

ISB light emission has been widely investigated in various III–V material systems, such as InGaAs/InAlAs, GaAs/AlGaAs or InAs/lnSb, in the mid-infrared spectral range and in the THz frequency domain. These intense research efforts translate into the demonstration of high-performance unipolar lasers relying on either current injection or optical pumping, such as the quantum cascade laser [Fais 94] or the quantum fountain laser (QFL) [Gaut 97]. III-nitride semiconductors, because of their large conduction-band offset, offer the potential to push the ISB emission to much shorter wavelengths, while offering prospects for high-performance lasers at THz frequencies, because of the large LO-phonon energy in this material system. ISB light emission has been investigated in III-nitride-based QWs or QDs using linear or non-linear optical processes. In the following we review the progress achieved so far.

12.6.1 *ISB light generation in GaN-based QWs through non-linear optical processes*

In semiconductor QWs, the second-order non-linear susceptibility is known to be strongly enhanced, when the waves are in close resonance with ISB transitions [Bouc 90]. However, since the second-order non-linear susceptibility is identically zero in a potential with inversion symmetry, an asymmetric potential is required to benefit from the ISB resonance enhancement. This is the case for hexagonal-phase GaN/AlGaN QWs grown along the *c*-axis which naturally exhibit an asymmetric saw-tooth potential. Large second-order non-linear susceptibilities were predicted by Liu *et al.* [Liu 00] in GaN/AlGaN QWs at near-infrared wavelengths due to ISB resonant enhancement.

The observation of room-temperature light emission by an ISB second-harmonic generation process (SHG) was first reported by Nevou *et al.* in lightly doped GaN/AlN QWs grown on AlN/*c*-sapphire templates [Nevo 06c]. The sample contained 200 periods of 10-ML (2.6 nm) thick GaN wells separated by 3-nm-thick AlN spacer layers. The QWs were designed to exhibit three bound states in the conduction band with almost identical e_1e_2 and e_2e_3 ISB transition energy corresponding to a wavelength of $\sim$2 μm. Under optical pumping by an optical parametric oscillator (OPO), light generation at a wavelength of $\sim$1 μm was observed. The harmonic power spectrum versus pump wavelength clearly showed a resonant enhancement of SHG at the pump wavelength of ~2 μm, which corresponds to the double-resonance condition with the e_1e_2 and e_2e_3 ISB transitions. The conversion efficiency was found to be sixteen times larger in the QW sample than in a reference bulk GaN sample, and could be enhanced further by increasing the doping concentration in the QWs.

12.6.2 *Intraband emission of GaN-based quantum dots via a resonant Raman process*

The emission properties of GaN/AlN QDs associated with intraband radiative recombinations have been studied by Nevou *et al.* [Nevo 08]. The investigated sample consisted of 200 periods of GaN/AlN QD layers with a dot height and diameter of 1.7 ± 0.1 nm and 6 ± 1 nm, respectively. The sample exhibited TM-polarized s-p_z intraband absorption peaking at 1.55 μm. Under TM-polarized optical excitation of the side facet at a wavelength of 1.37 μm, a back-scattered emission at 1.48 μm was observed at room temperature. The emission was found to be TM-polarized as expected from the s-p_z transition selection rules. This emission wavelength, made possible by the large conduction-band offset offered by GaN/AlN heterostructures, is the shortest value ever reported for an ISB emitter, whatever the material system.

The intraband emission was ascribed to a resonant Raman process involving zero-momentum LO phonons, as illustrated in Fig. 12.27. Such resonant Raman emission process is expected to be selective in terms of dot height, since it involves

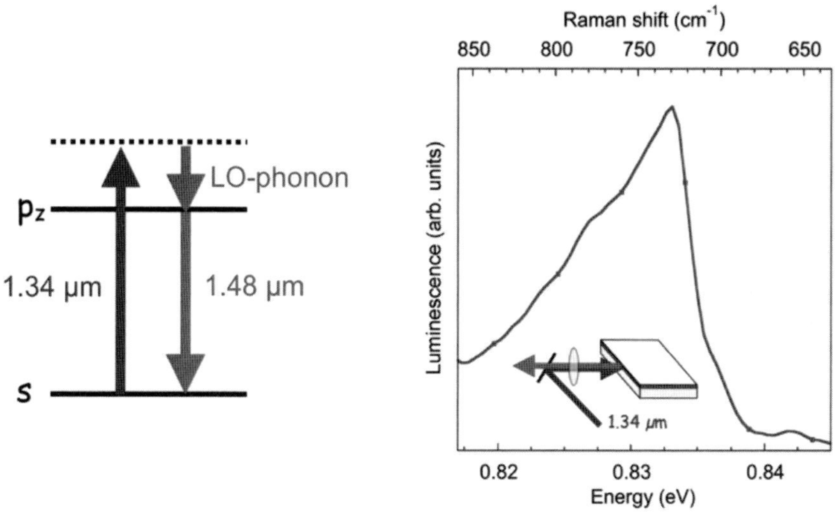

FIG. 12.27. Resonant Raman intraband emission at 1.48 μm from a GaN/AlN QD superlattice at room temperature. (After [Nevo 08].)

only the QDs within the size distribution, which can satisfy the energy conservation requirement. As a consequence, it was pointed out that the observed broadening of the emission, ~9 meV, should reflect the homogeneous linewidth of the p_z-s transition. It was also predicted that the resonant Raman process can provide population inversion and stimulated gain under intense optical excitation [Nevo 08]. In order to obtain further insight into the homogeneous broadening of the s-p_z intraband transition, non-linear hole burning measurements have been performed on the same sample [Nguy 10]. A value of the homogeneous linewidth of ~15 meV has been obtained at a temperature of 5 K, which is significantly smaller than the homogeneous broadening achieved in GaN/AlN QWs (~40 meV estimated from absorption measurements [Nevo 06b]).

12.6.3 *Intersubband luminescent devices*

It is well known that in contrast to its interband counterpart ISB luminescence is not an efficient process. The radiative time associated with ISB spontaneous emission is in the nanosecond range in the near-infrared, and in the tenths of a microsecond range in the THz domain. In contrast, non-radiative ISB scattering at short infrared wavelengths is dominated by electron–LO-phonon interactions and proceeds on a subpicosecond time-scale, while at long infrared wavelengths it is dominated by electron–electron or electron-impurity scattering as well as interface roughness or alloy disorder diffusion and proceeds on tenths of a picosecond. In both cases, this translates into a very weak ISB luminescence efficiency. It should be noted that this does not hinder the realization of high-performance ISB lasers, since, in the population inversion regime, short stimulated radiative time as well as large stimulated gain can be achieved because of the strong oscillator strength associated with ISB transitions.

Room-temperature ISB luminescence from GaN/AlN QWs at near-infrared wavelengths around 2 μm was first reported by Nevou *et al.* [Nevo 06a, Nevo 07b]. The investigated samples contained 200–250 periods of 2.1-nm (8-ML) thick GaN wells separated by 3-nm-thick AlN spacer layers. As shown in Fig. 12.28a, the QWs support three bound electron states. Under continuous-wave optical excitation at 0.98 μm wavelength in resonance with the e_1e_3 ISB absorption, a TM-polarized emission related to the e_3e_2 ISB transition was observed at room temperature, as shown in Fig. 12.28b. The peak wavelength was as short as 2.1 μm. The internal quantum efficiency was estimated to be ~0.3 μW/W of pump power. This low value is a consequence of the adverse effect of non-radiative electron scattering mechanisms. Accounting for the small collection efficiency, the actual power of the detected emission was around 10 pW. A large enhancement of the detected emission power (0.3 μW) could be achieved by Driscoll *et al.*, under intense pulsed optical excitation by an optical parametric oscillator [Dris 09]. The investigated sample contained 200 periods of Si-doped GaN/AlN QWs with a slightly smaller QW thickness (1.8 nm), leading to an ISB emission wavelength of 2.05 μm.

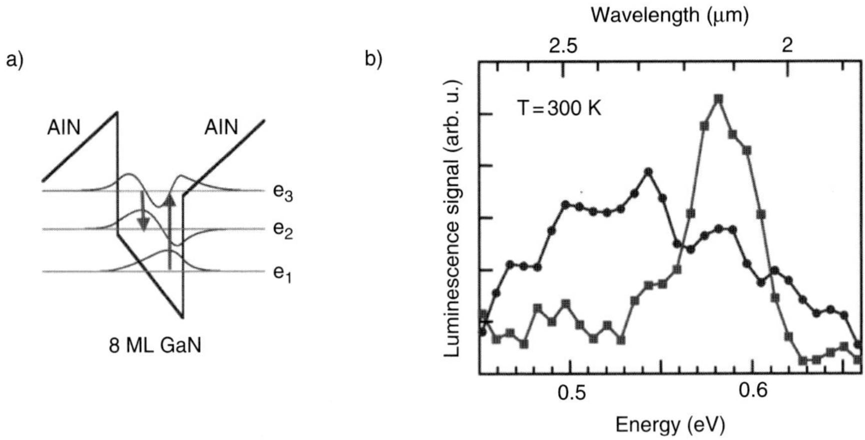

FIG. 12.28. a) Conduction-band profile and energy levels of 8-ML-thick GaN/AlN QWs. Optical pumping of the e_1e_3 ISB transition leads to e_3e_2 ISB light emission. b) Room-temperature intersubband luminescence spectrum of samples with 200 periods (squares) and 250 periods (circles) under excitation at 0.98 μm wavelength. (After [Nevo 07b].)

12.6.4 *Towards THz quantum cascade lasers*

Although ISB luminescence has been observed at near-infrared wavelengths, demonstration of lasing action in III-nitride based heterostructures through either electrical injection or optical pumping remains quite challenging in this spectral range, beside some theoretical propositions to make use of Raman processes [Sun 06]. One reason stems from the ultrafast non-radiative ISB scattering via LO-phonon emission, which is expected to impose very large injection-current or pumping-power density to achieve population inversion. It was pointed out in a few theoretical works that the situation would be much more favorable at long infrared wavelengths corresponding to the THz frequency domain [Jova 04, Sun 05, Bell 08, Bell 09]. In GaAs-based materials currently used for THz QCLs, there is a considerable slow-down of non-radiative scattering mechanisms when the ISB transition energy is not sufficient for allowing emission of optical phonons [Kohl 01].

It was also shown that GaN-based materials paves the way for THz QCLs operating at temperatures much above room temperature [Bell 08, Bell 09]. Indeed, one intrinsic reason limiting the operation of GaAs-based THz QCLs to cryogenic temperatures is the small energy of the LO phonon in GaAs (36 meV, 8.2 THz). As the temperature is increased, electrons in the upper lasing subband acquire sufficient thermal energy for activating fast (<1 ps) non-radiative relaxations via LO-phonon emission towards the lower lasing subband, hence ruining the population inversion. As illustrated in Fig. 12.29, GaN-based QCLs should not suffer

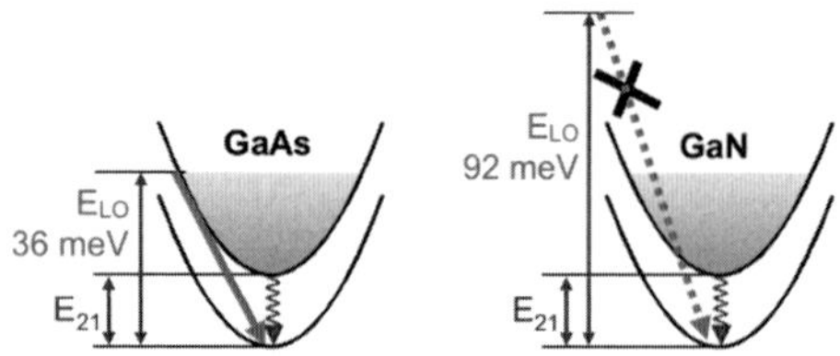

FIG. 12.29. Non-radiative scattering of thermal electrons by LO phonons between lasing subbands of THz QCLs in the case of GaAs contrary to GaN.

from these parasitic non-radiative channels because of the large energy of the LO phonons (~90 meV), almost three times higher than in GaAs. In addition, standard III–V semiconductor devices (such as GaAs-based QCLs) cannot be operated in the 4.6–11 THz frequency range because of their Reststrahlen band, i.e. the spectral region where the material is completely opaque due to absorption by optical phonons. The larger energy of optical phonons in GaN offers prospects for THz quantum cascade devices, which can operate in a much broader spectral range from 1 to 15 THz.

The first attempt to fabricate GaN-based QCLs at a target frequency of 7.5 THz was reported by Terashima and Hirayama using GaN/InAlGaN QWs and then GaN/ $Al_{0.2}Ga_{0.8}N$ QWs grown by RF MBE [Tera 09, Tera 10]. The authors claimed the observation of THz spontaneous emission at low temperatures, but the emitted power was too small to allow measurement of the emission spectrum. By growing a similar GaN/$Al_{0.2}Ga_{0.8}N$ QCL structure on a low-dislocation-density free-standing GaN substrate, a strong improvement of the emitted power was reported, as well as the observation of a quite sharp electroluminescent line at 1.37 THz [Tera 11]. All these designs were based on a resonant phonon scheme where a fast depopulation of the lower lasing state is provided by LO-phonon emission. However, it was shown by Yashida *et al.* that this resonant phonon scheme may not be the most appropriate design for providing sufficient gain for lasing action, because of the broadening of the subband levels caused by the very strong interaction between electrons and LO phonons in GaN [Yasu 12].

Observation of ISB electroluminescence at THz frequencies has been reported using in-plane transport of electrons at room temperature in GaN/AlGaN step QWs [Juli 11]. The device relied on forty periods of $Al_{0.1}Ga_{0.9}N$/GaN/$Al_{0.05}Ga_{0.85}N$ step QWs exhibiting ISB absorption at 2.1 THz [Mach 10]. Two ohmic contacts have been processed on the surface of the device to contact the 2D electron gas in the step QWs. Under application of a bias, some electrons in the ground subband with sufficient in-plane kinetic energy could be scattered into the first excited subband, giving rise to an ISB spontaneous emission peaking at 2 THz, as illustrated in Fig. 12.30. It should be noted that although this mechanism opens prospects for fast modulated THz light sources, it cannot provide population inversion.

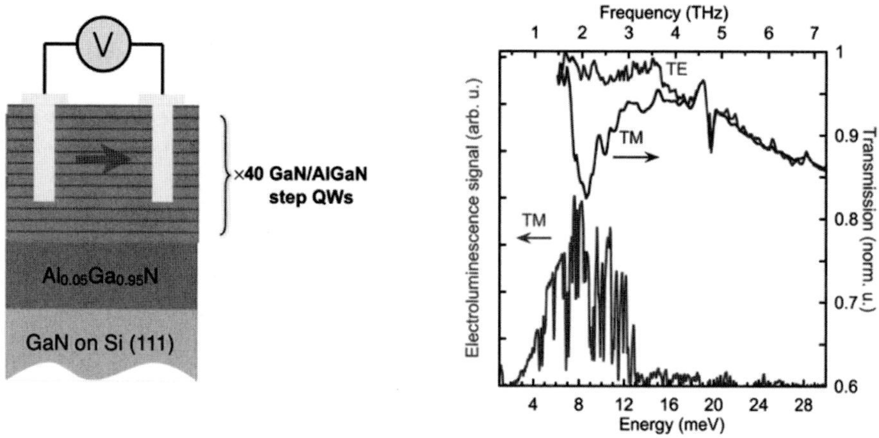

FIG. 12.30. Left: Scheme of the electroluminescent sample relying on in-plane transport. Right: TM-polarized electroluminescent spectrum at room temperature under 15 V applied bias (bottom curve) and transmission spectrum at 4 K for TM- and TE-polarized light (top curves).

One requirement for GaN-based QCLs is the control of vertical transport, which has proved to be elusive until recently. One main difficulty stems from the large densities of dislocations and other structural defects that are typically found in III-nitride materials, especially when grown on highly lattice-mismatched substrates. Such defects can act as strong scattering centers as well as leakage current paths. Negative differential resistance (NDR) has been observed at room temperature in GaN/AlGaN double-barrier structures and attributed to resonant tunneling [Kiku 02, Bely 04, Golk 06, Bayr 10]. However, the current–voltage characteristics exhibit hysteresis with two levels of current, and the NDR is not reproducible for positive and negative voltage sweeps as well as for multiple scans, and disappears at low temperatures. These anomalies have been related to the presence of activated filling of trap states in double-barrier resonant tunneling diodes [Sakr 11].

More convincing results have been achieved using GaN/AlGaN multiple QWs. In particular, sequential resonant tunneling has been reported in superlattices grown on free-standing GaN substrate containing GaN/AlGaN-coupled QW structures, and THz emission has been observed, though it could not be unambiguously ascribed to ISB emission [Sudr 10]. In a recent work, vertical electron transport has been investigated in a seven-period GaN/AlN MQW structure, leading to the observation at room temperature of five reproducible NDR features assigned to the successive resonant electron tunneling between the ground (e_1) and the first excited (e_2) states of adjacent QWs in a quantum cascade-like configuration [Sakr 12c].

12.7 Conclusions

In this chapter we have reviewed the field of ISB transitions in III-nitride semiconductors, which has emerged within the last fifteen years following the progress in terms of epitaxial growth of these materials with atomic-layer thickness control. The basic properties of ISB transitions in nitrides have been presented, with special emphasis on the effect of internal electric fields. We have also presented an overview of the insight gained on physical properties through spectroscopic investigations of ISB transitions in nitride QWs and QDs. Finally, we have described the progress towards ultrafast ISB devices at fiber-optics telecommunication wavelengths, as well as recent developments towards ISB devices operating in the THz frequency domain, which will certainly constitute a driving line of research in future years.

References

[Akim 01] R. Akimoto, Y. Kinpara, K. Akita, F. Sasaki, and S. Kobayashi. "Short-wavelength intersubband transitions down to 1.6 μm in ZnSe/BeTe type-II superlattices". *Applied Physics Letters*, Vol. 78, p. 580, 2001.

[Akim 02] R. Akimoto, K. Akita, F. Sasaki, and S. Kobayashi. "Short-wavelength (λ <2 μm) intersubband absorption dynamics in ZnSe/BeTe quantum wells". *Applied Physics Letters*, Vol. 80, p. 2433, 2002.

[Alle 76] S. Allen, D. Tsui, and B. Vinter. "Absorption of Infrared Radiation by Electrons in Semiconductor Inversion Layers". *Solid State Communications*, Vol. 20, No. 4, pp. 425–428, 1976.

[Ando 77] T. Ando. "Inter-subband optical transitions in a surface space-charge layer". *Solid State Communications*, Vol. 21, No. 1, pp. 133–136, 1977.

[Andr 00] A. Andreev and E. O'Reilly. "Theory of the electronic structure of GaN/AlN hexagonal quantum dots". *Physical Review B*, Vol. 62, No. 23, p. 15851, 2000.

[Andr 01] A. Andreev and E. O'Reilly. "Optical transitions and radiative lifetime in GaN/AlN self-organized quantum dots". *Applied Physics Letters*, Vol. 79, p. 521, 2001.

[As 09] D. As. "Cubic group-III nitride-based nanostructures: basics and applications in optoelectronics". *Microelectronics Journal*, Vol. 40, No. 2, pp. 204–209, 2009.

[Band 88] K. Bandara, D. Coon, B. O, Y. Lin, and M. Francombe. "Exchange Interactions in Quantum Well Subbands". *Applied Physics Letters*, Vol. 53, No. 20, pp. 1931–1933, 1988.

[Band 89] K. Bandara. "Correction". *Applied Physics Letters*, Vol. 55, No. 2, p. 206, 1989.

[Bast 81] G. Bastard. "Superlattice band structure in the envelope-function approximation". *Physical Review B*, Vol. 24, No. 10, p. 5693, 1981.

[Bast 82] G. Bastard. "Theoretical investigations of superlattice band structure in the envelope-function approximation". *Physical Review B*, Vol. 25, No. 12, p. 7584, 1982.

[Bast 88] G. Bastard. "Wave mechanics applied to semiconductor heterostructures". *Editions de Physique*, Paris, 1988.

[Baum 05a] E. Baumann, F. Giorgetta, D. Hofstetter, H. Lu, X. Chen, W. Schaff, L. Eastman, S. Golka, W. Schrenk, and G. Strasser. "Intersubband photoconductivity at 1.6 μm using a strain-compensated AlN/ GaN superlattice". *Applied Physics Letters*, Vol. 87, p. 191102, 2005.

[Baum 05b] E. Baumann, F. Giorgetta, D. Hofstetter, H. Wu, W. Schaff, L. Eastman, and L. Kirste. "Tunneling effects and intersubband absorption in AlN/GaN superlattices". *Applied Physics Letters*, Vol. 86, p. 032110, 2005.

[Baum 06a] E. Baumann, F. Giorgetta, D. Hofstetter, S. Golka, W. Schrenk, G. Strasser, L. Kirste, S. Nicolay, E. Feltin, J. Carlin, *et al.* "Near infrared absorption and room temperature photovoltaic response in AlN/ GaN superlattices grown by metal–organic vapor-phase epitaxy". *Applied Physics Letters*, Vol. 89, p. 041106, 2006.

[Baum 06b] E. Baumann, F. Giorgetta, D. Hofstetter, S. Leconte, F. Guillot, E. Bellet-Amalric, and E. Monroy. "Electrically adjustable intersubband absorption of a GaN/AlN superlattice grown on a transistor-like structure". *Applied Physics Letters*, Vol. 89, p. 101121, 2006.

[Bayr 10] C. Bayram, Z. Vashaei, and M. Razeghi. "AlN/GaN double-barrier resonant tunneling diodes grown by metal–organic chemical vapor deposition". *Applied Physics Letters*, Vol. 96, No. 4, 2010.

[Bell 08] E. Bellotti, K. Driscoll, T. Moustakas, and R. Paiella. "Monte Carlo study of GaN versus GaAs terahertz quantum cascade structures". *Applied Physics Letters*, Vol. 92, p. 101112, 2008.

[Bell 09] E. Bellotti, K. Driscoll, T. D. Moustakas, and R. Paiella. "Monte Carlo simulation of terahertz quantum cascade laser structures based on wide-bandgap semiconductors". *Journal of Applied Physics* , Vol. 105, No. 11, 2009.

[Bely 04] A. Belyaev, O. Makarovsky, D. Walker, L. Eaves, C. Foxon, S. Novikov, L. Zhao, R. Dykeman, S. Danylyuk, S. Vitusevich, M. Kappers, J. Barnard, and C. Humphreys. "Resonance and current instabilities in AlN/GaN resonant tunnelling diodes". *Physica E: Low-Dimensional Systems & Nanostructures*, Vol. 21, No. 2–4, pp. 752–755, 2004.

[BenD 66] D. BenDaniel and C. Duke. "Space-charge effects on electron tunneling". *Physical Review*, Vol. 152, No. 2, p. 683, 1966.

[Bern 97] F. Bernardini, V. Fiorentini, and D. Vanderbilt. "Spontaneous polarization and piezoelectric constants of III–V nitrides". *Physical Review B*, Vol. 56, pp. R10024–R10027, 1997.

[Bern 98] F. Bernardini and V. Fiorentini. "Macroscopic polarization and band offsets at nitride heterojunctions". *Physical Review B*, Vol. 57, pp. R9427–R9430, 1998.

[Blos 89] W. Bloss. "Effects of Hartree, exchange, and correlation energy on intersubband transitions". *Journal of Applied Physics*, Vol. 66, No. 8, pp. 3639–3642, 1989.

[Bouc 90] P. Boucaud, F. Julien, D. Yang, J. Lourtioz, E. Rosencher, P. Bois, and J. Nagle. "Detailed Analysis of 2nd-Harmonic Generation Near 10.6 μm in GaAs/AlGaAs Asymmetric Quantum-Wells". *Applied Physics Letters*, Vol. 57, No. 3, pp. 215–217, 1990.

[Braz 06] R. Brazis and R. Raguotis. "Monte Carlo modeling of phonon-assisted carrier transport in cubic and hexagonal gallium nitride". *Optical and quantum electronics*, Vol. 38, No. 4, pp. 339–347, 2006.

[Butt 07] R. Butté, J.-F. Carlin, E. Feltin, M. Gonschorek, S. Nicolay, G. Christmann, D. Simeonov, A. Castiglia, J. Dorsaz, H. J. Buehlmann, S. Christopoulos, G. B. H. von Hgersthal, A. J. D. Grundy, M. Mosca, C. Pinquier, M. A. Py, F. Demangeot, J. Frandon, P. G. Lagoudakis, J. J. Baumberg, and N. Grandjean. "Current status of AlInN layers lattice-matched to GaN for photonics and electronics". *Journal of Physics D: Applied Physics*, Vol. 40, No. 20, p. 6328, 2007.

[Cen 09a] L. Cen, B. Shen, Z. Qin, and G. Zhang. "Influence of polarization induced electric fields on the wavelength and the refractive index of intersubband transitions in AlN/GaN coupled double quantum wells". *Journal of Applied Physics*, Vol. 105, No. 9, p. 093109, 2009.

[Cen 09b] L. Cen, B. Shen, Z. Qin, and G. Zhang. "Near-infrared two-color intersubband transitions in AlN/GaN coupled double quantum wells". *Journal of Applied Physics*, Vol. 105, No. 5, p. 053106, 2009.

[Chak 05] A. Chakraborty, T. Baker, B. Haskell, F. Wu, J. Speck, S. Denbaars, S. Nakamura, and U. Mishra. "Milliwatt power blue InGaN/GaN light-emitting diodes on semipolar GaN templates". *Japanese Journal of Applied Physics Part 2: Letters & Express Letters*, Vol. 44, No. 28–32, pp. L945–L947, 2005.

[Cham 04] V. Chamard, T. Schülli, M. Sztucki, T. Metzger, E. Sarigiannidou, J. Rouviere, M. Tolan, C. Adelmann, and B. Daudin. "Strain distribution in nitride quantum dot multilayers". *Physical Review B*, Vol. 69, No. 12, p. 125327, 2004.

[Cong 07] G. Cong, R. Akimoto, K. Akita, T. Hasama, and H. Ishikawa. "Intersubband absorption with different sublevel couplings in [(CdS/ZnSe/BeTe)/(ZnSe/BeTe)] double quantum wells". *Applied Physics Letters*, Vol. 90, No. 18, p. 181919, 2007.

[Cywi 06] G. Cywiński, C. Skierbiszewski, A. Fedunieiwcz-Żmuda, M. Siekacz, L. Nevou, L. Doyennette, M. Tchernycheva, F. Julien, P. Prystawko, M. Kryśko, *et al.* "Growth of thin AlInN/GaInN quantum wells for applications to high-speed intersubband devices at telecommunication wavelengths". *Journal of Vacuum Science & Technology B: Microelectronics and Nanometer Structures*, Vol. 24, p. 1505, 2006.

[Cywi 08] G. Cywinski, C. Skierbiszewski, M. Siekacz, A. Feduniewicz-Zmuda, M. Krysko, M. Gladysiewicz, R. Kudrawiec, L. Nevou, N. Kheirodin, F. Julien, *et al.* "Enhancement of Intersubband Absorption in GaInN/AlInN Quantum Wells". *Acta Physica Polonica A*, Vol. 114, p. 1093, 2008.

[Daud 97] B. Daudin, F. Widmann, G. Feuillet, Y. Samson, M. Arlery, and J. Rouviere. "Stranski–Krastanov growth mode during the molecular beam epitaxy of highly strained GaN". *Physical Review B*, Vol. 56, No. 12, pp. 7069–7072, 1997.

[DeCu 07] E. DeCuir Jr, E. Fred, M. Manasreh, J. Schörmann, D. As, and K. Lischka. "Near-infrared intersubband absorption in nonpolar cubic GaN/AlN superlattices". *Applied Physics Letters*, Vol. 91, p. 041911, 2007.

[DeCu 08] E. DeCuir, M. Manasreh, E. Tschumak, J. Schörmann, D. As, and K. Lischka. "Cubic GaN/AlN multiple quantum well photodetector". *Applied Physics Letters*, Vol. 92, No. 20, p. 201910, 2008.

[Doye 05] L. Doyennette, L. Nevou, M. Tchernycheva, A. Lupu, F. Guillot, E. Monroy, R. Colombelli, and F. Julien. "GaN-based quantum dot infrared photodetector operating at 1.38 μm". *Electronics Letters*, Vol. 41, No. 19, pp. 1077–1078, 2005.

[Doye 06] L. Doyennette, A. Vardi, F. Guillot, L. Nevou, M. Tchernycheva, A. Lupu, R. Colombelli, G. Bahir, E. Monroy, and F. Julien. "Intraband photodetection at 1.3–1.5 μm in self-organized GaN/AlN quantum dots". *Physica Status Solidi (b)*, Vol. 243, No. 15, pp. 3993–3997, 2006.

[Dris 07] K. Driscoll, A. Bhattacharyya, T. Moustakas, R. Paiella, L. Zhou, and D. Smith. "Intersubband absorption in AlN/GaN/AlGaN coupled quantum wells". *Applied Physics Letters*, Vol. 91, No. 14, p. 141104, 2007.

[Dris 09] K. Driscoll, Y. Liao, A. Bhattacharyya, L. Zhou, D. Smith, T. Moustakas, and R. Paiella. "Optically pumped intersubband emission of short-wave infrared radiation with GaN/AlN quantum wells". *Applied Physics Letters*, Vol. 94, p. 081120, 2009.

[Dupo 93] E. Dupont, D. Delacourt, and M. Papuchon. "Mid-infrared phase modulation via Stark effect on intersubband transitions in GaAs/GaAlAs quantum wells". *IEEE Journal of Quantum Electronics*, Vol. 29, No. 8, pp. 2313–2318, 1993.

[Duss 10] A. Dussaigne, S. Nicolay, D. Martin, A. Castiglia, N. Grandjean, L. Nevou, H. Machhadani, M. Tchernycheva, L. Vivien, F. Julien, *et al.* "Growth of intersubband GaN/AlGaN heterostructures". *Proceedings of SPIE*, Vol. 7608, p. 76080H, 2010.

[Fais 94] J. Faist, F. Capasso, D. Sivco, C. Sirtori, A. Hutchinson, and A. Cho. "Quantum cascade laser". *Science*, Vol. 264, No. 5158, pp. 553–556, 1994.

[Fock 30] V. Fock. "Approximation method for the solution of the quantum mechanical multibody problems". *Zeitschrift Fur Physik*, Vol. 61, No. 1-2, pp. 126–148, 1930.

[Gaut 97] O. Gauthier-Lafaye, P. Boucaud, F. Julien, S. Sauvage, S. Cabaret, J. Lourtioz, V. Thierry-Mieg, and R. Planel. "Long-wavelength (15.5 μm) unipolar semiconductor laser in GaAs quantum wells". *Applied Physics Letters*, Vol. 71, p. 3619, 1997.

[Gend 04] L. Gendron, M. Carras, A. Huynh, V. Ortiz, C. Koeniguer, and V. Berger. "Quantum cascade photodetector". *Applied Physics Letters*, Vol. 85, p. 2824, 2004.

[Gior 07] F. Giorgetta, E. Baumann, F. Guillot, E. Monroy, and D. Hofstetter. "High frequency (f = 2.37 GHz) room-temperature operation of 1.55 μm AlN/GaN-based intersubband detector". *Electronics Letters*, Vol. 43, No. 3, pp. 185–186, 2007.

[Gior 09] F. Giorgetta, E. Baumann, M. Graf, Q. Yang, C. Manz, K. Kohler, H. Beere, D. Ritchie, E. Linfield, A. Davies, *et al.* "Quantum Cascade Detectors". *IEEE Journal of Quantum Electronics*, Vol. 45, No. 8, pp. 1039–1052, 2009.

[Gmac 00] C. Gmachl, H. Ng, S. Chu, and A. Cho. "Intersubband absorption at $\lambda \sim 1.55$ μm in well- and modulation-doped GaN/AlGaN multiple quantum wells with superlattice barriers". *Applied Physics Letters*, Vol. 77, p. 3722, 2000.

[Gmac 01a] C. Gmachl, S. Frolov, H. Ng, S. Chu, and A. Cho. "Sub-picosecond electron scattering time for $\lambda = 1.55$ μm intersubband transitions in GaN/AlGaN multiple quantum wells". *Electronics Letters*, Vol. 37, No. 6, pp. 378–380, 2001.

[Gmac 01b] C. Gmachl, H. Ng, and A. Cho. "Intersubband absorption in degenerately doped GaN/AlGaN coupled double quantum wells". *Applied Physics Letters*, Vol. 79, p. 1590, 2001.

[Golk 06] S. Golka, C. Pflugl, W. Schrenk, G. Strasser, C. Skierbiszewski, M. Siekacz, I. Grzegory, and S. Porowski. "Negative differential resistance in dislocation-free GaN/AlGaN double-barrier diodes grown on bulk GaN". *Applied Physics Letters*, Vol. 88, No. 17, 2006.

[Guil 06] F. Guillot, E. Bellet-Amalric, E. Monroy, M. Tchernycheva, L. Nevou, L. Doyennette, F. Julien, T. Remmele, M. Albrecht, T. Shibata, *et al.* "Si-doped GaN/ AlN quantum dot superlattices for optoelectronics at telecommunication wavelengths". *Journal of Applied Physics*, Vol. 100, p. 044326, 2006.

[Gunn 76] O. Gunnarsson and B. Lundqvist. "Exchange and correlation in atoms, molecules, and solids by the spin-density-functional formalism". *Physical Review B*, Vol. 13, No. 10, p. 4274, 1976.

[Hama 04] J. Hamazaki, S. Matsui, H. Kunugita, K. Ema, H. Kanazawa, T. Tachibana, A. Kikuchi, and K. Kishino. "Ultrafast intersubband relaxation and nonlinear susceptibility at 1.55 μm in GaN/AlN multiple-quantum wells". *Applied Physics Letters*, Vol. 84, p. 1102, 2004.

[Hama 05] J. Hamazaki, H. Kunugita, K. Ema, A. Kikuchi, and K. Kishino. "Intersubband relaxation dynamics in GaN/ AlN multiple quantum wells studied by two-color pump-probe experiments". *Physical Review B*, Vol. 71, No. 16, p. 165334, 2005.

[Harw 87] A. Harwit and J. Harris. "Observation of Stark shifts in quantum well intersubband transitions". *Applied Physics Letters*, Vol. 50, No. 11, pp. 685–687, 1987.

[Hebe 02] J. Heber, C. Gmachl, H. Ng, and A. Cho. "Comparative study of ultrafast intersubband electron scattering times at 1.55 μm wavelength in GaN/AlGaN heterostructures". *Applied Physics Letters*, Vol. 81, p. 1237, 2002.

[Hedi 71] L. Hedin and B. Lundqvist. "Explicit local exchange-correlation potentials". *Journal of Physics C: Solid State Physics*, Vol. 4, p. 2064, 1971.

[Helm 03a] A. Helman, F. Fossard, M. Tchernycheva, K. Moumanis, A. Lusson, F. Julien, B. Damilano, N. Grandjean, J. Massies, C. Adelman, *et al.*

"Intraband spectroscopy of self-organized GaN/AlN quantum dots". *Physica E: Low-Dimensional Systems and Nanostructures*, Vol. 17, pp. 60–63, 2003.

[Helm 03b] A. Helman, M. Tchernycheva, A. Lusson, E. Warde, F. Julien, K. Moumanis, G. Fishman, E. Monroy, B. Daudin, D. Dang, *et al.* "Intersubband spectroscopy of doped and undoped GaN/AlN quantum wells grown by molecular-beam epitaxy". *Applied Physics Letters*, Vol. 83, p. 5196, 2003.

[Helm 99] M. Helm. "The Basic Physics of Intersubband Transitions". In: H. Liu and F. Capasso, Eds., *Intersubband Transitions in Quantum Wells: Physics and Device Applications I*, Academic Press, 1999.

[Hira 94] Y. Hirayama, J. Smet, L. Peng, C. Fonstad, and E. Ippen. "Feasibility of 1.55 μm Intersubband Photonic Devices Using InGaAs/AlAs Pseudomorphic Quantum Well Structures". *Japanese Journal of Applied Physics Part 1: Regular Papers, Short Notes & Review Papers*, Vol. 33, No. 1B, pp. 890–895, 1994.

[Hofs 03] D. Hofstetter, S. Schad, H. Wu, W. Schaff, and L. Eastman. "GaN/AlN-based quantum-well infrared photodetector for 1.55 μm". *Applied Physics Letters*, Vol. 83, p. 572, 2003.

[Hofs 06] D. Hofstetter, E. Baumann, F. Giorgetta, M. Graf, M. Maier, F. Guillot, E. Bellet-Amalric, and E. Monroy. "High-quality AlN/ GaN-superlattice structures for the fabrication of narrow-band 1.4 μm photovoltaic intersubband detectors". *Applied Physics Letters*, Vol. 88, p. 121112, 2006.

[Hofs 07] D. Hofstetter, E. Baumann, F. Giorgetta, F. Guillot, S. Leconte, and E. Monroy. "Optically nonlinear effects in intersubband transitions of GaN/AlN-based superlattice structures". *Applied Physics Letters*, Vol. 91, No. 13, p. 131115, 2007.

[Hofs 10] D. Hofstetter, J. Di Francesco, P. Kandaswamy, A. Das, S. Valdueza-Felip, and E. Monroy. "Performance improvement of AlN–GaN-based intersubband detectors by using quantum dots". *Photonics Technology Letters, IEEE*, Vol. 22, No. 15, pp. 1087–1089, 2010.

[Holm 06] P. Holmström. "Electroabsorption modulator using intersubband transitions in GaN-AlGaN-AlN step quantum wells". *IEEE Journal of Quantum Electronics*, Vol. 42, No. 8, pp. 810–819, 2006.

[Holm 07] P. Holmström, X. Liu, H. Uchida, T. Aggerstam, A. Kikuchi, K. Kishino, S. Lourdudoss, T. Andersson, and L. Thylén. "Intersubband photonic devices by group-III nitrides". *Proceedings of SPIE*, Vol. 6782, p. 67821N, 2007.

[Hosh 02] K. Hoshino, T. Someya, K. Hirakawa, and Y. Arakawa. "Observation of intersubband transition from the first to the third subband (e1e3) in GaN/AlGaN quantum wells". *Physica Status Solidi (a)*, Vol. 192, No. 1, pp. 27–32, 2002.

[Hui 09] R. Hui and M. O'Sullivan. *Fiber Optic Measurement Techniques*. Academic Press, 2009.

[Iizu 00] N. Iizuka, K. Kaneko, N. Suzuki, T. Asano, S. Noda, and O. Wada. "Ultrafast intersubband relaxation ($\leq$150 fs) in AlGaN/GaN multiple quantum wells". *Applied Physics Letters*, Vol. 77, p. 648, 2000.

[Iizu 02] N. Iizuka, K. Kaneko, and N. Suzuki. “Near-infrared intersubband absorption in GaN/AlN quantum wells grown by molecular beam epitaxy”. *Applied Physics Letters*, Vol. 81, p. 1803, 2002.

[Iizu 04] N. Iizuka, K. Kaneko, and N. Suzuki. “Sub-picosecond modulation by intersubband transition in ridge waveguide with GaN/AlN quantum wells”. *Electronics Letters*, Vol. 40, No. 15, pp. 962–963, 2004.

[Iizu 05] N. Iizuka, K. Kaneko, and N. Suzuki. “Sub-picosecond all-optical gate utilizing an intersubband transition”. *Optics Express*, Vol. 13, No. 10, pp. 3835–3840, 2005.

[Iizu 06a] N. Iizuka, K. Kaneko, and N. Suzuki. “All-optical switch utilizing intersubband transition in GaN quantum wells”. *IEEE Journal of Quantum Electronics*, Vol. 42, No. 8, pp. 765–771, 2006.

[Iizu 06b] N. Iizuka, K. Kaneko, and N. Suzuki. “Polarization dependent loss in III-nitride optical waveguides for telecommunication devices”. *Journal of Applied Physics*, Vol. 99, p. 093107, 2006.

[Iizu 09] N. Iizuka, H. Yoshida, N. Managaki, T. Shimizu, S. Hassanet, C. Cumtornkittikul, M. Sugiyama, and Y. Nakano. “Integration of GaN/AlN all-optical switch with SiN/AlN waveguide utilizing spot-size conversion”. *Optics Express*, Vol. 17, No. 25, pp. 23247–23253, 2009.

[Jova 04] V. Jovanovic, D. Indjin, Z. Ikonic, and P. Harrison. “Simulation and design of GaN/AlGaN far-infrared ($\lambda \sim 34$ μm) quantum-cascade laser”. *Applied Physics Letters*, Vol. 84, No. 16, pp. 2995–2997, 2004.

[Juli 11] F. Julien, S. Sakr, H. Machhadani, M. Tchernycheva, R. Colombelli, E. Warde, J. Mangeney, Y. Kotsar, E. Bellet-Amalric, E. Sarigiannidou, and E. Monroy. “Latest developments towards GaN/AlGaN THz quantum cascade lasers”. *International Conference on Intersubband Transitions in Quantum Wells ITQW11, Badesi, Italy*, 11–17 September 2011.

[Juli 91] F. Julien, P. Vagos, J. Lourtioz, D. Yang, and R. Planel. “Novel all-optical 10 μm waveguide modulator based on intersubband absorption in GaAs/AlGaAs quantum wells”. *Applied Physics Letters*, Vol. 59, No. 21, pp. 2645–2647, 1991.

[Kamg 74] A. Kamgar, P. Kneschaurek, G. Dorda, and J. F. Koch. “Resonance spectroscopy of electronic levels in a surface accumulation layer”. *Physical Review Letters*, Vol. 32, pp. 1251–1254, 1974.

[Kand 08] P. Kandaswamy, F. Guillot, E. Bellet-Amalric, E. Monroy, L. Nevou, M. Tchernycheva, A. Michon, F. Julien, E. Baumann, F. Giorgetta, *et al.* “GaN/AlN short-period superlattices for intersubband opto-electronics: A systematic study of their epitaxial growth, design, and performance”. *Journal of Applied Physics*, Vol. 104, No. 9, p. 093501, 2008.

[Kand 09] P. Kandaswamy, H. Machhadani, C. Bougerol, S. Sakr, M. Tchernycheva, F. Julien, and E. Monroy. “Mid-infrared intersubband absorption in GaN/AlGaN superlattices on Si (111) templates”. *Applied Physics Letters*, Vol. 95, p. 141911, 2009.

[Kand 10] P. Kandaswamy, H. Machhadani, Y. Kotsar, S. Sakr, A. Das, M. Tchernycheva, L. Rapenne, E. Sarigiannidou, F. Julien, and

E. Monroy. "Effect of doping on the mid-infrared intersubband absorption in GaN/AlGaN superlattices grown on Si (111) templates". *Applied Physics Letters*, Vol. 96, p. 141903, 2010.

[Khei 08] N. Kheirodin, L. Nevou, H. Machhadani, P. Crozat, L. Vivien, M. Tchernycheva, A. Lupu, F. Julien, G. Pozzovivo, S. Golka, *et al.* "Electrooptical Modulator at telecommunication wavelengths based on GaN–AlN coupled quantum wells". *Photonics Technology Letters, IEEE*, Vol. 20, No. 9, pp. 724–726, 2008.

[Kiku 02] A. Kikuchi, R. Bannai, K. Kishino, C. Lee, and J. Chyi. "AlN/GaN double-barrier resonant tunneling diodes grown by rf-plasma-assisted molecular-beam epitaxy". *Applied Physics Letters*, Vol. 81, No. 9, pp. 1729–1731, 2002.

[Kish 02a] K. Kishino, A. Kikuchi, H. Kanazawa, and T. Tachibana. "Intersubband absorption at λ 1.2–1.6 µm in GaN/AlN multiple quantum wells grown by rf-plasma molecular beam epitaxy". *Physica Status Solidi (a)*, Vol. 192, No. 1, pp. 124–128, 2002.

[Kish 02b] K. Kishino, A. Kikuchi, H. Kanazawa, and T. Tachibana. "Intersubband transition in (GaN)/(AlN) superlattices in the wavelength range from 1.08 to 1.61 µm". *Applied Physics Letters*, Vol. 81, p. 1234, 2002.

[Kohl 01] R. Kohler, R. Iotti, A. Tredicucci, and F. Rossi. "Design and simulation of terahertz quantum cascade lasers". *Applied Physics Letters*, Vol. 79, No. 24, pp. 3920–3922, 2001.

[Kudr 08] R. Kudrawiec, M. Motyka, G. Cywiński, M. Siekacz, C. Skierbiszewski, L. Nevou, L. Doyennette, M. Tchernycheva, F. Julien, and J. Misiewicz. "Contactless electroreflectance spectroscopy of inter-and intersubband transitions in AlInN/GaInN quantum wells". *Physica Status Solidi (c)*, Vol. 5, No. 2, pp. 503–507, 2008.

[Kumt 07] C. Kumtornkittikul, T. Shimizu, N. Iizuka, N. Suzuki, M. Sugiyama, and Y. Nakano. "AlN waveguide with GaN/AlN quantum wells for all-optical switch utilizing intersubband transition". *Japanese Journal of Applied Physics*, Vol. 46, pp. L352–L355, 2007.

[Laho 07] L. Lahourcade, E. Bellet-Amalric, E. Monroy, M. Abouzaid, and P. Ruterana. "Plasma-assisted molecular-beam epitaxy of AlN (11-22) on *m* sapphire". *Applied Physics Letters*, Vol. 90, No. 13, p. 131909, 2007.

[Laho 08a] L. Lahourcade, P. Kandaswamy, J. Renard, P. Ruterana, H. Machhadani, M. Tchernycheva, F. Julien, B. Gayral, and E. Monroy. "Interband and intersubband optical characterization of semipolar (11-22)-oriented GaN/AlN multiple-quantum-well structures". *Applied Physics Letters*, Vol. 93, p. 111906, 2008.

[Laho 08b] L. Lahourcade, J. Renard, B. Gayral, E. Monroy, M. Chauvat, and P. Ruterana. "Ga kinetics in plasma-assisted molecular-beam epitaxy of GaN (11-22): Effect on the structural and optical properties". *Journal of Applied Physics*, Vol. 103, p. 093514, 2008.

[Leav 91] R. Leavitt. "Empirical two-band model for quantum wells and superlattices in an electric field". *Physical Review B*, Vol. 44, No. 20, p. 11270, 1991.

[Lei 08a] S. Lei, Z. Dong, B. Shen, and G. Zhang. "Intersubband transition in symmetric AlxGa1-xN/GaN double quantum wells with applied electric field". *Physics Letters A*, Vol. 373, No. 1, pp. 136–139, 2008.

[Lei 08b] S. Lei, Z. Dong, B. Shen, and G. Zhang. "Three-energy-level system in asymmetric AlxGa1-x N/GaN double quantum wells". *Journal of Physics D: Applied Physics*, Vol. 41, p. 065101, 2008.

[Lero 98] M. Leroux, N. Grandjean, M. Laügt, J. Massies, B. Gil, P. Lefebvre, and P. Bigenwald. "Quantum confined Stark effect due to built-in internal polarization fields in (Al,Ga)N/GaN quantum wells". *Physical Review B*, Vol. 58, pp. R13371–R13374, 1998.

[Levi 87] B. Levine, K. Choi, C. Bethea, J. Walker, and R. Malik. "New 10 μm infrared detector using intersubband absorption in resonant tunneling GaAlAs superlattices". *Applied Physics Letters*, Vol. 50, No. 16, pp. 1092–1094, 1987.

[Levi 93] B. Levine. "Quantum-well infrared photodetectors". *Journal of Applied Physics*, Vol. 74, No. 8, pp. R1–R81, 1993.

[Li 06a] B. Li, R. Akimoto, K. Akita, and T. Hasama. "λ 1.49–3.4 μm intersubband absorptions in (CdS/ZnSe)/ BeTe quantum wells grown by molecular beam epitaxy". *Applied Physics Letters*, Vol. 88, p. 221915, 2006.

[Li 06b] Y. Li and R. Paiella. "Intersubband all-optical switching based on Coulomb-induced optical nonlinearities in GaN/AlGaN coupled quantum wells". *Semiconductor Science and Technology*, Vol. 21, p. 1105, 2006.

[Li 07a] Y. Li, A. Bhattacharyya, C. Thomidis, T. Moustakas, and R. Paiella. "Nonlinear optical waveguides based on near-infrared intersubband transitions in GaN/AlN quantum wells". *Opt. Express*, Vol. 15, No. 9, pp. 5860–5865, 2007.

[Li 07b] Y. Li, A. Bhattacharyya, C. Thomidis, T. Moustakas, and R. Paiella. "Ultrafast all-optical switching with low saturation energy via intersubband transitions in GaN/AlN quantum-well waveguides". *Opt. Express*, Vol. 15, No. 26, pp. 17922–17927, 2007.

[Li 08] Y. Li, A. Bhattacharyya, C. Thomidis, Y. Liao, T. Moustakas, and R. Paiella. "Refractive-index nonlinearities of intersubband transitions in GaN/AlN quantum-well waveguides". *Journal of Applied Physics*, Vol. 104, No. 8, p. 083101, 2008.

[Liu 00] A. Liu, S. Chuang, and C. Ning. "Piezoelectric field-enhanced second-order nonlinear optical susceptibilities in wurtzite GaN/AlGaN quantum wells". *Applied Physics Letters*, Vol. 76, p. 333, 2000.

[Lupu 08] A. Lupu, F. Julien, S. Golka, G. Pozzovivo, G. Strasser, E. Baumann, F. Giorgetta, D. Hofstetter, S. Nicolay, M. Mosca, *et al.* "Lattice-matched GaN–InAlN waveguides at λ= 1.55 μm grown by metal–organic vapor phase epitaxy". *Photonics Technology Letters, IEEE*, Vol. 20, No. 2, pp. 102–104, 2008.

[Lupu 12] A. Lupu, M. Tchernycheva, Y. Kotsar, E. Monroy, and F. Julien. "Electroabsorption and refractive index modulation induced by intersubband transitions in GaN/AlN multiple quantum wells". *Optics Express*, Vol. 20, No. 11, pp. 12541–12549, 2012.

[Ma 07] B. Ma, W. Fan, Y. Dang, W. Cheah, W. Loke, W. Liu, D. Li, S. Yoon, D. Zhang, H. Wang, *et al.* "GaInNAs double-barrier quantum well infrared photodetector with the photodetection at 1.24 μm". *Applied Physics Letters*, Vol. 91, No. 5, p. 051102, 2007.

[Mach 09] H. Machhadani, P. Kandaswamy, S. Sakr, A. Vardi, A. Wirtmüller, L. Nevou, F. Guillot, G. Pozzovivo, M. Tchernycheva, A. Lupu, *et al.* "GaN/AlGaN intersubband optoelectronic devices". *New Journal of Physics*, Vol. 11, p. 125023, 2009.

[Mach 10] H. Machhadani, Y. Kotsar, S. Sakr, M. Tchernycheva, R. Colombelli, J. Mangeney, E. Bellet-Amalric, E. Sarigiannidou, E. Monroy, and F. Julien. "Terahertz intersubband absorption in GaN/AlGaN step quantum wells". *Applied Physics Letters*, Vol. 97, No. 19, p. 191101, 2010.

[Mach 11] H. Machhadani, M. Tchernycheva, S. Sakr, L. Rigutti, R. Colombelli, E. Warde, C. Mietze, D. As, and F. Julien. "Intersubband absorption of cubic GaN/Al (Ga) N quantum wells in the near-infrared to terahertz spectral range". *Physical Review B*, Vol. 83, No. 7, p. 075313, 2011.

[Mach 13] H. Machhadani, M. Beeler, S. Sakr, E. Warde, Y. Kotsar, M. Tchernycheva, M. P. Chauvat, P. Ruterana, G. Nataf, Ph. De Mierry, E. Monroy, and F. H. Julien. "Systematic study of near-infrared intersubband absorption of polar and semipolar GaN/AlN quantum wells". *Journal of Applied Physics*, Vol. 113, p. 143109, 2013.

[Mali 09] O. Malis, C. Edmunds, M. Manfra, and D. Sivco. "Near-infrared intersubband absorption in molecular-beam epitaxy-grown lattice-matched InAlN/GaN superlattices". *Applied Physics Letters*, Vol. 94, p. 161111, 2009.

[Miet 11] C. Mietze, M. Landmann, E. Rauls, H. Machhadani, S. Sakr, M. Tchernycheva, F. Julien, W. Schmidt, K. Lischka, and D. As. "Band offsets in cubic GaN/AlN superlattices". *Physical Review B*, Vol. 83, No. 19, p. 195301, 2011.

[Moty 06] M. Motyka, R. Kudrawiec, G. Cywiński, M. Siekacz, C. Skierbiszewski, and J. Misiewicz. "Energy difference between electron subbands in AlInN/ GaInN quantum wells studied by contactless electroreflectance spectroscopy". *Applied Physics Letters*, Vol. 89, p. 251908, 2006.

[Moum 03] K. Moumanis, A. Helman, F. Fossard, M. Tchernycheva, A. Lusson, F. Julien, B. Damilano, N. Grandjean, and J. Massies. "Intraband absorptions in GaN/AlN quantum dots in the wavelength range of 1.27–2.4 μm". *Applied Physics Letters*, Vol. 82, p. 868, 2003.

[Nels 87] D. Nelson, R. Miller, and D. Kleinman. "Band nonparabolicity effects in semiconductor quantum wells". *Physical Review B*, Vol. 35, No. 14, p. 7770, 1987.

[Neog 99] A. Neogi, T. Mozume, H. Yoshida, and O. Wada. "Intersubband transitions at 1.3 and 1.55 μm in a novel coupled InGaAs-AlAsSb double-quantum-well structure". *Photonics Technology Letters, IEEE*, Vol. 11, No. 6, pp. 632–634, 1999.

[Nevo 06a] L. Nevou, F. Julien, R. Colombelli, F. Guillot, and E. Monroy. "Room-temperature intersubband emission of GaN/AlN quantum wells at $\lambda = 2.3$ μm". *Electronics Letters*, Vol. 42, No. 22, pp. 1308–1309, 2006.

[Nevo 06b] L. Nevou, M. Tchernycheva, L. Doyennette, F. Julien, E. Warde, R. Colombelli, F. Guillot, S. Leconte, E. Monroy, T. Remmele, *et al.* "New developments for nitride unipolar devices at 1.3–1.5 μm wavelengths". *Superlattices And Microstructures*, Vol. 40, No. 4, pp. 412–417, 2006.

[Nevo 06c] L. Nevou, M. Tchernycheva, F. Julien, M. Raybaut, A. Godard, E. Rosencher, F. Guillot, and E. Monroy. "Intersubband resonant enhancement of second-harmonic generation in GaN/AlN quantum wells". *Applied Physics Letters*, Vol. 89, p. 151101, 2006.

[Nevo 07a] L. Nevou, N. Kheirodin, M. Tchernycheva, L. Meignien, P. Crozat, A. Lupu, E. Warde, F. Julien, G. Pozzovivo, S. Golka, *et al.* "Short-wavelength intersubband electroabsorption modulation based on electron tunneling between GaN/ AlN coupled quantum wells". *Applied Physics Letters*, Vol. 90, p. 223511, 2007.

[Nevo 07b] L. Nevou, M. Tchernycheva, F. Julien, F. Guillot, and E. Monroy. "Short wavelength ($\lambda = 2.13$ μm) intersubband luminescence from GaN/AlN quantum wells at room temperature". *Applied Physics Letters*, Vol. 90, No. 12, p. 121106, 2007.

[Nevo 08] L. Nevou, F. Julien, M. Tchernycheva, F. Guillot, E. Monroy, and E. Sarigiannidou. "Intraband emission at λ 1.48 μm from GaN/ AlN quantum dots at room temperature". *Applied Physics Letters*, Vol. 92, p. 161105, 2008.

[Nevo 09] L. Nevou, J. Mangeney, M. Tchernycheva, F. Julien, F. Guillot, and E. Monroy. "Ultrafast relaxation and optical saturation of intraband absorption of GaN/AlN quantum dots". *Applied Physics Letters*, Vol. 94, p. 132104, 2009.

[Ng 02] H. Ng, C. Gmachl, J. Heber, J. Hsu, S. Chu, and A. Cho. "Recent progress in GaN-based superlattices for near-infrared intersubband transitions". *Physica Status Solidi (b)*, Vol. 234, No. 3, pp. 817–821, 2002.

[Nguy 10] D. Nguyen, W. Wuester, P. Roussignol, C. Voisin, G. Cassabois, M. Tchernycheva, F. Julien, F. Guillot, and E. Monroy. "Homogeneous linewidth of the intraband transition at 1.55 μm in GaN/AlN quantum dots". *Applied Physics Letters*, Vol. 97, p. 061903, 2010.

[Nico 05] S. Nicolay, J. Carlin, E. Feltin, R. Butte, M. Mosca, N. Grandjean, M. Ilegems, M. Tchernycheva, L. Nevou, and F. Julien. "Midinfrared intersubband absorption in lattice-matched AlInN/ GaN multiple quantum wells". *Applied Physics Letters*, Vol. 87, p. 111106, 2005.

[Nico 06] S. Nicolay, E. Feltin, J. Carlin, M. Mosca, L. Nevou, M. Tchernycheva, F. Julien, M. Ilegems, and N. Grandjean. "Indium surfactant effect on AlN/ GaN heterostructures grown by metal–organic vapor-phase epitaxy: Applications to intersubband transitions". *Applied Physics Letters*, Vol. 88, p. 151902, 2006.

[Nico 07] S. Nicolay, E. Feltin, J. Carlin, N. Grandjean, L. Nevou, F. Julien, M. Schmidbauer, T. Remmele, and M. Albrecht. "Strain-induced interface instability in GaN/AlN multiple quantum wells". *Applied Physics Letters*, Vol. 91, No. 6, pp. 061927, 2007.

[Noda 93] S. Noda, T. Yamashita, M. Ohya, Y. Muromoto, and A. Sasaki. "All-optical modulation for semiconductor lasers by using three energy levels in n-doped quantum wells". *IEEE Journal of Quantum Electronics*, Vol. 29, No. 6, pp. 1640–1647, 1993.

[Paie 06] R. Paiella. *Intersubband Transitions in Quantum Structures.* McGraw-Hill Nanoscience and Technology Series, McGraw-Hill, 2006.

[Pere 09] N. Péré-Laperne, C. Bayram, R. McClintock, M. Razeghi, *et al.* "Tunability of intersubband absorption from 4.5 to 5.3 μm in a GaN/AlGaN superlattices grown by metal–organic chemical vapor deposition". *Applied Physics Letters*, Vol. 95, p. 131109, 2009.

[Pugh 99] S. Pugh, D. Dugdale, S. Brand, and R. Abram. "Electronic structure calculations on nitride semiconductors". *Semiconductor Science and Technology*, Vol. 14, p. 23, 1999.

[Ranj 03] V. Ranjan, G. Allan, C. Priester, and C. Delerue. "Self-consistent calculations of the optical properties of GaN quantum dots". *Physical Review B*, Vol. 68, No. 11, p. 115305, 2003.

[Rapa 03] R. Rapaport, G. Chen, O. Mitrofanov, C. Gmachl, H. Ng, and S. Chu. "Resonant optical nonlinearities from intersubband transitions in GaN/AlN quantum wells". *Applied Physics Letters*, Vol. 83, No. 2, pp. 263–265, 2003.

[Roma 06] A. E. Romanov, T. J. Baker, S. Nakamura, and J. S. Speck. "Strain-induced polarization in wurtzite III-nitride semipolar layers". *Journal of Applied Physics*, Vol. 100, No. 2, 2006.

[Sakr 10] S. Sakr, Y. Kotsar, S. Haddadi, M. Tchernycheva, L. Vivien, I. Sarigiannidou, N. Isac, E. Monroy, and F. Julien. "GaN-based quantum cascade photodetector with 1.5 μm peak detection wavelength". *Electronics Letters*, Vol. 46, No. 25, pp. 1685–1686, 2010.

[Sakr 11] S. Sakr, E. Warde, M. Tchernycheva, L. Rigutti, N. Isac, and F. H. Julien. "Origin of the electrical instabilities in GaN/AlGaN double-barrier structure". *Applied Physics Letters*, Vol. 99, No. 14, 2011.

[Sakr 12a] S. Sakr, E. Giraud, A. Dussaigne, M. Tchernycheva, N. Grandjean, and F. Julien. "Two-color GaN/AlGaN quantum cascade detector at short infrared wavelengths of 1 and 1.7 μm". *Applied Physics Letters*, Vol. 100, No. 18, p. 181103, 2012.

[Sakr 12b] S. Sakr, E. Giraud, M. Tchernycheva, N. Isac, P. Quach, E. Warde, N. Grandjean, and F. Julien. "A simplified GaN/AlGaN quantum cascade detector with an alloy extractor". *Applied Physics Letters*, Vol. 101, p. 251101, 2012.

[Sakr 12c] S. Sakr, Y. Kotsar, M. Tchernycheva, E. Warde, N. Isac, E. Monroy, and F. H. Julien. "Resonant tunneling transport in a GaN/AlN multiple-quantum-well structure". *Applied Physics Express*, Vol. 5, No. 5, 2012.

[Sakr 13] S. Sakr, Gacemi, P. Crozat, L. Vivien, M. Tchernycheva, N. Isac, Y. Kotsar, E. Monroy, and F. Julien."Ultrafast GaN/AlGaN quantum cascade detectors at 1.55 μm wavelength". *Applied Physics Letters*, Vol. 102, p. 011135, 2013.

[Sauv 97] S. Sauvage, P. Boucaud, F. Julien, J. Gerard, and J. Marzin. "Infrared spectroscopy of intraband transitions in self-organized InAs/GaAs quantum dots". *Journal of Applied Physics*, Vol. 82, No. 7, pp. 3396–3401, 1997.

[Shim 07] T. Shimizu, C. Kumtornkittikul, N. Iizuka, N. Suzuki, M. Sugiyama, and Y. Nakano. "Fabrication and measurement of AlN cladding AlN/GaN multiple-quantum-well waveguide for all-optical switching devices using intersubband transition". *Japanese Journal of Applied Physics*, Vol. 46, p. 6639, 2007.

[Sirt 94] C. Sirtori, F. Capasso, J. Faist, and S. Scandolo. "Nonparabolicity and a sum rule associated with bound-to-bound and bound-to-continuum intersubband transitions in quantum wells". *Physical Review B*, Vol. 50, No. 12, p. 8663, 1994.

[Skie 06] C. Skierbiszewski, G. Cywiński, M. Siekacz, A. Feduniewicz-Zmuda, L. Nevou, L. Doyennette, M. Tchernycheva, F. Julien, J. Smalc, P. Prystawko, *et al.* "Growth and characterization of AlInN/GaInN quantum wells for high-speed intersubband devices at telecommunication wavelengths". *Proceedings of SPIE: The International Society for Optical Engineering, Gallium Nitride Materials and Devices*, Vol. 6121, p. 612109, 2006.

[Sudr 10] F. Sudradjat, W. Zhang, K. Driscoll, Y. Liao, A. Bhattacharyya, C. Thomidis, L. Zhou, D. J. Smith, T. D. Moustakas, and R. Paiella. "Sequential tunneling transport characteristics of GaN/AlGaN coupled-quantum-well structures". *Journal of Applied Physics* , Vol. 108, No. 10, 2010.

[Sudr 12] F. Sudradjat, W. Zhang, J. Woodward, H. Durmaz, T. Moustakas, and R. Paiella. "Far-infrared intersubband photodetectors based on double-step III-nitride quantum wells". *Applied Physics Letters*, Vol. 100, No. 24, p. 241113, 2012.

[Sun 05] G. Sun, R. Soref, and J. Khurgin. "Active region design of a terahertz GaN/Al0.15Ga0.85N quantum cascade laser". *Superlattices and Microstructures* , Vol. 37, No. 2, pp. 107–113, 2005.

[Sun 06] G. Sun, J. Khurgin, and R. Soref. "Design of a GaN/AlGaN intersubband Raman laser electrically tunable over the 3-5 μm atmospheric transmission window". *Journal of Applied Physics* , Vol. 99, No. 3, 2006.

[Suzu 00] N. Suzuki, N. Iizuka, and K. Kaneko. "Intersubband transition in AlGaN/GaN quantum wells for ultrafast all-optical switching at communication wavelength". *Proceedings of SPIE*, Vol. 3940, p. 127, 2000.

[Suzu 03] N. Suzuki, N. Iizuka, and K. Kaneko. "Calculation of near-infrared intersubband absorption spectra in GaN/AlN quantum wells". *Japanese Journal of Applied Physics Part 1: Regular Papers Short Notes & Review Papers* , Vol. 42, No. 1, pp. 132–139, 2003.

[Suzu 96] M. Suzuki and T. Uenoyama. "Optical gain and crystal symmetry in III–V nitride lasers". *Applied Physics Letters*, Vol. 69, p. 3378, 1996.

[Suzu 97] N. Suzuki and N. Iizuka. "Feasibility study on ultrafast nonlinear optical properties of 1.55 μm intersubband transition in AlGaN/GaN quantum wells". *Japanese Journal of Applied Physics*, Vol. 36, pp. L1006–L1008, 1997.

[Suzu 98] N. Suzuki and N. Iizuka. "Electron scattering rates in AlGaN/GaN quantum wells for 1.55-μm inter-subband transition". *Japanese Journal of Applied Physics*, Vol. 37, pp. L369–L371, 1998.

[Suzu 99] N. Suzuki and N. Iizuka. "Effect of polarization field on intersubband transition in AlGaN/GaN quantum wells". *Japanese Journal of Applied Physics*, Vol. 38, pp. L363–L365, 1999.

[Tche 05] M. Tchernycheva, L. Nevou, L. Doyennette, A. Helman, R. Colombelli, F. Julien, F. Guillot, E. Monroy, T. Shibata, and M. Tanaka. "Intraband absorption of doped GaN/AlN quantum dots at telecommunication wavelengths". *Applied Physics Letters*, Vol. 87, No. 10, p. 101912, 2005.

[Tche 06a] M. Tchernycheva, L. Nevou, L. Doyennette, F. Julien, F. Guillot, E. Monroy, T. Remmele, and M. Albrecht. "Electron confinement in strongly coupled GaN/ AlN quantum wells". *Applied Physics Letters*, Vol. 88, p. 153113, 2006.

[Tche 06b] M. Tchernycheva, L. Nevou, L. Doyennette, F. Julien, E. Warde, F. Guillot, E. Monroy, E. Bellet-Amalric, T. Remmele, and M. Albrecht. "Systematic experimental and theoretical investigation of intersubband absorption in GaN/ AlN quantum wells". *Physical Review B*, Vol. 73, No. 12, p. 125347, 2006.

[Tche 10] M. Tchernycheva, L. Nevou, L. Vivien, F. Julien, P. Kandaswamy, E. Monroy, A. Vardi, and G. Bahir. "Intersubband optics in GaN-based nanostructures: physics and applications". *Physica Status Solidi (b)*, Vol. 247, No. 7, pp. 1622–1627, 2010.

[Tera 09] W. Terashima and H. Hirayama. "Design and fabrication of terahertz quantum cascade laser structure based on III-nitride semiconductors". In: *Physica Status Solidi C: Current Topics in Solid State Physics*, Vol 6, Suppl 2, pp. S615–S618, 2009.

[Tera 10] W. Terashima and H. Hirayama. "The Utility of Droplet Elimination by Thermal Annealing Technique for Fabrication of GaN/AlGaN Terahertz Quantum Cascade Structure by Radio Frequency Molecular Beam Epitaxy". *Applied Physics Express*, Vol. 3, No. 12, 2010.

[Tera 11] W. Terashima and H. Hirayama. "Spontaneous emission from GaN/ AlGaN terahertz quantum cascade laser grown on GaN substrate". In: *Physica Status Solidi C: Current Topics In Solid State Physics*, Vol 8, No 7-8, European Mat. Res. Soc. (E-MRS), 2011.

[Uchi 05] H. Uchida, S. Matsui, P. Holmström, A. Kikuchi, and K. Kishino. "Room-temperature operation of 1.55 μm wavelength-range GaN/AlN quantum well intersubband photodetectors". *IEICE Electronics Express*, Vol. 2, No. 22, pp. 566–571, 2005.

[Vald 08] S. Valdueza-Felip, F. Naranjo, M. González-Herráez, H. Fernández, J. Solis, F. Guillot, E. Monroy, L. Nevou, M. Tchernycheva, and F. Julien. "Characterization of the resonant third-order nonlinear susceptibility of Si-doped GaN–AlN quantum wells and quantum dots at 1.5 μm". *Photonics Technology Letters, IEEE*, Vol. 20, No. 16, pp. 1366–1368, 2008.

[Vard 06] A. Vardi, N. Akopian, G. Bahir, L. Doyennette, M. Tchernycheva, L. Nevou, F. Julien, F. Guillot, and E. Monroy. "Room-temperature demonstration of GaN/AlN quantum dot intraband infrared photodetector at fiber-optics communication wavelength". *Applied Physics Letters*, Vol. 88, No. 14, p. 143101, 2006.

[Vard 08a] A. Vardi, G. Bahir, F. Guillot, C. Bougerol, E. Monroy, S. Schacham, M. Tchernycheva, and F. Julien. "Near-infrared quantum cascade detector in GaN/AlGaN/AlN heterostructures". *Applied Physics Letters*, Vol. 92, No. 1, p. 011112, 2008.

[Vard 08b] A. Vardi, N. Kheirodin, L. Nevou, H. Machhadani, L. Vivien, P. Crozat, M. Tchernycheva, R. Colombelli, F. Julien, F. Guillot, *et al.* "High-speed operation of GaN/AlGaN quantum cascade detectors at $\lambda \sim 1.55$ μm". *Applied Physics Letters*, Vol. 93, p. 193509, 2008.

[Vard 09] A. Vardi, G. Bahir, S. E. Schacham, P. K. Kandaswamy, and E. Monroy. "Photocurrent spectroscopy of bound-to-bound intraband transitions in GaN/AlN quantum dots". *Physical Review B*, Vol. 80, p. 155439, 2009.

[Vard 10] A. Vardi, G. Bahir, S. Schacham, P. Kandaswamy, and E. Monroy. "Negative photoconductivity due to intraband transitions in GaN/AlN quantum dots". *Journal of Applied Physics*, Vol. 108, No. 10, p. 104512, 2010.

[Vard 11] A. Vardi, S. Sakr, J. Mangeney, P. Kandaswamy, E. Monroy, M. Tchernycheva, S. Schacham, F. Julien, and G. Bahir. "Femtosecond electron transit time characterization in GaN/AlGaN quantum cascade detector at 1.5 micron". *Applied Physics Letters*, Vol. 99, No. 20, p. 202111, 2011.

[Vodj 91] N. Vodjdani, B. Vinter, V. Berger, E. Bockenhoff, and E. Costard. "Tunneling assisted modulation of the intersubband absorption in double quantum wells". *Applied Physics Letters*, Vol. 59, No. 5, pp. 555–557, 1991.

[Vurg 03] I. Vurgaftman and J. Meyer. "Band parameters for nitrogen-containing semiconductors". *Journal of Applied Physics*, Vol. 94, p. 3675, 2003.

[Wada 00] O. Wada. "Femtosecond semiconductor-based optoelectronic devices for optical-communication systems". *Optical and quantum electronics*, Vol. 32, No. 4, pp. 453–471, 2000.

[Wang 05] Z. Wang, K. Reimann, M. Woerner, T. Elsaesser, D. Hofstetter, J. Hwang, W. Schaff, and L. Eastman. "Optical phonon sidebands of electronic intersubband absorption in strongly polar semiconductor heterostructures". *Physical Review Letters*, Vol. 94, No. 3, p. 37403, 2005.

[Wang 06] Z. Wang, K. Reimann, M. Woerner, T. Elsaesser, D. Hofstetter, E. Baumann, F. Giorgetta, H. Wu, W. Schaff, and L. Eastman. "Ultrafast hole burning in intersubband absorption lines of GaN/ AlN superlattices". *Applied Physics Letters*, Vol. 89, p. 151103, 2006.

[Webe 99] A. Weber, O. Gauthier-Lafaye, F. Julien, J. Brault, M. Gendry, Y. Desieres, and T. Benyattou. "Strong normal-incidence infrared absorption in self-organized InAs/InAlAs quantum dots grown on InP(001)". *Applied Physics Letters*, Vol. 74, No. 3, pp. 413–415, 1999.

[West 85] L. West and S. Eglash. "First observation of an extremely large-dipole infrared transition within the conduction band of a GaAs quantum well". *Applied Physics Letters*, Vol. 46, No. 12, pp. 1156–1158, 1985.

[Whit 81] S. White and L. Sham. "Electronic properties of flat-band semiconductor heterostructures". *Physical Review Letters*, Vol. 47, No. 12, pp. 879–882, 1981.

[Will 05] D. P. Williams, A. D. Andreev, E. P. O'Reilly, and D. A. Faux. "Derivation of built-in polarization potentials in nitride-based semiconductor quantum dots". *Physical Review B*, Vol. 72, p. 235318, 2005.

[Yasu 12] H. Yasuda, T. Kubis, I. Hosako, and K. Hirakawa. "Non-equilibrium Green's function calculation for GaN-based terahertz-quantum cascade laser structures". *Journal of Applied Physics*, Vol. 111, No. 8, 2012.

[Zain 10] N. Zainal, S. Novikov, C. Mellor, C. Foxon, and A. Kent. "Current-voltage characteristics of zinc-blende (cubic) AlGaN/GaN double barrier resonant tunneling diodes". *Applied Physics Letters*, Vol. 97, p. 112102, 2010.

[Zhou 03] Q. Zhou, J. Chen, B. Pattada, M. Manasreh, F. Xiu, S. Puntigan, L. He, K. Ramaiah, and H. Morkoç. "Infrared optical absorbance of intersubband transitions in GaN/AlGaN multiple quantum well structures". *Journal of Applied Physics*, Vol. 93, p. 10140, 2003.

13 The slow light in gallium nitride

Tatiana V. Shubina, Mikhail M. Glazov, Nikolay A. Gippius, and Bernard Gil

13.1 Introduction

Development of all-optical schemes of quantum communication and information processing implies a precise control of photon pulses—in particular, their spectral and temporal characteristics. For example, the operation of quantum logics, random-access memory, data synchronization, and network buffering need a definite delay of photons (Matsko *et al.*, 2001; Milonni, 2002). Therefore "slow light", propagating with a velocity much lower than the speed of light in vacuum, has been studied intensively in recent years. Related effects include "fast light" with the superluminal pulse velocity, and "stopped" light, when an optical pulse is trapped (or caught) (Bigelow *et al.*, 2003-1). The slowdown of light can be done by different methods. Many of them assume the use of dispersive media with a sharp change in optical parameters, e.g., a dielectric with a resonant absorption line (Loudon, 1970) or an atomic system with electromagnetically induced transparency (Harris, 1997). The factor of optical dispersion being a fundamental effect in any material is important for other ways to produce the slow light. In the following section we review briefly the history and methods of producing slow light, focusing basically on the effects of the optical dispersion.

GaN and related materials are very promising for different applications in nanophotonics and optoelectronics. This wide-gap semiconductor possesses strong excitonic resonances, and thus is suitable for light retardation due to optical dispersion. The value of longitudinal–transverse splitting, $\hbar\omega_{LT}$, about 1 meV promotes the emergence of exciton–polaritons, stable up to room temperature. Section 13.3 describes the basic mechanisms of the light transfer in GaN, and Section 13.4 considers experimental facts concerning slow light in GaN obtained by time-of-flight spectroscopy and supported by other methods.

Since the thesis by D. S. Rozhdestvensky, of 1912 (Frish, 1976), there has been a long tradition of using optical dispersion studies to determine the parameters of a resonant line. The so-called "Rozhdestvensky hooks" in the interference pattern near an atomic line hold information on its oscillator strength. It does not need to be justified that exact data on excitonic resonances are needed to design smart semiconductor devices. In Section 13.5 the advantages of time-of-flight spectroscopy for determining the parameters of bulk semiconductors, as well as investigating the fine structure of an excitonic spectrum, are demonstrated.

A light pulse consists of a large number of frequency components, each of those travels in a dispersive medium with its own phase velocity. This can cause a certain degree of pulse distortion, its broadening, and reshaping on an energy–time plane. Section 13.6 is devoted to a description of the distortion of the light pulses in GaN and ZnO.

In this chapter we restrict ourselves to relatively low-temperature and linearity in the intensity regimes, bearing in mind the limitations of our approach. Namely, strong interaction with phonons can significantly modify the light transfer at elevated temperatures. At high excitation powers it can be also affected by non-linear effects. These processes require detailed study.

We thank A. A. Toropov and B. Monemar for many fruitful discussions of the slow-light phenomenon in semiconductors, and the help of G. Pozina and J. P. Bergman in experimental studies is highly appreciated.

13.2 Slow light: history and recent studies

13.2.1 *Group velocity in a medium with optical dispersion*

In spite of very different mechanisms, accomplishing the slowdown of light, two systems—(i) dielectric with a resonant absorption line, and (ii) atomic system with electromagnetically induced transparency—have a common feature. This is the strong dispersion of optical parameters near a resonant frequency ω_0 (Fig. 13.1). Specifically, the variation of the refractive index $n \equiv n(\omega)$ results in the slowdown of group velocity $v_g(\omega)$.

The fundamental study of light propagation and its group velocity in a dispersive medium was initiated more than a century ago by A. Sommerfeld and L. Brillouin. In their works a number of interesting effects, such as the emergence of precursors, was predicted, and they drew attention to the variation of

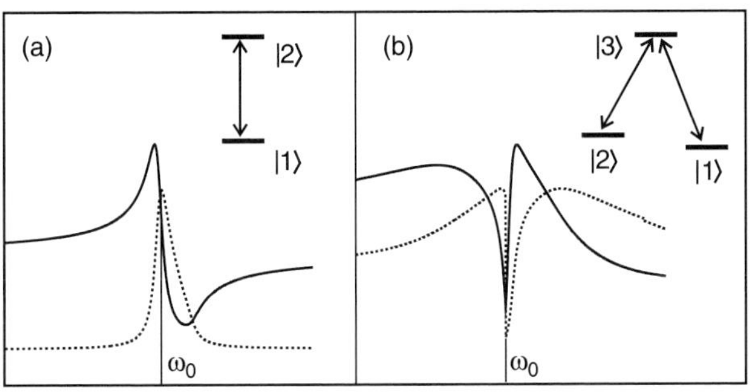

FIG. 13.1. Schematic representation of real (solid lines) and imaginary (dotted lines) parts of a complex dielectric function in (a) a dielectric with a resonant absorption line and (b) an atomic system with the window of electromagnetically induced transparency at frequency ω_0. The insets present the simplified schemes of levels.

the group velocity in a resonant system (Brillouin, 1960). While the phase velocity $v_{ph}(\omega) = c/n(\omega)$ is the speed at which the phase front of a monochromatic wave propagates through a medium with $n(\omega)$, the group velocity characterizes the motion of the envelope of the wave packet of a photon pulse. The commonly used expression for $v_g(\omega)$ as a function of frequency ω is written as

$$v_g(\omega) = \frac{\partial \omega}{\partial k} = \frac{c}{n(\omega) + \omega \frac{\partial n}{\partial \omega}} \tag{13.1}$$

Here, $k(\omega) = \omega/(c\sqrt{\varepsilon(\omega)})$ is a wavevector of light; $\varepsilon(\omega)$ is the dielectric function of a medium, c is the light speed in vacuum. Strictly speaking, eq. (13.1) is valid for transparent (non-absorbing) media only, otherwise the group velocity may become meaningless; see below. The denominator in eq. (13.1) represents the group refractive index, whose real part defines $v_g(\omega)$. If the spectrum of an optical pulse falls into the range of strong $n(\omega)$ dispersion, the factor $\partial n/\partial \omega$ starts to dominate the denominator. As a result, $v_g(\omega)$ decreases greatly. At the same time, the phase velocity is not influenced by this factor; its variation is much smaller. Both the group index and wavevector are complex in an absorbing dielectric; thus, the wave passing through such a medium is attenuated.

It was recognized long ago that in the medium with an absorption line the group velocity, given by eq. (13.1), may be superluminal in certain frequency ranges, may be negative in others, and may even equal zero. This intricate situation has been discussed in several textbooks (e.g., Born and Wolf, 1965). It was noted that the term of group velocity loses its strict physical meaning for an electromagnetic wave in the region of abnormal dispersion (Landau and Lifshitz, 1984). Recent analysis of the group velocity concept for absorbing media can be found in Sonnenschein *et al.* (1998). For a Gaussian light-pulse the possibility of the $v_g(\omega)$ variation from negative to superluminal values has been shown theoretically, along with the unchanged width of the pulse for many exponential absorption depths (Garrett and McCumber, 1970). It is worth mentioning that this abnormal $v_g(\omega)$ does not imply violation of the causality principle, but is instead a consequence of pulse-shape distortion. The superluminal group velocity of a pulse does not imply energy transfer (Loudon, 1970), which cannot be faster than the phase velocity of light in vacuum (Crisp, 1971). The same is true for the information, whose transfer rate is limited by c (Molotkov, 2010). This limitation is important for quantum cryptography, because it provides security in communication systems.

The passage of light-pulses through the resonant medium was intensively studied during the epoch of the development of communication systems (Vainshtein, 1976). The theory of energy transfer has been developed for spatially dispersive media with exciton-polaritons, both infinite and bounded (Bishop and Maradudin, 1976; Agranovich and Ginzburg, 1984). The remarkable decrease of energy transport velocity down to a velocity of 4×10^3 m·s^{-1} has been predicted for the exciton–polariton at a resonance in a semiconductor, e.g., GaAs (Puri and Birman, 1981). It has been demonstrated that a width of exciton-polaritons resonance Γ strongly affects this velocity. Its value may decrease by orders of magnitude with decreasing Γ/ω_0 value (Loudon *et al.*, 1997).

13.2.2 *Light propagation in semiconductors near excitonic lines*

The rise of lasers in the early 1960s drastically intensified experimental studies of the light velocity. First, such studies were carried out in gaseous systems. The velocity of an optical pulse in an Ne absorption cell placed inside a self-locked He–Ne laser has been found to be superluminal (Faxvog *et al.*, 1970). In a direct-gap semiconductor GaAs, the pioneer observation of the slow light propagation at an exciton-polariton resonance has been achieved using a cross-correlation technique (Ulbrich and Fehrenbach, 1979). Polariton group velocity in these experiments varied from $c/3.6$ down to $c/2000$, when the central frequency of a dye laser was tuned through the resonance in a 3.7-μm-thick GaAs crystal. The shape of the laser pulse on the time-intensity plane, registered by the cross-correlation technique, was almost unchanged during this tuning.

The results obtained in GaP:N crystals (Chu and Wong, 1982) seemed to verify the theoretical predictions (Garrett and McCumber, 1970). In this experiment a laser was tuned to a weak bound A exciton line. The samples have different doping and thicknesses up to 76 μm. The $v_g(\omega)$ values extracted from the experimental data exceeds c, tends to infinity, and even becomes negative. The shape of the pulses did not exhibit significant distortion, similar to that in the experiment by Ulbrich and Fehrenbach. It is worth noting that the cross-correlation technique used at that time did not allow the registration of the pulse spectrum on the energy–time plane, which remained unknown. The results of the dramatic $v_g(\omega)$ variation stimulated extensive theoretical investigations in order to exclude the causality violation. The model of the pulse propagation has been improved to take into account the finite thickness of the systems (Halevi and Fuchs, 1985).

A variety of time-resolved (TR) techniques were used to study the polariton effects in CuCl. It has been found that the propagation time in the 1.7-mm-thick sample was about 150 ps for the lower branch polariton and 700 ps for the higher one (Segawa *et al.*, 1978). A variation of the polariton lifetime in the bottleneck region was investigated by exploiting a narrowband tunable laser (Segawa *et al.*, 1979). TR non-degenerate four-wave mixing was used to determine the damping constant of the exciton-polaritons (Masumoto *et al.*, 1983). At low temperature, the value of this parameter is as small as 10 μeV with 20 meV detuning from the exciton resonance. It increases by an order of magnitude when approaching the transverse exciton energy. The temporal response of photoluminescence (PL) in this material was measured in both geometries—conventional "backward" and "forward", i.e., with registration from the opposite side of a sample (Kuwata *et al.*, 1988). The additional peak in the spectra near the Z_3 exciton of the backward PL is ascribed to the light reflection from the opposite boundary of a crystal. It was noticed that such a signal propagates at double distance in its passage inside a crystal. In view of the delay-time of this reflex, the group velocity smaller than $c/10^4$ has been found near the exciton resonance (Ikehara and Itoh, 1991). It was not a record, however, because in some materials, such as anthracene, the measured velocity of exciton-polaritons was below that of sound (Vidmont *et al.*, 1983).

The time-of-flight measurements using the pulsed lasers proved the polariton mechanism of the light transfer in a set of II–VI crystals. It has been recognized that the temporal evolution of the pulse cannot be described by the $v_g(\omega)$ of the wave packet, when it propagates very close to the exciton resonance in CdSe (Itoh *et al.*, 1981). To perfectly fit the data on the light propagation in CdZnTe it has been necessary to take into account all excitonic resonances, but not only the ground one (Sermage *et al.*, 1998). With increasing excitation power, non-linear effects such self-induced transparency were observed at the A exciton in CdSe (Nielsen *et al.*, 2001) and at the bound exciton in CdS (Jütte *et al.*, 1996). This transparency induces a noticeable increase in $v_g(\omega)$ in these materials. In Cu_2O, quantum beats were recorded by the time-resolved technique. They appear due to interference between the lower and upper polariton branches of the 1S quadrupole exciton (Frölich *et al.*, 1991).

Very demonstrative results on slow light in semiconductors have been obtained with the development of new detection systems. When the Hamamatsu streak camera was invented it became possible to record the optical precursors in CuCl (Sakai *et al.*, 2002), predicted long ago (Brillouin, 1960). PL propagation experiments at the donor-bound exciton (BX) resonance in single-crystal ZnO (Xiong *et al.*, 2005) have shown the curvature of the PL streak on the energy–time plane, not demonstrated previously. The group velocity at the resonance (3.36 eV) was measured to be 6×10^6 m·s^{-1}. The signal was attenuated by 95% on traversing a 0.5-mm-long crystal. The temperature-dependent measurements showed an absence of noticeable delay at 100 K, corresponding roughly to the thermal detrapping of the donor-bound excitons. Xiong *et al.* have explained their results using a disputable model of propagation of a bound-exciton polariton that hampers, in part, the timely recognition of their results.

In GaN, two mechanisms of the light transfer—namely, ballistic (polaritonic) and diffusive—has been found with the studies of both PL and photon-pulse propagations (Shubina *et al.*, 2008). The light velocity of 2.1×10^6 m·s^{-1} has been registered. The pulses of a tunable ps or fs laser do not excite PL when they propagate below the bound exciton lines. This allowed demonstration of the distortion of the photon pulses without PL admixture. The results of light propagation in GaN were considered as promising for realization of all-optical schemes (Wiersma, 2008).

Analysis of the time-of-flight data obtained in forward and backward experimental geometries using the developed model to fit the pulse shape on the energy-time plane has allowed determination of the low-temperature parameters of free and bound excitons in GaN (Shubina *et al.*, 2011-1) and ZnO (Shubina *et al.*, 2011-2). The homogeneous width of exciton-polaritons in these semiconductors at 2 K was found to be 0.01 and 0.003 meV, respectively, with measurements in the close vicinity of free exciton resonances. The minimal documented group velocity in ZnO was about $c/(1.5 \times 10^3)$, but the applicable delay value corresponds roughly to $c/100$, due to strong signal attenuation.

The time-of-flight studies of ps and fs laser pulses propagation through (ZnCd)Se (Godde *et al.*, 2010) have exhibited the purely polaritonic mechanism of light transfer in this compound. In a 745-μm-thick crystal, the time delay up to 350 ps has been recorded. This maximal delay decreases to 50 ps at 55 K, presumably due to the thermal quench of the exciton-polaritons. Time-resolved measurements in different crystal orientations have been attemped to check the valence-band ordering in ZnO (Chen *et al.*, 2011). Using the model developed in Shubina *et al.* (2008), Chen *et al.* have demonstrated that the diffusion process in ZnO is practically absent, due to low concentration of numerous types of donors, bounding excitons in the studied perfect crystals.

All these studies of the slow light in semiconductors have been performed at relatively low temperatures, while the resonant scattering by phonons, which is significant in polar semiconductors (Klingshirn, 2007), can modify and even prohibit the light retardation as the temperature rises. In particular, the output signal at room temperature can be the scattered light predominantly. The delay of this signal is much smaller than that needed due to the high group velocity at the energy of scattered photons (Shubina, 2012).

13.2.3 *Electromagnetically induced transparency and other effects*

In spite of the noticeable delay, the light slowing in the semiconductors due to optical dispersion was inevitably accompanied by dramatic attenuation of signal intensity. Therefore, the electromagnetically induced transparency (EIT) (Harris *et al.*, 1990; Boller *et al.*, 1991) was considered as a new approach to obtain the slow light. The EIT phenomenon features two previously incompatible characteristics: steepness of the $\partial n/\partial \omega$ dependence, and small attenuation in a certain energy range (see Fig. 13.1b). EIT exists in the presence of two optical fields in a medium possessing three quantum levels arranged in the so-called Λ-scheme (Fig. 13.1b, inset). A laser beam (linearly polarized) couples two states in the left arm of the scheme. For a weaker probe beam (circularly polarized), the ground state is a coherent superposition of the two lowest levels. Thus, quantum interference takes place due to two alternate excitation paths from this ground state to the excited state. To provide the interference, a frequency difference between the coupling and probe electromagnetic fields should be equal to the difference between the two lowest levels in the Λ-scheme. When the coupling beam has sufficient intensity it depletes the superposition state. This opens a narrow window (hole) in the spectrum, where attenuation of a probe beam is significantly suppressed. Thus, the probe beam can propagate almost without loss and with very slow speed.

The most impressive results in light retardation by EIT have been obtained using atomic systems. In a gas of sodium atoms cooled by a laser beam down to 50 nK, the velocity of an optical pulse of 17 $m{\cdot}s^{-1}$ has been achieved

(Hau *et al.*, 1999). Progress in this field presented opportunities for deterministic photon synchronization by integration of a single-photon source with the slow-light medium (Eisaman *et al.*, 2005; Hau, 2011). However, to provide the $\partial n/\partial \omega$ steepness required, the transparency window should be very narrow. In other words, the operating bandwidth is small. Due to the uncertainty principle, only the pulses of microsecond duration can propagate in such a narrow window, while shorter pulses would be strongly distorted.

For optical buffers there is one more limitation besides possible pulse-shape distortion. This is the ratio of the delay to the output pulse duration ν, which should be much higher than unity, while for EIT a best ratio of 4:1 has been reported. It was achieved in an isotopically pure lead vapor with a pulse velocity of $c/165$ and 55% intensity transmission for a 10-cm-long cell (Kasapi *et al.,* 1995). Some corroboration of the possibility for improving the ratio in a slow-light medium has been published (Boyd *et al.*, 2005), though more realistic considerations, which take into account the background absorption and signal broadening, were not so optimistic. On the other hand, certain improvements in the delay time–bandwidth product has been demonstrated (Tucker *et al.*, 2005; Matsko *et al.*, 2007). The reshaping procedure has been proposed to correct, in part, the shape of output pulses (Eastman *et al.*, 2004).

Besides EIT, various mechanisms have been exploited to realize the slowdown of light in dispersive media. These include coherent population oscillations (Bigelow *et al.*, 2003-2), stimulated Brillouin scattering (Okawachi *et al.*, 2005), stimulated Raman scattering (Dahan *et al.*, 2005), and so on. The relatively large ν value of about 7 with three-fold attenuation was realized in a spectral hole between two resonant lines in rubidium vapors at room temperature. However, the operating rate of such a system is even lower than at EIT. Some of these experiments have somewhat disputable explanations (Aleksandrov and Zapasskii, 2004).

The light slowdown can be achieved using semiconductor quantum-well/quantum-dot absorbers and optical amplifiers, where the process can be controlled electrically together with changing the pump laser intensity and wavelength (Chang-Hasnain and Chuang, 2006). In doped quantum wells, light slowing down to $c/40$ induced by the hyper-Raman process has been reported (Frogley *et al.*, 2006). Note that this value was found indirectly from analysis of the changes in absorption spectra. Another possible way to slow down light is the use of photonic crystals (Vlasov *et al.*, 2005), because higher-order photonic modes have a low group velocity.

A promising trend in slow-light formation is related to the solitons—single waves excited in a non-linear medium. The soliton can propagate at a long distance without noticeable attenuation and broadening (Zabusky and Kruskal, 1965). This is why the solitons, also called “optical bullets”, can be used for information transfer with controllable delay in all-optical schemes (Silberberg, 1990). Full stop of the light signal due to its transformation into a standing soliton has been observed in rubidium vapor (Bajcsy *et al.*, 2003). This stop took place because in atomic vapors and Bose–Einstein condensates the group velocity of the solitons decreases monotonically after the switching off of a pumping laser (Rybin *et al.*, 2005). These optical bullets can propagate in a system of optically coupled waveguides, when they are modulated in a specific way to suppress the light dispersion (Lederer and Biehlig, 1994; Sukhorukov and Kivshar, 2006; Minardi *et al.*, 2010). Solitons have been observed in semiconductor microcavities (Taranenko *et al.*, 2001; M. Sich *et al.*, 2012).

Progress in these studies is very fast, though control over the light velocity and overcoming the delay–bandwidth product problem is still a challenge.

13.2.4 *Light scattering and diffusion*

The multiple scattering process of photons is associated with intriguing phenomena, such as weak localization and coherent back-scattering of light (Van Albada and Lagendijk, 1985). The distinction between the group velocity and the velocity of energy transfer in the scattering media has been established during numerous discussions (Barabanenkov and Ozrin, 1992; Van Tiggelen *et al.*, 1993). Experimentally, photon scattering and diffusion has been studied in various disordered media, including dielectrics (Van Albada *et al.*, 1991), polymers (Garcia *et al.*, 1992), liquid crystals (Van Tiggelen *et al.*, 1996), semiconductors (Gamarts *et al.*, 1977, Ivchenko *et al.*, 1977), and even interstellar space, for which the theory of light propagation was developed for the first time (Chandrasekhar, 1960; Gnedin, Dolginov, 1963). In the scattering medium containing TiO_2 spheres of 220 nm diameter, which were considered as resonant Mie scatterers, the velocity of electromagnetic energy transfer differs by an order of magnitude from the phase velocity (Van Albada *et al.*, 1991). Note that this medium is optically inert, i.e. no reabsorption of phonons influences the energy transfer. The multiple scattered light has been also observed in ensembles of ZnS nanocrystals which, in this case, act as disordered media (Kurita *et al.*, 1999).

In semiconductors, where exciton-polariton modes can be supported, it was first demonstrated that scattering by neutral donors can affect the line shape of exciton-polariton emission. It has been observed in GaAs by means of measurements with temporal resolution in backward reflection geometry (Steiner *et al.*, 1986). Time-of-flight studies of perfect GaN crystals carried out in combination with far-field angular measurements have revealed the diffusion of the photons due to the elastic resonant scattering by neutral donor-bound exciton (BX) centers (Shubina *et al.*, 2008).

The principal issue for the semiconductors is that one needs to distinguish between three competing processes: i) ballistic (polariton-like) propagation with conservation of energy and wavevector $\boldsymbol{k}$; ii) elastic scattering with the conservation of energy, when the direction of the photon movement may be changed; and iii) inelastic relaxation or scattering with changing $\boldsymbol{k}$-vector and energy. It is difficult to distinguish the inelastic scattering and reabsorption/re-emission processes, which frequently imply exciton–phonon interaction (Ivchenko, 2005).

Both the relaxation process and the inelastic scattering, in contrast to the elastic one, are usually accompanied by spectral diffusion, i.e. change of the emission energy.

To discriminate between the different mechanisms one should take into account that the strength of the scattering is sensitive to the detuning from an exciton resonance, being different for elastic and inelastic types of scattering. The cross-section of photon–acoustic phonon scattering has a weaker (non-resonant) frequency-dependence in the vicinity of a donor-bound exciton line (Calleja and Cordona, 1977). Thus, the delay of light will vary more or less monotonically with energy. On the contrary, the spectral dependence of the delay due to the resonant diffusion should increase strongly with the approach of the bound exciton energy (Shubina *et al.*, 2008). The intensity quenching of the transmitted emission and its cut-off energy have to depend on concentration of the scattering centers. Besides, scattering by acoustic phonons should be strongly temperature-dependent (Wan *et al.*, 1993; Stanley *et al.*, 1997; Viswanath *et al.*, 1998). The spectrum of transmitted PL resembles a redshifted "Prussian helmet", when the exciton–acoustic-phonon interaction is substantial (Liptay *et al.*, 2007). Thus, the peculiarities of a pulse spectrum, its delay dependencies, and distortion can help to distinguish between the mechanisms.

13.3 Two mechanisms of light transfer in GaN: ballistic and diffusive

When considering light transfer in wide-gap semiconductors, two main physical mechanisms are important. First of all, optical dispersion results in the modification of phase and group velocities of light-propagation, and generally, delays optical pulse propagation compared with that in the empty space. Second, imperfections of structure result in the scattering of light, which may give rise to diffusive propagation of photons. The coexistence of these two mechanisms was demonstrated for GaN (Shubina *et al.*, 2008), and their schematic presentations are shown in Fig. 13.2. In the following, we outline basic theoretical concepts of light-propagation in crystals.

13.3.1 *Optical dispersion in a medium with several resonances of free and bound excitons*

The key quantity defining the optical properties of the system is the frequency ω and the wavevector $\boldsymbol{k}$-dependent dielectric susceptibility tensor $\hat{\epsilon}(\omega, \boldsymbol{k})$, which links the Fourier components of the displacement $\boldsymbol{D}$ and electric $\boldsymbol{E}$ fields inside the medium:

$$D_i(\omega, \boldsymbol{k}) = \epsilon_{ij}(\omega, \boldsymbol{k}) E_j(\omega, \boldsymbol{k}), \tag{13.2}$$

where $i, j = x, y$, or z are Cartesian coordinates. In what follows we consider for simplicity the following specific geometry of experiment: light propagates along the c-axis of the GaN, $\boldsymbol{k} \parallel z \parallel [0001]$, and the field is linearly polarized

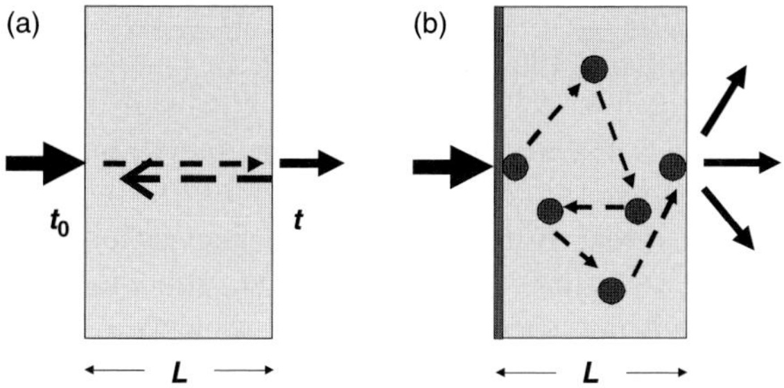

FIG. 13.2. (a) Ballistic and (b) diffusive light propagation through a crystal. The optical pulse impinges normally the sample surface at the time moment t_0 and goes out at time t. Thus, the delay time $T = t - t_0$.

in the xy plane, say, along the x axis. We neglect effects related to possible optical activity of the wurzite GaN. In this case, the vector nature of the fields can be disregarded, and the dielectric susceptibility tensor reduces to a scalar $\epsilon(\omega, \boldsymbol{k}) = \epsilon_{xx}(\omega, \boldsymbol{k}) \equiv \epsilon_{yy}(\omega, \boldsymbol{k})$.

For frequencies in the vicinity of the fundamental absorption edge of GaN, the main contribution to the dielectric susceptibility is provided by excitons. The dielectric polarization $\boldsymbol{P}$ of a single exciton resonance is related to its center-of-mass wavefunction, and satisfies the following equation of motion (Ivchenko, 1982)

$$\left(-\omega + \omega_0 + \frac{\hbar k^2}{2M} - \mathrm{i}\Gamma\right)\boldsymbol{P} = \frac{d^2}{\hbar}\boldsymbol{E}, \tag{13.3}$$

were ω_0 is the exciton resonance frequency, M is its effective mass, Γ is the non-radiative damping, and d is the optical dipole moment of the exciton. Hence, the displacement field $\boldsymbol{D} = \epsilon_b \boldsymbol{E} + 4\pi \boldsymbol{P}$, where ϵ_b is the background dielectric constant (whose frequency and spatial dispersion can, as a rule, be neglected), and the dielectric susceptibility in the presence of a single exciton resonance has the form:

$$\epsilon = \epsilon_b - \frac{f\omega_0}{\omega - \omega_0 - \frac{\hbar k^2}{2M} + \mathrm{i}\Gamma}, \tag{13.4}$$

where we have introduced the oscillator strength of the exciton transition $f = 4\pi d^2/(\hbar\omega_0)$. The longitudinal–transverse splitting of the bulk exciton is given by $\hbar\omega_{LT} = \hbar\omega_0 f/\varepsilon_b$. Hence, dielectric susceptibility has a pole at the frequency ω corresponding to the excitation frequency of the exciton, $\omega = \omega_0 + \hbar k^2/(2M) - \mathrm{i}\Gamma$. Equations (13.3) and (13.4) allow for simple classical interpretation where the optically active exciton state is treated as a classical dipole with a certain frequency, which is excited by the electromagnetic wave propagating in the system.

In real semiconductors there are several types of excitons contributing the susceptibility $\epsilon(\omega, k)$ (Hopfield, 1960; Lagois, 1977). In particular, in wurzite GaN there are three types of free excitons A, B, and C related to the corresponding subbands of the complex valence band. Moreover, inevitable impurities, such as Si and O donors, cause exciton localization and give rise to bound exciton (BX) states. Disorder in any real system also causes fluctuations of excitonic parameters, primarily the resonance frequency ω_0, i.e. inhomogeneous broadening. As a result, in the excitonic spectral region the dielectric susceptibility can be modeled as

$$\epsilon = \epsilon_b + \sum_j \int \frac{\exp(-\xi^2/\Delta_j^2)}{\sqrt{\pi}\Delta_j} \frac{f_j \omega_{0,j}}{\omega - \omega_{0,j} - \frac{\hbar k^2}{2M_j} - \xi + \mathrm{i}\Gamma_j} d\xi, \tag{13.5}$$

where the subscript j enumerates all relevant excitonic states and the convolution with the Gaussian function takes into account the inhomogeneous broadening, and Δ_j is the corresponding broadening of the resonance.

For completeness, we present here the expressions for the oscillator strengths of various exciton transitions. In semiconductors the dipole moment of an exciton is determined by the interband momentum matrix element and the wavefunction of an electron–hole pair at coinciding coordinates (Elliott, 1957). In particular, for free excitons we obtain

$$f_{free} = \frac{4}{\hbar\omega_0 a_B^3} \left(\frac{e|p_{cv}|}{m_0\omega_0} \right)^2, \tag{13.6}$$

where a_B is the excitonic Bohr radius, e is the electron charge, and p_{cv} is the interband momentum matrix element. If excitons are bound to donors with density N, the oscillator strength takes the form (Rashba and Gurgenishvili, 1962)

$$f_{bound} = \frac{4N}{\hbar\omega_0 a_B^3} \left(\frac{e|p_{cv}|}{m_0\omega_0} \right)^2 \left[\int \Psi(\boldsymbol{r}) d\boldsymbol{r} \right]^2, \tag{13.7}$$

where $\Psi(\boldsymbol{r})$ is the envelope function describing exciton center-of-mass localization on the donor. For an order-of-magnitude estimate one has $f_{bound} \sim f_{free} N V_{loc}$, where V_{loc} is the localization volume of the exciton.

Knowledge of the dielectric function $\epsilon(\omega, k)$ makes it possible to model the pulse propagation in the medium and extract all excitonic parameters for comparison of the delays and pulse shapes with available experimental data. Neglecting the spatial dispersion $M_j \to \infty$ in eq. (13.5), one can estimate the delay time T_B as a function of light frequency via the sample thickness L and the group velocity $v_g(\omega) = d\omega/dk$, where $k = k(\omega)(\omega/c)Re\{\sqrt{\epsilon(\omega)}\}$, as

$$T_B(\omega) = \frac{L}{v_g(\omega)}. \tag{13.8}$$

The more elaborate approach is based on the fitting of the transmitted pulse shape, assuming the linear regime of the light propagation. To that end, the spectrum of the impinging laser pulse $E_0(\omega)$ is approximated as a Gaussian with central frequency ω_0 and the envelope maximum at time $t = 0$. The amplitude and phase of the pulse in the frequency domain are given by the Fourier transform at a crystal boundary. We introduce the amplitude transmission coefficient of the sample $T(L, \omega)$, which can be calculated via $\epsilon(\omega, k)$ using the Pekar boundary conditions. It takes into account the multiple reflections of light at the crystal boundaries (Landau and Lifshitz, 1984; Chui, 1992), as well as its delay and attenuation during propagation inside the medium. This allows recasting of the transmitted field in the form (Vainshtein, 1976; Shubina *et al.*, 2011-2)

$$E(L, t) = \int_{-\infty}^{\infty} \frac{d\omega}{2\pi} E_0(\omega) T(L, \omega) e^{-i\omega t}, \tag{13.9}$$

where $E_0(\omega)$ is the Fourier transform of the pulse incident on the sample surface. For direct comparison with streak-camera images we use the Gabor transform, providing time-frequency representation of the transmitted intensity.

13.3.2 *Diffusive propagation of light and resonant photon scattering by bound excitons*

The light delay caused by excitonic contributions to the dielectric susceptibility equation (13.5) is a purely coherent process, which can be interpreted as a ballistic propagation of exciton–polariton whose group velocity differs from the speed of light in vacuum. Additionally, excitons can scatter light, resulting in diffusive propagation of photons inside the medium. This effect can be described in terms of the classical diffusion equation for the coordinate- and time-dependent concentration of photons $n_\omega(z, t)$ with the frequency ω:

$$\frac{\partial n_\omega(z,t)}{\partial t} = D_\omega \frac{\partial^2 n_\omega(z,t)}{\partial z^2} - \frac{n_\omega(z,t)}{\tau_\omega}, \tag{13.10}$$

where D_ω is the photon diffusion coefficient, and we have also introduced the phenomenological photon lifetime τ_ω in the system caused by real absorption processes.

The key quantity in eq. (13.10) is the photon diffusion coefficient related to the group velocity, $v_g(\omega)$, and the mean free path, $l_{tr}(\omega)$, as

$$D_\omega = \frac{v_g(\omega) l_{tr}(\omega)}{3}. \tag{13.11}$$

If the main contribution to the photon scattering is provided by the donor-bound excitons, then the photon mean free path can be expressed via the donor concentration density N and the transport cross-section, $\sigma_{tr}(\omega)$, as

$$l_{tr} = \frac{1}{N\sigma_{tr}(\omega)}. \tag{13.12}$$

The cross-section has a resonant character (Berestetskii *et al.*, 1971; Lang *et al.*, 2007):

$$\sigma_{tr}(\omega) \propto \frac{d_{BX}^4 \omega_{0,BX}^4}{(\omega - \omega_{0,BX})^2 + \Gamma^2}. \tag{13.13}$$

This equation should be averaged with the Gaussian distribution of $\omega_{0,BX}$ (similar to eq. (13.5)) to allow for inhomogeneous broadening in a real system. Obviously, the enhancement of the bound-exciton scattering cross-section results in significant reduction of the photon mean free path and, correspondingly, of the diffusion length if the photon energy approaches the bound-exciton resonance. This provides a significant delay in light propagation via imperfections in a crystal.

Interestingly, due to the interplay of the light diffusion and light absorption, the delay time has a non-trivial dependence on the sample length. If absorption processes are relatively weak, the delay time for the diffusive propagation is given by

$$T_D = \frac{L^2}{2D_\omega}, \tag{13.14}$$

and is quadratic in the sample length. However, if absorption is sufficiently strong ($\tau_\omega \ll T_D$), the delay time depends linearly on the sample thickness (Shubina *et al.*, 2008):

$$T'_D = \frac{L}{2}\sqrt{\frac{\tau_\omega}{D_\omega}}. \tag{13.15}$$

This regime can be realized in GaN epilayers in the vicinity of neutral donor bound-exciton resonances.

13.4 Time-of-flight spectroscopy of light propagating through GaN crystals

Studies in time-of-flight spectroscopy have been performed using high-quality free-standing GaN layers grown by hydride vapor phase epitaxy (HVPE), which have the width of neutral donor bound-exciton lines in the range 0.3–0.5 meV, pronounced A and B free exciton (FE) lines, and the lowest carrier concentration of $(5-8)\times 10^{15}$ cm^{-3}. The samples have different thicknesses L varying from 0.3 to 2 mm. A tunable Ti:sapphire femtosecond pulsed laser and a tunable picosecond laser (Mira-HP) were used as light sources. A Hamamatsu streak camera with a temporal resolution less than 20 ps was used for detection.

Two geometries of measurement have been exploited: i) the conventional backward one, and ii) forward transmission spectroscopy, when the light was detected from the opposite side of a sample with respect to the excitation. In the backward geometry with the 4.66 eV laser energy excitation, the doublet of BX lines related to the Si and O donors dominates the spectrum. The emission peaks of A and B FE are well resolved. In the transmission geometry, the PL signal quenches above a certain boundary, which is very close to the BX doublet. The sharp cut-off is at 3.470 eV in the 1-mm-thick sample, being smoother and redshifted in the 2-mm one. Below the cut-off energy the transmitted PL resembles that registered in the conventional geometry, with one remarkable exception. It appears with a delay in time, which increases gradually towards the cut-off, attaining 470 ps in the 1-mm sample (Fig. 13.3).

The sharp decrease in intensity of the transmitted PL near the BX lines is accompanied by the change of its decay curves (Fig. 13.4). The initial fast-decay characteristic for conventional backward PL disappears. It appears that related processes are terminated before the light exits the crystal, which is possible if their characteristic times are less than the time of the light-propagation. At slightly lower photon energies, where the delay is smaller, the fast component in the PL decay curves is recovered. The characteristic decay times of the transmitted PL increase slightly towards the BX lines, probably due to scattering as the photon passes through the sample. Note that this is in the energy range where the FE-BX scattering process has been suggested to be important (Korona, 2002). It can contribute to the observed difference in the decay of no-phonon PL lines and their phonon replicas, because the latter are in the field of relative transparency (Monemar *et al.*, 2010).

To study the propagation of the laser pulse solely, without superimposing the transmitted pulse with the PL excited by the laser, the central energy of an fs laser pulse was tuned to 3.465 eV. The bending of the pulse towards BX is observed in the samples of different thicknesses L, similar to PL. The delay time-dependence on L is very close to the linear ($T \sim L$), as is expected for polariton propagation. An additional curved streak ("reflex") has been recorded in the 1-mm sample of best quality (Fig. 13.5). It appears at times corresponding

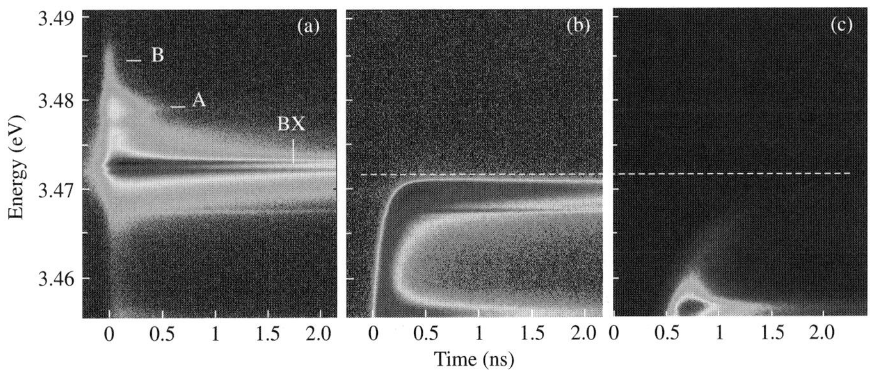

FIG. 13.3. (a) Backward and (b,c) forward TR PL images recorded in (a,b) 1-mm and (c) 2-mm-long samples with excitation by a 4.66-eV laser line at 2 K (logarithmic scale). A, B, and BX denote respective excitons, and the dashed line marks the light-transfer cut-off.

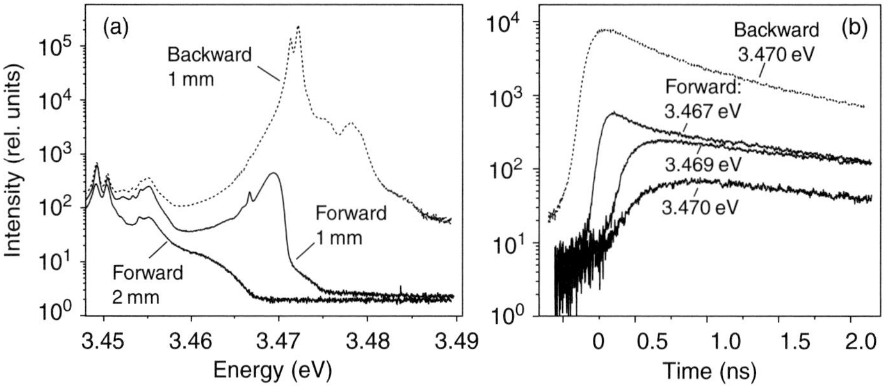

FIG. 13.4. (a) Integrated PL spectra measured with excitation by a 4.66-eV laser line at 2 K, using transmission (forward) geometry in 1-mm and 2-mm-long samples. The conventional (backward) PL measured in the 1-mm sample is also shown. (b) Decay curves of the forward PL measured at different energy in the 1-mm sample, shown together with the decay curve of the backward PL.

to the triple coherent passage of a wave packet, as a result of reflection from sample boundaries. The appearance of the reflex is an argument in favor of the predominant ballistic passage. From the point of common sense, a photon should lose any memory on an initial direction after a few acts of scattering. This conclusion has been proved by the modelling of the shapes of an optical pulse with the ballistic and diffusive propagations.

With a temperature rise, the free exciton resonances are shifted towards lower energies, following the band-gap variation. It results in the attenuation of the pulse, until its full disappearance. The reflex corresponding to the polariton propagation vanishes first from the time-resolved images, while certain decay of the optical signal, given by its curvature, exists up to 80–100 K. This temperature corresponds roughly to the delocalization of bound excitons. At higher temperatures one can anticipate the enhanced resonant interaction between excitons and phonons, whose population increases with a rise in temperature.

To gain additional information on the light transfer we have measured the far-field angular dependence of the transmitted light. We expected the ballistic propagation to yield a narrow angular diagram of the output intensity, while the diffusion could provide a much broader distribution, because the light should be scattered into the full solid angle. For this experiment we used a Xe lamp, whose light passed through a monochromator impinging normally to the sample surface. The measurement revealed a superposition of two components in the transmitted signal: relatively narrow and broad ones (Fig. 13.7). The intensity ratio between these two components depends on the photon energy. The ballistic

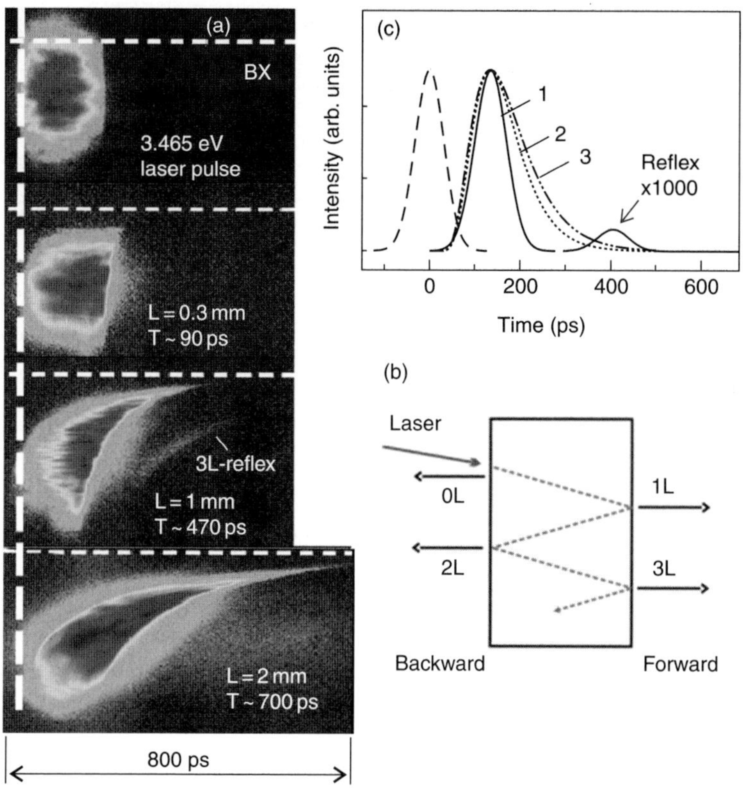

FIG. 13.5. (a) Variation of the shape of the 3.465-eV pulse of a fs laser passing through the GaN crystals of different thicknesses (logarithmic scale). (b) Scheme illustrating the light reflection from the crystal boundaries that provides the observed reflex. (c) Shape of the 3.465-eV pulse on an intensity-time plane calculated for ballistic (1) and diffusive propagations in a sample neglecting (2) and taking into account (3) the 97% light reflection at the crystal boundaries.

part dominates in the region of relative transparency, while the diffusive one prevails in the vicinity of the BX resonances.

The coexistence of two regimes of light propagation without significant broadening of the pulse implies that their delay times do not differ strongly. It is worth the reminding that for conventional diffusion the delay time would increase quadratically with the sample thickness. However, in the spectral range of strong absorption the diffusion also obeys the linear dependence (see Section 13.3 and eq. (13.15)). The photons that travel along a twisting trajectory, which differs significantly from a direct way of polariton propagation, simply cannot reach the opposite faces of the crystal. The applicability of the two models presented

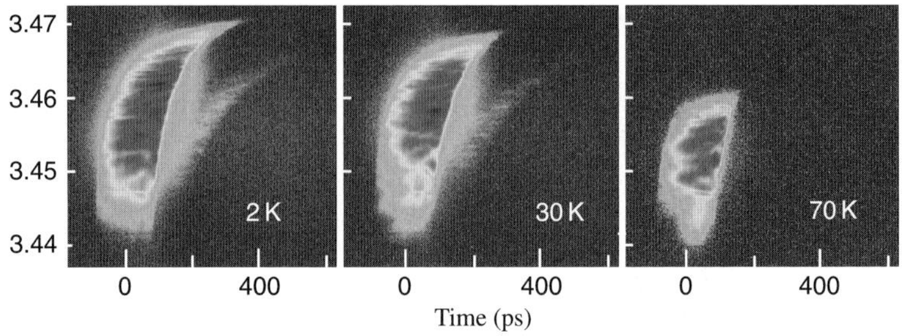

FIG. 13.6. Variation of a fs laser pulse shape with a temperature rise (logarithm scale). Figures adapted from Ref. (Shubina *et al.*, 2008).

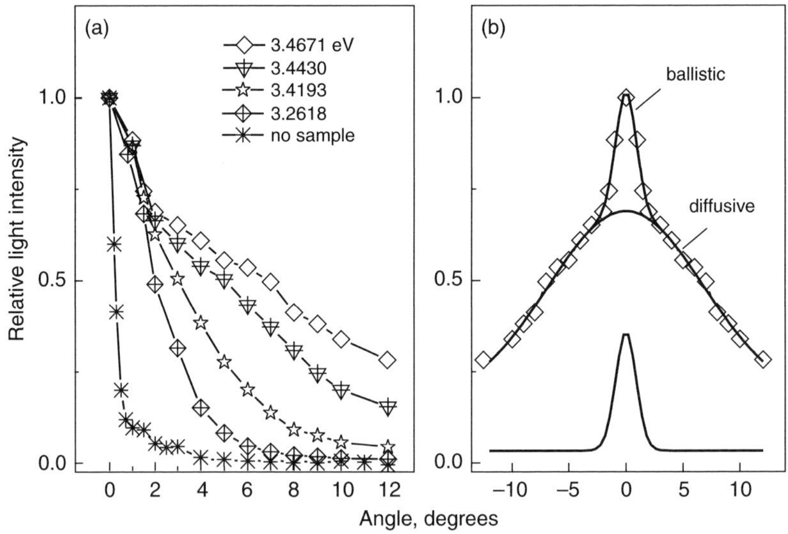

FIG. 13.7. (a) Angular diagrams of the intensity of light transmitted through the 1-mm GaN sample measured at different photon energies. The diagram without the sample is presented for comparison. (b) Deconvolution of the diagram of the transmitted light measured at 3.4671 eV into narrow and broad components. The broad diffusive component is distorted at large angles due to shading by cryostat windows. The width of the ballistic components is due to the light scattering by surface inhomogeneities. Figures adapted from Ref. (Shubina *et al.*, 2008).

in Section 13.3 to describe the light delay in GaN is confirmed by simulations carried out in the limiting cases of either ballistic or diffusive propagation.

The polariton model used has taken into account all excitonic resonances A, B, C, and BX (a single bound exciton state was considered for simplicity). The inhomogeneous broadening of the exciton line has been taken into account by means of convolution with the Gaussian with the width Δ_j. The free exciton energies were obtained from the fitting of reflection spectra. For each of these resonances, $\hbar\Gamma_j$ is about 10 μeV, which is comparable with the homogeneous linewidth of an exciton-polariton in other semiconductors (Masumoto *et al.*, 1983). The inhomogeneous width $\hbar\Delta_j$ is taken close to the average PL linewidth in PL spectra. The oscillator strength for the bound excitons $f_{BX} \approx 2 \cdot 10^{-5}$ is taken somewhat smaller than a value estimated according to eq. (13.7) for the measured electron concentration $\sim 5 \cdot 10^{15}$ cm^{-3}. The exciton localization radius is assumed to be approximately equal to the Bohr radius (Suffczynski and Wolniewicz, 1989). The oscillator strength for the A, B, and C excitons is found as a result of simultaneous fitting of cw transmission and delay-time spectra. In the next section a more precise way of determining the excitonic parameters will be described.

In the diffusive regime the photon mean free path $l_{tr}(\omega)$ and, respectively, the diffusion constant D_ω are determined by the light-scattering cross-section $\sigma(\omega)$; see eqs. (13.11)–(13.12). A reasonable fit of the delay dependencies has been obtained with $\gamma = 0.75$ meV and the same f_{BX} as in the ballistic model. Strictly speaking, eq. (13.13) describes scattering caused by individual BX; that is, it is valid only for the diluted donor concentration, i.e. when $N^{1/3}\lambda \ll 1$, where λ is a wavelength inside the crystal. Near BX, this is fulfilled for $N \sim 10^{15}$ cm^{-3}, while our best sample has $N \sim 5 \cdot 10^{15}$ cm^{-3}. However, two types of the donor are in GaN; namely, a neutral oxygen donor on the N site, and a neutral silicon donor on the Ga site. In PL spectra the lines at 3.4714 eV and 3.4723 eV are associated with an A exciton bound to these O and Si donors, respectively. With full width at half maximum (FWHM) for both lines about 0.3 meV (Monemar *et al.*, 2008), they are well separated. Thus, the scattering is dominated by the O donor centers, hence the actual N is smaller than the Hall electron concentration. Besides, the spatial distribution of donors is inhomogeneous. The donor clusters act as virtual scattering centers with effective concentration and localization radius.

These simulations show that the experimental data can be reasonably fitted in both models. The method of separating the polaritonic and diffusive components will be discussed in the next section. The calculated dependencies clarify the origin of the sharp cut-off of light propagation near BX resonances. It appears because the group velocity of light strongly decreases here, while the resonant absorption and scattering is enhanced.

13.5 Excitonic parameters of wide-gap semiconductors

Until now there is huge variation in reported values of the exciton oscillator strength and longitudinal–transverse splitting (ω_{LT}) of GaN, which is caused by various factors, such as different elastic strain (Gil *et al.*, 1996), the quality of samples (Reynolds *et al.*, 2002), and, certainly, by the specifics of various experimental techniques. A high value of ω_{LT} can be obtained if it is defined via

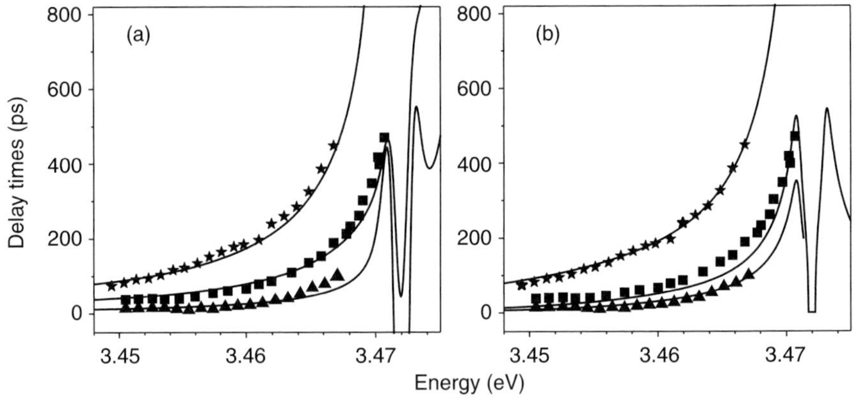

FIG. 13.8. Delay times measured in the 1-mm (squares), 2-mm (stars), and 0.3-mm (triangles) samples shown together with theoretical dependencies calculated using (a) polariton and (b) diffusion models.

a separation between photoluminescence (PL) peaks, which are influenced by a thermalization process (Gil *et al.*, 1998). To extract the free exciton parameters the modeling of reflectivity spectra are commonly used (Stepniewski *et al.*, 1997; Torii *et al.*, 1999). The uncertainty of this method is caused by a large number of fitting parameters. Besides, both PL and reflectivity techniques probe an area close to the sample surface rather than its volume, whereas the properties of these regions may differ (Lagois, 1981; Monemar *et al.*, 2010). As a result, the published data on the homogeneous width for an A exciton vary from 0.2 meV (Torii *et al.*, 1999) up to 2.1 meV (Toda *et al.*, 2005). This parameter is commonly considered as an empirical damping constant (Γ_{eff}), though it has a certain physical meaning. Its inverse value characterizes the dephasing/relaxation time (Takagahara, 1985).

To refine the excitonic parameters of bulk semiconductors we have proposed the following (Shubina *et al.*, 2011-1, 2011-2):

i) To use the data on processes taking place inside the crystal; namely, cw transmission and pulse propagation. When a sample is thick enough, the surface effects are negligible.
ii) To model the ballistic reflexes, because they do not contain any contribution of diffusive photons.

The typical images of the reflexes, emerging with the shift of the pulse energy from the BX lines towards a transparency region, are presented in Fig. 13.9. The delay and shape of these reflexes reproduce the variation of the group velocity, which in turn is controlled by the parameters of the exciton resonances. The light spot marked 0L is reflected straight from the front surface. Those marked as 2L and 4L correspond to the two-fold and four-fold passage of the light inside

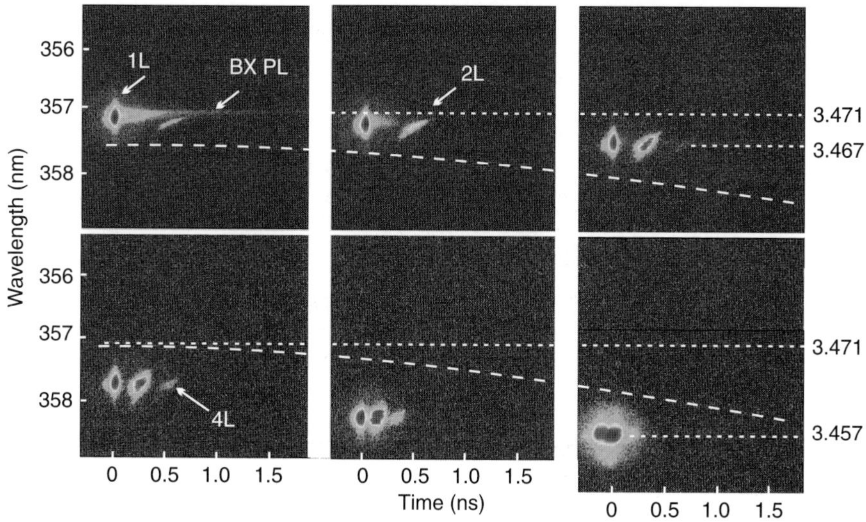

FIG. 13.9. Selected TR images of replicas of impinging pulses of a ps tunable laser, recorded in back-scattering configuration in the 1-mm-long GaN crystal (logarithmic scale). The notation $n{\cdot}L$ corresponds to the multiplicity of the light passage inside the crystal. The dashed lines are for guiding an eye.

the crystal, respectively. Besides the exclusion of the diffusion factor, the use of $n{\cdot}L$ replicas increases the accuracy of the methods, because their study is equivalent to the study of crystals with exactly the same parameters but of different thicknesses.

Figure 13.10a illustrates the importance of the elimination of a diffusion contribution. It shows the spectral dependencies of the delay times measured along the pulses of different energies for the same length of propagation; namely, through the 2-mm-long sample and for 2L replicas in the 1-mm-long sample. The delay time of the basic transmitted pulse in the 1-mm sample, multiplied by two in order to simulate the 2-mm length, is also shown. Obviously, the delay times per unit length are smallest for the 2L reflexes, when the photon scattering, enhancing the light path, is excluded. The difference in the dependencies for the 1-mm and 2-mm samples is related to different donor densities, which influence the scattering cross-sections. It can be used to estimate the contribution of the diffusion and its parameters.

The simulation shows that the influence of the neutral donor bound excitons is noticeable only at a distance less than 1–2 meV from their peaks. The oscillator strength and homogeneous width of the BX excitons for the 1-mm sample are found to be $5{\cdot}10^{-6}$ and 1 μeV, respectively. These parameters are assumed to be similar for both BX lines, related to the oxygen and silicon donors. A deviation from these values exceeding 10% does not allow to reproduce the shape of the

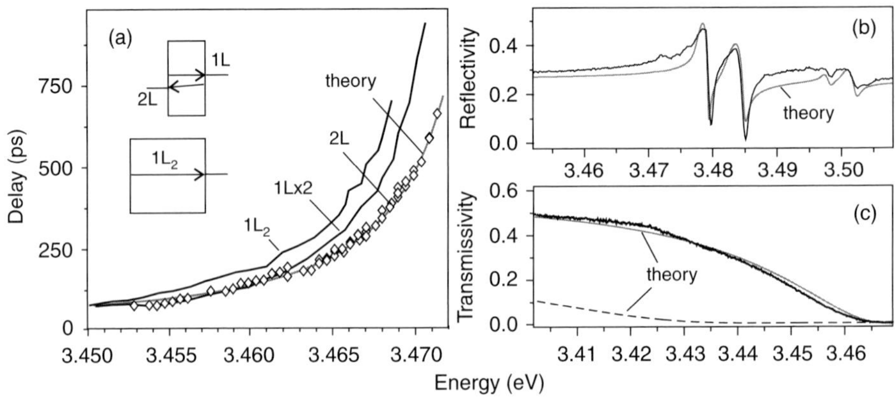

FIG. 13.10. (a) Experimental dependencies of delay times for: 1L, basic pulses propagating through the 1-mm-long sample, multiplied by two; 2L, reflexes in the same sample propagating the double path shown together with its fitting; and $1L_2$, basic pulses in the 2-mm-long sample. (b) Reflectivity and (c) transmittivity spectra measured in the 1-mm GaN sample, shown with their simulations. Fittings of the transmittivity neglecting the inhomogeneous exciton line-broadening is shown by a dashed line in (c).

reflexes shown in Fig. 13.9. The inhomogeneous width of the BX resonances is determined by consideration of the light attenuation in the close vicinity of these lines. A width of more than 35 μeV causes quenching of the higher-energy tail of the reflexes at the energies where the signal is actually observed.

A similar approach has been applied to study ZnO (Shubina *et al.*, 2011-2). The overall pulse delay in this material, which reached 1.6 ns at 3.374 eV with a 0.3-mm propagation path, is determined by the free exciton-polariton resonances. The BX lines influence locally; they provide dips, cutting the pulses into several parts, and induce extra retardation of light nearby. Because of the large number of donor types, the partial density of each of them is apparently not too high in the best ZnO samples. Therefore, the scattering and diffusion of photons is not pronounced. The delay times of the replicas in ZnO occur strongly proportional to the multiplicity of their passage. Both the directly transmitted signal and replicas can be fitted using the same values of the excitonic parameters.

The refined parameters of exciton-polaritons in GaN and ZnO are given in Table 13.1. (The data on distant C exciton are not presented here). These data demonstrate that the exciton–polaritons propagating within GaN at low temperature have an homogeneous width in the order of 10 μeV, and the oscillator strength is about 30% lower than the currently accepted values. In ZnO the $\hbar\omega_{LT} \approx 3$ meV turns out to be higher than the currently accepted 2 meV (Klingshirn *et al.*, 2010). The Γ value of 3 μeV at 2 K is the lowest among those

Table 13.1 Exciton–polariton parameters of A and B excitonic series in GaN and ZnO derived at 2 K, using time-of-flight spectroscopy.

Parameter	A exciton	B exciton
	GaN	
$\hbar\omega_0$ (eV)	3478.4 ± 0.0002	3483.4 ± 0.001
f	0.0022 ± 0.0002	0.0016 ± 0.001
ω_{LT} (meV)	0.8	~0.6
$\hbar\Gamma$ (μeV)	13	13
$\hbar\Delta$ (meV)	<0.85	<1
	ZnO	
$\hbar\omega_0$ (eV)	3.376 ± 0.0002	3.382 ± 0.001
f	0.0072 ± 0.0002	0.012 ± 0.001
ω_{LT} (meV)	3	5
$\hbar\Gamma$ (μeV)	3	3.5
$\hbar\Delta$ (meV)	<0.5	<1

published for polaritons in semiconductors. The unusual transparency of ZnO was noticed in the first spectroscopic studies of its excitonic spectrum, when a certain discrepancy was observed between the absorption data and the value deduced from the Kramers–Kronig analysis of reflectivity spectra (Thomas, 1960). Note that markedly higher values of Γ would result in full opaqueness in the transmission spectra in the range where the narrow BX lines can be clearly resolved. However, there is a certain uncertainty in the exact values caused by the determination of the parameter a bit below the exciton peak. Some increase of them can be predicted for the measurements directly at the resonance (Masumoto *et al.*, 1983).

Because even a very weak line causes noticeable absorption if a sample is sufficiently thick, the number of BX lines resolved by time-of-flight spectroscopy is higher than in the PL spectra. It allows us to reconstruct a fine structure of energy spectrum of bound excitons. The full list of BX lines and their parameters in ZnO recorded by this method can be found in Shubina *et al.* (2011-2).

13.6 Distortion of optical pulse near excitonic resonances

A light pulse generally contains a superposition of light-waves of different frequencies. Each individual frequency component of the pulse travels at its own phase velocity, which is determined by the refractive index of a material (Vainshtein, 1976). When entering a dispersive medium, the envelope of the pulse begins to change shape. The higher-energy tail of the pulse might move forwards or backwards with respect to the rest, depending on the sign of the group index (Wiersma, 2008). In order to provide high-speed data-processing photonic devices must operate with short pulses, fs or ps. The spectral width of such pulses can approach several meV. In the close vicinity of an exciton resonance, huge variation of all optical parameters takes place along the pulse spectrum. It results in the curvature of the pulse shape on the energy–time plane and its non-homogeneous attenuation. The following distortion with propagation of the short light-pulses can be anticipated:

1. Shift of the "center of gravity" of a pulse to lower energy.
2. Asymmetrical modification of the higher-energy and lower-energy fronts of the pulse.
3. Cutting of the pulses into separated parts by narrow absorption lines.
4. Extra retardation of an optical signal near these lines.
5. Quantum interference (beating) between separated parts of the signal of close energies.

These features can be found in experimental images of the laser pulses propagating through GaN and ZnO crystals (see, e.g., Shubina *et al.*, 2008, 2011-1, 2011-2). Note that there is a range of optimal thicknesses, where fine features such as the replicas and beating can be observed. For a distinct representation we have simulated the images of the fs laser pulses for two propagation lengths (Fig. 13.11), using the model described in Section 13.3 and the excitonic parameters extracted from the experimental data (Section 13.5). For direct comparison with the experimental images we use the Gabor transform with the temporal window of ~15 ps. With $L \approx 0.3$ mm, the images reproduce the modification of the pulse shapes on the energy–time plane induced by the narrow BX lines, further to the basic curvature and attenuation.

Figure 13.12 presents, for comparison, experimental images of fs and ps laser pulses propagating through the 0.3-mm long ZnO sample. In general, the tunable ps laser provides higher resolution at the time-of-flight spectroscopy, which is why it has been used to monitor reflexes in GaN and the fine spectrum of BX excitons in ZnO in our studies. By comparison with the image of the fs pulses, one can conclude that the smaller the pulse duration the less the disturbance of the pulse fronts. Thus, there is a chance for excluding the cutting of the pulse by the narrow BX lines with proper choice of the laser wavelength.

The observed modification of the optical pulses in the vicinity of exciton resonances can influence the threshold and frequency characteristics of a set of optoelectronic devices—high-frequency ones, or those where the light should propagate a sufficiently long way. Among them there are vertical cavity surface-emitting lasers, and polariton lasers with a semiconductor microcavity. It also concerns conventional laser heterostructures, where the light travelling through a planar waveguide passes through a distance of 1–2 mm.

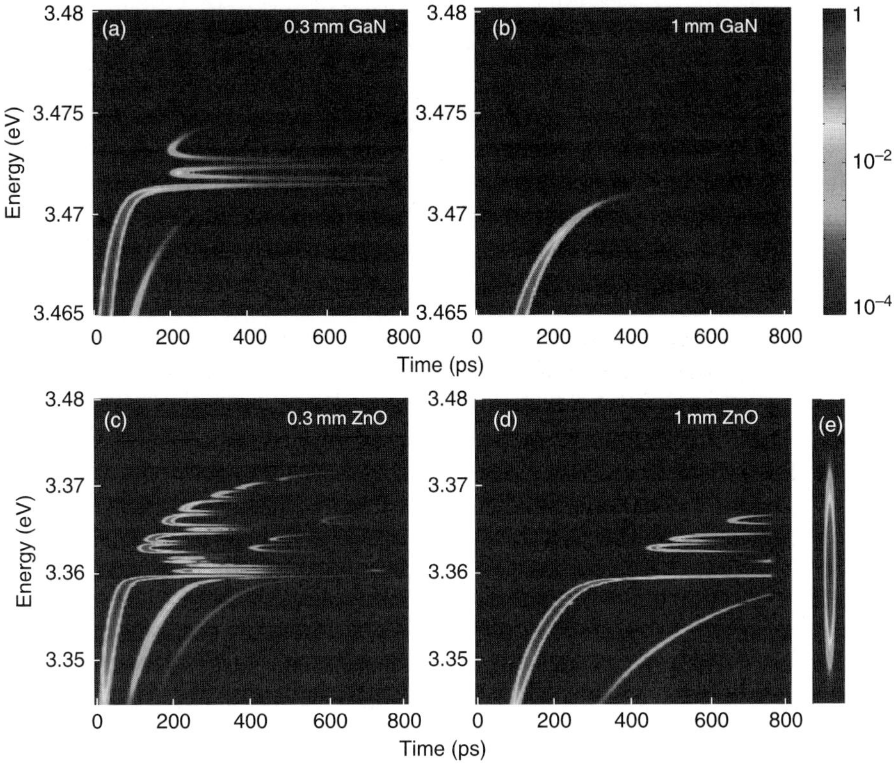

FIG. 13.11. Distortion of a Gaussian pulse of a fs laser on the energy–time plane after propagation through (a,b) GaN and (c,d) ZnO crystals of different thicknesses. The initial shape of the pulse is shown in (e).

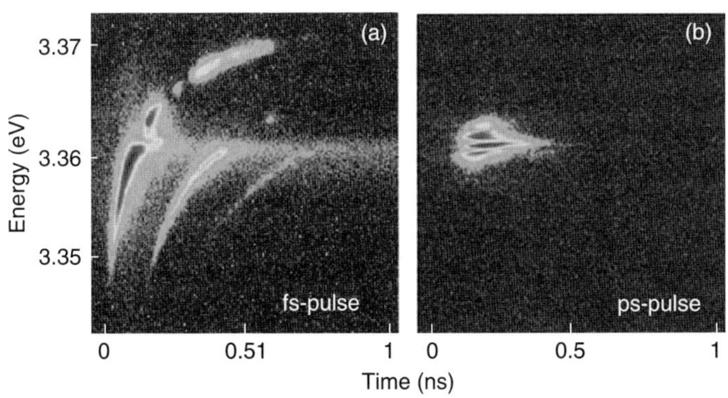

FIG. 13.12. Pulses of (a) femtosecond and (b) picosecond lasers possessing close central energies and propagating through a 0.3-mm ZnO crystal near the basic doublet of neutral donor-bound excitons (logarithm scale).

13.7 Concluding remarks

Our time-of-fight studies of wide-gap semiconductors GaN and ZnO have demonstrated that slow light transfer with a group velocity of $\sim c/100$ is possible in the spectral range of strong optical dispersion near excitonic lines. Simulation of the shape of laser pulses on the energy–time plane enables us to determine the excitonic parameters of bulk semiconductors at low temperatures, as well as to establish that the resonances in these materials must have a low damping and limited broadening to allow the light propagation near them. Besides, time-of-flight spectroscopy opens a new route for refining an excitonic spectrum of bound excitons.

As anticipated, slowing down the light velocity by excitons is accompanied by distortion of the light pulses and their strong attenuation. Moreover, a noticeable delay in studied semiconductors has been observed only at relatively low temperatures. This limits the immediate use of the results obtained for photonics and optoelectronics, which need in delay units operating at room temperature for random-access memory, data synchronization, pattern correlation, and network buffering. For all these purposes, it is necessary to have the delay of the light by several times exceeding the pulse duration with reasonable intensity attenuation. We believe that the time-of-fight studies of various semiconductors, both polar and nonpolar, over a wide temperature range will enable to make a final conclusion on the applicability of the optical dispersion in these materials to provide the slow light.

In general, we should conclude that an ideal approach to produce the slow light is currently absent, in spite of interesting results and certain progress in this field. It concerns both nonlinear effects and the optical dispersion at exciton resonances in semiconductors. Such a situation implies the necessity of further investigation of different mechanisms in various material systems, bearing in mind that the study of optical dispersion is a fundamental issue for any method of slow-light realization.

References

Agranovich V. M., and Ginzburg V. L. (1984) *Crystal optics with spatial dispersion, and Excitons* (Springer-Verlag Berlin and Heidelberg GmbH & Co. K).

Albada M. P. van, and Lagendijk A. (1985) *Phys. Rev. Lett.* **55**, 2692.

Albada M. P. van, Tiggelen B. A. van, Lagendijk A., and Tip A. (1991) *Phys. Rev. Lett.* **66**, 3132.

Aleksandrov E. B., and Zapasskii V. S. (2004) *Sov. Phys. Usp.* **174**, 1105.

Bajcsy M., Zibrov A. S., and Lukin M. D. (2003) *Nature* **426**, 638.

Barabanenkov Yu. N., and Ozrin V. D. (1992) *Phys. Rev. Lett.* **69**, 1364.

Berestetskii V. B., Lifshitz E. M., and Pitaevskii L. P. (1971) *Relativistic Quantum Theory* (Pergamon Press).

Bigelow M. S., Lepeshkin N. N., and Boyd R. W. (2003-1) *Science* 301, 200.

Bigelow M. S., Lepeshkin N. N., and Boyd R. W. (2003-2) *Phys. Rev. Lett.* **90**, 113.

Bishop M. F., and Maradudin A. A. (1976) *Phys. Rev. B* **14**, 3384.

Boller K.-J., Imamoğlu A., and Harris S. E. (1991) *Phys. Rev. Lett.* **66**, 2593.

Born M., and Wolf E. (1965) *Principles of Optics* (Oxford: Pergamon Press).

Boyd R. W., Gauthier D. J., Gaeta A. L., and Willner A. E. (2005) *Phys. Rev. A* **71**, 023801.

Brillouin L. (1960) *Wave Propagation and Group Velocity* (New York: Academic Press).

Calleja J. M., and Cordona M. (1977) *Phys. Rev. B* **16**, 3753.

Camacho R. M., Park M. V., and Howell J. C. (2006) *Phys. Rev. A* **73**, 063812.

Chang-Hasnain C. J., and Chuang S. L. (2006) J. *Lightwave Technology* **24**, 4642.

Chandrasekhar S. (1960) *Radiative Transfer* (Dover Publications, UK).

Chen S. L., Chen W. M., and Buyanova I. A. (2011) *Phys. Rev. B* **83**, 245212.

Chu S., and Wong S. (1982) *Phys. Rev. Lett.* **48**, 738.

Chui C. K. (1992) *Wavelet Analysis and Its Applications, Vol. 1: An Introduction to Wavelets* (Academic Press, New York).

Crisp M. D. (1971) *Phys. Rev. A* **4**, 2104.

Dahan D., and Eisenstein G. (2005) *Opt. Express* **13**, 6234.

Eisaman M. D., Childress L., André A., Massou F., Zibrov A. S., and Lukin M. D. (2004) *Phys. Rev. Lett.* **93**, 233602.

Eisaman M. D., André A. Massou F., Fleischhauer M., Zibrov A. S., and Lukin M. D. (2005) *Nature* **438**, 837.

Elliott R. J. (1957) *Phys. Rev.* **108**, 1384.

Faxvog F. R., Chow C. N. Y., Bieber T., and Carruthers J. A. (1970) *Appl. Phys. Lett* **17**, 192.

Frish C. E. (1976) *Physics-Uspekhi* **118**, 565.

Fröhlich D., Kulik A., Uebbing B., Mysyrowicz A., Langer V., Stolz H., and von der Osten W. (1991) *Phys. Rev. Lett.* **67**, 2343.

Frogley M. D., Dynes J. F., Beck M., Faist J., and Phillips C. C. (2006) *Nature Materials* **5**, 175.

Garcia N., Genask A. Z., and Lisyansky A. A. (1992) *Phys. Rev. B* **46**, 14475.

Garrett C. G. B. and McCumber D. E. (1970) *Phys. Rev. A* **1**, 305.

Gil B., Hamdani F., and Morkoç H. (1996) *Phys. Rev. B* **54**, 7678.

Gil B., Hoffmann A., Clur S., Eckey L., Briot O., and Aulombard R.-L. (1998) *J. Cryst. Growth* **189/190**, 639.

Gamarts E. M., Ivchenko E. L., Karaman M. I., Mushinskii V. P., Pikus G. E., Razbirin B. S., Starukhin A. N. (1977) *Zh. Eksp. Teor. Fiz.* **73**, 1113 [Engl. transl.: (1977) *Sov. Phys. JETP* **46**, 590].

Gnedin Yu. N., and Dolginov A. Z. (1963) *Zh. Exper. Teor. Fiz.* **45**, 1136 [Engl. transl: (1964) *Sov. Phys. JETP* 18, 784].

Godde T., Akimov I. A., Yakovlev D. R., Mariette H., and Bayer M. (2010) *Phys. Rev. B* **82**, 115332.

Ikehara T., and Itoh T. (1991) *Phys. Rev. B* **44**, 9283.

Itoh T., Lavallard P., Reydellet J., and Benoit à la Guillaume C. (1981) *Solid State Commun.* **37**, 925.

Ivchenko E. L., Pikus G. E., Razbirin B. S., and Starukhin A. N. (1977) *Zh. Eksp. Teor. Fiz.* **72**, 2230 [Engl. transl.: (1977) *Sov. Phys.* JETP **45**, 1172].

Ivchenko E. L. (1982) *Spatial dispersion effects in the exciton resonance region.* In: E. I. Rashba and M. D. Sturge, eds. *Excitons.* pp. 141–176 (North-Holland, Amsterdam).
Ivchenko E. L. (2005) *Optical Spectroscopy of Semiconductor Nanostructures* (Alpha Science, Harrow, UK).
Halevi P., and Fuchs R. (1985) *Phys. Rev. Lett.* **55**, 338.
Harris S. E. (1997) *Physics Today* 50, 36.
Harris S. E., Field J. E., and Imamoğlu A. (1990) *Phys. Rev. Lett.* **64**, 1107.
Hau L. V., Harris S. E., Dutton Z., and Behroozi C. H. (1999) *Nature* **397**, 594.
Hau L. V. (2011) *Nature Photonics* **5**, 197.
Hopfield J. J. (1960) *J. Phys. Chem. Solids* **15**, 97.
Jütte M., Stolz H., and von der Osen W. (1996) *J. Opt. Soc. Am. B* **13**, 1205.
Kasapi A., Jain M., Yin G.Y., and Harris S.E. (1995) *Phys. Rev. Lett.* **74**, 2447.
Klingshirn C., Hauschild R., Fallert J., and Kalt H. (2007) *Phys. Rev. B* **75**, 115203.
Klingshirn C., Fallert J., Zhou H., Sartor J., Thiele C., Maier-Flaig F., and Kalt H. (2010) *Phys. Status Solidi B* **247**, 1424.
Korona K. P. (2002) *Phys. Rev. B* **65**, 235312.
Kurita A., Kanematsu Y., Watanabe M., Hirata K., and Kushida T. (1999) *Phys. Rev. Lett.* **83**, 1582.
Kuwata M., Kuga T., Akiyama H., Hirano T., and Matsuoka M. (1988) *Phys. Rev. Lett.* **61**, 1226.
Lagois J. (1977) *Phys. Rev. B* **16**, 1688.
Lagois J. (1981) *Phys. Rev. B* **23**, 5511.
Landau L. D., and Lifshitz E. M. (1984) *Electrodynamics of Continuous Media* (Pergamon, Oxford).
Lang I., Korovin L., and Pavlov S. (2007) *Fiz. Tverd. Tela* **49**, 1304 [Engl. transl.: (2007) *Phys. Solid State* **49**, 1368].
Lederer F., and Biehlig W. (1994) *Electronics Letters* **30**, 1871.
Liptay T. J., Marshall L. F., Rao P. S., Ram R. J., and Bawendi M. G. (2007) *Phys. Rev. B* **76**, 155314.
Loudon R. (1970) *J. Phys. A* **3**, 233.
Loudon R., Allen L., and Nelson D. F. (1997) *Phys. Rev. E* **55**, 1071.
Masumoto Y., Shionoya S., and Takagahara T. (1983) *Phys. Rev. Lett.* **51**, 923.
Matsko A. B., Kocharovskaya O., Rostovtsev Y., Welch G. R., Zibrov A. S., Scully M. O. (2001) *Adv. At. Mol. Opt. Phys.* **46**, 191.
Matsko A. B., Strekalov D., Savchenkov A. A., and Maleki L. (2007) *Phys. Rev. A* **76**, 013806.
Milonni P. W. (2002) *J. Phys. B: At. Mol. Opt. Phys.* **35**, R31.
Minardi S., Eilenberger F., Kartashov Y. V., Szameit A., Röpke U., Kobelke J., Schuster K., Bartelt H., Nolte S., Torner L., Lederer F., Tünnermann A., and Pertsch T. (2010) *Phys. Rev. Lett.* **105**, 263901.
Molotkov S. N. (2010) *Pis'ma v Zh. Eksper. Teoret. Fiz.* **91**, 762.
Monemar B., Paskov P. P., Bergman J. P., Pozina G., Toropov A. A., Shubina T. V., Malinauskas T., and Usui A. (2010) *Phys. Rev. B* **82**, 235202.
Nielsen N. C., Linden S., and Kuhl J., Förstner J., Knorr A., Koch S. W., and Giessen H. (2001) *Phys. Rev. B* **64**, 245202.

Okawachi Y., Bigelow M. S., Sharping J. E., Zhu Z., Schweinsberg A., Gauthier D. J., Boyd R. W., and Gaeta A. L. (2005) *Phys. Rev. Lett.* **94**, 153902.

Puri A., and Birman J. L. (1981) *Phys. Rev. Lett.* **48**, 173.

Rashba E. L. and Gurgenishvili G. E. (1962) *Fiz. Tverd. Tela* **4**, 1029 [(1962) *Sov. Phys. Solid State* **4**, 759].

Reynolds D. C., Jogai B., and Collins T. C. (2002) *Appl. Phys. Lett.* **80**, 3928.

Rybin A.V., Vadeiko I. P., and Bishop A. R. (2005) *Phys. Rev. E* **72**, 026613.

Sakai M., Nakahara R., Kawase J., Kunugita H., Ema, K., Nagai M., and Kuwata-Gonokami M. (2002) *Phys. Rev. B* **66**, 033302.

Segawa Y., Aoyagi Y., Asuma K., and Namba S. (1978) *Solid State Commun.* **28**, 853.

Segawa Y., Aoyagi Y., and Namba S. (1979) *Solid State Commun.* **32**, 229.

Sermage B., Petiot S., Tanguy C., Dang Le Si, and André R. (1998) *J. Appl. Phys.* **83**, 7903.

Shubina T. V., Glazov M. M., Toropov A. A., Gippius N. A., Vasson A., Leymarie J., Kavokin A., Usui A., Bergman J. P., Pozina G., and Monemar B. (2008) *Phys. Rev. Lett.* **100**, 087402.

Shubina T. V., Toropov A. A., Pozina G., Bergman J. P., Glazov M. M., Gippius N. A., Disseix P., Leymarie J., Gil B., and Monemar B. (2011-1) *Appl. Phys. Lett.* **99**, 101108.

Shubina T. V., Glazov M. M., Gippius N. A., Toropov A. A., Lagarde D., Disseix P., Leymarie J., Gil B., Pozina G., Bergman J. P., and Monemar B. (2011-2) *Phys. Rev. B* **84**, 075202.

Shubina T.V., Gippius N. A., Toropov A. A., Pozina G., Bergman J. P., Disseix P., Leymarie J., and Monemar B. (2012) ICPS 2012, Zurich, Swiss, 3.7.

Sich M., Krizhanovskii D. N., Skolnick M. S., Gorbach A. V., Hartley R., Skryabin D. V., Cerda-Méndez E. A., Biermann K., Hey R., and Santos P. V. (2012) *Nature Photonics* **6**, 50.

Silberberg Y. (1990) *Opt. Lett.* **22**, 1282.

Sonnenschein E., Rutkevich I., and Censor D. (1998) *Phys. Rev. E* **57**, 1006.

Stanley R. P., Pau S., Oesterle U., Houdré R., and Ilegems M. (1997) *Phys. Rev. B* **55**, R4867.

Stepniewski R., Korona K. P., Wysmolek A., Baranovski J. M., Pakula K., Potemski M., Martinez G., Grzegory I., and Porowski S. (1997) *Phys. Rev. B* **56**, 15151.

Sukhorukov A. A., and Kivshar Y. S. (2006) *Phys. Rev. Lett.* **97**, 233901.

Suffczynski M., and Wolniewicz L. (1989) *Phys. Rev. B* **40**, 6250.

Takagahara T. (1985) *Phys. Rev. B* **31**, 8171.

Taranenko V. B., Ganne I., Kuszelewicz R., and Weisset C. O. (2001) *Appl. Phys. B* **72**, 377.

Tiggelen B. A. van, Lagendijk A., and Tip A. (1993) *Phys. Rev. Lett.* **71**, 1284.

Tiggelen B. A. van, Maynard R., and Heiderich A. (1996) *Phys. Rev. Lett.* **77**, 639.

Thomas D. G. (1960) *J. Phys. Chem. Solids* **15**, 86.

Toda Y., Adachi S., Abe Y., Hoshino K., and Arakawa Y. (2005) *Phys. Rev. B* **71**, 195315.

Torii K., Deguchi T., Sota T., Suzuki K., Chichibu S., and Nakamura S. (1999) *Phys. Rev. B* **60**, 4723.

Tucker R. S., Ku P.-C., and Chang-Hasnain C. J. (2005) *J. Lightwave Technology*, **23**, 4046.

Ulbrich R. G., and Fehrenbach G. W. (1979) *Phys. Rev. Lett.* **43**, 963.

Vainshtein L. A. (1976) *Sov. Phys. Usp.* **19**, 189.

Vidmont N. A., Maksimov A. A., and Tartakovskii I. I. (1983) *JETP Lett.* **37** 689 [(1983) *Pis'ma Zh. Eksp. Teor. Fiz.* **37**, 578].

Viswanath A. K., Lee J. I., Kim D., Lee C. R., and Leem J. Y. (1998) *Phys. Rev. B* **58**, 16333.

Vlasov Y. A., O'Boyle M., Hamann H. F., and McNab S. J. (2005) *Nature* 438, 65.

Wan J. S., Brebner J. L., Leonelli R., Zhao G., and Graham J. T. (1993) *Phys. Rev. B* **48**, 5197.

Wiersma D. S. (2008) *Nature* **452**, 942.

Xiong G., Wilkinson J., Ucer K. B., and Williams R. T. (2005) *J. Phys.: Condens. Matter* **17**, 7287.

Zabusky N. L., and Kruskal M. D. (1965) *Phys. Rev. Lett.* **15**, 240.

14 Nitride devices and their biofunctionalization for biosensing applications

Csilla Gergely

14.1 Introduction

The past two decades have witnessed considerable interest in biosensor development. The heart of these hybrid analytical devices is the immobilized biologically sensitive material integrated within a transducer. A biosensor ultimately converts the biological information into a quantitatively measurable signal, usually in the form of optical, acoustic, electrical, or magnetic responses. In the realm of biosensors, semiconductor-based devices constitute a promising alternative for molecular detection based on high-sensibility responses in electron transfer, adsorption, surface recombination, photoconductivity, and light-induced contact potential measurements.

Next-generation affinity-based biosensors will require significant improvements in sensitivity, specificity, and parallelism in order to meet future needs in a variety of fields, including *in vitro* medical diagnostics, pharmaceutical discovery, and drug and pathogen detection. The interface between biological molecules and semiconductor surfaces is a key issue in the development of efficient devices based on biomolecular recognition.

The use of high-sensitivity semiconductor materials such as nitrides can considerably enhance biosensing performances. For example, indium nitride (InN) has proven to have robust surface properties and unusually strong surface-electron accumulation, and gallium nitride (GaN) reveals outstanding chemical stability and is transparent, enabling easy optical readout, rendering nitrides as very promising substrates for biological detection. In terms of interfacing them with biological materials, nitrides also present the advantage of a relatively easy functionalization due to their affinity to various organic linkers. Nitrides can be selectively functionalized with peptides produced by a biotechnological method called "phage display", which can be crucial when the retention of contrast in multi-material components is necessary. Biomolecules' capturing can be electrically or optically monitored by metal semiconductor field-effect transistors or photonic crystal resonances, respectively. Photonic crystals provide strong light confinement and can be designed to localize the electric field in the low-refractive-index region, which makes the sensors extremely sensitive to a small change of refractive index produced by the biomolecular binding event. Moreover, their nanostructured design makes them ideal candidates for the development of miniaturized biosensors with very low requirements in terms of volume of biological material.

14.2 Configurations of nitride devices for sensing

Any biosensor system should include the analyte recognition, signal transduction, and readout processes. Sensors rely on measurable responses of the target molecule, such as chemical reactions with electron transfer, adsorption, surface recombination, photoconductivity, and change in contact potential with light. These external chemical stimuli can drastically alter semiconductor properties, and can be measured by monitoring surface current, potential, or impedance characteristics. Hence semiconductors are ideal matrices, including the sensor element and the transducer, for the development of a variety of chemical and biosensor systems. The development of such sensing devices is always based on the effect resulting from the specific or selective interaction of an analyte with the semiconductor surface. Such interaction will usually result in changes of the electrochemical characteristics of the surface. The most important of these parameters are the space charge, the field, and the potential. Once this interaction has been characterized, a variety of sensing schemes can be envisaged.

Hence, creating chemical sensors or biosensors from semiconductors requires precise chemical control of the surface chemistry—conditions necessary for optimizing the analytical characteristics of the sensor, such as selectivity, sensitivity, detection limit, response time, and signal stability. Since analyte recognition and detection result from the perturbation of the electro-optical properties of the semiconductor surface and subsurface layers, there must be a specific and reversible chemical interaction of the analyte with the semiconductor sensing element. During recent decades, tremendous advances in materials science have enabled excellent control of material growth processes, leading to high-quality factor devices for the development of sensors and biosensors.

When sensing via optoelectronic devices, an efficient and direct transfer of charge between semiconducting substrates and bioorganic layers is required. Typically semiconductors, such as GaN, InN, or AlGaN alloy systems, appear to be very promising for bioelectronic applications, due to the good match of their large, wide band-gap with the typical HOMO–LUMO gaps of organic molecules. Moreover, GaN and AlGaN are extremely stable chemically and are naturally biocompatible (Steinhoff *et al.* 2003).

Group III-nitride semiconductors present unique features for the development of photonics-based devices due to the direct band-gap electroluminescence in the blue and near-ultraviolet parts of their optical spectrum. Their significant electro-optic, piezoelectric, and non-linear coefficients render the wide-bandgap III-nitride epitaxial structures very attractive for the development of integrated optical devices and circuits operating at wavelengths throughout the blue/violet part of the visible spectrum and a substantial part of the near-ultraviolet spectrum. Nitrides containing gallium, aluminum, and their alloys, or those with low indium content, have been investigated widely within the last decades. InN with a band-gap lying between 0.6 and 0.7 eV is the least-known material of the nitride family (Davydov *et al.* 2002, Wu *et al.* 2002). InN alone, or combined with GaN or AlN, offers new perspectives in optoelectronics, solar cells, and terahertz applications. Efforts in recent years have provided crystalline-quality InN layers, exhibiting an electron accumulation layer at their surface. The application that can benefit of this unusually high surface sensitivity is sensing (Lu *et al.* 2004, Chen *et al.* 2006). It has also been demonstrated that InN samples grown by molecular beam epitaxy and metal organic vapor phase epitaxy exhibit an enhanced concentration of hydrogen near the surface (Darakchieva *et al.* 2010). However, although hydrogen is expected to behave like a donor in indium nitride (Limpijumnong *et al.* 2001), no correlation between the electron concentration and the hydrogen concentration, both located near the surface region, has been established yet. Wurtzite InN exhibits strong, intrinsic electron accumulation at its surface, which is very advantageous in the design of high-sensitivity sensors.

Due to its transparency in the visible regime and its convenient optical band-gap of 3.4 eV (at room temperature), GaN is a viable candidate for photonic-crystal biosensor applications. In a typical sensing setup, design of the resonant defect cavity in the photonic crystal is such that the resonant frequency of the cavity matches the emission frequency of the molecule. The emission frequency lies within the photonic band-gap, hence the mode is confined within the resonant cavity with little loss (high-quality factor). This resonant confinement leads to an increase in photon lifetime within the cavity. Moreover, the emission yield of the biomolecule is also increased due to the high density of optical modes within the defect region. This has enabled 25-fold amplification of fluorescence emission in the cavity, approaching the limits of detection at single-molecule concentrations (Dawson *et al.* 2008).

Miniaturization, keeping at the same time a large sensing area by assuring high specificity of the sensing surface, is the challenge of current research in biosensing. Nanostructured semiconductor substrates fulfil the miniaturization and large-sensing-area requirements, and are ideal candidates for photonic-based molecular detection, as they present unique possibilities for supporting device functionality that includes strongly confined and localized light emission. Fluorescence enhancement of molecules infiltrated in Bragg-mirror and microcavity structures has been reported previously (Palestino *et al.* 2007). Moreover, infiltrated protein acts as an internal two-photon-excited fluorescence emitter and second-harmonic generator (Martin *et al.* 2009). Significant enhancement in the linear and/or non-linear optical responses of biomolecules, when embedded in simple 1D photonic structures, compared to unstructured surfaces, has enabled molecular detection with extreme (femtomolar) sensitivity (Estephan *et al.* 2010).

Furthermore, (nitride) semiconductor-based sensors produce an electronic signal with much faster response times than conventional sensing methods. The electrical output signals from the sensor can be digitized and used for quantifying the concentration of the detected species. The device configuration for electric sensing is typically a transistor-type instrument.

A field-effect transistor (FET) is a very efficient device for converting a biological signal into an electrical signal, due to its sensitivity to changes in the surface potential. Due to their high pH sensitivity and electrochemical stability, aluminum–gallium nitride/gallium nitride (AlGaN/GaN) solution gate field-effect transistors are appropriate for monitoring enzymatic reactions (Steinhoff *et al.* 2003). Gate voltages induced by ions adsorbed onto the gate region modulate the source-drain currents, realizing the concept of the ion selective FET (ISFET).

AlGaN/GaN-based high-electron-mobility-transistors (HEMT) are ideally suited for sensing, since they have a high electron sheet concentration induced by spontaneous and piezoelectric polarization of the AlGaN and the GaN layer (Pearton *et al.* 2005, Kang *et al.* 2005). High-electron-mobility transistors of AlGaN/GaN have been employed as gas, chemical, biological, and medical sensors due to their excellent sensitivity (Kang *et al.* 2007, Wang *et al.* 2007).

14.3 Functionalization of nitrides

When used for sensing, device functionality can be seriously limited by complex phenomena at the solid/liquid interface, often resulting in a loss of activity of biological molecules due to unfolding upon immobilization (Norde 1986). On the other hand, surface modification presents great possibilities to define the sensing-material properties. Modification of material surfaces is usually achieved by chemical functionalization with a linker, (i) providing high biocompatibility, (ii) maintaining a low density of electronic defects at the material surface, (iii) providing a high density of docking sites to specifically attach biomolecules, (iv) lowering toxicity, (v) providing material stability (it inhibits oxidation), and (vi) converting hydrophobic materials into hydrophilic to enable their transfer into aqueous solution. The linker has to be chosen carefully for proper functioning of the material. The functional coupling groups in proteins are either available in the native form or are incorporated by chemical means or by genetic engineering.

Examples of materials coated with protein molecules through chemical interactions are shown in Fig. 14.1. Frequently-used groups for performing biomolecular coupling to the material surface are citrate, thiol, and phosphane groups (Katz *et al.* 2004). Exchange reactions with a stronger binding ligand can lead to direct protein linking with a metal surface. This is possible with a labile capping citrate layer, for example.

For covalent binding of molecules, complex chemistry such as thiolylation (Neves *et al.* 2006) and silanization (Babu *et al.* 2004) are to be applied, providing efficient coupling of molecules to a substrate. To reach biocompatibility, InN has to be functionalized with linkers, thereby reducing toxicity and ensuring the

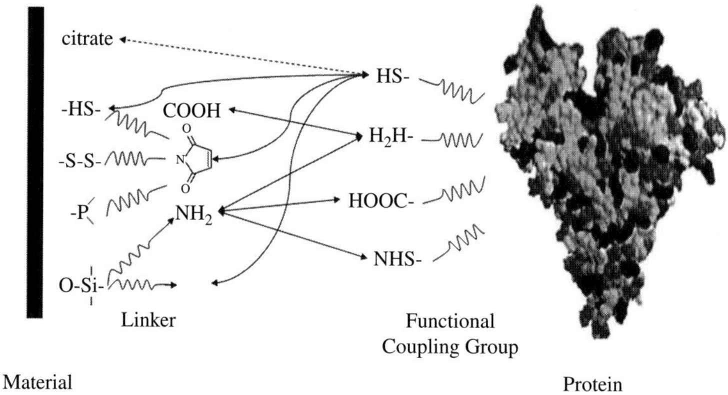

FIG. 14.1. Schematic representation of linkers and functional groups used to couple material with proteins.

materials' stability (Stutzmannn *et al.* 2006). The InN surface naturally forms a thin native oxide layer under ambient conditions, which enables direct coupling of aminosilane molecules onto its surface (Chen *et al.* 2006).

A very good overview of functionalization strategies for semiconductors is presented in Stutzmannn *et al.* (2006). Frequently, the biomolecules are covalently immobilized on the surface of the semiconductor substrate via linker molecules. The main role of the linker molecules is to provide a high density of docking sites for the attachment of biomolecules. The covalent immobilization is achieved via the silanization approach with different organosilanes such as aminopropyltriethoxymethylsilane (APTES) or octadecyl-trimethoxysilane (ODMTS) (Kang *et al.* 2005, Baur *et al.* 2005), taking advantage of the natural oxide of the nitride surfaces, or via photochemical functionalization with protected amines starting from the H-terminated GaN surface (Kim *et al.* 2006).

An alternative functionalization method is the selection of material-binding biomolecules from a random biological library, based on the natural recognition of proteins or peptides towards specific material (Brown *et al.* 1992, Whaley *et al.* 2000). We have reported on surface modification of the GaN and InN semiconductor by specific 12-mer peptides produced by the combinatorial phage display technique—an original technology for the identification and application of material-specific biomolecules (Estephan *et al.* 2008, Estephan *et al.* 2009, Estephan *et al.* 2011). Due to their small dimensions and unique surface recognition properties, peptides can be used further for controlled placement of biomolecules on patterned surfaces. The peptide functionalization method presents an easy-to-use alternative solution for the problems encountered in existing sensing methods, due to the unwanted unspecific interactions. These peptides and/or their derivatives are simply adsorbed from the liquid phase onto the surface of the semiconductor, due to their affinity for this material.

GaN is a semiconductor that reveals excellent chemical stability and biocompatibility (Steinhoff *et al.* 2003), rendering it very attractive for biomedical applications too. Interfacing semiconductors with complex biological systems require a physiological environment and a sufficient degree of biocompatibility of the substrate. This already precludes the use of many materials for this purpose, and/or requires the presence of specific buffer layers to increase the biocompatibility. Therefore, application of natural biomolecules such as peptides during functionalization could be desirable in many cases, such as the production of an implantable optical sensor.

Currently, there is intense demand for downscaling to micro-/nanometer scale when developing new sensing substrates, and they need also to present high binding selectivity and sensitivity for biological molecules. Miniaturization and a large sensing area can be assured by nanostructured substrates such as photonic crystals, and high specificity can be achieved by their proper functionalization, with short peptides produced by phage-display technology. Typically, a 12-mer specific peptide is extracted by means of several biopanning cycles in which a phage library (expressing a large number of peptides) is exposed to the target of interest, followed by washing, elution, and amplification of the eluted phages and then reexposure to the target (Whaley *et al.* 2000). After several biopanning rounds, a specific peptide presenting high affinity for the chosen target is obtained. We have successfully elaborated the peptide sequences for numerous semiconductors (GaN, InP, GaAs, ZnSe, InAs, Si, InN, and so on) (Estephan *et al.* 2008, Estephan *et al.* 2009, Estephan *et al.* 2011). Specific localization of peptides and reversibility of functionalization was demonstrated, and the interactions between the surfaces and its specific peptide have been measured.

As a proof of concept, for controlled placement of molecules via adhesion peptides we employed fluorescence. A nested square pattern was etched onto a GaN wafer (Fig. 14.2), and the pattern obtained was formed by 30-μm-thick GaN lines separated by 120-μm SiO_2 spacing. The pattern was exposed to biotinylated peptide, and then incubated with fluorescent streptavidin.

When the surface obtained was observed under a fluorescent microscope, the tagged peptides were seen as grey lines in Fig. 14.2, corresponding to GaN lines labeled by the GaN-specific peptide. The SiO_2 regions did not reveal any fluorescence; thus, this part of the pattern remained unbound by the specific peptide. In another application, similar selectivity was obtained when an indium phosphide (InP) star-structure within an InP/SiO_2 pattern, with a diameter of the inner circle of the star of 80 μm, was functionalized with the InP-specific peptide (Fig. 14.2).

Peptide route functionalization was found to preserve enzyme activity and its secondary structure while adsorbed on a ZnSe ATR-crystal (Saab *et al.* 2010). Peptides can resolve the toxicity problems of other promising semiconductors too, with a direct band-gap such as InP (Estephan *et al.* 2009), rendering them useful for optoelectronics, biomedical devices, and for fluorescence applications such as QDs.

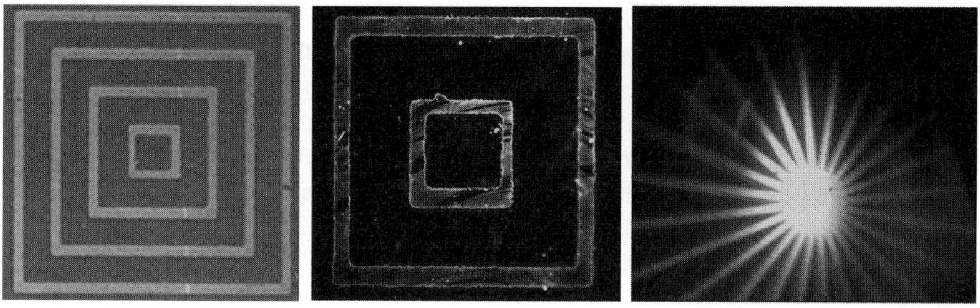

FIG. 14.2. Examples of selective functionalization via peptides, revealing high affinity for GaN (0001) and InP (001), respectively: a GaN net-structure within a GaN/SiO_2 pattern, and an InP star structure within an InP/SiO_2 pattern. The GaN lines are 30 μm thick, and the diameter of the inner circle of the star is 80 μm. Fluorescence is revealed by the labelled streptavidin molecules that are bound to the biotinilated peptide previously adsorbed on the semiconductor surface. (Reprinted with permission from Estephan *et al.* 2008, © 2008 by the American Chemical Society, and from Estephan *et al.* 2009, © 2009, with permission from Elsevier.)

Specific adhesion of these peptides was used for the controlled placement of biomolecules in porous silicon microcavities that were further used for optical biosensing (Estephan *et al.* 2011). Improved streptavidin sensing (twenty-one-times-lower detection limit) was obtained with peptide-functionalized porous silicon microcavities, compared to sensing performed with devices functionalized with the commonly used silanization method, suggesting that surface modification via specific peptides provides better interface layers for molecular detection. High-resolution atomic-force microscopy revealed the formation of ordered nanometer-sized molecular layers when peptide-route functionalization is performed. Hence, the obtained specific peptides proved to be ideal linkers for surface modification, opening the way for the elaboration of new, highly ordered and nanometer-sized organic/inorganic materials.

14.4 Sensing examples with nitride-based devices

The basic interest of using a GaN substrate is due to its light-emitting properties modified by exposure to a gas or liquid containing the molecule to be sensed for chemical or biological agent identification (Schalwig *et al.* 2002). Porous GaN was used to fabricate a Pd Schottky contact gas sensor for hydrogen detection (Yam *et al.* 2007). Hydrogen was measured as a current variation attributed to the unique microstructure at the Pd/porous GaN interface, which allowed a higher accumulation of hydrogen. This resulted in a stronger effect of H-induced dipole layer, and led to a significant change in the electrical characteristics of the porous sensor.

GaN Schottky diodes with catalytic platinum electrodes were used for detection of hydrogen and hydrocarbon species, and H_2, CO, and hydrocarbons increased the diode currents. High sensitivity was achieved, as a difference has been measured between adsorption of molecular and atomic hydrogen (Schalwig *et al.* 2002). Molecular adsorption produces molecular dipoles at the surface of the Schottky barrier device, thereby altering the apparent metal work function. Conversely, the atomic hydrogen penetrates into the Pd bulk, leading to an interfacial OH–dipole layer. However, hydrogen has a similar response on Ga-face or N-face Schottky diodes, indicating that the different polarization directions of the semiconductor material do not play a major role in gas sensing.

Dependence on sensitivity to certain gases of differently etched GaN possessing different morphologies has been reported (Popa *et al.* 2006). The main difference in the sensitivities to methane gas and alcohol vapors exhibited by the etched GaN lies in the morphology-induced differences in the dissociation of the gas molecules as reducing gas molecules implies their dissociation at the surface of grain boundaries in order to react chemically with the oxygen.

Applications of III-nitride semiconductors have been found in developing potentiometric anion sensors (Chaniotakis *et al.* 2004), or in elaborating GaN-based heterostructures for sensor applications monitoring adsorption of ions, wetting by polar liquids, and exposure to gases (Stutzman *et al.* 2002). GaN crystal grown on sapphire substrate is an excellent electroactive material for the development of anion-sensitive potentiometric sensors. The Ga atoms of the outer surface of the GaN (0001) crystal coordinate selectively and reversibly with anions in solution. The anion coordination generates interfacial potential, related to the activity of the anions (Chaniotakis *et al.* 2004).

Interaction of the wurtzite-type crystal of GaN (0001) with different gases such as hydrogen (H_2), methane, carbon monoxide (CO), acetylene (C_2H_2), and nitrous oxide (NO_2) has been studied, though still very little is known of the specific chemical interactions. A crystalline membrane of GaN was used as an anion selective potentiometric sensor based on the interaction of the basic ions with the acidic surface gallium atoms, while the lipophilicity of the anions has proven to be of minor importance (Alifragis *et al.* 2005). It was suggested that the double layer (inner and outer Helmholtz planes) produced by the specifically adsorbed charged species is responsible for the observed interfacial potential difference, measured under the potentiometric mode.

AlGaN/GaN ungated high electron mobility transistors (HEMT) exhibit extreme sensitivity towards polar liquids: the drain-source current dramatically changes due to bonding of polar liquid molecules in the gate region, altering the polarization-induced positive surface charge. Sensing with AlGaN/GaN heterostructures is based generally on the modulation of the carrier density in a surface near two-dimensional electron gas (2DEG) which is spontaneously created at the interface of these devices. The ion-sensitive field effect transistor (ISFET) is another design of III-nitride-based sensors used for the detection of negative or positive ions via the strong modulation of the source-drain current. GaN or AlGaN/GaN heterostructures grown on sapphire substrates can easily provide metal oxide gate layers in transistor structures, at the same time benefiting from the high chemical stability of the III-nitrides. Hence, GaN-based ISFETs allow simultaneous electronic and microscopic detection of processes on the device surface, due to the optical transparency of both the thin semiconducting film and the sapphire substrate. These devices reveal a high pH sensitivity, providing the possibility of measuring H-concentration in the electrolyte.

Sensing with an ultrathin (10-nm) InN-based ISFET has been investigated (Lu *et al.* 2008). In such devices (Fig. 14.3a) the current channel, composed of the accumulated surface electrons located just below the gate surface, is highly sensitive to the variation of adsorbed gate charges. The variation of adsorbed ion concentration in the gate region is immediately transduced into the device current output.

Figure 14.3b depicts the dynamic response of the InN ISFET for potassium chloride (KCl) solution with a concentration ranging from 10^{-5} M to 10^{-1} M when the voltage between the drain and the source VDS was fixed at 0.5 V. Ion detection is indicated as an abrupt current change in the InN ISFET when exposed to different ion concentrations. The current became smaller as the chlorine ion concentration was increased. The estimated response time of one decadechlorine ion concentration change is less than 10 s.

The main advantages of the AlGaN/GaN heterostructures include their chemical stability and the presence of a polarization-induced two-dimensional electron gas (2DEG), which responds to any change of charge on the free unpassivated gate surface (Eickhoff *et al.* 2003), while realizing a low noise level (Steinhoff

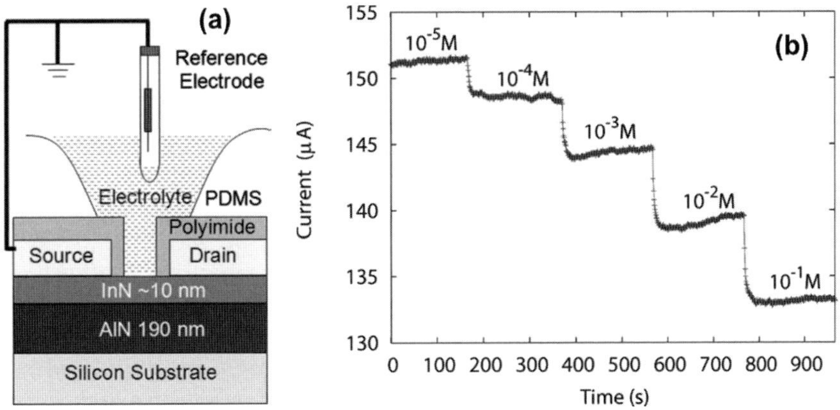

FIG. 14.3. (a) Schematic representation of an ultrathin InN ISFET device. InN is used as the sensing-gate material in direct contact with the sensing environment. Adsorption of anions to InN surface builds up a gate potential with respect to the reference electrode, resulting in a reduced current output. (b) Dynamic response of InN ISFET in a KCl solution with varying KCl concentration from 10^{-5} M to 10^{-1} M at VDS = 0.5 V. (Reprinted with permission from Lu *et al.* 2008, © 2008 American Institute of Physics.)

et al. 2005). Moreover, AlGaN/GaN sensors on sapphire are transparent for visible light, which allows for the implementation of additional optical techniques such as spectrophotometric detection or simple microscopic observation of attached cells.

Recently, online monitoring of living nerve cells with an AlGaN/GaN electrolyte gate field effect transistor (EGFET) has been reported (Gebinoga *et al.* 2012). The nerve cells attached to the transistor structure were provoked by an inhibitor, creating prolonged ion flux through the protein channels. This ionic current flows along the thin film of electrolytes and generates an extracellular potential change between cell and chip that modulates the 2DEG density and thus the induced current. By measuring this current versus time and comparing it to the sensor response of a pure medium with the same inhibitor, information about the ion channel activity of the cells at the interface between transistor and cells was obtained.

A GaN-based biosensor for the detection of penicillin has been reported (Baur *et al.* 2006). The penicillinase enzyme was covalently grafted on an AlGaN/GaN heterostructure, leading to an enzyme-modified field-effect transistor (EnFET). When the enzyme catalyses the transformation of penicillin to penicilloic acid, protons are released. The resultant modification in the pH value was sensed by the two-dimensional electron gas at the AlGaN/GaN interface, and was quantified as shifts in the surface potential.

Using zinc-oxide nanorods integrated into an Al-GaN/GaN HEMT nanomols of glucose has been detected successfully (Kang *et al.* 2007) for diagnosis of diabetes. Diagnosis of breast cancer has been also performed via functionalization of the HEMT sensor with antibodies against a breast-cancer biomarker in human saliva (Chen *et al.* 2008)

Diagnosing bacterial infections or genetic diseases requires biosensing of deoxyribonucleic acid (DNA) via oligomer hybridization to form duplex DNA using one immobilized strand on a surface. Most of the methods used to detect DNA hybridization through measuring changes in mass, optical properties, or electrochemical characteristics require prelabeling of the DNA target with special isotopes or fluorophores. Electronic measurements using semiconductor-based field-effect transistors to detect base pairs in DNAs rapidly and with high accuracy are very promising sensing methods (Han *et al.* 2005, Xuan *et al.* 2005).

Gold (Au)-gated AlGaN/GaN HEMT structures were functionalized in the gate region to bind oligomers for sensing via hybridization of matched target DNA (Kang *et al.* 2006). The HEMT drain-source current showed a clear decrease of 115 μA when the probe matched the target DNA on the surface of the device (Fig. 14.4). Importantly, the repeated use of the nitride HEMT sensor was also demonstrated, after rinsing the device with hot water, indicating the highly reproducible sensing surface.

GaN and InN wurtzite nanocolumns processed by molecular beam epitaxy can be used both as enzyme immobilization matrices and as highly sensitive transducers (Sofikiti *et al.* 2009). Due to their stabilizing nano-environment within

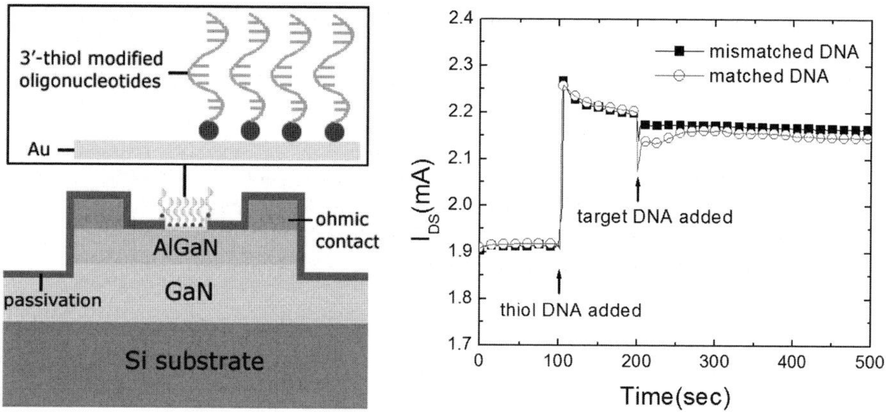

FIG. 14.4. Schematic representation of a functionalized AlGaN/GaN HEMT and sensing via hybridization of DNA. (Reprinted with permission from Kang *et al.* 2006, © 2006 American Institute of Physics.)

which urease retains its catalytic activity over a prolonged period, these electrochemically active nanocolumns were successfully used as potentiometric urea biosensors (Fig. 14.5a). The urease enzyme was physically adsorbed on both nanocolumnar and flat GaN and InN surfaces. The enzymatic reaction of urease alters the pH of the media onto the semiconductor surface, rendering possible the potentiometric detection of urea. The calibration curve to urea, presented in Fig. 14.5b for both planar and nanocolumnar GaN and InN-based structures, indicates that the biosensors based on GaN and InN nanocolumns exhibit a much higher sensitivity to urea, compared to that of the corresponding planar substrates. Since both planar and nanocolumnar sensors exhibit similar pH sensitivity, this result is attributed to the fact that nanostructures can adsorb higher amounts of enzyme, due to increased surface-to-volume ratios. Moreover, when evaluated over time, the authors found that an enzyme can be stabilized against denaturation if encapsulated within nanocavities, which have an average diameter two to six times larger than that of the biomolecule. In the case of these InN and GaN nanostructures, the space between the nanocolumns is approximately 80 nm, whereas the average diameter of urease is around 13 nm.

Finally, very promising biomedical applications appeared recently with complex lab-on-chip devices as next-generation implantable biosensors. Other than the transducer, there is a need for new light-sources with advanced characteristics. Recently, a flexible GaN light-emitting diode (LED) fabricated on a liquid-crystal polymer substrate has been successfully elaborated (Lee *et al.* 2012). These GaN LEDs are mechanically and optically stable, and integrated on flexible substrates might constitute future light-sources. The measurement module constituted of the flexible GaN LED for excitation, an electroluminescence detector, and a reaction chip successfully detected prostate-specific antigens.

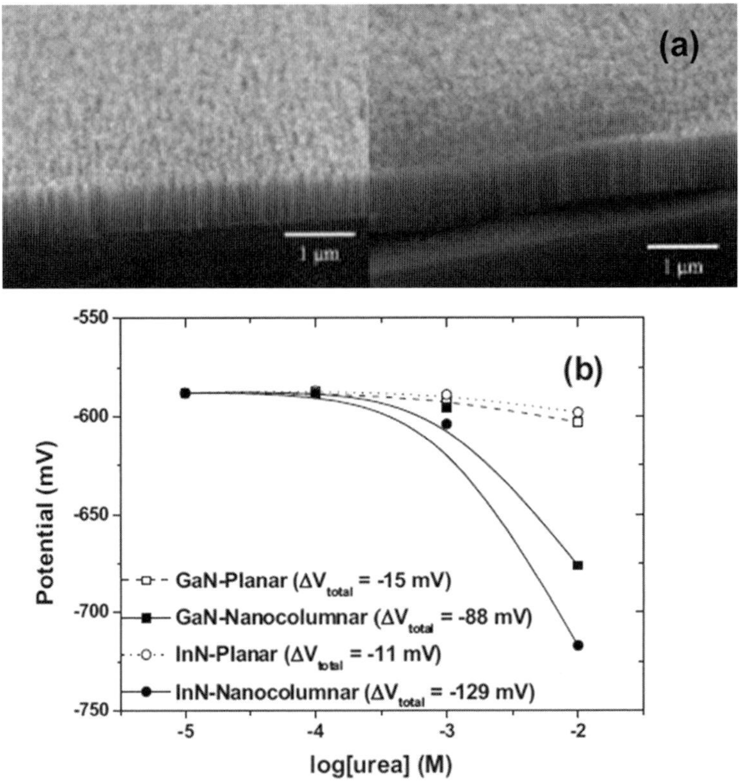

FIG. 14.5. (a) Scanning electron microscope images of InN (left) and GaN (right) nanocolumns; (b) Calibration curves to urea of GaN and InN nanocolumnar and planar biosensors. (Reprinted with permission from Sofikiti *et al.* 2009, © 2009 American Institute of Physics.)

References

Y. Alifragis, G. Konstantinidis, A. Georgakilas, and N. Chaniotakis, *Electroanalysis* 17, 5 (2005).

V. R. S. Babu, M. A. Kumar, N. G. Karanth, and M. S. Thakur, *Biosens. Bioelectron.* 19, 1337 (2004).

B. Baur, G. Steinhoff, J. Hernando, O. Purrucker, M. Tanaka, B. Nickel, M. Stutzmann, and M. Eickhoff, *Appl. Phys. Lett.* 87, 263901 (2005).

B. Baur, J. Howgate, H.-G. Von Ribbeck, Y. Gawlina, V. Bandalo, G. Steinhoff, M. Stutzmann, and M. Eickhoff, *Appl. Phys. Lett.* 89, 183901 (2006).

S. Brown, *Proc. Natl. Acad. Sci. USA* 89, 8651 (1992).

N. A. Chaniotakis, Y. Alifragis, G. Konstantinidis, and A. Georgakilas, *Anal. Chem.* 76, 5552 (2004).

C. F. Chen, C. L. Wu, and S. Gwo, *Appl. Phys. Lett.* 89, 252109 (2006).

K. H. Chen, B. S. Kang, H. T. Wang, T. P. Lele, F. Ren, Y. L. Wang, C. Y. Chang, S. J. Pearton, J. W. Johnson, P. Rajagopal, J. C. Roberts, E. L. Piner, and K. J. Linthicum, *Appl. Phys. Lett.* 92, 192103 (2008).

V. Darakchieva, K. Lorenz, N. P. Barradas, E. Alves, B. Monemar, M. Schubert, N. Franco, C. L. Hsiao, L. C. Chen, W. J. Schaff , L. W. Tu, T. Yamaguchi, and Y. Nanishi, *Appl. Phys. Lett.* 96, 81907 (2010).

V. Y. Davydov, A. A. Klochikhin, V. V. Emtsev, S. V. Ivanov, V. V. Vekshin, F. Bechstedt, J. Furthmuller, H. Harima, A. V. Mudryi, A. Hashimoto, A. Yamamoto, J. Aderhold, J. Graul, and E. E. Haller, *Phys. Status Solidi B*, 230, 787 (2002).

J. M. Dawson, J. R. Nightingale, R. P. Tompkins, X. Cao, T. H. Myers, L. A. Hornak, and D. Korakakis, *Mater. Res. Soc. Symp. Proc.* Vol. 1040 (2008).

M. Eickhoff, J. Schalwig, G. Steinhoff, O. Weidemann, L. Görgens, and R. Neuberger, *Phys. Stat. Sol. C,* 0(6), 1908 (2003).

E. Estephan, M.-B. Saab, V. Agarwal, F. J. G. Cuisinier, C. Larroque, and C. Gergely, *Adv. Funct. Mat.* 21 (11), 2003 (2011).

E. Estephan, M.-B. Saab, M. Martin, C. Larroque, F. J. G. Cuisinier, O. Briot, S. Ruffenach, M. Moret, and C. Gergely, *J. of Peptide Science* 17, 143 (2011).

E. Estephan, D. Bajoni, M.-B. Saab, T. Cloitre, R. Aulombard, C. Larroque, L. C. Andreani, M. Liscidini, A. M. Malvezzi, and C. Gergely, *Langmuir* 26(12), 10373 (2010).

E. Estephan, C. Larroque, N. Bec, P. Martineau, F. J. G. Cuisinier, T. Cloitre, and C. Gergely, *Biotechnology & Bioengineering* 104 (6), 1121 (2009).

E. Estephan, C. Larroque, F. J. G. Cuisinier, Z. Bálint, and C. Gergely, *J. Phys. Chem.* B, 112, 8799 (2008).

E. Estephan, M.-B. Saab, C. Larroque, M. Martin, F. Olsson, S. Lourdudoss, and C. Gergely, *J. Coll. Int. Sci.* 337, 358 (2009).

M. Gebinoga, P. Maia, M. Donahuea, M. Kittler, I. Cimalla, B. Lübbers, M. Kletta, V. Lebedev, L. Silveira, S. Singh, and A. Schober, *J. Neurosci. Meth.* 206, 195 (2012).

D. I. Han, D. S. Kim, J. E. Park, J. K. Shin, S. H. Kong, P. C. Choi, J. H. Lee, and G. Lim, *Jpn. J. Appl. Phys* 1 (44), 5496 (2005).

B. S. Kang, F. Ren, L. Wang, C. Lofton, W. W. Tan, S. J. Pearton, A. Dabiran, A. Osinsky, and P. P. Chow, *Appl. Phys. Lett.* 87, 023508 (2005).

B. S. Kang, S. J. Pearton, J. J. Chen, F. Rena, J. W. Johnson, R. J. Therrien, P. Rajagopal, J. C. Roberts, E. L. Piner, and K. J. Linthicum, *Appl. Phys. Lett.* 89, 122102 (2006).

B. S. Kang, F. Ren, L. Wang, C. Lofton, Weihong Tan, S. J. Pearton, A. Dabiran, A. Osinsky, and P. P. Chow, *Appl. Phys. Lett.* 87, 023508 (2005).

B. S. Kang, H. T. Wang, F. Ren, S. J. Pearton, T. E. Morey, D. M. Dennis, J. W. Johnson, P. Rajagopal, J. C. Roberts, E. L. Piner, and K. J. Linthicum, *Appl. Phys. Lett.* 91, 252103 (2007).

B. S. Kang, H. T. Wang, T. P. Lele, F. Ren, S. J. Pearton, J. W. Johnson, P. Rajagopal, J. C. Roberts, E. L. Piner, and K. J. Linthicum, *Appl. Phys. Lett.* 91, 112106 (2007).

B. S. Kang, H. T. Wang, F. Ren, B. P. Gila, C. R. Abernathy, S. J. Pearton, D. M. Dennis, J. W. Johnson, P. Rajagopal, J. C. Roberts, E. L. Piner, and K. J. Linthicum, *Electrochem. Solid State Lett.* 11, J19 (2007).

E. Katz and I. Willner, *Angew. Chem. Int. Ed.* 43 (45), 6042 (2004).

H. Kim, P. E. Colavita, K. M. Metz, B. M. Nichols, B. Sun, J. Uhlrich, X. Wang, Th. F. Kuech, and R. J. Hamers, *Langmuir* 22, 8121 (2006).

S. Y. Lee, K. I. Park, C. Huh, M. Koo, H. G. Yoo, S. Kim, C. S. Ah, G. Y. Sung, and K. J. Lee, *Nano Energy* 1, 145 (2012).

S. Limpijumnong and C. G. Van de Walle, *Phys. Stat. Sol. B* 228, 303 (2001).

H. Lu, W. J. Schaff, and L. F. Eastman, *J. Appl. Phys.* 96, 3577 (2004).

Y.-S. Lu, C.-L. Ho, J. A. Yeh, H.-W. Lin, and S. Gwo, *Appl. Phys. Lett.* 92, 212102 (2008).

M. Martin, G. Palestino, V. Agarwal, T. Cloitre, L. Zimányi, and C. Gergely, *Appl. Phys. Lett.* 94, 223313 (2009).

M. T. Neves-Petersen, T. Snabe, S. Klitgaard, M. Duroux, and S. B. Petersen, *Protein Sci.* 15, 343 (2006).

W. Norde, *Adv. Colloid Interface Sci.* 25, 267 (1986).

G. Palestino, M. B. de la Mora, J. A. del Rio, C. Gergely, and E. Perez, *Appl. Phys. Lett.* 91, 12, 121909 (2007).

S. J. Pearton, B. S. Kang, S. Kim, F. Ren, B. P. Gila, C. R. Abernathy, J. Lin, and S. N. G. Chu, *J. Phys. Condens. Matter* 16, R 961 (2004).

V. Popa, I. M. Tiginyanu, V. V. Ursaki, O. Volcius, and H. Morkoc, *Semicond. Sci. Technol.* 21, 1518 (2006).

M.-B. Saab, E. Estephan, T. Cloitre, C. Larroque, and C. Gergely, *J. Phys. Chem. C* 114 (43), 18509 (2010).

J. Schalwig, G. Muller, M. Eickhoff, O. Ambacher, and M. Stutzmann, *Materials Sci. Eng.* B93, 207 (2002).

N. Sofikiti, N. Chaniotakis, J. Grandal, M. Utrera, M. A. Sanchez-Garcia, E. Calleja, E. Iliopoulos, and A. Georgakilas, *Appl. Phys. Lett.* 95, 113701 (2009).

G. Steinhoff, O. Purrucker, M. Tanaka, M. Stutzmann, and M. Eickhoff, *Adv. Funct. Mater.* 13, 841 (2003).

G. Steinhoff, M. Hermann, W. J. Schaff, L. F. Eastman, M. Stutzmann, and M. Eickhoff, *Appl. Phys. Lett.* 83, 177 (2003).

M. Stutzmann, J. A. Garrido, M. Eickhoff, and M. S. Brandt, *Phys. Stat. Sol. (a)* 203, 14, 3424 (2006).

M. Stutzmann, G. Steinhoff, M. Eickhoff, O. Ambachera, C. E. Nebela, J. Schalwigb, R. Neuberger, and G. Muller, *Diam. Rel. Mat.* 11, 886 (2002).

H. T. Wang, B. S. Kang, F. Ren, S. J. Pearton, J. W. Johnson, P. Rajagopal, J. C. Roberts, E. L. Piner, and K. J. Linthicum, *Appl. Phys. Lett.* 91, 222101 (2007).

S. R. Whaley, D. S. English, E. L. Hu, P. F. Barbara, and A. M. Belcher, *Nature* 405, 665 (2000).

J. Wu, W. Walukiewicz, K. M. Yu, J. W. Ager, E. E. Haller, H. Lu, W. J. Schaff, Y. Saito, and Y. Nanishi, *Appl. Phys. Lett.* 80, 3967 (2002).

G. Xuan, J. Kolodzey, V. Kapoor, and G. Gonye, *Appl. Phys. Lett.* 87, 103903 (2005).

F. K. Yam and Z. Hassan, *Appl. Surf. Sci.* 253, 9525 (2007).

15 Heterovalent ternary II-IV-N_2 compounds: perspectives for a new class of wide-band-gap nitrides

Walter R. L. Lambrecht and Atchara Punya

15.1 Introduction

The predominant method for adjusting materials properties in materials science, and, in particular, semiconductor band structures, is isovalent alloying. For example, by mixing the various group-III cations (Al, Ga, In) one can more or less interpolate between the properties of the binary group-III nitride compounds. Of course, these interpolations need not be linear. For example, the band-gap variation is usually described by means of a bowing coefficient, b. For an $A_{1-x}B_xC$ alloy, the band gap is written as

$$E_g(x) = E_g(A)(1-x) + E_g(B)x - bx(1-x) \tag{15.1}$$

and a positive b indicates the usual downward bowing. Finer control of the band structure can be obtained by nanostructuring, in the form of heterojunctions, superlattices, quantum wires, and quantum dots, and this approach is usually combined with alloying, so as to avoid building up too much lattice strain. On the other hand, it is important to remember that all these structures are essentially thermodynamically metastable. They can be grown only by means of non-equilibrium growth techniques. Also, the alloys are for the most part disordered systems in which the two types of cation are distributed randomly, or at least in a disordered way, on the cation sublattice. Although ordering can be observed in homovalent semiconductor alloys, they do not form truly stable compounds, and such ordering is now understood to result from a quasi-equilibrium of the surface layers during growth (1).

There is, however, another paradigm for adjusting electronic structure of tetrahedrally bonded semiconductors: one can replace two of the four group-III cations surrounding each anion by a group II element, and two by a group-IV element. For example, GaN becomes $ZnGeN_2$. We call this concept: *heterovalent ternaries.* This possibility is well known for other group-V compounds, which have the zincblende structure. It leads to the family of chalcopyrite structure compounds (2), such as $ZnGeP_2$, $CdGeAs_2$, etc. These materials have received significant attention because of their favorable non-linear optical properties (3, 4). We have studied some of their band structures in (5–8). In particular, the symmetry lowering from the cubic to tetragonal structure leads to birefringence, which allows one to do phase-matching and hence make efficient frequency doubling and tunable frequency sum or difference generating materials (9). The same concept can also be applied to II-VI materials by replacing the group-II with a group-I and group-III element. For example, ZnSe becomes $CuInSe_2$. The importance of Cu(In,Ga)Se_2 (better known as CIGS) as a thin-film solar cell material is widely appreciated.

It is much less well known that this same concept can also be applied to the more ionic semiconductors that exhibit the wurtzite structure. As in the zincblende-to-chalcopyrite case, a well-defined crystal structure forms with a lower symmetry; in this case, orthorhombic. This crystal structure, of which β-$NaFeO_2$ is the prototype, applies to a whole new class of II-IV-N_2 compounds, which form the subject of this chapter. The same concept can also be applied to ZnO, and in that case leads to, e.g., $LiGaO_2$. That material is in fact familiar to the nitride community because it has been proposed as a suitable substrate for GaN growth in view of its close lattice match. It is, however, a material of considerable intrinsic interest in its own right. We refer the interested reader to some of our recent work on this material, because it falls outside the scope of this review on nitrides (10–13).

An important difference with isovalent alloying is that heterovalent substitution leads to new thermodynamically stable ordered compounds. It is not *a priori* evident why this is so. One can in principle construct different and even disordered structures which locally obey the octet rule of pairing two group-II with two group-IV elements around each anion. Thus, order–disorder transitions are a possibility. In fact, in chalcopyrite semiconductors, such order–disorder transitions have been observed. Nonetheless, we expect that deviations from the local octet rule, requiring each anion to be surrounded by exactly two cations of group-II and two cations of group-IV, will locally destroy charge neutrality and are thus expected to be energetically unfavorable. This issue was studied for $LiGaO_2$ in (13) by comparing structures with only Li_2Ga_2 tetrahedrons surrounding oxygen to a structure with a combination of $LiGa_3$ and Li_3Ga tetrahedra. The latter were found to be about 1 eV less stable per formula unit. This is a more severe restriction for more ionic materials. It is well known that in GaN, antisite defects are less common than in GaAs. This could severely limit the possibilities for disordering the lattice, since atom-switches need to occur in pairs, preserving the local octet rule. Thus II-IV-N_2 compounds may be expected to have an energetic incentive to be well-ordered. Even in $ZnGeP_2$ the predominant defects are not cation antisites, as one might have expected on the basis of order–disorder considerations, but rather Zn-vacancies, related to the high vapor pressure of Zn (14–16). In reality, of course, this remains one of the important questions to be studied for the nitrides.

From a band structure and defect engineering point of view, one may anticipate that heterovalent compounds provide a new degree of flexibility. One has two different cation sublattices with different valence to play with. Thus, slight deviations from stoichiometry might already intrinsically dope the materials. This could be viewed as a challenge, but also as an opportunity. It also means that one has two different valence cations which can be replaced by different types of dopant to achieve n- or p-type doping. For example, one might envision doping $ZnGeN_2$ p-type by placing group-I elements, such as Na, Cu, on the Zn site, or group-III elements on the Ge site. Again, this presents an opportunity but also the challenge of how to control the site selectivity of the dopants.

At first, it may appear that with this new form of band-gap engineering, the octet rule limits us to exactly 50% compounds and is thus less flexible than the usual isovalent alloying. However, alloying elements of the same valence can still be applied to these compounds. For example, one might envision $ZnGe_{1-x}Si_xN_2$ alloys with a continuously varying x. The fact that we have different elements for both the group-II and group-IV sites at our disposal means that we can in principle obtain several new well-defined compounds in the (band gap)–(lattice constant) space. Because alloying away from the well-defined compounds invariably leads to strain, this means that we have new anchor points available around which we can fine-tune the properties of the band gap while minimizing the strain.

It may even be possible to grow mixed (III-N)(II-IV-N_2) systems. This has already been explored for the analogous ZnO-$LiGaO_2$ systems, where continuous alloys were grown (17, 18). This is an interesting concept, because a group-III element would dope the II-IV-N_2 n-type if it occurs on the group-II site, and p-type if it occurs on the group-IV site. But if one is capable of adding the group-III so that it occupies both elements equally, we may change from a perfectly compensated semiconductor to a new type of alloy. From a heterojunction point of view, GaN and $ZnGeN_2$ have very similar band gaps, as we will see below, so one expects a type-II band offset situation. This would mean that in alternating $ZnGeN_2$, GaN layers one would expect electrons and holes to localize in the different regions. This separation of charges is in principle of interest for photovoltaics. At present we do not yet know which way the band-offset would go, but because of the presence of the Zn-3d bands in $ZnGeN_2$ one expects the VBM to be higher in $ZnGeN_2$ than in GaN, so holes would localize in $ZnGeN_2$ and electrons in GaN.

Another important consideration is that the new crystal structures have lower symmetry. As we will show later, this leads to additional splittings of the valence bands, giving rise to interesting anisotropies in optical properties. It also leads to a much more complex phonon spectrum than in the wurtzite structure. These additional anisotropic properties are of fundamental as well as practical interest.

At present, we are only at the start of exploring this vast new materials system. We are where the group-III nitrides were in the early 1970s. The literature on these materials can still be summarized in a few pages. In the remainder of this chapter we first give a brief review of the literature in Section 15.2. In particular, this describes some of the growth methods that have been used for these materials. Next, we describe the crystal structure and symmetry in Section 15.3. We follow this with some considerations on thermodynamic stability, which are relevant to crystal growth, and more precisely the energies of formation as calculated from first principles in Section 15.4. Next, we discuss the electronic band structure in Section 15.5. In particular, we focus on our own latest results using the quasiparticle self-consistent *GW* method. This method, which came of age only recently, allows one to overcome the long-lasting shortcomings of the band-gap underestimate for semiconductors occurring in the local density approximation, used in most of the early work on band structures of these materials. After a brief discussion of the method, we present an overview at the large energy scale of the band structures. We then focus on the band-gap region and describe the effective masses and effective mass Hamiltonians, including the valence-band splittings, which will be useful in future work for investigating nanostructures. Next, we turn to lattice dynamics in Section 15.6. While one might think at first that these would be closely related to the well-known phonon spectra in wurtzite by a simple Brillouin zone folding scheme, we will show that the differences in II-N and IV-N bond force constants and the lower symmetry lead to much more complex phonon spectra. Comparison with the few experiments available will be discussed. Various tensorial properties related to lattice deformation are presented next in Section 15.7, including elastic constants and piezoelectric constants. Spontaneous polarization and its relevance for heterojunctions is discussed in Section 15.8. Optical properties in the visible and UV range, related to the band structure, have not yet been investigated in detail, but we summarize what is known today in Section 15.9. This also includes a discussion of the non-linear optical properties. On defects, we have not many results to present yet. We have only just started studying these. But we will present some preliminary considerations on why interesting changes from the III-nitrides are expected, in Section 15.10. We finish with an outlook toward future work in Section 15.11.

15.2 History and crystal growth

The first reported synthesis of a II-IV-N_2 compound appears to be of $CaSiN_2$ (19–21). Several other compounds with the same crystal structure $CaGeN_2$ (22), $MgSiN_2$ (23, 24), $MgGeN_2$ (24, 25), $BeSiN_2$ (26, 27), $MnGeN_2$ (28, 29), $MnSiN_2$ (29), and $SrSiN_2$ (30), were synthesized by groups in France and Germany, in the late 1960s and early 1970s, but for the most part the only properties investigated were the crystal structure and the chemical stability. A general similarity in properties to group-III nitrides was pointed out.

Most of the research published to date has been on the Zn-IV-N_2 compounds. The first reported synthesis of $ZnGeN_2$ in 1970 (31) used a reaction of NH_3 with Zn_2GeO_4, the latter being separately obtained from reacting GeO_2 with ZnO:

$$3Zn_2GeO_4 + 8NH_3 \rightarrow 3ZnGeN_2 + N_2 + 3Zn + 12H_2O \qquad (15.2)$$

as well as the direct reaction of Zn with Ge_3N_4:

$$2Zn + Ge_3N_4 \rightarrow 2ZnGeN_2 + Ge \qquad (15.3)$$

It discusses the difficulties of having to work in a narrow temperature-range of 800–810°C to avoid forming Ge_3N_4 (above 810°C) and the inevitable traces of unreacted Ge that remain and gradually become nitrided. Below 800°C, an oxynitride is found to form. Such oxynitrides may be of interest in their own right, and were recently exlored for photocatalytic water splitting (32). The initial reports on the crystal structure showed it to have a structure related to that of wurtzite, but with a lower, monoclinic symmetry. Because of the similarity in electron density of Zn and Ge and hence their X-ray and electron-scattering factors, it was impossible using X-ray diffraction to determine whether the Zn and Ge atoms were ordered or disordered on the cation lattice. A neutron-diffraction study (33) determined the ordering of the Zn and Ge atoms, and thus established the orthorhombic structure, discussed in Section 15.3.

The stability of $ZnGeN_2$, Zn_3N_2 and Ge_3N_4 in partial mixtures of NH_3 and its dissociation products H_2 and N_2, was investigated by Zykov *et al.* (34). The difficulty of growing these nitrides from ammonia is that at high temperature ammonia dissociates and the resultant H_2 can reduce the nitride, e.g.,

$$ZnGeN_2 + 3H_2 \rightarrow ZnGe_{alloy} + 2NH_3 \tag{15.4}$$

They find that $ZnGeN_2$ is the more stable of the three nitrides and persists up to high dissociation rates α of NH_3 of up to 0.58 at 600°C and 0.62 at 800°C. By contrast, Ge_3N_4 already dissociates at $\alpha = 0.39 - 0.41$ at temperatures of 720–840°C, and Zn_3N_2 already starts to dissociate at $\alpha = 0.01$ at 600°C.

Larson *et al.* (35) were the first to use a vapor-growth method for $ZnGeN_2$ using cation chlorides. They started from elemental Zn and Ge which was first reacted with a HCl–N_2 mixture at temperatures of 550°C and 950°C respectively, and subsequently this gas mixture was mixed with NH_3–N_2. The reaction is thus assumed to be

$$ZnCl_2 + GeCl_4 + 2NH_3 \rightarrow ZnGeN_2 + 6HCl \tag{15.5}$$

They deposited the films on c-plane and r-plane sapphire at 850°C. They reported wurtzite structure with $a = 3.1993$ Å and $c = 5.187$ Å and an absorption onset at 2.67 eV, which is now known to be an underestimate resulting from defects.

High-pressure synthesis of $ZnGeN_2$ and $ZnSiN_2$ starting from mixtures of Zn_3N_2 and Si_3N_4 or Ge_3N_4 was reported by Endo *et al.* (36) in Japan in 1992. High pressure had been used previously to grow $MgSiN_2$ by Whitney and Giese (23). Endo *et al.* (36) reported a band gap of 3.64 eV for $ZnSiN_2$ based on the optical absorption onset. They reported their $ZnGeN_2$ crystals to be opaque and black, indicating a high degree of contamination.

In 1999, metal-organic chemical vapor deposition (MOCVD) was used by Zhu *et al.* (37) to synthesize thin films of $ZnGeN_2$. They used diethylzinc and germane as starting gases and grew on either r-plane or c-plane sapphire with a GaN buffer layer. N was supplied via NH_3. The r-plane sapphire was found to lead to finer and more uniform grains. They concluded that their $ZnGeN_2$ exhibited disorder on the cation sublattice. The same group also reported $ZnSiN_2$ growth and alloy growth of $ZnSi_{1-x}Ge_xN_2$ growth, and these materials were investigated by a number of other collaborators determining optical properties (38, 39), transistor devices on SiC (40), and even their suitability as magnetic semiconductor host by implantation of Mn (41). Muth *et al.* (38) determined band gaps from optical absorption data as a function of alloy composition in $ZnSi_{1-x}Ge_xN_2$, and found it to vary between 3.1 and 3.2 eV for $ZnGeN_2$ to 4.4 eV in $ZnSiN_2$. Cook *et al.* (39) determined the indices of refraction, and Mintairov *et al.* (42) determined the infrared reflection relating to the vibrational spectrum.

A few other groups, both in Japan and in France, have used chemical-vapor deposition-type methods. Misaki *et al.* (43–45) used remote plasma-enhanced MOCVD for thin film growth of $ZnGeN_2$ directly on sapphire without GaN buffer layer, and reported a band gap of 3.3 eV at room temperature. They determined the temperature-dependence of the band-gap luminescence and fitted it to the Varshni equation

$$E_g(T) = E_g(0) - \frac{\alpha T^2}{\beta + T} \tag{15.6}$$

with $E_g(0) = 3.36$ eV, $\alpha = 4.54 \times 10^{-3}$ eV/K, and $\beta = 4850$ K. They also reported optical reflectivity in the UV up to 20 eV. Most of this work used r-plane sapphire as the substrate. Cloitre *et al.* reported $ZnSiN_2$ growth by MOVPE on c-plane sapphire (46). Kikkawa and Morisaka (47) reported RF-sputter deposition of $ZnGeN_2$ thin films on Si and glass substrates, and estimated the gap to be about 3.1 eV. Viennois *et al.* (48) performed the first Raman-spectroscopy study on powder samples of $ZnGeN_2$, synthesized by a similar approach as used by Maunaye and Lang.

Recently a vapor–liquid–solid synthesis method was used by Du *et al.* (49) using pure Zn and Ge, and NH_3, at growth temperatures between 750° C and 900° C. Photoluminescence on these materials indicate a gap of 3.40±0.01 eV and a high ratio of band-gap versus defect luminescence. This growth technique uses the reaction of NH_3 with a Zn-saturated Ge liquid as well as an extra source of Zn-vapor. It leads to polycrystalline material but also fairly large micron needle-shaped crystals of several microns in diameter and of order 100 microns in length.

$ZnSnN_2$ has very recently been synthesized for the first time (50). The authors used a remote microwave N_2 plasma as source of the nitrogen in order to keep the temperature low enough to prevent dissociation. Depending on growth conditions, either Sn_3N_4, Zn_3N_2 or $ZnSnN_2$ can form. Polycrystalline samples of $ZnSnN_2$ were obtained and analyzed with powder X-ray diffraction (XRD), Auger electron spectroscopy (AES), Raman spectroscopy, photoluminescence (PL), and photoluminescence excitation (PLE) spectroscopy.

The first band-structure calculations of $ZnGeN_2$ were presented in 1999 by Limpijumnong *et al.* (52) using the local density approximation and linearized muffin-tin orbital method. They also investigated the linear and non-linear optical properties such as the index of refraction and second harmonic generation coefficient. Other groups also presented band-structure calculations with pseudopotential plane-wave approaches (53, 54), and included some other members of this family of II-IV-N_2 compounds.

Recently, Lambrecht *et al.* (55–59) presented a series of papers focusing mainly on the lattice dynamics of $ZnSiN_2$, $ZnGeN_2$, and $ZnSnN_2$. As part of this work, the electronic band structures were also calculated using a pseudopotential plane-wave approach (60). All these previous works on the band structure, however, suffer from the limitations of the local density approximation. The band structure of Zn-IV-N_2 compounds with IV = Si, Ge, Sn were recently calculated using a much more accurate GW approach (61, 62), which will be described in Section 15.5.1. Most recently, the method was also used to calculate $CdGeN_2$ and $ZnGe_{1-x}Sn_xN_2$ alloys (63).

One might also wonder about the possibility of carbonitrides. Lambrecht and Segall (64) proposed $BeCN_2$ to be a direct wide-band-gap semiconductor in the chalcopyrite structure. Because the parent compound c-BN has the zincblende structure, this makes sense, but the alternative wurtzite-based structure was not considered. Petukhov *et al.* (65) subsequently considered $BeCN_2$ and $MgCN_2$ as well as the corresponding siliconitrides in the chalcopyrite structure, and predicted both carbonitrides to be direct-gap semiconductors. These materials have not been synthesized as far as we can tell. An important problem is that carbon and nitrogen can form a CN_2^{2-} cyanamide ion, which is a linear molecular ion. Cyanamides of group II elements such as Mg, Sr, and Ca are indeed known, but form layered structures (66). In other words, for synthesizing tetrahedrally bonded carbonitrides one would have to overcome the problem that C and N prefer to use π bonding in a linear molecule, which then leads to totally different crystal structures. This problem is in that sense similar to the problem of the relative stability of graphite versus diamond in carbon.

15.3 Crystal structure and symmetry

15.3.1 *Symmetry*

The prototype for the crystal structure of the II-IV-N_2 compounds is β-$NaFeO_2$, which is actually a I-III-VI_2 compound. This crystal structure can be viewed as a $2 \times \sqrt{3}$ superlattice of wurtzite along ortho-hexagonal axes, as illustrated in Fig. 15.1, showing a projection on the c-plane.

It has sixteen atoms per unit cell. The b/a and c/a can both adjust as well as the internal positions x, y, z of each of the four types of atom, the group II, group IV, and two inequivalent N positions, N_{II} being on top of the group II and N_{IV} being on top of the group IV element. We choose the a axis to be $2a_w$ and $b \approx \sqrt{3}a_w$ unlike the commonly found choice in the crystallography literature, which interchanges a and b from ours.

The space group is $Pbn2_1$ (space group No. 33, or C_{2v}^9), meaning that there is a two-fold screw axis along the z direction with translation $\frac{1}{2}\mathbf{c}$, a diagonal

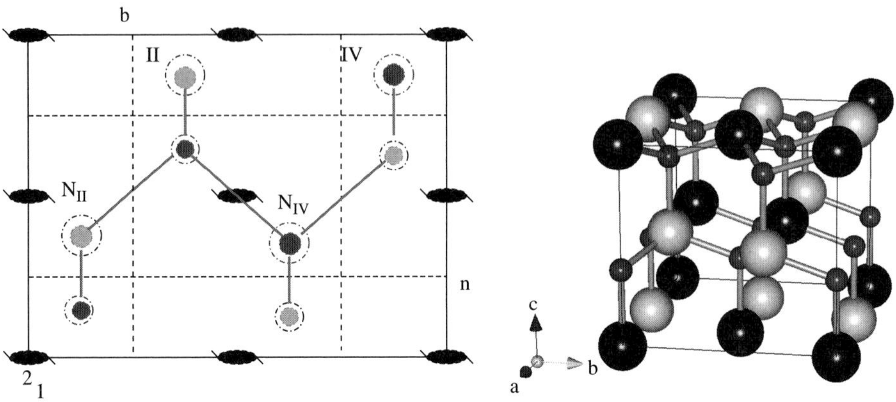

FIG. 15.1. Left: Projection of crystal structure of Zn-IV-N_2 compounds on c-plane, indicating symmetry elements. Large open circles indicate cations in bottom plane, (light grey group-II, darker grey group IV), small open circles in top plane, filled circles are N atoms above them, as indicated. The symmetry elements are indicated and chosen so that the 2_1 axes passes through the origin. From (62). Right: 3D view of the crystal structure: small light grey spheres are N, large spheres are cations, dark grey: group-II, lightest grey: group-IV. From (10).

glide plane n perpendicular to $\mathbf{b}$ with translations $\frac{1}{2}(\mathbf{a}+\mathbf{c})$, and an axial glide plane perpendicular to the a-axis with translation $\frac{1}{2}\mathbf{b}$. These symmetry elements are indicated in Fig. 5.1. We note that the space-group denomination in the International Tables of Crystallography (ITC) is $Pna2_1$. Our choice has the a and b axes interchanged. Most of the experimental papers in the literature adopt the ITC choice of axes, but we prefer the interchange of a and b axes because it makes the relation to wurtzite easier to see. This means that in our case, $a = 2a_w$ is the longest lattice constant, $b \approx \sqrt{3}a_w$ is the next longest, and $c = c_w$, where the subscript indicates the wurtzite structure. In this setting the origin is chosen to fall on the two-fold screw axis. This simplifies the non-primitive translation vector of the screw axis to be just $c/2$.

The point group is C_{2v}. The character table for this group is given in Table 15.1. The meaning of this table is easy to understand in this case, even to readers not familiar with group theory. The characters of the irreducible representations (irrep for short) listed as rows, simply describe whether the functions corresponding to this irrep are odd or even under the corresponding symmetry operations in the column headings. The irreducible representations a_1, b_1, and b_2 correspond to basis functions transforming like z, x, and y, with x along a, y along b, and z along the c-axes. The a_2 irreducible representation is even under the two-fold rotation, but odd under both mirror-planes, and corresponds to an xy basis function. These character tables are useful for classifying the vibrational modes at Γ as well as the electronic states.

Table 15.1 Character table for the group C_{2v}: first column, basis functions; second column, irreducible representations; first row, classes.

		E	C_{2z}	σ_x	σ_y
z	a_1	1	1	1	1
x	b_1	1	−1	−1	1
y	b_2	1	−1	1	−1
	a_2	1	1	−1	−1

15.3.2 *Lattice constants*

The lattice parameters and atomic positions for the II-IV-N_2 materials studied in our group were optimized using first-principles calculations. The details of the method are given in Section 15.5.1. The results, given in Table 15.2, show excellent agreement with experimental data where available. There is still significant uncertainty on the experimental lattice constants. For $ZnGeN_2$ a discussion of the lattice constants found in literature is given in Du *et al.* (49). For the last two materials, $ZnSnN_2$ and $CdGeN_2$, our calculations were predictions, when we first performed them, though $ZnSnN_2$ has recently been grown for the first time (50). The internal positions are given in Table 15.3. All atoms occur in 4(a) Wyckoff positions: (1) x, y, z (2) $-x, -y, z + \frac{1}{2}$, (3) $x + \frac{1}{2}, -y + \frac{1}{2}, z$, and (4) $-x + \frac{1}{2}, y + \frac{1}{2}, z + \frac{1}{2}$. The average bond lengths are summarized in Table 15.4.

In the process of minimizing the energy per cell as a function of cell volume, we also determined the bulk modulus and its pressure derivative by fitting an energy-versus-volume curve to the Vinet equation of state (51), given by

$$E(V) = E_0 + \frac{2B_0V_0}{(B_0^{'} - 1)^2}\Big\{2 - [5 + 3B_0^{'}\left((V/V_0)^{1/3} - 1\right) \\ -3(V/V_0)^{1/3}]e^{-\frac{3}{2}(B_0^{'}-1)[(V/V_0)^{1/3}-1]}\Big\} \quad (15.7)$$

For each volume the energy is first minimized as a function of internal coordinates, but we kept the ratios of the b/a and c/a fixed to those determined by minimization for the starting volume. The bulk moduli are given in Table 15.5.

15.4 Thermodynamic stability

An important question for crystal growth is the thermodynamic stability. While the full ternary phase diagram as a function of composition and temperature for these ternaries is still not well known, we present some results we have obtained on the energies of formation at zero temperature.

The energies of formation are defined with respect to the elements in their equilibrium phase at room temperature and ambient pressure, but here are

Table 15.2 Lattice parameters a, b, and c (Å), lattice volume V (Å^3), average error with respect to experiment ($\sigma = [(\delta V/V) - 1]/3$), and lattice constant ratio ($a_w = a/2$) in II-IV-N_2.

Compound		LDA	GGA	LDA-ABINIT[a]	Expt.[b]
$ZnSiN_2$	a	6.08	6.16	6.01	6.18
	b	5.27	5.41	5.28	5.35
	c	5.02	5.11	4.98	5.05
	V	160.85	170.29	158.02	166.97
	$\frac{\delta V}{V}$	0.96	1.02	0.95	
	σ	−0.012	0.007	−0.018	
	b/a_w	1.733	1.756	1.757	1.731
	c/a_w	1.651	1.659	1.657	1.634
$ZnGeN_2$	a	6.38	6.42	6.33	6.44
	b	5.45	5.54	5.36	5.45
	c	5.22	5.27	5.11	5.19
	V	181.50	187.44	173.38	182.16
	$\frac{\delta V}{V}$	0.996	1.03	0.952	
	σ	−0.001	0.009		
	b/a_w	1.708	1.726	1.693	1.693
	c/a_w	1.636	1.641	1.614	1.612
$ZnSnN_2$	a	6.59	6.70	6.76	6.753 ± 0.006
	b	5.70	5.80	5.85	5.842 ± 0.007
	c	5.41	5.53	5.58	5.462 ± 0.003
	V	203.21	214.89	220.67	215.48
	$\frac{\delta V}{V}$	0.943	0.997	1.024	
	σ	−0.019	−0.001	0.008	
	b/a_w	1.730	1.731	1.731	1.730
	c/a_w	1.642	1.651	1.651	1.618
$CdGeN_2$	a	6.74			
	b	5.71			
	c	5.47			
	V	210.10			
	b/a_w	1.694			
	c/a_w	1.622			

[a] From (58).
[b] For $ZnSiN_2$ from (42), for $ZnGeN_2$ from (33), and for $ZnSnN_2$ from Quayle *et al.* (50).

Table 15.3 Wyckoff 4(a) positions (reduced coordinates) in the unit cell.

Compound	Atoms	x	y	z
$ZnSiN_2$	Zn	0.623	0.089	0.000
	Si	0.126	0.074	0.000
	N_{Si}	0.102	0.057	0.346
	N_{Zn}	0.649	0.109	0.402
$ZnGeN_2$	Zn	0.625 (0.620)[a]	0.085 (0.083)	0.000 (0.000)
	Ge	0.126 (0.125)	0.079 (0.083)	0.001 (0.000)
	N_{Ge}	0.113 (0.115)	0.069 (0.070)	0.360 (0.365)
	N_{Zn}	0.637 (0.640)	0.096 (0.095)	0.389 (0.385)
$ZnSnN_2$	Zn	0.626	0.084	0.000
	Sn	0.127	0.083	0.002
	N_{Sn}	0.125	0.081	0.377
	N_{Zn}	0.622	0.085	0.372
$CdGeN_2$	Cd	0.620	0.070	0.000
	Ge	0.127	0.054	0.002
	N_{Ge}	0.098	0.047	0.284
	N_{Cd}	0.651	0.101	0.336

[a] For $ZnGeN_2$, the numbers in parentheses are the experimental values from Wintenberger *et al.* (33).

Table 15.4 Average bond lengths in Å.

Compound	II-N (Å)	IV-N (Å)
$ZnSiN_2$	2.02	1.74
$ZnGeN_2$	2.03 (2.01)[a]	1.88 (1.88)
$ZnSnN_2$	2.04	2.02
$CdGeN_2$	2.24	1.88

[a] Experimental values from Wintenberger *et al.* (33) in parentheses.

calculated at zero pressure and temperature. In other words, for $ZnGeN_2$ it corresponds to the reaction energy of the hypothetical reaction,

$$\mathrm{Zn}(hcp) + \mathrm{Ge}(dia) + \mathrm{N}_2(g) \rightarrow \mathrm{ZnGeN}_2(orth) \tag{15.8}$$

between Zn in the hexagonal close packed (hcp) structure, Ge in the diamond structure, and N_2 gas molecules to the orthorhombic crystal structure of $ZnGeN_2$. It is important here to include the spin-polarization contributions of the spin-polarized atoms in determining the individual cohesive energies. These properties

Table 15.5 Bulk moduli and their pressure derivatives of II-IV-N_2 compounds obtained from fit to Vinet equation of state.

Compound	B_0 (GPa)		B'_0	
	LDA	GGA	LDA	GGA
$ZnSiN_2$	234 (228)[a]	208	4.4 (4.4)	4.4
$ZnGeN_2$	207 (197)	178	4.8 (4.4)	4.9
$ZnSnN_2$	177 (184)	150	4.9 (4.8)	4.8
$CdGeN_2$	128		4.7	

[a] Values in parentheses from Paudel *et al.* (58).

were calculated using density functional theory (DFT)(67, 68) in the local density approximation (LDA) (69) and generalized gradient approximation (GGA) (70, 71), and using the full-potential linearized muffin-tin orbital (FP-LMTO) method (72, 73). More details on our computational method are given in Section 15.5.1.

The energies of formation and cohesive energies of the elements involved in their calculation are summarized in Table 15.6. These results were first reported

Table 15.6 Energies of formation (eV/per formula unit), cohesive energies (without zero-point motion correction), and bonding energy of the N_2 molecule (in eV/formula unit).

		LDA	GGA	Expt
ΔE_f	$ZnSiN_2$	−4.68	−4.84	
	$ZnGeN_2$	−1.98	−2.43	
	$ZnSnN_2$	−1.68	−2.32	
	$CdGeN_2$	−1.19		
	GaN	−1.67	−1.90	-1.70 ± 0.16[a]
$-E_{coh}$	Zn	−1.86	−1.08	−1.34[b]
	Cd	−1.51		−1.16
	Si	−5.95	−5.40	−4.65
	Ge	−5.19	−4.49	−3.84
	Sn	−4.45	−3.74	−3.12
	Ga	−3.55	−2.86	−2.77
$-E_B$	N_2	−10.83	−8.45	−9.91[c]

[a] Peshek *et al.* (74).
[b] Gshneider *et al.* (75).
[c] Huber and Herzberg (76).

in (61) for the Zn-IV-N_2 compounds. Here we added $CdGeN_2$. For comparison, we also show results for GaN where we can compare with experimental data (74). The cohesive energies of the elements involved in this calculation help to ascertain the accuracy of this calculation. We can see that LDA typically overestimates the cohesive energy by about 30% for all the solids and about 10% for the molecule. This is because primarily the LDA error occurs for the free atom and molecule. There is a large degree of compensation of errors when subtracting the energies. We find that all four compounds have exothermic energies of formation, including $ZnSnN_2$ and $CdGeN_2$.

Table 15.6 does not include zero-point motion corrections, which we now estimate. The latter were calculated from the phonons for all three Zn-IV-N_2 compounds in our previous work (58), and can be estimated as $9/8k_B\Theta_D$ from the Debye temperature for the other solids. The Debye temperatures were taken from an interactive periodic table, online at http://www.infoplease.com/periodictable.php.

For the N_2 molecule we use, simply, $h\nu/2 = hc/2\lambda$ with $1/\lambda = 2360$ cm^{-1}, the vibrational frequency of the N_2 molecule, obtained by Wathelet *et al.* (77) in the LDA and the harmonic approximation. The important point here is that the zero-point motion contribution of the gas molecule is significantly larger than that of the solids. The zero-point motion corrections on the cohesive energy and molecular binding energy are given in Table 15.7. We can see, however, that the zero-point motion correction to the energies of formation amount to only −0.13, −0.12, and −0.12 eV for Zn-IV-N_2 with IV = Si, Ge, and Sn respectively, and they help to make the energies of formation more negative.

We caution that the energies of formation here are not sufficient to establish the thermodynamic stability of these compounds. It is necessary to also investigate the energies of formation of competing compounds such as Zn_3N_2, Si_3N_4, Ge_3N_4, Sn_3N_4, and other possible IV-N_x compounds. Using these, we can construct maps of the available chemical potential region where the different compounds would be stable. These types of maps are usually used to determine the allowed ranges of the chemical potentials in the context of impurity calculations.

Table 15.7 Zero-point motion contribution to the cohesive energy (in meV).

	E_{zpm} (meV)		
Zn	20	Ga	21
N_2	146		
Si	54	$ZnSiN_2$	87
Ge	31	$ZnGeN_2$	78
Sn	15	$ZnSnN_2$	63

We explain the procedure for $ZnSnN_2$, for which the chemical stability is of the largest concern, since until recently, no reports have been found of synthesis of this material.

For $ZnSnN_2$ the chemical potentials of the elements involved must obey

$$\mu_{\rm Sn} + \mu_{\rm Zn} + 2\mu_{\rm N} = E_f({\rm ZnSnN_2}). \tag{15.9}$$

Here the chemical potentials are *excess* chemical potentials. This means they are in principle Gibbs free energies relative to their value in the reference states, bulk Sn, bulk Zn, and a N_2 gas. We ignore here the pressure and temperature-dependence of the N_2 gas and focus only on the energy of a N atom in a N_2 molecule. In other words, it is half the binding energy of a N_2 molecule. The energy of formation $E_f(\mathrm{ZnSnN_2})$ is the cohesive energy of $ZnSnN_2$ minus the cohesive energies of Zn and Sn per atom (in their solid form) and the binding energy of the N_2 molecule. Thus, when two of the chemical potentials are chosen, the third is defined by eq. (15.9). We choose $\mu_{\rm Sn}$ and $\mu_{\rm Zn}$ as the basic variables in Fig. 15.2. The diagonal line from the upper left to the lower right in Fig. 15.2 corresponds to eq. 15.9 when $\mu_{\rm N} = 0$. The chemical potential of N, μ_N, must lie above this line, otherwise the system becomes unstable to evolution of nitrogen gas. On the other hand, the chemical potentials of Sn and Zn must both be negative. If they became positive it would be favorable to precipitate bulk Sn and bulk Zn. However, more severe restrictions apply if other compounds can form between Sn and N or Zn and N. Zn is known to form a compound Zn_3N_2, while for Sn and N one might expect Sn_3N_4, but reports of other stoichiometries

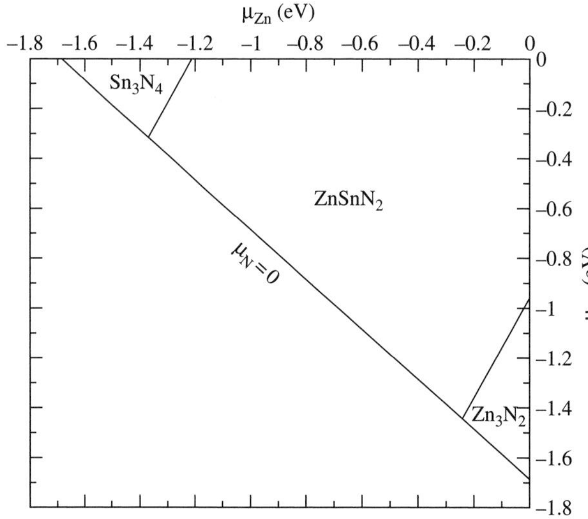

FIG. 15.2. Chemical potential diagram for stability of $ZnSnN_2$. (For discussion, see the text.)

SnN_x with $0.9 < x < 3.3$ exist in literature (78). Most interesting is the report (79) that Sn_3N_4 has its lowest energy in a spinel structure, which for Si_3N_4 only exists as a high-pressure phase. Using, this stoichiometry, we require

$$3\mu_{\rm Zn} + 2\mu_{\rm N} \leq E_f({\rm Zn_3N_2})$$
$$3\mu_{\rm Sn} + 4\mu_{\rm N} \leq E_f({\rm Sn_3N_4}) \quad (15.10)$$

These define two more lines and the region where $ZnSnN_2$ has a chance to be stable. We have calculated the energy of formation of Zn_3N_2 in LDA to be −0.72 eV per formula unit, and that of Sn_3N_4 in the spinel structure to be −0.95 eV, using experimental structural data. These may have to be adjusted further by relaxation, and may thus be slight overestimates. With the present estimates it appears that there is a wide range of stability of the chemical potentials where $ZnSnN_2$ would be stable. We can see that if $\mu_{\rm Zn}$ becomes close to zero (Zn-rich) and $\mu_{\rm Sn}$ is below a certain value (Sn-poor) and above the $\mu_{\rm N} = 0$ line, (nitrogen-rich) Zn_3N_2 would prefer to form. Similarly, if we are Sn-rich and Zn-poor, Sn_3N_4 might form in N-rich conditions. If the borderline between $ZnSnN_2$ and Sn_3N_4 would shift down and to the right, conceivably it could move past the Zn_3N_2–$ZnSnN_2$ borderline. If this occurred it would mean that $ZnSnN_2$ would have no stability region at all. However, we see that this would require a much more negative energy of formation for Sn_3N_4. Similarly, if the Zn_3N_2 line were to move up and to the right, the range of allowable chemical potentials required for the stability of $ZnSnN_2$ could be reduced. Similar conclusions hold for $ZnGeN_2$ and $ZnSiN_2$.

15.5 Electronic structure

As already mentioned in the historical overview of the literature in Section 15.2, there have been various electronic structure calculations since the pioneering work in 1999 by Limpijumnong *et al.* (52). The emphasis of most electronic structure calculations in the early works was to take them as a starting point to discuss the total energy properties of binding, structural optimization, etc. Most of this work used either the LDA or GGA. The main drawback of these methods, however, is that they severely underestimate the band gaps. While one may attempt to correct these by making the assumption that the corrections should be similar in similar materials and therefore could be derived from the known experimental band gap and LDA calculated band gaps in the parent III-N compounds (58), the predictive capability of such an approach is limited. The uncertainty on the band gaps is exacerbated by the fact that most early attempts at growing a new material resulted in rather defective samples. Absorption can then be strongly affected by point defects, leading to subband-gap absorption tails. Thereby, the band gaps are usually underestimated. For example, the earliest band-gap determinations for $ZnGeN_2$ by absorption suggested a gap of 2.67 eV, while later measurements based on photoluminescence (49) gave a gap of 3.40 eV.

The good news is that due to the development of the quasiparticle self-consistent *GW* method, we can now make reliable predictions of the band gaps. In this section, we first describe the computational method, both standard DFT in the LDA which are used to determine the structural parameters and the *GW* method. Next we describe the electronic band-structure and use partial densities of states to determine the atomic orbital character of the bands in different energy ranges. We then zoom in to the band-gap region, describe various corrections required to determined the band gaps accurately, and then zoom in further to determine the valence-band maximum fine structure by means of effective mass Hamiltonians.

15.5.1 *Computational methods*

The basic framework of first-principles electronic structure theory is density functional theory (DFT) (67, 68). These methods provide a "mean-field" prescription for determining the effective potential felt by an electron due to all the other electrons and the nuclei in the solid. Based on the Hohenberg and Kohn theorems (67), these potentials should be completely determined by the electron density. In other words, they are not orbital-dependent. Unfortunately, the exact functional form for the energy and potential as a function of density are not known, and one has to approximate them. The most common approximations are the local density approximation (LDA) (68, 69) and the generalized gradient approximation (GGA)(71). In these methods one assumes that the exchange and correlation energy per particle at any point in space depends only on the density at that point in space (or also on its gradient in GGA), and, furthermore, that these dependencies are the same as in a free electron gas of the same density. Thereby, the results from many-body theory on the exchange and correlation energy in the electron gas, i.e. a system with constant density, can be transferred straight forwardly to the case of varying density. This methodology, however, is focused on calculating the total energy properties, and is in principle valid only for the ground state of the system. It is used to optimize the lattice constants and atomic positions inside the unit cell, and also used to calculate the force constants, from which one derives the vibrational modes. More details are presented in Section 15.6.1.

The method, however, does not apply strictly to excited-state properties such as the optical properties, or even to calculating the so-called quasiparticle excitations, i.e. the energy to extract an electron from the occupied bands, or add one into the empty bands. For the latter, the *GW* method provides a better approach.

In any case, to solve the DFT Kohn–Sham equations self-consistently, one has a choice between various band-structure methods. The latter differ mostly in how they make, or do not make, further approximations to the potential, and in what basis set they are utilized. For the phonon calculations in the later section, we utilize a pseudopotential approach and plane-wave basis set, because these methods facilitate calculation of the derivatives of the total energy. The most accurate methods, however, are free from the ambiguities of how to separate the electrons in valence and core states. They are the so-called all-electron methods. Even in these methods the core-electrons are calculated somewhat differently from the valence electrons. They are calculated as for free atoms, assuming the wavefunctions go to zero at the sphere boundary of spherical regions around each atom. However, their charge density is incorporated fully in the total charge density, and so is the potential they give rise to. In pseudopotential methods, on the other hand, one replaces the problem of the valence electrons by a problem of pseudowavefunctions, which lack the fast oscillations of the wavefunctions near the core required by orthogonality to the core states, but are nonetheless expected to produce the same eigenvalues. They hence rely on transferability of the pseudopotential, representing the smoothed-out interaction with the core-electrons from one chemical environment to another, say from the free atom to that in a solid.

The all-electron calculations are usually performed in the so-called linearized method, in which the basis set is shown to be accurate over a given energy range. Such methods include the linearized augmented plane-wave (LAPW) method, as well as the linearized muffin-tin orbital (LMTO) methods. While early versions of these methods used to make additional "shape" approximations, i.e. spherical potentials, were assumed within spheres, and a constant potential in the region between spheres, this is no longer required, and methods without shape approximation are labeled as being full-potential. We use here the full-potential linearized muffin-tin orbital (FP-LMTO) method as described in (72, 73). This method uses basis functions obtained from smoothed Hankel function envelope functions augmented inside muffin-tin spheres. The Hankel functions describe the overall decay of the functions centered on atomic sites for each type of spherical harmonic. The augmentation means that these functions outside the spheres are matched continuously and with continuous derivative to radial functions inside the spheres which are exact solutions of the Schrödinger equation of the self-consistent potential in that sphere at some given energy, and its energy derivative. Sometimes, additional freedom to the basis set is added by including further "local orbitals" defined only inside the sphere, or "floating orbitals", which are orbital only outside the sphere. The former are used to describe, for example, semi-core states which are not exactly zero yet at the sphere boundary. The latter are mainly useful to better describe highly-lying conduction band-states and improve the basis set for the GW calculations in an efficient manner. The convergence of the basis set can be tested by adding augmented plane-waves. The method, in fact, is now a hybrid of LAPW and LMTO. We emphasize once more that although muffin-tin potentials are used in the construction of the basis set, no shape approximations are made to the potential or charge density.

We use a $4 \times 4 \times 4$ k-point sampling for the Brillouin zone integration in the self-consistent LDA or GGA calculations. The basis set contains two sets of smoothed Hankel function decay parameters κ, and smoothing radii, R_{sm}, and includes *spd* functions in the first and the second set for Zn and includes *spd* functions in the first and *sp* in the second set for Si, Ge, Sn, and N. In addition, we add Zn-4d as local orbitals. Finally, we point out that scalar relativistic effects are included by solving the scalar Dirac equation instead of the Schrödinger equation. The effects of spin–orbit coupling can also be included.

The quasiparticle band structure is calculated using the quasiparticle self-consistent GW approach implemented in the same FP-LMTO method, as described in (80–82). Hedin's GW approximation (83, 84) is the first term in a series expansion of the self-energy, in terms of the screened Coulomb interaction. Schematically, $\Sigma(12) = iG(12)W(1_+2)$, with $G(12)$ the one-electron Green's function and $W(1_+2)$ the screened Coulomb interaction—written in position, spin, time coordinates $1 = \{\mathbf{r}_1, \sigma_1, t_1\}$, with $1_+ = t_1 + \delta$. Applying a Fourier transform to the frequency (or energy) domain, the following equations are used to calculate these quantities:

$$\Pi(\mathbf{r}, \mathbf{r}', \omega) = -iG_0(\mathbf{r}, \mathbf{r}', \omega) \otimes G_0(\mathbf{r}', \mathbf{r}, -\omega) \tag{15.11}$$

$$\epsilon^{-1}(\mathbf{r}, \mathbf{r}', \omega) = \delta(\mathbf{r} - \mathbf{r}') + \int d\mathbf{r}'' v(\mathbf{r} - \mathbf{r}'')\Pi(\mathbf{r}'', \mathbf{r}', \omega) \tag{15.12}$$

$$W(\mathbf{r}, \mathbf{r}', \omega) = \int d\mathbf{r}''\, \epsilon^{-1}(\mathbf{r}'', \mathbf{r}', \omega) v(\mathbf{r}'' - \mathbf{r}') \tag{15.13}$$

$$\Sigma(\mathbf{r}, \mathbf{r}', \omega) = \frac{i}{2\pi} \int d\omega'\, G(\mathbf{r}, \mathbf{r}', \omega + \omega') W(\mathbf{r}, \mathbf{r}', \omega') e^{i\delta\omega'} \tag{15.14}$$

Here, $\otimes$ denotes convolution over frequencies, W is calculated as $W = (1 - v\Pi)^{-1}v$ and $\Pi(12) = -iG(12)G(21)$ in the random phase approximation (RPA) and GW is usually applied as a single-shot correction to LDA within first-order perturbation theory. In other words, the Green's functions $G = G^0 = G^{LDA}$, calculated from the eigenstates of the LDA Hamiltonian. One may now extract a new static (but non-local) exchange correlation potential

$$V_{xc}^{QSGW} = \frac{1}{2} \sum_{nm} |\Psi_n\rangle \Re[\Sigma_{nm}(E_n) + \Sigma_{nm}(E_m)] \langle \Psi_m| \tag{15.15}$$

from the Σ self-energy. Here, Ψ_n are the one-electron Kohn–Sham eigenstates and $\Re$ denotes taking the Hermitian part. This new exchange correlation potential can then be used as a new starting point for a new H^0, and iterated to convergence. This approach (called QSGW) is designed so that the eigenvalues of the Kohn–Sham independent particle equations $\epsilon_{n\mathbf{k}}$

$$H^0 \Psi_{n\mathbf{k}}(\mathbf{r}) = \left[-\frac{1}{2}\nabla^2 + v_N(\mathbf{r}) + v_H(\mathbf{r}) + v_{xc}(\mathbf{r}) \right] \Psi_{n\mathbf{k}}(\mathbf{r}) = \epsilon_{n\mathbf{k}} \Psi_{n\mathbf{k}}(\mathbf{r}) \tag{15.16}$$

approach the quasiparticle energies $E_{n\mathbf{k}}$, given by

$$\left[-\frac{1}{2}\nabla^2 + v_N(\mathbf{r}) + v_H(\mathbf{r})\right]\Phi_{n\mathbf{k}}(\mathbf{r}) + \int d^3r'\Sigma(\mathbf{r},\mathbf{r}',E_{n\mathbf{k}})\Phi_{n\mathbf{k}}(\mathbf{r}') = E_{n\mathbf{k}}\Phi_{n\mathbf{k}}(\mathbf{r}) \tag{15.17}$$

in the GWA. Here we use atomic units: v_N is the interaction with the nuclei, and v_H is the Hartree potential.

An important distinguishing feature of the present *GW* and QS*GW* implementation is that it is an all-electron implementation, not relying on pseudopotentials and thereby properly including core-valence exchange. It is implemented in the full-potential LMTO method. A mixed plane-wave and product basis set (85) is used to express all two-particle operators. For details, we refer the reader to the above-mentioned papers by the van Schilfgaarde group.

Let us here just remark that this method provides very accurate results for a wide range of solids, including all standard III-V and II-VI semiconductors (80). Most importantly, the method has highly systematic small remaining errors: it gives a slight overestimate of the gaps compared to experiment. This remaining discrepancy is thought to arise from the under-screening by the RPA. It can be corrected by scaling the final $\Delta V_{xc} = V_{xc}^{QSGW} - v_{xc}^{LDA}$ by about 80%, as obtained empirically by comparing QSGW with experiment for a wide variety of semiconductors (80). We refer to this as the 0.8Σ approximation. Within this approximation, most band gaps are obtained to better than 0.1 eV. The ΔV_{xc} represented in a muffin-tin orbital basis set can be Fourier-transformed to real space, and then Bloch summed for arbitrary k-points on a finer mesh than the one on which the time-consuming *GW* calculation is carried out. This allows us to obtain accurate effective masses and band plots along symmetry lines.

As explained in (82), the $\Delta V^{xc}_{\tilde{n}\tilde{m}}(\mathbf{k})$ in the basis set of the LDA Hamiltonian eigenstates (as indicated by the $\tilde{}$ over the indices) can be approximated to be linear in the LDA eigenvalue: $a\epsilon^{LDA}_{\tilde{n}\mathbf{k}} + b$ for energy bands above some cut-off energy E_{cut}. We find it important to take this cut-off sufficiently high for the nitrides, say $E_{cut} = 3$ Ryd above the Fermi level, to obtain well-converged results. The gaps for $E_{cut} = 2.5$ Ryd were still underestimated by 5–10 %. As a **k**-point sampling set for the *GW* self-energy we use a $2 \times 2 \times 2$ mesh corresponding to the sixteen-atom unit cell. This is more or less equivalent to an $8 \times 8 \times 4$ sampling for wurtzite.

15.5.2 *Band-structure overview*

The electronic band structures of Zn-IV-N_2, for IV = Si, Ge, Sn calculated using the QSGW method, were reported in (62). The band structure of $CdGeN_2$ was presented at the ICSCRM2011 meeting, and is reported in (63). As an example, the band structure and density of states of $ZnGeN_2$ are shown in Fig. 15.3.

Since there are sixteen atoms in the unit cell, eight nitrogens, four Zn, and four Ge, there are 104 valence electrons per unit cell or 52 occupied bands. We

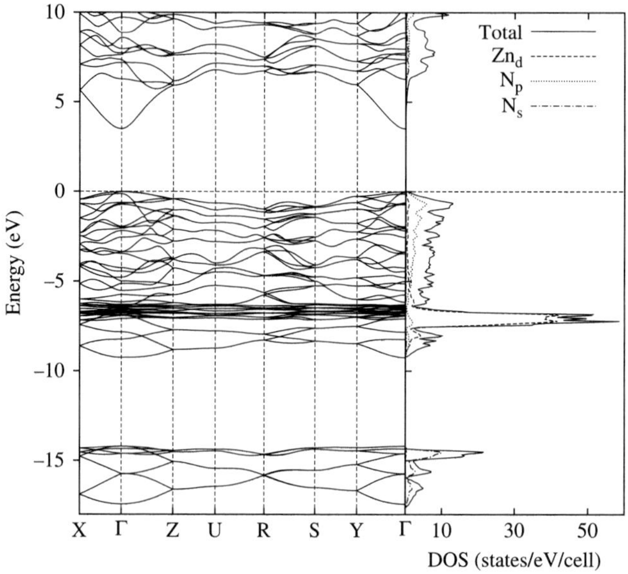

FIG. 15.3. Left: Band structure of $ZnGeN_2$, right: density of states and partial densities of states, both calculated in the QS*GW* approximation.

can see that the eight bands near −15 eV are predominantly N-2*s*-like bonding with Zn and Ge. The twenty-four bands between −5 and 0 eV are the N-2*p*-like bands, and the conduction bands have mixed Ge and Zn-*s* character. The twenty Zn-3*d* bands are found near the bottom of the N-2*p* band in a narrow range. A similar ordering of states occurs in the other II-IV-N_2 nitrides. For further details on the partial densities of states, showing also those with smaller contributions, we refer the reader to (62). A similar general ordering of the bands is applicable to the other II-IV-N_2 compounds.

The effects of *GW* compared to LDA can be seen in Fig. 15.4 for $ZnSiN_2$. The gap is increased by 2.1 eV, with 1.6 eV of the shift occurring in the conduction band and −0.5 eV in the valence band. In $ZnGeN_2$, the gap change is 1.8 eV, with 1.7 eV in the conduction band and only −0.1 eV in the valence band. In $ZnSnN_2$, both the valence and conduction bands shift up relative to the average electrostatic potential (by 0.2 eV and 1.8 eV respectively), and the net gap shift is 1.6 eV. In $CdGeN_2$, the gap shift is also primarily in the conduction band and amounts to 1.55 eV. The absolute shifts of the band structure relative to the average electrostatic potential are important for band offsets at interfaces and also play a role in defects and impurities. A discussion of this issue can be found in (62). We can also see that the Zn-3*d* shift down in *GW*. This shift amounts to −2.3 eV in $ZnSiN_2$, −2.0 eV in $ZnGeN_2$, −1.5 eV in $ZnSnN_2$, and −1.8 eV in $CdGeN_2$. The N-2*s* bands also

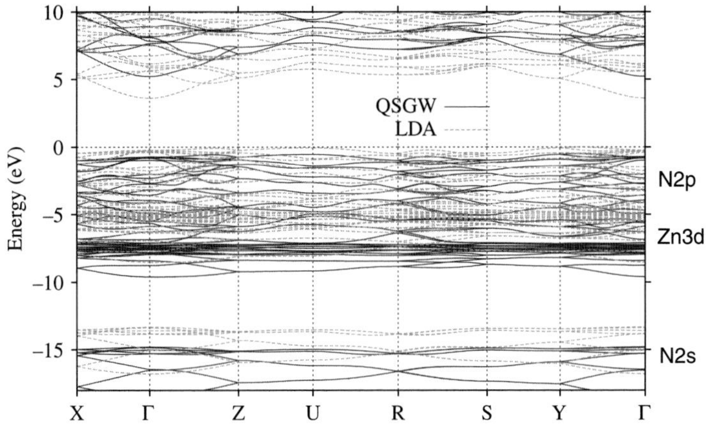

FIG. 15.4. Comparison of LDA (dashed lines) and QSGW (solid lines) band structure of $ZnSiN_2$. The zero of energy is chosen at the VBM of the LDA calculation. (From (62).)

shift down significantly more than the top of the valence band, as expected, by about 1 eV.

Having understood the overall band structure at a scale of about a Rydberg or so above and below the VBM, we now zoom in on the region near the gap. We can see that in all materials the conduction band minimum is clearly at Γ. The conduction band dispersion is high, implying a small electron effective mass. The region near the valence bands is shown in Fig. 15.5. The valence-band maximum is also at Γ except in $ZnSiN_2$. In $ZnSiN_2$ one of the valence bands is seen to disperse upwards both in the $\Gamma - Z$ and $\Gamma - Y$ directions, and the actual VBM occurs along the $\Gamma - Y$ direction. Thus, of the four II-IV-N_2 nitrides studied here, we find three direct-gap and one indirect-gap semiconductors. Even, so the direct gap is not very different from the indirect gap in $ZnSiN_2$ because the dispersion of the upper valence bands is relatively small. One may expect that optical absorption will still be dominated by the lowest direct transition at Γ, but in photoluminescence one might expect the lower gap to show up at low temperature because the holes would thermalize to their lowest energy, i.e. the VBM.

Next, we discuss various corrections to the band gap and their comparison to experimental data. First of all, we have already included a correction to strict QSGW by including the 0.8 reduction factor of Σ_{xc}. Second, we should include the effects of spin–orbit coupling. Since these split the valence bands, they slightly reduce the gap. This effect, discussed in the following, is extremely small in Zn-IV-N_2 semiconductors.

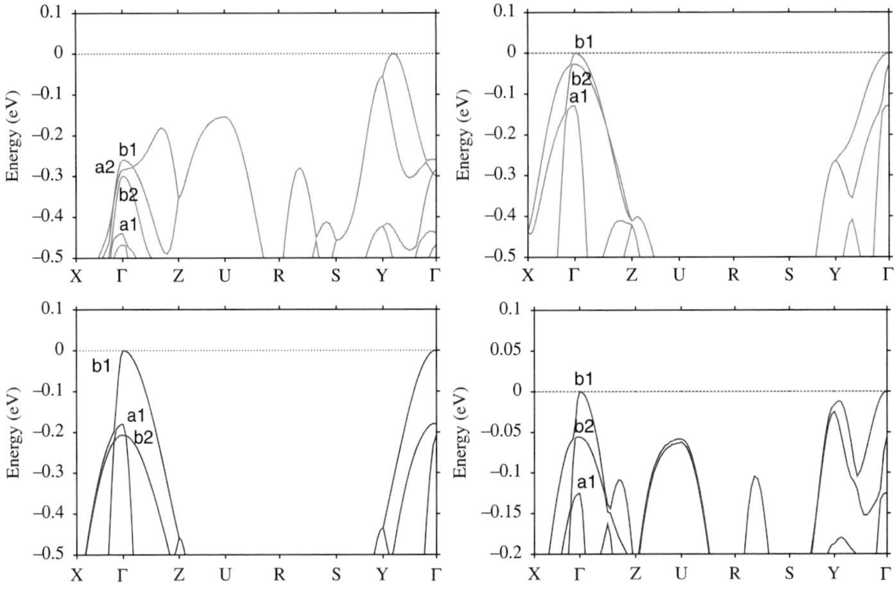

FIG. 15.5. Zoom-in on the band structure near the valence-band maximum of II-IV-N_2 compounds: top row, $ZnSiN_2$ and $ZnGeN_2$; bottom row, $ZnSnN_2$ and $CdGeN_2$. The levels at Γ are labeled according to the irreducible representations. The results for Zn-IV-N_2 were previously published in (62), and the one for $CdGeN_2$ in (63).

Third, there is a zero-point motion correction. The electron–phonon effects on the band structure are discussed in a review article by Cardona and Thewalt (86). While at high temperature the gap varies more or less linear with temperature, in part because of the thermal expansion and hence smaller gaps (because the covalent interactions are reduced), one typically finds a curving over of the gap at low temperatures. If one were to extrapolate the gap toward zero temperature based on the slope and value at room temperature, one would overestimate the band gap. This remaining discrepancy is due to the zero-point motion effect on the electron–phonon interaction. It is tabulated in (86) for an extensive list of semiconductors and in particular for the III-nitrides. Examining these values for various closely related families as a function of the gap itself, one can see that they are approximately linearly related. Plotting the values for the III-nitrides, we find that the zero-point motion correction can be approximated by

$$\Delta(0) = -50\text{meV} - 31 \times 10^{-3} E_g \tag{15.18}$$

Assuming that the same slope and offset hold for the II-IV-N_2 nitrides, we obtain sizable gap corrections of −227, −161, −126, and −128 meV in $ZnSiN_2$, $ZnGeN_2$,

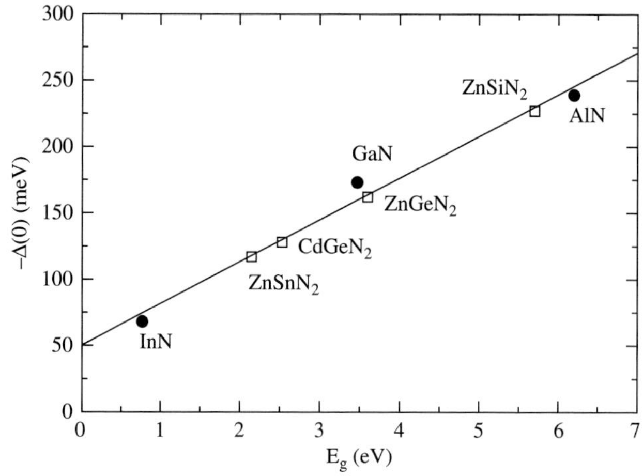

FIG. 15.6. Zero-point motion correction $\Delta(0)$ to the band gap as a function of the band gap: the data for group-III nitrides (filled circles) from Cardona and Thewalt (86) are used to estimate the linear interpolation for the II-IV-N_2 compounds, indicated by open squares.

$ZnSnN_2$, and $CdGeN_2$ respectively. These are shown in Fig. 15.6, along with the III-N from which they are derived.

Next, we need to consider the exciton binding energies, estimated from

$$E_{xb} = \frac{\mu}{\kappa_0^2}\text{Rydberg} \tag{15.19}$$

where we use the reduced mass $\mu = m_c m_v/(m_c + m_v)$ with m_c, m_v a directional averaged conduction and valence band hole mass as obtained in the next section. Here we neglect the interaction of the different valence bands and take only the highest VBM into account (which is justified by the fairly large crystal field splittings), and average the inverse masses over direction. We use the static dielectric constants κ_0, calculated using DFPT in (58) and discussed later. We have not yet carried out such calculations for $CdGeN_2$. We therefore used an estimate of the dielectric constant for $CdGeN_2$, assuming that within this family of materials it is proportional to the gap. The use of static dielectric constants is consistent with the fact that the resultant exciton binding energies are significantly smaller than the optical phonon frequencies, so one may assume that lattice distortion contributes to the screening of the Coulomb interaction between electron and hole. The effective Bohr radius of the exciton is given by $(\kappa_0/\mu)a_0$, with a_0 the Bohr radius of hydrogen: 0.529 Å. The results are given in Table 15.8. We note that for $ZnSiN_2$ and $ZnGeN_2$ our values are close to that for GaN: 25 meV (87).

Table 15.8 Exciton binding energy and related parameters: conduction band mass m_c, valence band mass m_v, reduced mass μ, static dielectric constant κ_0, exciton binding energy (meV), and effective Bohr radius (Å).

	m_c	m_v	μ	κ_0	E_{xb} (meV)	a_* (Å)
$ZnSiN_2$	0.293	0.520	0.188	9.33	29	26
$ZnGeN_2$	0.185	0.530	0.136	9.70	20	38
$ZnSnN_2$	0.141	0.392	0.104	12.71	9	65
$CdGeN_2$	0.166	0.527	0.126	11.84	12	50

Table 15.9 Band gaps (in eV) of II-IV-N_2 compounds in various approximations.

Compound		LDA	GGA	QSGW	0.8Σ	$0.8\Sigma + \Delta(0) + E_{xb}$ [a]	Expt.
$ZnSiN_2$	indirect	3.60	3.23	6.01	5.70	5.44	
	direct	3.84	3.45	6.26	5.92	5.66	4.46 [b]
$ZnGeN_2$	direct	1.66	1.57	3.99	3.60	3.42	3.40±0.01 [c]
$ZnSnN_2$	direct	0.55	0.35	2.64	2.15	2.02	1.7±0.1 [d]
$CdGeN_2$	direct	1.02		2.92	2.53	2.38	

[a] $\Delta(0)$ is the estimated zero-point motion renormalization correction (see text), and E_{xb} is the exciton binding energy.
[b] From Osinksy *et al.* (40).
[c] From Du *et al.* (49).
[d] From Quayle *et al.* (50).

Putting the various corrections together we obtain the results of Table 15.9. We can see that both LDA and GGA significantly underestimate the gaps. Their difference is mostly related to the difference of lattice constant obtained in these methods. GGA produces slightly larger lattice constants, and hence a smaller gap. The full QS*GW* method overestimates the gap typically by about 20%. When finally adding the various corrections mentioned, we obtain extremely good agreement with experiment for $ZnGeN_2$. To be honest, one should keep in mind that the 0.8 factor is a bit of a fudge factor, and various empirical approximations were made previously to estimate the other corrections. We therefore should still put a conservative error bar of 0.1 eV on the gap. Also, notice that we used the LDA lattice constants, which are somewhat small, and therefore might overestimate the gaps. Thus, we might actually need a smaller correction factor than 0.8 if we use experimental lattice constants and explicitly include all the other corrections mentioned.

For $ZnSiN_2$ our value appears to be a significant overestimate by about 1 eV. We should note, however, that far less experimental work has been done on

$ZnSiN_2$. Significant uncertainty clearly exists. The work from Endo *et al.* (36) produces a larger lattice constant and an even lower gap of only 3.64 eV. The first gap-determination for $ZnGeN_2$ by means of optical absorption (35) produced a gap estimate of 2.67 eV, also about 1 eV lower than the recent more accurate results. The value of 4.46 eV for $ZnSiN_2$ was obtained by absorption on a series of $ZnSi_xGe_{1-x}N_2$ alloys (40). The temperature effect on the gap is expected to be at most 0.2 eV, and cannot explain this discrepancy. We thus think that this results from defect band-gap tailing and the intrinsic difficulty in determining an accurate gap from transmission measurements. Since it is determined from absorption and not from photoluminescence, we should clearly compare it with the direct gap, but even our indirect gap is only a few 0.1 eV smaller. The slightly smaller indirect than direct gap was noted in previous LDA calculations by Shaposhnikov *et al.* (54), and our LDA results are close to theirs, with the small discrepancies attributable to differences in the band-structure method or the lattice constants used.

$ZnSnN_2$ has only very recently been grown in polycrystalline form. Its gap was determined, by photoluminescence excitation spectroscopy (PLE), to be 1.7 ± 0.1 eV at room temperature (50). Our overestimate of the calculated gap (2.02 eV) can in large part be explained by the change in the lattice constant. Our LDA lattice constant at zero temperature is about 6% smaller than the measured room-temperature lattice constant. Note that this is probable not only because of the thermal expansion but also because of the LDA underestimate of the lattice constant, even at 0 K. Using our calculated band-gap deformation potential $dE_g/d\ln V$ of -3.89 eV, this reduces the gap by about -0.2 eV, giving a gap of 1.8 ± 0.1 eV. Thus, theory and experiment are consistent to within about 0.1 eV.[1] For $CdGeN_2$ there are no experimental data available.

We note that the direct gaps calculated here correspond to wavelengths of 218 nm, 362 nm, 617 nm, and 469 nm for $ZnSiN_2$, $ZnGeN_2$, $ZnSnN_2$, and $CdGeN_2$. These values span the red-to-ultraviolet region, with $CdGeN_2$ very close to the green region. The band gaps versus lattice constants are summarized in Fig. 15.7. Clearly, this materials family has a significant potential for optoelectronic applications, similar to the group-III nitrides. While the known III-nitrides extend to smaller gaps of 0.7 eV in InN and a larger gap of 6.2 in AlN, we have not yet explored the full set of possibilities in the II-IV-N_2 family. For example, one might expect $CdSnN_2$ to have a smaller gap closer to that of InN. Hg instead of Cd would also be expected to lead to smaller gaps. Mg instead of Zn might extend the gaps on the UV side. On the other hand, the materials included in Fig. 15.7 have a closer lattice constant match among each other than the III-nitrides. They are somewhat closer clustered together, and thus interpolation between them by alloying would be expected to result in lower strain and

[1] We have recently become aware of two recent papers on $ZnSnN_2$ grown by molecular beam epitaxty (MBE) (88, 89), which indicates a gap of about 2.1 eV. A larger gap may result from n-type doping and the Moss–Burstein effect.

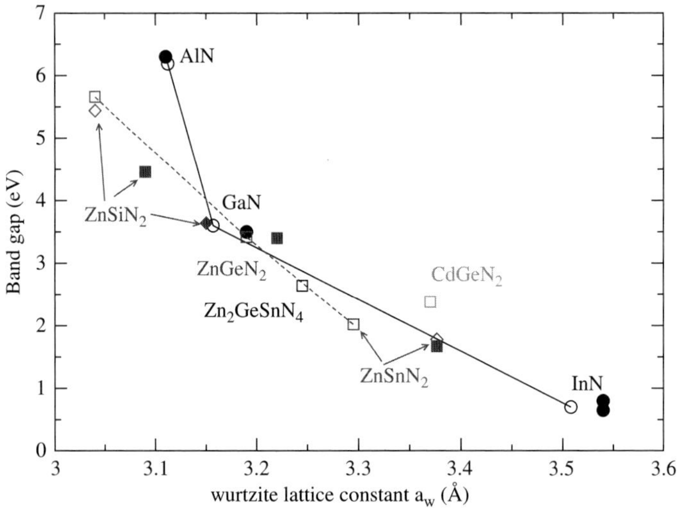

FIG. 15.7. Band gaps as a function of lattice constant in III-N and II-IV-N_2 nitrides. Open symbols: LDA lattice constants and QS*GW* gaps; filled symbols: experimental values. For $ZnSiN_2$, the square indicates the direct gap, and the diamond the indirect gap. The lower experimental gap value for $ZnSiN_2$ is from (36), and is not mentioned if it is direct or indirect. The higher value is from (40). The gap value of $ZnGeN_2$ is from (49). The experimental gap for $ZnSnN_2$ is from (50), and the open diamond for $ZnSnN_2$ is estimated by correcting for the change in the lattice constant, using the LDA gap-deformation potential.

more flexibility in achieving specific target band-gaps. For example, we picked $CdGeN_2$ with the specific aim of testing whether it could fill the so-called green gap. Unfortunately, $ZnSiN_2$ is found to be an indirect gap material. At present, we do not yet know up to what concentration of Si the gap remains direct in $ZnSi_xGe_{1-x}N_2$. However, interpolating the indirect gap "near Y to Γ" and the direct gap at Γ linearly between $ZnGeN_2$ and $ZnSiN_2$, one would predict the gap to stay direct up to about 51% Si and to reach up to a gap of about 4.56 eV. Similarly, in $ZnSi_xSn_{1-x}N_2$, the gap is expected to stay direct up to almost 58% and to reach a maximum direct gap of about 4.15 eV. This is shown in Fig. 15.8. Because Zn, Si, and Sn are all abundant elements, in contrast to Ga and In, this is important from a sustainability point of view, especially for applications in photovoltaics and solid-state lighting, which require a potentially large amount of material if widely deployed.

In (63) we also made a first estimate of the band gap of the $ZnSn_{0.5}Ge_{0.5}N_2$ alloy. For this purpose we used only one structural model in which the first layer contains Sn and the second Ge in the usual II-IV-N_2 unit cell. This further

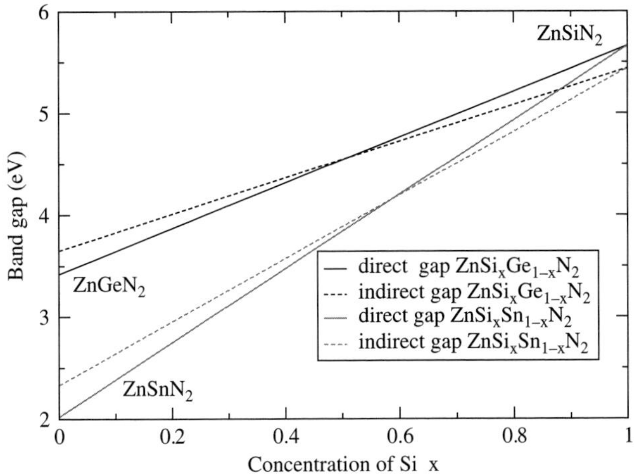

FIG. 15.8. Crossing of direct (at Γ) and indirect ($Y - \Gamma$) gaps in $ZnSi_xGe_{1-x}N_2$ and $ZnSi_xSn_{1-x}N_2$, assuming linear interpolation.

reduces the symmetry. We caution that disorder on the cation sublattice is not yet taken into account. Band-gap corrections beyond LDA were estimated by interpolating between those of $ZnGeN_2$ and $ZnSnN_2$. We extract a value of the band-gap bowing parameter of 0.3 eV. The band gap as a function of concentration according to Eq. 15.1 is shown in Fig. 15.9 This predicts a gap corresponding to a 469-nm wavelength in the blue region of the spectrum.

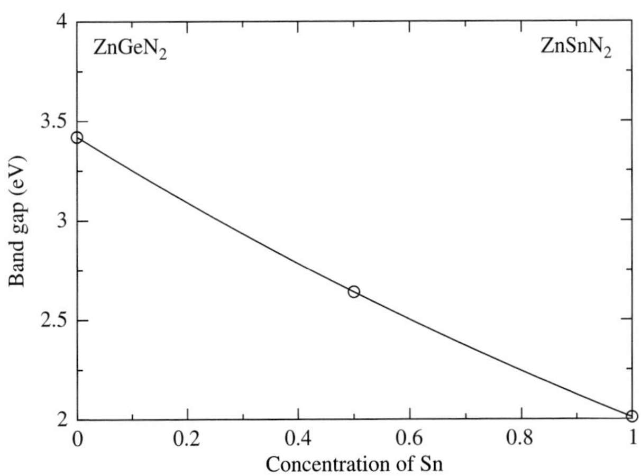

FIG. 15.9. Band gaps in $ZnGe_{1-x}Sn_xN_2$ alloys.

15.5.3 *Effective-mass Hamiltonians*

In this section we zoom further in to the top of the valence band, and discuss its fine structure. This is shown for all four materials on which we focus here in Fig. 15.5. We first notice that the valence bands at Γ are already split into three, even without including spin–orbit coupling. This is a result of the orthorhombic symmetry. By inspecting the eigenvectors we determined their symmetry labeling as indicated in Fig. 15.5. The conduction-band minimum at Γ has a_1 symmetry in all cases. This implies that optical transitions are only dipole allowed between a_1 and a_1 for $\mathbf{E} \parallel \mathbf{c}$ between b_1 and a_1 for $\mathbf{E} \parallel \mathbf{a}$, and between b_2 and a_1 for $\mathbf{E} \parallel \mathbf{b}$. In other words, the optical absorption and luminescence should display additional anisotropy compared to the wurtzite group-III-nitrides. Even for normal emission from the basal plane, the light emission should be anisotropic. This is potentially of use for polarization-sensitive detectors.

The valence-band splittings in the absence of spin–orbit coupling can be described in terms of two crystal field splitting parameters (62):

$$
\begin{aligned}
E_{a_1} &= 0 \\
E_{b_2} &= \Delta_{1c} + \Delta_{2c} \\
E_{b_1} &= \Delta_{1c} - \Delta_{2c}
\end{aligned}
\tag{15.20}
$$

The values of these parameters are given in Table 15.11. Based on symmetry there are three different spin–orbit coupling parameters. The crystal field and spin–orbit coupling Hamiltonian of the VBM can be written as

$$
\begin{aligned}
H^{C_{2v}}(\Gamma) = {} & \Delta_{1c} {L_z}^2 + \Delta_{2c}({L_x}^2 - {L_y}^2) \\
& + \Delta_{1so} L_z \sigma_z + \sqrt{2}(\Delta_{2so} L_x \sigma_x + \Delta_{3so} L_y \sigma_y)
\end{aligned}
\tag{15.21}
$$

Using a basis of $L = 1$ spherical harmonics and including the two-fold degeneracy because of spin, the 6×6 Hamiltonian breaks into two equivalent 3×3 Hamiltonians of the form

$$
\begin{pmatrix}
\Delta_{1c} + \Delta_{so1} & \Delta_{so2} - \Delta_{so3} & \Delta_{2c} \\
\Delta_{so2} - \Delta_{so3} & 0 & \Delta_{so2} + \Delta_{so3} \\
\Delta_{c2} & \Delta_{so2} + \Delta_{so3} & \Delta_{c1} - \Delta_{so1}
\end{pmatrix}
\tag{15.22}
$$

The order of the basis function is $m = 1 \uparrow$, $m = 0 \downarrow$, and $m = -1 \uparrow$, or equivalently, $m = -1 \downarrow$, $m = 0 \uparrow$, and $m = 1 \downarrow$. The eigenvalues are indeed doubly degenerate in spin because of Kramer's theorem, related to time-reversal symmetry. If we assume that the spin–orbit coupling parameters are essentially isotropic, $\Delta_{so1} = \Delta_{so2} = \Delta_{so3} = \Delta_s$, which is usually a good approximation, because spin–orbit coupling arises from the inner part of the atom where the potential is close to spherically symmetric, we can make some progress by first applying the unitary transformation that diagonalized the matrix without spin–orbit coupling. This leads to

$$\begin{pmatrix} \Delta_{c1}+\Delta_{c2} & \sqrt{2}\Delta_s & \Delta_s \\ \sqrt{2}\Delta_s & 0 & -\sqrt{2}\Delta_s \\ \Delta_s & -\sqrt{2}\Delta_s & \Delta_{c_1}-\Delta_{c2} \end{pmatrix} \tag{15.23}$$

If we assume, furthermore, that the spin–orbit parameter is small compared to the crystal field splitting of the levels, then in second-order perturbation theory we obtain

$$\begin{aligned} E_{a1} &= -\frac{2\Delta_s^2}{\Delta_{c1}+\Delta_{c2}} \\ E_{b2} &= \Delta_{c1}+\Delta_{c2}+\frac{2\Delta_s^2}{\Delta_{c1}+\Delta_{c2}}+\frac{\Delta_s^2}{2\Delta_{c2}} \\ E_{b1} &= \Delta_{c1}-\Delta_{c2}+\frac{2\Delta_s^2}{\Delta_{c1}-\Delta_{c2}}-\frac{\Delta_s^2}{2\Delta_{c2}} \end{aligned} \tag{15.24}$$

Strictly speaking, all levels have symmetry Γ_5 if we use the double-group notation, but we still label them by the single-group label from which they are derived. We carried out calculations including spin–orbit coupling for the Zn-IV-N_2, and found that the splittings were essentially indistinguishable from those without spin–orbit coupling. This indicates that the spin–orbit coupling parameter is very small in these compounds, and makes it at this point impractical and at the same time unimportant to try to extract these small parameters.

The reason for this small spin–orbit coupling is worth some discussion. For the parent compounds, GaN, we find it is already rather small, because (1) the N atom is a low-Z element for which relativistic effects are small, (2) there is a negative contribution from the Ga-3d states lying below the VBM, but which are nevertheless somewhat hybridized with the VBM. In $ZnGeN_2$ this negative contribution, which now derives from the Zn-3d, must be even stronger. In fact, we know that in ZnO the spin–orbit splitting becomes negative (90). One expects it here to be intermediate between GaN and ZnO, because only half the cation sites contribute. The Ge-3d or Sn-4d states lie significantly deeper, so their contribution must be significantly smaller. Apparently, the 3d contribution must almost perfectly cancel the N-2p contribution. The same is true in $CdGeN_2$ with the 4d level.

In order to describe the valence-band maximum manifold of states for **k**-points close to the Γ-point, it is useful to derive an effective mass Hamiltonian (EMH) by means of the theory of invariants. In zincblende crystals this is well known as the Kohn–Luttinger Hamiltonian (91, 92), and in wurtzite it is known as the

Rashba–Sheka–Pikus Hamiltonian (93, 94). In Punya *et al.* (62) we generalized this for the orthorhombic symmetry. The Hamiltonian for this case is of the form:

$$\begin{aligned} H^{C_{2v}} = {} & \Delta_{1c}{L_z}^2 + \Delta_{2c}({L_x}^2 - {L_y}^2) + \Delta_{1so}L_z\sigma_z + \sqrt{2}(\Delta_{2so}L_x\sigma_x + \Delta_{3so}L_y\sigma_y) + \\ & \left[A_1 + A_2{L_z}^2 + A_3\left({L_x}^2 - {L_y}^2\right)\right]{k_z}^2 + \left[B_1 + B_2{L_z}^2 + B_3\left({L_x}^2 - {L_y}^2\right)\right]\left({k_x}^2 + {k_y}^2\right) + \\ & \left[C_1 + C_2{L_z}^2 + C_3\left({L_x}^2 - {L_y}^2\right)\right]\left({k_x}^2 - {k_y}^2\right) + \\ & D_1\{L_x, L_y\}k_xk_y + D_2\{L_z, L_x\}k_zk_x + D_3\{L_z, L_y\}k_zk_y + E_1L_xk_y + E_2L_yk_x \end{aligned} \tag{15.25}$$

where $\{L_i, L_j\}$ are anti-commutators. This includes all terms up to second order that can be formed from the operators L_i, k_i, and σ_i that are invariant (i.e. belong to the a_1 representation). At $\mathbf{k} = 0$ we have two crystal field splittings and three spin–orbit splitting parameters. The remaining terms are inverse effective masses, and the parameters E_1 and E_2 are linear in k_i. We have neglected purely relativistic linear terms involving σ_i and k_i. If $\Delta_{2c} = 0$, $\Delta_{2so} = \Delta_{3so}$, $A_3 = 0$, $B_3 = 0$, $C_1 = C_2 = 0$, $C_3 = 2D_1$, $D_2 = D_3$, and $E_2 = -E_1$, the above Hamiltonian is reduced to the RSP Hamiltonian valid for C_{6v} symmetry. If we denote the RSP Hamiltonian parameters as defined in Kim *et al.* (95) by $\tilde{A}_i$, we have the relations:

$$\begin{aligned} A_1 &= \tilde{A}_1 \quad A_2 = \tilde{A}_3 \\ B_1 &= \tilde{A}_2 \quad B_2 = \tilde{A}_4 \\ C_3 &= \tilde{A}_5 \quad D_1 = 2\tilde{A}_5 \\ D_2 &= D_3 = \sqrt{2}\tilde{A}_6 \\ E_1 &= -E_2 = \sqrt{2}\tilde{A}_7 \end{aligned} \tag{15.26}$$

For arbitrary $\mathbf{k}$, the 3×3 Hamiltonian matrix without spin–orbit coupling terms is of the form

$$\begin{pmatrix} \Delta_{1c} + (B_1 + B_2 + C_1 + C_2)k_x^2 + (B_1 + B_2 - C_1 - C_2)k_y^2 + (A_1 + A_2)k_z^2 & (D_2k_zk_x - iD_3k_zk_y)/\sqrt{2} + E_1k_y - iE_2k_x & \Delta_{2c} + (B_3 + C_3)k_x^2 + (B_3 - C_3)k_y^2 + A_3k_z^2 - iD_1k_xk_y \\ (D_2k_zk_x + iD_3k_zk_y)/\sqrt{2} + E_1k_y + iE_2k_x & (B_1 + C_1)k_x^2 + (B_1 - C_1)k_y^2 + A_1k_z^2 & (-D_2k_zk_x + iD_3k_zk_y)/\sqrt{2} + E_1k_y - iE_2k_x \\ \Delta_{2c} + (B_3 + C_3)k_x^2 + (B_3 - C_3)k_y^2 + A_3k_z^2 + iD_1k_xk_y & (-D_2k_zk_x - iD_3k_zk_y)/\sqrt{2} + E_1k_y + iE_2k_x & \Delta_{1c} + (B_1 + B_2 + C_1 + C_2)k_x^2 + (B_1 + B_2 - C_1 - C_2)k_y^2 + (A_1 + A_2)k_z^2 \end{pmatrix} \tag{15.27}$$

If we now consider the bands in each of the orthogonal directions, e.g. ($k_x = k_y = 0$), we can diagonalize the Hamiltonian in the same way as for $\mathbf{k} = 0$ and read off the hole inverse masses. We will use the notation m_α^λ, where λ denotes the irreducible representation of the level at Γ, and α denotes the Cartesian direction x, y, z. Thus, we obtain the following relations for the inverse hole masses:

$$\begin{aligned}
-(m_z^{a_1})^{-1} &= A_1 \\
-(m_z^{b_2})^{-1} &= A_1 + A_2 + A_3 \\
-(m_z^{b_1})^{-1} &= A_1 + A_2 - A_3 \\
-(m_x^{a_1})^{-1} &= B_1 + C_1 \\
-(m_x^{b_2})^{-1} &= B_1 + B_2 + B_3 + C_1 + C_2 + C_3 \\
-(m_x^{b_1})^{-1} &= B_1 + B_2 - B_3 + C_1 + C_2 - C_3 \\
-(m_y^{a_1})^{-1} &= B_1 - C_1 \\
-(m_y^{b_2})^{-1} &= B_1 + B_2 + B_3 - C_1 - C_2 - C_3 \\
-(m_y^{b_1})^{-1} &= B_1 + B_2 - B_3 - C_1 - C_2 + C_3
\end{aligned} \tag{15.28}$$

In Table 15.10 we summarize the various VBM and CBM effective masses, and in Table 15.11 we summarize the A_i, B_i, C_i parameters. The remaining parameters D_i enter only if we look at bands in directions intermediate between in-basal-plane and out-of-basal-plane. Assuming the crystal is not too far from hexagonal, which we could call a quasi-hexagonal approximation, one would have the relation $D_1 = C_3/2$. Furthermore, within the quasi-cubic relation, one has the approximate relation $\tilde{A}_6 = (4\tilde{A}_5 - \tilde{A}_3)/\sqrt{2}$ in the wurtzite case. This leads here to the relation $D_2 = D_3 = 4C_3 - A_2$. Thus the D_i parameters can be obtained approximately from those already determined. On the other hand, to obtain these parameters more accurately we can consider a direction in k-space intermediate between the principal axes directions. We find that they affect only the bands farther away from Γ. Compared to our previous results (62) we here refitted the data over a larger region up to k about 0.1 in units of $2\pi/a$. The E_i linear in k-parameters are related to the anti-crossing behavior of the bands one can observe, for example, for the b_2 and a_1 bands along $\Gamma - Y$ in $ZnGeN_2$. They are determined by fitting the bands close to these anti-crossing points.

We may note that the parameters for GaN, which were recently determined by us from QS*GW* calculations (96), when described as an orthorhombic material, are not too far from those of $ZnGeN_2$. The crystal field splittings, however, are significantly larger in the II-IV-N_2. This may have important consequences for shallow acceptor levels. We found recently that in III-N materials the shallow acceptor levels in the effective mass approximation depend sensitively on the crystal field splitting. This would indicate deeper levels in the II-IV-N_2 materials. These Hamiltonians are also useful for describing size quantization effects in various nanostructures, 2D quantum wells, cylindrical nanowires, and spherical nanocrystals.

Table 15.10 Effective masses (in units of free electron mass m_e).

		$ZnSiN_2$	$ZnGeN_2$	$ZnSnN_2$	$CdGeN_2$
CBM at Γ	m_x^c	0.30	0.22	0.17	0.18
	m_y^c	0.47	0.21	0.16	0.17
	m_z^c	0.21	0.16	0.13	0.15
VBM at Γ	$m_x^{a_1}$	4.80	1.92	1.92	1.89
	$m_y^{a_1}$	0.80	2.00	2.17	3.70
	$m_z^{a_1}$	0.21	0.16	0.12	0.16
	$m_x^{b_1}$	0.37	0.21	0.16	0.21
	$m_y^{b_1}$	4.80	2.50	1.92	2.13
	$m_z^{b_1}$	2.97	2.22	2.08	2.17
	$m_x^{b_2}$	0.31	2.94	2.17	3.03
	$m_y^{b_2}$	1.06	0.22	0.17	0.26
	$m_z^{b_2}$	1.18	2.44	2.27	2.78
actual VBM	m_x^v	0.55			
in $ZnSiN_2$	m_y^v	0.33			
	m_z^v	1.09			

15.6 Lattice dynamics

15.6.1 *Computational method*

In order to calculate the vibrational modes we need the force-constants, which are second derivatives of the total energy versus the atomic displacements. Unlike forces, which can be obtained via the force theorem from the charge density of the unperturbed system, this requires knowledge of the first-order corrections to the wavefunctions. The reason why the unperturbed density can be used to estimate the forces is that errors in the charge density lead only to second-order errors in the total energy, and forces are first-order changes in the energy. The appropriate theoretical framework for calculating force constants is density-functional perturbation theory (DFPT), or linear-response theory. This method differs from "textbook" perturbation theory in the sense that the first-order corrections of the wavefunction are calculated self-consistently with the first-order changes in charge density which they produce within the framework of density-functional theory. Once the charge density has only second-order errors, the corresponding total energy has only fourth-order errors, and hence, up to third-order derivatives of the total energy can be computed. This is known as the $2n + 1$-theorem (97). The beauty of linear-response theory is that if we impose a first-order change in the external potential, say a displacement wave with wavevector $\mathbf{q}$, then the

Table 15.11 Parameters of effective Hamiltonian: inverse mass parameters (A_i, B_i, C_i, D_i) in $\hbar^2/2m_e$, E_i parameters in $e^2/2$, and energy splittings Δ_{ci} (meV), $i = 1, 2$.

parameter	$ZnSiN_2$	$ZnGeN_2$	$ZnSnN_2$	$CdGeN_2$	GaN
A_1	−4.71	−6.39	−8.23	−6.3	−5.98
A_2	4.11	5.96	7.77	5.90	5.44
A_3	−0.26	0.02	0.02	0.05	
B_1	0.73	−0.51	−0.49	−0.35	−0.58
B_2	−1.93	−2.01	−2.80	−1.94	−2.46
B_3	0.04	0.05	0.11	0.25	
C_1	−0.52	−0.01	−0.03	−0.13	
C_2	−1.26	−0.01	−0.05	−0.05	
C_3	−0.32	2.15	2.80	1.95	2.53
D_1	−0.64	4.3	5.7	3.89	5.06
D_2	−5.41	4.1	5.5	4.90	−2.19
D_3	−5.41	4.1	5.5	3.2	−2.19
E_1	0	0.015	0	0.007	0.021
E_2	0	−0.005	−0.005	−0.005	−0.021
Δ_{1c}	160	115	82	97	12
Δ_{2c}	−20	−14	−94	−28	0

response to first order also involves only Fourier components of wavevector **q**. By rewriting the perturbation theory in terms of the periodic parts of the Bloch functions, one can obtain the relevant force constants for any arbitrary **q** from calculations of the standard primitive cell. In other words, there is no need to construct supercells with sizes commensurate with the imposed perturbation wavevector periodicity, as in the frozen-phonon methods (98, 99).

The long wavelength limit corresponding to $\mathbf{q} \to 0$ presents its own special problems for compounds with partially ionic bonding. As is well known, there are LO-TO splittings related to the long-range Coulomb forces. These are treated analytically in the DFPT by considering separately the response to a static electric field. The latter are treated as a separate type of perturbation. The derivative of the total energy versus the electric field is the polarization, and although it also breaks the crystal periodicity, it can be calculated using the modern theory of polarization (100, 101) in terms of a Berry phase. The polarization itself is not uniquely defined, but adiabatic changes in polarization, which maintain the insulating phase along the entire adiabatic transition, are. We may then consider various mixed second-derivatives:

$$K_{R\tau\alpha,\ R'\tau'\alpha'} = \frac{\partial^2 U}{\partial \tau_{R\alpha} \partial \tau'_{R'\alpha'}}, \tag{15.29}$$

$$Z^*_{\tau\alpha,\ \beta} = \frac{\partial^2 U}{\partial \tau_\alpha(\mathbf{q}=0)\partial E_\beta} = \frac{\partial P_\beta}{\partial \tau_\alpha(\mathbf{q}=0)} = \frac{\partial F_{\tau\alpha}}{\partial E_\beta}, \tag{15.30}$$

$$\chi_{\alpha,\ \beta} = \frac{\partial^2 U}{\partial E_\alpha \partial E_\beta} \tag{15.31}$$

Here, $\boldsymbol{\tau}_\mathbf{R} = \boldsymbol{\tau} + \mathbf{R}$ is the position of atom $\boldsymbol{\tau}$ in a unit cell, with lattice vector $\mathbf{R}$ and τ_α indicating its Cartesian components. $F_{\tau\alpha}$ is the long-range force on atom τ in the α direction, and P_β is the polarization in the β direction. The first of these equations gives the short-range contribution to the force constants, the second represents the Born-effective charge tensor, and the third is the high-frequency electronic susceptibility. The latter corresponds to below-band-gap frequencies, because we include only static electric fields, but do not include the phonon contributions. For a general wavevector $\mathbf{q}$, we form the Bloch sum,

$$K_{\tau\alpha,\ \tau'\alpha'} = \sum_{\mathbf{R}'-\mathbf{R}} K_{R\tau\alpha,\ R'\tau'\alpha'} e^{i\mathbf{q}\cdot(\mathbf{R}'-\mathbf{R})} \tag{15.32}$$

and find the phonon frequencies $\omega_{m\mathbf{q}}$ from the generalized eigenvalue problem:

$$K_{\tau\alpha,\ \tau'\alpha'}(\mathbf{q}) u^{m\mathbf{q}}_{\tau'\alpha'} = \omega^2_{m\mathbf{q}} M_\tau u^{m\mathbf{q}}_{\tau\alpha} \tag{15.33}$$

with normalization

$$[u^{m\mathbf{q}}_{\tau\alpha}]^* M_\tau u^{n\mathbf{q}}_{\tau\alpha} = \delta_{mn} \tag{15.34}$$

Summation over repeated indices is understood here. For $\mathbf{q} = 0$ this gives the transverse phonon frequencies ω_n and eigenvectors $u^n_{\tau\alpha} \equiv u^{n\mathbf{q}=0}_{\tau\alpha}$. Using the Born effective charges and the eigenvectors of the above diagonalization problem, the dielectric function can be written:

$$\varepsilon_{\alpha\alpha}(\omega) = \varepsilon^\infty_{\alpha\alpha} + \frac{4\pi}{\Omega} \sum_n \frac{S_{n,\alpha\alpha}}{\omega_n^2 - \omega^2 - i\Gamma_n\omega} \tag{15.35}$$

where ω_n, Γ_n are the mode frequencies and damping factors respectively, and Ω is the volume of the unit cell. The damping factors are added empirically. The oscillator strengths are given by

$$S_{n,\alpha\alpha} = \sum_{\tau\alpha} |Z^*_{\tau,\ \alpha\alpha} u^n_{\tau\alpha}|^2 \tag{15.36}$$

Here we consider that in view of the orthorhombic symmetry, the effective charge tensor is diagonal in the Cartesian indices. The zeros of this dielectric function give the longitudinal modes, while the poles give the transverse modes. Thus the LO-TO splittings of each mode that is subject to long-range forces can be calculated.

As a byproduct, we obtain the dielectric constant tensor at high frequency, $\varepsilon^{\infty}_{\alpha\alpha}$, from which we can obtain the indices of diffraction, discussed in Section 15.9. Furthermore, the imaginary part of the above dielectric function gives us explicitly the infrared absorption spectrum due to the phonons. Using the real part, we can also obtain the infrared reflectivity and the loss function, $\Im\{-\varepsilon^{-1}_{\alpha\alpha}(\omega)\}$, whose peaks correspond to the LO modes. As already mentioned, we can also obtain the phonons for finite wavevector $\mathbf{q}$ and hence the full phonon spectrum, as well as the phonon density of states. By taking various integrals of these functions one obtains the lattice-dynamical contributions to the thermal properties, such as the specific heat, and the vibrational entropy.

Finally, we turn to Raman spectroscopy. The Raman intensity is proportional to $|\mathbf{e}_i \cdot \mathbf{R}^m \cdot \mathbf{e}_o|^2$ where $\mathbf{e}_i$ and $\mathbf{e}_o$ are the incident and scattered light polarization vectors and $\mathbf{R}$ is the second rank Raman susceptibility tensor for mode m, given by

$$\mathbf{R}^m_{\alpha\beta} = \sqrt{\Omega}\sum_{\tau\gamma} \mathbf{T}_{\alpha\beta,\ \tau\gamma} u^m_{\tau\gamma} \tag{15.37}$$

This involves the third-rank tensor,

$$\mathbf{T}_{\alpha\beta,\ \tau\gamma} = \frac{\partial\chi_{\alpha\beta}}{\partial\tau_\gamma} = \frac{\partial^3 U}{\partial E_\alpha \partial E_\beta \partial\tau_\gamma} \tag{15.38}$$

calculated from a third derivative of the total energy. As already mentioned, this can still be obtained from the first-order corrections to the wavefunctions, because of the $2n+1$ theorem.

Now we turn to some specific group-theoretical considerations for the materials at hand. The modes subject to LO-TO splitting are those that transform like the components of a vector, i.e. b_1, b_2, a_1 for electric fields along x, y, z or a, b, c crystallographic axes. These modes are infrared-active. So, the only mode which is not infrared-active is a_2.

All modes are Raman active but still different selection rules apply. For a_1 symmetry the tensor is diagonal:

$$a_1: \begin{pmatrix} a & \cdot & \cdot \\ \cdot & b & \cdot \\ \cdot & \cdot & c \end{pmatrix}, \tag{15.39}$$

which means that incident and scattered light must have parallel polarization but different values will be obtained for the x, y and z direction. The scattering geometry is fully specified by giving $\mathbf{k}_i(\mathbf{e}_i\mathbf{e}_o)\mathbf{k}_o$. If the scattering wavevector $\mathbf{q} = \mathbf{k}_o - \mathbf{k}_i$ is parallel to the polarization of the mode, then the longitudinal modes are excited, otherwise the transverse ones. For example, $x(yy)\bar{x}$ will measure the b component of the a_{1T} modes, while $z(yy)\bar{z}$ will measure the b for the a_{1L} modes. The remaining modes correspond to off-diagonal matrix elements of the Raman tensor, as follows:

$$a_2: \begin{pmatrix} \cdot & d & \cdot \\ d & \cdot & \cdot \\ \cdot & \cdot & \cdot \end{pmatrix} \quad b_1: \begin{pmatrix} \cdot & \cdot & e \\ \cdot & \cdot & \cdot \\ e & \cdot & \cdot \end{pmatrix} \quad b_2: \begin{pmatrix} \cdot & \cdot & \cdot \\ \cdot & \cdot & f \\ \cdot & f & \cdot \end{pmatrix} \tag{15.40}$$

15.6.2 *Vibrational modes at the zone center*

Since there are sixteen atoms per cell, there are forty-eight degrees of freedom. These break up in twelve modes of each symmetry, a_1, b_1, b_2, and a_2. The three modes of a_1, b_1, and b_2 of zero frequency correspond to translations of the lattice and are omitted. This leaves eleven modes of these symmetries each, but these are subject to LO-TO splitting, so we list twenty-two modes for each of these. This amounts to seventy-eight modes in total. They are listed in Table 15.12 for $ZnSiN_2$ from (56), $ZnGeN_2$ from (55), and $ZnSnN_2$ from (58). They are all calculated using the DFPT described above and in the local-density approximation.

Compared to the wurtzite crystals, the first thing to notice is the multitude of modes and the much smaller LO-TO splittings. Because the unit cell of the II-IV-N_2 crystal is doubled in two directions compared to the wurtzite unit cell, one might think that the modes would be closely related to folded modes from the Brillouin zone boundary of wurtzite. The latter, in turn, are often interpreted as folded zone modes of the zincblende cell. For example, the E_2 modes and B_1 silent modes can be interpreted as L-point modes of zincblende folded onto Γ. This concept was explored in some detail in (55) for $ZnGeN_2$. While it holds to some degree for the acoustic branches, which extend more or less between the E_2^{low} and the B_1^{low} modes of wurtzite GaN, the optic-mode range shows a clear splitting in two peaks when viewed as a histogram or Gaussian broadened spectrum. This is shown in Fig. 15.10, reproduced from (55). This reflects the bimodal bond distribution of Zn-N and Ge-N bonds, which have clearly different frequencies as well as the LO-TO splitting. However, the situation is actually more complex. While in wurtzite, A_1 and B_1 symmetries allow motion only in the z direction, and E_1, E_2 symmetry allows motion only in the basal plane, this separation of out-of-plane and in-plane motion no longer holds in the orthorhombic structure.

Figure 15.11 illustrates the displacements corresponding to each symmetry. Thus, for example, an a_1 symmetry mode involves motion of the four equivalent Wyckoff position atoms in the same direction in the z-motion, and can also involve x and y motions with the appropriate symmetry pattern that leave the orthorhombic symmetry unchanged. This means that much more interaction between the wurtzite-derived modes is allowed once we break the symmetry

Table 15.12 Phonon frequencies in Zn-IV-N_2 at the Γ point in cm^{-1}.

a_2	b_{1T}	b_{1L}	b_{2T}	b_{2L}	a_{1T}	a_{1L}
			$ZnSiN_2$			
139	164	165	147	148	135	135
154	178	178	169	169	168	169
181	258	279	224	241	242	266
300	399	403	357	358	287	287
350	493	501	422	423	443	443
422	507	509	466	471	456	456
522	583	584	538	597	506	528
538	630	634	625	648	584	586
569	656	713	697	697	624	702
678	784	812	776	784	766	766
802	819	903	856	921	818	923
831						
			$ZnGeN_2$			
130	167	168	129	129	133	133.
163	194	194	166	166	168	168.
183	238	239	208	210	192	192.
201	308	310	266	266	224	227
266	324	324	340	340	300	301
341	512	538	482	491	483	484
471	550	557	506	561	485	527
555	598	651	600	628	586	623
568	652	702	673	673	623	668
654	767	794	775	790	760	763
754	815	823	791	860	764	825
820						
			$ZnSnN_2$			
109	134	135	107	107	108	108
127	153	153	131	131	124	124
130	178	180	155	159	146	147
153	203	211	191	191	163	167
185	224	224	237	238	206	206
235	368	401	355	355	344	359
338	429	435	370	440	359	389
424	516	539	496	525	527	477
483	541	579	560	562	528	554
543	673	708	675	692	666	666
668	713	720	700	739	677	716
713						

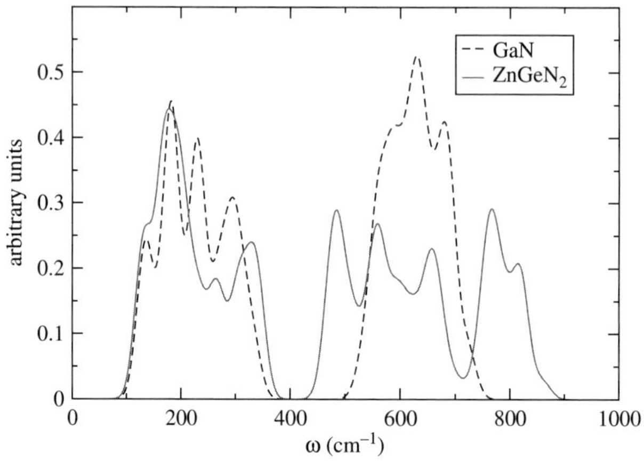

FIG. 15.10. Phonon mode distribution at Γ for $ZnGeN_2$ compared to GaN, displayed as a Gaussian broadened spectrum. From (55).

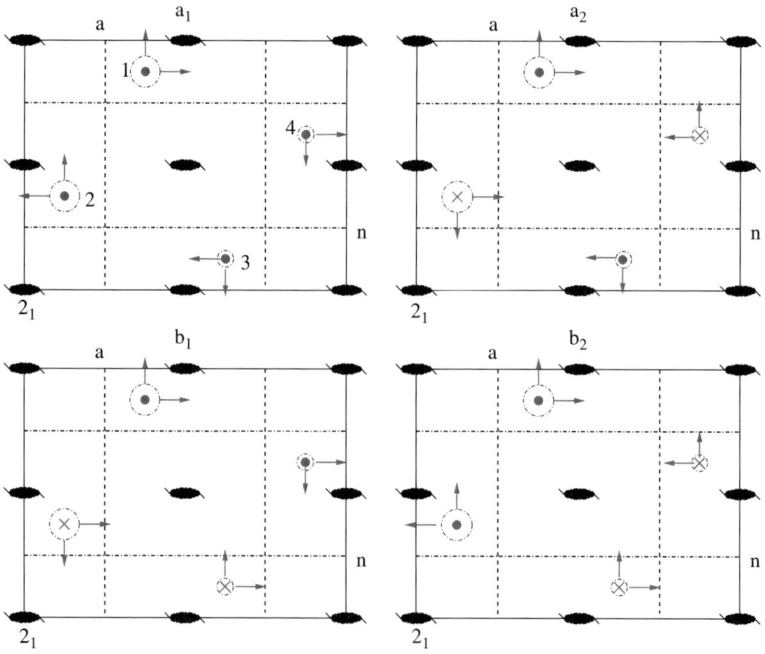

FIG. 15.11. Vibrational mode patterns in II-IV-N_2 material. For each type of atom, the 4 equivalent atoms indicated in the cell move in x, y, z direction according to the indicated pattern relative to each other. Smaller circles lie in the upper plane, larger ones in the lower plane. From (10).

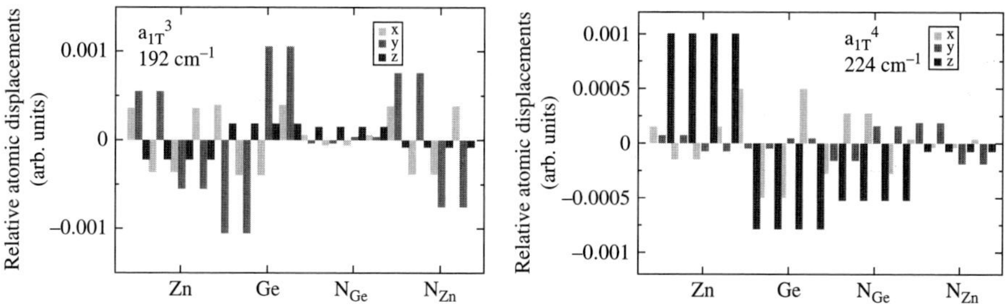

FIG. 15.12. Vibrational mode pattern of a_{1T}^3 (left) and a_{1T}^4 modes: x, y and z displacements are shown in light grey, darker and even darker grey bar graphs for each of the 4 atoms of each type as indicated.

to C_{2v}. This also allows for modes of fairly low frequency to already have a significant oscillator strength.

Low-frequency modes correspond primarily to motions of groups of atoms, as illustrated for an example mode a_{1T}^3 at 192 cm^{-1} and a_{1T}^4 at 224 cm^{-1} in Fig. 15.12, in which we indicate the pattern of the motion by showing the x, y, and z displacements of each of the atoms. One can see that the z-motions of all atoms of a given type are in unison, but one may also see that the Zn and the N_{Zn} and the Ge and N_{Ge} move in the same direction. This is why it is a low-frequency mode. However, the displacements of these atoms are not equal, so there is some stretching of the bonds accompanying it, and hence this mode is weakly optically active. At the same time, the atoms also exhibit in-plane vibrations because these are also allowed by a_1 symmetry. In this particular mode we can see that the x and y displacements are actually larger than the z-motions. We can see that the Zn and N_{Zn}, as well as Ge and N_{Ge}, move as a group in the x and y as well as the z direction. The mode a_{1T}^4 has a significant oscillator strength, even though it is still a low-frequency mode. We can see that it has a significant Zn–N bond stretch, but Zn has a larger displacement than the N to which it is bonded. On the other hand, Ge and N_{Ge} move in the same direction.

We examine various other a_{1T} modes in Figs. 15.14 and 15.15. The a_1 symmetry modes for Zn-IV-N_2 are shown in Fig. 15.13 from (58). From this figure and the following mode patterns one can see which modes ultimately have a strong oscillator strength in the infrared spectrum.

We may notice that for the higher-frequency modes, the cations and anions in a bond move in opposite directions. Also, the displacements of the lighter N atoms are much larger than those of the cations. Mode a_{1t}^9 is of special interest, because we show in the following that it is the strongest mode in the Raman spectrum. The upshot is that the modes are complex combinations of the wurtzite modes, and symmetry is not of much use to unravel them.

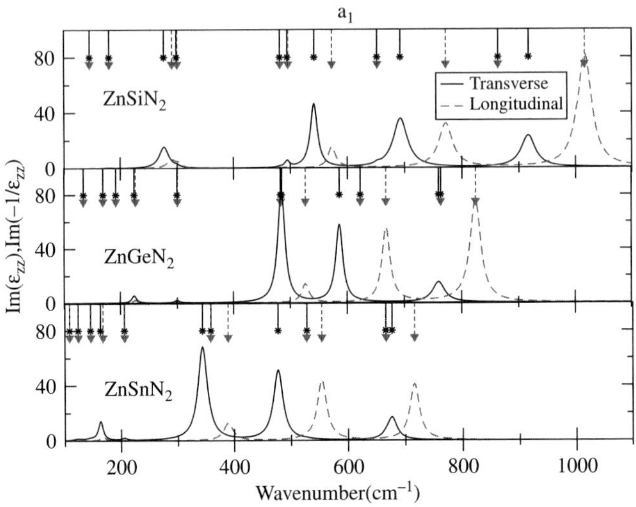

FIG. 15.13. Vibrational modes of a_1 symmetry. The bars at the top in black and dashed grey show the TO and LO modes respectively. The solid line spectra show the $\text{Im}\{\varepsilon(\omega)\}$ whose peaks are the TO modes, and grey dashed line $\text{Im}\{-\varepsilon^{-1}(\omega)\}$ whose peaks are the LO modes. From (58).

15.6.3 *Phonon dispersions and density of states*

We have already indicated in the previous section that the vibrational spectrum of the II-IV-N_2 is much more complex than one would expect on the basis of zone-folding. In fact, in $ZnSiN_2$ even the distinction between acoustic branch modes and optic branch modes becomes somewhat blurred. In wurtzite, the six upper modes (A_1, B_1^{high}, E_1 doubly degenerate, E_2^{high} doubly degenerate) can be viewed as truly optic modes, with the lower six bands folded acoustic bands. Because of the quadrupling of the number of atoms in the cell, one thus expects twenty-four optic and twenty-four acoustic-type vibrational bands. In $ZnGeN_2$ and $ZnSnN_2$ there is indeed a clear gap between the lower and upper twenty-four bands. However, in $ZnSiN_2$, only the lower twelve bands are separated by a small gap from the upper thirty-six bands. This is most clearly seen in Fig. 15.16, showing the full phonon band structure and phonon density of states, taken from (58).

From the phonon densities of states, one can also determine various thermal properties such as the vibrational specific heat and entropy contributions. We refer the interested reader to (58).

15.6.4 *Infrared spectroscopy*

We now turn to a discussion of the infrared spectra as a measurement of the vibrational modes. As already mentioned, the b_1, b_2, and a_1 modes are optically

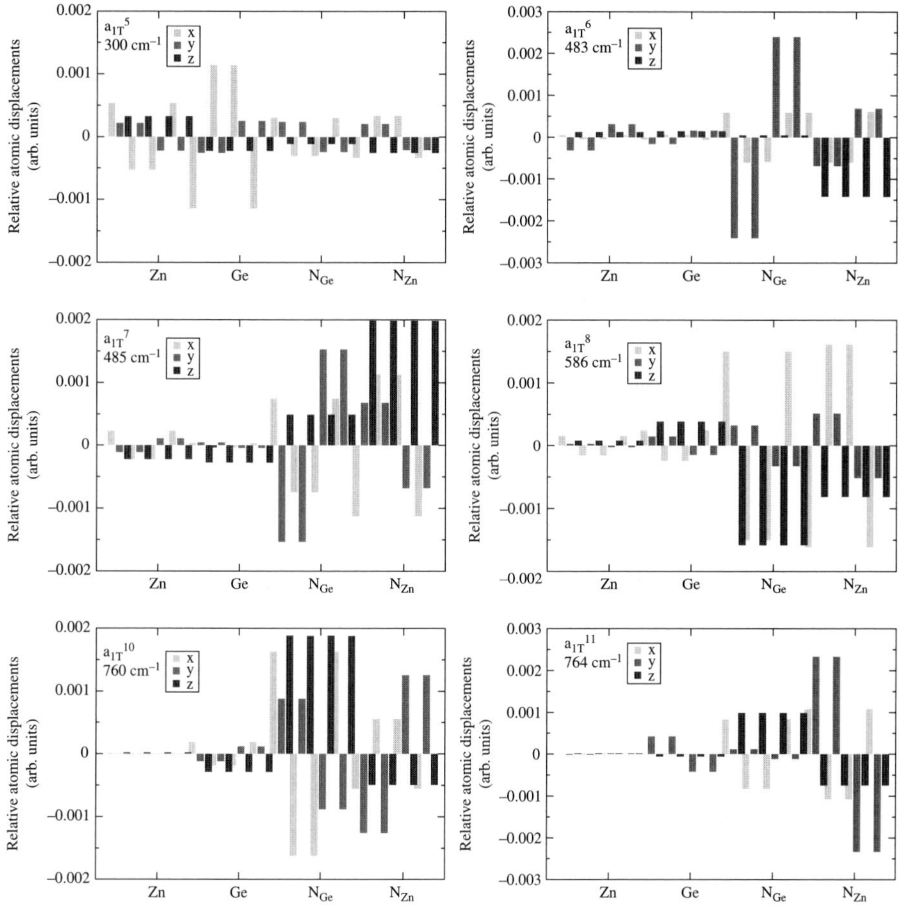

FIG. 15.14. Vibrational mode patterns of a_{1T} symmetry.

active for polarizations of the light along x, y, z respectively. Full sets of predicted spectra can be found in (58). Here we focus on the only case for which experimental data are available: $ZnSiN_2$. Even in this case, only the modes for one symmetry were determined experimentally: namely, the b_1 symmetry (56, 42). The experiments were carried out on thin films grown on r-plane sapphire by Mintairov *et al.* (42) The results are shown in Fig. 15.17.

While there is fair agreement between the calculated and experimental results, we emphasize that there is a significant difference in the interpretation of the results. One would expect that five modes are of the folded acoustic type and would have low oscillator strength, and six are true optic and have strong oscillator strength and correspondingly strong LO-TO splitting. Thus, the experimentalists fitted their measured reflectivity spectrum to six modes, and these

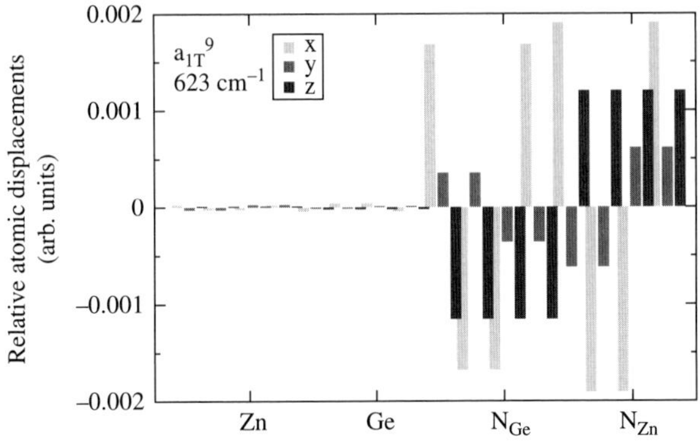

FIG. 15.15. Vibrational mode pattern of a_{1T}^{9}

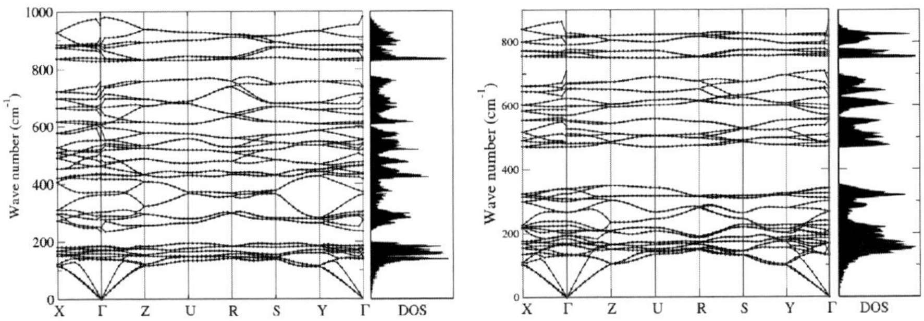

FIG. 15.16. Phonon dispersions and density of states in $ZnSiN_2$ and $ZnGeN_2$. From (58).

correspond to the six peaks one sees in the Im$\varepsilon(\omega)$ and $-$Im$\varepsilon^{-1}(\omega)$ spectra, extracted from the fit. However, as we can see from the calculation, the highest two energy peaks actually correspond to two individual frequencies. On the other hand, some of the lower-frequency modes have significant oscillator strength. This was discussed in the previous section.

The broadening of the modes was adjusted in our calculation to obtain the best fit to the experiment. This broadening may arise from the anharmonic coupling to modes with wavevector $\mathbf{q} \neq 0$. For example, the sharpest mode at 417 cm^{-1} occurs in a region where the density of phonon states integrated over the whole Brillouin zone is low, while the broader peaks correspond to frequencies where the DOS is high. This can be seen in Fig. 15.16 This may in particular explain why the lowest-frequency peak at about 300 cm^{-1} is not seen at all in the experiment. Additional complications in extracting the frequencies and experimental oscillator strengths occur because of the interference of modes with those from the sapphire substrate. The peak at 417 cm^{-1} appears to be much

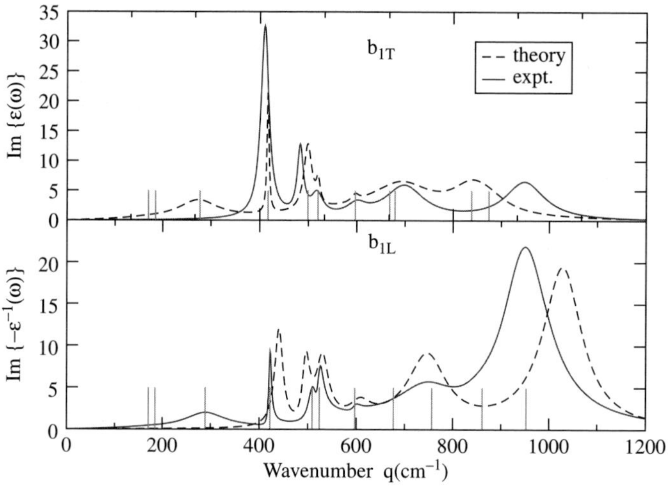

FIG. 15.17. Calculated (dashed line) and experimental (grey solid line) transverse (top panel) and longitudinal (bottom panel) infrared spectra. The vertical green lines indicate the actual calculated frequencies of the eleven modes of b_1 symmetry. Adapted from (56).

stronger in the experiment than in the theory, but happens to lie close to a peak in the substrate reflection. Furthermore, the samples may be expected to be under some strain, which could lead to peak shifts. On the other hand, LDA also somewhat overbinds and may thereby overestimate the phonon frequencies. The LDA overestimate of dielectric screening, on the other hand, may lead to an underestimate of the LO-TO splittings. More details can be found in (56).

Clearly, much more work needs to be done to fully determine the infrared spectra for other symmetries and for the other II-IV-N_2 nitrides.

15.6.5 *Raman spectroscopy*

For $ZnGeN_2$ there are two experimental studies of Raman data: one on polycrystalline powder samples (48), and one on single-crystal needles of about 3 μm in diameter, as well as polycrystalline material consisting of sub-micron-size crystallites (57). These spectra are quite different. For the samples of Viennois *et al.* (48) the Raman spectrum looks like a phonon density of states. This would make sense if the crystal momentum conservation law is broken, e.g. by the small grain size. Although the crystallite sizes were not discussed by Viennois, we proceed with the assumption that they are small, or other severe disorder breaks the crystalline symmetry. A comparison of their spectrum with phonon density of states is shown in Fig. 15.18. One can see the importance of the second harmonics in this spectrum. The peaks observed above 900 cm^{-1} are all harmonics,

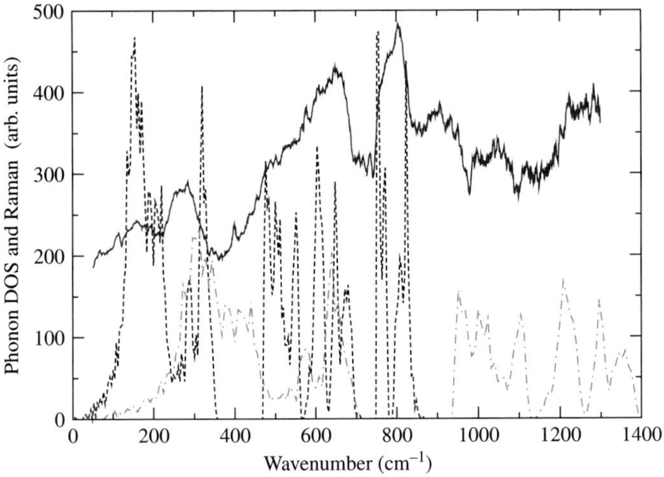

FIG. 15.18. Comparison of powder Raman spectrum, from Viennois *et al.* (48) (solid curve) with phonon density of states for $ZnGeN_2$ (dashed). The dash–dotted curve is the phonon density of states, but with ω replaced by 2ω to mimic the second harmonic spectrum. Adapted from (61).

and the largest peaks occur where harmonics of lower modes coincide with the fundamentals of the spectrum.

In contrast, for the single-crystal needles one obtains a very sharp Raman spectrum in which the $\mathbf{q} = 0$ allowed modes are clearly observed. For polarization of the incident and reflected light both parallel to the c-axis, these are the a_{1T} modes, which show the zz component of the Raman tensor. This spectrum is shown in Fig. 15.19. However, one can still see some peaks which correspond to the phonon density of states, showing that there is an additional disordered Raman scattering contribution.[2] The nature of this disorder is not entirely clear. Point defects may be responsible. The second harmonics of the latter appear much weaker, but could be responsible for the peak at 400 cm^{-1}—a region where there are no fundamental modes, because it lies in the gap between folded acoustic and optic modes. Additional information on the polarization-dependence of this spectrum are given in (57). In particular, it was possible to determine some changes in the Raman tensor for the a_{1T} modes near 200 cm^{-1} when the

[2]An alternative interpretation for these peaks is that they correspond to forbidden LO modes, excited by the Fröhlich interaction. As the LO bands are flat, LO modes at Γ, and peaks in the density of states arising from the Brillouin zone boundary, are close to each other. Which of the two interpretations, disorder or LO phonons, is correct will require further study. LO phonons would be strongly enhanced in resonant Raman scattering, while disorder-induced modes should be reduced in samples of improved quality.

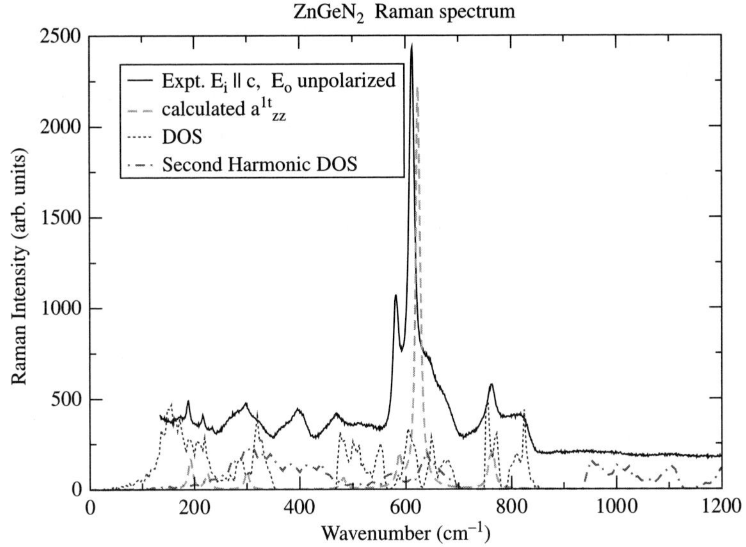

FIG. 15.19. Raman spectrum for a needle single crystal for $E_i \parallel c$, $E_o \parallel c$, from Peshek *et al.* (57) (solid black line), compared with the zone-center-calculated Raman spectrum for a_{1T}^{zz} Raman tensor component (dashed line) and phonon density of states (dotted line) and its second harmonic (dash–dotted line). The experimental spectrum is offset vertically for better viewing, but goes to zero intensity at high frequencies.

polarization was perpendicular to the c-axis, which is the long axis of the needle crystals. This is shown in Fig. 15.20.

Comparison with the calculated spectra allowed us to determine that the light polarization corresponds to the y, i.e. b direction of the crystal. This means that we could use Raman spectroscopy to determine how the needle crystal was lying on the microscope slide used in these micro-Raman studies! This was further confirmed by the measurements with crossed-polarization, which then showed some evidence of the b_2 modes. See (57) for further details.

The occurrence of sharp peaks supports the high crystalline quality, which was placed in evidence by XRD and the clear facetting of the crystals. It also provides evidence for the cation ordering. However, the experimentally observed polarization changes in the Raman spectrum were far smaller experimentally than expected from theory. In (57) we analyzed the effects of the focusing of the beam into a small spot, which leads to some angular spread in the sample, and found that these effects could not fully explain the effect, though they could explain partially why the a_{1T} spectrum is not completely suppressed under crossed-polarizer conditions.

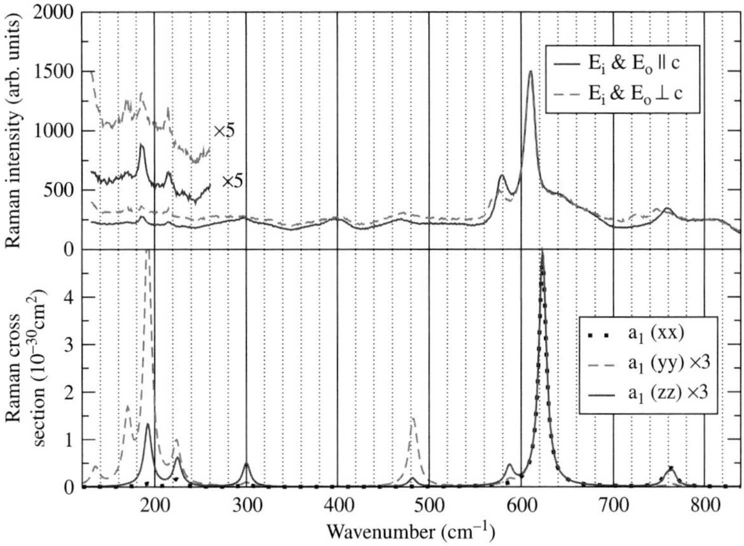

FIG. 15.20. Polarization-dependence of the measured Raman spectrum in $ZnGeN_2$, from (57). Both spectra for $E_i \parallel E_o \parallel c$ and $E_i \parallel E_o \perp c$ measure the same a_{1T} modes, but correspond to a different Raman tensor element. The calculated spectra for each Raman tensor component are shown in the bottom panel. The occurrence of three peaks instead of two peaks at low frequency is consistent with the yy component of the Raman tensor.

The question of cation disorder is somewhat subtle. In fact, if the cations were fully disordered on their wurtzite sublattice, one would expect to see modes more similar to those in wurtzite crystals. In wurtzite crystals the E_2^{high} mode is often the dominant mode visible in Raman spectrocopy. The strongest mode in the a_{1T} Raman spectrum happens to coincide with a mode that is somewhat similar in its displacement pattern to an E_2 wurtzite-type mode. In fact, E_2 of the C_{6v} point group decomposes into a_1, and a_2 of C_{2v}. This means that it is compatible with a_1 symmetry. Inspecting the a_1 pattern in Fig. 15.11 one can see that corresponding atoms in the upper and lower plane move opposite to each other in the x and y directions. This is precisely what an E_2 mode implies in wurtzite. In other words, as far as its in-plane motion is concerned, an a_1 mode is like an E_2 mode in wurtzite. The same is true for an a_2 mode, but in this case the z motions are like a B_1 of wurtzite. One can also see, from Fig. 15.15, which is the strong Raman mode, that the x displacements (in-plane) are large for both the N_{Zn} and N_{Ge}. This in-plane motion corresponds to E_2^{high}-like bond stretches. The fact that this E_2^{high}-like mode stays strong even for other polarizations may indicate that the sample responds in part like a wurtzite crystal.

This could mean that there is partial disorder on the cation sublattice. At present, we cannot tell how good the cation-ordering is. X-ray diffraction or electron diffraction is not capable of determining it in $ZnGeN_2$ because of the closeness of Zn and Ge electron densities. The presence of the other sharply defined modes indicates that there is at least partial order, but the robustness of the E_2^{high}-like mode may indicate that the spectrum is also compatible with a wurtzite-like component in the spectrum; in other words, some cation sublattice disorder. This antisite disorder may also in part be responsible for the density-of-states-like features in the Raman spectrum.

Nonetheless, there is clearly a large change between the powder spectra of Viennois *et al.* (48) and the needle Raman spectra. Even for the polycrystalline regions of the samples studied in (57), and whose growth is described by Du *et al.* (49), the Raman spectrum showed rather sharp features more similar to those of the needles than to the broad density-of-states-like spectrum of Viennois *et al.* (48). This might be indicative of a larger grain size than in Viennois *et al.* (48).

Again, much work remains to be done. Raman on thin-film samples with the c-plane as surface would allow one possibly for studying other vibrational modes. No Raman spectra have yet been reported for $ZnSiN_2$. For $ZnSnN_2$ the Raman spectrum agreed well with the calculated density of phonon states (50). The sharp $\mathbf{q} = 0$ allowed lines were not observed, and the origin of the larger disorder is presently unknown.

15.7 Elastic and piezoelectric tensors

15.7.1 *Theoretical considerations*

Using an extension of the DFPT method developed by Hamann and Wu (102, 103), strains, i.e. changes of the shape and size of the unit cell, can also be treated as perturbations. This opens the way toward calculating a number of other interesting lattice-dynamical properties. The second derivatives

$$c_{ij} = \frac{\partial^2 U}{\partial \eta_i \partial \eta_j} \tag{15.41}$$

give the elastic constants. Here the index $i = 1 \dots 6$ uses the Voigt notation. It is a shortcut for two Cartesian indices $\alpha\beta$, using the fact that the strain tensor is symmetric, $\eta_{xx} \to \eta_1$, $\eta_{yy} \to \eta_2$, $\eta_{zz} \to \eta_3$, $\eta_{yz}, \eta_{zy} \to \eta_4$, $\eta_{xz}, \eta_{zx} \to \eta_5$, $\eta_{xy}, \eta_{yx} \to \eta_6$. The stress tensor is $\sigma_i = \partial U/\partial \eta_i$. Using electric field and strain as basic variables, the direct and converse piezoelectric effect can be written:

$$P_\alpha = e_{\alpha j}\eta_j + \chi^\eta_{\alpha\beta} E_\beta \tag{15.42}$$

$$\sigma_i = c^E_{ij}\eta_j - e_{\alpha i} E_\alpha \tag{15.43}$$

The first equation tells us that a polarization is induced not only by an electric field but also by a strain. The electric susceptibility $\chi^{\eta}_{\alpha\beta}$ is specified to be measured at fixed strain, say zero strain. Thus,

$$e_{\alpha j} = \left.\frac{\partial P_\alpha}{\partial \eta_j}\right|_E = -\left.\frac{\partial \sigma_i}{\partial E_\alpha}\right|_\eta = \frac{\partial^2 U}{\partial E_\alpha \partial \eta_i} \tag{15.44}$$

Here the electric field must be kept fixed while taking the strain derivative, for example at zero value, in the first equation, and the strain must be kept fixed in the second equality, while the third shows us that this is just another second derivative of the total energy. Instead of polarization P one can also use the dielectric displacement instead, and write

$$D_\alpha = e_{\alpha j}\eta_j + \varepsilon^{\eta}_{\alpha\beta} E_\beta \tag{15.45}$$

Instead of the susceptibility, the dielectric constant appears in this case. From a computational point of view it is most easy to prescribe the strain and calculate the strain-derivative of the polarization using the Berry phase approach. From an experimental point of view, however, one prefers to fix the stress. Switching to stress as a basic variable, one can write the converse piezoelectric field as

$$\eta_i = s^E_{ij}\sigma_j + d_{\alpha i} E_\alpha \tag{15.46}$$

Here, s^E_{ij} is the compliance tensor, which is the inverse of the elastic stiffness tensor, measured at constant electric field. The second type of piezoelectric tensor $d_{\alpha i}$ is thus given by

$$d_{\alpha i} = \left.\frac{\partial \eta_i}{\partial E_\alpha}\right|_\sigma \tag{15.47}$$

In other words, one measures the strain induced by an electric field at constant stress. It is therefore called the piezoelectric strain tensor. On the other hand, $e_{\alpha i}$ is called the piezoelectric stress tensor, because according to eq. (15.43) it gives the stress produced by an electric field. Using the direct equation but with stress as basic variable, one writes

$$P_\alpha = d_{\alpha i}\sigma_i + \chi^{\sigma}_{\alpha\beta} E_\beta \tag{15.48}$$

Setting the electric field equal to zero and using $\sigma_i = c^E_{ij}\eta_j$, we obtain $P_\alpha = d_{\alpha i} c^E_{ij}\eta_j$ and hence

$$e_{\alpha j} = d_{\alpha i} c^E_{ij} \tag{15.49}$$

Thus, given the elastic constants measured at fixed (or zero electric field) we can convert the piezoelectric strain tensor to the piezoelectric stress tensor. Similarly,

$$d_{\alpha i} = e_{\alpha j} s^E_{ji} \tag{15.50}$$

The number of independent components of these tensors is determined by the macroscopic or point-group symmetry of the crystal. For example, in wurtzite, with C_{6v} symmetry, $c_{11} = c_{22}$, $c_{13} = c_{23}$, $c_{44} = c_{55}$, while in the orthorhombic C_{2v} case, these tensor components can be different.

15.7.2 *Values for nitrides*

In Table 15.13 we show the elastic constants of the Zn-IV-N_2 compounds compared to those of the III-N compounds taken from (59).

The values for the III-nitrides were calculated by Wright *et al.* (104), using a more direct approach using finite strains and numerical derivatives. The values for GaN in the second GaN column were calculated by Boonchun and Lambrecht (10) using the same DFPT approach as used for the Zn-IV-N_2 materials. For GaN we also give the experimental values determined by Brillouin scattering from Polian *et al.* (105). This gives some idea of the uncertainties of the different computational methods compared to experiment. The trends are similar to those in the nitrides. The elastic constants decrease as we go from Si to Ge to Sn, just as they do from Al to Ga to In. Using the values calculated by the same DFPT method for GaN and $ZnGeN_2$, their elastic constants are within 10% of each other. This is within the error bar of uncertainty, given the deviations from experiment and other computational methods. $ZnSnN_2$ appears to have systematically somewhat larger lattice constants than InN.

The piezoelectric stress constants $e_{\alpha i}$ are given in Table 15.14, again compared with those of the III-nitrides. Note that in wurtzite, $e_{15} = e_{24}$ and $e_{31} = e_{32}$.

Table 15.13 Elastic constants of Zn-IV-N_2 compounds, C_{ij}, in Voigt notation in units of 100 GPa = 1 Mbar, compared to the corresponding ones in the III-N. Values for III-N are taken from Wright (104).

Components	$ZnSiN_2$	$ZnGeN_2$	$ZnSnN_2$	AlN	GaN			InN
					Wright	DFPT[a]	Expt.[b]	
C_{11}	3.83	3.41	2.72	3.96	3.67	3.45	3.90	2.23
C_{22}	4.08	3.58	2.90					
C_{33}	4.63	4.01	3.06	3.73	4.05	3.83	3.98	2.24
C_{44}	1.04	0.86	0.64	1.16	0.95	0.91	1.05	0.48
C_{55}	1.10	0.95	0.67					
C_{66}	1.24	1.05	0.74	1.29	1.16	1.08	1.22	0.54
C_{12}	1.46	1.36	1.28	1.37	1.35	1.30	1.45	1.15
C_{13}	1.17	1.03	1.00	1.08	1.03	0.95	1.06	0.92
C_{23}	1.05	0.98	1.05					

[a] From (10).
[b] From (105).

Table 15.14 Piezoelectric tensor $e_{\alpha j}$ in (C/m^2) for Zn-IV-N_2 and III-N materials. The values for III-N are from Bernardini *et al.* (106), except for GaN, where they are from Boonchun *et al.* (10), with the value in parenthesis from Bernardini *et al.* Values in square brackets are experimental values, as discussed in the text.

Components	$ZnSiN_2$	$ZnGeN_2$	$ZnSnN_2$
e_{15}	−0.31	−0.27	−0.41
e_{24}	−0.30	−0.35	−0.44
e_{31}	−0.45	−0.43	−0.59
e_{32}	−0.44	−0.49	−0.59
e_{33}	0.80	0.73	1.09
	AlN	GaN	InN
e_{15}	[−0.42]	−0.30 [−0.29]	
e_{33}	1.46 [1.5]	0.70 (0.73) [1.12]	0.97
e_{31}	−0.60 [−0.6]	−0.42 (−0.49) [−0.55]	−0.57

Experimental values (107) for bulk GaN are $e_{33} = 1.12$ Cm^{-2}, $e_{31} = -0.55$ Cm^{-2}, while for AlN (108) they are 1.5 and −0.6 C/m_2. Experimentally, the measured quantities are the $d_{\alpha j}$ coefficients. We converted them to $e_{\alpha j}$ using the above-mentioned relationship eq. (15.49) and the elastic constants from Wright (104). Specifically, this means

$$\begin{aligned} e_{15} &= d_{15}c_{44} \\ e_{31} &= d_{31}(c_{11} + c_{12}) + d_{33}c_{13} \\ e_{33} &= 2d_{31}c_{13} + d_{33}c_{33} \end{aligned} \tag{15.51}$$

Experimental data are available for d_{15} in GaN, and AlN by Muensit *et al.* (108), and using the elastic constants from Wright *et al.* (104) convert to $|e_{15}| = 0.29$ Cm^{-2} for GaN and 0.42 Cm^{-2} for AlN. This value for GaN is close to our calculated value for $ZnGeN_2$, as expected. The sign of these coefficients depends on how the Cartesian axes are oriented versus the crystal structure polarity (positive from anion to cation along its axial bond), and it is not clear from the experimental data if the sign was actually determined.

Overall, we can see that the values for the Zn-IV-N_2 compounds are comparable to the corresponding III-N; in particular, for $ZnGeN_2$ and GaN, although the values for $ZnSiN_2$ are smaller than those in AlN. AlN has been considered for surface-acoustic-wave applications because of its high piezoelectric constants. Thus the Zn-IV-N_2 compounds are also suitable for piezoelectric applications.

Table 15.15 Piezoelectric strain tensors (in pm/V).

	$ZnSiN_2$	$ZnGeN_2$	$ZnSnN_2$
d_{31}	−1.5	−1.1	−2.9
d_{32}	−1.2	−1.1	−2.7
d_{33}	2.4	1.8	5.4
d_{15}	−2.8	−2.3	−6.2
d_{24}	−2.9	−3.1	−6.9

For convenience, we also provide the $d_{\alpha i}$ coefficients for the Zn-IV-N_2 compounds in Table 15.15, which can be calculated from the $e_{\alpha i}$ and the compliances given in (59) using eq. (15.50). In the present case this reduces to

$$\begin{pmatrix} d_{31} \\ d_{32} \\ d_{33} \end{pmatrix} = \begin{pmatrix} s_{11} & s_{12} & s_{13} \\ s_{12} & s_{22} & s_{23} \\ s_{13} & s_{23} & s_{33} \end{pmatrix} \begin{pmatrix} e_{31} \\ e_{32} \\ e_{33} \end{pmatrix}$$

$$d_{15} = e_{15}s_{55}$$

$$d_{24} = e_{24}s_{24} \tag{15.52}$$

As a byproduct of our investigation into lattice dynamics, we obtain the dielectric constants. These are calculated at zero strain. In Table 15.16 we give the high-frequency as well as the static dielectric constants, including the phonon contribution. There is also a piezoelectric contribution to the static dielectric constant, given by

$$\chi^{piezo}_{\alpha\beta} = e_{\alpha i}d_{\beta i} \tag{15.53}$$

This is because at zero stress an electric field causes a strain $\eta_i = d_{\beta i}E_\beta$, and this strain in turn creates a polarization $P_\alpha = e_{\alpha i}\eta_i = e_{\alpha i}d_{\beta i}E_\beta \equiv \chi^{piezo}_{\alpha\beta}E_\beta$. We define $\varepsilon^s_{\alpha\beta} = \varepsilon^0_{\alpha\beta} + \chi^{piezo}_{\alpha\beta}$. To obtain the relative dielectric constant κ we need to divide this by ε_0, the dielectric permeability of vacuum.

Table 15.16 High-frequency (κ^∞), static (κ^0) relative dielectric tensor components, including phonon contributions, and finally also including piezoelectric contributions (κ^s).

Compound	κ^∞_{xx}	κ^∞_{yy}	κ^∞_{zz}	κ^0_{xx}	κ^0_{yy}	κ^0_{zz}	κ^s_{xx}	κ^s_{yy}	κ^s_{zz}
$ZnSiN_2$	4.842	4.867	5.290	8.704	9.278	10.015	8.801	9.374	10.460
$ZnGeN_2$	5.166	5.239	5.725	9.222	9.276	10.610	9.292	9.399	10.872
$ZnSnN_2$	5.853	5.931	7.239	11.151	11.895	15.088	11.936	12.186	16.130

15.8 Spontaneous polarization

The II-IV-N_2 materials, like the wurtzite structure group-III nitrides, are polar materials but also pyroelectric. This means that they can have spontaneous polarization. Spontaneous polarizations have played an important role in group-III nitrides. They lead to a surface charge density at the interface between two nitrides along the basal plane $\sigma = \Delta\mathbf{P} \cdot \hat{\mathbf{n}}$, with $\Delta\mathbf{P}$ the difference in polarization between the two materials (2-1) and $\hat{\mathbf{n}}$ the normal to the interface pointing from 1 to 2. In turn, this leads to an electric field in quantum wells made of different III-N semiconductors, such as GaN sandwiched between $Al_xGa_{1-x}N$. This can have beneficial as well as detrimental effects, depending on the type of device pursued. In optoelectronic devices it leads to a separation of electron and holes on two different ends of the quantum well, which reduces the overlap of their wavefunctions and hence the optical oscillator strength of excitons. To avoid this, there is now an effort to grow III-nitrides on non-polar and semipolar planes. On the other hand, the electric field may provide confinement in a two-dimensional electron gas (2DEG) at the heterojunction between two nitrides. Such high-density electron gases are of interest in field-effect transistors, and can provide doping without extrinsic impurities. This in particular occurs in radial heterojunction quantum wires, where much larger concentration differences between the two materials can be tolerated than in planar structures.

The spontaneous polarization, however, is not a uniquely defined quantity for a bulk material. Strictly speaking, only adiabatic changes in the polarization are well-defined. However, in the modern theory of polarization (100, 101) it can be calculated from a Berry phase in k-space. The so-calculated $\mathbf{P}$ is only defined *modulo a quantum* $2e\mathbf{R}/\Omega$ where $\mathbf{R}$ is a lattice vector and Ω the unit cell volume, and in the present case (59) only to within $2e(n + \frac{1}{4})c/\Omega$. However, this arbitrary quantum depends only on the cross-sectional area, and at a coherent interface this stays the same. So, it plays no role in determining the $\Delta\mathbf{P}$ at a heterojunction. As in wurtzite, the only non-zero component of the polarization is along the c direction.

The spontaneous polarizations of the Zn-IV-N_2 nitrides are given in Table 15.17. One can see that the difference between them are much smaller than those between the III-nitrides, and that all values are close to that of GaN. This means that one may expect the spontaneous polarization effects between GaN

Table 15.17 Spontaneous polarizations in Zn-IV-N_2 and III-N materials, in C/m^2. Values for III-N are taken from Bernardini *et al.* (106).

$ZnSiN_2$	$ZnGeN_2$	$ZnSnN_2$
−0.022	−0.023	−0.029
AlN	GaN	InN
−0.090	−0.029	−0.042

and the various Zn-IV-N_2 materials to be much less of a problem in optoelectronic devices than in III-N heterojunctions.

15.9 Optical properties

Optical properties in the infrared and Raman spectroscopy have already been discussed in the section on lattice dynamics. Here we focus on optical properties related to the electronic interband transitions.

15.9.1 *Indices of refraction*

The indices of refraction are given by $n = \sqrt{\varepsilon^\infty}$ and are shown in Table 15.18. These values are obtained from the high-frequency dielectric constant tensor κ^∞ given previously. By "high-frequency" we mean high relative to the phonon frequencies but low with respect to the band-gap energy.

One can see that the x and y components are closer to each other than to the z component. This is expected in view of the close similarity of the structure to wurtzite. On the other hand, Limpijumnong *et al.* (52) obtained the different result that $n_z \approx n_y \neq n_x$. These results were obtained in a different way, using the RPA equation for the imaginary part of the dielectric function as a function of frequency above the band gap and then calculating the real part of $\varepsilon(\omega)$ for $\omega \to 0$. This does not include local field effects, whereas the Berry phase calculation does. On the other hand, it is probable that this discrepancy results from a not-fully-relaxed crystals structure in (52). This matter requires further investigation. We can now calculate the frequency (or wavelength dependence) in the region below the band gap using *GW* calculations, and plan to do so in the near future. In any case, we believe the more recent result to be more accurate.

Misaki *et al.* (44) measured the UV reflectivity of $ZnGeN_2$ films. Although the thin films showed significant interference effects at low frequency below the gap, they were able to perform a Kramers–Kronig analysis, and hence extract indices of refraction, and extinction coefficients, as well as the real and imaginary part of the dielectric function up to about 100 eV. They found a larger n_z than $n_\perp$, but did not distinguish between different directions in the c-plane. From ellipsometry they obtained a $n_z = 2.46 \pm 0.12$ at 632.8 nm. This is in quite good agreement with the data shown in Table 15.18.

Table 15.18 Indices of refraction.

Compound	n_x	n_y	n_z
$ZnSiN_2$	2.200	2.206	2.300
$ZnGeN_2$	2.273	2.289	2.393
$ZnSnN_2$	2.419	2.435	2.690

The indices of refraction were measured for $ZnSiN_2$ by Cook *et al.* (39), and for $ZnSi_xGe_{1-x}N_2$ alloys by Muth *et al.* (38). They discuss the indices of refraction corresponding to TE and TM-guided modes in the thin film on top of r-plane sapphire with a buffer layer of GaN between, which complicates the analysis. If we assume that their TM modes correspond to n_y, their TE_{min} to n_x and TE_{max} to n_z, then their values for the longest wavelength measured (676.4 nm) are $n_z = 2.205$, $n_x = 2.179$, and $n_y = 2.180$ for $ZnSiN_2$. This agrees with the notion that $n_z > n_x \approx n_y$. The values are a little smaller than our calculated values. Our overestimate is probably due to the use of the LDA. LDA underestimates the gap, hence overestimates ε and the index of refraction. For $ZnGeN_2$, Muth *et al.* (38) give an index of refraction of 2.34 for the TM mode at 632.8 nm. We assume this is n_y. This is indeed smaller than the n_z measured and reported by Misaki *et al.* (44) at the same wavelength, and would indicate a birefringence of about 0.12. This is consistent with our calculations, which give a $\Delta n = n_z - n_y \approx 0.1$. In this case, our calculated values are a little smaller than the experimental values. This is perhaps because we have not yet properly taken into account the wavelength dispersion.

In summary, the calculated indices of refraction are in reasonable agreement with the rather incomplete experimental data available on $ZnSiN_2$ and $ZnGeN_2$. It is safe to say that the materials are approximately birefringent, with $n_z > n_x$. In other words, the materials are positive birefringent with $n_e > n_o$, although strictly speaking the materials are biaxial.

15.9.2 *UV dielectric functions*

The imaginary part of the dielectric function in the UV region has already been calculated by Limpijumnong *et al.* (52), but a full analysis of the dielectric function in terms of specific interband transitions has not yet been carried out. Overall, it showed a $\varepsilon_2(\omega)$ function similar to that of GaN, with several peaks between 7 and 12 eV. After the initial onset at the gap, it shows a sharp rise at just above 5 eV and decays rapidly above 13 eV.

This is in basic agreement with the measurements of Misaki *et al.* (44) mentioned previously. Misaki *et al.* (44) performed their own LDA calculations but apparently only for $\mathbf{E} \perp \mathbf{c}$, and without explicitly telling which direction in the plane. They obtained a reasonable agreement between peak positions in theory and experiment, and also in agreement with the results of Limpijumnong *et al.* (52). However, as already mentioned for the indices of refraction, there are discrepancies between Limpijumnong *et al.* polarization dependence and experiment also for the higher frequency $\varepsilon_2(\omega)$. Misaki *et al.* mostly just note a slight shift to lower frequencies for $\mathbf{E} \parallel \mathbf{c}$ for peaks at 4.5 and 6.5 eV, while the peaks at 7.3 eV and 10.7 eV stayed the same. Clearly, more work is needed to analyze these spectra.

15.9.3 *Non-linear optics*

Because chalcopyrite crystals have found their main applications as non-linear optical-frequency conversion materials, it is of interest to also examine these properties in the II-IV-N_2 materials. Limpijumnong *et al.* (52) calculated the second harmonic generation coefficients. They find $\chi^{(2)}_{\alpha\beta\gamma}$ coefficients of -0.800 pm/V for the zzz or 33 component, -0.293 pm/V for the zyy or 32, and -0.682 pm/V for the zxx or 31 components. These are quite small compared with typical chalcopyrite semiconductors such as $ZnGeP_2$, which are of order 100 pm/V (3). This is not surprising, because $\chi^{(2)}$ scales like E_g^{-3}. Thus, wide band-gap materials are not expected to exhibit large $\chi^{(2)}$. Nevertheless, these are even smaller than those in GaN by an order of magnitude. Limpijumnong *et al.* (52) thus concluded that $ZnGeN_2$ is not particularly suitable as non-linear optical material.

Recently, the static values of the second harmonic generation coefficient were recalculated by Paudel and Lambrecht (59), using the Berry phase DFPT approach. The second-order susceptibilities, $\chi^{(2)}_{\alpha\beta\gamma} = \partial^3 U/\partial E_\alpha \partial E_\beta \partial E_\gamma$, can indeed be viewed as a third-order derivative. Their values are significantly higher than Limpijumnong's, and are reasonably close to those of GaN, as shown in Table 15.19.

The values for AlN and GaN were calculated by Chen *et al.* (109), using a different approach but also including local field-effects. The values by Rashkeev *et al.* (112), which do not include local field-effects, are almost a factor 2 smaller. Even so, it appears that the results by Limpijumnong *et al.* (52) are significantly in error. This may be related to the errors noted previously on the polarization dependence. There appear to be some ambiguities about the sign of the $\chi^{(2)}$ for AlN. We note that the sign of $\chi^{(2)}$ depends on how one chooses the polarity of the axes pointing from N to Al or from Al to N along the bonds, and these details are not fully specified in all previous work.

Overall, the second harmonic generation coefficient $\chi^{(2)}_{33}$ appear to be similar in magnitude to those of the group-III nitrides, and increase from $ZnSiN_2$ to $ZnGeN_2$ to $ZnSnN_2$ as the gaps decrease.

Table 15.19 Second-order non-linear optical tensor coefficients $\mathbf{d}_{\alpha j} = \frac{1}{2}\chi^{(2)}_{\alpha j}$ in pm/V. The results for AlN and GaN are from calculations by Chen *et al.* (109), and the experimental value in parentheses from Fujii *et al.* (110) for AlN, and Miragliotta *et al.* (111) for GaN.

Components	$ZnSiN_2$	$ZnGeN_2$	$ZnSnN_2$	AlN	GaN
$\mathbf{d}_{15} = \mathbf{d}_{31}$	-2.19	-2.63	-0.99	$-0.24(\leq \|0.26\|)$	-3.45 (2.66)
$\mathbf{d}_{24} = \mathbf{d}_{32}$	-2.09	-3.17	-2.96		
$\mathbf{d}_{33}$	5.60	6.98	9.27	-4.61 (-6.3 ± 3.5)	5.76 (5.35)

The ratio $\chi^{(2)}_{31}/\chi^{(2)}_{33} \approx -1/2$ is expected on the basis of the relation between zincblende and wurtzite. Furthermore, one expects $\chi^{(2)}_{31} \approx \chi^{(2)}_{32}$ based on the relation between the orthorhombic $ZnGeN_2$ structure and wurtzite. The actual ratios $\chi^{(2)}_{31}/\chi^{(2)}_{33}$ are -0.39, -0.38, and -0.10 for $ZnSiN_2$, $ZnGeN_2$, and $ZnSnN_2$ respectively. The $\chi^{(2)}_{32}/\chi^{(2)}_{33}$ ratios are -0.37, -0.45, and -0.31. Thus the structural deviations of the orthorhombic from wurtzite and non-ideal wurtzite relative to cubic material significantly affect these ratios.

Another non-linear effect considered by Paudel *et al.* (59) is the electro-optic effect of Pockel's effect. This gives the change in index of refraction induced by a static electric field. More specifically, it is defined by the change in the inverse dielectric constant at optical frequencies by a low-freqency electric field. It is defined by

$$\Delta(\varepsilon^{-1})_{\alpha\beta} = r_{\alpha\beta\lambda}E_\lambda \tag{15.54}$$

in which we use summation convention as usual, but additionally, use indices λ for the static field and α, β for the optic-frequency electric fields. Depending on the mechanical boundary conditions, it may contain three contributions: a purely electronic one, which is directly related to the non-linear optic coefficient, an ionic contribution, and a piezoelectric contribution. These are respectively given by

$$\begin{aligned} r^{el}_{\alpha\beta\lambda} &= \frac{-8\pi}{n^2_\alpha n^2_\beta}\chi^{(2)}_{\alpha\beta\lambda} \\ r^{ion}_{\alpha\beta\gamma} &= -\frac{4\pi}{\sqrt{\Omega}n^2_\alpha n^2_\beta}\sum_n \frac{\mathbf{R}^n_{\alpha\beta}p^n_\gamma}{\omega^2_n} \\ r^{piezo}_{\alpha\beta\gamma} &= p_{\alpha\beta i}d_{\gamma i} \end{aligned} \tag{15.55}$$

in which the Cartesian axes are chosen along principle axes of the crystal so that the indices of refraction are diagonal, and with

$$p^n_\alpha = \sum_\tau Z^*_{\tau\alpha\alpha}u^n_{\tau\alpha} \tag{15.56}$$

gives the mode polarity for vibrational mode n, with $Z^*_{\tau\alpha\alpha}$ the Born effective charges, and $u^n_{\tau\alpha}$ the mode eigenvectors at $\mathbf{q} = 0$, as defined in the section on lattice dynamics. The ionic contribution requires also the Raman tensor $\mathbf{R}^n_{\alpha\beta}$ for each mode. The piezoelectric contribution involves the elasto-optic coefficients

$$\begin{aligned} p_{\alpha\beta i} = &\frac{-1}{n^2_\alpha n^2_\beta}\frac{\partial\varepsilon_{\alpha\beta}(\tau_0, \eta_0)}{\partial\eta_i} \\ &+ \frac{4\pi}{n^2_\alpha n^2_\beta\sqrt{\Omega}}\sum_n\sum_{\tau\gamma}\frac{\mathbf{R}^n_{\alpha\beta}}{\omega^2_n}\frac{\partial^2 U}{\partial\eta_i\partial\tau_\gamma}u^n_{\tau\gamma} \end{aligned} \tag{15.57}$$

Here the first term gives the direct effect of the strain on the electronic dielectric constant keeping the internal coordinates fixed, while the second gives the

Table 15.20 Clamped (A) and unclamped (B) linear electro-optical coefficients (pm/V).

Components	$ZnSiN_2$		$ZnGeN_2$		$ZnSnN_2$	
	A	B	A	B	A	B
r_{15}	0.40	0.40	0.45	0.48	0.25	0.25
r_{24}	0.31	0.31	0.52	0.46	0.50	0.50
r_{31}	0.17	0.18	0.20	0.47	−0.16	−0.15
r_{32}	0.12	0.21	0.09	0.12	−0.14	0.09
r_{33}	−0.66	−0.59	−0.77	−1.15	−0.77	−0.60

indirect effect of the strain on the internal coordinates, which in turn affect the electronic susceptibility. Strictly speaking, the energy here should be replaced by the electrical enthalpy in the presence of a finite electric field. For details, we refer the reader to Veithen *et al.* (113). The "clamped" value refers to the value measured under strain-free conditions, and includes the electronic and ionic contribution but not the piezoelectric contribution. In practice, this means that it corresponds to electric fields oscillating above about 100 MHz, so that the ions can relax (low compared to the phonon frequencies), but high enough that the crystal shape does not have time to relax to the electric field via the converse piezoelectric effect. The "unclamped" value refers to the value measured under stress-free conditions, and includes the additional piezoelectric contribution arising from the changes in the unit cell shape in response to the electric field. In practice, this is below 1 MHz. The values for the clamped and unclamped electro-optic tensor are given in Table 15.20.

No experimental data on these coefficients are available in Zn-IV-N_2 or III-N materials, to the best of our knowledge. They have not even been determined for the group-III nitrides.

15.10 Defects

Very little is known about the defect physics in II-IV-N_2 compounds. Here we present merely some speculative ideas based on comparison with $ZnGeP_2$ and general intuition.

From a basic perspective there is a richer variety of defects than in group-III nitrides. On general grounds one may expect cation vacancies of two types V_{II} and V_{IV} to have different properties. One might expect cation antisites such as II_{IV} and IV_{II}. Based on electron and state counting rules, one might expect a V_{II} to be a double acceptor and V_{IV} to be a quadruple acceptor. In a II-IV-N_2 material, one may think of the N bonding to a group-II to share 3/2 of its electrons in the II-N bond, while a N bonding to a Ge to share 1 electron in a IV-N bond. Since each N is surrounded by two II and two IV elements, this amounts to $3 + 2 = 5$ electrons per N. The group-IV element shares 1 electron in each bond, and the group-II shares 1/2 an electron. Now consider a $V_{\rm II}$. The four N-dangling bonds surrounding it will contain each 3/2 electron, or six electrons in total. The dangling bonds in a tetrahedral environment form an a_1 level and a t_2 level. Usually, the a_1 level lies below the VBM, and it will take the two electrons. The t_2 level lies in the gap and will take four electrons, so it has room for two more or will be a double acceptor. One may expect a Jahn–Teller distortion to D_{2d}, and hence a splitting of the t_2 into a_1 and e. If the a_1 level lies lowest and possibly also below the VBM, the e level is still only half filled and may be expected to split further to C_s symmetry. If the e-level is lowest, one expects an a_1 empty level in the gap.

For a group-IV vacancy the level splittings are expected to be similar, but now there are only four electrons in total in the dangling bonds, so one has in principle a quadruple acceptor. Whether all four charge states are allowed in the gap is unclear without explicit calculations. In $ZnGeP_2$ one finds that the $V_{\rm Ge}$ is unstable toward (14)

$$V_{\rm Ge} \rightarrow {\rm Zn}_{\rm Ge} + V_{\rm Zn} \tag{15.58}$$

In other words, a nearby Zn, hops into the Ge vacancy and creates a ${\rm Zn}_{\rm Ge}$ antisite $V_{\rm Zn}$ complex (114). The $V_{\rm Zn}$ in this case is found to be the dominant defect in as-grown crystals because of the high vapor pressure of Zn. Both the $V_{\rm Zn}$ and ${\rm Zn}_{\rm Ge}$ are acceptors. Thus, if Zn-IV-N_2 behave somewhat like chalcopyrites, one may expect acceptor-type defects to be more prominent than in GaN. In contrast, in GaN the $V_{\rm Ga}$ is the highest energy of formation-defect, except in very n-type material (115).

The nitrogen vacancy $V_{\rm N}$ is nominally a triple donor, because removing a N atom removes eight states and only five electrons. N-vacancies were long thought to be responsible for a residual n-type doping in GaN, until it was found by calculations that the $V_{\rm N}$ has actually a fairly deep defect level. In GaN, it was ultimately found that either O-impurities or ${\rm Si}_{\rm Ga}$ impurities were responsible for the n-type background. In Zn-IV-N_2 materials, Si impurities (arising, for example, from the quartz of the vapor growth chamber) would be isovalent impurities and would not dope the material. So, overall, it appears somewhat more promising to achieve p-type doping in II-IV-N_2 materials. On the other hand, ${\rm Ge}_{\rm Zn}$ antisites have been observed in $ZnGeP_2$, and form donor levels. So, Ge-rich material might be n-type if ${\rm Ge}_{\rm Zn}$ antisites have a lower energy of formation than $V_{\rm Zn}$ defects. This is not the case in $ZnGeP_2$, but remains to be seen in $ZnGeN_2$.

Next, let us consider doping opportunities. To make p-type material, one has two natural choices: group-I elements on group-II site, or group-III elements on the group-IV site. The former is tricky, because Cu in ZnO, for example, tends to give deep levels (in part because of the much higher lying $3d$ levels), and group-I elements from the Ia column of the periodic table, such as Li, Na, tend to diffuse easily because of their small ionic radius. For group-III elements the crucial question is the site selectivity. If the group-III goes on a group-II site it is expected to behave as a donor; if it goes on the group-IV site it is expected to behave as an acceptor. One might try to influence the site selectivity by controlling the stoichiometry or chemical potentials. To promote Ga to go on the Ge site, for example, one would grow Ge-poor material. On the other hand, Ge-poor material will likely promote Zn_{Ge} antisites, which are also acceptors. So, instead of inducing compensating donors one would expect additional acceptors to form. Which of these levels are deep or shallow remains to be seen by actual calculations. Also, the actual energies of formation need to be calculated carefully as a function of chemical potential. Still, the outlook is promising for achieving p-type material. Nonetheless, the only measurements of background doping in $ZnGeN_2$ were indicative of n-type doping. Whether this is due to N-vacancies or O-impurities or other contamination is presently not known. It is also not known *a priori* what the role of interstitials might be in the case of deviations from stoichiometry.

From an effective-mass-approximation (EMA) point of view, the similar inverse-effective-mass parameters, as in GaN but significantly larger crystal field splitting, predicts that the shallow impurity levels would be somewhat deeper than in GaN.

15.11 Outlook

At this point in time, a significant amount of fundamental properties of the II-IV-N_2 materials have been predicted, and some of them have been confirmed by experiment. There is still a large amount of work to be done to explore the full richness of the whole class of materials. From a general perspective they present equally promising properties for optoelectronics as did the III-N materials half a century ago. Of course, for their further development it will be crucial that more people start working on them. At present, the III-nitrides are much farther along and are a fierce competitor to the II-IV-N_2 nitrides because, inevitably the latter will lag behind. Some specific advantages and opportunities for the II-IV-N_2 start emerging, which we hope will encourage researchers to pursue them further. First, they present a richer variety of well-defined, strain-free thermodynamically stable compounds. We have already identified $CdGeN_2$ as a material with a predicted gap corresponding to green light emission. This is a valuable opportunity, given the technical difficulties in pushing toward green in the $In_xGa_{1-x}N$ system. Second, we have found that the materials exhibit smaller differences in spontaneous polarization among each other than in the group-III nitrides. Their piezoelectric tensors are similar, and thus the smaller lattice constant differences also generally will lead to smaller strain-induced polarization fields at heterojunctions. Their lower symmetry may create additional opportunities for polarization-sensitive detectors and other niche applications. Last, and most importantly, but at this point purely speculatively, they may well find their greatest benefits compared to the III-N in a different point-defect physics, which looks promising for more easy p-type doping. We consider that if more researchers become interested in these materials, they may have a bright future. Our hope is that this chapter will be a step in that direction.

Acknowledgments

I would like to thank the students and post-doctoral students who have worked with me on these materials: Sukit Limpijumnong did the early work on $ZnGeN_2$ in 1999; Sergey Rashkeev developed and performed the non-linear optics calculations at that time; Kwiseon Kim (a former student who worked with me on elastic constants and effective masses in the group-III nitrides) hosted me at NREL during my sabbatical in 2005, where we began work on the phonon calculations; Tula Ram Paudel did most of the work on the lattice dynamics; Atchara Punya (co-author of this chapter) performed the *GW* band-structure calculations and energy-of-formation calculations. I also would like to thank my colleagues Prof Kathleen Kash and Jie Shan, and various students who worked on these materials: Timothy Peshek, Paul Quayle, Eric Blanton, and Keliang He, for discussions on the growth and characterization of these materials. Parts of this work were sponsored by the National Science Foundation. Calculations made use of the Ohio Supercomputer Center and the High Performance Computing Resource in the Core Facility for Advanced Research Computing at Case Western Reserve University.

References

(1) James E. Bernard, S. Froyen, and Alex Zunger. Spontaneous surface-induced long-range order in $Ga_{0.5}In_{0.5}P$ alloys. *Phys. Rev. B*, 44:11178–11195, Nov 1991.

(2) J. L. Shay and J. H. Wernick. *Ternary Chalcopyrite Semiconductors: Growth, Electronic Properties, and Application.* Pergamon, New York, 1975.

(3) Sergey N. Rashkeev, Sukit Limpijumnong, and Walter R. L. Lambrecht. Second-harmonic generation and birefringence of some ternary pnictide semiconductors. *Phys. Rev. B*, 59(4):2737–2748, Jan 1999.

(4) Sergey N. Rashkeev and Walter R. L. Lambrecht. Second-harmonic generation of I-III-VI_2 chalcopyrite semiconductors: Effects of chemical substitutions. *Phys. Rev. B*, 63(16):165212, Apr 2001.

(5) Sukit Limpijumnong, Walter R. L. Lambrecht, and Benjamin Segall. Electronic structure of $ZnGeP_2$: A detailed study of the band structure near the fundamental gap and its associated parameters. *Phys. Rev. B*, 60(11):8087–8096, Sep 1999.

(6) Sukit Limpijumnong and Walter R. L. Lambrecht. Band structure of $CdGeAs_2$ near the fundamental gap. *Phys. Rev. B*, 65(16):165204, Apr 2002.

(7) Chandrima Mitra and Walter R. L. Lambrecht. Large band-gap bowing in $Cu_{1-x}Ag_xGaS_2$ chalcopyrite semiconductors and its effect on optical parameters. *Phys. Rev. B*, 76(3):035207, 2007.

(8) Chandrima Mitra and Walter R. L. Lambrecht. Band-gap bowing in $AgGa(Se_{1-x}Te_x)_2$ and its effect on the second-order response coefficient and refractive indices. *Phys. Rev. B*, 76(20):205206, 2007.

(9) Walter R. L. Lambrecht and Xiaoshu Jiang. Noncritically phase-matched second-harmonic-generation chalcopyrites based on $CdSiAs_2$ and $CdSiP_2$. *Phys. Rev. B*, 70:045204, Jul 2004.

(10) Adisak Boonchun and Walter R. L. Lambrecht. First-principles study of the elasticity, piezoelectricity, and vibrational modes in $LiGaO_2$ compared with ZnO and GaN. *Phys. Rev. B*, 81(23):235214, Jun 2010.

(11) A. Boonchun and W. R. L. Lambrecht. Electronic structure, doping, and lattice dynamics of $LiGaO_2$. *Proc. SPIE 7940*, page 79400N, 2011.

(12) S. N. Rashkeev, S. Limpijumnong, and W. R. L. Lambrecht. Theoretical evaluation of $LiGaO_2$ for frequency upcoversion to ultraviolet. *J. Opt. Soc. Am. B*, 16:2217–2222, 1999.

(13) S. Limpijumnong, W. R. L. Lambrecht, B. Segall, and K. Kim. Band structure and cation ordering in $LiGaO_2$. *Mater. Rese. Soc. Symp. Proc.*, 449:905–910, 1997.

(14) Xiaoshu Jiang, M. S. Miao, and Walter R. L. Lambrecht. Theoretical study of cation-related point defects in $ZnGeP_2$. *Phys. Rev. B*, 71(20):205212, 2005.

(15) N. C. Giles, Lihua Bai, M. M. Chirila, N. Y. Garces, K. T. Stevens, P. G. Schunemann, S. D. Setzler, and T. M. Pollak. Infrared absorption bands associated with native defects in $ZnGeP_2$. *J. Appl. Phys.*, 93(11):8975–8981, 2003.

(16) L. E. Halliburton, G. J. Edwards, M. P. Scripsick, M. H. Rakowsky, P. G. Schunemann, and T. M. Pollak. Electron-nuclear double resonance of the zinc vacancy in $ZnGeP_2$. *Appl. Phys. Lett.*, 66(20):2670–2672, 1995.

(17) Takahisa Omata, Keizo Tanaka, Atsushi Tazuke, Katsuhiro Nose, and Shinya Otsuka-Yao-Matsuo. Wide band gap semiconductor alloy: $x(LiGaO_2)_{1/2}$-$(1-x)$ ZnO. *J. Appl. Phys.*, 103(8):083706, 2008.

(18) Takahashi Omata, Kita Masao, Katsuhiro Nose, Kosuke Tachibana, and Shinya Otsuka-Yao-Matsuo. Zn_2LiGaO_4, Wurtzite-derived wide band gap oxide. *Jpn. J. Appl. Phys.*, 50:031102, 2011.

(19) L. Wöhler and O. Bock. Das Silicium-Analogon des Kalkstickstoffs. *Z. anorg. allg. Chem.*, 134:221, 1924.

(20) Heinrich Frank and Victor Louis. Calciumsilicide und Stickstoff. *Z. anorg. allg. Chem.*, 242:128–137, 1939.

(21) Yves Laurent. Contribution à l'étude des systèmes calcium-azote et calcium-silicium-azote. *Rev. Chim. Min.*, 5, 1968.

(22) J. Guyader and J. Lang. Étude du système Ca_3N_2-Ge_3N_4. *Rev. Chim. Min.*, 4:937, 1967.

(23) E. D. Whitney and R. F. Giese Jr. Preparation of a new ternary lithium silicon nitride, $LiSi_2N_3$ and the high-pressure synthesis of magnesium silicon nitride, $MnSiN_2$. *Inor. Chem.*, 10:1090–1092, 1971.

(24) J. David. Study on Mg_3N_2 and some of its compounds. *Rev. Chim. Min.*, 9:717, 1972.

(25) Jean David, Yves Laurent, and Jean Lang. Structure de $MgSiN_2$ et $MgGeN_2$. *Bull. Soc. Fr. Mineral. Cristallogr.*, page 253, 1979.

(26) P. Eckerlin, A. Rabenau, and H. Nortmann. Darstellung und Eigenshaften von $BeSiN_2$. *Z. anorg. allg. Chemie*, 353:113–121, 1967.

(27) P. Eckerlin. Die Kristallstruktur von $BeSiN_2$. *Z. anorg. allg. Chem.*, 353:225–336, 1967.
(28) M. Wintenberger, J. Guyader, and M. Maunaye. Étude cristallographique et magnétique de $MnGeN_2$ par diffraction neutronique. *Solid State Commun.*, 11:1485, 1972.
(29) M. Maunaye, R. Marchand, J. Guyader, Y. Laurent, and J. Lang. Structure de $MnSiN_2$ et $MnGeN_2$. *Bull. Soc. Fr. Mineral. Cristallogr.*, 94:561, 1971.
(30) J. Gaude and J. Lang. Le système Si_3N_4-Sr_3N_2. *C. R. Acad. Sc., C*, 268:1785, 1969.
(31) M. Maunaye and J. Lang. Préparation et Proprietés de $ZnGeN_2$. *Mater. Res. Bull*, 5:793, 1970.
(32) Yungi Lee, Hiroaki Terashima, Yoshiki Shimodaira, Kentaro Teramura, Michikazu Hara, Hisayoshi Kobayashi, Kazunari Domen, and Masatomo Yashima. Zinc germanium oxynitride as a photocatalyst for overall water splitting under visible light. *The Journal of Physical Chemistry C*, 111(2):1042–1048, 2007.
(33) M. Wintenberger, M. Maunaye, and Y. Laurent. Groupe spatial et ordre des atomes de zinc et de germanium dans $ZnGeN_2$. *Mat. Res. Bull.*, 8:1049, 1973.
(34) G. A. Zykov, F. F. Grekov, and D. M. Demidov. Conditions for the formation and reduction of Ge_3N_4, Zn_3N_2, and $ZnGeN_2$, in a stream of partially dissociated ammonia. *Zh. Pirkl. Khim.*, 44:1700, 1971. (English translation in *J. Appl. Chem. USSR*).
(35) W. L. Larson, H. P. Maruska, and A. Stevenson. Synthesis and properties of $ZnGeN_2$. *J. Electrochem. Soc.*, 121:1683, 1974.
(36) T. Endo, Y. Sato, H. Takizawa, and M. Shimada. High pressure synthesis of new compounds, $ZnSiN_2$ and $ZnGeN_2$ with distorted wurtzite structure. *J. Mater. Sci. Lett.*, 11:424, 1992.
(37) L. D. Zhu, P. H. Maruska, P. E. Norris, P. W. Yip, and L. O. Bouthillette. Epitaxial growth and structural characterization of single crystalline $ZnGeN_2$. *MRS Internet J. Nitride Semicond. Res.*, 4S1:G3.8, 1999.
(38) J. Muth, A. Cai, A. Osinsky, H. Everitt, B. Cook, and I. Avrutsky. Optical properties of II-IV-N_2 semiconductors. In Christian Wetzel, Bernard Gil, Masaaki Kuzuhara, and Michael Manfra, editors, *GaN, AlN, InN, and Their Alloys*, volume 831 of *Mater. Res. Soc. Symp. Proc.*, page E11.45.1. MRS, Pittsburgh, 2005.
(39) B. P. Cook, H. O. Everitt, I. Avrutsky, A. Osinsky, A. Cai, and J. F. Muth. Refractive indices of $ZnSiN_2$ on r-plane sapphire. *Appl. Phys. Lett.*, 86(121906), 2005.
(40) A. Osinsky, V. Fufluigin, L. D. Zhu, L. D. Goulakov, J.W. Graff, and E. F. Schubert. New concepts and preliminary results for SiC bipolar transistors: $ZnSiN_2$ and $ZnGeN_2$ heterojunction emitters. In *2000 IEEE/Cornell Conference on High Performance devices*, pages 168–172. IEEE, 2000.
(41) S. J. Pearton, M. E. Overberg, C. R. Abernathy, N. A. Theodoropoulou, A. F. Hebard, S. N. G. Chu, A. Osinsky, V. Fuflyigin, L. D. Zhu, A. Y. Polyakov, and R. G. Wilson. Magnetic and structural characterization of Mn-implanted, single-crystal $ZnGeSiN_2$. *J. Appl. Phys.*, 92:2047, 2002.

(42) Alexander Mintairov, James Merz, Andri Osinsky, Vladimir Fuflyigin, and L. D. Zhu. Infrared spectroscopy of $ZnSiN_2$ single-crystalline films on r-sapphire. *Appl. Phys. Lett.*, 76(18):2517, 2000.

(43) T. Misaki, K. Tsuchiya, D. Sakai, A. Wakahara, H. Okada, and A. Yoshida Growth and characterization of $ZnGeN_2$ by using remote-plasma enhanced metalorganic vapor phase epitaxy. *Phys. Stat. Solidi (c)*, 0:188–191, 2000.

(44) T. Misaki, A. Wakahara, H. Okada, and A. Yoshida. Optical properties of $ZnGeN_2$ epitaxial layer. *Phys. Stat. Solidi (c)*, 0:2890–2893, 2003.

(45) T. Misaki, A. Wakahara, H. Okada, and A. Yoshia. Epitaxial growth and characterization of $ZnGeN_2$ by metalorganic vapor phase epitaxy. *J. Cryst. Growth*, 260:1049, 2004.

(46) T. Cloitre, A. Sere, and R. L. Aulombard. Epitaxial growth of $ZnSiN_2$ single-crystalline films on sapphire substrates. *Superlatt. and Microstruc.*, 36, 2004.

(47) S. Kikkawa and H. Morisaka. RF-sputter depostiion of Zn-Ge nitride thin films. *Solid State Commun.*, 112:513–515, 1999.

(48) R. Viennois, T. Taliercio, V. Potin, A. Errebbahi, B. Gil, S. Charar, A. Haidoux, and J.-C. Tédenac. Prospective investigations of orthorhombic $ZnGeN_2$: synthesis, lattice dynamics and optical properties. *Mater. Sci. Eng. B*, 82:45–49, 2001.

(49) K. Du, C. Bekele, C. C. Hayman, J. C. Angus, P. Pirouz, and K. Kash. Synthesis and characterization of $ZnGeN_2$ grown from elemental Zn and Ge sources. *J. Cryst. Growth*, 310:1057, 2008.

(50) P. C. Quayle, Keliang He, J. Shan, and K. Kash. Synthesis, lattice structure, and band gap of $ZnSnN_2$. *MRS Communications*, FirstView:1–4, 2013.

(51) P. Vinet, J. H. Rose, J. Ferrante, and J. R. Smith. Universal features of the equation of state of solids. *J. Phys.: Condens. Matter*, 1:1941, 1989.

(52) Sukit Limpijumnong, Sergey N. Rashkeev, and Walter R. L. Lambrecht. Electronic structure and optical properties of $ZnGeN_2$. *MRS Internet J. Nitride Semicond. Res.* 4S1:G6.11, 1999.

(53) Takao Misaki, Xiaoping Wu, Akihiro Wakahara, and Akira Yoshida. Theoretical analysis of multinary nitride semicondutors by density functional theory. In *Proceedings of the International Workshop on Nitride Semiconductors*, volume 1 of *IPAP Conference Series*, pages 685–688. Institute of Pure and Applied Physics, Tokyo, 2000.

(54) V. L. Shaposhnikov, A. V. Krivosheeva, F. Arnaud D'Avitaya, J.-L. Lazzari, and V. E. Borisenko. Structural, electronic and optical properties of II-IV-N_2 compounds, (II=Be, Zn; IV=Si, Ge). *Phys. Stat. Solidi (b)*, 245:142–148, 2008.

(55) Walter R. L. Lambrecht, Erik Alldredge, and Kwiseon Kim. Structure and phonons of $ZnGeN_2$. *Phys. Rev. B*, 72(15):155202, 2005.

(56) Tula R. Paudel and Walter R. L. Lambrecht. First-principles calculation of the zone center phonons in $ZnSiN_2$: Comparison with infrared data. *Phys. Rev. B*, 76(11):115205, 2007.

(57) Timothy J. Peshek, Tula R. Paudel, Kathleen Kash, and Walter R. L. Lambrecht. Vibrational modes in $ZnGeN_2$: Raman study and theory. *Phys. Rev. B*, 77(23):235213, 2008.

(58) Tula R. Paudel and Walter R. L. Lambrecht. First-principles study of phonons and related ground-state properties and spectra in Zn-IV-N_2 compounds. *Phys. Rev. B*, 78(11):115204, 2008.
(59) Tula R. Paudel and Walter R. L. Lambrecht. First-principles calculations of elasticity, polarization-related properties, and nonlinear optical coefficients in Zn-IV- N_2 compounds. *Phys. Rev. B*, 79(24):245205, Jun 2009.
(60) Tula Ram Paudel. *Structure, Phonons and related Properties in Zn-IV-N_2 (IV=Si, Ge, Sn), ScN and Rare-Earth Nitrides.* PhD thesis, Case Western Reserve University, 2008.
(61) A. Punya, T. R. Paudel, and W. R. L. Lambrecht. Electronic and lattice dynamical properties of II-IV-N_2 semiconductors. *Phys. Stat. Solidi C*, 8:2492–2499, 2011.
(62) Atchara Punya, Walter R. L. Lambrecht, and Mark van Schilfgaarde. Quasiparticle band structure of Zn-IV-N_2 compounds. *Phys. Rev. B*, 84:165204, Oct 2011.
(63) Atchara Punya and Walter R. L. Lambrecht. $CdGeN_2$ and $ZnGe_0.5Sn_0.5N_2$: Two new nitride semiconductors with band gaps in the blue-green. In T. P. Chow, R. P. Devaty, and M. Dudley, editors, *Proceedings of the International Conference on Silicon Carbide and Related Materials, 2011*, Stafa-Zrich, 2012. Trans Tech Publications.
(64) W. R. L. Lambrecht and B. Segall. Electronic structure of $becn_2$: A proposed nearly direct wide-band-gap semiconductor. *Phys. Rev. B*, 45:1485–1487, Jan 1992.
(65) A. G. Petukhov, W. R. L. Lambrecht, and B. Segall. Electronic structure of wide-band-gap ternary pnictides with the chalcopyrite structure. *Phys. Rev. B*, 49:4549–4558, Feb 1994.
(66) U. Berger and W. Schnick. Syntheses, crystal structures, and vibrational spectroscopic properties of $MgCN_2$, $SrCN_2$ and $BaCN_2$. *J. Alloys and Compounds*, 206:179–184, 1994.
(67) P. Hohenberg and W. Kohn. Inhomogeneous electron gas. *Phys. Rev.*, 136(3B):B864–B871, Nov 1964.
(68) W. Kohn and L. J. Sham. Self-consistent equations including exchange and correlation effects. *Phys. Rev.*, 140(4A):A1133–A1138, Nov 1965.
(69) U. von Barth and L. Hedin. A local exchange-correlation potential for the spin polarized case. *J. Phys. C: Solid State Phys.*, 5:1629, 1972.
(70) John P. Perdew and Yue Wang. Accurate and simple analytic representation of the electron-gas correlation energy. *Phys. Rev. B*, 45:13244–13249, Jun 1992.
(71) John P. Perdew, Kieron Burke, and Matthias Ernzerhof. Generalized gradient approximation made simple. *Phys. Rev. Lett.*, 77(18):3865–3868, Oct 1996.
(72) M. Methfessel, M. van Schilfgaarde, and R. A. Casali. A full-potential LMTO method based on smooth hankel functions. In H. Dreyssé, editor, *Electronic Structure and Physical Properties of Solids. The Use of the LMTO Method*, volume 535 of *Lecture Notes in Physics*, p. 114. Berlin Springer Verlag, 2000.
(73) Takao Kotani and Mark van Schilfgaarde. Fusion of the LAPW and LMTO methods: The augmented plane wave plus muffin-tin orbital method. *Phys. Rev. B*, 81(12):125117, Mar 2010.

(74) T. J. Peshek, J. C. Angus, and K. Kash. Experimental investigation of the enthalpy, entropy, and free energy of formation of GaN. *J. Cryst. Growth*, 311:185–189, 2008.

(75) K. A. Gschneider. Physical properties and interrelationships of metallic and semimetallic elements. In F. Seitz and D. Turnbull, editors, *Solid State Physics, Advanced in Research and Applications*, volume 16, pp. 275–426. Academic Press, New York, 1964.

(76) K. P. Huber and G. Herzberg. *Molecular Structure and Molecular Spectra. IV. Constants of Diatomic Molecules.* Van Nostrand Reinhold, New York, 1979.

(77) V. Wathelet, B. Champagne, D. H. Mosley, J.-M. André, and S. Massida. Vibrational frequencies of diatomic molecules from Car and Parrinello molecular dynamics. *Chem. Phys. Lett.*, 275:506–512, 1997.

(78) R. S. Lima, P. H. Dionisio, and W. H. Schreiner. Magnetron sputtered tin nitride. *Solid State Commun.*, 79:295–398, 1991.

(79) M. Huang and Y. P. Feng. Theoretical prediction of the structure and properties of Sn_3N_4. *J. Appl. Phys.*, 96(7):4015–4017, 2004.

(80) M. van Schilfgaarde, Takao Kotani, and S. Faleev. Quasiparticle self-consistent GW theory. *Phys. Rev. Lett.*, 96(22):226402, Jun 2006.

(81) Mark van Schilfgaarde, Takao Kotani, and Sergey V. Faleev. Adequacy of approximations in GW theory. *Phys. Rev. B*, 74(24):245125, 2006.

(82) Takao Kotani, Mark van Schilfgaarde, and Sergey V. Faleev. Quasiparticle self-consistent GW method: A basis for the independent-particle approximation. *Phys. Rev. B*, 76(16):165106, 2007.

(83) Lars Hedin. New method for calculating the one-particle Green's function with application to the electron-gas problem. *Phys. Rev.*, 139(3A):A796–A823, Aug 1965.

(84) Lars Hedin and Stig Lundqvist. Effects of electron–electron and electron–phonon interactions on the one-electron states of solids. In F. Seitz, D. Turnbull, and H. Ehrenreich, editors, *Solid State Physics, Advanced in Research and Applications*, volume 23, pp. 1–181. Academic Press, New York, 1969.

(85) F. Aryasetiawan and O. Gunnarsson. Product-basis method for calculating dielectric matrices. *Phys. Rev. B*, 49(23):16214–16222, Jun 1994.

(86) Manuel Cardona and M. L. W. Thewalt. Isotope effects on the optical spectra of semiconductors. *Rev. Mod. Phys.*, 77(4):1173–1224, Nov 2005.

(87) A. V. Rodina, M. Dietrich, A. Göldner, L. Eckey, A. Hoffmann, Al. L. Efros, M. Rosen, and B. K. Meyer. Free excitons in wurtzite GaN. *Phys. Rev. B*, 64(11):115204, Aug 2001.

(88) N. Feldberg, B. Keen, J. D. Aldous, D. O. Scanlon, P. A. Stampe, R. J. Kennedy, R. J. Reeves, T. D. Veal, and S. M. Durbin. $ZnSnN_2$: A new earth-abundant element semiconductor for solar cells. In *38th IEEE Photovoltaic Specialists Conference (PVSC)*, page 002524, 2012.

(89) Lise Lahourcade, Naomi C. Coronel, Kris T. Delaney, Sujeet K. Shukla, Nicola A. Spaldin, and Harry A. Atwater. Structural and optoelectronic characterization of RF sputtered $ZnSnN_2$. *Adv. Mater.*, 28:2562, May 2013.

(90) Walter R. L. Lambrecht, Anna V. Rodina, Sukit Limpijumnong, B. Segall, and Bruno K. Meyer. Valence-band ordering and magneto-optic exciton fine structure in ZnO. *Phys. Rev. B*, 65(7):075207, Jan 2002.

(91) J. M. Luttinger. Quantum theory of cyclotron resonance in semiconductors: General theory. *Phys. Rev.*, 102:1030–1041, May 1956.

(92) J. M. Luttinger and W. Kohn. Motion of electrons and holes in perturbed periodic fields. *Phys. Rev.*, 97:869–883, Feb 1955.

(93) G. L. Bir and G. E. Pikus. *Symmetry and Strain-Induced Effects in Semiconductors*. Wiley, New York, 1974.

(94) Yu. M. Sirenko, J.-B. Jeon, K. W. Kim, M. A. Littlejohn, and M. A. Stroscio. Envelope-function formalism for valence bands in wurtzite quantum wells. *Phys. Rev. B*, 53:1997–2009, Jan 1996.

(95) Kwiseon Kim, Walter R. L. Lambrecht, Benjamin Segall, and Mark van Schilfgaarde. Effective masses and valence-band splittings in gan and aln. *Phys. Rev. B*, 56:7363–7375, Sep 1997.

(96) Atchara Punya and Walter R. L. Lambrecht. Valence band effective-mass Hamiltonians for the group-III nitrides from quasiparticle self-consistent *GW* band structures. *Phys. Rev. B*, 85:195147, May 2012.

(97) Xavier Gonze. First-principles responses of solids to atomic displacements and homogeneous electric fields: Implementation of a conjugate-gradient algorithm. *Phys. Rev. B*, 55(16):10337–10354, Apr 1997.

(98) Stefano Baroni, Paolo Giannozzi, and Andrea Testa. Green's-function approach to linear response in solids. *Phys. Rev. Lett.*, 58(18):1861–1864, May 1987.

(99) Xavier Gonze and Changyol Lee. Dynamical matrices, Born effective charges, dielectric permittivity tensors, and interatomic force constants from density-functional perturbation theory. *Phys. Rev. B*, 55(16):10355–10368, Apr 1997.

(100) David Vanderbilt and R. D. King-Smith. Electric polarization as a bulk quantity and its relation to surface charge. *Phys. Rev. B*, 48(7):4442–4455, Aug 1993.

(101) Raffaele Resta. Macroscopic polarization in crystalline dielectrics: the geometric phase approach. *Rev. Mod. Phys.*, 66:899–915, Jul 1994.

(102) D. R. Hamann, Xifan Wu, Karin M. Rabe, and David Vanderbilt. Metric tensor formulation of strain in density-functional perturbation theory. *Phys. Rev. B*, 71(3):035117, 2005.

(103) Xifan Wu, David Vanderbilt, and D. R. Hamann. Systematic treatment of displacements, strains, and electric fields in density-functional perturbation theory. *Phys. Rev. B*, 72(3):035105, 2005.

(104) A. F. Wright. Elastic properties of zinc-blende and wurtzite AlN, GaN and InN. *J. Appl. Phys.*, 82:2833–2839, 1997.

(105) A. Polian, M. Grimsditch, and I. Grzegory. Elastic constants of gallium nitride. *J. Appl. Phys.*, 79(6):3343–3344, 1996.

(106) Fabio Bernardini, Vincenzo Fiorentini, and David Vanderbilt. Spontaneous polarization and piezoelectric constants of III-V nitrides. *Phys. Rev. B*, 56(16):R10024–R10027, Oct 1997.

(107) I. L. Guy, S. Muensit, and E. M. Godbys. Extensional piezoelectric coefficients of gallium nitride and aluminum nitride. *Appl. Phys. Lett.*, 75, 1999.

(108) S. Muensit, E. M. Goldbys, and I. L. Guy. Shear piezoelectric coefficients of gallium nitride and aluminum nitride. *Appl. Phys. Lett.*, 75:3965, 1999.

(109) Jian Chen, Zachary H. Levine, and John W. Wilkins. Calculated second-harmonic susceptibilities of BN, AlN, and GaN. *Applied Physics Letters*, 66(9):1129–1131, 1995.

(110) Y. Fujii, S. Yoshida, S. Misawa, S. Maekawa, and T. Sakudo. Nonlinear optical susceptibilities of AlN film. *Applied Physics Letters*, 31(12):815–816, 1977.

(111) J. Miragliotta, D. K. Wickenden, T. J. Kistenmacher, and W. A. Bryden. Linear- and nonlinear–optical properties of GaN thin films. *J. Opt. Soc. Am. B*, 10:1447, 1993.

(112) Sergey N. Rashkeev, Walter R. L. Lambrecht, and Benjamin Segall. Efficient *ab initio* method for the calculation of frequency-dependent second-order optical response in semiconductors. *Phys. Rev. B*, 57:3905–3919, Feb 1998.

(113) M. Veithen, X. Gonze, and Ph. Ghosez. Nonlinear optical susceptibilities, Raman efficiencies, and electro-optic tensors from first-principles density functional perturbation theory. *Phys. Rev. B*, 71(12):125107, 2005.

(114) Xiaoshu Jiang, M. S. Miao, and Walter R. L. Lambrecht. Theoretical study of cation-related point defects in $ZnGeP_2$. *Phys. Rev. B*, 71:205212, May 2005.

(115) Sukit Limpijumnong and Chris G. Van de Walle. Diffusivity of native defects in GaN. *Phys. Rev. B*, 69:035207, Jan 2004.

16 Terahertz emission in polaritonic systems with nitrides

Oleksandr Kyriienko, Ivan A. Shelykh, and Alexey V. Kavokin

16.1 Introduction

The problem of creation of an efficient THz source is of high importance in modern physics and technology. High-power emitters with frequency tunable in the terahertz range are required for applications in information communication technologies, medicine, and security. However, while for optical spectrum and radio-wave frequencies there are numerous efficient sources of radiation, the THz range of spectrum remains the least explored. None of the existing THz emitters universally satisfies the application requirements. The problem is challenging, since the efficiency of semiconductor THz sources is limited due to the short life-time of involved electronic states (typically, fractions of a nanosecond) compared to a long typical time of spontaneous emission of a THz photon (typically milliseconds).

The existing terahertz sources are based on various operation principles (1). The standard solid-state oscillators such as the Gunn diode or tunnelling diode are restricted only to generation in the lower boundary of the THz frequency window. The group of emitters based on the free electron lasers are bulky, and cannot be integrated easily in the optoelectronic schemes. Finally, the semiconductor sources of THz radiation are represented by several groups of emitters. First, exploiting the intersubband transition of the quantum well, together with electron tunnelling through the structure, one can realize quantum cascade lasers (QCL) (2). In such waveguide structures one can achieve comparably high output power (up to 50 mW), but overall efficiency remains low due to the small spontaneous emission rate. Another drawback of the system is the ultra-low operating temperature of the device, and current research is focused on its raise. Alternatively, the semiconductor THz emission source can be done using downshift optical to THz range conversion. This can be realized if a semiconductor with a large difference of electron and hole mobility is illuminated by a short laser pulse. The resultant density oscillations with THz frequency last for a few periods, and lead to classical dipole emission. However, the power of such an emitter is limited to nanowatt range and can not be easily tuned.

In the following section we review different mechanisms which were proposed for THz emission generation using cavity polaritons, and discuss their advantages over previous schemes. First, we present a broad introduction of the physics of excitons in semiconductor materials. Then, we introduce strong-light matter coupling and discuss properties of emergent polaritonic states. Next, we describe currently existing proposals of THz emitters based on polaritons. Finally, we conclude that realization of presented proposals with nitrides will lead to a significant step towards realization of efficient room-temperature THz sources, including lasers.

16.1.1 *Excitons*

An important property of a semiconductor material is its optical absorption spectrum. It contains the information about energy-band structure, the quality of the structure (impurities), and collective excitations in the system. At room temperature the optical properties of material are dominated by interband absorption due to optical excitation of the electron from valence into conduction band. The behavior of the system in this case depends drastically on the dimensionality of the system, and is governed by the density of states. However, with the lowering of the temperature of the system, many-body effects come into play and alter the absorption spectrum of the material.

In most semiconductors at low temperature the formation of excitons—bound electron-hole pairs—takes place. An exciton represents a solid-state analog of a hydrogen atom. If the electron–hole interaction is not screened, a tightly-bound Frenkel exciton state is formed. It has a radius of the order of the lattice constant, and large binding energy in the range 0.1–1 eV. However, this type of exciton is usually absent in typical solid-semiconductor materials, and is important mostly in organic structures. In the case of screened interaction, another type of exciton appears. It has a radius of order of tens lattice constants with typically small binding energy (0.01 eV), and is referred to as Wannier–Mott exciton (3).

Wannier–Mott excitons can be described conveniently using effective-mass approximation. The Hamiltonian of such a two-body problem is written as

$$\hat{H} = \hat{H}_e + \hat{H}_h + V_{eh}(r) = -\frac{\hbar^2}{2m_e}\triangle_e - \frac{\hbar^2}{2m_h}\triangle_h - \frac{e^2}{4\pi\epsilon\epsilon_0|\mathbf{r}_e - \mathbf{r}_h|}, \tag{16.1}$$

which consists of kinetic energy operators for an electron and a hole, where $m_{e,h}$ denotes the mass of electron and hole, respectively, and $\triangle$ denotes the Laplace operator. ϵ and ϵ_0 correspond to the dielectric permittivity of medium and vacuum. The last term in (16.1) represents an attractive Coulomb interaction between particles, and leads to the formation of the bound state. It depends on the relative distance between an electron and a hole, $r = |\mathbf{r}_e - \mathbf{r}_h|$. The Hamiltonian can be rewritten in the center-of-mass (CM) coordinates, with a separation of CM and relative motion of carriers. The problem can be solved similarly to the hydrogen-atom case, and consequent minimization of the energy (16.1) gives the spectrum of the exciton. It has the structure of absorption lines corresponding to ground state $1s$, first excited $2s$ and $2p$ states, and states of higher order. The binding energy E_b of the ground-state exciton, which is given by the symmetric wavefunction and exciton Bohr radius a_B, is written as (in the 3D case) (4):

$$E_b^{3D} = \frac{e^4 \mu}{8h^2 \epsilon_0^2 \epsilon^2}, \quad a_B^{3D} = \frac{h^2 \epsilon \epsilon_0}{\mu \pi e^2}, \tag{16.2}$$

where $\mu = m_e m_h/(m_e + m_h)$ is reduced exciton mass, and h is the Planck constant. Similar calculation in the two-dimensional (2D) case gives the exciton parameters in the form $E_b^{2D} = 4E_b^{3D}$ and $a_B^{2D} = a_B^{3D}/2$. The excited states are separated from 1s by the energy $\Delta E = 8E_B^{2D}/9$ (2D).

Let us consider the properties of an exciton in a two-dimensional quantum well (QW). The corresponding wavefunctions of 1s, 2s, and 2p states read (in the 2D case) (4):

$$\psi_{1s}(r) = \sqrt{\frac{2}{\pi}\frac{1}{a_B}} \exp(-r/a_B) \tag{16.3}$$

$$\psi_{2s}(r) = \sqrt{\frac{4}{3\pi}\frac{1}{a_B}} \left(1 - \frac{2r}{3a_B}\right) \exp(-r/3a_B) \tag{16.4}$$

$$\psi_{2p}(r, \phi) = \sqrt{\frac{4}{3\pi}} \frac{r}{(3a_B)^2} \exp(-r/3a_B) \exp(\pm i\phi) \tag{16.5}$$

where we denote $a_B \equiv a_B^{2D}$ and use polar coordinates (r, ϕ).

One can note that 1s and 2p states have different parity. The symmetry of the wavefunction of a 2p exciton state does not allow its direct optical excitation with a single photon, and the 2p state is usually referred to as a *dark* exciton. However, it can be created using two-photon absorption. Additionally, one can see that while 2s–1s optical transition is forbidden, the radiative transition between 2p and 1s excitonic states is allowed in the xy polarization plane. These will be important for further consideration of the 2p–1s THz generation proposal.

Another type of exciton appears if it is formed not by a ground-state electron–hole pair, but with a hole situated in the second quantization subband (e1hh2). The spontaneous emission rate of such an exciton is greatly reduced, but it can radiatively decay to a ground-state exciton e1hh1. This transition is allowed due to the different parity of envelope functions in the growth direction, and results in the emission of a photon with z polarization.

Next, let us discuss the spin properties of an exciton. An exciton is a composite boson with a spin defined by the spin of an electron (e) and a hole (h). Typically, in semiconductors one has to distinguish light-hole (lh) and heavy-hole (hh) states, which have spin $S_{lh} = \pm 1/2$ and $S_{hh} = \pm 2$, respectively. In most semiconductor nanostructures the main role is played by e–hh exciton. It has four possible spin projections: +1, −1, +2, and −2. An important issue is the optical selection rules of the system, which define both the quasiparticle lifetime and the possibility of efficient observation of the excitons. In particular, if the total spin projection of the exciton is not unity, it is not possible to excite the exciton with absorption of a single photon. This comes from the fact that angular momentum should be conserved. Simultaneously, radiative transitions between ± 2 and ± 1 spin states are forbidden, since they have the same parity. Therefore, excitons with spin ± 2 correspond to dark excitons, due to reduced interaction with light. Their energy usually is slightly detuned from the bright exciton resonance. However, under certain conditions they can still play some role in the physics of excitons; for instance, if spin–orbit interaction is taken into account.

Formation of the exciton results in the appearance of narrow absorption peaks lying below the interband continuum, with the energy distance given by the binding energy of the exciton. Due to the large oscillator strength of excitonic transition it represents an important mechanism of absorption at low temperatures. The critical temperature of exciton dissociation is defined by the value of the binding energy. While for narrow and medium band-gap semiconductors it is typically low (up to 100 K in GaAs), in wide band-gap semiconductors such as GaN or ZnO, excitons can survive at room temperature. This is extremely important for the successful realization of devices based on the optics of excitons.

An important issue connected to the existence of excitons is excitation conditions. In particular, they exist only in the low-intensity excitation regime where the density of photo-generated electrons and holes is low. If the density of electrons and holes exceeds the critical value, called the Mott density, the system drives into the electron–hole plasma regime, where formation of excitons is suppressed. Moreover, previously we discussed that excitons have integer spin and thus are composite bosons. The validity of the application of bosonic statistics is defined by the density of electrons and holes, and the corresponding condition is

$$na_B^D \ll 1 \tag{16.6}$$

where n is the concentration, a_B is a Bohr radius of the exciton corresponding to its size in the real space, and D is a dimensionality of the system ($D = 1, 2, 3$).

The properties of exciton gas depend strongly on the interactions between the excitons. Formally, the problem of exciton interaction could be solved as a two-body interaction with four possible processes (5): direct Coulomb interaction between electrons and holes, exchange of electrons between excitons, hole exchange, and simultaneous electron and hole exchange. Since an exciton is a dipole, one expects that the interaction between excitons is short-range. Moreover, the conventional direct exciton does not have a dipole moment pointing in a single direction, and the resultant interaction will be not dipole–dipole. It was shown that direct interaction between excitons is zero for small exchanged momenta, and that the main contribution for short-range point-like interaction comes from electron and hole exchange. Finally, in terms of exciton parameters, it reads (5, 6):

$$V_{XX} \approx \frac{6E_b a_B^2}{A} \tag{16.7}$$

where A denotes the area of the sample. However, one should note that similarly to 2D Fourier transform of Coulomb interaction, area does not enter into the equations as an independent parameter. Finally, the exchange nature of interaction leads to the spin-anisotropy of excitons. In particular, this question will be important in the case of polaritons, where spin-singlet interaction is suppressed by one order of magnitude, compared to triplet interaction.

The bosonic statistics of excitons, together with efficient exciton–exciton interactions, make them a solid-state candidate for observation of Bose–Einstein condensation. It was prescribed theoretically by Keldysh and Kozlov more than fifty years ago (7) and since then has became a holy grail for experimentalists. The effective mass of the excitons is about five orders of magnitude smaller than those of cold atoms, which shifts the critical temperature of BEC from nano-Kelvins to about 1 K. This represents a major step forward, but still does not allow any technological implementation of the effect. However, the condensation of excitons in the semiconductor structure has not yet been observed. This fact is connected to the high requirements of the quality of the sample set by the coherence length. The breakthrough was made by combining excitons with photons, which leads to the formation of hybrid quasiparticles known as exciton polaritons, which are considered in the next section.

16.1.2 *Exciton-polaritons*

New horizons for bringing quantum collective phenomena into device applications were opened in the field of quantum microcavities which lies at the intersection of condensed-matter physics and optics (8).

A semiconductor microcavity is a photonic microstructure designed to enhance light–matter interaction (see Fig. 16.1). The cavity photons are confined between two mirrors, and resonantly interact with the excitonic transition of a semiconductor quantum well or a quantum dot (QD), which makes possible the achievement of the strong coupling regime, where the elementary excitations, cavity polaritons, have a hybrid half–light half–matter nature. In planar cavities with embedded quantum wells, polaritons are two–dimensional bosonic

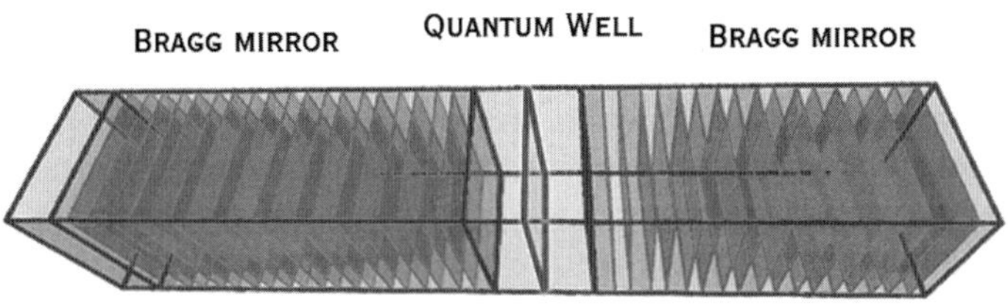

FIG. 16.1. The geometry of a semiconductor microcavity formed by two dielectrical Bragg mirrors (DBR), which confine the photonic mode. In the action region it couples to the excitons due to the strong light–matter interaction. In the present case, the active structure is a quantum well (QW).

quasi-particles, free in the plane of the structure and confined in the growth direction.

Mathematically, the formation of cavity polaritons can be described in the following way. The Hamiltonian of the coupled exciton–photon system can be written as

$$H = \sum_{\mathbf{k}} E_X(\mathbf{k}) b^\dagger_{\mathbf{k}} b_{\mathbf{k}} + \sum_{\mathbf{k}} E_C(\mathbf{k}) a^\dagger_{\mathbf{k}} a_{\mathbf{k}} + \sum_{\mathbf{k}} \frac{V_R(\mathbf{k})}{2} \left(a^\dagger_{\mathbf{k}} b_{\mathbf{k}} + a_{\mathbf{k}} b^\dagger_{\mathbf{k}} \right), \qquad (16.8)$$

where the operators $b^\dagger_{\mathbf{k}}, b_{\mathbf{k}}, a^\dagger_{\mathbf{k}}, a_{\mathbf{k}}$ are creation and annihilation operators for excitons and photons with in-plane momenta **k**. We suppose that condition (16.6) is satisfied, and both excitonic and photonic operators obey the rules of bosonic algebra. The first and second terms correspond to free excitons and photons, with $E_X(\mathbf{k})$, $E_C(\mathbf{k})$ being the dispersions of the free excitons and photons, for which effective-mass approximation can be normally used:

$$E_{X,C}(k) = \frac{\hbar^2 k^2}{2m_{X,C}} + E_{X,C}(0), \qquad (16.9)$$

with $m_{X,C}$ being the effective masses of the exciton and cavity photon, respectively, and $E_{X,C}(0)$ being energies of the modes at k = 0 point. Finally, the third term in (16.8) corresponds to the light–matter interaction, which is given by the exciton–photon coupling constant, $V_R/2$.

The non-zero effective mass of the cavity photon comes from the confinement of the electromagnetic field in the cavity growth direction z, and can be estimated as

$$m_C = \frac{\hbar n_c k_z}{c} \qquad (16.10)$$

where n_c is a refractive index of the cavity, $k_z = 2\pi n_c N/L$ is the photon wavevector in the cavity growth direction, L is an effective cavity length, and integer N is a number of the cavity mode. In realistic cavities, an effective mass of a photon estimated by this formula is extremely small, and normally lies in the range 10^{-4}–10^{-5} of the mass of free electron m_e, while the effective mass of the exciton is of the same order as m_e. Thus, if the value of the *detuning* parameter $\Delta = E_C(0) - E_X(0)$ is negative, the dispersion curves of the free excitons and photons cross at the point $k_0 = 2\Delta(m_X - m_C)/\hbar^2 m_X m_C$, while for positive detuning the photon dispersion lies everywhere above the exciton one.

The third term in eq. (16.8) corresponding to a coupling between photons and excitons is described by the constant V_R, which is determined by the parameters of the microcavity (9):

$$V_R(k) = \frac{E_X(k)}{\hbar c} \sqrt{\frac{2\pi\hbar c}{n_c L\left(k^2 + k_z^2\right)}} F_k(0) d_{cv} I \qquad (16.11)$$

where $F_k(r)$ is a wavefunction of the relative motion of the electron and hole, d_{cv} is a dipole matrix element of the excitonic transition, k_z denotes normal-to-plane momentum, and factor $I \leq 1$ is determined by the geometry of the QW with excitonic resonance. One should note that the coupling constant $V_R(k)$ is a weak function of momentum at small in-plane polariton momenta $k \ll k_z$, and E_X is almost constant.

The Hamiltonian (16.8) can be diagonalized easily by a linear transformation of the operators (10):

$$p_{L,\mathbf{k}} = C_{L,\mathbf{k}} a_{\mathbf{k}} + X_{L,\mathbf{k}} b_{\mathbf{k}} \tag{16.12}$$

$$p_{U,\mathbf{k}} = C_{U,\mathbf{k}} a_{\mathbf{k}} + X_{U,\mathbf{k}} b_{\mathbf{k}} \tag{16.13}$$

The transformed Hamiltonian reads:

$$H = \sum_{\mathbf{k}} E_L(\mathbf{k}) p^\dagger_{L,\mathbf{k}} p_{L,\mathbf{k}} + \sum_{\mathbf{k}} E_U(\mathbf{k}) p^\dagger_{U,\mathbf{k}} p_{U,\mathbf{k}} \tag{16.14}$$

where

$$E_{U,L}(k) = \frac{E_C(k) + E_X(k)}{2} \pm \frac{1}{2}\sqrt{(E_C(k) - E_X(k))^2 + V_R^2} \tag{16.15}$$

are the eigenenergies of the upper and lower polariton branch, respectively.

Operators (16.12) and (16.13) describe the upper and lower polariton—new eigenmodes of the coupled QW-cavity system. According to the above expression, a polariton is a composite particle containing photonic and excitonic fractions. The percentage of these fractions is given by the square of Hopfield coefficients $C_{U,L}$, $X_{U,L}$, which can be determined as

$$C_{U,\mathbf{k}} = X_{L,\mathbf{k}} = \frac{E_U(\mathbf{k}) - E_X(\mathbf{k})}{\sqrt{V_R^2/4 + [E_U(\mathbf{k}) - E_X(\mathbf{k})]^2}} \tag{16.16}$$

$$X_{U,\mathbf{k}} = -C_{L,\mathbf{k}} = \frac{V_R/2}{\sqrt{V_R^2/4 + [E_U(\mathbf{k}) - E_X(\mathbf{k})]^2}} \tag{16.17}$$

The coupling between photonic and excitonic modes in the sample with negative detuning leads to the phenomenon of the anti-crossing of the polariton branches: at the point corresponding to k_0 they do not cross as dispersions of bare excitons and photons, but are separated by distance V_R, corresponding to the *Rabi splitting* in the system. It should be noted that in realistic samples, where excitons and cavity photons have a finite lifetime, their energies can be considered as complex functions with imaginary parts describing their decay, $E_{X,C} = E_{X,C} + \hbar^2 k^2/2m_{X,C} - i\gamma_{X,C}$. In this case, expression (16.15) also become complex. Their real parts correspond to dispersions of polariton modes, while imaginary parts give polariton lifetimes. The Rabi splitting in this case is determined as a distance between the real parts of the polariton states in the anti-crossing point:

$$\hbar\Omega_R = \begin{cases} \sqrt{V_R^2 - (\gamma_X - \gamma_C)^2}, V_R^2 > (\gamma_X - \gamma_C)^2 \\ 0, V_R^2 < (\gamma_X - \gamma_C)^2 \end{cases} \tag{16.18}$$

The transition to the strong coupling regime when exciton and photon damping is overcome by their coherent interaction and anti-crossing is observed, is thus given by the condition

$$V_R^2 > (\gamma_X - \gamma_C)^2 \tag{16.19}$$

The plot in Fig. 16.2 shows dispersion of upper and lower polaritons in the strong coupling regime in GaAs microcavity. One can note the following. First, the effective mass of the lower polariton has the same order of magnitude as the effective mass of the cavity photon. Second, the dispersion curve of the lower polaritons is highly non-parabolic, and has an inflection point in the anti-crossing region. Both these facts have important consequences, which we consider in the next section, and make quantum microcavities in a strong coupling regime a unique laboratory for the implementation of quantum collective phenomena for the design of optoelectronic devices of the new generation (11). In particular, differently from the case of bare excitons, the formation of polariton Bose–Einstein condensate (12, 13) and onset of superfluidity (16) were reported experimentally.

Due to the fact that Rabi splitting depends on the exciton properties, the material properties play an important role in light–matter coupling. For instance, it was shown that in large band-gap semiconductors GaN and ZnO the binding energy of excitons is several times higher than in the conventional GaAs case. This leads to the large Rabi splitting in nitride materials and the strongest light–matter coupling over other semiconductor systems (15). Therefore, accounting

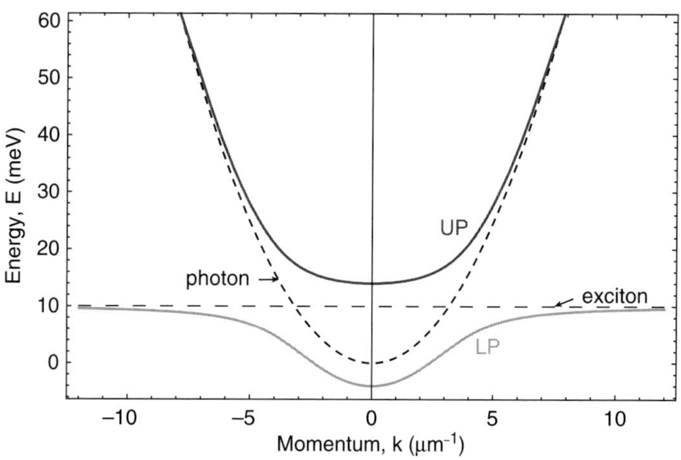

FIG. 16.2. The dispersion of upper and lower polariton modes (UP and LP solid lines) plotted for negative detuning $\delta = -10$ meV and $V_R = 15$ meV. The bare exciton and cavity modes are shown by dashed lines.

for the high critical temperature of exciton dissociation, nitride semiconductors represent an excellent research base for exciton-polariton studies at room temperature (16).

16.2 Polariton-based terahertz emitters

In the previous section we introduced the phenomenon of strong coupling between a microcavity photon and exciton in a solid, and discussed its possible applications. Recently, the different polariton systems were proposed as an effective radiation emitters in the terahertz (THz) range of frequency. Here we overview currently proposed schemes and compare them with existing THz emitters. The basic idea is that the possibility of the polariton BEC opens a way for drastic enhancement of the rate of the THz emission due to the effects of the final-state bosonic stimulation.

16.2.1 *Upper-to-lower polariton transition*

The distance between upper and lower polariton lies in the range of milielectronvolts, and can be changed with the material and quality of the resonator. If the scattering from the upper to lower polariton branch was accompanied by emission of THz photons, polariton lasers would emit THz radiation, and this emission would be stimulated by the population of the lowest-energy polariton state. However, at first glance this process is forbidden: an optical dipole operator cannot directly couple the polariton states formed by the same exciton state. Once it is overcome, the polariton system allows for generation of THz radiation with quantum efficiency, which can be tuned to a great extent by means of the optical pumping.

The transition between upper and lower polariton modes is allowed if one of the polariton states of interest is mixed with an exciton state of a different parity; for instance, the e1hh2 exciton state formed by an electron at the lowest energy level in a quantum well (QW) and a heavy hole at the second energy level in the QW. This state typically lies a few meV above the exciton ground state e1hh1. By a proper choice of the QW width and exciton–photon detuning in the microcavity, it can be brought into resonance with the lowest-energy upper polariton state. Being resonant, the two states can be easily hybridized by any weak perturbation, such as a built-in or external electric field. Therefore, the optical transition between such a hybridized state and the lowest e1hh1 exciton polariton state becomes allowed.

The geometry of the system considered in the original proposal is shown in Fig. 16.3 (17). It represents a QW with excitons embedded in a planar microcavity in the strong coupling regime. Additionally, this system can be placed inside a THz cavity, which enhances the rate of the emission of the THz photons. Together with the waveguiding effect of microcavity structure, adding the lateral THz cavity aims for an effective three-dimensional confinement of the THz mode, leading to enhancement of the spontaneous emission rate via the Purcell effect

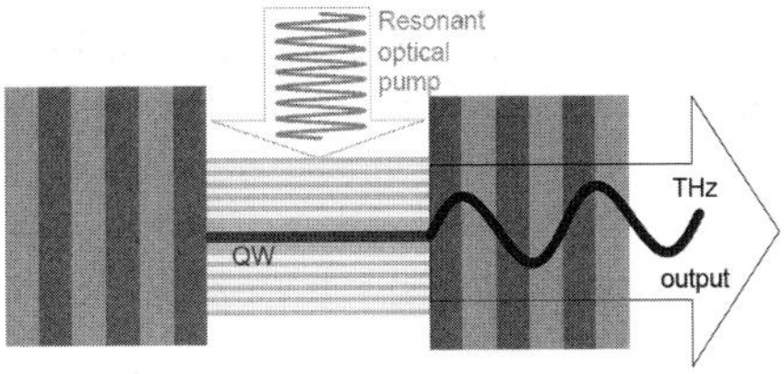

FIG. 16.3. Scheme of the polariton terahertz emitter (not shown to scale). A planar microcavity is embedded in the lateral terahertz cavity.

(18). The Hamiltonian of the system written in the basis of the bright exciton, dark exciton, and cavity photon states is written as

$$\widehat{H} = \begin{pmatrix} E_{1X} & \delta/2 & \hbar\Omega_R/2 \\ \delta/2 & E_{2X} & 0 \\ \hbar\Omega_R/2 & 0 & E_C \end{pmatrix} \tag{16.20}$$

Here, E_{1X}, E_{2X}, and E_C are the energies of bright and dark excitons and the cavity photon mode respectively. The off-diagonal element $\hbar\Omega_R$ denotes Rabi splitting of bright exciton and cavity photon coupling, and δ is a parameter describing the mixture between bright and dark exciton states due to the electric field ε normal to the QW plane, $\delta \approx e\varepsilon \int_{-\infty}^{+\infty} u_{hh1}(z) z u_{hh2}(z) dz$, with $u_{hh1}(z)$ and $u_{hh2}(z)$ being the confined wavefunctions of the ground and first excited states of the heavy hole in a QW. One can easily estimate that in order to have δ_ϵ larger than the upper polariton linewidth, which is typically of the order of 0.1 meV, it is enough to apply an electric (piezoelectric) field of 3 kV/cm in a quantum well with a width of 10 nm.

The diagonalization of the Hamiltonian (16.20) permits the finding of new eigenenergies and eigenstates of the system. The scheme in Fig. 16.4 shows the spectrum of the system with sketched processes which occur in the system. In particular, for the case of $|E_{1X} - E_{2X} - \hbar\Omega_R/2| \ll \hbar\Omega_R$ and zero detuning between bright exciton and photon modes ($E_{1X} = E_C = E_1$), the eigenstates can be written as

$$|L\rangle \approx \frac{1}{\sqrt{2}} \left(|C\rangle - |1X\rangle \right), \tag{16.21}$$

$$|U_+\rangle \approx \frac{1}{\sqrt{1+b_-^2}} \left[\frac{1}{\sqrt{2}} \left(|C\rangle + |1X\rangle \right) + b_- |2X\rangle \right] \tag{16.22}$$

$$|U_-\rangle \approx \frac{1}{\sqrt{1+b_+^2}} \left[\frac{1}{\sqrt{2}} \left(|C\rangle + |1X\rangle \right) + b_+ |2X\rangle \right] \tag{16.23}$$

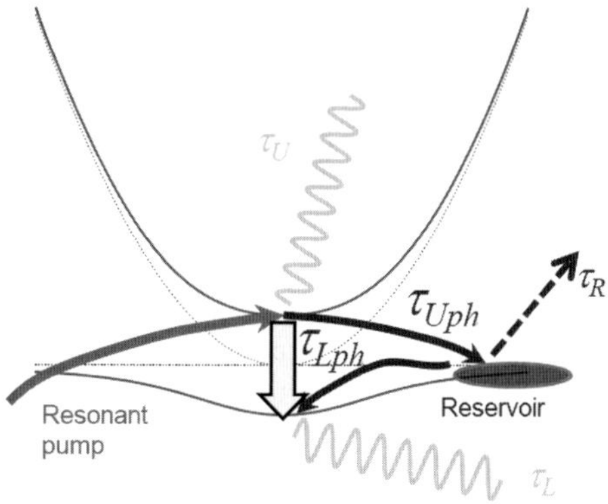

FIG. 16.4. Scheme illustrating possible transitions in the system. The vertical axis is energy, and the horizontal axis is the in-plane wavevector.

with

$$b_{\pm} = \delta^{-1} \left[E_{2X} - \frac{\hbar\Omega_R}{2} - E_1 \pm \sqrt{\left(E_{2X} - \frac{\hbar\Omega_R}{2} - E_1 \right)^2 + \delta^2} \right] \tag{16.24}$$

Therefore, together with the lower polariton mode L, two upper branches U_+ and U_- are formed due to the mixture between the bright upper polariton $|U\rangle = (|1X\rangle - |C\rangle)/2$ and the dark exciton $|2X\rangle$. One should note that if $\delta \neq 0$, both upper polariton eigenstates $|U_-\rangle$ and $|U_+\rangle$ contain fractions of the bright exciton and photon and of the dark exciton. This allows the radiative transition between the states $|U_{\pm}\rangle$ and the lower polariton state $|L\rangle$ accompanied by emission of the THz photon and their direct optical excitation. The rate of the spontaneous emission of THz radiation can be estimated from the Planck formula:

$$W_{\pm} \approx \frac{\omega^3 |d_{\pm}|^2 n}{3\pi\epsilon_0 \hbar c^3} F_P = W_0 F_P \tag{16.25}$$

where n is the refractive index of the cavity, ω denotes the frequency of radiation, and c is the speed of light. The dipole matrix element of the optical transition with emission of a THz photon in the plane of the cavity is written as

$$d_{\pm} = e\langle U_{\pm}|z|L\rangle \approx \frac{16 b_{\pm} e L_z}{9\pi^2 \sqrt{2\left(1 + b_{\pm}^2\right)}} \tag{16.26}$$

where the last equality holds for a QW of width L_z with infinite barriers. The coefficient F_P is the Purcell factor, which describes the enhancement of the rate of the spontaneous emission of THz photons due to the presence of the THz cavity. The principal effect of the cavity is to increase the electric-field operator by a factor of $\sqrt{Q}$ within the frequency band $\Delta\omega \approx \omega_0/Q$ around the cavity resonance frequency ω_0, where Q is the quality factor of the cavity. In the case of a narrow spectral width of the electronic oscillator coupled to the cavity, this results in the Purcell formula (18):

$$F_P = \frac{3}{4\pi^2}\frac{\lambda_c^3}{n^3}\frac{Q}{V_{eff}} = \xi Q \tag{16.27}$$

where Q is a quality factor of the cavity, V_{eff} is an effective cavity volume, and λ_c is the working wavelength. The equation can be written in a simplified way using a dimensionless geometric factor ξ ($\xi \approx 1$ for $V \approx (\lambda_c/2)^3$). In our case, the electronic resonance is broadened due to short lifetimes of the initial ($U_\pm$) and final (L) states. Applying the Fermi golden rule to the cavity tuned in resonance with a transition between homogeneously broadened levels characterized by lifetimes τ_i and τ_f, one obtains the following generalization of the Purcell formula:

$$F_P = \xi\left[Q^{-1} + (\omega_0\tau_i)^{-1} + (\omega_0\tau_f)^{-1}\right]^{-1} \tag{16.28}$$

Estimates show that for lifetimes of the electronic states of the order of 10 ps, F_P cannot be made much larger than approximately 50, however high Q might be.

Finally, above, we described the system of coupled-bright-exciton, dark-exciton, and cavity-photon modes. The simple model allows one to estimate the rate of spontaneous THz photon emission due to upper polariton to lower polariton transition. However, to describe the dynamics of the system one needs to account properly for interaction between polaritons, acoustic phonon relaxation mechanisms, and coherent polariton–THz photon interaction.

Dynamics of the system. The model Hamiltonian (16.20) can be rewritten in the second quantization formalism, and generalized for accounting for interactions and decoherence in the system (19). Written in terms of the creation and annihilation operators for upper polaritons ($a_U^\dagger, a_U$), lower polaritons ($a_L^\dagger, a_L$), THz photons ($c^\dagger, c$), reservoir states ($a_{R\mathbf{k}}^\dagger, a_{R\mathbf{k}}$), and acoustic phonons ($b_{\mathbf{k}}^\dagger, b_{\mathbf{k}}$), it recasts as a sum of four terms:

$$H = H_0 + H_{pol-pol} + H_T + H_R \tag{16.29}$$

The first term

$$H_0 = \epsilon_L a_L^\dagger a_L + \epsilon_U a_U^\dagger a_U + \epsilon_T c^\dagger c + \sum_k \epsilon_{\mathbf{k}} a_{R\mathbf{k}}^\dagger a_{R\mathbf{k}} \tag{16.30}$$

corresponds to the energy of the uncoupled upper polariton ϵ_U and lower polariton ϵ_L states, the THz mode with energy ϵ_T, and the polariton reservoir with dispersion relation given by $\epsilon_{\mathbf{k}}$. Here, hybridized upper polariton modes are assumed to be degenerate, $|U_\pm\rangle \approx |U\rangle$.

The second term

$$H_{pol-pol} = U_{LL}a_L^\dagger a_L^\dagger a_L a_L + U_{UU}a_U^\dagger a_U^\dagger a_U a_U + \\ + 2U_{UL}a_U^\dagger a_U a_L^\dagger a_L + \sum_k \left(U_{LR}a_L^\dagger a_L + U_{UR}a_U^\dagger a_U\right) a_{R\mathbf{k}}^\dagger a_{R\mathbf{k}} \quad (16.31)$$

describes polariton–polariton interaction. The interaction constants can be estimated as $U_{ij} = X_i^2 X_j^2 U$, where U denotes the bare interaction between excitons given by eq. (16.7), and X_i^2 denotes excitonic fraction in each polarition state, using the Hopfield coefficients written in eq. (16.16). The reservoir is supposed to be purely excitonic, $X_R = 1$.

It is well known that strong polariton–polariton interactions in microcavities yield observations of pronounced non-linear effects for the intensities of the pump orders of magnitude smaller than in other non-linear optical systems. Among them are polariton superfluidity (14), bistability and multistability (20, 21), soliton formation (22), and others. One can expect that polariton–polariton interactions will also strongly affect the process of THz emission.

The third term

$$H_T = V_T \left(a_U^\dagger a_L c + a_U a_L^\dagger c^\dagger\right) \quad (16.32)$$

describes radiative THz transion between the upper and lower polariton states. The matrix element of the THz emission can be estimated using a standard formula for the coupling constant of the dipole transition with confined electromagnetic mode, $V_T = \omega^2 d\sqrt{\hbar n/2\pi^3\epsilon_0 c^3}$, where d is the matrix element of the radiative transition, and n is the refractive index of the terahertz cavity (23).

Interaction of the upper and lower polaritons with an incoherent reservoir is described by the fourth term:

$$H_R = H_R^+ + H_R^- = D_1 \sum_{\mathbf{k}} \left(a_U a_{R\mathbf{k}}^\dagger b_{\mathbf{k}}^\dagger + a_U^\dagger a_{R\mathbf{k}} b_{\mathbf{k}}\right) \\ + D_2 \sum_{\mathbf{k}} \left(a_L a_{R\mathbf{k}}^\dagger b_{\mathbf{k}} + a_L^\dagger a_{R\mathbf{k}} b_{\mathbf{k}}^\dagger\right) \quad (16.33)$$

where $b_{\mathbf{k}}^\dagger$ and $b_{\mathbf{k}}$ are creation and annihilation operators of acoustic phonons with wave vector $\mathbf{k}$, and D_1, D_2 denote the polariton-phonon interaction constants.

The sketch in Fig. 16.5 shows the addition of processes which were not accounted fore in Fig. 16.4. Here, an incoherent polariton reservoir coupled with upper and lower polaritons and polariton–polariton interaction appear.

In general, the dynamics of a quantum system which has both coherent and incoherent interaction parts can be described using the Lindblad master equation for the density matrix (24, 25). The dynamics of the coherent part (H_0, $H_{pol-pol}$, H_T) can be written in Liouville–von Neumann form, while phonon-assisted interaction with reservoir (H_R) is described by the Lindblad term. The equation for the density matrix can be written as

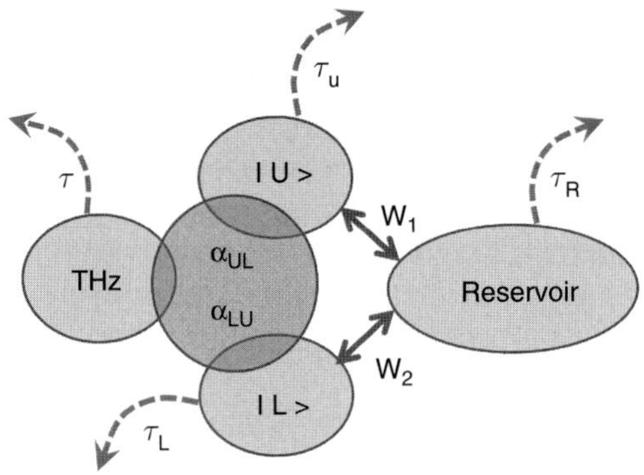

FIG. 16.5. The scheme of transitions in THz emitting cavity. The upper polariton is mixed with dark exciton state due to the application of the gate voltage V_g. The radiative transition between the upper hybrid state $|U\rangle$ and lower polariton state $|L\rangle$ thus becomes possible. Upper hybrid and lower polariton states are also coupled with an incoherent reservoir of the polaritons via phonon-assisted processes.

$$\begin{aligned}\frac{\partial \rho}{\partial t} &= \frac{i}{\hbar}\left[\rho; H_0 + H_{pol-pol} + H_T\right] + \frac{\delta_{\Delta E}}{\hbar}\Big\{2(H_R^+ \rho H_R^- + H_R^- \rho H_R^+) \\ &\quad - (H_R^+ H_R^- + H_R^- H_R^+)\rho - \rho(H_R^+ H_R^- + H_R^- H_R^+)\Big\} + \frac{1}{2\tau_L}\widehat{L}_{a_L} \\ &\quad + \frac{1}{2\tau_D}\widehat{L}_{a_D} + \frac{1}{2\tau_R}\widehat{L}_{a_R} + \frac{1}{2\tau}\widehat{L}_c + \frac{P}{2}\widehat{L}_{a_U^\dagger} + \frac{I}{2}\widehat{L}_{c^\dagger}\end{aligned} \tag{16.34}$$

where $\widehat{L}_A$ is the Lindblad superoperator, defined as $\widehat{L}_A = 2A\rho A^\dagger - A^\dagger A\rho - \rho A^\dagger A$ for each operator of type A. τ_L, τ_U, τ_R and τ are lifetimes of lower polaritons, upper polaritons, polaritons in the reservoir, and THz photons. P and I are the pumping intensities of the upper polariton state and the terahertz mode. The delta function $\delta_{\Delta E}$ denotes the conservation of energy in the process of phonon-scattering. The first term in the RHS accounts for the coherent processes in the system, and the second, third, and fourth terms placed in curly brackets correspond to the phonon-assisted coupling with the incoherent reservoir of the polaritons. The last six terms account for the decay of quasiparticles and pump.

The equations for the populations of polariton states and terahertz photons can be obtained by writing the Lindblad equation for particle-number operators:

$$\partial_t n_i = \mathrm{Tr}\left(\widehat{n_i}\frac{\partial \rho}{\partial t}\right) \tag{16.35}$$

To find a solution of eq. (16.35) one can use the mean-field approximation and derive the closed system of the dynamic equations for the occupancies $n_L = \langle a_L^\dagger a_L \rangle$, $n_U = \langle a_U^\dagger a_U \rangle$, $n_{R\mathbf{k}} = \langle a_{R\mathbf{k}}^\dagger a_{R\mathbf{k}} \rangle$, and $n = \langle c^\dagger c \rangle$, connected by the correlators $\alpha_{LU} = \langle a_L^\dagger a_U c^\dagger \rangle$, $\alpha_{UL} = \langle a_L a_U^\dagger c \rangle = \alpha_{LU}^*$, which reads:

$$\begin{aligned} \partial_t n_L = & -2\frac{V_T}{\hbar}\mathrm{Im}\,(\alpha_{UL}) - \frac{n_L}{\tau_L} \\ & + W_2 \sum_{\mathbf{k}} \left\{ (n_L + 1)\, n_{R\mathbf{k}} \left(n_{\mathbf{k}}^{ph} + 1 \right) - n_L \left(n_{R\mathbf{k}} + 1 \right) n_{\mathbf{k}}^{ph} \right\} \end{aligned} \tag{16.36}$$

$$\begin{aligned} \partial_t n_U = & \, 2\frac{V_T}{\hbar}\mathrm{Im}\,(\alpha_{UL}) - \frac{n_U}{\tau_U} + P \\ & + W_1 \sum_{\mathbf{k}} \left\{ (n_U + 1)\, n_{R\mathbf{k}} n_{\mathbf{k}}^{ph} - n_U \left(n_{R\mathbf{k}} + 1 \right) \left(n_{\mathbf{k}}^{ph} + 1 \right) \right\} \end{aligned} \tag{16.37}$$

$$\begin{aligned} \partial_t n_{R\mathbf{q}} = & - n_{R\mathbf{q}}/\tau_R + W_1 \sum_{\mathbf{k}} \left\{ n_U \left(n_{R\mathbf{k}} + 1 \right) \left(n_{\mathbf{k}}^{ph} + 1 \right) - (n_U + 1)\, n_{R\mathbf{k}} n_{\mathbf{k}}^{ph} \right\} \\ & + W_2 \sum_{\mathbf{k}} \left\{ n_L \left(n_{R\mathbf{k}} + 1 \right) n_{\mathbf{k}}^{ph} - (n_L + 1)\, n_{R\mathbf{k}} \left(n_{\mathbf{k}}^{ph} + 1 \right) \right\} \end{aligned} \tag{16.38}$$

$$\partial_t n = -2\frac{V_T}{\hbar}\mathrm{Im}\,(a_{UL}) - \frac{n}{\tau} + I \tag{16.39}$$

$$\begin{aligned} \partial_t \alpha_{UL} = & \frac{i}{\hbar} \left(\tilde{\epsilon}_U - \tilde{\epsilon}_L - \epsilon_T \right) \alpha_{UL} - \frac{a_{UL}}{\tau_{corr}} + \\ & + i\frac{V_T}{\hbar} \left\{ (n_U + 1)\, n_L n - n_U \left(n_L + 1 \right) (n + 1) \right\} + \\ & + \left\{ W_1 \sum_{\mathbf{k}} \left(-n_{R\mathbf{k}} - n_{\mathbf{k}}^{ph} - 1 \right) + W_2 \sum_{\mathbf{k}} \left(n_{R\mathbf{k}} - n_{\mathbf{k}}^{ph} \right) \right\} a_{UL}. \end{aligned} \tag{16.40}$$

In the above expressions, $\tau_{corr}^{-1} = \tau_L^{-1} + \tau_U^{-1} + \tau_R^{-1} + \tau^{-1}$, $V_T \approx 1\ \mu eV$ is a coupling constant between polaritons and terahertz photons, and $W_{1,2} \approx 2\ \mathrm{ps}^{-1}$ are transition rates between the reservoir and upper/lower polariton states determined by polariton–phonon interaction constants ($W_{1,2} \sim |D_{1,2}|^2$). Note that the characteristic time of terahertz photon emission is about three orders of magnitude smaller than the characteristic time of the scattering with acoustic phonons. However, THz emission is dramatically enhanced by bosonic stimulation, and becomes the dominant mechanism for a sufficiently strong pump.

$n_{\mathbf{k}}^{ph}$ gives the occupancies of the phonon mode determined by Bose distribution function. The reservoir can be considered as N identical states, for simplicity.

Quasiclassical approximation. First, let us discuss the case where coherent interaction is not important. It can be switched off by equating $d\alpha_{UL}/dt = 0$, and the resultant system corresponds to the following set of Boltzmann equations (17):

$$\begin{aligned}
\partial_t n_U &= P - \left(\tau_U^{-1} + \tau_{UR}^{-1}\right) n_U + W\left[n_L n(n_U+1) - n_U(n_L+1)(n+1)\right] \\
\partial_t n_L &= -\tau_L^{-1} n_L + \tau_{LR}^{-1} n_R + W\left[n_U(n_L+1)(n+1) - n_L n(n_U+1)\right] \\
\partial_t n &= -\tau^{-1} n + W\left[n_U(n_L+1)(n+1) - n_L n(n_U+1)\right] \\
\partial_t n_R &= -\tau_R^{-1} n_R - \tau_{LR}^{-1} n_R(n_L+1) + \tau_{UR}^{-1} n_U
\end{aligned} \tag{16.41}$$

where we introduced the rates of acoustic phonon-assisted transitions $\tau_{UR,LR}$ between the upper/lower polariton modes and the reservoir, which are inversely proportional to the $W_{1,2}$ matrix elements. Here, W is a rate of THz emission given by eq. (16.25) for a three coupled oscillator Hamiltonian, and is similar to the $V_T/\hbar$ matrix element. Thus, for resonantly optically excited hybridized upper states $|U\rangle$, polaritons relax to the lower polariton states, either directly, emitting THz photons, or via a cascade of $k \neq 0$ states of the lower polariton mode (acoustic phonon-assisted relaxation).

In the stationary regime the occupation number of the THz mode can be found from the solution of the above set of equations, putting $\partial_t n_U = \partial_t n_L = \partial_t n = \partial_t n_R = 0$.

Figure 16.6 shows the dependence of the quantum efficiency parameter $\beta = n/P\tau$ on the pumping rate P. The parameters used in the calculation are $\tau_U = \tau_L = 20$ ps, $\tau_R = 100$ ps, $\tau_{UR} = \tau_{LR} = 10$ ps, $\tau/Q = 10$ ps, and $W/F_p = W_0 = 10^{-9}$ ps^{-1}. One can see that for the realistic choice of parameters corresponding to the existing polariton lasers, the quantum efficiency achieves $\beta = 1 - 1.5\%$. Note that the mechanism of THz emission considered here is qualitatively different from the one used in quantum cascade lasers embedded in THz cavities. Effectively, here the THz emission is stimulated by the population of the lowest-energy polariton state, which provides quite high quantum efficiency compared with the best commercially available quantum cascade lasers. Semiconductor microcavities in the regime of polariton lasing may be used as efficient sources of THz radiation, having a quantum efficiency exceeding 1%, according to estimates.

Full quantum dynamics. Second, we describe the situation where all interactions and coherence in the system should be taken into account. In (19) it was shown that for the interactions between polaritons present in the system, new qualitative effects appear.

An important consequence of polariton–polariton interaction is emergence blueshift, which modifies energies of the upper and lower polariton states. The renormalized energies read

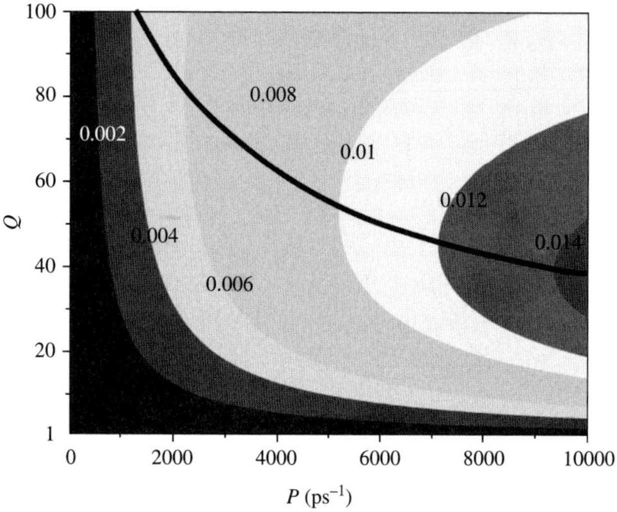

FIG. 16.6. Quantum efficiency of the terahertz emitter as a function of the pump intensity P and a quality factor of the terahertz cavity Q. The black line shows the dependence of the optimum value of Q on the pump intensity.

$$\tilde{\epsilon}_U = \epsilon_U + 2\left(U_{UU}n_U + U_{UL}n_L + U_{UR}\sum_{\mathbf{k}} n_{R\mathbf{k}}\right) \quad (16.42)$$

$$\tilde{\epsilon}_L = \epsilon_L + 2\left(U_{LL}n_L + U_{UL}n_U + U_{LR}\sum_{\mathbf{k}} n_{R\mathbf{k}}\right) \quad (16.43)$$

Due to the difference of the Hopfield coefficients for the upper and lower polariton states, the difference $\tilde{\epsilon}_U - \tilde{\epsilon}_L$ depends on the polariton concentrations, and thus is determined by the intensity of the pump P. This dependence can have important consequences, allowing for the onset of bistability in the system.

The calculation of dynamics of the system corresponding to eqs. (16.36)–(16.40) was performed in (19). Authors considered a planar GaAs microcavity in a strong coupling regime, with Rabi splitting Ω_R between upper and lower polariton modes equal to 16 meV (≈ 4 THz), which was embedded into THz cavity with an eigenfrequency slightly different from Ω_R, and having a quality factor $Q = 100$. The initial state was assumed to be unpopulated. When the constant non-resonant pump of the upper polariton state is switched on, the number of THz photons n starts to increase until it reaches some equilibrium level defined by the radiative decay of polaritons and escape of THz radiation from the cavity, as shown in Fig. 16.7.

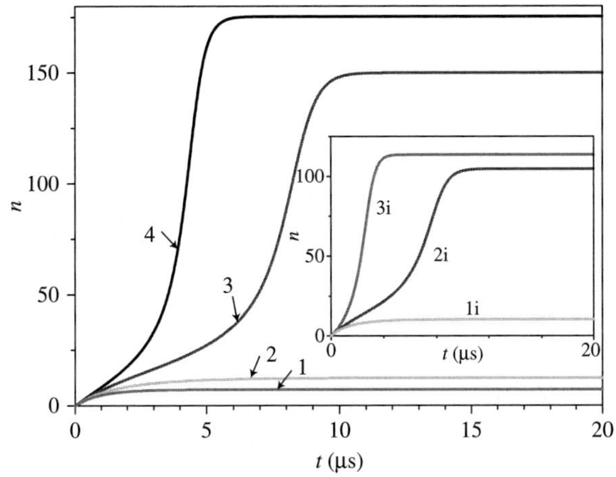

FIG. 16.7. Time-evolution of the terahertz photons number at zero temperature for different pumps: 4500 ps^{-1} (curve 1), 5000 ps^{-1} (curve 2), 5500 ps^{-1} (curve 3) and 6000 ps^{-1} (curve 4). The reservoir consists of $N = 3 \cdot 10^5$ identical particles. The inset shows evolution of the THz photons number for the constant pump $P = 6 \cdot 10^3$ ps^{-1} for different temperatures: 1 K (curve 1i), 10 K (curve 2i) and 20 K (curve 3i).

The equilibrium value of the THz population n as a function of pumping P demonstrates threshold-like behavior. For sufficiently high temperatures, the THz photon occupation number depends weakly on the pump in the subthreshold region. When pumping reaches the certain threshold value, polariton condensate is formed in the lower polariton state, and radiative THz transition is amplified by bosonic stimulation. Thus, the occupancy of the THz mode increases superlinearly together with the occupancy of lower the polariton state n_L (Fig. 16.7, curve 3). This behavior is qualitatively the same as in the approach operating with semiclassical Boltzmann equations. However, the decrease of temperature leads to the onset of the bistability and hysteresis in the dependence $n(P)$. The bistable jump occurs when the intensity of the pump tunes $\tilde{\epsilon}_U - \tilde{\epsilon}_L$ into the resonance with the cavity mode ϵ_T. The parameters of the hysteresis loop strongly depend on the temperature (Fig. 16.8). It is very pronounced and broad for low temperatures, narrows with an increase in the temperature, and disappears completely at $T \approx 20$ K.

The coherent nature of the interaction between excitons and THz photons makes possible the periodic exchange of energy between polariton and photon modes and oscillatory dependence of the THz signal in time. Figure 16.8 shows the temporal evolution of the occupancy of the THz mode after excitation of the upper polariton state by a short pulse having a duration of about 2 ps. One can see that the occupancy of the THz mode reveals a sequence of the short

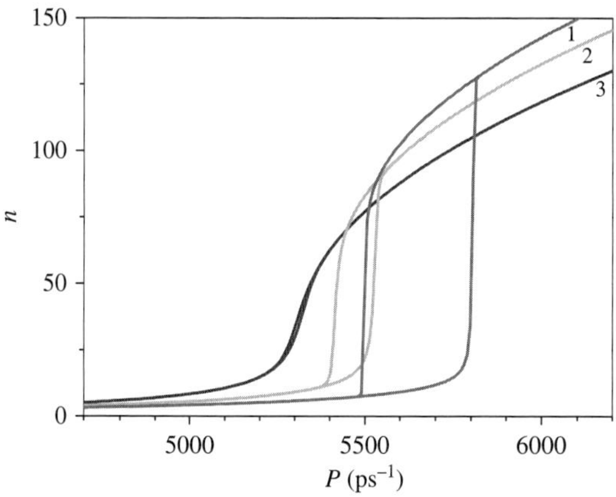

FIG. 16.8. Dependence of occupancy of the THz mode on pump in an equilibrium state for different temperatures: 1 K (curve 1), 10 K (curve 2), and 20 K (curve 3).

pulses having a duration of dozens of ps with an amplitude decaying in time due to escape of THz photons from a cavity, and radiative decay of polaritons. The period of the oscillations is sensitive to the number of the injected polaritons, N_0, and decreases with increasing N_0. If the lifetime of the polaritons is less than the period of the oscillations, single-pulse behaviour can be observed, as shown in the inset of Fig. 16.9. Appropriate choice of the parameters can lead ultimately to a generation of THz wavelets composed of one or several THz cycles, which makes a polariton-THz system suitable for application in short-pulse THz spectroscopy.

If the system of coupled THz photons and cavity polaritons is in the state corresponding to the lower branch of the S-shaped curve in the bistability region, illumination of the system by a short THz pulse $I(t)$ can induce its switching to the upper branch, as demonstrated in Fig. 16.10. One sees that the response of the system is qualitatively different for different values of the pump P. If P lies outside the bistability region, the application of a THz pulse leads to a short increase in n, but subsequently the system relaxes to its original state (curves 1 and 3). However, when the system is in the bistability regime, the switching occurs. The latter effect has quantum origin, and cannot be described using the set of semiclassical Boltzmann equations (16.41).

16.2.2 *THz emission by 2p-exciton to 1s-polariton transition*

In the previous section we considered the transition between the upper polariton state hybridized with the dark exciton and lower polariton mode. The estimated efficiency of such a process is comparable to standard quantum cascade laser

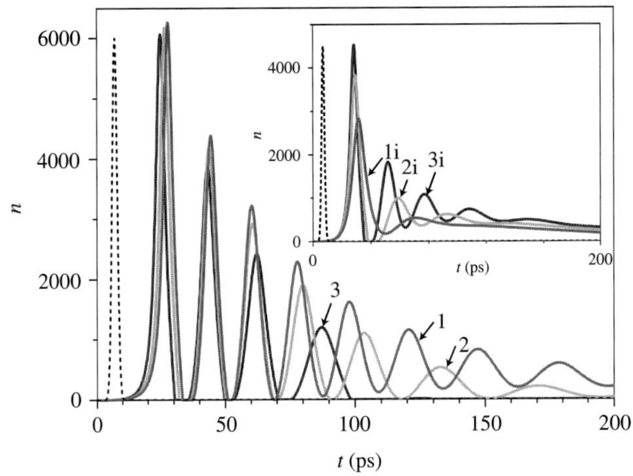

FIG. 16.9. The temporal dependence of the terahertz-mode occupancy. The background pump is switched off, and the polariton lifetime is taken as $\tau_L = \tau_U = 50$ ps. The temperatures are 1 K (curve 1), 10 K (curve 2), and 20 K (curve 3). The inset shows the calculation for $T = 1$ K and different lifetimes of polariton states: 15 ps (curve 1i), 20 ps (curve 2i) and 25 ps (curve 3i).

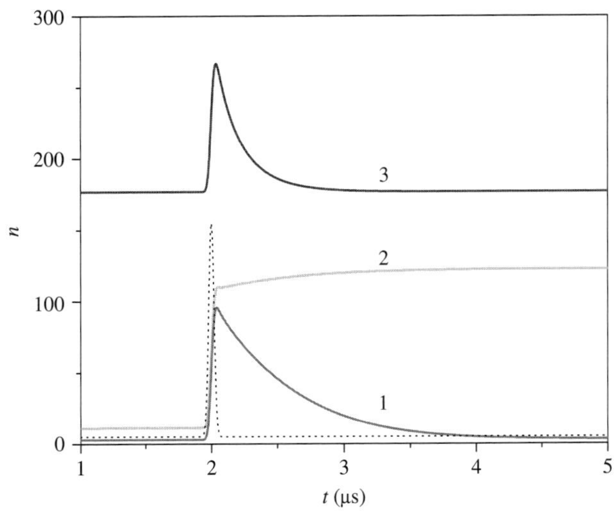

FIG. 16.10. Response of the system to a short single pulse (FWHM= 40 ps, black dotted curve) for different values of the background pump: 4600 ps^{-1} (curve 1), 5750 ps^{-1} (curve 2), 6500 ps^{-1} (curve 3). The switching occurs only when the background pump corresponds to the bistable region (compare with Fig. 16.7).

structures, but is still low due to the much smaller upper polariton lifetime compared to the THz photon emission rate. The situation can be improved if the initial state of THz emission is optically dark and therefore has long radiative lifetime.

The dark $2p$ exciton state is split from the bright $1s$ state by several meV. Usually, it is not populated and is not considered in the strong-coupling case. We remind that single-photon optical generation of the dark exciton is prohibited due to angular momentum conservation. It requires absorption of two photons, and usually such processes are suppressed in the microcavity. However, one can achieve sufficient population of $2p$ excitons using resonant two-photon pumping (see Fig. 16.11). Such excitons have a very long radiative lifetime, and can radiatively decay to the lower exciton-polariton mode formed by the $1s$ exciton and cavity photon. This transition is accompanied by emission of a THz photon. Such a process was proposed for the construction of an efficient THz emitter based on $2p$–$1s$ transition (26). Additionally, macroscopic occupation of the lowest-energy polariton state leads to stimulation of THz photon emission. Ideally, in the polariton lasing regime, the cavity would emit one THz photon for each optical photon emitted by the polariton laser. This allows a considerable increase in the quantum efficiency of the THz emitter.

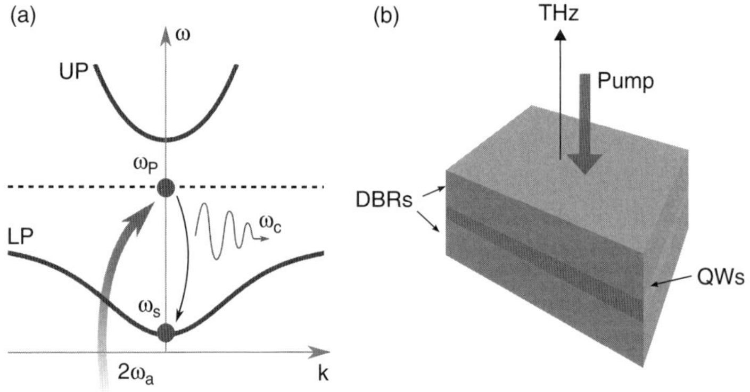

FIG. 16.11. (a) Energy spectrum of the system showing the polariton dispersion relation with the lower polariton (LP) and upper polariton (UP) branches, as well as the $2p$ exciton state with frequency ω_p. The two-photon pump process is shown schematically. The energy $\hbar\omega_a$ is a half of the $2p$ exciton state energy, $\hbar\omega_p$. (b) Sketch of the considered structure consisting of an active layer containing quantum wells (QWs) sandwiched between two distributed Bragg reflectors (DBRs) which form a semiconductor microcavity. The structure is pumped vertically, i.e. in the direction perpendicular to the microcavity plane, and the resultant THz emission from the cavity is in the same direction.

The design of the laser and the involved energy levels are illustrated schematically in Fig. (16.11). This design has two crucial advantages with respect to that previously considered: it allows for operation with an optically allowed THz transition, and it provides vertical emission of THz photons. Moreover, due to the dark nature of the initial state, no waveguides or supplemental THz cavities are needed, and the structure remains microscopic. This latter point is also a significant advantage with respect to the quantum cascade laser, which operates in the waveguide geometry.

The dynamics of the system can be studied using the quantum model of the laser based on the Lindblad master-equation. Written for the density matrix describing the $2p$ exciton state, the lower polariton mode, optical and THz photons, it accounts correctly for both statistics of the particles and their decay.

The model includes coherent processes of $2p$ excitons and $1s$ exciton-polariton interaction with an external electromagnetic field, and incoherent interaction with phonons in a semiconductor microcavity. Therefore, the total Hamiltonian can be divided into two parts:

$$H = H_c + H_d \tag{16.44}$$

where the part H_c includes all types of coherent processes in our system and part H_d describes the decoherence in the system due to exciton interactions with acoustic phonons, modelled as a classical reservoir (24, 25).

The equation for the density matrix is written as

$$\frac{d\varrho}{dt} = \frac{i}{\hbar}\left[\varrho, H_c\right] + \hat{L}\varrho \tag{16.45}$$

where $\hat{L}$ denotes the Lindblad superoperator. The incoherent part of the Hamiltonian can be divided into two parts, $H_d = H^+ + H^-$, where H^+ creates an excitation in the quantum system (and thus annihilates an excitation in the classical reservoir), and H^-, conversely, annihilates an excitation in the quantum system and creates an excitation in the classical reservoir.

The Lindblad terms can be then written as (25)

$$\begin{aligned}\hat{L}\varrho = \frac{\delta(\Delta E)}{\hbar}\{&\left(H^+\rho H^- + H^-\varrho H^+\right) - \\ &- \left(H^+H^- + H^-H^+\right)\varrho - \varrho\left(H^+H^- + H^-H^+\right)\}\end{aligned} \tag{16.46}$$

where $\delta(\Delta E)$ accounts for energy conservation, and in realistic calculations should be taken as an average inverse broadening of the states in our system, $\delta(\Delta E) \to \zeta^{-1}$. In the particular system which we consider, the coherent and incoherent parts of the Hamiltonian can be written as

$$H_c = \epsilon_p p^\dagger p + \epsilon_s s^\dagger s + \epsilon_a a^\dagger a + \epsilon_T c^\dagger c \tag{16.47}$$

$$H_d = H^+ + H^- \tag{16.48}$$

where

$$H^+ = gp^\dagger a^2 + Gp^\dagger sc \tag{16.49}$$
$$H^- = gpa^{\dagger 2} + Gps^\dagger c^\dagger \tag{16.50}$$

and ϵ_p, ϵ_s, ϵ_a and ϵ_T are the energies of $2p$ exciton, $1s$ exciton-polariton, pump photon ($2\epsilon_a = \epsilon_p$), and THz photon, $\epsilon_T = \epsilon_p - \epsilon_s$, respectively. Here, p and s denote annihilation operators of the $2p$ exciton and the lowest-energy $1s$ polariton states, a is the annihilation operator for laser photons exciting the $2p$ exciton state, and c is an annihilation operator for THz photons produced by the $2p \to 1s$ transition. Constants g and G define the strengths of the two-photon absorption and $1s \to 2p$ radiative transitions, respectively. They are defined by corresponding dipole matrix elements. The upper polariton state is assumed to lie higher in energy than the $2p$ exciton state, and thus can be excluded from consideration. The classical reservoir consists of the photons of the external laser-light and THz photons. Note that the coherent part of the Hamiltonian (16.47) describes only the free modes of the system, and thus is irrelevant to the kinetic equations.

The occupancy of the $2p$ exciton mode can be calculated and written in the form

$$\begin{aligned}\frac{dN_p}{dt} &= \mathrm{Tr}\left\{p^\dagger p \frac{d\varrho}{dt}\right\} = \frac{2}{\hbar\zeta}\mathrm{Re}\left\{\mathrm{Tr}\left(\varrho[H^-;[p^\dagger p;H^+]]\right)\right\} = \\ &= W_g\left[\frac{1}{2}\langle a^{\dagger 2}a^2\rangle - \langle p^\dagger p(2a^\dagger a+1)\rangle\right] + \\ &\quad + W_G\left[\langle s^\dagger s c^\dagger c\rangle - \langle p^\dagger p(s^\dagger s + c^\dagger c + 1)\rangle\right]\end{aligned} \tag{16.51}$$

where $\langle ...\rangle = \mathrm{Tr}\{...\varrho\}$ denotes averaging with the appropriate density matrix ϱ, $W_g = 4g^2/\hbar\zeta$, $W_G = 2G^2/\hbar\zeta$. The equations for the polariton mode occupancy $N_s = \langle s^\dagger s\rangle$, and the terahertz mode occupancy $N_c = \langle c^\dagger c\rangle$, are analogous to those for N_p. The occupancy of the pumping mode $N_a = \langle a^\dagger a\rangle$ is defined by the intensity of the external pump, and an independent dynamic equation for it is not necessary. The same holds true for the higher-order correlators involving pump operators, e.g., $\langle a^{\dagger 2}a^2\rangle$. It follows from eq. (16.51) that the dynamic equations for the occupancies of the modes contain quantum correlators of fourth order, such as $\langle s^\dagger s c^\dagger c\rangle$. For them, one can also write the dynamic equations analogous to eq. (16.51), which would contain correlators of sixth order. Proceeding further, one would obtain an infinite chain of coupled equations for the hierarchy of correlators.

In the easiest form the system of equations for correlators can be solved using mean-field approximation. Therefore, fourth-order correlators are truncated into the products of second-order ones. One can approximate $\langle p^\dagger p a^\dagger a\rangle \approx \langle p^\dagger p\rangle\langle a^\dagger a\rangle = N_p N_a$, $\langle s^\dagger s c^\dagger c\rangle \approx N_s N_c, \ldots$ etc.

A particular analysis was performed for the truncation of the correlator $\langle a^{\dagger 2}a^2\rangle$, containing four operators corresponding to the pumping mode (26). It can be represented as

$$\langle a^{\dagger 2}a^2\rangle = g^{(2)}(0)N_a^2 \tag{16.52}$$

where we use the definition of the second-order coherence $g^{(2)}(0)$ function.

Finally, the closed set of equations of motion describing the dynamics of the system is written as

$$\begin{aligned}\frac{dN_p}{dt} &= -\frac{N_p}{\tau_p} + W_g\left[\frac{g^{(2)}(0)}{2}N_a^2 - N_p(2N_a+1)\right] + \\ &\quad + W_G\left\{N_sN_c\left(N_p+1\right) - N_p\left(N_s+1\right)\left(N_c+1\right)\right\} \end{aligned} \tag{16.53}$$

$$\frac{dN_s}{dt} = -\frac{N_s}{\tau_s} - W_G\left\{N_sN_c\left(N_p+1\right) - N_p\left(N_s+1\right)\left(N_c+1\right)\right\} \tag{16.54}$$

$$\frac{dN_c}{dt} = -\frac{N_c}{\tau_c} - W_G\left\{N_sN_c\left(N_p+1\right) - N_p\left(N_s+1\right)\left(N_c+1\right)\right\} \tag{16.55}$$

where τ_p denotes the non-radiative lifetime of the $2p$ exciton state, and the lifetimes of the lower polaritons and THz photons are τ_s and τ_c, respectively.

Let us analyze in more detail the kinetic equation for the $2p$ state occupation N_p. The terms describing incoming and outcoming scattering rates for this state can be written as

$$(1+N_p)\left\langle {a^+}^2a^2\right\rangle - N_p\left\langle a^2{a^+}^2\right\rangle$$

As already mentioned in eq. (16.52), the incoming rate is proportional to $g^{(2)}(0)N_a^2$. The outgoing rate can be rewritten as $\propto \langle a^2(a^+)^2\rangle = g^{(2)}(0)N_a^2 + 4N_a + 2$. Hence, all terms containing $N_a^2N_p$ cancel each other, and eq. (16.53) holds. Note that for processes of non-degenerate two-photon absorption or emission neglected here, two-photon emission processes where photon frequencies are different can be included in τ_p, the decay rate of the $2p$ state. The scattering rates have the standard form $(1+N_p)N_{a_1}N_{a_2} - N_p(N_{a_1}+1)(N_{a_2}+1)$, independent of the photon field statistics. Therefore, one can assume that $N_a \gg 1$, and it is possible to neglect all other modes.

The key feature of eq. (16.53) is that the pumping term $W_g g^{(2)}(0)N_a^2$ contains the second-order coherence of the pump, and thus is strongly dependent on its statistics. For a laser pump corresponding to the coherent state this statistic is Poissonian, and $g^{(2)}(0) = 1$. On the other hand, in principle other values of $g^{(2)}(0)$ are possible. For example, for a thermal pump, $g^{(2)}(0) = 2$. Different values of the second-order coherence function lead to the different lasing thresholds in the system. In Fig. 16.12, the steady-state solution for N_p and N_s is plotted. The assumption for calculations is that the terahertz mode occupation N_c is zero for the cases of coherent and thermal pumps. For a qualitative understanding, we

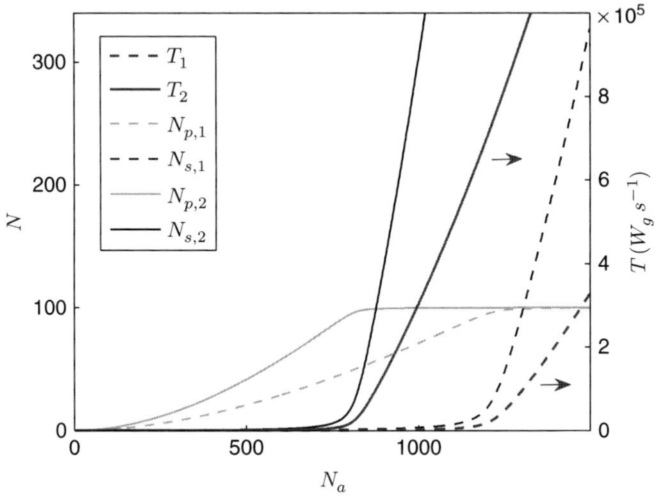

FIG. 16.12. Occupation of the $2p$ and $1s$ states in the steady-state (left axis) and terahertz generation rate (right axis), as a function of pumping-mode occupancy. The subscript 1 or 2 refers to the value of $g^{(2)}(0)$, assumed in the calculations. $W_g = 1, W_G = 10, 1/\tau_s = 1000, 1/\tau_p = 5000$, in units of W_g.

note that the threshold is reached if $N_s \sim 1$, or equivalently when $N_p \sim 1/\tau_s W_G$, since this signals the onset of Bose stimulation of the transition from the $2p$ state to the $1s$ lower polariton, accompanied by terahertz emission. The terahertz generation rate is given by $T = W_G N_p(N_s + 1) \approx 1/\tau_s N_s$, and therefore shows the same threshold behaviour with pumping intensity as N_s, as can also be seen in Fig. 16.12. We observe that the threshold is higher in the case of a coherent pump, which may be understood in terms of the enhanced losses from the $2p$ state due to stimulated two-photon emission.

Finally, the proposed mechanism of a $2p - 1s$ terahertz emitter is a promising tool for the creation of a vertical-cavity surface-emitting terahertz lasers (Fig. 16.11(b)). For practical realization of compact terahertz lasers operating at room temperature, microcavities based on wide band-gap semiconductors (GaN or ZnO) seem to be the most advantageous. The resonant two-photon pumping of $2p$ exciton states in such a cavity may be assured by a conventional vertical-cavity light-emitting diode (VCLED) emitting in red. This prospective structure would consist of a GaN or ZnO microcavity grown on the top of a GaAs-based VCLED structure. As emission of coherent terahertz light coexists with the polariton lasing in the optical frequency range in this scheme, one can use the visible blue or green light produced by the polariton laser as a marker for the terahertz light-beam, which may be important for applications in medicine, security control, and other areas.

16.2.3 *Dipolariton THz emission*

In the introductory section we discussed several types of exciton. The excitons of the first type, referred to as direct Wannier–Mott excitons, can be created optically in single quantum wells and be strongly coupled to a microcavity photon mode. The second type of exciton appears in coupled quantum wells, where for sufficiently high electric fields the pairing between an electron and a hole located in different QWs occurs (27). These are spatially indirect excitons. They can be cooled to lattice temperature due to their long lifetime, which is why they are frequently referred to as cold excitons. Radiative lifetimes of indirect excitons are usually much longer than those of direct excitons due to lower electron–hole overlap. Another important feature of indirect excitons is their large dipole moments in the normal to QW plane direction, resulting in strong exciton–exciton interactions.

Recently, it has been shown that exciton polaritons and spatially indirect excitons can be intermixed in the biased semiconductor microcavities with embedded coupled QWs (28, 29). In this structure, new quasiparticles, being linear superposition of cavity photon (C), direct exciton (DX) and indirect exciton (IX), appear. They form three exciton-polariton modes; namely, the upper polariton (UP), middle polariton (MP), and lower polariton (LP) modes. Therefore, hybrid dipolariton quasiparticles emerge—cavity exciton-polaritons characterized by large dipole moments.

The dynamics of a three-coupled oscillators system has a rich phenomenology. In particular, the strong coupling between levels leads to the so-called Rabi flopping process, where quasiparticle oscillate between two coupled states with characteristic Rabi frequency. For the tunnel coupling between direct and indirect excitons it results in coherent oscillation of dipole moment in the system, with the tunnelling frequency lying in the terahertz frequency range. This leads to classical radiation of the THz signal by the Hertz dipole array (30). Moreover, due to the strong coupling of direct exciton with a cavity mode the system exhibits superradiant behavior (31). Rabi oscillations and consequent THz signal can be sustained by the cavity for several tens of picoseconds. This strongly improves the quantum efficiency of the emitter.

The considered structure represents two QWs separated by a barrier which is sufficiently thin to allow for resonant electron tunneling (Fig. 16.13). The electron wavefunction is shared between two QWs in this case. The optical microcavity is tuned to the wider QW exciton resonance (the left QW in Fig. 16.13), while the right QW remains decoupled from the optical pump. With the pulsed pump one can create electron–hole pairs which form direct excitons. Furthermore, the strong light–matter coupling between direct exciton in the left QW and cavity photon can be achieved. Then, the application of an electric field in the growth direction induces mixing of direct and indirect exciton states, which leads to the appearance of dipolariton modes.

The system can be described effectively by three coupled oscillators Hamiltonian, which successfully describe the behavior of dipolaritons measured

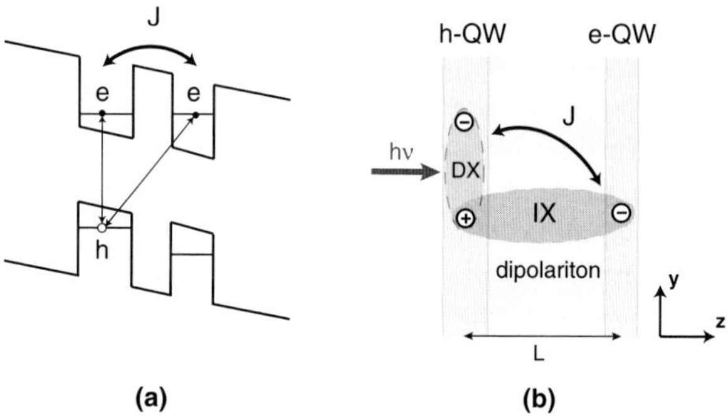

FIG. 16.13. (a) Sketch of the system which shows double quantum well (DQW) heterostructure with resonantly coupled electron levels and a hole in the left quantum well (QW). The right QW has a larger band gap than the left QW tuned by the presence of indium alloy or width of the well. (b) Electron–hole bilayer with a schematic picture of coupled spatially indirect and direct excitons which form dipolariton.

experimentally (29). It can be written in matrix form, together with the dipolariton wavefunction, as

$$H = \hbar \begin{pmatrix} \omega_{IX} & J/2 & 0 \\ J/2 & \omega_{DX} & \Omega/2 \\ 0 & \Omega/2 & \omega_C \end{pmatrix}, \ |\Psi\rangle = \begin{pmatrix} \Psi_{IX} \\ \Psi_{DX} \\ \Psi_C \end{pmatrix}, \tag{16.56}$$

where $\hbar\omega_C$, $\hbar\omega_{DX}$, and $\hbar\omega_{IX}$ denote cavity mode, direct exciton, and indirect exciton energies, respectively. The coupling constant between photon and direct exciton is $\hbar\Omega$ (Rabi energy), while the tunneling rate corresponding to DX–IX coupling is $\hbar J$. The wave-function $|\Psi\rangle$ of a dipolariton can be expanded on the basis of uncoupled states Ψ_C, Ψ_{DX} and Ψ_{IX}, being field operators for cavity photon, direct exciton, and indirect exciton, respectively. The evolution of the $|\Psi(t)\rangle$ state after illumination of the sample by a short optical pulse describes a time-dependent response of the system. To separate the rapidly oscillating part of dynamics, the rotating frame transformations are used: $\Psi_{IX}(t) = \psi_{IX}(t)e^{-i\omega_{IX}t}$, $\Psi_{DX}(t) = \psi_{DX}(t)e^{-i\omega_{DX}t}$, and $\Psi_C(t) = \psi_C(t)e^{-i\omega_C t}$ (8). In the following, all modes are assumed to be macroscopically occupied, and fields $\psi_{C,DX,IX}$ are treated as classical. This allows description of the dynamics of the system by a closed set of time-dependent Schrödinger equations:

$$\frac{\partial \psi_{IX}(t)}{\partial t} = -i\frac{J}{2}\psi_{DX}(t)e^{i\delta_J t} - \frac{\gamma_{IX}}{2}\psi_{IX}(t) \tag{16.57}$$

$$\frac{\partial \psi_{DX}(t)}{\partial t} = -i\frac{J}{2}\psi_{IX}(t)e^{-i\delta_J t} - i\frac{\Omega}{2}\psi_C(t)e^{-i\delta_\Omega t} - i\frac{\gamma_{DX}}{2}\psi_{DX}(t) \tag{16.58}$$

$$\frac{\partial \psi_C(t)}{\partial t} = -i\frac{\Omega}{2}\psi_{DX}(t)e^{i\delta_\Omega t} - iP_0(t)e^{-i\Delta t} - \frac{\gamma_C}{2}\psi_C(t) \tag{16.59}$$

where the photonic detuning $\delta_\Omega = \omega_C - \omega_{DX}$ for the cavity photon coupled to the direct exciton, and energy distance between spatially indirect and direct excitons $\delta_J = \omega_{IX} - \omega_{DX}$, are introduced. The frequency of the resonant optical pumping $P_0(t)$ is detuned from the photon mode by $\Delta = \omega_P - \omega_C$ (the second term in the right-hand side of eq. (16.57)). Additionally, the damping rates for each mode $\gamma_i = 2\pi/\tau_i$, $i = C,\ DX,\ IX$ were taken into account, the lifetimes of the modes being typically $\tau_C \approx 3$ ps, $\tau_{DX} \approx 1$ ns, and $\tau_{IX} \approx 100$ ns.

The numerical calculation of dipolariton system dynamics subjected to a picosecond pulsed optical pumping was performed by the authors of (30). The presence of mixing terms between different mode implies the oscillating behavior similar to Rabi flopping in the classical model of a two-level system imposed on a time-varying field. It is important to note that this system has several characteristic frequencies. They are governed by the exciton–photon coupling strength Ω and detuning δ_Ω, the IX-DX coupling strengths J, and detuning δ_J. Additionally, the pumping frequency ω_P governs the efficiency of the pump. Varying these characteristic frequencies one can control the frequency, magnitude, and damping rate of indirect exciton density oscillations.

While the coupling constants Ω and J are dependent on the geometry of the structure and can hardly be tuned for a given sample, the detunings between modes δ_J and δ_Ω are strongly sensitive to the applied electric field and the incidence angle of the cavity pump. Tuning of these parameters can bring the oscillating dipole system to different regimes. If the cavity mode is far-detuned from IX-DX anti-crossing, the light–exciton coupling is weak, which was referred to as the regime I. If the detuning δ_Ω is small, the strong intermixing of IX, DX, and C modes takes place, which corresponds to regime II.

The behavior of the system in regime I is shown in Fig. 16.14a for the detuning $\delta_\Omega = -10$ meV. The antiphase oscillations of IX and DX densities with decaying amplitudes can be observed. The inset in Fig. 16.14a shows a zoom of longstanding phase-locked oscillations which last for several tens of picoseconds. For the electric field F approximately equal to the resonant field of IX–DX modes anti-crossing F_0, the frequency of oscillations is given by $\nu \approx J/2\pi = 1.45$ THz. The time-dependence of the indirect exciton density oscillation can expressed as

$$n_{IX}(t) = n_{IX}^0 \cos^2(\omega t/2)e^{-t/\tau} \tag{16.60}$$

where $\omega = 2\pi\nu$ is the frequency of oscillations, and n_{IX}^0 denotes the magnitude of the oscillations, which decrease in time with the damping rate τ^{-1}. Tuning the electric field, the frequency of generated oscillations changes in the range

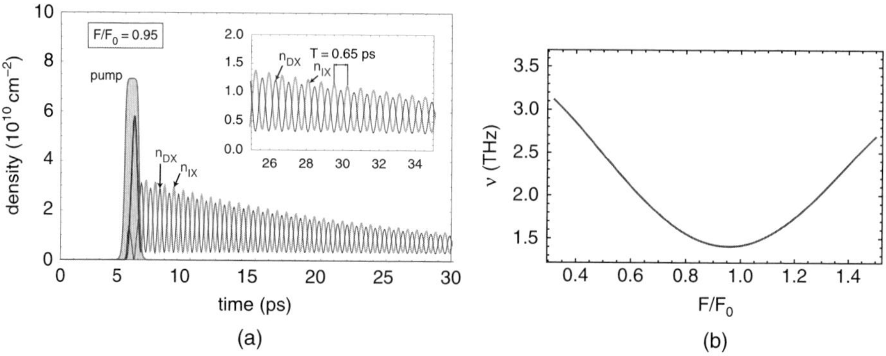

FIG. 16.14. (a) Dynamics of the dipolariton system subjected to pulsed optical pumping, which shows oscillations of indirect exciton (curve n_{IX}) and direct exciton (curve n_{DX}) density. The coupling constants are equal to $\hbar J = \hbar\Omega = 6$ meV, and cavity mode is detuned from direct exciton resonance by $\delta_\Omega = -10$ meV. The pumping frequency is chosen as $\hbar\Delta = 11.5$ meV, and we plot oscillations for the electric field $F = 0.95F_0$, where F_0 is the field corresponding to IX-DX resonance. The gray area schematically represents the optical pulse with duration $\Delta\tau = 1$ ps and pulse edge defined by 0.1 ps transient time (pump, scaled intensity). The inset shows the long-term term oscillations of $n_{IX} = |\psi_{IX}|^2$ and $n_{DX} = |\psi_{DX}|^2$ density, which are in antiphase. The period of oscillations is equal to $T = 0.65$ ps. (b) Frequency of indirect exciton density oscillations as a function of a dimensionless electric field calculated for the parameters defined above.

of several THz due to its IX–DX detuning dependence (Fig. 16.14(b)). Other important parameters of the system, such as the magnitude of oscillations and dimensionless oscillation quality factor ξ, can be tuned by pumping conditions as well as by the applied electric field. The parameter ξ can be defined as the ratio of the magnitude to the decrement of oscillations.

Taking $\delta_\Omega = 2$ meV, one can bring the system to regime *II*. In this regime the IX density oscillations are observed as well, while their quality factor is different. Strong interactions of the cavity mode with the IX–DX resonance result in a higher magnitude of n_{IX} oscillations than in regime *I*. The damping rate of these oscillations is also larger. Clearly, the regime of strong coupling between all modes is advantageous for the pulsed pumping regime, and allows for high-power generation. On the other hand, regime *I* is preferential for the longstanding signal generation providing a higher-quality factor ξ, with lower amplitude of oscillations.

It has been shown that due to the coupling between modes IX and DX, densities oscillate with a THz frequency. This infers that a dipolariton is an oscillating dipole, with a dipole moment in the z direction d_z changing periodically from value $d_z = d_0$, corresponding to an indirect exciton (IX) to $d_z = 0$, corresponding to a direct exciton (DX). One can define the total dipole moment of the system as $D_z = N_{IX} d_z$, where $N_{IX} = n_{IX} A$ is the number of indirect excitons within the area A illuminated by the pumping light. Since we have shown previously that after initial transient regime the density of indirect excitons $n_{IX}(t)$ is an harmonic function of time, the total dipole moment of the system can be found as

$$D_z(t) = d_0 n_{IX} A \cos^2(\omega t/2) e^{-t/\tau} \tag{16.61}$$

Here, n_{IX} is the maximum density of indirect excitons.

The mechanism of THz radiation by an oscillating density of indirect excitons is similar to the photo-Dember effect, where oscillations of electron–hole plasma on a surface of semiconductor generate a THz signal and other laser-driven terahertz emitters. However, the system of dipolaritons has some important advantages over previously studied systems; namely, (1) the better tunability of the system, allowing for fast modulation of the THz emission frequency, (2) improved spectral characteristics of the THz signal, which can be controlled by using applied voltage and pumping conditions, and (3) the possibility of achieving a high output power. For a typical laser-driven THz source, the output power lies in the range of nanowatts, with its upper boundary in the microwatt range. The power of a dipolariton emitter is calculated as follows.

The total intensity of the far-field radiation emitted by a classical Hertz dipole can be found from (32):

$$I = \frac{\ddot{D}_z^2}{6\pi\epsilon_0 c^3} \tag{16.62}$$

where D_z is the total dipole moment of the dipolariton array, ϵ_0 is the vacuum permittivity, and c is the speed of light. For the particular case of an array of harmonic dipole oscillations, the intensity is

$$I = N_{IX}^2 \frac{d_0^2 \omega^4}{3\pi\epsilon_0 c^3} \tag{16.63}$$

where $d_0 = eL$ is a dipole moment of indirect exciton, with L being the separation between centers of QWs. Similarly to the conventional case of an elementary dipole emitter, the polar pattern is given by $I_\theta \sim \sin^2\theta$, where θ is an angle between the direction of radiation and the axis of the structure (see Fig. 16.15).

It is important to note that the total emitted intensity is proportional to the square of the density of indirect excitons, and is non-linearly dependent on the pump intensity. This is a manifestation of the *superradiance* effect (31): due to the interference of coherent in-phase oscillations of elementary dipoles, the output power is enhanced. This effect is sensitive to the quality factor of the cavity: the longer that Rabi oscillations persist, the stronger the amplification effect. Superradiance is a specific feature of the dipolariton THz emitter, which makes it more efficient than any existing laser-to-THz convertor.

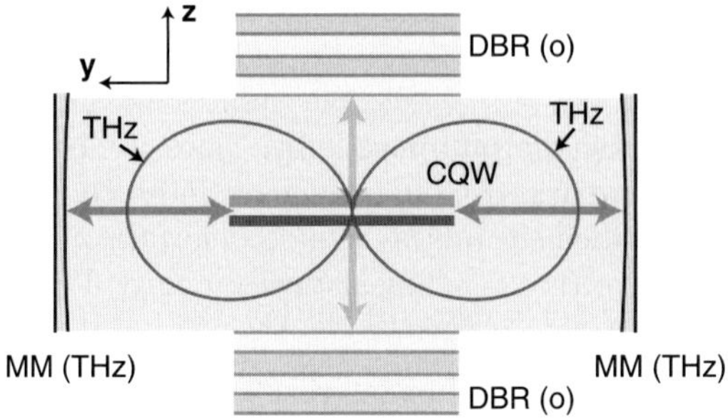

FIG. 16.15. Sketch of the geometry where the power of THz emission is increased due to the Purcell effect (top view). Here, DBR (o) represents optical cavity mirrors, while MM (THz) denotes a metallic mirror cavity tuned to reflect a signal of THz frequency. Curves with label THz show the peculiar polar pattern of a dipolariton THz emitter.

For $\hbar J = 6$ meV and the typical distance between QWs of $L = 12$ nm, the power of the THz emission of one elementary dipole formed by a dipolariton is $I_0 = 1.8 \times 10^{-17}$ W $= 18$ aW. The total power emitted is given by this quantity multiplied by the number of elementary dipoles N_{IX}. For the typical concentration of indirect excitons achievable experimentally of $n_{IX} = 10^{10}$ cm^{-2} and a 60-μm-diameter of the pumping spot, one obtains $I_{tot} \approx 1.4\ \mu$ W. By embedding a stack of double quantum wells, $n_{DQW} =$ four in each microcavity, and using several sets of cavities on one chip one can obtain the output power similar to one of quantum cascade lasers.

Furthermore, to improve the efficiency of THz radiation, the system can be placed in an external cavity tuned to the THz frequency. It can be created using metallic mirrors or an inductor capacitor circuit. A sketch of the system is shown in Fig. 16.15. The efficiency of the emission would be further increased due to the Purcell factor of the external cavity (8). For the chosen frequency, 1.45 THz, the experimentally measured Purcell factor of an inductor–capacitor cavity is $F_P = 17$. However, strictly speaking, the THz cavity is not necessary, and the described physics holds for the sample without it.

16.3 Conclusion

In this chapter we have presented an overview of recent proposals for radiation sources in the terahertz range of frequencies, based on microcavity polaritons. The Rabi-splitting of polariton modes in microcavities is usually in the THz frequency range, which allows for different scenarios of realization of THz sources based on microcavities. An additional advantage can be taken from the bosonic nature of exciton-polaritons. Stimulation of THz emission by the polariton condensate population leads to the bosonic laser effect. Bosonic lasers are expected to show a higher quantum efficiency than existing fermion lasers—in particular, quantum cascade lasers. Stimulated emission of THz radiation in these lasers does not require a THz cavity. Bosonic terahertz lasers may operate both in horizontal and vertical geometry. In the latter case, they can be assembled into matrices. The crucial factor in these proposals is the use of a wide band-gap semiconductor (in particular, GaN), which opens the way to realization of room-temperature THz emitters. We believe that existing optically pumped polariton lasers based on GaN, combined with a pumping LED, may be used for the creation of a room-temperature vertical-cavity surface-emitting terahertz laser.

Acknowledgments

Alexey Kavokin acknowledges support from the Marie-Curie Initial Training network "INDEX", and thanks the editor of this volume, Prof Bernard Gil, for many enlightening discussions.

References

(1) P. H. Siegel, *IEEE Trans. Microw. Theory Techn.* **50**, 910 (2002).
(2) J. Faist, F. Capasso, D. L. Sivco, C. Sirtori, A. L. Hutchinson, and A. Y. Cho, *Science* **264**, 553 (1994).
(3) R. S. Knox, *Theory of Excitons* (Academic Press Inc., 1963).
(4) H. Haug and S. W. Koch, *Quantum Theory of the Optical and Electronic Properties of Semiconductors* (World Scientific, Singapore, 1990).
(5) C. Ciuti, V. Savona, C. Piermarocchi, A. Quattropani, and P. Schwendimann, *Phys. Rev. B* **58**, 7926 (1998).
(6) F. Tassone and Y. Yamamoto, *Phys. Rev. B* **59**, 10830 (1999).
(7) L. V. Keldysh and A. N. Kozlov, *Sov. Phys. JETP* **27**, 521 (1968).
(8) A. V. Kavokin, J. J. Baumberg, G. Malpuech, and F. P. Laussy, *Microcavities* (Oxford University Press, Oxford, 2007).
(9) V. Savona, Z. Hradil, A. Quattropani, and P. Schwendimann, *Phys. Rev. B* **49**, 8774 (1994).
(10) H. Deng, H. Haug, and Y. Yamamoto, *Rev. Mod. Phys.* **82**, 1489 (2010).
(11) T. C. H. Liew, I. A. Shelykh, and G. Malpuech, *Polaritonic Devices, Physica E* **43**, 1543 (2011).
(12) J. Kasprzak, M. Richard, S. Kundermann, A. Baas, P. Jeambrun, J. M. J. Keeling, F. M. Marchetti, M. H. Szymanska, R. Andre, J. L. Staehli, V. Savona, P. B. Littlewood, B. Deveaud and Le Si Dang, *Nature* **443**, 409 (2006).
(13) R. Balili, V. Hartwell, D. Snoke, L. Pfeiffer, and K. West, *Science* **316**, 1007 (2007).
(14) A. Amo, D. Sanvitto, F. P. Laussy, D. Ballarini, E. del Valle, M. D. Martin, A. Lemaître, J. Bloch, D. N. Krizhanovskii, M. S. Skolnick, C. Tejedor, and L. Viña, *Nature* **457**, 291–295 (2009).
(15) B. Gil and A. V. Kavokin, *Appl. Phys. Lett.* **81**, 748 (2002).

(16) S. Christopoulos, G. Baldassarri Höger von Högersthal, A. Grundy, P. G. Lagoudakis, A. V. Kavokin, J. J. Baumberg, G. Christmann, R. Butté, E. Feltin, J.-F. Carlin, and N. Grandjean, *Phys. Rev. Lett.* **98**, 126405 (2007).
(17) K. V. Kavokin, M. A. Kaliteevski, R. A. Abram, A. V. Kavokin, S. Sharkova, and I. A. Shelykh, *Appl. Phys. Lett.* **97**, 201111 (2010).
(18) E. M. Purcell, *Phys. Rev.* **69**, 681 (1946).
(19) I. G. Savenko, I. A. Shelykh, and M. A. Kaliteevski, *Phys. Rev. Lett.* **107**, 027401 (2011).
(20) A. Baas, J.-Ph. Karr, M. Romanelli, A. Bramati, and E. Giacobino, *Phys. Rev. B* **70**, 161307(R) (2004).
(21) N. A. Gippius, I. A. Shelykh, D. D. Solnyshkov, S. S. Gavrilov, Yuri G. Rubo, A. V. Kavokin, S. G. Tikhodeev, and G. Malpuech, *Phys. Rev. Lett.* **98**, 236401 (2007).
(22) O. A. Egorov, D. V. Skryabin, and F. Lederer, *Phys. Rev. B* **82**, 165326 (2010).
(23) M. O. Scully and M. A. Zubairy, *Quantum Optics* (Cambridge University Press, 1997).
(24) E. B. Magnusson, H. Flayac, G. Malpuech, and I. A. Shelykh, *Phys. Rev. B* **82**, 195312 (2010).
(25) I. G. Savenko, E. B. Magnusson, and I. A. Shelykh, *Phys. Rev. B* **83**, 165316 (2011).
(26) A. V. Kavokin, I. A. Shelykh, T. Taylor, and M. M. Glazov, *Phys. Rev. Lett.* **108**, 197401 (2012).
(27) Yu. E. Lozovik and V. I. Yudson, *Sov. Phys. JETP* **44**, 389 (1976).
(28) G. Christmann, A. Askitopoulos, G. Deligeorgis, Z. Hatzopoulos, S. I. Tsintzos, P. G. Savvidis, and J. J. Baumberg, *Appl. Phys. Lett.* **98**, 081111 (2011).
(29) P. Cristofolini, G. Christmann, S. I. Tsintzos, G. Deligeorgis, G. Konstantinidis, Z. Hatzopoulos, P. G. Savvidis, and J. J. Baumberg, *Science* **336**, 704 (2012).
(30) O. Kyriienko, A. V. Kavokin, and I. A. Shelykh, arXiv:1211.0688v1.
(31) R. H. Dicke, *Phys. Rev.* **93**, 99 (1954).
(32) L. D. Landau and E. M. Lifshitz, *The Classical Theory of Fields* (Butterworth-Heinemann, 1980).

Index

Note: Page numbers in *italics* indicate figures and page numbers in **bold** indicate tables. '*n*' following a page number indicates a footnote.

J

K

L

P